Diagnostics for Experimental Thermonuclear Fusion Reactors 2

Diagnostics for Experimental Thermonuclear Fusion Reactors 2

Edited by

Peter E. Stott

JET Joint Undertaking
Abingdon, Oxfordshire, England

Giuseppe Gorini

Paolo Prandoni and

Elio Sindoni

University of Milan
Milan, Italy

Plenum Press • New York and London

Library of Congress Cataloging-in-Publication Data

On file

Proceedings of the International School of Plasma Physics "Piero Caldirola" Workshop on Diagnostics for Experimental Fusion Reactors, held September 4–12, 1997, in Varenna, Italy

ISBN 0-306-45835-7

A Division of Plenum Publishing Corporation
233 Spring Street, New York, N.Y. 10013

http://www.plenum.com

10 9 8 7 6 5 4 3 2 1

Printed in the United States of America

PREFACE

This book of proceedings collects the papers presented at the workshop on "Diagnostics for Experimental Fusion Reactors" held at Villa Monastero, Varenna (Italy) September 4-12, 1997. This workshop was the seventh organized by the International School of Plasma Physics "Piero Caldirola" on the topic of plasma diagnostics and the second devoted to the diagnostic studies for the International Thermonuclear Experimental Reactor (ITER). The proceedings of the first workshop on ITER diagnostics were published by Plenum Press in 1996 with the title "Diagnostics for Experimental Thermonuclear Fusion Reactors". While many of the ideas and studies reported in the first workshop remain valid, there has been substantial progress in the design and specification of many diagnostics for ITER. This motivated a second workshop on this topic and the publication of a new book of proceedings.

ITER is a joint venture between Europe, Japan, Russia and USA in the field of controlled thermonuclear fusion research. The present aim of ITER is to design an experimental fusion reactor that can demonstrate ignition and sustained burn in a magnetically confined plasma. To achieve this goal, a wide range of plasma parameters will have to be measured reliably. It is also anticipated that diagnostics will be used much more extensively as input to control systems on ITER than on present fusion devices and this will require increased reliability and long-term stability.

The workshop opened with a series of papers on general issues concerning ITER diagnostics. Other oral sessions dealt with the application to ITER of magnetic diagnostics, reflectometry and ECE, interferometry and polarimetry, Thomson scattering, passive and active spectroscopy, fusion products and divertor diagnostics. Special attention was paid to issues such as radiation effects in diagnostic components and control applications of diagnostic systems. The workshop was attended by 81 participants from the ITER parties with strong participation from the USA, Japan and Russia as well as Europe. There were 50 invited talks and 30 contributed papers presented in 2 poster sessions. Five discussion sessions were dedicated to neutron diagnostics, Thomson scattering, reflectometry, radiation effects and spectroscopy. The 75 papers collected in these proceedings were produced by the authors either before or shortly after the conference. The papers have not been refereed and authors remain responsible for their own views.

Thanks are due to the institutions which have contributed financially to the organisation of this Workshop, in particular the Istituto di Fisica del Plasma "Piero Caldirola" of CNR, the DG XII of the European Commission, the Department of Physics of the University of Milan, the Province of Lecco and the Camera di Commercio of Lecco.

P. E. Stott, G. Gorini, P. Prandoni, E. Sindoni

CONTENTS

THE ITER PROJECT

REFLECTOMETRY AND ECE

INTERFEROMETRY, POLARIMETRY AND THOMSON SCATTERING

RADIATION EFFECTS

PASSIVE SPECTROSCOPY AND NPA

ACTIVE SPECTROSCOPY

BOLOMETRY

FUSION PRODUCTS

OTHER DIAGNOSTICS

DIAGNOSTICS OF OTHER FUSION EXPERIMENTS

Diagnostics for Experimental Thermonuclear Fusion Reactors 2

ITER PHYSICS BASIS, MACHINE DESIGN AND DIAGNOSTIC INTEGRATION

ITER Joint Central Team and Home Teams
G. Janeschitz and D. Boucher, T. Burges, K. Ioki, H. Pacher, R. Parker, D. Post, R. Thome, C. Walker

ITER Garching Joint Worksite
Max Planck-Institut für Plasma Physik
Boltzmannstrasse 2
85748 Garching bei München, Germany

INTRODUCTION

In the last Diagnostic Workshop, which took place at the mid-point of the ITER EDA, the ITER machine design and physics performance was presented [1] based on the Interim Design Report (IDR). Whereas the design of several major machine components (e.g. TF-coils) were changed considerably compared to the earlier Outline Design (OD), it was pointed out at the time that the overall objectives and the plasma performance remained in line with the requirements stated by the Special Working Group(SWG 1).

Since then the design of all machine components as well as of the auxiliary systems and buildings evolved further, as documented in the Detailed Design Report (DDR) in December 1996. However, the main machine parameters and the general design concepts for the various components have remained unchanged compared to the IDR.

In addition to the more detailed design available now, a considerable increase in knowledge concerning the physics basis for ITER has been gained through the work of the Expert groups. On the basis of this knowledge a physics basis document is presently being written by the 7 Expert Groups which is planned to be published in early 1998. A brief summary of the present understanding and the resulting predictions for the performance of ITER will be given in this paper concentrating on areas where changes occurred compared to the last workshop.

The goal for the design work after the DDR is to develop an integrated design of the ITER machine and of the auxiliary systems with an emphasis on the various interfaces and to optimise the design of the machine components further. The result of this work will be documented in the Final Design Report (FDR) to be issued in January 1998. An update of the machine design including the tokamak building and the pit layout at the present time, the mid-point between DDR and FDR, will be given in this paper.

A major evolution of the design since the IDR and even since the DDR has occurred in the area of diagnostic integration. This relatively late progress is due to the fact that the boundary conditions for diagnostic integration, namely the design of the major machine components, have to be sufficiently mature before diagnostics can be integrated. From the machine design point of view the diagnostic integration is performed in 4 major areas, namely the in-vessel sensors, the diagnostics in the divertor cassettes, the diagnostics in the equatorial ports and the diagnostics in the top ports.

Whereas only a brief overview of the diagnostic integration will be given here, a more detailed discussion of it can be found in [2].

PHYSICS BASIS

During the interval between the last ITER Diagnostics Workshop and today, substantial progress has been made in the physics basis of ITER through the physics R and D program coordinated by the ITER Expert Groups.

Energy Confinement:

Three methods have been used to predict the energy confinement requirements and plasma performance for ITER: statistical analysis of the dependence of the confinement time on the parameters of present experiments, theoretically developed models for energy confinement, and scaling from experiments with similar dimensionless parameters as ITER. Based on these scalings the sensitivity of the performance of ITER to uncertainties in the confinement projections has been quantified using the 1 1/2 dimensional transport code PRETOR. A simple model for supplying edge boundary conditions for PRETOR has been developed and coupled to the code. The results of this model indicate that the pumping provided in the ITER divertor is sufficient to reduce the estimated central helium fraction during ignited burn from previously ~ 15% to ~ 10%, resulting in a reduction of the required operating density from 1.3 10^{20} m^{-3} to ~ 0.90 10^{20} m^{-3} ($n_e/n_{Greenwald}$ ~ 1.1) for the ITER93-H scaling. The coupled model also showed that it is possible to control the fusion power and divertor heat loads by simultaneous feedback control of the density, heating power and impurity puffing.

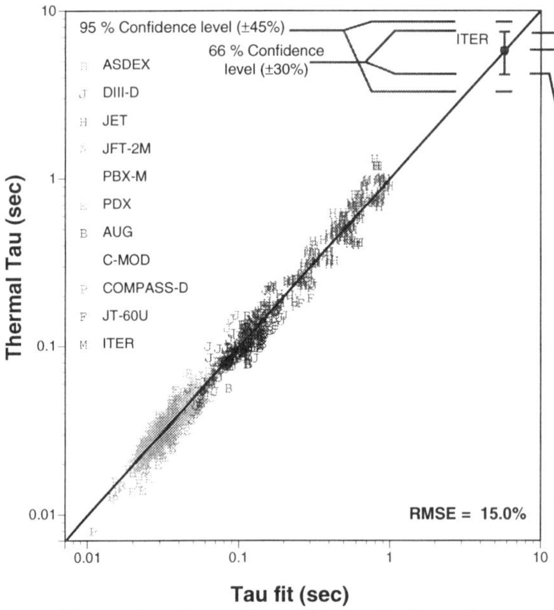

Fusion Power (GW) and Q				
H_H	21 MA		24 MA	
	$n \geq n_{GW}$	$n=n_{GW}$	$n \geq n_{GW}$	$n=n_{GW}$
1.3	1.5 ignited	1.5 ignited	1.5 ignited	1.5 ignited
1.0	1.5 ignited	1.4 Q =15	1.5 ignited	1.5 ignited
0.7	1.5 Q = 75	0.8 Q = 8	1.5 ignited	1.3 Q=15

Normalized Beta: b_N = <b> aB/I				
H_H	21 MA		24 MA	
	$n \geq n_{GW}$	$n=n_{GW}$	$n \geq n_{GW}$	$n=n_{GW}$
1.3	2.7 ignited	2.7 ignited	3.5 ignited	3.5 ignited
1.0	1.5 ignited	2.4 Q =15	2.7 ignited	2.7 ignited
0.7	1.5 Q = 75	1.5 Q = 8	2.0 ignited	2.0 Q=15

Figure 1: Comparison of H-mode thermal energy confinement time with the scaling expression in equation (1) for ELMy data in the ITER H-mode database version DB3[1] and a summary of the ITER performance for variations of the confinement for ± 1 σ from the nominal prediction. The performance projections were calculated with the PRETOR code using a transport model constrained to reproduce the ITER97 scaling.

Recently the ITER Confinement Database and Modelling Expert Group has developed an improved H-mode confinement database (ITERH.DB3) to correct some of the deficiencies of the previous databases (ITERH.DB1,2). The newer database includes data from a wider range of experiments and experimental conditions and heating methods. Analysis of the new database results in a new scaling (Eq. 1, Fig. 1) [3]. This new scaling satisfies the Kadomtsev and high β constraints[4] and roughly has a gyro-Bohm scaling.

$$\tau_{th}^{ELMy} = 0.0286 I^{0.90} B^{0.20} P^{-0.66} n_{19}^{0.40} M^{0.20} R^{2.03} \varepsilon^{0.19} \kappa^{0.91} \quad (1)$$

The new scaling is better conditioned than prior scalings (e.g. ITER93-H) and has been developed explicitly for ELMy conditions rather than for the ELM-free conditions used for ITER93-H. The prediction for the ITER confinement time is 5.8 s, and the scaling gives slightly worse performance than ITER93-H (e.g. $\tau_E^{ELMy} = 0.85 \times \tau_{E-ITER93H}^{ELM-Free}$.) for the 95% confidence level. A key determining factor for the plasma performance is the maximum plasma density. Constraining the density to be below the "Greenwald" limit [5] reduces the fusion power (Fig. 1). For $H_H = 1$ ($\tau_E = H_H \times \tau_{E-ITER97}^{ELMy}$), $n_e/n_{Greenwald} = 1.3$ for 1.5 GW fusion power. Additional margin is available by operating ITER with a plasma current of 24 MA (with a reduced pulse length) (Fig. 1).

Local transport models have also been used to predict the performance of ITER [3]. These models include the most widely accepted theoretical and semi-empirical local transport coefficients. The performance of these models is roughly consistent with the variations of the H_H factor illustrated in Fig.1. Many of these models are very sensitive to the temperature inside the edge transport barrier, which must be specified independently of the models. All of the models predict fusion powers of 1.2 GW or greater for edge transport barrier temperatures of 4 keV, but the IFS/PPPL and GLF23 models predict fusion powers as low as 100 MW for 1 keV edge temperatures. These two models are based on gyro-fluid calculations of the Ion Temperature Gradient (ITG) turbulence. There are preliminary indications from potentially more accurate gyro-kinetic calculations that the gyro-fluid calculations strongly overestimate the turbulent transport. In addition, extrapolations to ITER of the emerging measurements of the edge pedestal conditions on present tokamaks indicate that the edge pedestal temperature on ITER will likely be above 4 keV (e.g.[6]) for which all of the models predict ignited burn. The validation and development of these models is still in an early stage, and their present extrapolations have large uncertainties.

The third approach to predicting the performance of ITER is to scale directly from present tokamak discharges using the dimensionless parameters, ρ^*, ν^*, β. Similar ν^* and β can be obtained in present experiments, but ρ^* in ITER is at least a factor of 2 to 3 smaller than present experiments. The scaling in ρ^* is $\Omega_i \tau_E = F(\rho^*, \nu^*, \beta) \propto (\rho^*)^{-2-\alpha}$ (for fixed ν^*, β), where Ω_i is the ion gyro-frequency, and α is 0 for "Bohm"-like scalings and 1 for "gyro-Bohm" scalings. Scaling experiments on JET, DIII-D, ASDEX Upgrade, Alcator C-Mod, COMPASS-D and JT-60/U support the general conclusion that $\Omega_i \tau_E$ is constant for constant ρ^*, ν^*, and β. For ELMy H-mode discharges, the general result is that α is close to 1, and that the confinement time in ITER will roughly scale as $a^3 B^4 T^{-3/2}$[3]. Since a and B in ITER are 2.8 m and 5.7 T in ITER and 1 m, 1.1 T and 0.56 m, 2.1 T for JET and DIII-D, respectively, the ρ^* scaling is quite strong. The predicted confinement time for ITER based on these scaling experiments on JET and DIII-D ranges from 6.4 s (JET) to 21 s (DIII-D) (Table 1).

Table 1. Experimental Determination of the ρ^*-Scaling Exponent α.

Tokamak	β_N	ρ^*/ρ^*_{ITER}	α	$\delta\alpha$	$\tau_{th,ITER}$ (s)	$\delta\tau$ (σ)
DIII-D	2.0	7.3	1.15	±0.44	21	±12
JET	2.2	5.5	0.7	±0.32	6.4	±3

Reduction of the uncertainties requires experiments with a greater range in $\rho^* \propto I_p^{-2/3} R^{-1/6}$,. Experiments with the widest size range, JET and COMPASS-D, will be very important.

H Mode Power Threshold Requirements:

A new scaling for the H-mode threshold power in terms of global parameters has been developed from the updated H-mode threshold database (Equation 2, Fig. 2)[7]. Like the previous scaling[8], it is not consistent with the dimensionless parameter constraints, indicating that non-plasma processes, such as atomic physics, probably play a strong role in determining the threshold.

$$P_{thr} = 0.65\bar{n}_e^{0.93} B_T^{0.86} R^{2.15} \qquad [\text{MW, } 10^{20}\ \text{m}^{-3}\text{, T, m}] \qquad (2)$$

The lowest threshold powers (about 60% of the scaling in Equation 2) were obtained with single null divertors, grad B drift direction toward the divertor, low wall recycling conditions, and improved neutral retention in the divertor. The ITER divertor and first wall are designed to achieve these conditions, so that the threshold power in ITER will likely be consistent with the lower threshold powers achieved (Fig. 2). In addition, there are indications that the H-mode power threshold with early DT experiments on JET is lower than with D alone (consistent with the lower threshold with D operation compared to H). The next JET DT campaign will quantify the threshold power reduction. Both this scaling (Equation 2) and prior scalings have significant uncertainties for projecting to ITER.

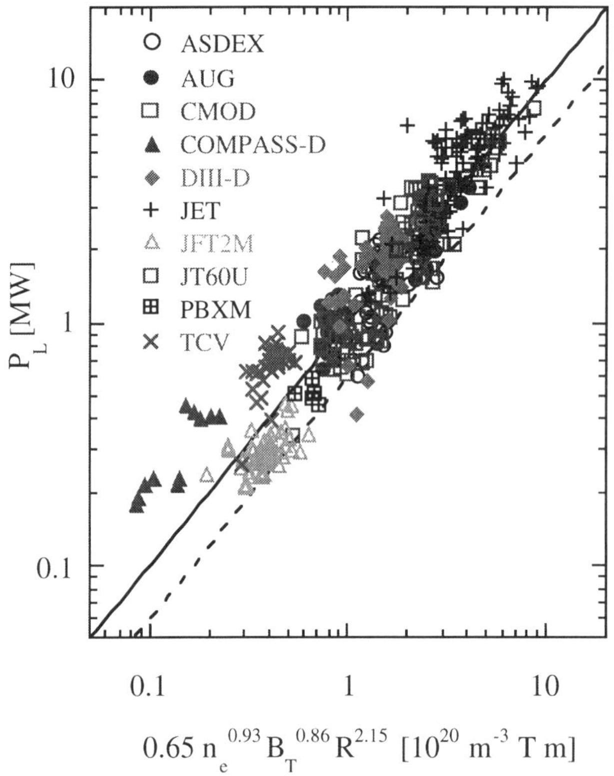

$P_{threshold}$ (MW)

n_e	new	60%
0.25	75	43
0.5	137	82
1.0	260	156

Figure 2 : Measured H-mode Power Threshold vs. scaling expression. The dashed line is 0.6 time, the scaling expression [7].

The uncertainties have not yet been examined in detail yet for this scaling but are expected to be similar to those of the old scaling. For the old scaling, they can be expressed in terms of the dependence on n and R using the exponent α in Equation (3):

$$P_{thr} = 0.45\bar{n}_e^{0.75} B_T R^2 \left(0.6nR^2\right)^{\alpha} \qquad [\text{MW, } 10^{20}\ \text{m}^{-3}\text{, T, m}] \qquad (3)$$

Taking α to be either -0.25 or 0.25 results, for $n_e \approx 0.5\ 10^{20}\ \text{m}^{-3}$ in a threshold power of 50 and 200 MW, respectively, and for $n_e \approx 1.0\ 10^{20}\ \text{m}^{-3}$ in a power of 70 and 425 MW, respectively.

Whereas further versions of the global database will reduce the uncertainties in these global scalings, there are strong indications from experiments that the L to H mode transition requirements depend on the local edge parameters, such as the local edge temperature and edge pressure just inside the separatrix(e.g. [9]) rather than the global parameters. The ITER Divertor Modelling and Database Expert Group is assembling a database of edge pedestal parameters to characterise the dependence on the local edge parameters, and several groups are developing models for L to H transition in terms of those parameters.

Edge Temperature:

An initial analysis of the published edge data and the data being assembled by the ITER Divertor Modelling and Database Expert Group indicates that the temperature at the top of the H-mode transport barrier will be in the 4 keV range or higher [10]. A projection of the edge pedestal width Δ_{ITER} for ITER using a scaling roughly proportional to size gives $\Delta_{ITER} \approx 19$ cm and a weaker size scaling gives $\Delta_{ITER} \approx 9$ cm. The assumption that the limiting pressure gradient is determined by local ballooning stability results in edge pedestal temperatures of 4 to 8 keV at $n_{e\text{-}Ped} \sim 6.\ 10^{19}\ m^{-3}$ [11]. This is likely to be a conservative assumption since the edge pedestal electron pressure in DIII-D and JET is consistent with local ballooning stability based on the total local plasma pressure.

Density Limits:

The ITER97H scaling indicates that $n_e/n_{Greenwald} \sim 1.3$ is required to reach 1.5 GW fusion power. In addition to the difficulty of achieving $n_e/n_{Greenwald} \geq 1$, energy confinement is degraded in many experiments with the density close to the Greenwald limit. Whereas this poses a potential difficulty for ITER operation, there is strong evidence that $n_e/n_{Greenwald}$ is not a universal constraint on the plasma density. The density limit in present tokamaks appears to be a limit on the edge density (e.g.[9]). A number of experiments have exceeded the Greenwald limit with gas puffing and with pellet fuelling. In shallow fuelling pellet experiments, the fuelling efficiency is low and the confinement is often significantly degraded. However, high velocity pellet injection in DIII-D allowed the achievement of $n_e/n_{Greenwald}$ of up to 1.5 without significant confinement degradation [12]. Deep fuelling techniques, such as high field side injection in ASDEX-Upgrade [13], can be used to increase the central density with minimal increases in the edge density. Furthermore, the recent compact toroid fuelling experiments on TdeV offer the promise of another central fuelling technique [14].
Edge parameter measurements indicate that the effective density limit is reached in H-mode by gas puffing, which reduces the edge temperature sufficiently for the plasma to make a transition to the L-mode. As a consequence the particle flux increases and a disruption often follows. For ITER, the temperature for the H-L mode transition is likely to be substantially below the edge pedestal temperature, leading to adequately high edge densities (~ 5—7 10^{19} m-3).

Power and Particle Control:

Experiments on divertor and limiter tokamaks have identified new phenomena and have provided increased understanding of divertor operation. This gives added confidence that most of the heating power in ITER can be radiated to the main chamber and divertor chamber walls resulting in acceptable peak heat loads on the divertor plates without unacceptably high confinement degradation.

All of the divertor experiments have been able to reduce the peak heat flux on the divertor plates by a factor of five or more by a combination of gas puffing and impurity injection. The detailed understanding of divertor physics has benefited greatly from recent advances in divertor diagnostics. Two-dimensional temperature and density measurements have allowed the validation of divertor models to a much greater extent than before [15]. In addition, impurity injection experiments on ASDEX Upgrade, DIII-D, Alcator C-Mod and JET have identified the significance of hydrogen recombination for detachment (Fig. 3)[13]. Whereas the confinement often degrades as P_{rad}/P_{tot} approaches 1, ASDEX

Upgrade and DIII-D have been able to achieve good confinement for high values of P_{rad}/P_{tot}. Good confinement appears to require that the plasma remain partially attached, i.e. detached near the strike point and attached further out in the SOL.

Two-dimensional divertor models have been validated with the help of the improved diagnostic data, and are being used to project potential divertor operational regimes for ITER. B-2 EIRENE calculations for ITER including a moderate degree of Ne puffing and a source of carbon from chemical and physical sputtering from the divertor plates indicate that most of the SOL power can be radiated from the divertor by acceptable levels of Ne, carbon, He, and D/T with Zeff < 1.8 [16], [17]. The calculations also show that these regimes are compatible with He exhaust requirements. This operating point is therefore favourable in several respects: low radiated power in main plasma, low neutral density in main chamber, well-distributed radiation, low peak power on the plates, and adequate He pumping.

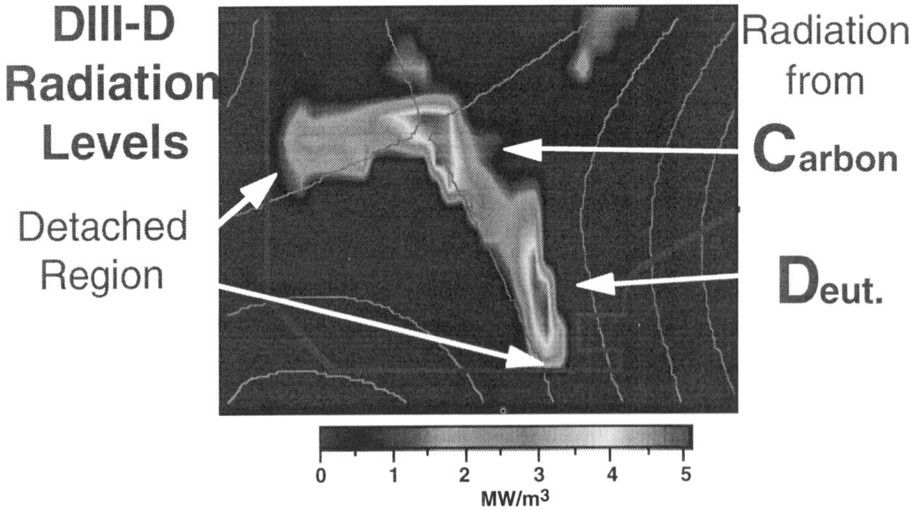

Figure 3: Bolometer measurements of the radiated power from DIII-D with gas puffing. The radiation near the X point is due to carbon and the radiation in the lower region is due to hydrogen, particularly hydrogen recombination[15].

While the present expectation is that between 100 to 200 MW power conducted across the separatrix will be needed to sustain the H-mode and maintain a high temperature edge pedestal, the capability to radiate most of the power from the main plasma without excessive levels of plasma dilution and central radiation losses would offer additional margin. Recent experiments on TEXTOR and TFTR with limiters, as well as divertor experiments on ASDEX Upgrade, showed that losses of 80% or more of the heating power from the main plasma can be sustained without substantial loss of confinement, and tokamak transport calculations indicate that the fuel dilution and central radiation losses from such regimes are acceptable.

Disruptions:

Experiments on many of the major tokamaks have led to improved characterisation of the magnitude and toroidal symmetry of the halo currents during vertical displacement events. This, in turn, has led to improved (and somewhat more severe) design guidelines for disruption forces on the ITER internal components. The forces from halo currents are among the most severe mechanical requirements for the design of the internal components. During the past two years, there has been mounting evidence that their toroidal distribution is not uniform (e.g.[18]).

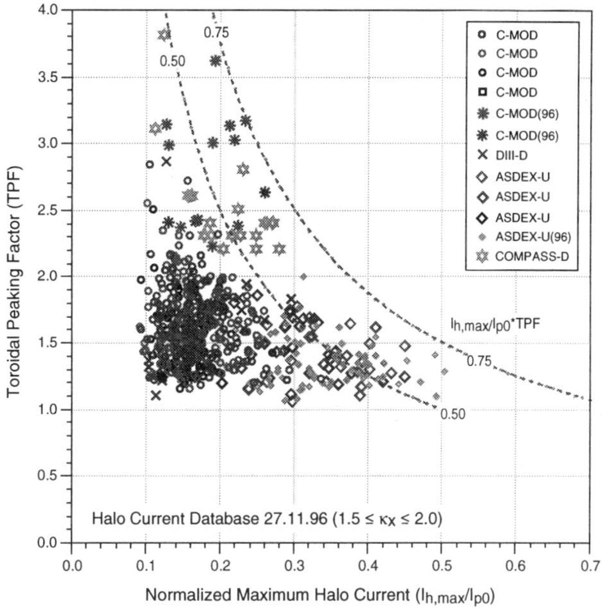

Figure 4 Toroidal peaking of halo currents in various tokamaks, for plasmas with $1.5 \leq k_X \leq 2.0$. High peaking factors occur only at low halo current fraction. The hyperbolic curves show limiting bounds for the data.

A database of the measured halo currents and their toroidal peaking has been assembled by the MHD Stability and Disruption ITER Expert Group (Fig. 4). Ratios of I_{halo}/I_o as large as 0.5 and toroidal peaking factors up to a factor 4 have been observed. Fortunately, the highest values of these two factors do not occur simultaneously, and a design guideline of $I_{halo}/I_o \times$ TPF (toroidal peaking factor) of 0.58 appears consistent with all but the most severe conditions. There is no data from the largest divertor experiments in the database (JET and JT-60/U), but preliminary data from JET not yet in the database indicates that $I_{halo}/I_o \times$ TPF is lower than 0.5.

ITER OVERALL MACHINE DESIGN

As already mentioned in the introduction no major change in the layout or in the parameters of the ITER machine occurred since the last workshop. The emphasis in this section will therefore be on an overview of design improvements and of new engineering solutions for the various machine components and the status of the major R&D supporting the design solutions. All the improvements or changes were driven by simplifications, cost (seek reductions), by maintenance considerations, by major interfaces or by safety issues. Fig. 1 shows a cross section through the present ITER device including the pit, the cryostat, the magnet system, the VV, and the in-vessel components.

Magnet System:

The magnet system is described in [20] and [21]. The magnet system including the central solenoid and the VV support constitutes a closed mechanical system where all forces except gravity are short circuited [22]. This closed system is then supported on the cryostat through a ring of laminated gravity support columns which are attached to the lower outer intercoil structure. The lower outer intercoil structure has been extended downwards and provides the connection to these gravity supports (Fig. 5). The advantage of this approach is that all disruption forces are short-circuited in a very compact mechanical system. This removed the concern about relaying e.g. halo current forces onto the VV through two different gravity support rings and the cryostat into the magnet system (IDR-design). Because the gravity supports now have to support only the weight of the machine and related seismic loads a greatly simplified design has become possible. Thus the new gravity supports consist of Inconel plates which provide the required flexibility in the radial direction (thermal contraction during cool down).

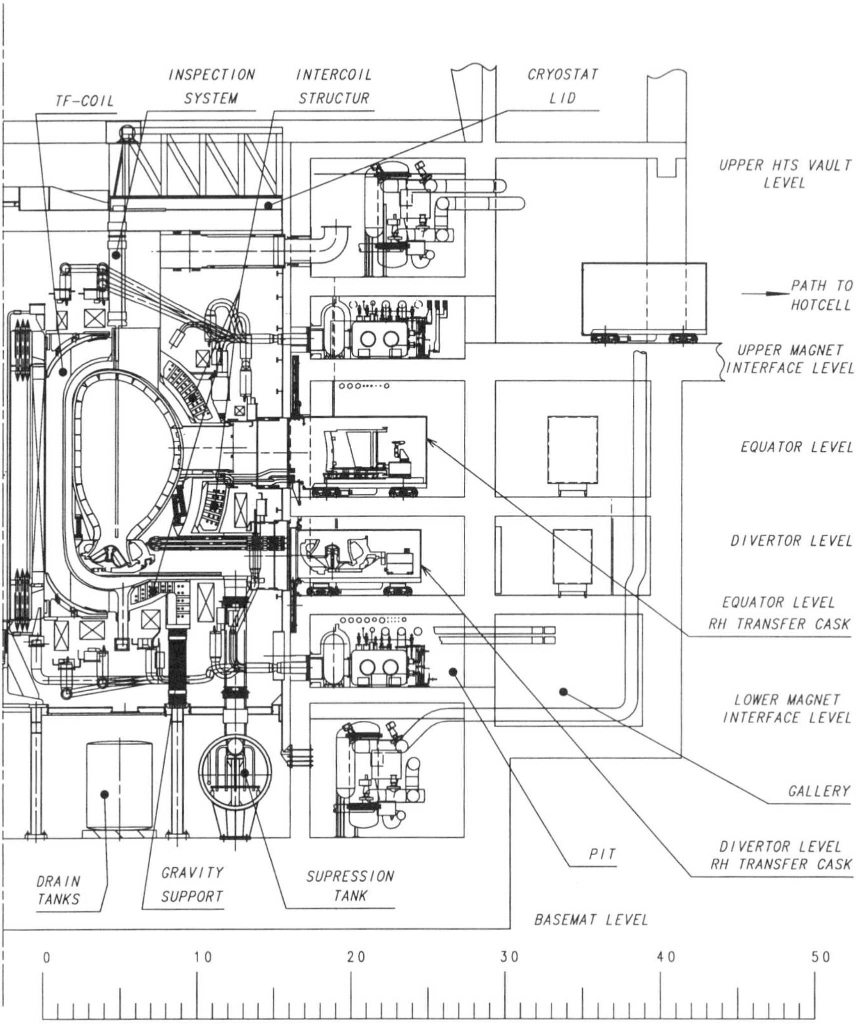

Figure 5: Cross section through the ITER machine the cryostat and the tokamak pit

As mentioned in [1] a segmented central solenoid has been considered, which would result in a reduction of manufacturing risks (smaller sub components), would allow building all other PF coils with NbTi superconductors (reduced cost, improved fabrication schedule, and feasible under-machine replacement if required) and would provide greater flexibility for plasma shape control. In these studies a fully segmented solenoid as well as a so called hybrid solution with a shorter central section (~ 9m high) and two additional coils on top and bottom were investigated. The fully segmented solenoid had to be ruled out because joints between pancake to pancake which would work in a 13 T field were not considered practical. Therefore a fully segmented solenoid would require adoption of a wedged instead of a bucking approach to support the TF coils (with significant increase of the major radius) in order to bring the joints outside the solenoid into an area with lower field.

A hybrid solution, however, seemed to be very promising initially. It would not reduce the manufacturing risks considerably (9 m high layer-wound CS) but all the PF coils could be manufactured from NbTi and it would improve shape control as well as allow higher triangularity, which has considerable benefits for high density operation in H-mode. In this approach a gap of ~ 1.5 m between the central solenoid and the additional coils would provide room for the joints resulting in a field at the joints of ~ 6.5 T which would be considered acceptable. However, more detailed stress analysis showed that due to the field penetrating through these gaps a significant increase of the out of plane forces occurred for the TF coils, resulting in a reduction of the fatigue lifetime of the TF cases by a factor of 2. In addition it was found that the shear stresses in the central solenoid coil itself were unacceptably high . Consequently, this approach had to be abandoned.

The solution adapted for the FDR design therefore retains the monolithic 12m high central solenoid. However, to reduce cost, improve manufacturing schedule and in order to allow a possible replacement of the PF coils trapped below the machine, it was considered important to manufacture all PF coils with NbTi which limits the maximum field in these coils to 5 T. In order to retain at least the IDR shaping capability it was necessary to split the PF2 and the PF7 coils into two. These new PF coils have to be supported by clamps which penetrate the upper and lower crowns. The resulting PF system consists of 9 PF coils (Fig. 5), with all but the CS coil manufactured using NbTi superconductors. It has similar shaping capabilities as that of the IDR while allowing somewhat higher average triangularity (δ_{95}~0.26).

Because all PF coils are manufactured with NbTi a replacement of the trapped PF coils is now feasible should it become necessary. The procedure envisages that a ring below the damaged PF coils is opened in the now flat cryostat floor (see below) and the damaged PF coil is lowered into the ~ 7.5 m high room below the cryostat. This coil is then cut into pieces and removed. A winding machine will be installed in this room after fitting a temporary shield into the opening of the cryostat floor and a new PF coil will be manufactured. The new coil will then be lifted into its place and the cryostat floor will be rewelded.

A concern for ITER are the error fields due to manufacturing and installation errors in the magnet system. Data on the scaling of the onset of locked modes in present machines shows an unfavourable scaling, indicating that locked modes in ITER are expected for error fields exceeding 2.10^{-5} B_T. Whereas tight manufacturing and installation tolerances will be applied and an attempt will be made to measure the error fields early on, it is anticipated that error fields smaller than several 10^{-4} B_T will not be achievable. Therefore three sets of superconducting correction coils are foreseen, one set at the top of the machine inside the PF coils, one around the equator outside of the PF coils and one below the machine. These coil sets will allow to reduce the most important error field components (n:m = 1:1, 2:1, 3:1) to < 2.10^{-5} B_T.

Another concern for ITER operation was the relatively high ripple in the outer upper region of the plasma, which causes considerable alpha losses in particular for advanced tokamak regimes. Therefore ferromagnetic inserts are foreseen in the VV segments at locations under the TF coils between the equatorial ports and the vertical ports. Calculations show a factor of 2.6 reduction in the ripple can be achieved at full field, from 1.6% to 0.62% at the plasma edge, resulting in acceptable alpha losses for all envisaged equilibria. Other detailed changes and improvements to the magnet system are described elsewhere [20].

Beside the design effort two large R&D projects are underway for the magnet system, one aiming at manufacturing a model CS coil by the US HT & JA HT and one manufacturing a model TF coil by the EU HT. However, all 4 HTs are deeply involved in the various manufacturing processes of both model coils. Presently all strands for the CS coil and 75% of the strands for the TF coil have been fabricated. A 300 m jacketing line and a 1000 m jacketing line is in operation for the CS type and the TF type of conductors respectively. Winding of 12 out of 18 layers of the model CS has been performed and four layers were heat treated, forging of the TF coils radial plates and trial windings and heat treatments are underway. Test facilities in JA for the CS and in EU for the TF coil are either complete or well advanced.

Cryostat:

The cryostat, which consisted of a double wall cylinder and two double wall domes on top and bottom, has been completely redesigned (Fig. 5). In order to make room below the cryostat for several services such as drain tanks and the pressure suppression system as well as for the possible replacement of a PF coil without increasing the overall height of the machine and thus its cost, a flat floor was designed. The gravity supports for the machine which penetrate the cryostat floor and the gravity supports used for the CS installation, which are radially further inboard, support a double wall flat floor which is thus able to sustain the atmospheric pressure (Fig. 1).

The cylindrical walls of the cryostat are supported at the bottom on the bio-shield and consist of 60 mm (in some areas 50 mm) thick SS plates strengthened by toroidal and vertical ribs resulting in a simpler and thus cheaper solution.

The main problem for the new cryostat design was to design a flat cryostat lid which also should have a central flange big enough to allow the removal of the CS as well as of the two innermost PF coils without having to lift the whole lid. This allows a changeout of a damaged TF coil.

In order to be able to sustain the atmospheric pressure the single wall 40 mm thick lid is supported by 60 mm thick and ~ 1m high toroidal ribs and a bridge like structure (Fig. 5). The previously separate bio-shield lid is now integrated into the cryostat lid where part of the shielding is performed by shielding blocks on the lid itself and part of the shielding rests on top of the bridge structure. The bridge structure is only present above every second port and above the two adjacent TF coils providing good access to the area of the 10 vertical diagnostic ports. The large central flange also consists of a 40 mm thick lid supported by 60 mm thick and up to 2 m high radial ribs.

Tokamak Building and Hot Cell:

The general layout of the Tokamak building has not changed considerably since the IDR but the pit and also the Hot Cell building has evolved considerably. The tokamak pit now consists of a 60 m diameter cylinder which can be separated by a seismic gap and by shock absorbers from the rest of the building if required. All major machine components including the primary heat transfer systems (PHTS) are inside this cylinder (Fig. 5). There are two cylindrical HTS vaults on top and bottom of the pit connected by 20 vertical shafts which also contain the divertor and part of the blanket cooling pipes. The top vault contains 10 HTS loops for the blanket and the bottom vault 8 HTS loops, four for the divertor and four for the blanket/ limiter. The water holdup in each loop is such that a major leak would not exceed the pressure design limit for the two connected HTS vaults (safety requirement).

Outside the pit a 10 m wide ring of galleries is foreseen providing space for toroidal services routing and for movements of transfer casks which are foreseen to transport RH equipment and machine components to and from the Hot Cell building. These casks move on a railway system which is connected to all equatorial and divertor ports. Within the gallery there is a lift shaft allowing the casks to be moved up to the -9m level from which a tunnel leads to the Hot Cell building.

The Hot Cell building consists of a lower level with docking ports for the various casks as well as storage areas for divertor cassettes, for diagnostic and heating systems, VV plug assemblies, as well as for RH tools. The upper level houses several repair sections where divertor cassettes and diagnostic plugs can be remotely refurbished and thus can be reused in order to minimise radioactive waste. It also contains a waste processing area mainly foreseen for compacting and detritiating waste before shipping it off site. A storage area for all the RH casks is foreseen in the basement of the laydown hall in the Tokamak building.

Vacuum Vessel:

Only changes in details occurred in the VV design [19] whereas the general concept of a double-walled structure constructed from two toroidal shells (40 - 60 mm thick) and joined by ribs cut from 40 - 60 mm plate is the same as reported in [1]. The space between shells is filled with Boron doped SS shield plates directly cooled by water. As already mentioned the vessel is attached on the bottom directly to the magnet structure by supports which permit differential radial displacement. In addition toroidal supports joining the vessel equatorial ports with the TF structure limit possible differential horizontal displacements, as might occur due to seismic or vertical displacement events (VDE's). Because of the positions of these supports, the out of plane forces acting on the TF coils cause a toroidal rocking of the vessel in the region of the horizontal port by ± 25 mm in the course of a discharge.

No direct deposition of the plasma thermal energy on the vessel is possible. However, body forces exerted on the in-vessel systems must ultimately be reacted by the vessel and its supports. The most troublesome load in this respect originates from Vertical Displacement Events (VDE's). An axisymmetric VDE with a limiting plasma safety factor of $q = 1.5$, yields a net vertical force of $F_V \leq 150$ MN. Toroidal peaking of the halo currents as described in the physics section above can cause the force per unit area to exceed the average value locally. Experimental data also suggest that net horizontal forces can arise. In the absence of a sufficient database, these have been evaluated by a model assuming a tilted plasma, yielding ~ 50 MN net horizontal force. For the bounding case of a non-axisymmetric VDE the maximum membrane stress in the vessel due to the vertical load is 90% of allowable (131 MPa), whereas for the specified horizontal component of the load the maximum stress, occurring in the area of the equatorial ports, is 60 % of allowable. The maximum stresses resulting from SL-1 and SL-2 seismic loads are also well within allowable stresses.

An R&D task aiming at building a full VV sector is under way in the JA HT. Two different manufacturing methods are used for each toroidal half of this sector: one using a conventional TIG+MAG welding approach and closure plates to manufacture the double wall structure and one using an advanced electron beam welding technique, for which the welding is performed through the outer shell into the ribs. Two thirds of the VV sector parts are already finished and await assembly.

Blanket:

The blanket modules, their attachment to the backplate and the backplate itself were changed considerably since the IDR and even since the DDR. The changes were necessary to solve a number of problems, discovered by detailed analysis, when joining two strong structures (modules and backplate) with different thermal time constants leading to transient differential thermal expansions. While this requires flexibility of the attachment, the modules (and the backplate) are subject to very high electromagnetic loads and thus additional locking devices are required to sustain shear loads between modules and backplate or between adjacent modules.

In addition it was found that the hoop stresses arising on the backplate from either radial disruptions (~ 75 MPa) or asymmetric VDE's (115 MPa) require a minimum material thickness for SS of 120 mm in order to be below the allowable membrane stress (131 MPa). A single plate design with this thickness had to be ruled out due to cooling problems leading to a solution with a double wall backplate which also constitutes the cooling manifold for the blanket modules. The double wall approach with a gap of 40 mm between two plates (inner 50 mm, outer 70 mm) also improves the resilience against bending stresses induced at the bottom of the blanket during VDEs. The new manifolding allows to concentrate the cooling pipes which connect the blanket with the HTS in 10 vertical ports thus freeing the other 10 ports for diagnostic access and off-line in-vessel viewing. As an additional benefit, the increased manifold volume raises the module cooling water residence time in the machine thus reducing the activation caused by ^{17}N decay (produced by $^{17}O(n,p)^{17}N$ reactions)with $T_{1/2}$.= 4.17 s [23] as well as radiation loads caused by ^{16}N.

In port areas, reinforcement must be used to reduce peak stresses and to limit distortions arising from high local load. These arise because the presence of the port interrupts electrical continuity, modifying the current path, and because of the missing material in the port area, reducing the resistance to the hoop stress. These reinforcements consist of toroidal and poloidal ribs which are introduced in the space between backplate and VV in case of the horizontal ports and in form of rectangular extensions into the vertical port.

As in previous designs, the modules have a first wall consisting of a 20 mm thick copper alloy mat on top of which (facing the plasma) is a 10 mm thick array of Be tiles. The copper mat is bonded to a water cooled stainless steel block, which serves as a neutron shield, and is cooled by water flowing through an embedded array of stainless steel tubes (typical ID = 10 mm, thickness = 1 mm, and pitch = 22 mm). This configuration accommodates the design heat flux of 0.5 MW/m^2 with substantial margin.

As can be seen from Figure 6, the modules are oriented with their long side in the toroidal direction. They are attached to the backplate by means of 4 flexible connections, which are slotted inconel cylinders approximately 200 mm long and 150 mm in diameter which can bend or flex in the direction transverse to their cylindrical axis in order to permit differential expansions of order 1 mm between the backplate and the module for 10^4 cycles. Each "flexible" can be bolted or welded to the module and is attached to the backplate by means of a single M45 high strength bolt. The flexible fits in a hole recessed in the backplate. Adjustment for tolerance and backplate deformations is accommodated by an insert which is machined after survey of the positions and orientations of the backplate holes. Four 30 mm holes are situated in the module to permit access to the bolts from the plasma side.

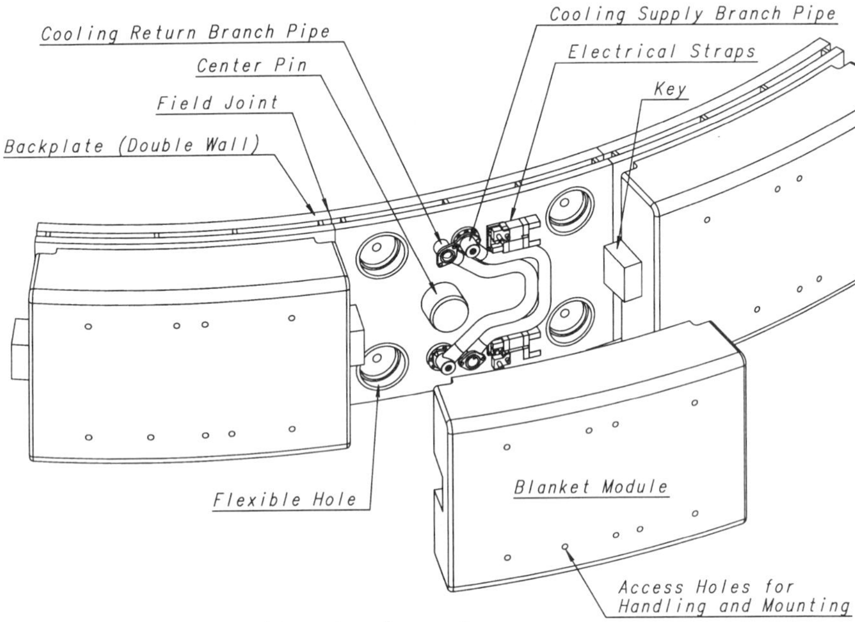

Figure 6: Blanket modules and attachment scheme

The pipes connecting the backplate manifolds with the water cooling circuits in the module are sufficiently flexible to allow ± 2 mm displacements when connected to the flanges. The new backplate design, the concern about the reliability of welding the branch pipes from the modules to the cooling manifold with a crawler device introduced into the manifold as well as the new attachment scheme resulted in an approach where all mechanical, hydraulic and electrical connections from the module to the backplate are made from the plasma side. This resulted in nine 30 mm diameter access holes, 4 for the attachment, 2 for the branch pipe welding, 2 for attaching two electrical connectors and 1 RH hole needed for gripping in combination with 3 of the existing holes. Analysis shows that the reduction of shielding as well as nuclear heating (bolts) and He production (branch pipe weld) due to these holes at the level of the backplate is acceptable.

The electromagnetic loads on the blanket system from centred disruptions induce a current in the blanket and vacuum vessel. Most of the current appears in the backplate, reaching 1.2 MA/m near the inside equator. Such current densities also appear in the adjacent equatorial modules, flowing in the toroidal direction in the first wall and returning again in the reverse toroidal direction in the rear surface of the module. Since there is no net toroidal current in the module (the modules have an electrical connection at only a single toroidal location) , the radial pulling force on the module is negligible. Instead, this force (a radial pressure of 1 MPa) is applied directly to the backplate.

Although the net force on the modules is negligible, there are strong moments created by the induced currents as they cross the toroidal field near the toroidal ends of the modules. On the inside equator the radial torque (poloidal forces) reaches 1.3 MNm. This torque is reacted by (insulated) keys linking each module with its toroidally adjacent neighbour which experiences the same torque. Whereas the radial torque diminishes in modules above and below the equator, such modules experience a poloidal torque (radial forces) whose origin is less obvious than that of the radial torque just described. It arises from poloidal currents crossing the toroidal field, also at the toroidal ends of the module, which are associated with penetration of radial magnetic field through the poloidal gaps between modules. EM modelling indicates for example that in the number 8 module, counting from the lowest module on the inside, the poloidal current reaches 110 kA and produces a poloidal torque of 0.7 MNm depending on the location [24]. This torque must be reacted by the flexibles and produces their dominant axial load.

In addition to the standard blanket modules there are special areas with higher heat load and / or neutral CX fluxes. Two of these areas are the toroidal belts of modules in the vicinity of the X-point forming the baffle. A compromise between neutrals back-streaming from the divertor and direct neutral recycling on these modules results in arranging their FW along the 6 cm flux surface (~ 6 power e-folding lengths outside the separatrix). The relatively high neutral density near these modules is expected to produce a significant flux of low energy CX neutrals ($\leq$ 100 eV). To mitigate the erosion from these neutrals, tungsten has been selected as the first wall material for the lower row of baffle modules while beryllium is retained for the upper row. The peak heat fluxes on the baffle first wall are expected to be ~3 MW/m^2, ~ 0.5 MW/m^2 from CX neutrals and up to 2 MW/m^2 in cases where MARFEs develop at the X-point while the surface-averaged heat load is only 1 MW/m^2.

Another special area is the startup limiter. While previous designs included a belt limiter below the equatorial ports consisting of modified blanket modules, two or three local limiters situated in up to three dedicated RH ports are foreseen for the FDR design. The reason for this change is that the alignment requirements for which a toroidal belt limiter becomes attractive (< 1 mm absolute to the field lines) are too stringent for a machine like ITER. This would furthermore require a first wall designed for 5-10 MW/m^2, of which only a small fraction (a few %) would be used at this level, resulting in a very inefficient cooling scheme.

The design of the discrete port limiters follows a similar approach as the divertor vertical target/albeit with a Be FW. The heat load capability is in the range of 8 to 10 MW/m^2 and will thus allow limiter plasmas with ~ 20 MW conducted into the SOL (accounting for some additional heating) if two (10 MW/m^2) or three limiters (8 MW/m^2) are utilised. In order to utilise more than one limiter, again a minimum alignment tolerance of 1 mm of the limiter modules relative to the limiting surface has to be achieved. In contrast to the belt limiter, the alignment for these local port limiters is feasible through adjustments which can be performed after initial operation. Additional advantages are the relatively fast replacement possibility should the necessity arise and the more localised cooling required for this limiter system.

A comprehensive R&D programme for the blanket led by the EU HT is presently under way albeit somewhat delayed. However, many small and medium size mockups addressing the manufacturing of modules and of the FW have been made and were successfully tested. Among them was a 5 t module mock up produced by powder HIP and many small FW mockups demonstrating the feasibility of hipping the Be-Cu-SS FW mat. One full size mockup will be fabricated within the EDA, but testing will be done during the post-EDA period.

Divertor:

The general design approach of the divertor has not changed since it was presented in [1] and [25]. It still consists of 60 cassettes onto which all HHF components are mounted. Also, the choice for plasma facing materials (C at strike zones on the vertical target (VT) and short dump target and W everywhere else) has not changed. The expected lifetime due to erosion in particular of the VT is of the order of 3000 full power discharges. Based on R&D results, the reference concept adopted for the lower part of the VTs and the short dump targets consists of CFC monoblocks joined to Cu pipes while rectangular Cu heat sinks clad with 4x4 mm 10 mm thick W tiles are use in the upper part of the VT. The monoblocks and the Cu heat sinks are supported by SS plates (slices) which form a backplate when all 18 (outboard) or 13 (inboard) slices of a VT are assembled. Swirl tube technology is used to enhance the heat transfer coefficient in the lower part of the VT and the short dump targets. Critical heat fluxes (CHF) of up to 27 MW/m^2 have been achieved in this geometry with ITER-relevant thermal hydraulic conditions in R&D performed by the EU Home Team [26] giving sufficient margin to the expected maximum transient heat load of 20 MW/m^2. HHF tests on mockups with CFC monoblocks and others with the above mentioned W cladding survived > 1000 cycles of up to 30 MW/m^2 (CFC) and 10 MW/m^2 (W) respectively.

Minor changes have been also made in many other areas (Fig. 7): a slightly bigger dome to optimise neutral baffling, the design of maintenance-friendly attachments for the HHF components, a cassette body design considering manufacturing constraints, a dome body design following the modular slice approach of the VT as well as the increase of the heat load capability of the short dump target to ~ 16 MW/m^2 are among them.

Recently the concern about C-T codeposition and about the related cleaning techniques together with the adoption of semi-attached operation as the reference, reducing the need for the wings in the private region, has resulted in a design option without wings (Fig. 7). The driver for this new design was to try to keep the surfaces in the private region sufficiently hot (> 450°C) in order to prevent- or to reduce the T trapped with C in codeposited layers. In addition it was attempted to design smooth surfaces in order to allow local ECH discharges in oxygen, which, beside baking in air, is one of the most promising techniques to remove C-T codepositied layers.

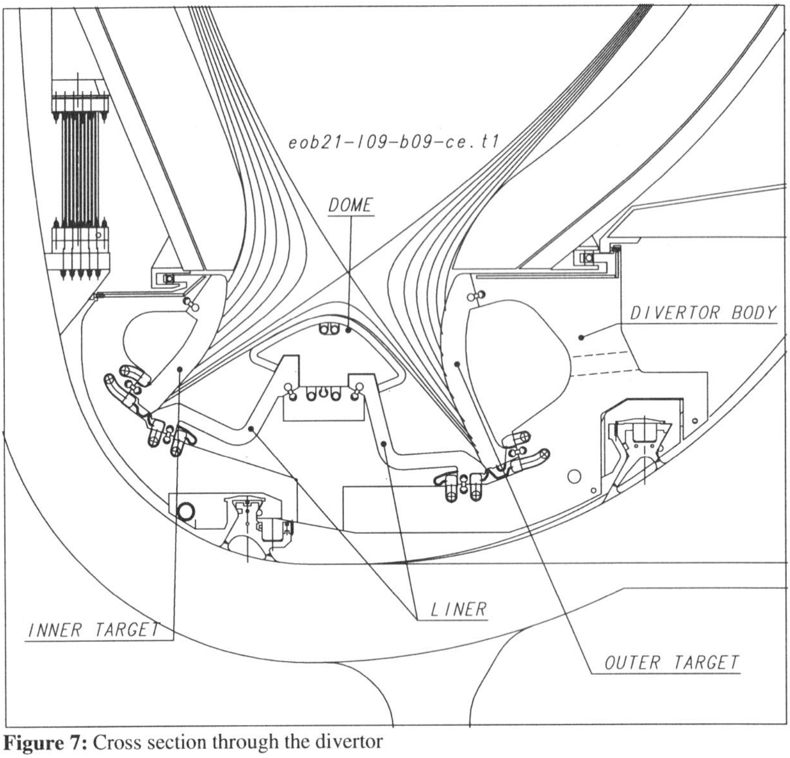

Figure 7: Cross section through the divertor

Due to these requirements and in order to standardise the manufacturing and attachment of HHF components, again a similar design approach as for the vertical target (VT) was adopted for this new liner albeit with a different kind of W cladding and with pumping slots between the slices forming the liner. In order to keep the surface temperature sufficiently high, special W tiles with a narrow neck (limits heat conduction) which are attached by active metal casting to the Cu heat sink are envisaged. The design is such that the W temperature near the surface will be in the range between 450°C and 1200°C for radiation heat loads calculated to be between 0.3 and 1.0 MW/m^2. In order to provide a smooth surface for cleaning purposes the pumping holes are fully covered by this new liner. However, due to the modular design consisting of slices, pumping slots between each subsection of this HHF component together with a 50 mm gap between cassette body and liner provide enough conductance to retain ~ 90% of the previous pumping speed. Further physics analysis applying a Monte Carlo code for neutrals including n-n collisions and as well as discussions with industry about this concept are presently ongoing in order to enable a final judgement on this new approach by the end of October.

The main electromagnetic loads in the divertor are experienced during downward VDEs. The resulting halo currents and eddy currents lead to ~ 4 MPa loads on the HHF components and a maximum downward force of 4.5 MN on each cassette body and its supports. The cassette body as well as the HHF components are designed to withstand these loads.

R&D coordinated by the US HT and aiming to build close to full size HHF components and a toroidal half of a full size cassette is well under way. All HTs participate in this task, in particular in the development and manufacturing of HHF components. Many results which are already available from small and medium size mockups (see above) support the present design choices and provide also several viable backup options for the HHF components as well as for the cassette body. Further results from close to full size components are expected during 1998.

Fuelling and Pumping System:

The fuelling and pumping systems have evolved considerably during the last two years but it would go beyond the scope of this paper to describe in particular the various pumping systems in detail. Therefore only a brief description of the fuelling system and some aspects of the torus pumping system will be given here. The fuelling / pumping system is designed to provide a range of fuelling rates based on the following considerations: replacement of the DT burnup requires 50 Pa-m^3/sec assuming 10% He concentration in the core plasma, a He enrichment in the divertor of 0.2 (experimental; B2-Eirene gives 0.15), and a He production rate of 2 Pa-m^3/sec for 1.5 GW of fusion power. The upper limit chosen for the fuelling rate is 200 Pa-m^3/s: a compromise between the need to generate some SOL flow for impurity entrainment, and the pumping / T processing capability of the ITER plant. Up to 500 Pa-m^3/s are foreseen for up to 50 s for density ramp up as well as divertor and burn control actions. Control of the DT mix and of the divertor target power load requires injection of up to three hydrogenic gases (D, T, DT) and of up to three impurity gases (e.g. Ne, Ar, He) at the same time. To mitigate local erosion of the first wall (0.2 mm Be after 10000 discharges) from CX sputtering, five evenly spaced injection locations are provided around the torus at the top of the machine and in the divertor. An injection time response of ~ 0.5 sec, required for divertor and burn control, allows pipes 25 m long (10 mm diameter) from the injection valve to the plasma chamber.

The gas fuelling system consists therefore of 5 valve containers (6 injection valves each) evenly distributed around the torus inside the upper Heat Transfer vault and of 5 valve containers in the divertor level pit. The valves inside each container are arranged for independent injection of 6 fuelling gases via a common 10 mm dia; discharge manifold (~20 m long) connected to the plasma chamber and the divertor volume, respectively. Two safety isolation valves controlled by the safety system can isolate the fuelling system from the torus in case of a pressure excursion, and can also stop fuelling of the plasma instantly to shut down the fusion power within ~ 10 sec. Two additional puffing valves also controlled by the safety system can bypass the flow controllers of two of the impurity injection lines to provide strong impurity puffing for an even faster fusion power shutdown (FPSS).

In addition to gas injection, a pellet fuelling system is provided to inject up to 100 Pa-m^3/s of hydrogenic species (3 to 8 mm pellets with 1.5 km/sec) from the high field side. This pellet injection system consists of: (a) two centrifuge pellet injector drivers, (b) three extruders per injector, (c) a feed manifold for the gas supply (6 gas lines and a pumping/ flushing line), (d) a continuous regenerating cryopump per injector, (e) a flight tube routed through a divertor port below the cassettes behind the inner blanket penetrating the backplate between two modules, (f) a diagnostic, control & data acquisition system. The two pellet injector assemblies are installed inside a transport cask (6 m long x 4 m high x 3 m wide) located at the divertor level.

The primary pumping system for ITER comprises two major subsystems: the torus roughing system and the high vacuum system (cryo-pumps). The former must perform two primary functions: to evacuate the torus to < 50 Pa from atmosphere in less than 60 hours with the torus back-filled with air, helium or nitrogen, and to regenerate the primary cryopumps to < 5% of their initial inventory in < 75 s during plasma burn. The latter requirement is the design driver for the roughing system, (two mechanical pump sets, one for redundancy) located in the vacuum pump room [27].

The high vacuum system consists of 16 batch regenerating cryosorption pumps designed to provide the required ~ 200 m^3/sec pumping speed in the divertor private region. The minimum gas throughput (D, T and He) of 50 Pa-m^3/s can thus be provided at ~ 0.3 Pa pressure in the divertor private region whereas the maximum throughput (200 Pa-m^3/s) requires 0.8 Pa. These or even higher pressures are readily observed during detached or semi-detached divertor operation in present day machines and when modelling the ITER divertor.

The cryo-pumps are independently controlled to allow individual pumps to be regenerated, shut down in the event of failure, or to regulate the pumping speed. They consist of a cylindrical SS housing 1500 mm OD x 1500 mm long into which charcoal coated cryopanels (8 m^2 / pump) and an 80 K radiation shield are mounted. A pneumatically driven inlet valve is used to control the pumping speed and to isolate the pump from the torus during regeneration. Each cryopump is mounted on the inside of the vacuum vessel port closure flange within the divertor port duct.

During the plasma discharge 12 pumps are in pumping mode and 4 pumps are in various states of regeneration, namely rapid warm-up to ~80K, gas desorption and evacuation, and rapid cool down. Thus each cryopump will pump for 900 s followed by a 300 s regeneration cycle. The tritium content in the 16 cryopumps reaches a maximum steady state value of 170 g after 900 sec. The total 4.5 K refrigeration required to operate the cryopumps is an estimated 10 kW and depends on the flow rate required for rapid cool down [27].

REMOTE MAINTENANCE APPROACH FOR IN-VESSEL COMPONENTS

It would go beyond the scope of this paper to describe the Remote Maintenance approach for ITER in detail (see [28]). However, a brief description of the general concepts for maintaining the divertor, the blanket and the equipment inside the equatorial and the vertical ports will be given here. Generally all maintenance operations are performed by docking so called transfer casks to divertor or equatorial ports (Fig. 5) which then insert the various maintenance tools into the machine. These casks also transport components and tools to and from the Hot Cell [29]. A slight deviation of this approach is the Blanket maintenance where a mixture of cask operated tools and permanently installed RH tools are used. The maintenance approach for the vertical ports is not as far developed as the one for the other areas in the machine but a concept using crane operated vertical casks is presently envisaged.

While it is planned to store the casks during plasma operation in the basement of the laydown hall, the RH tools are stored and maintained in the lower level of the Hot Cell building. During RH operations the tools will be picked up as needed by docking casks to this area and inserting or removing tools from the casks. As already mentioned above, all spare or damaged in-vessel components of the divertor, the blanket, the heating systems and diagnostics will be also stored in the Hot Cell building and can be accessed by cask dockings.

Divertor:

The cassette solution is fundamental to the remote maintenance concept [30], which is based on installation of a small number of integrated components inside the vessel to provide short RH times and high reliability. Connections to the cooling system and the locking mechanism are in an area of low neutron flux to permit frequent rewelding and bolting. The cassettes are locked on toroidal rails from the side (the last one from the port), avoiding access holes through the HHFC's.

Remote maintenance ([28]) is performed in parallel through four dedicated RH ports (one per quadrant) serving 15 cassettes each. A Cassette Toroidal Mover (CTM) handles the cassettes between their in-vessel location and the central position in front of the

RH port. Not larger than a cassette, it is introduced in the vessel once the central cassette (in front of the RH port) and the second cassette (next to the central one) have been removed. It uses a dextrous manipulator arm which locks (bolting tool) the supports and mounts the earth-straps. The CTM is guided by the toroidal rails and driven by two rack-and-pinion assemblies. It handles the cassette using three grips and two wheeled lifting forks inserted between toroidal rail grooves and the cassette bottom (Fig 8).

A Radial Tractor (RT), driven by a rack-and-pinion and combined with a set of rolling carriers, moves the cassettes and the CTM along the RH port radial rail, between the vessel and a transport cask which is docked to the cryostat flange to confine T and radioactive dust [29]. The RT includes a manipulator arm for RH operations inside the port (pipe handling, connections) and is supplied from the cask (Fig. 8). This cask is also used to transport cassettes and tools between the Hot Cell and the machine.

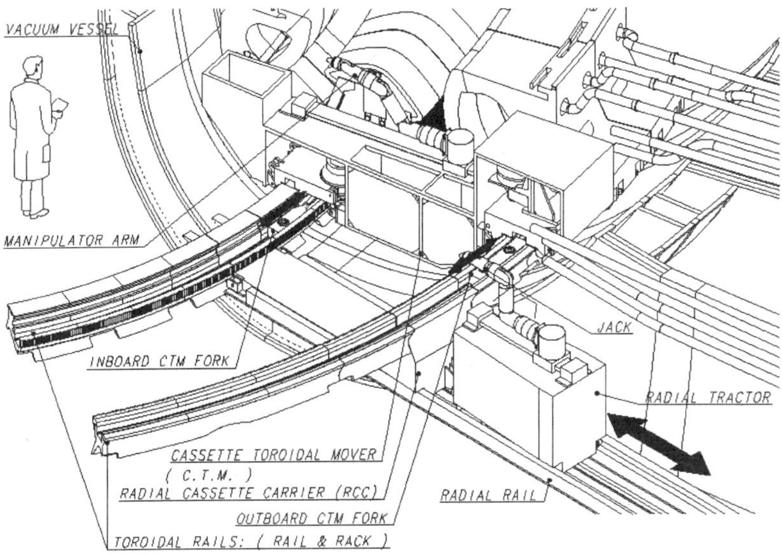

Figure 8: Divertor Remote maintenance showing TCM and radial tractor

Cutting and welding of the cassette water feeds (6 pipes per port connecting three cassettes) is performed in a parallel process . A set of hands-on assisted RH pipe tools cuts, welds and inspects the pipe end caps outside the bio-shield (orbital tools) and the joints located inside the vessel (bore tools) between the cassette water-feeds and the fixed straight pipes inside the 20 ducts (Fig. 6). Due to radioactive corrosion products inside the pipework, the tools are confined in glove boxes or casks.

A preliminary assessment of replacement times for the whole divertor and for a single faulty cassette yielded less than 6 months and less than 3 month, respectively.

Beside the design effort an R&D programme led by the EU HT [31] is underway aimed on building a full size 70° mockup of the ITER divertor area including 4 ports. Full size cassette and mover mockups will allow assessment of the divertor RH concept. First operation of this RH platform is expected in early 1998. In addition to the in-vessel RH a so-called refurbishment platform will test the Hot Cell operations required to repair damaged divertor cassettes. In the post-EDA phase the scope of this refurbishment platform will be extended and will also include diagnostic cassette refurbishment trials.

Blanket:

While the backplate constitutes a semipermanent structure (RH class 3) the blanket modules have to be exchangeable in a reasonable time frame for repair and also for installing a breeding blanket for the Extended Performance Phase (EPP). The modular concept of the blanket together with an in-vessel transporter are fundamental for the ITER blanket maintenance concept. The in-vessel transporter when installed consists of a toroidally continuous rail located at the centre of the plasma chamber supported by four beams extending from the four dedicated RH ports (Fig. 9). The rail itself consists of 2 times 8 segments connected by hinges and a locking clamp. On this rail up to four so-called vehicles can move toroidally. Each of these vehicles has an extendable arm with 6 degrees of freedom and is therefore able to reach every position at the FW. It has a net lifting capability of 4.3 t and thus limits the weight of the blanket modules. The end effectors on these arms are capable of performing all operations for module attachment to the backplate (see above).

The two halves of the rail are stored during plasma operation in the form of two storage rings inside dedicated RH cells in the pit in front of two of the RH ports. The rail is installed by first inserting two support beams from casks at the RH ports located 90° from the ones where the rail is stored. Then the support arms in the handling cells on which a vehicle and the first segment of the rail is fixed are extended into the vessel pulling the rail from the circular storage ring. Once the support beams are extended the vehicle pulls the rail further into the vessel directing it into the toroidal direction. At each hinge a special bolting tool mounted on the vehicle closes the locking clamp thus producing a stiff rail. Once both rail halves have been fully installed they are connected to form the above described rigid toroidal structure. The two vehicles are then released from the support arms and can now move along the rail. Two more vehicles as well as all the blanket modules are introduced / removed through the RH ports located 90^0 from the RH cells.

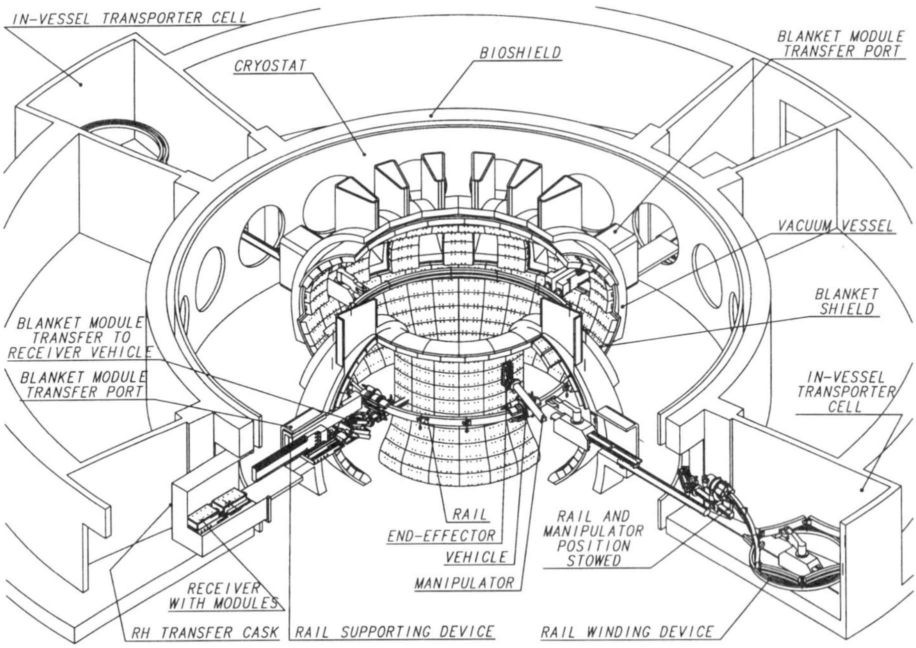

Figure 9: In-vessel transporter and RH cells

In addition to its function for blanket maintenance the in-vessel transporter will be used for several ancillary operations such as leak checking and maintenance of the in-vessel diagnostic sockets mounted on the backplate [2].

The JA HT leads an R&D task [32] to manufacture a full size 180° rail system and a full size 20° blanket segment. The task is well advanced and test results from this full size system can be expected in 1998. Previous mockups with reduced size have already shown satisfactory operation and thus support the design choices made for the blanket, its attachment, and the RH scheme. An additional subtask within this effort is the manufacture and test of a transport cask and its docking to the cryostat flange. This test also includes a double door mechanism which is a general feature of all the casks envisaged for ITER to avoid contaminated external surfaces after undocking the casks.

Equatorial ports:

Nearly all the equatorial ports including the RH ports are equipped with standard plug assemblies (Fig. 10) and are maintained by standardised maintenance tools. An exception is the NBI ports and port 7 which is obstructed by the NBI where different RH schemes are used.

The integrated plug assembly (weight up to 90 t) consists of a blanket section of the size of the port (back plate opening 1.6 x 2.6 m x 0.5 m deep, port opening 1.8 x 2.8 x 4.0 m) mounted on a VV shielding block which also includes the closure plate. The whole assembly is supported by rollers on the floor of the port and is mechanically fixed onto the VV flange. The vacuum seal is provided by a welded lip seal. These integrated assemblies contain the front ends of heating systems or diagnostics or constitute the port limiter assembly (in 2 to 3 RH ports). On the bottom of the port below the VV flange 4 blanket cooling pipes exit the vessel and are routed in the port interspace through a platform mounted on the cryostat which constitutes shielding (N_{16}) and provides support for RH tools and for shielding blocks in case this is required.

Maintenance of such a port is performed by removing the bio-shield plug with hands-on assistance, inserting tools to cut and remove the VV plug cooling lines, and remove the cryostat closure plate. Then a cask is docked to the port remove the shielding assembly in the port interspace (rolling it onto a platform extended from the cask) where it exists. In a subsequent cask docking operation a RH platform supported by the above mentioned floor is inserted into the port interspace and a vehicle moves forward to cut the lip seal and to remove the mechanical locking on the VV plug assembly (Fig. 10). The same vehicle then grips the VV plug and pulls it ~ 1.5 m forward then a floor segment from the inserted platform is lifted to support the plug behind the VV flange, allowing to pull the assembly fully onto the platform with subsequent floor segments lifted for additional support. The platform is then retracted into the cask and the VV plug assembly is subsequently transported to the Hot Cell where detailed maintenance and testing will be performed. The concept of handling inside the machine (only a few big assemblies), and the standardisation, permit fast maintenance times to be achieved, so that the downtime of the machine can be minimised if spare VV plug assemblies are available.

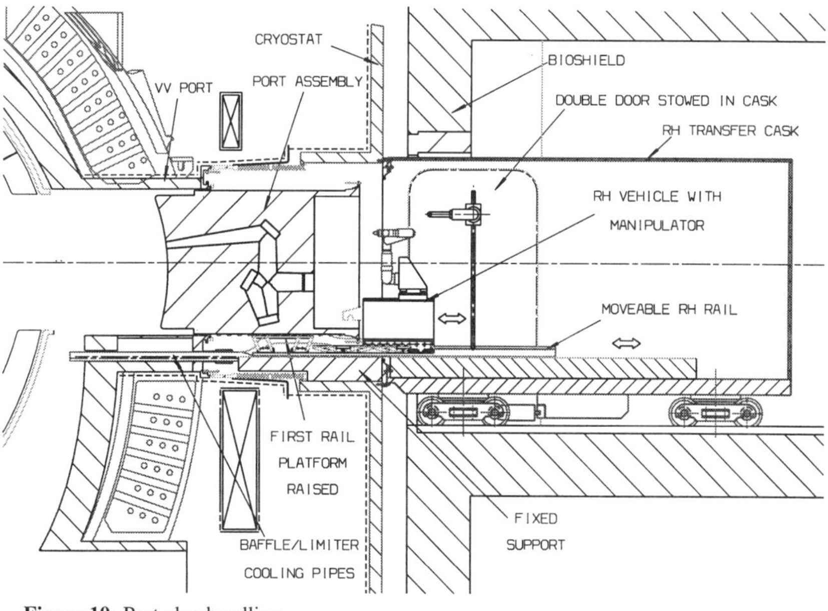

Figure 10: Port plug handling

The maintenance of the port 7 area is very similar, but the NBI results in more limited space and mandates the use of curved rails. Therefore, an integrated plug assembly is not possible in this environment [33] and more docking operations are required. The NBI maintenance will be performed by special casks docking at the back of the beam line aiming to maintain mainly the filaments of the ion sources.

Vertical Ports:

As already mentioned there are two types of vertical ports namely 10 ports dedicated to blanket cooling pipes and 10 diagnostic ports. While for the cooling pipe ports no frequent maintenance is foreseen (RH class 3), regular maintenance is expected on the 10 diagnostic ports. These 10 ports provide space for diagnostic access but have to incorporate also the in-vessel inspection (viewing) system consisting of insertable probes and the also insertable glow discharge electrodes. Access to the plasma is provided through a 150 mm wide slot over the whole length of the port. On the blanket either special modules with side wall cooling locking into keys on a frame like extension of the backplate or an integrated blanket plug inserted through the port will be used (decision pending). Behind this slot in the blanket custom made VV plug assemblies will provide shielding and will include the VV closure plate and its vacuum seal. This shielding plug incorporates a number of cylindrical holes, one for the inspection system (~ 150 mm wide) at the radially innermost area of the port and either four 300 mm holes or two 300 mm holes plus one 500 mm hole. A re-entry structure of the size of the vertical port protrudes from the cryostat lid to within 150 mm of the VV flange. Bellows connecting the holes mentioned above with this structure provide vacuum tight continuity thus avoiding a vent of the cryostat for maintenance of the diagnostic subassemblies. A closure flange at the level of the cryostat lid provides secondary confinement.

The in-vessel inspection systems are located in 9 ports. They require an opening to the plasma of ~ 150 mm diameter and a space above the VV flange inside the re-entry structure of ~ 500 mm diameter. These systems are maintained by closing valves located at the bottom of the re-entry structure and then using a special cask and tools to open the vacuum seals and mechanical attachments and finally lifting the whole assembly into the cask. This cask is then moved into a special area where hands on maintenance in protective suits is possible.

The rest of the port space is occupied by diagnostics and by glow discharge electrodes which are inserted as cylindrical assemblies into the holes mentioned above and then sealed to the flanges on the bottom of the re-entry structure. Maintenance of these components also makes use of a crane operated cask holding all required RH tools. However, these components are transported into the Hot Cell for refurbishment.

Maintenance or modification of the RH class 3 components (VV plug assembly, blanket plug) needs venting of the cryostat and will therefore be performed only a few times in the life of ITER.

R&D on the remote maintenance of the above described port assemblies (vertical and equatorial ports) will be performed in the post EDA phase.

GENERAL APPROACH FOR DIAGNOSTIC INTEGRATION

Some aspects of diagnostics integration have already been mentioned throughout the paper when describing machine components or the maintenance approach. The detailed engineering aspects of the diagnostic integration are the subject of a separate paper in this conference [2] and will be therefore not be addressed here. In this section a short description of the general approach and the present distribution of diagnostics on the machine will be given.

Diagnostic integration in ITER takes place in four generic areas, namely the blanket and VV mounted diagnostics (in-vessel diagnostics), the divertor diagnostics, the equatorial port diagnostics and the diagnostics at the vertical ports.

In-Vessel Diagnostics:

The majority of the in-vessel diagnostics will be mounted on the plasma facing side of the backplate while only a few flux and voltage loops as well as a few magnetic pick-up coils in the divertor region will be mounted on the VV, serving as a back-up for the much more extensive magnetic diagnostics on the backplate.

There are two generally different types of in-vessel diagnostics which pose different kinds of problems for diagnostic integration. They can be divided into localised diagnostics (coils, bolometers...) and into diagnostics consisting of flux loops or toroidal voltage loops [34]. The problem for installation of toroidal voltage loops is that due to the assembly procedure of the machine, toroidally unimpeded access to the VV walls and also to the blanket backplate is not possible (installation of backplate and pre-installation of blanket modules). In addition, provisions have had to be found for reconnecting toroidal loops after the possible replacement of a TF coil which requires the removal of a complete VV - blanket sector. The solution found is that on every sector joint a connection bridge is foreseen. It is spot welded into special sockets on each side and a spare connection is provided for a possible repair action [2]. This approach allows flexibility and reliable installation of voltage loops for both assembly and repair. Another special case are waveguides for high field side reflectometer systems which are routed through the vertical ports and then poloidally downward on the plasma facing side of the backplate similar to the cable looms decribed below.

All the other types of diagnostics (magnetic pick up coils, bolometers, SX detectors, and microfission chambers) are mounted in standardised cylindrical inserts in the backplate which are very similar to those for the flexible attachments of the modules. These inserts are located in the middle of a poloidal gap between blanket modules allowing bolometers and SX diagnostics to view the plasma. Magnetic pickup coils are hidden behind one module while a removable plug connecting the diagnostic in the insert to a cable loom is protected by the adjacent module. Each of these inserts can be exchanged against a new one if the two blanket modules on top of it are removed [35].

A row of these inserts is integrated into each blanket sector served by a cable loom routed poloidally on the plasma facing side of the backplate behind modules. The cable looms penetrate the backplate below the equatorial ports and are then routed along the blanket supports to the cryo-pump ports. All cabling for electrically connected in-vessel sensors exits the machine at the divertor level through the cryo-pump ports [2] and is terminated in cubicles situated in the pit at the divertor level.

Divertor Diagnostics:

These are described in [36]. Diagnostic access into the divertor is provided at the 4 RH ports. The so-called central cassettes in front of these ports are modified to incorporate optical and microwave diagnostics. While the microwave diagnostic is integrated into cooled plates attached on the side of the cassette body, the optical diagnostics use a mirror rack located in the space between the cassette and the VV (200 mm high) viewing the divertor targets through slots in the cassette body and liner [2]. A ~ 20 mm wide slot between the central cassette and its two neighbours is used as a viewing slot for e.g. the X-point LIDAR system. Routing of the microwaves as well as of the optical signals between the cassettes and the VV closure plate is performed by means of four dedicated diagnostic blocks which are handled by the same tool as the central cassette. They are used as platforms for waveguides as well as for optical assemblies (VUV spectrometers, LIDAR system, relay optics...). All connections to these blocks are designed for fast removal in order to minimise the time penalty for the divertor maintenance .

The standard cassettes on each side of these central cassettes are modified to incorporate electrically connected sensors such as magnetic coils, bolometers, pressure gauges and Langmuir probes. All of these sensors are incorporated in plates similar to those of the microwave diagnostic, mounted on the two sides of the cassette body. The cables of these diagnostics end in a plug which automatically connects to its counterpart on the VV during cassette installation. Cables from these diagnostics are also routed through the cryo-pump ports.

Diagnostics in Equatorial Ports:

Up to seven equatorial ports each 1.6 x 2.6 m wide are fully or partially available for diagnostic integration and thus provide the main access to the ITER plasma (Fig. 11) [2]. As already mentioned in the RH section all equipment in these ports has to be incorporated in standard integrated VV plug assemblies (Fig 10). Due to the size of these ports and the numerous diagnostic demands several diagnostic systems have to share a port. The diagnostic components will be assembled into such a VV plug inside the Hot Cell using a multi-purpose repair street. Included in this street will be a testing area where all functions of these diagnostic components have to be checked. Only after a detailed test has been passed will such a plug be installed on the machine.

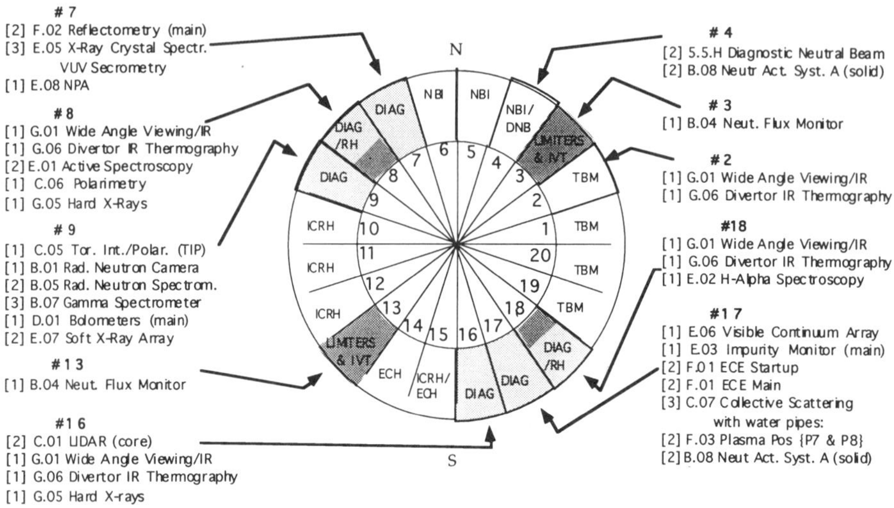

Figure 11: Diagnostic access at the equatorial level. diagnostics labeld with [1] belong to the control set, the ones with [2] belong to the advanced control set and ones labeld with [3] belong to the set with physics evaluation

As can be seen in Fig. 11 four ports are dedicated to diagnostics, of which one is somewhat obstructed by the NBI. Due to the fact that this port can be included in the NBI vault it is used for vacuum extended diagnostics [33]. In addition to the dedicated diagnostic ports the two RH ports without a handling cell can be used for diagnostic equipment which allows fast removal (e.g. optically coupled diagnostics). The number of RH ports which can be shared with diagnostics will depend on the number of port limiters needed and will be decided at a later stage. An additional opportunity for diagnostic integration is to use or share one of the Test Blanket Module (TBM) ports at least for a limited time. Small diagnostics like neutron flux monitors which do not need a direct view of the plasma could be mounted in the ports where the limiters are situated. By sharing two RH ports and one TBM port all diagnostic systems envisaged so far can be integrated into the machine, however, a final decision on the number of ports available for sharing can only be made in the medium term future when the boundary conditions will be better known.

Diagnostics in Vertical Ports:

The general approach of diagnostic integration in the vertical ports has already been briefly explained in the RH section. Therefore the point remaining to be made here is that regardless of spatial limitations in the 10 vertical diagnostic ports due to the inspection system and the glow discharge electrodes, all diagnostics envisaged have been accommodated so far. While 9 of the ports are configured as described in the RH section one is specially modified to incorporate the vertical neutron camera [37] and thus neither houses an inspection system nor a glow discharge electrode. Vacuum extended diagnostics (X-ray, VUV- spectrometers) are foreseen in two ports. The boundary condition here is that all the diagnostics have to be placed below the cryostat lid for reasons of secondary contfinement.

SUMMARY

The physics basis as well as the machine design have evolved considerably since the last workshop but as already pointed out above the major machine parameters as well as the performance predictions for ITER have not changed. Due to the excellent collaboration of the fusion physics community and its coordination through the ITER Expert Groups, it has been possible to establish a physics basis for a reactor class machine such as ITER: While some uncertainties remain which ultimately will have to be resolved by operating ITER the present physics knowledge allows one to be confident that ITER will reach its performance goals (1.5 GW fusion power for 1000 sec). In addition, evolution in divertor physics has made a divertor design with acceptable lifetime and He removal rates possible. A disruption database based on improved input from many divertor tokamaks is instrumental for defining the design requirements in particular for the in-vessel components (blanket and divertor).

During the last two years the design of all the major machine components was optimised for performance, assembly schedule, maintenance, safety and if possible cost reductions. An R&D program focusing on 7 Large Projects which aim to manufacture and test full size or representative sized mockups of the major machine components and the RH tools is well under way. Preliminary results from this programme have influenced design solutions and further optimisation is expected on the basis of results to come in 1998. However, it is not expected that these will result in major changes of design solutions but rather in further refinements taking into account manufacturing processes and their cost.

Due to the well-advanced machine design diagnostic integration has progressed significantly, in particular during the last year. Already now it becomes clear that it is feasible to achieve an extensive diagnostic coverage also in a reactor class machine like ITER despite the many constraints imposed by radiation and safety issues. This diagnostic coverage together with its performance capability will make ITER the most interesting physics and technology tool for the world-wide fusion community once it is operational.

REFERENCES

[1] R. Parker, et.al, Proceedings of Diagnostic Workshop, Varenna, September 1995

[2] C. Walker, et.al. this conference

[3] J. G. Cordey, *Energy Confinement Scaling and the Extrapolation to ITER,* Plasma Physics and Controlled Fusion to appear (EPS meeting, 1997) (1997).

[4] J. W. Connor and J. B. Taylor, , Nuclear Fusion 17, 1047 (1977).

[5] M. Greenwald, J. Terry, S. Wolfe *et al.*, *A New Look at Density Limits in Tokamaks,* Nuclear Fusion 28, 2199-2207 (1988).

[6] A. Loarte and et al., *Experimental Edge Results and Multi-machine Comparison,* Contributions to Plasma Physics to appear (1998).

[7] J. Snipes et al, Proceedings of the EPS Conference on Plasma Physics and Controlled Fusion, Berchesgarten, June, 1997.

[8] F. Ryter and H-Mode Database Working Group, *H-mode Power Threshold Database for ITER,* Nuclear Fusion 36, 1217 (1996).

[9] M. Kaufmann, J. Schweinzer, M. Albrecht *et al.*, "Overview of ASDEX Upgrade Results," presented at the 16th IAEA Fusion Energy Conference, Montreal, Canada, 1996 IAEA-CN-64/O2-5.

[10] A. Loarte, et al, "Experimental Edge Results and Multi-machine Comparison", to appear in Contributions to Plasma Physics, 1998.

[11] G. Janeschitz, A. Hubbard, Yu. Igitkhanov *et al.*, "ITER operation space in terms of Te and ne at the plasma edge," presented at the European Physical Society, Berchesgarten, Germany, 1997.

[12] M. Mahdavi, R. Maingi, R. LaHaye *et al.*, "Recent H-Mode Density Limit Experiments on DIII-D," presented at the 1997 EPS Conference on Plasma Physics and Controlled Fusion, Berchesgarten, Germany, 1997.

[13] V. Mertens, A. Herrmann, A. Kallenbach *et al.*, "Edge and Divertor Physics in ASDEX Upgrade with emphasis on density limit characteristics," presented at the Sixteenth IAEA Fusion Energy Conference, Montreal, Canada, 1996 IAEA-CN-64/A4-4.

[14] R. Raman, F. Martin, B. Quirion *et al.*, *Experimental Demonstration of*

Nondisruptive, Central Fuelling of a Tokamak by Compact Toroid Injection, Physical Review Letters 73 (23), 3101-3104 (1994).
[15] S. Allen, D. Hill, T. Carlstrom *et al., Measurements of Electron Temperature and Density with Divertor Thomson Scattering in Radiative Divertor Discharges on DIII-D,* Journal of Nuclear Materials to appear (1996).
[16] A. Kukushkin, H. D. Pacher, V. Abramov, *et al.*, "Analysis of The Performance of The ITER Divertor And Analysis of The ITER Tokamak Edge Parameter Database," presented at the Sixteenth IAEA Fusion Energy conference, Montreal, Canada, 7-11 October, 1996 IAEA-CN-64/FP-27.
[17] A. Kukushkin, H. D. Pacher *et al.*, 24th EPS Conference on Controlled Fusion and Plasma Physics, Berchtesgaden, Germany, 9–13 June, 1997. Paper P3.013
[18] R.S. Granetz, , Nuclear Fusion 36, 545 (1996).
[19] K.Ioki et al, "ITER Vessel and Blanket", presented at the Sixteenth IAEA Fusion Energy conference, Montreal, Canada, 7-11 October, 1996 IAEA-CN-64/FP-7.
[20] Thome, R., Barabaschi et al, "Recent Progress on ITER Magnet Systems, Proc. SOFE 1997
[21] M. Huguet., "ITER Magnets and Plasma Control", presented at the Sixteenth IAEA Fusion Energy conference, Montreal, Canada, 7-11 October, 1996 IAEA-CN-64/FP-11
[22] C. Sporchia et al, "Mechanical Structrues for the ITER Magnet Systems", Proc. 19th SOFT, Lisboa, 1996
[23] V. Khripunov and R. T. Santoro, Radionuclide production in the ITER coolant water, ITER Report, Garching Joint Work Site, G 30 RI 1 96-07-10, 1996.
[24] M. Roccella, Private communication.
[25] Janeschitz, G. et al, , Proc. 4th ISFNT, Tokyo, 1997.
[26] J. Schlosser, J. Boscary, F. Escourbiac, et. al., Comparison between various thermal hydraulic tube concepts for the ITER divertor, 19th Symp. on Fusion Tech., Lisbon (1996)
[27] Janeschitz, G. et al, "ITER Divertor, Pumping, and Fuelling System Designs, presented at the Sixteenth IAEA Fusion Energy conference, Montreal, Canada, 7-11 October, 1996 IAEA-CN-64/FP-2
[28] T. Burgess et al, Remote Maintenance of In-Vessel Components for ITER, Proc. Fourth International Symposium on Fusion Nuclear Technology, Tokyo, Japan, April 1997, to appear.
[29] R. Haange et al, Radiation and Safety Aspects of Removal of Activated Components from the ITER Vacuum Vessel, Proc. Fourth International Symposium on Fusion Nuclear Technology, Tokyo, Japan, April 1997, to appear.
[30] E. Martin et al, Design of the ITER Divertor Remote Handling System, Proc. 19th Symposium of Fusion Technology, Lisbon, Portugal, Oct. 1996, to appear.
[31] D. Maisonnier et al, Status of Development of Remote Maintenance of ITER Divertor Cassettes, Proc. 16th IAEA Fusion Energy Conference, Montreal, Canada, Oct. 1996, to appear.
[32] E. Tada et al, Development of Remote Maintenance Equipment for ITER Blankets, Proc. Fourth International Symposium on Fusion Nuclear Technology, Tokyo, Japan, April 1997, to appear
[33] P. Edmonds, et al., this conference
[34] B. de Kock, et al, this conference
[35] P. Barabaschi, Private communication.
[36] G. Janeschitz, B. de Kock et al, this conference
[37] L. Johnson et al, this conference

Acknowledgement
This report is an account of work undertaken within the framework of the ITER EDA Agreement. The views and opinions expressed herein do not necessarily reflect those of the Parties to the ITER Agreement, the IAEA or any agency thereof. Dissemination of the information in this paper is governed by the applicabable terms of the ITER EDA Agreement.

ROLE AND REQUIREMENTS FOR PLASMA MEASUREMENTS ON ITER

V.S. Mukhovatov,[1] R. Bartiromo,[2] D. Boucher,[1] A.E. Costley,[1] L. de Kock,[3] K. Ebisawa,[1] P. Edmonds,[3] Yu. Gribov,[4] G. Janeschitz,[3] L. C. Johnson,[1] S. Kasai,[5] E. Marmar,[6] A. Nagashima,[5] M. Petrov,[7] D.E. Post,[1] P.E. Stott,[8] V.S. Strelkov,[9] G. Vayakis,[1] C.I. Walker,[3] J. Wesley,[1] S. Yamamoto,[3] K.M. Young,[10] V.S. Zaveriaev,[9] and the ITER Joint Central Team and Home Teams

[1]ITER JCT, San Diego JWS, 11025 North Torrey Pines Rd, La Jolla, CA 92037, USA
[2]Instituto Gas Ionizzati del CNR, Padova, Italy
[3]ITER Joint Central Team, Garching JWS, Germany
[4]ITER Joint Central Team, Naka JWS, Japan
[5]JAERI, Naka, Ibaraki-ken, Japan
[6]MIT Plasma Fusion Center, Cambridge, USA
[7]Ioffe Institute, St. Petersburg, Russia
[8]JET Joint Undertaking, Abingdon, Oxfordshire, UK
[9]NFI, RRC Kurchatov Institute, Moscow, Russia
[10]Princeton Plasma Physics Laboratory, Princeton, USA

INTRODUCTION

Measurement of plasma and key first wall parameters will have three main roles on ITER. Some of the measurements will be used in real time to prevent the on-set of conditions which could potentially damage the first wall and other in-vessel components (machine protection); others will be used in real-time feedback control loops to control the value of key parameters at values required for specific plasma performance (plasma control); while others will be used to evaluate the plasma performance and to provide information on key phenomena which may limit ITER performance (physics studies). The measurements of some parameters may contribute to all three roles although the requirements on the measurements (accuracies, resolutions etc.) may be different depending on the role.

At the last diagnostic workshop in Varenna, the first work on the role and requirements for plasma measurements on ITER was presented.[1] Since then both topics have developed further and the demands for plasma measurements are now more clear. The current requirements are presented in this paper.

MEASUREMENTS FOR MACHINE PROTECTION

The stored thermal and magnetic energies in the ITER plasma are high (~1 GJ each) as is the power flow through the separatrix (~200-300 MW). Abnormal and uncontrolled behavior of the plasma can cause severe damage of ITER in-vessel components. Therefore, timely

Diagnostics for Experimental Thermonuclear Fusion Reactors 2
Edited by Stott *et al.*, Plenum Press, New York, 1998

information about such an events must be available to activate protective techniques. The major parameters that should be measured for machine protection are shown in Table 1 with approximate limits that must be avoided for safe machine operation.

Table 1. Parameters to be measured for machine protection

Parameter	Limit
Separatrix/wall gaps	>5 cm
First wall surface temperature	≤400°C (hot spots)
Divertor plate surface temperature	≤800°C (hot spots)
Fusion power	≤1.8 GW (20% above nominal value)
Locked modes	Avoidance
Runaway electron current	≤0.1 I_p
Type I (Giant) ELMs (at full parameters)	Avoidance
Specific impurity emission	To be defined
Line-averaged density (NBI shine-through)	≥0.3x10^{20} m^{-3} for D^o beams

The precision of the separatrix positioning relative to the first wall is critical for keeping first wall power load below the allowable limit (0.5 MW/m^2). Even a relatively short (≥0.1-1 s) contact of the plasma separatrix with the first wall can result in localized surface melting (~1 mm deep for Be wall). Similarly, an uncontrolled rise of the fusion power and/or a full divertor attachment can increase the thermal load onto the divertor plates above the tolerable level (5 MW/m^2 at steady state).

Plasma disruptions can deposit up to 100 GJ/m^2 of energy on the divertor plates during a thermal quench (1-10 ms) with consequential significant plate erosion.[2] Also, high mechanical stresses can arise in the blanket/shield structure and divertor cassettes due to high poloidal 'halo' currents through these components during the vertical displacement event (VDE). In addition, the formation of runaway electrons during the disruption increases the potential for damage to the plasma facing components. Since the disruptions will probably be a major factor defining the life-time of the divertor plates, the number of disruptions should be kept small and/or their effects mitigated. Experiments show that locked (non-rotating) modes, especially those induced by the resonant m/n=2/1 error field component, often lead to disruptions. An early and reliable detection of locked modes is a prerequisite for their unlocking (e.g., through a plasma rotation induced by neutral beam injection) or a controlled plasma termination. Frequently, a clear disruption precursor is absent or observed too late to be useful for disruption mitigation. An artificial neural network, combining signals from a large number of plasma diagnostics, is a promising approach for predicting the disruption boundary in a real time, as was shown recently for high-β disruptions in DIII-D.[3]

The edge localized mode (ELM) is beneficial for impurity and helium expulsion from the plasma edge but Type I ('Giant') ELMs can lead to a substantial pulse heat load (i.e. ~25 MJ for ~0.1-1 ms every 1 s) to the divertor plates. This heat load is very close to, or above, the ablation threshold and may increase significantly the erosion rate of the divertor plates. The measurements of ELMs is therefore included as a machine protection parameter.

A sudden and strong increase in the emission of a particular impurity, e.g., Be, can be a clear indication of increased plasma-wall interaction. Measurement of specific impurity emission is therefore included.

The 50MW/1MeV neutral beam injection in the absence of plasma will produce a localized heat load to the first wall up to 30 MW/m^2, i.e., 60 times higher than design value. A relatively high density plasma, i.e., $\bar{n}_e$≥0.3 10^{20} m^{-3} for deuterium beams and $\bar{n}_e$≥0.4 10^{20} m^{-3} for hydrogen beams, must be maintained during the neutral beam injection to reduce beam shine through to acceptable level. Hence, reliable measurement of the line-averaged electron density is required for machine protection.

MEASUREMENTS FOR PLASMA CONTROL

Many of the measurement requirements for plasma control in ITER are identical to those successfully employed in the present generation of divertor tokamaks. However, operation with a burning DT plasma results in additional plasma control requirements which are inherent in a reactor-grade plasma. In particular, a simultaneous control of the fusion power, divertor heat load and helium-ash density represents a new challenge that can not be fully studied in present experiments.

Plasma control assumes coupling of a measurement of the parameter to be controlled with an actuator that can modify the measured parameter in a required way. The concept of plasma control adopted for ITER is represented schematically by a matrix shown in Fig. 1. The vertical structure of the matrix reflects the division of the plasma control system into four hierarchical layers, i.e., operation scenario, control of magnetic configuration, kinetic control and fast shutdown.

Operation Scenario

The operation scenario is pre-set and defines a sequence in which the plasma progresses from initiation in a circular configuration near the outboard limiters through current ramp up and plasma channel expansion and elongation, divertor formation, current flat top during auxiliary heating and plasma burn to current ramp down and termination. It provides the operational framework within which three other control levels effect their actions.

Control of Magnetic Configuration

This encompasses the dynamic control of plasma current, and plasma position and shape, including the control of selected separatrix-wall clearance gaps, and the divertor configuration. The essential requirement is to maintain the magnetic configuration suitably positioned ideally independent of the changes in kinetic plasma parameters. This control must provide feedback stabilization of vertical displacement of the plasma and keep the plasma-wall clearance gaps within acceptable tolerance during the 'normal' operation and at typical plasma disturbances, i.e., ELMs, sawteeth and minor disruptions. It leads to requirements for measurements of the plasma current, critical separatrix-to-first wall clearance gaps and position of divertor strikepoints.

Fig. 2 shows simulation results for a time behavior of six representative plasma-wall gaps in ITER after a minor disruption with relatively large drops in the internal inductance, Δl_i =-0.1, and the poloidal beta, $\Delta \beta_p$=-0.2.[4] Time behavior of the poloidal field coil power is also shown. One can see that the control system restores the plasma position within 10-15 s after initial large (10-12 cm) shift. The time scale for initial changes in the gap widths is ~1s which translates into a required time resolution for the gap measurements of (0.1-0.3) s.

Plasma Kinetic Control

Plasma kinetic control in ITER addresses the problem of controlling simultaneously the fusion power and semi-detached divertor conditions. The ITER divertor is designed for a steady-state peak power load on the divertor plate $\hat{P}_{div,ss} \leq 5MW/m^2$ with power transients up to $\hat{P}_{div,tr} \leq 20MW/m^2$ for $\leq 10\ s$. This means that under the normal conditions only a small fraction (≤12-16%) of the total heating power ($P_{tot} = P_\alpha + P_{aux}$= 300-400 MW where P_α is the alpha particle power and P_{aux} is the auxiliary heating power) can be removed through the divertor plates, and more than 80% of P_{tot} must be radiated from the main and divertor plasma. It requires an admixture of a controlled amount of a specific impurity or a combination of impurities (e.g., Ne, Ar) to the divertor and scrape-of-layer (SOL) plasma while maintaining the bremsstrahlung power loss and plasma dilution in the core at the level acceptable for the plasma burn. The kinetic control must also keep the plasma away from the β and density limits and provide a sufficient power flow through the separatrix to insure the H-mode of plasma operation. It is clear that a sophisticated multi-input, multi-actuator feedback control scheme is required for the successful operation of ITER in the basic ignited regime. This leads to requirements for a wide range of plasma measurements including fusion power, divertor plate temperature, radiative power loss from plasma core, SOL, X-point region and from divertor, plasma density, β, n_T/n_D ratio, locked and rotating MHD modes and a degree of divertor detachment, i.e., 'ionization front' position and/or T_e and n_e at the divertor plate. For advanced scenarios, e.g., that with the negative central shear, a careful current profile tailoring and control is required along with some level of pressure profile control in addition to the basic kinetic control. For this, measurements of the current density profile (q profile) and the electron and ion density and temperature profiles are required.

Fig. 3 shows the results of a simulation of simultaneous control of fusion power and heat load on the divertor plates using the PRETOR transport code with simplified SOL and divertor model.[4] In the simulations, three actuators are used, i.e., argon injection to limit the divertor target power to an allowable value (50 MW); DT injection to keep fusion power at preset level (1.5 GW); and auxiliary heating power - in order of decreasing priority - to (i) maintain the plasma in H-mode, (ii) maintain the plasma density below a maximum value and (iii) assist the control of fusion power. To test a control scheme, auxiliary power P_{RF}=100 MW is intentionally added for 50 s during the sustained burn. As the figure shows, a satisfactory control is obtained. The high-frequency oscillations of alpha-particle power (P_α) and divertor target load (P_{target}) are likely simulation artefacts. Such simulations will help in determining the optimum accuracies and resolutions for kinetic control. Further work on this topic is in progress.

Fast Plasma Shutdown

Fast plasma shutdowns are necessary to limit damage to the first wall and divertor plates which could arise from a loss of position control or positive fusion power excursions in situations where the burn control is inadequate. These events will usually be followed by VDEs and disruptions. Fast plasma termination may be desirable also in the case of a locked mode development which usually leads to disruption. The proposed method[5] provides conditions for a radiative transfer of a major part of the thermal and magnetic plasma energy onto the first wall for ~300 ms. Theoretically, this can be achieved by a fast increase of plasma density (by up to two orders of magnitude) using a massive (~50 g) injection of hydrogen or deuterium. Injection of impurities ('killer' pellets) instead of hydrogen may provoke the generation of a dangerous amount of the relativistic runaway electrons carrying a large fraction of the plasma current. This leads to requirements of reliable measurements of the parameters listed in Table 1.

MEASUREMENTS FOR PLASMA PERFORMANCE EVALUATION AND PHYSICS STUDIES

The ITER plasma will be the first in which there is significant alpha heating so the experimental program will have an extensive physics component.[6] Key topics to be investigated include confinement physics; operational limits; high-current plasma disruptions; physics of high power radiative divertor; alpha-particle effects; and steady-state burn. An extensive set of plasma measurements is required to support these programs.

Confinement Physics in Reactor-Scale Plasma

One can expect a certain change in confinement physics in a burning DT plasma compared to the present-day experiments. In particular, a strong dependence of the heating power on the ion density and temperature, $P_\alpha \propto n_D n_T T^2$, will affect the temperature and pressure profiles which in turn can influence the stability and transport properties of a reactor plasma. Spontaneous formation and destruction of internal transport barriers may lead to thermal stability and control issues. Besides, the heating source in a reactor is coupled with internal injection of impurity (helium) that will affect the plasma performance due to the plasma dilution and increase in the bremsstrahlung power loss. A change in the confinement physics may be associated with a large step (a factor of ~5) in the non-dimensional machine size, $1/\rho^* = a/\rho_{ci}$ where a is minor plasma radius and ρ_{ci} is the ion gyroradius.[7] There is a chance that in a burning plasma, new modes of operation will be discovered which are more favorable for sustained ignition compared to the ELMy H-mode.

The principal parameters that must be measure for confinement studies include the ion and electron temperature and density profiles, radiation power profiles in the plasma core and at the edge, alpha particle source profile, and density profiles of major impurities, i.e., He, Be, C and W, and extrinsic impurities (Ne, Ar) injected deliberately for the divertor head load control. Measurement of parameters of edge pedestal, edge and internal transport barriers and plasma rotation profile are also be required.

PLASMA CONTROL MATRIX

● = Major direct effect

◎ = Appreciable secondary effect

○ = Possible secondary effect

blank = No appreciable effect or not applicable

	Control Action (Controllable Parameter or System)																				
		Scenario and Magnetics						Fueling and Exhaust							Auxiliary Heating and Current Drive (options)						Shutdown
	Measurable Quantity	TF field (static, 0-5.7 T)	PF currents (CS, 2-8)	PF voltages (CS, 2-8)	Prefill pressure (D_2)	Startup EC (ω, P, t, Δt)	Error field compensation current	DT fueling (gas, into SOL)	DT fueling (shallow pellet, ~1 km/s)	DT fueling (deep pellet, ~5 km/s)	Impurity fueling (Ne, Ar...to SOL)	DT divertor fueling (gas)	Impurity divertor fueling (Ne, Ar, ...)	Pumping speed (0-300 m^3/s)	NBI power (0-max)	ICH power (0-max, ω, mode)	ECH power (0-max, launch angle)	FWCD (power, radial location)	ECCD (power, radial location)	LHCD (power, $n_{\|\|}$)	Shutdown Pellet
1: Scenario 2: Magnetics	Plasma current, q_{edge}	●	●																		
	Plasma shape (R, a, κ, δ)		●																		
	Plasma shape (FW gaps)		●																		
	m=2 mode							◎	◎	◎	◎	○	◎					◎	◎	◎	
	Locked modes		◎				●								●						
3(a): Core Kinetics	Plasma density							●	●	●	●	●	○	◎	○						
	Fusion power							●	●	●	◎	●	◎	◎	◎	◎	◎				
	He density							●	●	●	◎	◎	○	●							
	Core D/T ratio							●	●	●		●									
	Zeff										●		◎	◎							
	Core radiation power							●	●	●	●		◎	●							
	Core plasma rotation (f_{rot})														●						
	β_N							●	●	●	◎	●	◎	◎	◎	◎	◎	◎	◎	○	
	q profile														●			●	●	●	
	Sawtooth period														◎			◎	◎	○	
3(b): Edge Kinetics	ELM period, magnitude							●	◎		●										
	n_{edge}							●	◎		◎			◎							
	Impurity exhaust							●	◎		◎			●							
	X-point and SOL radiation										●		◎								
3(c): Divertor	In-divertor radiation (x,y)										◎	◎	●	◎							
	Target plasma (n,T)							◎	◎	◎	●	●	●	◎	◎	◎	◎				
	Target temperature							◎	◎	◎	●	●	●	●	◎	◎	◎				
	Divertor neutral pressure							◎	◎	◎	●	●	●	●							
	Divertor He density							◎	◎	◎	●	◎	●	●							
4: Shutdown	P_{fus} and I_p																				●

Figure 1. ITER plasma control matrix for measurable parameters and control actions.

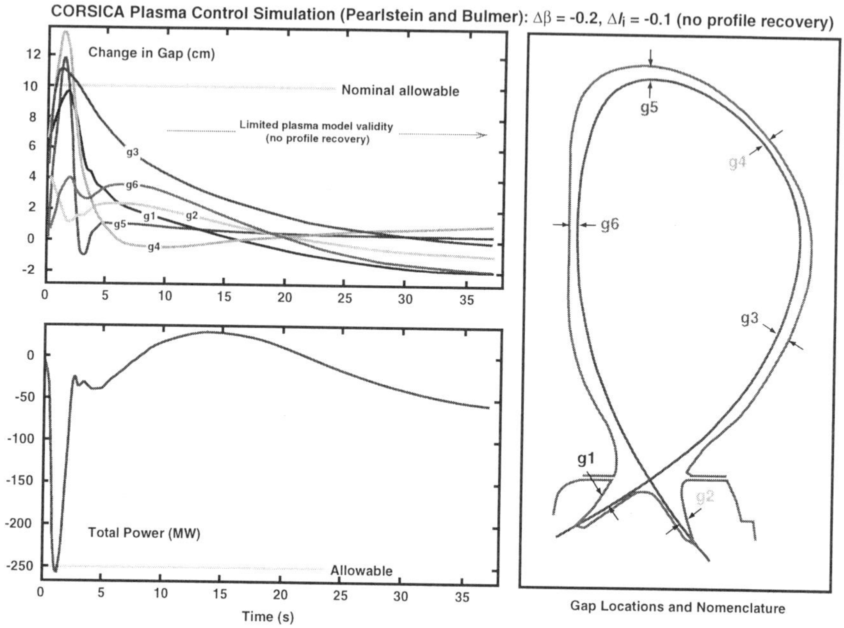

Figure 2. Results of simulations of time behavior of the separatrix/wall gap widths after a minor disruption in ITER using the CORSICA code. Simultaneous drops in poloidal β and l_i by 0.2 and 0.1 respectively have been assumed at t=0. Also shown are the positions of the gaps g1 - g6 in the ITER poloidal cross-section and time derivative of magnetic energy of poloidal magnetic field coil system (marked as total power).

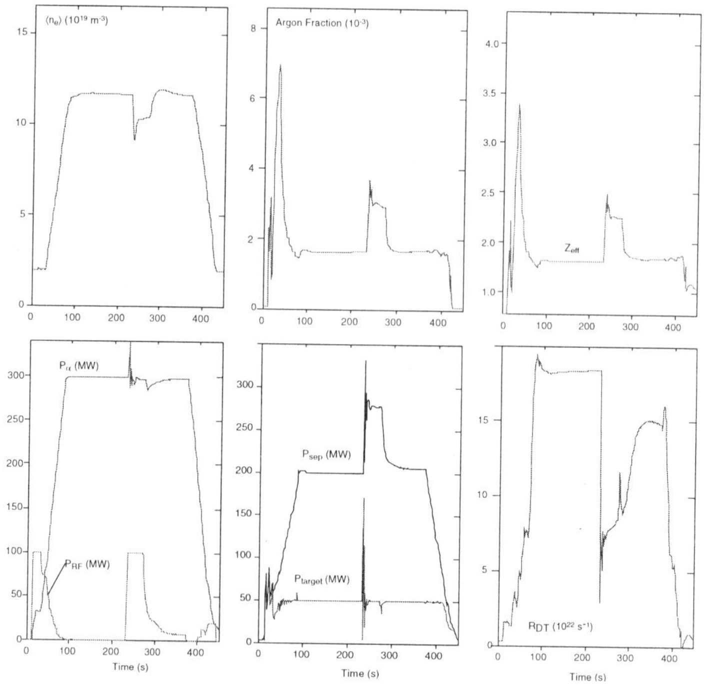

Figure 3: Simulation of simultaneous burn and divertor control. Additional heating power P_{aux}=100 MW is intentionally added at t=230 s to test the response of control loops.

Operational Limits

ITER will operate at the β and Greenwald density limits and close to H-mode power threshold. While the β limit is defined by stability of the plasma core, the density limit and L-H transition are essentially edge phenomena, affected by atomic processes, although the high beta physics seems to be involved in these phenomena as well. Due to a difference in edge and core plasma physics, a plasma with an ITER-like core (i.e., the same normalized beta β_N an collisionality $\nu*$) in present-day machines must necessarily be well below the Greenwald limit and well above the H-mode power threshold, and a plasma with an ITER-like edge will necessarily have a cold, collisional core.[8] This means that the integrated demonstration of reactor-like core and reactor-like edge can be made only in an ITER-scale experiment.

The β limit is a key parameter for a demonstration reactor (DEMO) design, and experiments on ITER should address it. The 'neoclassical' magnetic islands with n=1 that are expected to be a major β limiting factor in long-pulse discharges at low $\nu*$, have a rather long growth time, i.e., 10-30 s, and experiments aiming at increasing the β-limit by the electron cyclotron current drive (ECCD) stabilization of the islands are planned on ITER. Poloidal magnetic field fluctuations (f ~ 500 Hz) associated with rotating islands will be used as input signal for the feedback ECCD control system. Measurement of the island size will also be necessary for this program.

Studies of L-H transition and density limit will require careful measurements of T_e, T_i and n_e at the plasma edge, i.e. in the region of the edge transport barrier and pedestal, with a space resolution of 0.5 cm because of steep gradients expected there. Also, measurements of neutral density between plasma and the first wall, and the radiation power loss within the separatrix (to determine the power flow through the separatrix), are required.

High-Current Plasma Disruptions and Runaway Electrons

The physics of a plasma disruption in a reactor is expected to be essentially the same as in present-day tokamaks although the consequences of disruptions in a reactor can be much more severe as discussed above. Also, due to high electric field produced at disruption and substantial population of high-energy electrons created in the plasma by Compton scattering of gamma-rays emitted by the wall, a high runaway current can be produced through knock-on avalanche multiplication of the fast electrons.

To document the physics phenomenology of disruptions, fast (~0.1 ms) time and space resolved measurements of major plasma parameters, i.e., T_e, T_i and n_e are essential. Also, time and space resolved measurements of energy deposition on the first wall, divertor entrance baffle and divertor target are required for documenting energy partitioning in disruptions. Magnetic measurements of both plasma current and configuration evolution and in-vessel component halo current flow will be required to document VDE and halo current characteristics. These measurements and the associated plasma equilibrium reconstruction methods must be capable of accurately resolving configuration and in-vessel halo current evolution on a 10-ms or faster time scale. Measurement of the toroidal distribution of the in-vessel currents is required. Measurement of synchrotron radiation from the plasma and hard X-ray emission from the wall will be necessary to characterize the onset and evolution of the runaway electron current.

Physics of Radiative Divertor

The radiative divertor concept is based on reduction of power reaching the divertor plates due to radiation of impurities introduced into the divertor chamber.[9] The main issue is a compatibility of partially or fully detached radiative divertor with low impurity level in main plasma and high-confinement operation mode.

Several measurements are needed for understanding the divertor physics. In particular, (i) parallel and perpendicular particle and energy fluxes in the two divertor legs; (ii) gradients in the plasma parameters perpendicular to the flux surfaces with scale lengths of mm's; (iii) 2-D profiles of electron and ion temperatures and densities in the divertor and edge region; (iv) concentrations and fluxes of the impurities; (v) partial pressures of the constituents of the gas in the divertor chamber; (vi) edge plasma profiles during transient phenomena (e.g., MARFE's and ELMs). These data are essential to calibrate the divertor models which are required for extrapolation to the DEMO. The role of the diagnostic measurement in the ITER divertor is discussed in greater detail by deKock et al.[10]

Alpha Particle Effects

The classical losses of the alpha-particles in ITER will be caused mainly be the ripples of the toroidal magnetic field and should be low (≤1%) in the 21-MA plasma. They are expected to be low also at reduced currents (e.g., in a 12-MA reversed-shear steady state mode) because of the planned installation of ferromagnetic inserts which will reduce the field ripple. Therefore in the absence of anomalous alpha-particle loss, the fusion power in an ignited plasma will constitute the heating power, and the power deposition profile will coincide closely with the neutron emission profile.

Several candidates for the anomalous loss of the alpha-particles have been identified. The toroidicity-induced Alfvèn eigenmode (TAE) is considered to be the most important one.[11] Calculations show that low-n TAE modes observed in present-day experiments will be stable in ITER. However, modes with 10≤n≤50 can be unstable and their effect on alpha particle confinement is not yet known. Alpha-particle transport can also be affected by the sawtooth crash and low frequency MHD modes such as internal kinks, fishbones and kinetic ballooning modes. On the other hand, fast alpha particles can stabilize sawteeth for a long period to ~100 s on ITER. Experiments on ITER should determine the importance of anomalous loss of the alpha-particles in reactor-scale plasma. Detection of these instabilities and the measurement of associated alpha-particle losses are required for this program.

Studies of Ignition Domain

The size of the ignition domain in a reactor is defined by the range of plasma parameters which can provide sufficient alpha-particle heating to compensate the transport and radiation power loss from the plasma. The plasma parameters must be compatible with the operational limits, i.e., $\beta_N < \beta_N^{max}$, $n_e < n_e^{max}$, $q_{95\%} > q_{95\%}^{min}$, $P_{tr}^{sep} / P_{thr}^{L-H} > 1$, and $P_{tr}^{sep} - P_{rad}^{div} \leq$, P_{plate}^{max} where P_{tr}^{sep} is the power flux through the separatrix, P_{thr}^{L-H} is the H-mode power threshold and P_{rad}^{div} is the radiation and charge exchange loss from the SOL and divertor plasma. This program is similar to studies of the operational limits with an emphasis on the energy balance of the main plasma. Measurements required for this program are those for plasma confinement and operational limits.

Steady-State Burn and Advanced Operation Scenarios

Demonstration of steady-state burn of the D-T plasma is an ultimate goal for ITER. Because of low efficiency of non-inductive current drive, a true steady-state mode at a limited auxiliary power can be achieved in ITER only at high bootstrap current and, hence, at high $\beta_N > 3$. It is presently expected that requirements for steady-state operation of ITER can be met by reversed (or weak) shear modes with the internal transport barrier.[12] Such regimes with high performance are at present transient, and a high level of plasma profile control seems to be necessary to extend them to steady state.

The principal parameters that must be measured for understanding and optimizing the steady-state burn are radial profiles of the safety factor $q(r)$ and plasma pressure $p(r)$ with a good spatial resolution (~10 cm), and profiles of the poloidal and toroidal plasma rotation and radial electric field. Measurements of plasma shape, β_N, escaping alpha particles and key divertor parameters are also required.

SELECTED MEASUREMENTS AND THEIR ROLES

The above requirements imply a need for a large number of measurements. Table 2 classifies the measurements according to their major role, i.e., those that are required for machine protection and basic control (group 1a); those that could potentially be used for advanced plasma control (group 1b); and additional measurements that provide data for plasma evaluation and physics studies (group 2). The separation of control measurements between *basic* and *advanced* is somewhat arbitrary but it recognizes the fact that some parameters will have to be under real-time control for every ITER pulse, for example the plasma shape and position, plasma current and line-averaged electron density, while others will only have to be controlled for specific programs. Examples of the latter are the plasma rotation and *q* profile which may have to be under real-time control for specific high confinement modes of operation but are not necessarily controlled on every ITER pulse. Some parameters will have multiple roles: for example, measurements of the plasma density are required for machine protection, plasma control and for physics studies. Many parameters required for physics studies will be measured for plasma control, although as a rule, better space and time resolutions will be needed.

Table 2: Measurements for machine protection and plasma control and for performance evaluation and physics studies

Group 1a Measurements for machine protection and basic control	Group 1b Measurements for advanced control	Group 2 Additional measurements for performance evaluation and physics studies
Plasma shape and position, separatrix-wall gaps	Neutron and α-particle source profile	Confined α-particles
Plasma current, q(a), q(95%)	Helium density profile (core)	TAE Modes, fishbones
Fusion power	Plasma rotation (toroidal and poloidal)	T_e profile (edge)
$\beta_N = \beta_{tor}(aB/I)$	Current density profile (q-profile)	n_e, T_e profiles (X-point)
Line-averaged electron density	Electron temperature profile (core)	T_i in divertor
Impurity and D,T influx (divertor & main plasma)	Electron density profile (core and edge)	Plasma flow (divertor)
Surface temperature (divertor plates)	Ion temperature profile (core)	$n_T/n_D/n_H$ (edge)
Surface temperature (first wall)	Radiation power profile (core, X-point, divertor)	$n_T/n_D/n_H$ (divertor)
Runaway electrons	Z_{eff} profile	T_e fluctuations
'Halo' currents	Helium density (divertor)	n_e fluctuations
Radiated power (main plasma, X-point, divertor)	Heat deposition profile (divertor)	Radial electric field and field fluctuations
Divertor detachment indicator (J_{sat}, n_e, T_e at divertor plate)	'Ionization front' position in divertor	Edge turbulence
Disruption precursors (locked modes, m=2 mode)	Impurity density profiles	MHD activity in plasma core
H/L mode indicator	Neutral density between plasma and first wall	Pellet ablation
Z_{eff} (line-averaged)	Alpha-particle loss	
n_T/n_D in plasma core	Low m/n MHD activity	
ELMs	Sawteeth	
Gas pressure (divertor & duct)	Net erosion (divertor plate)	
Gas composition (divertor & duct)	Neutron fluence	
Loop voltage		
Toroidal magnetic field		

PARAMETER RANGES, TARGET MEASUREMENT RESOLUTIONS AND ACCURACIES

The parameter ranges, target measurement resolutions and accuracies required for plasma control, evaluation and physics studies have been presented at the 1995 Varenna Diagnostic

Table 3: Parameter ranges, target measurement resolutions and accuracies*

1) Plasma Current

Parameter	Parameter range	Time resolution	Rate of change	Accuracy
Plasma current	0.1-28 MA	1 ms	40-5x10^5 As^{-1}	1% (for I_p>1 MA)
		0.1 ms	<5 x 10^9 As^{-1} †	~30%

† *At current quench at a disruption*

2) Plasma Position and Shape

Parameter	Time resolution	Rate of change	Accuracy*
Δ_{in}, Δ_{out}, Δ_{top}	10 ms	10 cm s^{-1}	1 cm (for I_p>2 MA)
	10 ms	1 m s^{-1}†	2 cm
Divertor channel	10 ms	10 cm s^{-1}	1 cm
positions (r-direction)	10 ms	1 m s^{-1}†	2 cm

Δ is a width of gap between separatrix and first wall

† *At current quench at a disruption*

* *For full bore plasma*

3) Loop Voltage

Voltage range	Number of loops	Time resolution	Rate of change	Accuracy
0.01 - 30 V	≥4	1 ms	5 V s^{-1}	0.005 V
< 500 V †		1 ms	5 kV s^{-1}	10%

† *At current quench at a disruption*

4) Plasma Beta

Beta range	Time resolution	Rate of change	Accuracy
.01<β_p<3	1 ms	0.05 s^{-1}	5% at β_p~1
	1 ms	< 1 x 10^5 s^{-1} †	~30%

† *At thermal quench of a disruption.*

5) Radiated Power from Main Plasma, X-point and Divertor

	Power Range	Time resolution	Accuracy
Main plasma	≤ 0.6 GW	10 ms	10%
	<100 GW †	3 ms	20%
X-point/MARFE	≤ 0.6 GW	10 ms	10%
Divertor	≤ 0.6 GW	10 ms	10%

† *At current quench at a disruption.*

6) Line-Averaged Electron Density

Density range	Time resolution	Accuracy
10^{18}- 4x10^{20} m^{-3}	1 ms	1%
up to 2x10^{22} m^{-3} (after killer pellet injection)	1 ms	100%

*Time resolutions shown in this table are for plasma performance evaluation and optimization; time resolution of ~100 ms is estimated for plasma position and shape control.

7) Total Neutron Flux and Emission Profile

Parameter	Parameter range	Spatial resolution	Time resolution	Accuracy
Total neutron flux	10^{14}-10^{21} n s^{-1}	integral	1 ms	10%
Neutr./α source	10^{14}-4×10^{18} $s^{-1}m^{-3}$	30 cm	1 ms	10%
Fusion power	≤2 GW	integral	1 ms	10%
Fusion power density	≤10 MWm^{-3}	30 cm	1 ms	10%

8) Locked Modes

Parameter	Parameter range	Mode number	Time resolution	Accuracy
$\tilde{B}_r / B_p$	10^{-4}-10^{-2}	m/n=2/1	1 ms	30%

9) m=2 MHD Mode, Sawteeth, Disruption Precursors

	Parameter	Frequency	Spatial resolution	Accuracy
m=2 Mode	$\tilde{B}_r / B_p$	0-3 kHz	-	10%
Sawteeth	$\tilde{T}_e / T_e$	0-3 kHz	10 cm	10%
Disruption precursors	Combination of parameters (to be defined)			

10) Toroidal and Poloidal Plasma Rotation

Parameter	Parameter range	Spatial resolution	Time resolution	Accuracy
$v_{toroidal}$	1-200 km/s	5 cm	10 ms	30%
$v_{poloidal}$	1-50 km/s	5 cm	10 ms	30%

11) n_T/n_D Ratio in Plasma Core

Parameter	Parameter range	Spatial resolution	Time resolution	Accuracy
n_T/n_D	0.1-3	30 cm	100 ms	20%

12) Impurity Species Monitoring

Impurity	Influx (particles/s)	Content	Time resolution	Accuracy
Be	4×10^{16} - 2×10^{19}	10^{-4} - 5×10^{-2}	10 ms	10%(rel.)
C	4×10^{16} - 2×10^{19}	10^{-4} - 5×10^{-2}	10 ms	10%(rel.)
Cu	4×10^{15} - 2×10^{18}	10^{-5} - 5×10^{-3}	10 ms	10%(rel.)
W	4×10^{14} - 2×10^{17}	10^{-6} - 5×10^{-4}	10 ms	10%(rel.)
Extrinsic (Ne, Ar, Kr)	4×10^{16} - 8×10^{18}	10^{-4} - 2×10^{-2}	10 ms	10%(rel.)

13) Z_{eff} (Line-Averaged)

Z_{eff} range	Time resolution	Accuracy
1 - 5	10 ms	20%

14a) ELMs

Parameter	Time resolution	Accuracy
D_α bursts	0.1 ms	-
n_e fluctuations	0.1 ms	10%
T_e fluctuations	0.1 ms	10%

14b) L-H Transition Indicator

Parameter	Time resolution	Space resolution
D_α emission	0.1 ms	-
Edge pedestal (n_e, T_e)	0.1 ms	1 cm

15) Runaway Electrons

Parameter	Parameter range	Time resolution	Accuracy
Max. energy	1-100 MeV	10 ms	20%
Runaway current	$(0.05\text{-}0.7)I_p{}^{\dagger}$	10 ms	30% rel.

† *After thermal quench at a disruption*

16a) Divertor Plate Temperature

Parameter	Parameter range	Time resolution	Accuracy
Max. surface plate temperature	200-2500°C	2 ms	10%

16b) Net Divertor Plate Erosion

Parameter	Parameter range	Spatial resolution	Time resolution	Accuracy
Net plate erosion in real time	0 - 3 mm	1 cm	100 s (steady state) 1 s (transients)	10%

16c) Divertor Gas Pressure

Parameter	Parameter range	Spatial res.	Time res.	Accuracy
Divertor gas pressure	10^{-4} - 20 Pa	Several points	50 ms	20% (during shot)

16d) Divertor Gas Composition

Parameter	Parameter range	Spatial res.	Time res.	Accuracy
Divertor gas composition	A=1-100; $\Delta A \leq 0.5$	Several points	1 s	20% (during shot)

16e) 'Ionization Front' Position in Divertor (detachment indicator)

Parameter	Parameter range	Spatial res.	Time res.	Accuracy
'Ionization front' position	0-2 m	~10 cm	1 ms	10%

17) First Wall Visible Image & Wall Temperature

Parameter	Parameter range	Spatial res.	Time res.	Accuracy
First wall visible image	-	10 cm	100 ms	-
Wall surf. temper.	200-1500°C	-	10 ms	20°C

18) Gas Pressure and Composition in Main Chamber

Parameter	Parameter range	Time res.	Accuracy
Gas pressure	10^{-4} - 20 Pa	1 s	20%
Gas composition	A=1-100; $\Delta A \leq 0.5$	10 s	50% (during shot)

19) Gas Pressure and Gas Composition in Duct

Parameter	Parameter range	Time res.	Accuracy
Gas pressure	$\leq 5 \times 10^{-2}$ Torr	100 ms	20%
Gas composition	A=1-100; $\Delta A \leq 0.5$	1 s	20% (during shot)

20) In-Vessel Inspection (under responsibility of the Remote Handling Group)

	Inspected surface	Spatial resolution
First wall	100%	1 mm
Divertor	100%	1 mm

21) Halo Currents

Parameter	Parameter range	Time res.	Accuracy
Poloidal current (10 toroidal locations)	$\leq 0.2 \times I_p$	1 ms	20%

22) Toroidal Magnetic Field

Parameter	Parameter range	Time res.	Rate of change	Accuracy
B_T	3-6T	1s	steady state	0.1%

23) Electron Temperature Profile

	Temp. range	Spatial res.	Time res.	Accuracy
Core (r/a≤0.9)	0.5-30 keV	10 cm	10 ms	10%
Edge (r/a≥0.9)	0.05-10 keV	0.5 cm	10 ms	10%

24) Electron Density Profile

	Density range	Spatial res.	Time res.	Accuracy
Core (r/a≤0.9)	$(0.3\text{-}3)10^{20}m^{-3}$	10 cm	10 ms	5%
Edge (r/a≥0.9)	$(0.05\text{-}3)10^{20}m^{-3}$	0.5 cm	10 ms	5%

25) q(r)

Range of q	Spatial res.	Time res.	Accuracy
0.5-10	15 cm	10 ms	10%

26) Z_{eff} Profile

	Z_{eff} range	Spatial res.	Time res.	Accuracy
Steady state	1 - 5	30 cm	100 ms	10%
Transients		30 cm	0.1 ms	20%

27) Fishbones, TAE Modes

Type of perturbation	Frequency range	Mode numbers
Fishbones ($\tilde{B}_p/B_p, \tilde{T}_e/T_e, \tilde{n}_e/n_e$)	0.1 - 10 kHz	m/n=1/1
TAE modes ($\tilde{B}_p/B_p, \tilde{T}_e/T_e, \tilde{n}_e/n_e$)	30 - 300 kHz	n=10-50

28) Ion Temperature Profile

	Temp. range	Spatial res.	Time res.	Accuracy
Core (r/a≤0.9)	0.5-50 keV	30 cm	100 ms	10%
Edge (r/a≥0.9)	0.05-10 keV	TBD	100 ms	10%

29) Helium Density in Plasma Core (r/a≤0.9)

n_{He}/n_e range	Spatial res.	Time res.	Accuracy
1-20%	30 cm	100 ms	10%

30) Confined Alphas

Parameter	Parameter range	Spatial res.	Time res.	Accuracy
Energy spectrum	0.1-3.5 MeV	30 cm	100 ms	20%
Density profile	$(0.1\text{-}2)10^{18}$ m^{-3}	30 cm	100 ms	20%

31) Escaping Alphas

	Flux on first wall	Spatial res.	Time res.	Accuracy
Steady state	≤ 2 MW/m^2	30 cm	100 ms	10%
Transients	≤ 20 MW/m^2	-	10 ms	30%

32) Impurity Density Profile

Impurity	Content range	Spatial res.	Time res.	Accuracy
$Z \leq 10$	0.5-20%	30 cm (core)	100 ms	20%
		5 cm (edge)	100 ms	20%
$Z > 10$	0.01-0.3%	30 cm (core)	100 ms	20%
		5 cm (edge)	100 ms	20%

33) Edge n_T/n_D, n_H/n_D

Parameter	Parameter range	Spatial res.	Time res.	Accuracy
n_T/n_D	0.1-10	-	100 ms	20%
n_H/n_D	0.01-0.1	-	100 ms	20%

34) Long-Term Neutron Fluence

	Fluence range	Spatial res.	Time res.	Accuracy
First wall	0 - 3 MWy/m^2	~10 locations	10 s	10%

35) Impurity and D,T Influx in Divertor

Parameter	Parameter range	Spatial res.	Time res.	Accuracy
Γ_{Be}, Γ_C, Γ_W	10^{17} - 10^{22} at/sec	1 cm	1 ms	30%
Γ_D, Γ_T	10^{19} - 10^{25} at/sec	1 cm	1 ms	30%

36) Ion Saturation Current, n_e and T_e at Divertor Target

Parameter	Parameter range	Spatial res.	Time res.	Accuracy
Γ_i	10^{19} - 10^{25} ion/sec	0.3 cm	1 ms	30%
n_e	10^{19} - 10^{22} m^{-3}	0.3 cm	1 ms	30%
T_e	1 eV - 1 keV	0.3 cm	1 ms	30%

37) Radiation Profile [see also 5]

Region	Rad. power range	Spatial res.	Time res.	Accuracy
Plasma core	0.01-1 MWm^{-3}	20 cm	10 ms	20%
X-point/MARFE	≤ 300 MWm^{-3}	20 cm	10 ms	20%
Divertor	≤ 100 MWm^{-3}	5 cm	10 ms	30%

38) Heat Loading Profile in Divertor

Parameter	Parameter range	Spatial res.	Time res.	Accuracy
Surface temperature	200 - 2500 ^oC	0.3 cm	2 ms	10%
Power loading	≤ 25 MWm^{-2}	0.3 cm	2 ms	10%
(calc. from temp.)	< 5 $GWm^{-2\dagger}$		0.1 ms	20%

† *During disruption*

39) Divertor Helium Density

Parameter	Parameter range	Spatial res.	Time res.	Accuracy
n_{He}	10^{17}-10^{21} m^{-3}	-	1 ms	20%

40) n_T/n_D, n_H/n_D in Divertor

Parameter	Parameter range	Spatial res.	Time res.	Accuracy
n_T/n_D	0.1-10	-	100 ms	20%
n_H/n_D	0.01-0.1	-	100 ms	20%

41a) Electron Density in Divertor

Parameter	Parameter range	Spatial res.	Time res.	Accuracy
n_e	10^{19}-10^{22} m^{-3}	10 cm (along legs) 0.3 cm (across legs)	1 ms	20%

41b) Electron Temperature in Divertor

Parameter	Parameter range	Spatial res.	Time res.	Accuracy
T_e	0.3 - 200 eV	10 cm (along legs) 0.3 cm (across legs)	1 ms	20%

42) Ion Temperature in Divertor

Parameter	Parameter range	Spatial res.	Time res.	Accuracy
T_i	0.3 - 200 eV	10 cm (along legs) 0.3 cm (across legs)	1 ms	20%

43) Divertor Plasma Flow

Parameter	Parameter range	Spatial res.	Time res.	Accuracy
v_p	$\leq 10^5$ m/s	10 cm (along legs) 0.3 cm (across legs)	1 ms	20%

44) n_H/n_D Ratio in Plasma Core

Parameter	Parameter range	Spatial res.	Time res.	Accuracy
n_H/n_D	0.01-0.1	30 cm	100 ms	20%

45) Neutral Density between Plasma and First Wall

Parameter	Parameter range	Spatial res.	Time res.	Accuracy
D/T influx in main chamber	10^{18}-10^{20} $at{\cdot}m^{-2}s^{-1}$	Several points in poloidal and toroidal directions	100 ms	30%

Workshop.[1] Since then some changes in the requirements have been made resulting mainly from more detailed analysis of the requirements by the ITER Physics Expert Groups, especially the ITER Diagnostic Expert Group,[13] and by the ITER Joint Central Team. Parameters not normally measured, e.g., the neutral pressure between the plasma and first wall, and poloidal plasma rotation, have been identified as potentially important for transport barrier formation in ITER. Recent studies of the advanced regimes lead to requirement of better space resolutions (≤10 cm) for core plasma measurements. Some requirements are still under discussion, in particular those for space resolutions in the divertor. The present status of the target measurement requirements is shown in Table 3.

CONCLUSIONS

The critical plasma and first wall parameters that must be measured for machine protection have been identified and operational limits specified. Further, the parameters that must be measured for plasma control and for evaluating the plasma performance have been revised on the basis of a more detailed consideration of the role of the measurements in the experimental programme. The target measurement resolutions and accuracies have been updated. Further work on the simulation of feedback schemes for simultaneous control of plasma burn and divertor conditions is needed and work on this topic is in progress.

ACKNOWLEDGEMENT

This report has been prepared as an account of work performed under the Agreement among the European Atomic Energy Community, the Government of Japan, the Government of the Russian Federation, and the Government of the United States of America on Cooperation in the Engineering Design Activities for the International Thermonuclear Experimental Reactor ("ITER EDA Agreement") under the auspices of the International Atomic Energy Agency (IAEA).

REFERENCES

1. A.E. Costley, R. Bartiromo, L. de Kock, *et al.*, Requirements for ITER diagnostics, in: *Diagnostics for Experimental Thermonuclear Fusion Reactors*, P.E. Stott et al., eds., Plenum Press, New York (1996).
2. J. Wesley, *et al.*, Disruption, vertical displacement event and halo current characterization for ITER, 16th IAEA Fusion Energy Conference (Montreal, Canada, 1996) , F1-CN-64/FP-25.
3. D. Wroblevski, G.L. Jahns, J.A. Leuer, Tokamak disruption alarm based on neural network model of the high-β limit, *Nucl. Fusion*, 37:725 (1997).
4. Plasma operation scenario and control, in: *Technical Basis for the ITER Detailed Design Report, Cost Review and Safety Analysis (DDR)*, Chapter III, 31 January 1997, unpublished.
5. M. Rosenbluth, *at al.*, Runaway electrons and fast plasma shutdown, F1-CN-64/FP-26, in Ref. 2.
6. V. Mukhovatov, A. Costley, R. Bartiromo, *et al.*, ITER physics program and implications for plasma measurements, *Rev. Sci. Instrum.*, 68:1250 (1997).
7. The ITER Confinement Database and Modeling Working Group presented by J.G. Cordey, Energy confinement scalings and the extrapolation to ITER, to be published in *Control. Fusion Plasma Phys.* (Proc. 24th Eur. Conf. Berchtesgaden, 1997) Paper RL08.
8. F. Perkins, *et al.*, ITER operational limits, F1-CN-64/FP-24, in Ref. 2.
9. A. Kukushkin *et al.*, Analysis of the performance of the ITER divertor and analysis of the ITER tokamak edge parameter database, F1-CN-64/FP-27, in Ref. 2.
10. L. deKock, G. Janeschitz, A. Kukushkin, *et al.*, Diagnostic requirements for the ITER divertor, these proceedings.
11. C. Cheng, *et al.*, Energetic particle physics issues for ITER, F1-CN-64/FP-23, in Ref. 2.
12. D. Boucher, *et al.*, ITER scenarios including non-inductive steady state operation, F1-CN-64/FP-22, in Ref. 2.
13. K.M. Young, A.E. Costley, T. Matoba, D. Orlinski, P.E. Stott, The ITER diagnostic programme, in Ref. 1.

OVERVIEW OF THE ITER DIAGNOSTIC SYSTEM

A E Costley[1], K Ebisawa[1], P Edmonds[2], G Janeschitz[2], L C Johnson[1], S Kasai[3], L deKock[2], V Mukhovatov[1], P E Stott[4], G Vayakis[1], C Walker[2], S Yamamoto[2], K M Young[5] and V Zaveriaev[6] and the ITER Joint Central and Home Teams.

1 ITER Joint Central Team, San Diego, USA
2 ITER Joint Central Garching, Germany
3 JAERI, Naka, Japan
4 JET, Abingdon, UK
5 PPPL, Princeton, USA
6 Kurchatov Institute, Moscow, Russia

ABSTRACT

The individual diagnostics which constitute the ITER Diagnostic System are outlined and the present state of development of the designs is summarised. The results of an assessment of the overall performance of the System are presented and the areas where the probable performance falls below the target specifications are identified. The design and R&D plans which are in place to address the shortcomings are outlined.

INTRODUCTION

Since the last diagnostics workshop in Varenna [1], the ITER Diagnostic System has become better defined: in particular, the individual systems which together make up the System have been chosen. The choice is influenced by a number of factors including the anticipated reliability of the diagnostic in the application on ITER, the size, cost and complexity of the diagnostic hardware, the number of ports available for diagnostics, the difficulties of access and maintenance, the difficulties and cost of providing adequate shielding for the diagnostic penetrations, and the effect of radiation on materials relevant to the diagnostic construction. In all cases, the choice is guided by the need to satisfy the role and detailed requirements for plasma measurements on ITER [2].

In this paper, we outline the result of the selection: we describe briefly the individual systems which constitute the ITER Diagnostic System. More details are given in several accompanying papers which are referenced. The systems are being designed by an ongoing combined effort which involves the diagnostic staff within the ITER Joint Central Team and diagnostic specialists in the four Home Teams. We present the status of the designs and the plans for the remainder of the EDA. Recently, an assessment of the probable performance of the ITER Diagnostic System has been made and we summarise the result of this exercise. We outline the R&D plans which are in place to address the shortcomings identified.

THE ITER DIAGNOSTIC SYSTEM

The ITER Diagnostic System comprises about 40 individual measurement systems which can be conveniently grouped into seven generic groups:

- magnetic diagnostics
- neutron diagnostics
- optical/IR systems
- bolometric system
- spectroscopic and neutral particle analyser (NPA) systems
- microwave diagnostics
- plasma facing component diagnostics and operational systems

The full list of diagnostics is shown in Table 1.

Magnetic Diagnostics

The magnetic diagnostics [3] will provide the plasma shape and position, plasma current, loop voltage, plasma energy and the magnitude of the 'halo' currents in some key support structures.

The diagnostic consists of several individual sub-systems: pick-up coils and voltage loops mounted on the inner wall of the vacuum vessel; coils, saddle loops and voltage loops mounted on the plasma side of the back-plate; coils mounted in special divertor diagnostic cassettes; four continuous poloidal (Rogowski) loops mounted on the outside of the vacuum vessel; four poloidal diamagnetic loops; and small Rogowski loops on the earth straps between the back-plate and divertor and around the supports of the blanket/shield modules for the measurement of 'halo' currents.

The in-vessel equipment has to fulfill requirements on vacuum integrity, radiation tolerance, temperature excursions and extreme reliability. This is solved by the choice of materials and method of construction and by using shielded locations.

The pick-up coils mounted on the inner wall of the vacuum vessel are coils of mineral insulated (MI) cable wound on a stainless steel form with a protective cover. They are cooled by conduction. The voltage loops are also MI cables with a bridge at every sector joint which is repairable after sector exchange. The coils and saddle loops mounted on the back-plate and in the divertor cassettes will provide the plasma shape and position on a fast time-scale (~ 10 ms) and are not affected by axi-symmetric eddy currents. Special coils with high frequency response mounted near the poloidal gap between shield modules will give information on MHD modes up to ~ 200 kHz. These coils are made from ceramic insulated wire wound on a ceramic former. The coils, the MI cable and other ceramics are shielded by placing them in the shadow of the shield modules so that their life time is comparable or longer than the life time of ITER.

Neutron Diagnostics

The principal neutronic systems are a radial neutron camera, a vertical neutron camera, neutron spectrometers, neutron flux monitors, and a neutron activation system [4].

The radial neutron camera consists of a fan-shaped array of flight tubes, viewing the ITER plasma through slots in the blanket/shield, intersecting at a common aperture defined by a specialized shielding plug in a midplane port, and penetrating the vacuum vessel, cryostat, and biological shield through thin stainless steel "windows" [5]. Each flight tube culminates in a set of detectors chosen to provide the required range of sensitivity and temporal and spectral resolution. The detectors are housed in an auxiliary shielded enclosure outside the biological shield, and additional shielding along and between flight tubes provides necessary collimation. All electronics are situated outside the biological shield. The vertical neutron camera viewing from above has a similar configuration but is installed on a top port. The cameras provide the measurement of the total neutron flux and emission profile from which the fusion power and power density and the alpha particle source profile are derived.

Several types of neutron detectors and spectrometers are being considered. These range in size from compact natural diamond detectors to magnetic proton recoil (MPR) and time of flight spectrometers. The present design of the radial neutron camera is sufficiently flexible to accommodate many possible instruments and this is the preferred arrangement for the installation of the spectrometers. From the measurements, the ion temperature would be determined and possibly the tritium-to-deuterium ion density ratio in the core, n_T/n_D.

The neutron flux monitors consists of a number of fission chambers containing ^{235}U or other isotopes, situated at various locations around the tokamak inside the biological shield [6,7]. The system measures the global neutron source strength from which the total fusion power is obtained.

The neutron activation system uses pneumatic transfer methods to place a sample of material close to the plasma where it is irradiated by neutrons. The system will provide a robust relative measure of fusion power and allows an absolute calibration of fusion power production.

The most critical design area is the access for the radial and vertical neutron cameras. The cameras need a wide angle view in order to be able to measure the neutrons from the full plasma cross-section and this has to be obtained simultaneously with maintaining high levels of radiation and thermal shielding of the tokamak. Modifications to the standard midplane and vertical ports are being designed, and additional shielding will be installed outside the biological shield and on the ceiling above the tokamak.

Optical/Infrared Systems

The principal optical systems are two multipulse Thomson scattering systems and an equatorial plane interferometer. One of the Thomson scattering systems is optimised for making measurements in the core region while the other is optimised for making measurements in the edge region. The core system operates on the time-of-flight (LIDAR) principle [8]. Light from a high power laser is transmitted to the plasma using a folded mirror system inside a shielded labyrinth located in a mid-plane port. The plasma-facing mirror is an actively cooled metal mirror. The scattered radiation returns along the same path to spectrometers and detectors sited remotely. The system provides the spatial profile of the electron density and temperature.

A LIDAR type system will not provide the spatial resolution necessary for measurements in the edge region where high gradients are expected. Hence, the design of the edge system is being based on a conventional Thomson scattering arrangement [9].

A vibration-compensated interferometer employing Faraday rotation techniques will be used to measure the line-integrated density for use in the feedback control of the plasma density [10]. The measurements will be made in the equatorial plane. The probing beams will have wavelengths of 10.6 μm and 3.39 μm. The radiation will be transmitted to and from the plasma through shielding labyrinths in an equatorial port and small (~ 100 mm diameter) retroreflectors mounted in other equatorial ports will return the radiation. The lasers and detectors will be sited remotely.

Additional optical systems under investigation are Thomson scattering (LIDAR type) systems for the X-point and divertor regions [11], a polarimeteric system [12], and a collective scattering system. The polarimeteric system would provide measurements of the current profile while the collective scattering system would provide measurements of the confined alpha particle population.

The most critical design issue is the survivability of the first mirrors: these must maintain a good optical quality in the presence of the nuclear heating, plasma radiation and neutral particle bombardment. For the Thomson scattering systems, the mirrors also have to handle the high level of laser power. The mirrors will be actively cooled, metallic, mirrors possibly protected with shutters. Key tests on candidate mirror materials are being pursued in the diagnostic R&D programme.

Bolometric System

The bolometric system aims to provide the spatial distribution of the radiated power in the edge of the main plasma and in the divertor with a resolution of 10-15 cm, and in the centre of the main plasma with a resolution of about 50 cm . The proposed method, as used on many contemporary tokamaks, is sparse data tomography. The total number of lines of sight (LOS) is approximately 300 [13].

The bolometer arrays will be installed in the equatorial and vertical ports, in the specially instrumented divertor diagnostic cassettes, and in selected locations on the back-plate viewing through the gaps between adjacent blanket/shield modules. From each of these locations, several arrays of LOS observe the plasma.

In the equatorial and vertical ports, the bolometer arrays are integrated rigid units with all wiring attached mounted in a port shield plug. In the divertor, the bolometer has many small heads assembled in a rigid conduit with the wiring and attached to the side wall of the instrumented divertor cassette. The wiring runs to an automatic remote handling connector near the exit (pumping) port. The viewing is through the gap between cassettes, nominally 1 cm. The bolometers on the back-plate are housed in removable carriers that are plugged into permanent sockets in the back plate.

Designs of potentially suitable bolometers exist but further development is required to ensure and validate that they are radiation hard.

Spectroscopic and Neutral Particle Analysis Systems

An extensive array of spectroscopic instrumentation will be installed on ITER covering the X-ray to visible range: both passive and active measurement techniques will be employed. The four main regions of the plasma - the core, the radiation mantle, the scrape off layer (SOL) and the divertor - will be probed. The measurements can potentially provide impurity species identification, ion densities, ion temperature, electron temperature, Zeff, plasma rotation, internal magnetic field, impurity transport coefficients, and information on MHD phenomena.

The specific instruments to be employed have not yet been finalized. However, for the X-ray region it is probable that there will be two spectrometers: a medium resolution survey instrument providing full coverage in the wavelength range 0.1 - 10 nm and a high resolution (crystal) instrument with narrow coverage in the range between about 0.1 - 0.2 nm [14, 15]. For the UV and XUV region the spectrometers and detectors will have to be directly coupled to the tokamak vacuum because it is not possible to use windows. However, mirrors can be used at low angles of reflection and so it is possible to shield the detectors from neutron and gamma radiation by using labyrinthine optical systems imbedded in shielding blocks. Passive spectral measurements in the visible wavelength range are of limited value for probing the core because of the high plasma temperatures. However, active measurements employing charge exchange recombination spectroscopy with beams of energetic neutrals are a rich source of information [16, 17]. For several of the important measurements, the optimum beam energy is ~ 100 keV/amu which is significantly below the energy of the heating beams (1 MeV). This generates a requirement for a dedicated diagnostic neutral beam (below). The beams are viewed through labyrinthine optical systems imbedded in shielding blocks. Motional Stark Effect measurements of the current profile, however, benefit from the higher energy of the heating beams and so an optical system to view one of these beams is also planned [18].

Measurements in the divertor region will be made through folded optical paths in special divertor diagnostic cassette modules which will be located at each of the four divertor remote handling (RH) ports [19]. The wavelength range is presently limited to wavelengths > 200 nm due to absorption in the quartz windows. The light will be transported with mirrors, lenses and optical fibres to the spectrometers sited outside the biological shield. Mirror damage due to the deposition of target plate material during a major disruption is a potentially serious problem which is presently being investigated.

The spectroscopic diagnostics will be supplemented by two neutral particle analyzers, both viewing radially through an equatorial port along the same line of sight [20]. One will be used for monitoring the n_T/n_D ratio in the plasma edge (r/a > 0.7) and the other will provide the fast alpha particle distribution function in the energy range 0.5-4 MeV.

Microwave Diagnostics

The principal microwave diagnostics will be a system to measure the electron cyclotron emission (ECE) from the main plasma, and three reflectometry systems for probing the main plasma, the divertor plasma, and for measuring the plasma position. Additional systems under study are an electron cyclotron absorption system for use in the divertor region, a fast wave reflectometry system and two microwave scattering systems.

The ECE system consists of an array of antennas, a system for transporting the radiation, and spectrometers and detectors for analysing the emission [21]. The antennas are mounted in a radial port plug. The transmission system carries the radiation through the vacuum vessel, cryostat and shield and on to the spectrometers and detectors which are located remote from the tokamak in the diagnostic hall. Calibration of the ECE system is an important process and special equipment and procedures are employed. The system will measure the electron temperature with good spatial and temporal resolutions.

The reflectometer for the main plasma has three sub-systems: (i) a system which utilizes the upper cutoff (extraordinary mode) from the low-field side of the plasma to provide measurements of the density profile in the scrape-off layer (SOL); (ii) a system which utilizes the plasma frequency cutoff (ordinary mode) from both the low-field and high-field sides of the plasma to provide the inboard and outboard density profile in the gradient region; and (iii) a system which utilizes the lower cutoff (extraordinary mode) launched from the high-field side of the plasma to provide the core profile [22].

The plasma position reflectometer provides the location of particular density layers which occur in the vicinity of the separatrix [23]. This system is intended as a stand-by reference to the plasma position and shape control system for very long ITER pulses (> 3000 s), so reliability is of the highest priority. The divertor reflectometer is intended to provide density profiles across the divertor plasma.

For all systems, dog-leg structures in the transmission lines reduce neutron streaming outside the vacuum vessel and bioshield. Distinct vacuum windows of fused quartz directly bonded to metal structures provide pressure boundaries. The second window is inclined at the Brewster angle for the appropriate polarisation. For the low field side components, a third window at the Brewster angle is used at the exit of the NB cell.

Plasma Facing Component Diagnostics and Operational Systems

Several diagnostics will be dedicated to monitoring the condition of the high heat flux components in the main chamber and the divertor. The principal system will be a wide-angle optical system which will give views of the in-vessel components in the infrared and visible. The design specification is for a 2 mm resolution at 10 m and the intention is to use several systems to achieve a high coverage (> 80%) of the area of the first wall. The design of the system is based on the design of a similar system installed on JET. On JET the first elements in the system are lenses but on ITER because of the high level of neutron radiation the first components will be metal mirrors. The image is transmitted through a rigid labyrinth using a series of plane mirrors to a catadioptric flattening array immediately before the vacuum window. From here the image is transmitted by lenses to the cameras which will be mounted on the shielded side of the biological shield.

Several of the divertor diagnostics will be installed in special divertor cassettes. These have special provisions for diagnostic sensors, for example Langmuir probes, and have optical and microwave transmission lines.

Finally, several diagnostics will be installed to aid the protection and operation of the tokamak: for example, pressure gauges, residual gas analysers, hard X-ray monitors and 'halo' current monitors.

Diagnostic Neutral Beam

The optimum beam energy for the diagnostic neutral beam is ~ 100 keV [24]. This is considerable below the energy of the heating beams (1 MeV) and so a dedicated beam is required. To minimise the cost, a beam system is being designed which as far as possible utilises the same components and can be maintained using the same tools and RH procedures that are being developed for the heating beams. In order to achieve the required measurement performance, the beam would have to have a power of about 5 MW, a narrow beam divergence, a small beam footprint and be capable of modulation at frequencies of a few Hz. An alternative approach is to use a very short pulse (1 μs), intense (5 GW) beam which can be pulsed at a repetition rate of about 100 Hz [25]. Such a beam would have an average beam power requirement of about 500 kW, would be significantly smaller than the conventional beam and would give an improved measurement performance. However, it is at a much earlier state of development and is a subject of on-going R&D.

Diagnostic Integration

Diagnostic components will be installed inside the vacuum vessel - for example on the back-plate - and in equatorial ports, upper ports and divertor ports [26]. Outside the biological shield, they will be installed in the pit and in the remote diagnostic areas.

The ports available for diagnostics have been provisionally allocated. At the mid-plane four ports are dedicated to diagnostics. In addition, diagnostics share two ports with Remote Handling. During operation these ports are fully available for diagnostics but there is some limitation on the type of systems that can be installed. Specifically, only optical systems that can be removed in a short time can be installed on these ports. A provisional allocation of diagnostics to the available ports has been made [26].

At the divertor level, special diagnostic cassettes are installed at four locations equispaced toroidally. At each location, there are two cassettes instrumented with 'electrical' diagnostics and one which carries the optical and microwave transmission lines. At the top of the machine, diagnostics can be installed in ten of the 20 ports. However, the top diagnostic ports also have to accommodate the In-vessel Inspection System and other equipment such as the Glow Discharge Cleaning System. Substantial progress has been made with the integration of diagnostics into the available ports [27].

STATUS OF DIAGNOSTIC DESIGN

Activity in diagnostic design started shortly after the beginning of the ITER Engineering Design Activities (EDA) in 1992 but the early work concentrated on designs for generic diagnostic access, establisihng the R&D programme on irradiation effects on materials relevant to diagnostic design, and on determining the role and requirements that plasma measurements must meet in the ITER programme. Work on specific system designs did not start in a substantial way until late 1994.

The design of individual systems is essentially being taken through four phases. Phase I is an initial feasibility assessment in which the parameters to be measured are determined; initial estimates are made of the parameter ranges, space and time resolutions, possible signal/noise and accuracy; the system is outlined and the key areas of design and required R&D are identified. In Phase II the concept and scientific design is developed. The characteristics of the principal system components are determined, the port and space requirements are estimated, the design work that must be completed during the EDA is identified, and a first estimate of the system cost are made. Phase III is the engineering design. In this phase it is necessary to demonstrate that there are engineering solutions to (a) the critical system components (eg first mirrors, antennas...), (b) the transmission systems through the diagnostic ports and into the pit without causing unacceptable nuclear heating of the TF coils, and (c) the principal interfaces with the major machine components - shield blanket modules, vacuum vessel cryostat etc. It is also necessary to demonstrate that the installation and maintenance are feasible. The cost estimates are also improved. Phase IV is the integration phase. In this phase the system is integrated into the available ports, space in the pit, buildings etc. The systems are taken sequentially through Phases I through III and only progress if satisfactory results are maintained. Phase IV is carried out in parallel with Phases II and III.

The current status of the designs is shown in Table 1. A phase completed is indicated with a X, a phase in progress is indicated with a (X), and an outstanding phase which it is intended to complete before the end of the EDA is indicated with a O. The principal party carrying out the design work is also shown.

Table I also contains other information. In particular, some of the diagnostics are marked with an asterix. These are the diagnostics in the 'Start-up Set'. It is neither necessary nor desirable to construct all of the diagnostics during the machine construction phase: some diagnostics will not be required until later in the operational programme and a phased installation of the diagnostics will permit the most advanced techniques and technologies to be used. However, it will be necessary to construct all the interfaces for all the diagnostics and to meet all the diagnostic requirements for space and services during the machine construction phase in order to avoid expensive modification costs later. The combination of the construction of the diagnostics required for first machine operation, and the interface and service work, constitutes the work that must be done during the construction phase.

The design of most of the systems is based on experience gained with similar systems on present-day tokamaks. However, for a few of the required measurements existing techniques are not applicable and new approaches are needed. The systems involved are termed 'New Concept' diagnostics and marked N/C in Table I.

Finally, the plasma measurement(s) that each diagnostic is intended to address are also shown. The number of the measurements corresponds to those in the accompanying paper by Mukhovatov et al [2].

Table 1: Status Of Diagnostc Design: September 1997

DIAGNOSTIC SYSTEM	Measure-ments	Principal Party	I	II	III	IV
Magnetic Diagnostics						
Ves. Wall Sens (Coils and Loops)*	*1, 2, 3, 8*	JCT	X	X	(X)	(X)
Back Plate Sens (Coils and Loops)*	*1, 2, 3, 8, 9, 14a, 27*	JCT	X	X	(X)	(X)
Divertor Coils*	*1, 2, 14a*	JCT	X	X	O	O
Continuous Rogowski Coils*	*1*	JCT	X	X	O	O
Diamagnetic Loop*	*4*	JCT	X	(X)	O	O
'Halo' Current Sensors*	*21*	JCT	X	(X)	O	O
Neutron Diagnostics						
Radial Neutron Camera*	*7, 28*	EU/JCT	X	X	(X)	(X)
Vertical Neutron Camera*	*7*	US/JCT	X	(X)	O	O
Microfis. Chambers (In-Ves.)* N/C	*7, 34*	JA	X	O		
Neutron Flux Mons (Ex-Vessel)*	*7, 28, 34*	US	X	(X)	O	O
Gamma-Ray Spectrometers	*30*	RF	X	(X)		
Activation System	*7*	US	X	(X)	O	
Lost Alpha Detectors* N/C	*31*	US	O	O	O	
Knock-on Tail Neutron Spect. N/C	*30*	US	(X)			
Optical/IR Systems						
Thomson Scattering (Core)*	*6, 23, 24*	EU	X	X	(X)	(X)
Thomson Scattering (Edge)	*23, 24*	US	X	(X)	O	O
Thomson Scattering (X-Point)		RF	X	(X)	O	
Thomson Scattering (Divertor)*	*41*	RF	(X)	O		
Toroidal Interf/Polarim. System*	*6, 24*	US	X	X	(X)	(X)
Polarim. Sys (Pol. Mag Field Meas.)	*25*	EU	(X)	O	O	O
Collective Scattering System N/C	*30*	EU/US	(X)	O		

Table 1: Status Of Diagnostc Design: September 1997 (Continued)

DIAGNOSTIC SYSTEM	Measure-ments	Principal Party	I	II	III	IV
Bolometric System						
Bolom. Array For Main Plasma*	*5, 37*	EU	X	(X)	O	O
Bolom. Array For Divertor*	*5, 16e, 37*	EU	X	(X)	O	O
Spectroscopic and NPA Systems						
Active Spectroscopy	*10, 25, 28, 29, 30, 32*	US	X	(X)	O	O
H Alpha Spectroscopy*	*14, 33, 35, 40, 45*	RF	(X)	O	O	O
Impurity Monitor (Main Plasma)*	*12, 32*	EU	(X)	O	O	O
Impurity Monitor (Divertor)*	*12, 16b, 16e, 35, 39, 41, 42, 43*	JA	X	X	(X)	O
X-Ray Crystal Spectrometers	*10, 12, 28, 32*	EU	X	O	(X)	O
Visible Continuum Array*	*6, 13, 26, 29*	US	X	(X)	O	O
Soft X-Ray Array	*8, 9, 27*	RF	(X)	O		
Neutral Particle Analyzers	*11, 33, 44*	RF	X	X	(X)	
Two Photon Ly-Alpha Fluo. N/C	*45*	EU	(X)			
Laser Induced Fluorescence N/C	*39*	RF	(X)			
Microwave Diagnostics						
ECE Diagnostics for Main Plasma*	*8, 9, 14a, 15, 23, 27*	EU	X	X	(X)	(X)
Reflectometers for Main Plasma*	*14, 24, 27*	EU/RF	X	X	(X)	O
Reflectometers for Plasma Position	*2*	US	X	X	O	O
Reflectometers for Divertor Plasma	*14, 16e, 41a,*	EU	X	O	O	O
ECA for Divertor Plasma	*41a, 41b*	EU	O	O		
Microwave Scatt. (Main Plasma)			O			
Fast Wave Reflectometry N/C	*11, 29, 44*	US	(X)	O		
Microwave Scatt. (Divertor) N/C	*43*		O			
Plasma-Facing Components and Operational Diagnostics						
IR Cameras, Visible/IR TV*	*16a, 17, 31, 38*	EU/RF	X	(X)	O	O
Thermocouples*	*16a*	JCT/RF	X	O	O	O
Pressure Gauges*	*16c, 18, 19, 45*	EU	X	(X)	O	O
Residual Gas Analyzers*	*16d, 19*		X	O		
Hard X-Ray Monitor*	*15*	JCT	O			
Langmuir Probes/ Tile Shunts*	*21, 36, 43*	JCT	X	X	(X)	(X)
Diagnostic Neutral Beam	*10, 28, 29,*	US/JCT	X	O	O	O

ASSESSMENT OF THE MEASUREMENT CAPABILITY

Although the design of the diagnostic systems is still in progress, it is useful to try to judge the performance that the systems will have at the end of the design process relative to the specified measurement requirements. Such an exercise should identify any areas where there are serious difficulties so that appropriate adjustments in the design and R&D work can be made. With this motivation, a detailed assessment was carried out earlier this year. The probable performance of each of the diagnostics was assessed and the results expressed in terms of the capability to meet the target measurement requirements [28]. The results were combined into a summary table (Table 3) which groups the parameters into three categories: a) those which appear to be reasonably well diagnosed and for which we expect to meet the target measurement requirements; b) those which have some difficulties and for which more information is required before it is clear what the final outcome will be; and c) those for which serious difficulties exist and it is already obvious that it will not be possible to meet the target measurement requirements. Not surprisingly, most parameters are in the middle category at this stage. However, there are some in the first category but there are also some in the third and a few of these are potentially important.

A considerable cause of uncertainty in the assessments comes from uncertainty of the performance and lifetime of the first mirrors of the optical systems and from the lack of definition of the interfaces of the diagnostic components with the blanket and backplate. These are subjects on which current design work is concentrating but this exercise highlights their importance. Naturally the resolution of these difficulties are high priority items in the current work.

R&D PLANS

An extensive R&D programme supports the design work. A radiation effects programme investigates the effects neutron and gamma radiation has on materials relevant to diagnostic components. Materials suitable for insulators, cables, windows, optical fibres, mirrors, etc, are examined [29, 30]. A parallel programme aims to develop prototype diagnostic components and to develop the new concept diagnostic techniques where these are essential to meet an identified need in the measurement capability. A brief summary of the ongoing work in this programme is given in Table 2. Finally, much relevant development work is ongoing in the laboratories of the ITER Parties and the work directly related to the needs of ITER diagnostics is approved by the ITER Physics Committee and coordinated by the Diagnostics Expert Group [28].

For the post EDA phase, it is intended that this programme should continue but with a change in emphasis towards developing and testing prototype diagnostic components. The development of the new concept diagnostics will also continue.

Table 2: R&D in progress for Diagnostic Components and New Concept Diagnostic Techniques (September 1997)

Task	Party
Development of radiation hard bolometers	EU
Development of radiation hard calibration source for ECE measurements	EU
Development of in-vessel microwave components for reflectometry	EU
Development and Testing of Pressure Gauges	EU
Development of plasma viewing system	EU
Development of neutron diagnostics (fluid flow activation concept, concept of compact fibre detectors)	EU
Development of vacuum seals for diagnostic windows (fused quartz, sapphire, ZnSe)	JA
Development of optical fibre feedthroughs	JA
Development of steady state magnetic sensor	JA
Development of synthetic diamond detectors for neutron spectrometry	JA
Development of particle detection system for Neutral Particle Diagnostic with reduced sensitivity to neutron and gamma radiation	RF
Fabrication and study of high temperature multipin connectors for iter backplate	RF
Feasibility assessment of threshold bubble chamber for neutron tails measurements in ITER	RF
Development of technology for fabrication of first mirrors for optical systems and study of their properties	RF
Development of natural diamond detectors for neutron spectrometry	RF
Testing of a prototype magnetic coil.	US
Demonstration of Fast Wave Reflectometry	US
Demonstration of knock-on tail alpha diagnostic	US
Feasibility assessment for measurement of confined alpha particles using CO_2 Collective Scattering Technique	US

CONCLUSIONS

The ITER Diagnostic System is now well defined: in particular, the individual systems which constitute the System have been chosen. The design of many of the systems is well advanced and the integration of the systems onto the tokamak is in progress. An assessment of the integrated measurement capability has been made. This confirms that many of the required measurements can be made to the target specifications but there are some important shortfalls. Current and planned design and R&D work is addressing these difficulties. For the post EDA phase, it is intended that the design and R&D should continue along the same lines but with an emphasis towards the development of key diagnostic components.

Table 3: Results of Assessment of ITER Measurement Capability

a) Parameters which appear to be reasonably-well diagnosed

Meas no.	Parameter	Purpose	Candidate Diagnostics	Comment
1	Plasma current	Current control	Rogowski coils Magnetic probes	Magnetic probes mounted on the plasma side of the backplate are required for disruption studies. Accuracy may be a problem for $t_{pulse} \geq 1000s$.
2	Plasma position and shape	Poloidal field control	Position loops & coils, reflectometry (plasma position)	Requires loops and coils on the plasma side of the backplate, connectors. Long term radiation effects. Back-up (reflectometry) may be necessary for long pulses.
3	Loop voltage	l_i, startup	Magnetic flux loops	Requires loops on the plasma side of the backplate.
6	Line averaged electron density	Plasma operation, disruption avoidance, NBI protection	Tor. interferometer/polarimeter, LIDAR, visible continuum	The key issue of installation of retroreflectors for interferometer/polarimeteris is not yet solved. Calibration difficult for LIDAR.
9	m=2 MHD modes, Sawteeth	Disruption avoidance	Mirnov loops, ECE, reflectometry, neutrons, soft X-ray array	Not all these techniques will work, especially soft X-rays. Determination of n-number requires multiple toroidal locations.
12	Impurity species monitor	Impurity control, wall protection	Visible spectrometry, VUV, X-ray spectrom.	Plasma viewing system will also be useful.
14a	ELMs	Optimization control, divertor protection	H_a spectroscopy, ECE, reflectometry, magnetics, IR camera, Langmuir probes	At least some of these will work.
14b	L/H mode indicators	H-mode realization	H_a spectroscopy, reflectometry (fluctuations)	Both look promising but need further study.
16c	Divertor gas press.	Divertor operation	Pressure gauges	Expect to meet requirements.
17	First wall vis. image, wall temp.&erosion	Startup, position & hot spot monitor	Visible & IR TV	Present design of plasma viewing system looks promising but needs many diagnostic ports
18	Base pressure	Tokamak readiness	Pressure gauges	As 16e
19	Gas pressure (duct)		Pressure gauges	As 16e
22	Toroidal magn. field		Current shunts, Rogowski coils	
23	Electron temperature profile	Profile. control with auxiliary heating, confinement studies, plasma energy	ECE, LIDAR, Thomson scattering (edge)	Core: Expect to meet measurement requirements: ECE for outer half of plasma, LIDAR for mid-plane profile. However, there will be problems if the operating scenario use plasmas that are not on the mid-plane. Edge: May meet requirements. Depends on implementation of Thomson scattering system on top port.
24	Electron density profile	Transport, fueling optimization, density & beta control	LIDAR, reflectometry, interferometry/polarimetry, Thomson scattering (edge)	LIDAR and Thomson scattering (edge) need calibration, reflectometry probably OK but needs high field side antenna for inner edge, interferometry/polarimetry have limited sight lines. Problems, as for T_e profile out of mid-plane.

Table 3: Results of Assessment of ITER Measurement Capability (continued)

a) Parameters which appear to be reasonably-well diagnosed (continued)

Meas no.	Parameter	Purpose	Candidate Diagnostics	Comment
33	Edge n_T/n_D, n_H/n_D	Fueling optimization, plasma dilution	NPA, H_a spectrosc., fast wave reflectom., two phot. Ly-a flour.	Expect to meet measurement requirements.
34	Long term neutron fluence		Activation transfer system	Essential for absolute neutron calibration. Expect to meet requirement but implementation has not yet been studied.
45	Neutral density between plasma and wall	L-H transition, particle source	H-alpha spectroscopy, pressure gauges	Looks possible, need multiple locations.

b) Parameters which appear to have some difficulties

Meas no.	Parameter	Purpose	Candidate Diagnostics	Comment
4	Beta	Disruption avoidance	Diamagnetic loops	Cancellation of spurious signal is very difficult. Long term effects. Accuracy is problematic for long pulses. Kinetic measurements will be necessary.
5	Radiated power (core, X-point and divertor)	Disruption avoidance	Bolometer arrays	Requires full array (see #37) to characterize new configurations. Development of radiation-hard bolometers, connectors, cables needs substantial R&D
7	Total neutron flux & emission profile	Burn control, fueling control	Neutron flux monitors, 2-D neutron camera	Calibration difficult. Needs neutron profile to accurately calculate total flux. Viewing angles of neutron cameras are limited.
8	Locked modes	Disruption avoidance	Magnetic flux loops, ECE, reflectometry, soft X-ray array	All have difficulties to positively identify locked modes but extended saddle loops should see large modes. ECE, reflectometry and X-ray array only see slowing down of modes.
13	Z_{eff} (line averaged)	Impurity control, He accumulation	Visible/IR Bremmsstrahlung	Calibration may be a problem. Plasma viewing system should be investigated.
15	Runaway electrons	Runaway avoidance and measurements after disruptions	Tang. view X-ray monitor, ECE, synchrotron radiation	Absolute measurement difficult for all techniques. Radiation background may be a problem for X-rays. Pitch angle distribution of runaways is required for synchrotron radiation.
16a	Divertor plate temperature	Divertor protection	IR cameras, thermocouples	Thermocouples too slow and unreliable. Viewing in IR through divertor optical cassette can give full poloidal coverage at fixed toroidal location but survivability of optics is unknown. IR plasma viewing cameras give good spatial coverage though 100% are hardly possible.

Table 3: Results of Assessment of ITER Measurement Capability (continued)

b) Parameters which appear to have some difficulties (continued)

Meas no.	Parameter	Purpose	Candidate Diagnostics	Comment
5	Divertor radiated power	Divertor operation	Bolometers	Total rad power requires the measurement of the radiated power profile (see #37)
16e	Divertor ionization front position	Divertor optimization	Visible spectrometry, bolometry, reflectom.	Needs sensor arrays in divertor region.
20	In-vessel inspection	Viewing the damage from the preceding pulse (when requested)	Viewing camera	#17 would probably function also as in vessel inspection system and would have the benefit of working also in real time. Requires addition of internal illumination source.
21	Halo currents	Monitoring of forces on vessel structure	Magnetic coils on structures	Needs a'priori knowledge of locations.
26	Zeff profile	Impurity transport	Visible continuum array	see #13
27	Fishbones, TAE modes	Burn optimization, beta limit indication	Mirnov loops, reflectometry, ECE	Needs high frequency magnetic probes.
28	Ion temperature profile	Burn optimization, transport, reaction rate control	X-ray crystal spectr., neutron spectrometer, neutron flux, CXRS	None of these methods will be sufficient by itself. Measurement hardly possible during H (and probably DD) phases. CXRS which gives best spatial resolution needs DNB and looks marginal.
29	He density in core	Burn optimization & transport, reaction rate control	CXRS with DNB, vis. Bremsstrahlung, fast wave reflectometry	CXRS should be able to meet requirements if DNB installed. Measurements in core ($r/a<0.3$) will be difficult/impossible in highest density range.
32	Impurity density profile	Tokamak condition, necessary if high Z used for radiative loss	CXRS, VUV spectrometry, X-ray crystal spectrometry	Visible spectrometry has serious problems of quantitative interpretation in the core. X-ray and VUV spectroscopy requires vacuum extension beyond the cryostat.
35	Impurity & D,T influxes in divertor with spatial resolution	Divertor optimization & control	Impurity monitors (div), High resolution H_a spectroscopy (div)	Spatially resolved spectroscopy below the divertor might have problems with mirrors. Plasma TV from radial ports looks the best bet. Needs some UV spectroscopy for quantitative measurements.
36	Ion saturation current, n_e, T_e at divertor plate	Divertor optimization & control	Langmuir probes, tile shunts	Will Langmuir probes survive, do they give T_e correctly?
37	Radiated power profiles in core, edge, divertor (combine with #5)	Divertor optimization & control	Bolometer arrays	Requires many (~100 sight lines), especially viewing the divertor. Requires a robust radiation-hard bolometric detector. Serious worries about in-vessel cabling, connectors and lead-throughs. Needs to be in Category (i).
38	Heat deposition profile in divertor	Divertor optimization & control	IR camera	Needs several camera positions to cover significant fraction of the divertor. Full coverage will not be possible.
39	Divertor He density	Divertor optimization He pumping physics	Impurity monitors, laser induced fluorescence (LIF)	Measurement capability is not yet clear. LIF looks virtually impossible in ITER divertor.
40	n_T/n_D, n_H/n_D in divertor	Fueling optimization, plasma dilution	H_a spectroscopy	H, D and T lines overlap. Needs design and modeling to assess measurement capability.
44	n_H/n_D in core	Plasma dilution	NPA, Fast wave reflectometry	Probably cannot meet measurement requirements for $r/a \leq 0.8$. Fast wave reflectometer in testing.

Table 3: Results of Assessment of ITER Measurement Capability (continued)

c) Parameters which appear to have some serious difficulties

Meas no.	Parameter	Purpose	Candidate Diagnostics	Comment
10	Toroidal & poloidal plasma rotation	Disruption avoidance confinement optimization	CXRS with diagn. NB, X-ray crystal spectrometry	Both techniques have difficulties. Orthogonal views needed to determine both rotations. Velocity shear may be important but needs good spatial resolution & coverage.
11 & 44	n_T/n_D, n_H/n_D, in core	Fueling control	NPA, Fast wave reflectometry	NPA can not see core, fast wave reflectometry needs demonstrating in principle and application to ITER needs study.
16c	Net divertor plate erosion	Plate erosion	Impurity monitoring, reflectometry Plasma viewing	Impurity monitoring is too insensitive and too late for localized erosion. Reflectometry needs developing, also would not cover all vulnerable area. Plasma viewing & radar offers some possibilities.
25	q (r) profile	Perform. evaluation, plasma stability, transport, adv. modes	MSE with heating NB, polarimetry	MSE may have operational conflict with occasional brief modulating the heating beams. Polarimetry needs further design and will have constraints on sight lines.
30	Confined alphas	Alpha transport & confinement, fusion heating	Collective scattering, CXRS/NPA with DNB or pellets, neutrons via knock on tail effect, gamma spectrometer	None of these techniques is working convincingly in a present day tokamak and all would have serious problems in ITER.
31	Escaping alphas	Performance evaluation	Lost alpha detectors, IR TV	Development of lost-alpha detectors is an R&D issue.
41	n_e, T_e in divertor	Divertor optimizat. & control	Visible/UV spectrosc. reflectometry, ECA, Thomson scattering	Expect to be able to make measurements of line-averaged and peak density. Measurements of T_e look difficult but only a preliminary investigation of measurement possibilities has been carried out.
42	T_i in divertor	Divertor optimizat. & control	Spectroscopy, CX	
43	Plasma flow in divertor	Divertor optimizat. & control	Spectroscopy	

ACKNOWLEDGEMENT

This report has been prepared as an account of work performed under the Agreement among the European Atomic Energy Community, the Government of Japan, the Government of the Russian Federation, and the Government of the United States of America on Cooperation in the Engineering Design Activities for the International Thermonuclear Experimental Reactor ("ITER EDA Agreement") under the auspices of the International Atomic Energy Agency (IAEA).

REFERENCES

[1] *Diagnostics for Experimental Thermonuclear Fusion Reactors*, edited by P E Stott, G Gorini and E Sindoni, Plenum Press, New York (1996).

[2] V S Mukhovatov et al, 'Role and Requirements for Plasma Measurements on ITER', these proceedings.

[3] L deKock et al, 'Magnetic Diagnostics for ITER', these proceedings.

[4] L C Johnson et al, 'Overview of Fusion Product Diagnostics for ITER', these proceedings.

[5] F B Marcus et al, 'A Radial Neutron Camera Design for ITER with Integral Shielding', these proceedings.

[6] C W Barnes et al, 'Design Considerations for Neutron Activation and Neutron Source Strength Monitors for ITER", these proceedings.

[7] T Nishitani et al, 'In-Vessel Neutron Monitoring using Microfission Chanbers for ITER', these proceedings.

[8] H Salzmann et al, 'ITER Core Plasma LIDAR System', these proceedings.

[9] B Grek and D Johnson, 'Edge Thomson Scattering for ITER', these proceedings.

[10] T N Carlstrom et al, 'Application of Interferometry for Density Measurements on ITER', these proceedings

[11] G T Razodabarin, 'Status of LIDAR Thomson Scattering Project for the ITER X-Point and the Adjacent Plasma Regions', these proceedings.

[12] A J H Doneé et al, 'Polarimetry for Poloidal Field Measurements', these proceedings.

[13] R Reichle et al, 'Progress of the Reference Design for ITER Bolometers and Development of a High Performance Alternative', these proceedings.

[14] P H Edmonds et al, 'Integration of Vacuum Coupled Diagnostics', these proceedings.

[15] R Barnsley et al, 'X-ray Spectrometry Systems for ITER', these proceedings.

[16] D M Thomas et al, 'Prospects for Core Helium and Related Measurements on ITER using Active Charge Exchange', these proceedings.

[17] R W T Konig et al, Neutral Beam Based Spectroscopic Diagnostics for ITER Extraploated from Recent JET Results', these proceedings.

[18] F M Levinton, 'The Motional Stark Effect Diagnostic Study for ITER', these proceedings.

[19] T Sugie et al, 'Divertor Impurity Monitor for ITER', these proceedings.

[20] A I Kislyakov et al, 'NPA at ITER: Physical Bais and Conceptual Design', these proceedings.

[21] G Vayakis et al, 'The ITER ECE Diagnostic Front End Design', these proceedings.

[22] G Vayakis et al, 'Overview of the Reflectometry Diagnostic Systems', these proceedings.

[23] N L Bretz et al, 'ITER Position Control Reflectometry - Conceptual Design' these proceedings.

[24] E S Marmar, 'Active Spectroscopy' pp 281 - 290, in *Diagnostics for Experimental Thermonuclear Fusion Reactors*, edited by P E Stott, G Gorini and E Sindoni, Plenum Press, New York (1996).

[25] H A Davis et al, 'Progress Toward a Microsecond Duration, Repetitively Pulsed, Intense Ion Beam for Active Spectroscopy Measurements on ITER', pp 332 - 335, Rev. Sci. Instrum., 68,, 1997.

[26] G Janeschitz et al, 'ITER Physics Basis, Machine Design and Diagnostic Integration', these proceedings.

[27] C Walker et al, 'Diagnostic Access for ITER', these proceedings.

[28] Minutes of the Sixth Meeting of the ITER Diagnostics Expert Group, held in Naka, 10 - 11 March, 1997 (S CX MI 12 97-04-23 F1)

[29] S Yamamoto et al, 'Irradiation Tests on ITER Diagnostic Components', these proceedings.

[30] E R Hodgson, 'Radiation Problems and Testing of ITER Diagnostic Components', these proceedings.

DIAGNOSTICS ACCESS FOR ITER

C.I.Walker[1], T.Ando[1], A.Costley[2], L.deKock[1], K.Ebisawa[2], P.Edmonds[1], G.Janeschitz[1], L.Johnson[2], V.Mukhovatov[2], G.Vayakis[2], S.Yamamato[1]

1 - ITER, Joint Central Team, Garching, Germany
2 - ITER, Joint Central Team, San Diego, USA

INTRODUCTION

The design of the major diagnostic systems for ITER has now progressed from generic approaches[1] to the engineering and design integration phases. The design of relevant parts of the ITER machine, in particular the access points for diagnostics such as ports and blanket modules, has developed significantly[2]. Diagnostic equipment is still found in the three main port areas (equatorial, vertical and divertor) and in the blanket back plate, but no longer in the blanket modules. Diagnostic access at these sites is considered in terms of the access to the plasma through the first wall and the implementation of the signal transmission paths away from the Tokamak. The arrangements for the installation and maintenance of the diagnostic components are described.

Access to the plasma is dominated by the requirement to minimise the penetrations through the first wall. Nuclear aspects (radiation shielding, irradiation effects and removal of nuclear heating) of the diagnostic components and of the torus systems influence the design. Many systems are required to be integrated into limited positions with some degree of standardisation for handling and access. The core plasma is probed through holes in special blanket modules at horizontal and vertical ports, and through the gaps between the blanket modules. Diagnostic sensors are mounted on the blanket back plate. In the divertor area some cassettes are specially adapted for visible, IR and UV viewing. Waveguides are mounted on the sides of these cassettes. Electrical sensors are located on standard cassettes observing the plasma through existing gaps.

Signal transmission from diagnostics introduces further considerations of the relative movements of machine components, requirements for safety and reliability and the general layout configuration of ITER boundaries and buildings. In the ports the transmission path is often labyrinthed through shielding but the peculiar requirement of straight beams has been considered. Alignment considerations predominate in the design of these and of optical systems. The strict requirements of integrity, for normal and foreseen fault conditions, of components on the vacuum and tritium confinement barriers have been met.

Maintenance interfaces are developed to minimise the time and equipment requirements for access. Depending on the location and time history of the various components, maintenance will be performed from hands-on to entirely by remote handling tools. The need to remove diagnostic equipment without venting the cryostat and the modularity of the installed components stem from these considerations. Standardised designs and procedures, with a small number of well developed solutions, are applied to a majority of the diagnostic interfaces. Railed transfer arrangements are provided for the removal of large components from the horizontal ports. These components, and the diagnostic equipment within them, are refurbished and tested in the hot cell. After removal of blanket modules, some diagnostic components on the back plate are accessible for remote replacement.

Diagnostics for Experimental Thermonuclear Fusion Reactors 2
Edited by Stott *et al.*, Plenum Press, New York, 1998

Equatorial Port Diagnostics

The equatorial (or mid-plane) ports are the preferred access for diagnostics: in total 9 are used. Three ports are assigned solely to diagnostics (Sectors 9, 16 and 17) for the larger systems such as the radial neutron camera, the toroidal interferometer, and the LIDAR. Smaller systems are integrated around these primary diagnostics. The port adjacent to the neutral beams (Sector 7), with its limited access but pressurised vault beyond the bio-shield, is used by directly coupled systems[3] such as NPA, X-ray spectrometers, VUV spectrometers. The diagnostic neutral beam (DNB) shares a port (Sector 4) with a heating neutral beam. Diagnostic systems have been assigned to ports allocated to remote handling for the removal of in-vessel equipment (Sectors 8 & 18) where they can be easily disconnected, for example the plasma viewing system[4]. The in-vessel transport system of the remote handling (in Sectors 3 & 13) is normally retracted in an enclosure in the access cell and is deployed when in-vessel maintenance is required. The unused spaces behind the two start-up limiters in these, diametrically opposed, ports make ideal sites for the Neutron Flux Monitors.

Diagnostics in equatorial ports are incorporated in a modular port plug that comprises the blanket shield plug, all of the diagnostic equipment in shielding (referred to as the diagnostic block) and the primary vacuum flange. This weighs ~90t, mainly from shielding, and EM analysis indicates that the radial force on it is 0.5MN towards the plasma with a net moment about a radial axis of 13 MN-m during survivable disruptions (VDE).

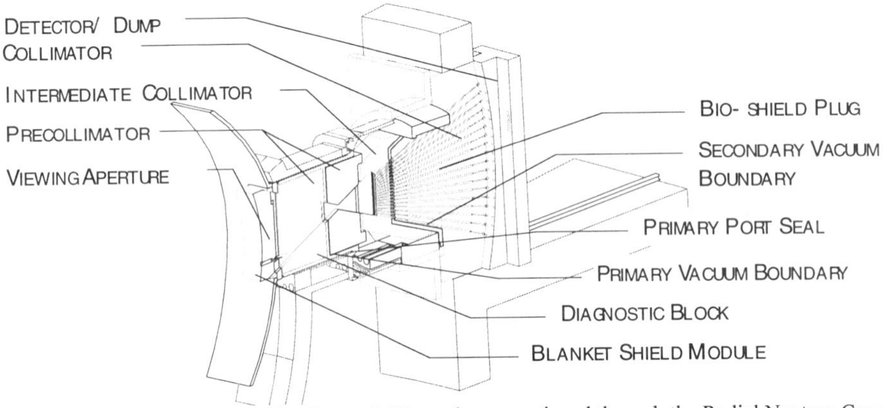

Figure 1. Port Plug installed at the Equatorial Port, shown sectioned through the Radial Neutron Camera.

The arrangement for Sector 9 (Fig.1) is used to illustrate the typical access considerations at the equatorial ports. Here the neutron camera, interferometer, bolometers and possibly soft X-ray cameras are combined. The radial neutron camera[5] consists of a fan-shaped array of flight tubes, viewing the plasma through a slot in the blanket shield module and precollimating shielding in the diagnostic block. To economise on blanket penetrations, this view of the plasma is shared by modular arrays containing bolometers[6] and soft X-ray cameras sensors. The toroidal interferometer[7] requires a separate, horizontal, slot in the blanket to acquire the best toroidal view while keeping the first mirror as deep within the shielding as possible (Fig. 2). The second mirrors are in a slice of the diagnostic block on the other side of the neutron collimator.

The number and size of holes in the first wall is kept to a minimum and the sides of the holes are cooled when they are larger than ~100mm. A number of alternative construction techniques under consideration for the blanket shield plugs have been shown to be modifiable for diagnostic holes. The first wall aperture is further minimised by attaching the blanket shield plug to the diagnostic block guaranteeing alignment.

The diagnostic block is the first part of the port plug with active elements. It carries the sensors, mirrors and shutters in modules within the massive shielding which is assembled to most easily access these components. Shutter modules have simple plug-in connectors for mechanical actuation. The relative merits of various designs is presently under investigation. First mirror designs (cooled by the 4.0 MPa water of the vacuum vessel cooling) have been developed further concentrating on surface and substrate materials. Mirror mounts, with fixed adjustment features are under design. The construction of the equatorial port plugs facilitates their remote handling, removal and maintenance (Fig.2). The plugs are constructed on a substantial chassis that also acts as shielding. This is supported on linear roller bearings from the vacuum vessel port. The diagnostic block is built up on this chassis from the component segments of shielding, cooling lines are attached with standard welded connections and the block is completed with a top cover for structural and shielding purposes. The cover is tailored to fit local variations in the ports. Transmission of signal from electrical sensors is by mineral insulated cables routed in channels to the back of the diagnostic block.

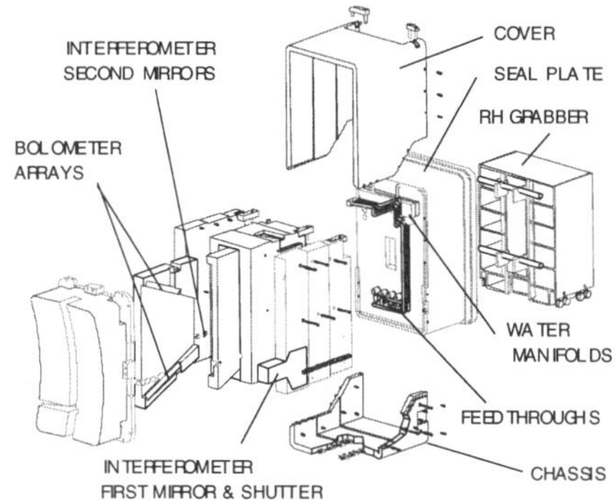

Figure 2. Assembly of an Equatorial Port Plug.

The vacuum vessel seal plate is the main vacuum boundary component of the port plug incorporating the main port seal, windows, electrical and water feedthroughs. All components are rated at 0.5MPa on this primary boundary. Aluminium diffusion bonded windows and alloy solder materials (e.g. AgPbSn) have been shown to be adequate for all ITER parameters[8] with the exception of the anticipated radiation effects which are the subject of further R&D. A new bond technique using a spring seal on the glass edge is showing promising early results[9]. More development work is expected to add this to the acceptable techniques. Beryllium foil windows supported on a honeycomb frame have been shown to survive to 1.7 MPa. Sealing to Zinc Selenide still remains to be demonstrated within the current window development programme. The electrical feedthroughs on the vacuum boundary are robust constructions of mineral insulated cables brazed into stainless steel bulkheads. The irradiation characteristics of the terminations are presently under investigation. Collimating shielding is installed in this port interspace. The total shielding thickness in the port plug has been shown to be adequate for a number of individual diagnostic penetrations[10]. The neutron streaming from the integration of many diagnostic penetrations in one block has to be confirmed as acceptable when the final designs are fixed. The interspace shielding thus can be used to increase the shielding if required without increasing the weight of the port plug.

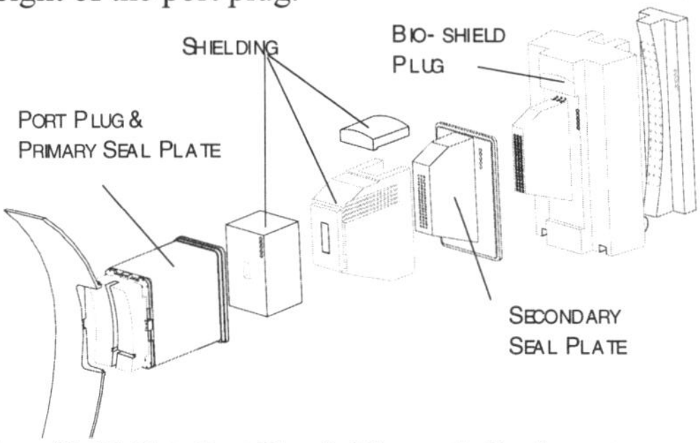

Figure 3. Installation of Mid-Plate Port Plug & Diagnostic Equipment.

The equatorial port plug will be handled as one unit. It is removed in a contamination containing transport cask running on rails into the port, which docks to the cryostat flange or a complementary flange on the hot cell. It is dismantled and refurbished in the hot cell, where the modular sub-assemblies can be replaced and tested. Reuse and refurbishment of the port plugs is important to minimise the accumulation of activated waste. It is also crucial that diagnostic commissioning and testing is completed before port plugs are transported for installation. The diagnostics in port plugs are finally commissioned at the hot cell in a clean room in diagnostic test stands with dedicated equipment. The second vacuum boundary and the removable shield (that sits in the port) re installed by similar techniques to the port plug. The bio-shield port plug collimator, detector array and any necessary local cabling and cubicles are manually installed behind the cryostat.

An example of an equatorial port where many optical diagnostics are integrated is sector 16 with the Thomson Scattering LIDAR, Wide-angle Plasma and Divertor Viewing, and Hα spectroscopy (Fig.4).

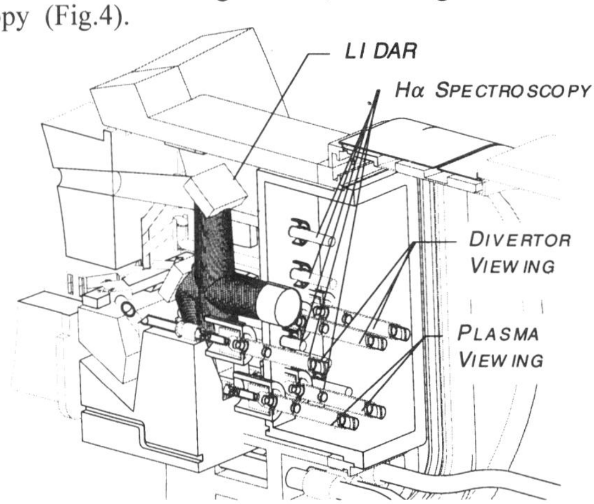

Figure 4. Optical Diagnostics in an Equatorial Port Plug.

The LIDAR system[11] is positioned on the port centre line for optimal shielding, and the Hα spectroscopy fan view array is on a side slice of the diagnostic block. For these systems mirror mounts are adjustable through the cover and are removable without a complete disassembly of the block.

Windows are in tubes on the vacuum seal plates and there are intermediate mirrors in the port interspace located on the rail system use by the RH tools. Laser and collected light is guided to the system through vacuum light tubes from the diagnostic hall. In the wide angle plasma viewing system and the divertor viewing system[4], the front end mirrors (probably of tungsten) need to be very close to the plasma on an extension of the diagnostic block. The small entrance aperture mitigates the problems with charge exchange sputtering. Image transmission is by reflective optics and lenses are only used near the vacuum window. A close alignment of the front mirrors relative to the window optics is achieved with cooled flight tube structures fixed only at the front of the port plug. Five of these tube structures are shown, for two opposed wide angled views and for three views of the divertor leg. The cameras are located outside the biological shield.

The only special handling requirements are for port 7, adjacent to the neutral beam injection. Here vacuum extensions are provided for close coupled diagnostics such as X-ray crystal and VUV spectrometry and neutral particle analysers. Reflectometer waveguides[12] are labyrinthed around these through windows on the vacuum boundaries.

Diagnostics in the Vertical Ports

The vertical ports give views of the core and edge plasma and some of the first wall and divertor that are otherwise inaccessible. The main blanket cooling pipes are now routed through alternate vertical ports, no diagnostics are foreseen on these ports. Thus the ten even numbered vertical ports are assigned to diagnostics. It is necessary to share these ports with the RH insertable in-vessel viewing (IVV) probe (required on all ten ports), the retractable glow discharge cleaning (GDC) probes (required on at least 8 ports) and the gas injection valves and pipes (required on 5 ports). Allocation of diagnostics to the remaining space is mainly dependant on diagnostic access factors, for example, completeness of the plasma and divertor viewing, NPA sight on to DNB, gas-free bolometer viewing, etc.

All vertical ports for diagnostics are standard in concept, they contain a range of holes for the diagnostic views with special components in diagnostic plugs. Generally there will only be radial and vertical viewing; shielding and replacement requirements determine the design and effectively preclude toroidal viewing.

There is a poloidal gap in the back plate, 150mm wide, across the radial width of the port. The dimensions of this slot are limited both by stresses in the back plate and by the neutron loading on vacuum vessel joints and on the P2 poloidal field coil.

Optical systems, (such as edge spectroscopy and divertor viewing) can access the plasma core, divertor and upper plasma using front end mirrors. Direct line of sight systems have a much more limited access to the plasma by virtue of the other machine systems. This can be extended somewhat by locating the detectors lower in the cryostat.

The method of completing the shielding blanket in this area under the port is still under review. One approach is with normal blanket shield modules (but with side wall cooling) leaving a 150m gap. An intermediate yoke is placed in the back plate hole to react against the blanket torques. In the other design there is a slotted blanket shield plug (complete with first wall armour and shielding) that is inserted from above the port. There are significant implications on remote handling access to these two variants, but from the point of view of the diagnostic access there is little to differentiate them. The latter lends itself to wider angle cones and slightly non-radial viewing; for these reasons this design is presently selected.

The gap in the shield plug is blocked where it is not required for access by the replacement shielding, a simple water cooled, stainless steel, block shaped internally for each local diagnostic requirement. There are no active or sensitive components here so access will only be required for repair or altered diagnostic requirements. The diagnostic shield block above this is again passive stainless steel shielding with a range of access holes. This block may be attached to the primary vacuum seal plate which is welded to the port above the level of the toroidal field coils. This plate is also considered to be permanent with no normal maintenance requirements, (RH Class 3), but it is the access point for maintainable systems.

There are three types of arrangement at the port seal plate: i) discrete vertical diagnostic plugs, ii) straight vacuum extensions and iii) fanned collimation.

i) Discrete vertical diagnostic plugs are used for optical diagnostics (such as the divertor viewing periscopes, impurity and Hα monitoring), and electrical diagnostics (such

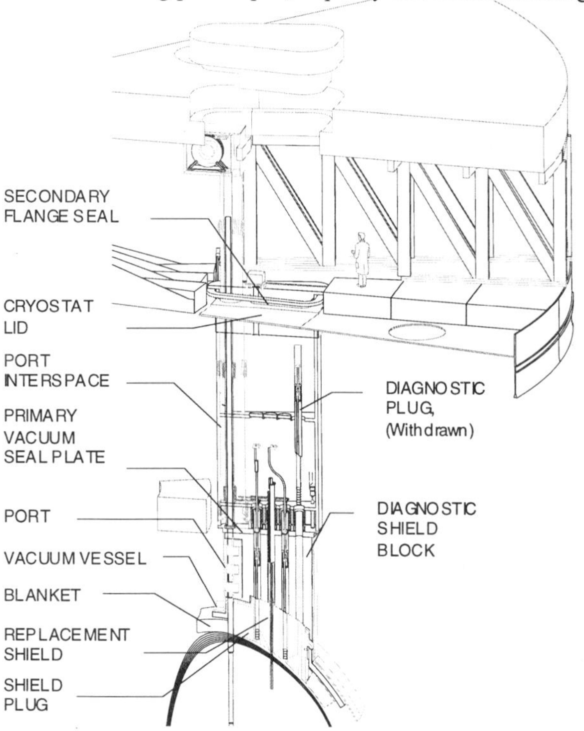

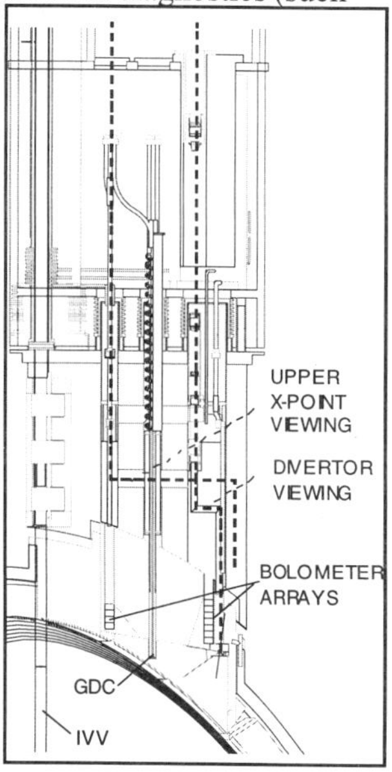

Details of In-Vessel Components for another Port

Figure 5. Vertical Port Arrangement for a typical diagnostic sector.

as bolometry). Windows, relay optics, sensors, wiring and feedthroughs are all integrated with the shielding of a diagnostic plug which serves as structure and vacuum seal. Most ports are of this type, either with four plugs (each 300mm diameter) or with three plugs (2x300mm, 1x500mm). The GDC probes also use these ports. The 500mm ports are provided, for example, for the viewing periscopes which require close dimensional stability of the head mirror to the windows, while including a dog-legged labyrinth. It can be seen that a significant part of the large plug is unused by a single diagnostic system. Other systems, especially electrical sensors, will be integrated in the plug. The plugs are sealed to tubes in the seal plate by automatic welding, the only RH class 2 component on the primary vacuum boundary.

ii) The straight vacuum extension arrangement is used for directly coupled diagnostics (such as NPAs, VUV and X-ray spectrometry), where there is little requirement for replaceable equipment in the port shielding. The vacuum flange is extended with bellowed vacuum flight tubes to the diagnostic detectors. Two ports of this type are anticipated.

iii) Fanned collimation is required for the vertical neutron camera[13], again with no requirements to remove the plug but with some form of permanent collimating shielding in the secondary vacuum volume. Only one of this arrangement is foreseen.

Micro-waveguides are brought through remotely handleable windows mounted in tubes offset from the centre of the port flange. These appear on a number of ports.

Above the primary vacuum seal is one large port interspace, providing the secondary containment. It is connected to the primary seal plate by small bellows and is supported by the cryostat lid. The biological shielding is split, up to 1m lies directly on the top of the cryostat lid. The remaining 1.5m is above the open frame structure which supports both the cryostat vacuum and pressure loads and the biological shield weight. No diagnostic features may extend beyond this biological shield into the non-radiation area above. Diagnostic detectors and services are situated either in the port interspace, above the secondary seal plate or in the plenum between the two layers of shielding. In all design solutions the top of bio-shield is kept clear for other uses.

Replacement of all diagnostic equipment is from the top of the port. The port interspace enclosure is an important feature of the maintenance procedures, allowing the port to be accessed and plugs withdrawn without venting the cryostat. Maintenance at the top of the biological shield will generally be manual down to the secondary vacuum seal plate. For convenience the diagnostic equipment in the port interspace is made modular. As soon as the interspace is opened maintenance will be with standard remote handling tools because of the small aperture and long distance access as well as because of the radiation levels. A vertical plug containment flask is introduced through the bio-shield and docked to the secondary seal flanges.

Divertor Port

The central divertor cassettes in front of each of the four remote handling ports are specially adapted to meet the diagnostic requirements. Space for viewing and components has to be found within these diagnostic cassettes in areas that can use, as much as possible, the natural gaps between the cassettes. The cassette structural integrity and cooling provisions must be preserved. Channels are provided in the central plane of the diagnostic cassette for optical access. Microwave access is along the sides using the gaps between the target plates (10-25 mm). The two cassettes either side of a diagnostic cassette are instrumented with diagnostic sensors, such as magnetic pick up coils, bolometers and pressure gauges.

The divertor impurity monitor diagnostic[14] (Fig.6) illustrates many of the access features of optical diagnostics. Here the divertor area is probed radially by intersecting arrays of viewing chords for each leg. Mirror assemblies are located under the dome and a space under the cassette left by the removal of RH tools. A diagnostic tray is located here that is used for relay mirror elements. Mirror damage due to the deposition of target plate material during a major disruption is a main consideration in this region. Mirrors are protected as much as possible by collimating apertures in the structure. The concept of direct viewing between the cassettes has been developed. For this approach no mirrors are required in the divertor, but the side of the cassette must be locally modified and alignment is especially critical. During a pulse the port movement relative to the cryostat is 28mm in a radial-vertical plane (from relative thermal expansions) and 23mm toroidally (from magnetic movement of the vessel).

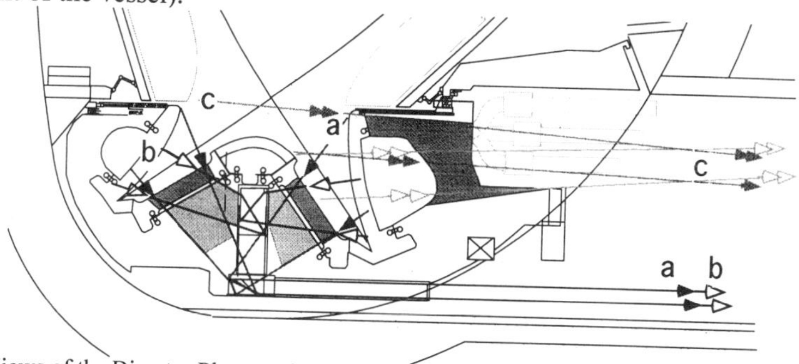

Figure 6. Views of the Divertor Plasma with Mirrors, a) on the Diagnostic Tray and b) in the Dome. Direct Views are given between cassettes (c).

The optical relay continues (Fig.7), if necessary, through intermediate optical elements in a diagnostic block located in the port, to windows in the vacuum boundaries. Mirror labyrinths or optical fibres bring light through the bio-shield plug. Outside the bio-shield the light is focused onto the different channels of the optical fibre array leading into the spectrometers. Some spectrometers are set directly outside the biological shield on the transport cask rails in the pit. Others (for longer wavelengths) are remote in the diagnostic hall, fed by long optical fibre bundles. These allow greater freedom of access to the spectrometers and help to minimise the access time to the divertor access ports.

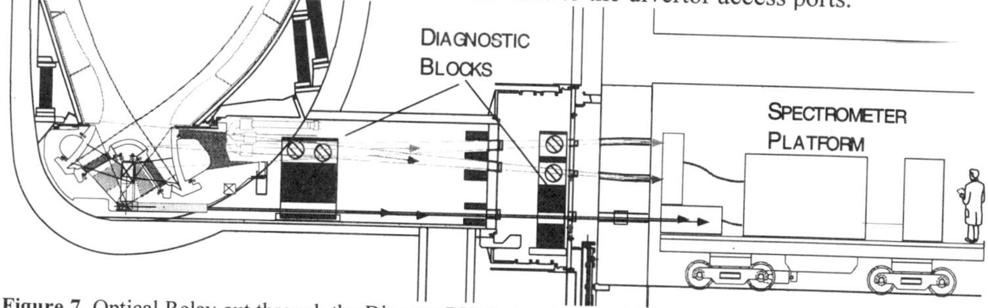

Figure 7. Optical Relay out through the Divertor RH Port, using the Diagnostic Tray under the Cassette and Diagnostic Blocks in port and in the interspace.

The divertor reflectometry system[15] and the divertor ECA each use a diagnostic cassette (sectors 8 & 18). The reflectometry system illustrates the solutions found for microwave diagnostic access. Special water cooled side plates are bolted on to the cassette (Fig.8), with the antennas and wave-guides running from the plasma viewing area to the outboard end of the cassette. There is space here for conversion and coupling boxes. A set of waveguides is mounted on a trolley within the port (Fig.9). These are articulated to make a narrow assembly for installation that is opened up for remote coupling to the back side of the divertor cassette. The outboard end of these waveguides is expanded on to waveguides and windows in the fixed part of the port flange. Within the port interspace there are modules of flexibly jointed waveguides.

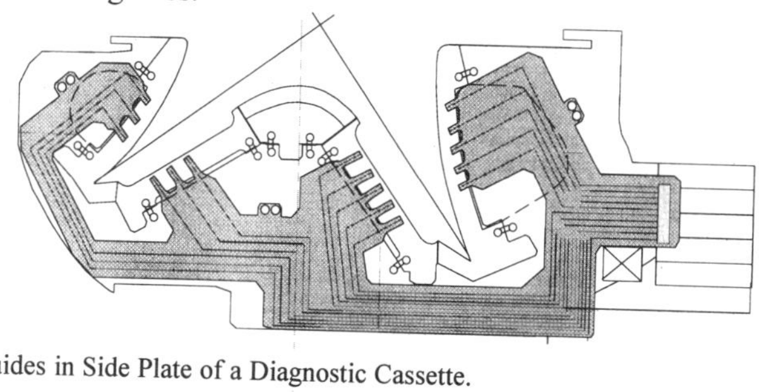

Figure 8. Waveguides in Side Plate of a Diagnostic Cassette.

The concept of side plates is also used on instrumented divertor cassettes, for electrical sensors including Langmuir probes, bolometer cameras, magnetic pick up coils, pressure gauges and Rogowski coils around earth straps. Wiring looms are routed to a connector at the back. During the installation of an instrumented cassette an electrical connector automatically engages with a socket mounted on the vacuum vessel wall. Permanent looms of mineral insulated cables are conduited to the diagnostic wiring in the cryo-pumping ports either side of the divertor RH ports. The divertor cassette handling operations are unaffected by the presence of diagnostic equipment. The remote maintenance of systems on the cassettes is entirely performed in the hot cell. The modular side plates are removed as the initial operation in the refurbishment of a cassette.

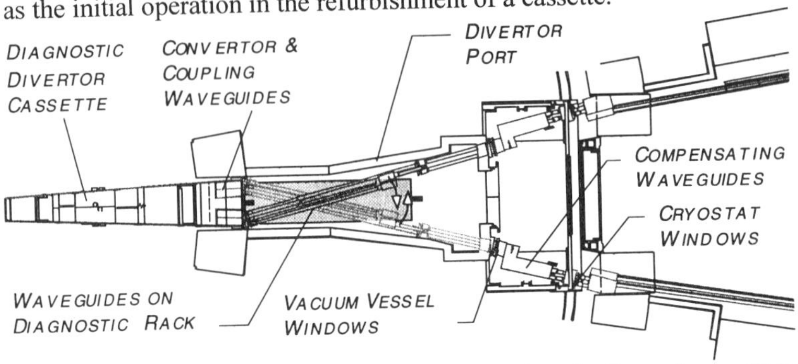

Figure 9.Waveguide relay for the Divertor Reflectometry System, from the diagnostic cassette to the pit.

As well as viewing the divertor plasma from within the cassettes, as much use as possible is made of the views from other parts of the Torus. The divertor IR thermography system, mounted in the equatorial ports, probes both the inner leg from a straight radial view and the outer leg by an oblique toroidal view. The vertical ports give a general view of the divertor dome and some special arrangements, used for example by the bolometry system[5], can allow the plasma in the inner leg to be viewed. The further requirement of bolometery, to probe the divertor plasma other than through the main plasma, has led to the development of a view from a downward viewing diagnostic socket in the back plate. The nominal gap between the cassettes is closed on the top outside by a gas tight seal.

Systems Installed In-vessel (Blanket & Back Pate)

Use is made of the 20mm gaps (both toroidal and poloidal) between the blanket modules, to give views of the plasma that cannot be attained from the other locations. Diagnostic sensors and components that are possibly degradable are made remotely removable by installing them in Diagnostic Plugs (Fig.10a & b) mounted in sockets in the back plate. Socket mounted diagnostics include the magnetic pick-up coils[16], bolometers[6], soft X-ray cameras, micro-fission chambers[17], irradiation foils, interferometer-retroreflectors, and polarimeter mirrors or retroreflectors.

In each sector of the back plate a row of diagnostic sockets are permanently welded under the poloidal gaps between blanket modules. The sockets are centred on poloidal gaps but slightly offset from toroidal gaps to give well shielded locations with plasma views and with the fastest response to field changes. Because of the blanket arrangement, some sockets (on the inner wall) are located on the back plate sector joints. During the construction of the back plate, groups of up to four sectors are assembled before installation in the Tokamak pit. These groups are then joined together in the pit. Some (12) of these rows of diagnostic locations are, therefore, better suited to sensors that require installation, connection and commissioning before back plate completion. The other toroidal locations are better used for systems, such as waveguide runs and neutron activation lines, installed later.

The magnetic pick-up coils are the most demanding in terms of position and quantity. Four orthogonal poloidal arrays of coils are provided with a further set for higher modes and redundancy. Bolometers are the heaviest users of cabling and the limitations for any site are given by wiring and connector space, at least 5 back plate sectors are used for these. In the diagnostic carriers a number of cable types (mineral insulated, steel and glass braided) and configurations (twisted pairs, coaxial, multi-axial etc.) are under consideration.

The diagnostic plug is installed, when the two covering blanket modules are removed, with a simple quarter turn bayonet action that both locates the plug and provides good thermal contact to the cooled socket. Remote removal is possible with a single dedicated tool mounted on the In-Vessel Transport system (IVT).

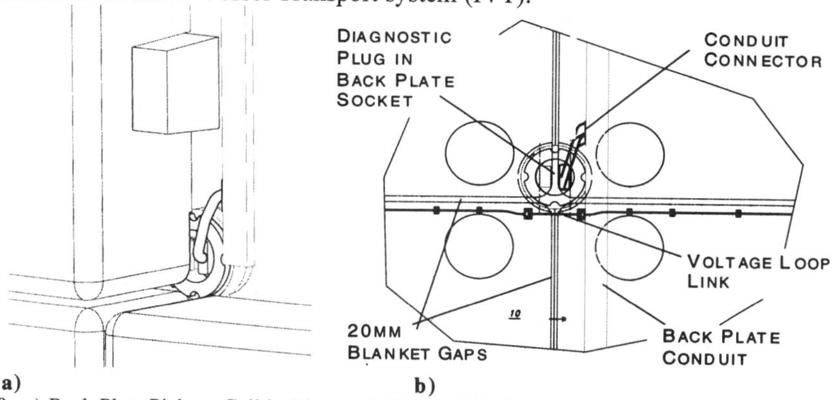

Fig 10. a) Back Plate Pick-up Coil in Diagnostic Plug and b) Voltage Loop behind Blanket Modules.

As well as these discrete sensors there is a need to provide for extensive diagnostics on the back plate. The voltage and saddle loops of the magnetic system are attached directly to the back plate in regions protected by the blanket modules. Loops are provided with joints that are made permanent on assembly of the adjacent sectors. They can be replaced with secondary links in the event of a possible sector replacement but are generally considered to be permanent. Reflectometry waveguides and Neutron Activation sample lines[18] run poloidally to the vertical port in sectors without electrical sensors using the conduit grooves in the blanket.

Most of these back plate diagnostics require a wiring connection. Diagnostics in carriers can be directly wired (e.g. some magnetic pick up coils) or incorporate a connector in the carrier (e.g. Microfission chambers). Special provisions are made for bolometer camera arrays, because of their large quantity of wires, with connector sockets on the sides of the conduits. The back plate wiring is permanent and so only mineral insulated wires (with stainless steel sheaths and Alumina insulation) are considered.

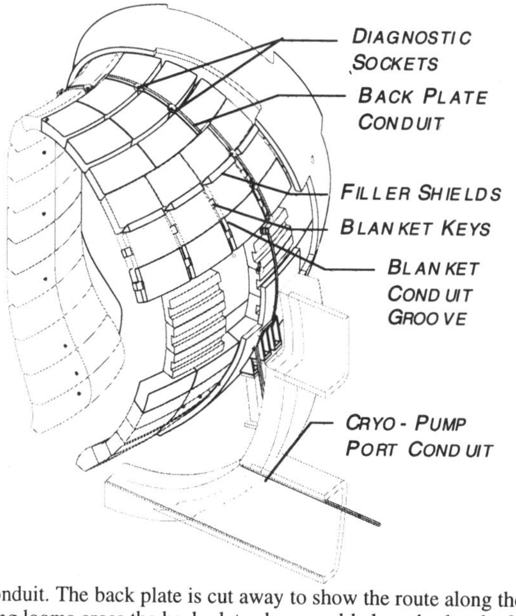

Fig 11 Back Plate Conduit. The back plate is cut away to show the route along the sides of the blanket modules, wiring looms cross the back plate above and below the level of the equatorial port holes

These will be vacuum sealed, even where they do not cross vacuum boundaries, to minimise the outgassing and to guarantee the integrity while in store prior to installation. Throughout its length, there will be a number of joints in the cabling, militated by the assembly sequence of the back plate and vacuum vessel. Where possible these joints and the electrical connections will be 'fritted' (with a high current pulse) or spot welded to ensure their permanence. The cabling of each sector will be tested before it is installed in the pit. A conduit system is welded to the plasma side of the back plate. The conduit is situated under the protection of blanket modules, it runs generally poloidally within one sector with short toroidal steps where the blanket segmentation changes (Fig.11). There is a complementary groove in the back of the blanket module over its full poloidal length. The wiring in the conduits exits through holes in the back plate in the region near the equatorial ports. Outside the back plate, the conduit is continued down the support legs to the vacuum vessel wall. Here the wiring joins that from loops and coils of the vacuum vessel diagnostics and from the divertor. Vacuum Vessel Loops also can be replaced with secondary links in the event of a possible Torus sector replacement. Final exit from the vacuum vessel is made along the divertor cryo-pumping ports with standard vacuum feedthroughs at the two sealing flanges. These ports have been selected as they provide the best distribution of almost complete toroidal access with a high level of standardisation and infrequent RH access requirements.

CONCLUSIONS & SUMMARY

Design solutions have been developed to give specific diagnostics viewing, signal transmission and maintenance access. Most of the planned diagnostics for ITER have been integrated in the available locations. The extreme radiation environment, life and safety requirements have been taken into consideration. Some design and integration issues still remain to be resolved. For the equatorial ports the handling of the port plugs must be developed in more detail. For vertical port access an important issue is the inability to view the full plasma and the design of the direct coupled diagnostics. Other developments required include blanket shield plug compatibility with the main blanket assembly, and definition of the allowable slot dimensions in the back plate. The absence of IVV on the vertical neutron camera port must be compensated for and maintenance of it and its shield lock remain to be developed. In the divertor region, for viewing through natural gaps, innovative solutions are required for compensation of the differential movements for direct port viewing. The thermal considerations of each back plate diagnostic require further analysis.

Acknowledgements

This report is an account of work undertaken within the framework of the ITER EDA Agreement. The views and opinions expressed herein do not necessarily reflect those of the Parties to the ITER Agreement, the IAEA or any agency thereof. Dissemination of the information in this paper is governed by the applicable terms of the ITER EDA Agreement.

References

1. C.I.Walker, L.deKock, *ITER Plasma Diagnostics Generic Access*, Varenna 1995
2. G.Janeschitz et al., *ITER Design and Diagnostic Integration*, these proceedings.
3. P.Edmonds et al., *Integration of Vacuum Coupled Diagnostics*, ibid.
4. P.Stott et al., *Plasma Viewing System*, ibid.
5. F.Marcus et al, *A Radial Neutron Camera Design for ITER*, ibid.
6. R.Reichle, et al,. *Reference Design for ITER Bolometers*. ibid.
7. R.Snider et al,. *Application of Interferometry and Plasma Rotation Techniques for Density Measurements on ITER*:, Varenna. 1995.
8. C.Walker, *Vacuum Window Seals for Diagnostic Windows*, (ITER Task D19),1995
9. A.Nagashima .*Development of New Vacuum Window Seal* ...., these proceedings.
10. G.Shatalov, *Neutronics For ITER Diagnostic Systems and Ports*, ibid..
11. H.Salzmann et al,. *ITER Core Plasma LIDAR System*, ibid.
12. G.Vayakis et al, *Overview of The ITER Reflectometry Diagnostic Systems*, ibid.
13. L.Johnson et al,., *Overview of Fusion Product Diagnostics*: ibid.
14. T.Sugie, *Divertor Impurity Monitor for ITER*, ibid.
15. M.Manso et al., *Reflectometry in the ITER Divertor*, ibid.
16. L.deKock et al., *Magnetic Diagnostics for ITER*, ibid.
17. T.Nishitani, *Micro-Fission Chamber Neutron Monitor*, ibid.
18. C.Barnes, *Design of Neutron Activation & Monitors*, ibid.

DIAGNOSTIC REQUIREMENTS FOR THE ITER DIVERTOR

L. de Kock[1], G. Janeschitz[1], A. Kukushkin[1], H. D. Pacher[3], T. Ando[1], A. Costley[2], K. Ebisawa[2], M. Sugihara[1], G. Vayakis[2], C. Walker[1]

1. ITER JCT, Boltzmannstrasse 2, D-85748 Garching, Germany
2. ITER JCT, 11025 N. Torrey Pines Rd, La Jolla, CA 92037, USA
3. The NET Team, Boltzmannstrasse 2, D-85748 Garching, Germany

1. INTRODUCTION

The requirements for the divertor diagnostics have originally been determined by the examination of the diagnostic experience on present large divertor tokamaks. In the past years the design of most diagnostic systems inclusive of those for the divertor has progressed beyond the conceptual stage so that their expected performance can now be predicted using this design information and the results of 2D modelling of the divertor plasma in a regime close to the desired operating point. The profiles of plasma parameters and the line integrals of spectral emission, which would be measured along the chords of the planned diagnostics, have been calculated for 3 cases for which the 2D model predicts sustainable heat loads on the divertor target. The results are assessed according to their capability to fulfil the requirements for protection, plasma control and physics understanding.

The paper describes in detail the preferred plasma configuration as obtained from the 2D model, the reference case, and the role of the diagnostic systems in kinetic control and physics evaluation and optimisation. It also shows a selection of relevant diagnostic profiles and line integrals for the 3 cases in the relevant parameter range. In the last section it assesses each diagnostic system in detail and concludes with recommendations.

2. 2D MODELLING OF ITER DIVERTOR PARAMETERS

ITER divertor operation has been described most recently in [1,2,3,4]. It is dominated by the necessity to exhaust around 200 MW via the scrape-off layer (SOL), assuring simultaneously good core confinement, adequate helium pumping, and peak power loads in the order of 5 MW/m^2 on the divertor targets. A strong reduction of both energy and particle flux near the strike point is needed to meet the latter requirement. Therefore, fully attached operation of the divertor plasma at the plates is unacceptable and the plasma must be either partially detached (detached near the strike point, attached further out in the SOL) or fully detached. In order to attain this partially or fully detached state, a large fraction of the SOL power must be radiated by impurities, either intrinsic (He plus C and/or Be) or seeded (e.g.

neon). Many present-day experiments (e.g. JET, ASDEX Upgrade) have shown that, as the radiation fraction and/or density increases and the divertor plasma proceeds toward full detachment, the core confinement is degraded, an effect which could be related to cooling of the edge of the main plasma, either by an increase in the radiation from closed flux surfaces at the plasma edge inside the separatrix or by the penetration of neutrals towards the core plasma. For fully detached operation, radiation is located near the X-point because the ionisation front and the radiation peak move close to the X point, cooling the edge plasma, and the amount of neutrals in the main chamber is expected to increase because the plasma plugging decreases and the divertor neutral pressure increases. A partially detached operating point for ITER, which avoids strong radiation from the area around the X-point, is therefore preferred.

Such an operation point has been found by 2D modelling for ITER [4]. Carbon physical and chemical sputtering (using a simplified expression) at the target plate provide a self-consistent influx of light impurities, a small amount (0.2%) of Ne seeding is added, and 10% concentration of He is assumed at the core-edge interface. Under these conditions, 140 MW are radiated from the divertor and only about 60 MW are transported to the plates by plasma and neutrals. The plasma is more detached on the inner divertor leg than on the outer, but remains partially attached for both. The plasma state can be clearly seen in Fig. 1, giving the electron temperature and density. The electron temperature drops strongly toward the plates, but the peak electron temperature remains above 10 eV in the divertor SOL region for both divertor legs, so the plasma remains partially attached. However, the temperature near the separatrix strike point is much lower and the pressure in the flux tubes close to the separatrix drops by an order of magnitude toward the plates, thus the plasma is detached on those flux tubes. The low electron temperature near the strike point favours penetration of neutrals (both DT and He) produced by recycling on the outer part of the vertical target toward the strike point and the private flux region, increasing the neutral density near the strike point to several 10^{20} m^{-3}. Neutrals from the private flux region can penetrate into the SOL plasma upstream because of the strong radial gradients on the private flux side (Fig. 1), enhancing somewhat the radial pressure profile upstream. Moreover, the high pressure in the private flux region favours He pumping.

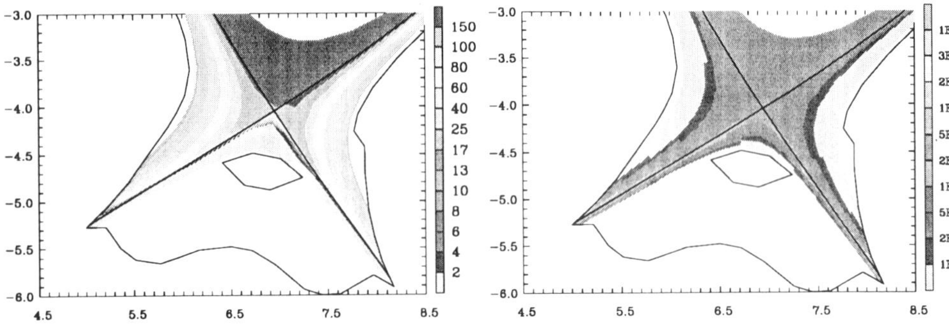

Fig. 1: 2D profile of electron temperature (eV) and density (m^{-3}) in the divertor region for partially attached case

In order to avoid strong radiation from the edge plasma inside the separatrix, and also strong local radiation loads near the strike point, it is important that the radiation source be well-distributed along the entire divertor leg. For the partially detached case presented here, the distribution of radiation power per unit volume for carbon (which provides most of the radiation) is shown in Fig. 2, demonstrating this favourable poloidal distribution.

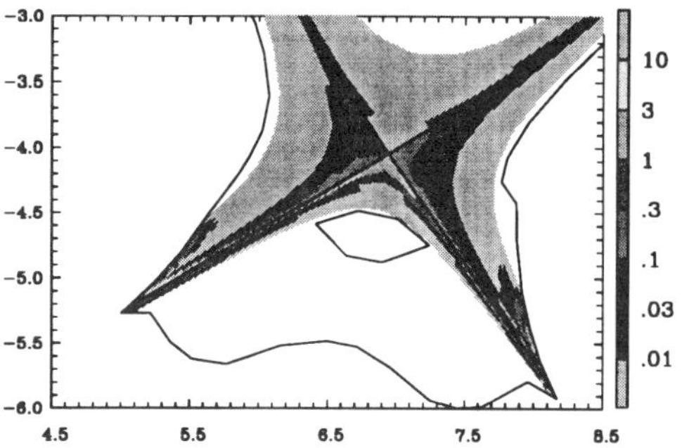

Fig. 2: 2D profile of radiation power per unit volume (MW/m^3) from all carbon charge states for partially detached case.

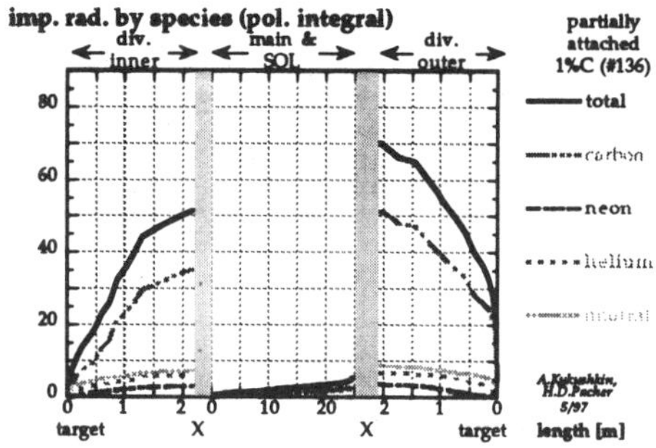

Fig. 3: Cumulative poloidal integral of impurity radiation in inner divertor, edge region above X-point, and outer divertor, for partially detached case.

This distribution is shown also in Fig. 3. Here, the cumulative poloidal integral of the power radiated between a reference point (the targets for the divertor legs, the X point for the edge) and the point under consideration, is shown. (Radial integration is performed first along the cells of the computational grid, which is skewed so as to be parallel to the targets at the strike point and orthogonal to the surfaces at the X point; poloidal distance is along the separatrix). It is seen that essentially all the radiation comes from the divertor legs (only 5 MW from the edge plasma), and that, except for 30 MW radiated close to the outer target, the remaining 110 MW of radiated power is well spread out along the divertor legs.

The partially detached case shown here as the preferred operating point results in a Zeff of 1.8 at the core boundary and peak power loads on the target of about 5 MW/m^2. Furthermore, an average He partial pressure in the private flux region of about 0.1 Pa is predicted. This is adequate to exhaust the He produced, 2 Pa-m^3/s, at the design pumping speed of 200 m^3/s with some margin. This operating point is therefore favourable in several respects: low radiated power in the main plasma, low neutral density in the main chamber, well-distributed radiation, low peak power on the plates, and adequate He pumping.

3. ROLE OF DIAGNOSTIC MEASUREMENTS IN THE DIVERTOR

To bring the divertor plasma into a regime in which the heat load to the target can be sustained, and to keep it in this condition, is one of the most challenging tasks in ITER, both in terms of control and physics understanding.

The requirements for the measurements for control and for physics understanding are quite different. For control, i.e. as part of a feedback loop, we require robust and simple signals without complicated interpretation, where robustness is achieved by redundancy and back up with related diagnostics. The control signals must show the essential features of the parameter to be controlled e.g. the degree of detachment. In [5] the measurements for control and protection and their specifications have been identified in category 1. For physics evaluation we require detailed information on the essential physics aspects of the divertor. These are mainly the profiles of density and temperature along flux tubes close to the separatrix, the impurity content as distributed in the divertor legs, and the localised He density. The profiles are generally very steep perpendicular to the flux surfaces leading to a demand for high spatial resolution. The measurements have been specified in [5] as category 2. The status of the design and the assessment of the capability of the proposed diagnostic systems is given in [6].

Control and Protection

The main diagnostic for the protection of the target plates is IR thermography of the target. The diagnostic purpose is to prevent too high target temperatures by a feedback loop linked to a suitable actuator. A detailed power deposition profile is not required, but small areas of excessive surface temperature must be distinguished. Ultimately a controlled shut-down action may be necessary. This measurement needs to be supported by an impurity influx monitor confirming the large impurity influx due to excessive local power load.

For kinetic control of the divertor plasma it will be necessary to provide the diagnostic tools to keep the plasma in a partially detached regime in both divertor legs, the regime shown in detail in section 2. This depends very sensitively on where the power is radiated, on where the plasma looses contact with the target plates, and on the plasma temperature on and near the target. One of the diagnostics having the required potential are the Langmuir probes mounted between some of the cassettes. These take the form of triple probes giving instantaneous measurements of density, temperature and floating potential. The saturation current as measured by the Langmuir probes is a well tested relative indicator for detachment. The obvious disadvantage is that Langmuir probes rely on plasma contact and are therefore prone to excessive erosion from disruptive events, limiting the lifetime. The other readily accessible signals are the line emission profiles along the separatrix. For well chosen lines these represent the local temperatures and have the required characteristics. From the same viewing chord the impurity influx along the target plate can be obtained giving similar information on the degree of detachment, but without indicating where the flame front is. Other signals which have the required robustness and capability are the bolometer measurements, which indicate where the power is radiated.

The divertor control system will be a multiple input/multiple output system. The actuators can only be gas feeds of various kinds and the heating systems. The kinetic control loops for divertor operation are being developed in the major divertor tokamaks and a substantial improvement in performance is to be expected with a concurrent increase in diagnostic input. Therefore the ITER diagnostic set must be designed to have sufficient capability to expand beyond today's standards. Because of the complexity of the control problem the (kinetic) control of the divertor of ITER is being modelled numerically to gain insight in the possibilities and limitations [5].

Physics Evaluation and Optimisation

For physics evaluation and understanding, the requirements on the diagnostic systems are very demanding. The amount of information would ideally need to be of comparable quality to that given by the 2D divertor model. This requires high spatial resolution of the 2D profiles expanding the requirements far beyond those for control. This paper endeavours to quantify some of these requirements.

For the divertor to operate in the desired parameter regime with partially detached plasmas in both legs a delicate internally consistent balance has to be struck between local density and temperature, the amount of power radiated by the different impurity species and their localisation, and the ability to pump the He ash. In order to understand the physics mechanisms the requirements for the diagnostic system for the local parameters n_e and T_e show high spatial resolution (10 cm parallel and 0.3 cm perpendicular to the flux lines). However, for good understanding we need a similar resolution for the local impurity species. Ideally this should consist of a system with measurements in the visible and VUV in both divertor legs separately. Imaging systems with observation points from perpendicular directions would be needed to achieve the required spatial resolution. Such a system would provide a tomographic 2D profile of the impurity line radiation, which needs to interpreted using the local plasma parameters. It is clear that this ideal situation can only partly be achieved in the ITER environment with its restricted access and harsh environment. For the physics evaluation and understanding it means that we need to combine the measurements of the various available diagnostics and use them to our best interpretation capability. Other measurements providing essential information on the degree of detachment are all observations on the target. They need to show the power deposition (from IR thermography), recycling (from measurements in visible light) and the target parameters (from Langmuir Probes) with good (1 cm) spatial resolution.

This overview of the requirements on the divertor diagnostic in the light of the protection, kinetic control and the physics evaluation and understanding shows that in practice there will be a hierachy of measurements. Some are very basic, both from an interpretational and technical point of view, others seem to be ambitious either because of interpretation difficulties or the technical implementation. The scientific and technical design effort endeavours to resolve these difficulties.

4. PROFILES AND LINE INTEGRALS ALONG DIAGNOSTIC CHORDS FROM MODELLING

In order to assess the capabilities of the proposed divertor diagnostics, observation chords have been taken from the present diagnostic designs. The 2D divertor model B2-EIRENE has been used to calculate the relevant plasma parameter profiles and line integrals of spectral emission along these chords. Where no satisfactory diagnostic design seemed to exist, explorative profiles and integrals have been produced along chords of a possible diagnostic system to show the kind of information that is needed for an evaluation of the detailed physics of the divertor plasma. This was the case for the divertor Thomson Scattering where only a single chord has been considered as an annex to the X-point LIDAR system [8].

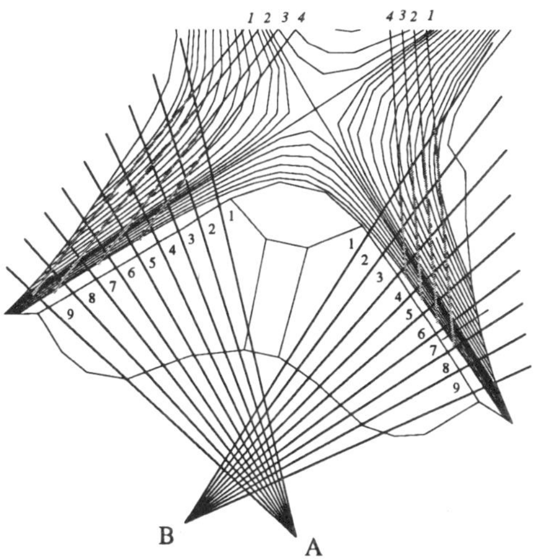

Fig.4 a: Spectroscopic observation channels in inner (A) and outer (B) divertor. Input beams of a possible Thomson Scattering system are show in inner and outer divertor

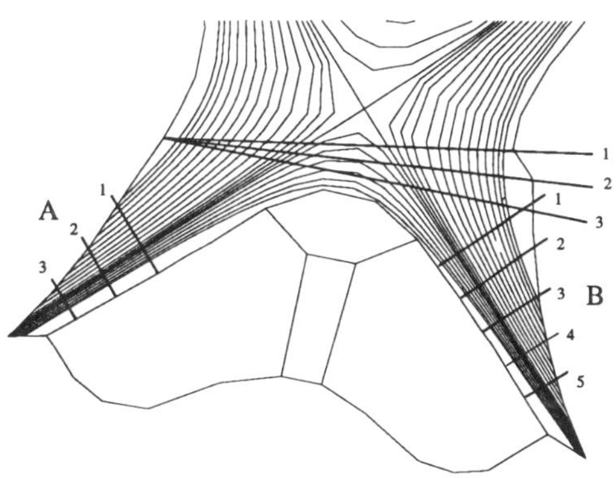

Fig. 4 b: Chords of the O-mode reflectometers in inner (A1,2,3) and outer (B1,2,3,4,5) divertor. X-point LIDAR chords go to the dump on the left. Channel 2 is existing. Channel 1,3 are not foreseen now.

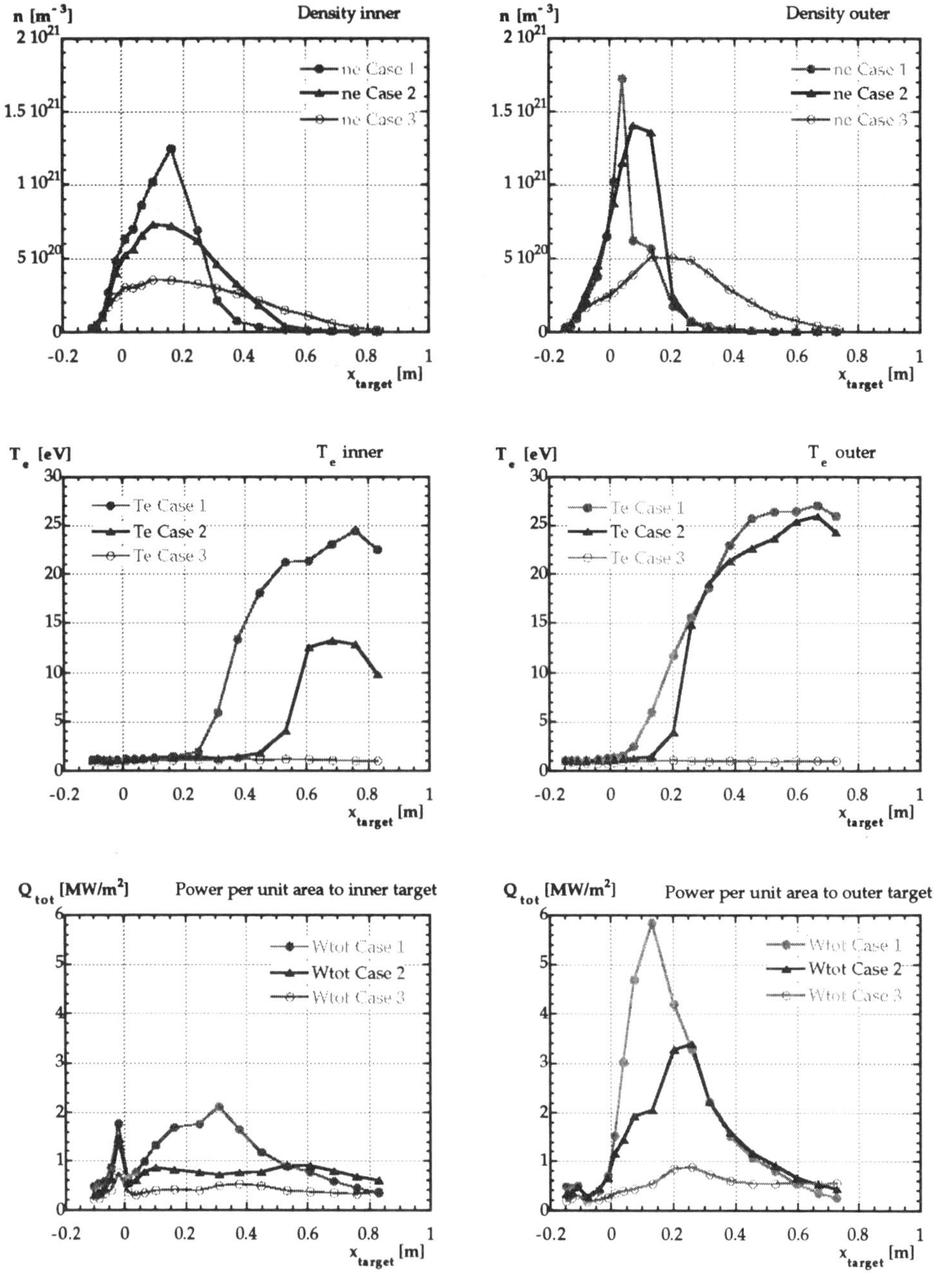

Figure 5: Electron density and temperature along the target plates for 3 cases (1-semi-detahed reference, 2-inner leg starts to detach, 3-fully detached). Also shown is the deposited power including that from optical (localised) radiation. Note that on the horizontal axis zero is the strike point.

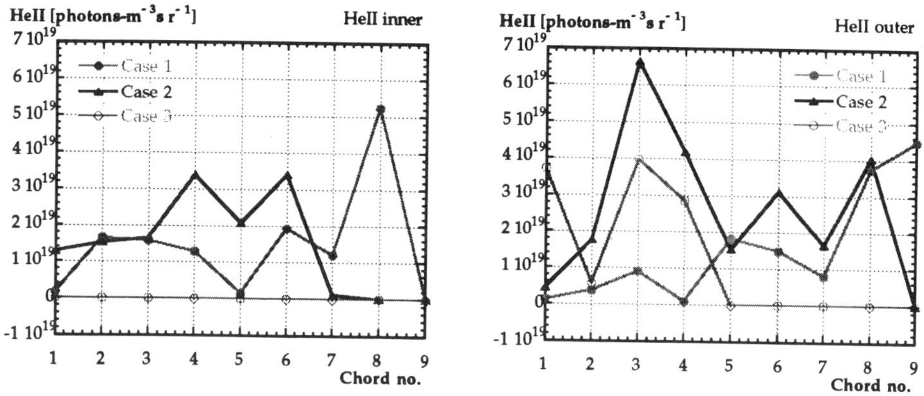

Figure 6: Line integrated He II radiation as seen by the arrays A and B for the 3 cases

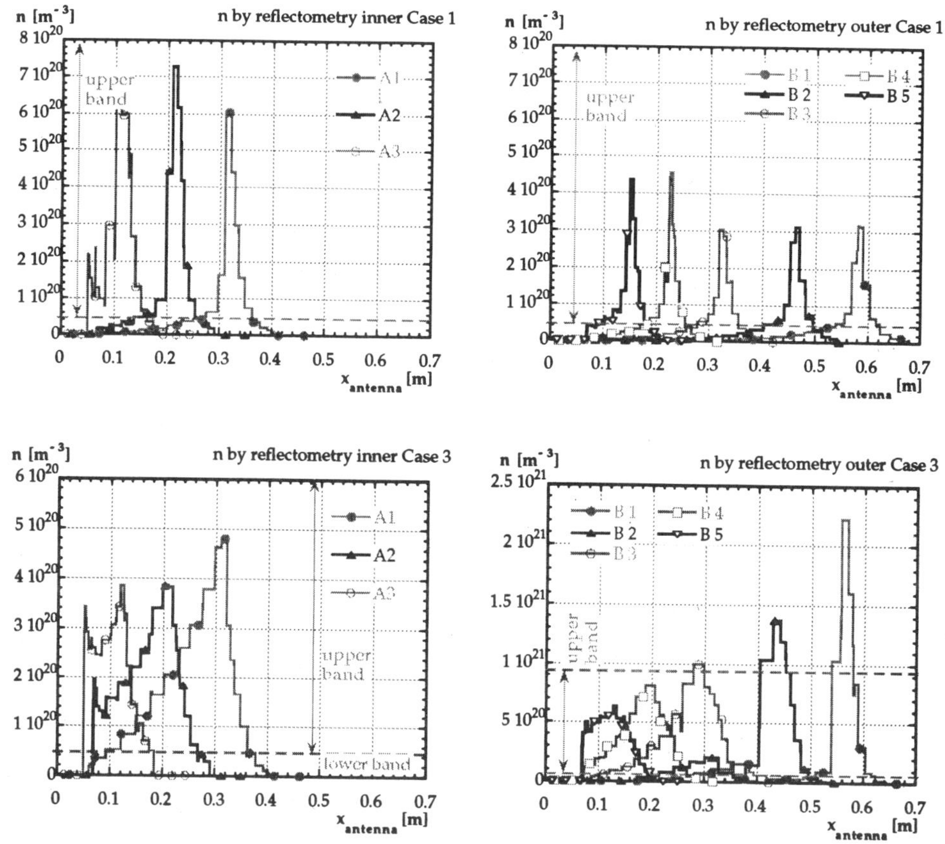

Figure 7: Density profiles along the chords of the O-mode reflectometer. Note that on the horizontal axis zero represents the mouth of the antenna. Moving from the top to the strike point the distance to the density peak becomes gradually smaller. Except in channels B1 and 2 the onset of the density profile is the vertical target plate.

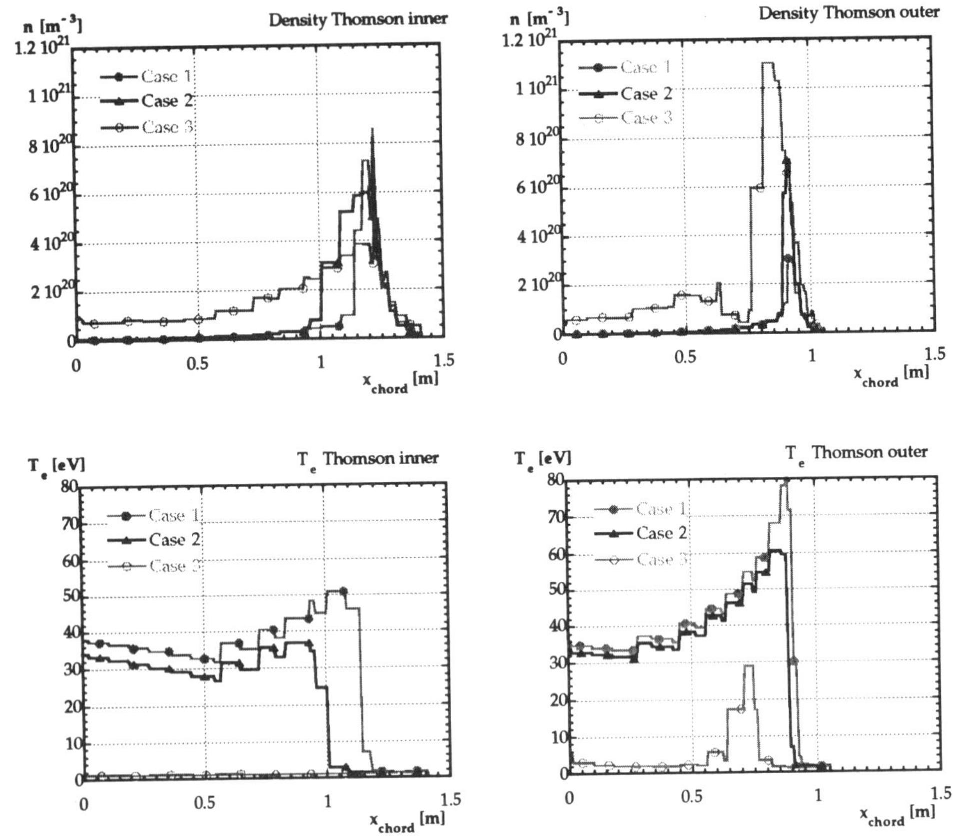

Figure 8: Electron density and temperature profiles as seen by the spectroscopic arrays A and B along chord 2 of the laser input beam of the Thomson Scattering system for the 3 cases.

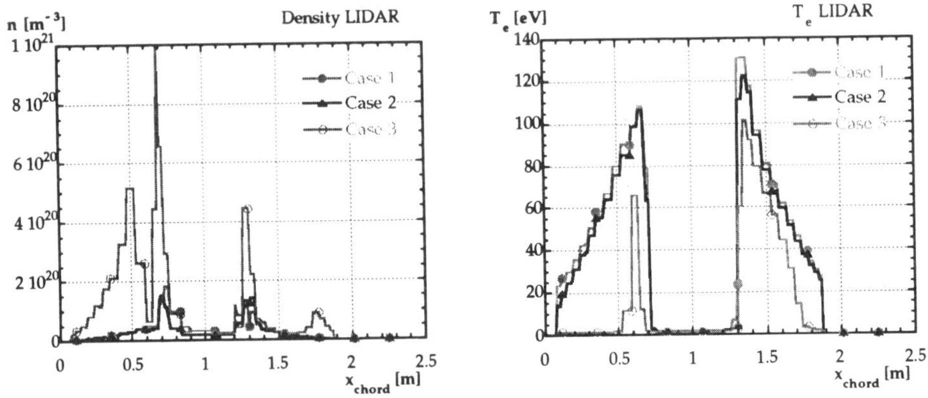

Figure 9: Electron density and temperature along chord 2 of the X-point LIDAR system for the 3 cases.

Profiles have been calculated along the targets and along the chords for reflectometry, a Thomson Scattering system in the divertor legs, and the X-point LIDAR system. Line integrals of the radiation from impurity species (He II, Ne II, Ne III, etc, CII, CIII, etc.) have been evaluated along the optical observation chords of the impurity monitoring system. The calculations have been performed for 3 cases:

1. The semi-attached reference case of section 2.
2. The case for which the inner leg starts to detach.
3. The case for which both legs are fully detached.

The layout of the diagnostic chords is shown in Figure 4a and b. A selection of representative examples, chosen to show a clear distinction between the 3 cases, are shown in Figures 5, 6, 7, 8, 9.

Figure 5 shows the n_e, T_e and power deposition profiles for the 3 cases along the target. From the deposited power and the temperature profiles the differences can be clearly distinguished. Note that the power deposition from optical (not from nuclear) radiation is included.

Figure 6 shows the profiles of the He II luminosity (line integral) across the spectroscopic arrays viewing the plasma from the bottom mirror in the diagnostic cassette. The views from the mirrors under the dome are not significantly different (not shown here).

Figure 7 shows the density profiles along the reflectometer viewing chords in one inner and one outer chord with the indications of the limits of the lower (20-60 GHz) and higher (60-280 GHz) O-mode bands. Along the horizontal axis zero corresponds to the antenna mouth and for most cases the onset of density corresponds to the target surface (except for B1 and B2). The fully detached outer leg shows double-humped profiles that will make interpretation difficult. It is clear that the lower O-mode band (4.8×10^{18}-4.3×10^{19} m^{-3}) is important in the phase leading up to the desired operating point.

Figure 8 shows the n_e and T_e profiles along selected chords in the inner and outer leg with observation points taken from the spectroscopic arrays representing a Thomson Scattering arrangement. The laser input beam is represented by 4 parallel chords. Profiles are given as function of the chord distance along chord 2 of the input beam between crossings of array A and B. Since the input beam closely follows the poloidal flux lines, the T_e profile is broad and the n_e profile shows the characteristic peak at the location of the ionisation front. There is no sharp qualitative distinction between attached and detached regimes, but the numerical values are different. A good spatial resolution of n_e and T_e, if achieved in this way, will, in combination with a sufficiently detailed coverage of spectroscopic observation chords, be essential to unfold the spectroscopic signals in terms of local impurity density.

Figure 9 shows the n_e and T_e profiles along the chord of the X-point LIDAR system [8]. The input laser beam has a dump between the cassette and the lower edge of the back plate, determining its location. The data from this diagnostic would be the basis for obtaining an impurity species distribution around the X-point if the spectroscopic measurements were also available.

5. ASSESSMENT OF THE DIAGNOSTIC CAPABILITY

With the predicted plasma parameter ranges and the spatial profiles derived from the numerical modelling the capability of the diagnostic systems to fulfil the requirements for control and physics understanding can now be examined. However, it is important to recognise that this analysis reflects the present state of the design and the modelling and that this will develop further, which may change the conclusions of this assessment. Moreover, a major question, and a challenge for the diagnostics, is the stability of the partially detached solution. For example, JET [8] finds, for L-mode partially detached plasmas, oscillations of the divertor plasma corresponding to a movement of the peak radiation zone from the target to the X-point and back on a millisecond time scale. Whether such oscillations are to be expected for the ITER conditions and divertor length remains to be investigated. These effects have not been considered here and have consequences for the required time resolution of the diagnostics and the response time of the actuators.

In the assessment either for control or physics understanding we will find that the diagnostic systems have to be considered an interrelated system and not as single individual systems.

Control and protection

We assume that the magnetic configuration has been established independently by the PF system and kept under feedback control. For the protection of the target the specification of the IR Thermography covers the range and resolution adequately and the impurity influx monitor will support this measurement. In particular it will help to distinguish hot spots.

For the kinetic control of the divertor we assume that the desired operating point has been reached. Then the main task is to keep the plasma in the semi-detached state. The profiles along the target (figure 5) indicate that the clearest signal showing the degree of detachment is the plasma temperature profile. This is difficult to measure with the Langmuir probes but temperature effects are clearly seen in the intensity of the impurity line radiation. This is shown in figure 6 for He II. Similar profiles can be found for other impurities such as C III (not shown here). The line integrals of He II as seen by the arrays A and B give a very sensitive signature of the degree of detachment with adequate spatial resolution. Important supporting evidence is provided by the IR Thermography showing large variations of the deposited power.

The role of the bolometers for control is to ensure that most power is radiated from the divertor channels and not near the X-point. This will be accomplished by selecting a limited number of channels which characterise these areas clearly.

Examining the profiles obtained for Thomson Scattering, reflectometers and X-point LIDAR it appears that there is no clear and simple signature of the degree of detachment. These measurements have a supporting role in the kinetic control of the operating point and some may be essential in the pulse scenario leading from initialisation to ignition. An example is the divertor density from the interferometer function of the reflectometer higher O-mode system.

In the overall picture there seems adequate capability to control the operating point of the divertor using the proposed diagnostic systems. How these diagnostics will be integrated into the control algorithm remains a point of further study.

Physics Evaluation and Understanding

The reference frame for the evaluation and understanding is the magnetic configuration. This has an estimated absolute accuracy of ±2 cm. Since the diagnostic systems are referenced to the external frame, i.e. the structural elements of the machine, the relation between the two has to be established with the spatial accuracy of the diagnostics (±3 mm). A good reference point will be the ridge of electron density on the separatrix (from reflectometer measurements) and the magnetic configuration should be adjusted to match this. This method should just achieve the required accuracy.

The Thomson Scattering system will provide the basis for the 2D plasma parameter profile in the form of an array of discrete measurement points. The design of this system has not yet advanced sufficiently to be confident that it can be built. The obstacles to overcome are alignment between input and output, the low signal to noise ratio due to the restricted access and the limitations in input power, and the calibration system. The Thomson Scattering on its own provides the 10 cm resolution along the flux surfaces but cannot match the 0.3 cm perpendicular resolution. The reflectometers can in principle provide these high resolution profiles at a limited number of locations along the leg. These profiles, together with the discrete measurement points of the scattering system, should help reconstruct the spatial distribution of n_e. A combination of low and high frequency O-mode reflectometers in the cm/mm bands (<300 GHz) can measure over most of the profile, including the lower density profiles immediately after the X-point formation (figure 7). Theoretically the profiles can be measured with the required accuracy. For fully detached cases (channels B1 and B2) one can also encounter double-humped profiles, which increase reconstruction errors[12]. The practical success of this measurement depends on the quality of the wave guide routing, as (negative) experience on other tokamaks shows.

The impurity measurement system has a limited number of channels in the array but just sufficient to match the parallel resolution. The measurements shown in this paper are from mirrors in location A and B. There is also a set of mirrors higher up closer to the X-point. Profiles from there show a very similar behaviour casting doubt on the ability of this system to provide the perpendicular resolution. Ideally one would need a view looking into the divertor channel from the mid plane. A 2D tomographic luminosity measurement of the different impurity lines would need to be obtained to be able to produce the interpretation in terms of localised impurity densities. Such a requirement is not quoted in the diagnostic specifications [5] and will certainly be difficult to achieve. However, other diagnostic systems will help in the interpretation. The bolometer system will have cameras specially adapted to the divertor and the number and location suggest a best spatial resolution of 5 cm [7].

At the target very detailed measurements can be obtained. The Langmuir probes have a pitch of 1 cm with tips of 3 mm wide. This allows an unrestricted profile measurement of n_e and T_e on the target. The same is expected for the surface temperature and power depostion profile from the IR Thermography. The range of temperatures is adequate, but estimates of the error bars indicate that the required accuracy cannot easily be obtained at low surface temperature [11]. The impurity measurement in the visible part of the spectrum will quantify the impurity influxes from the target and so support the overall picture of the impurities.

Near the X-point the flux expansion allows the application of a LIDAR system to obtain the parameter profile upstream of the divertor [12]. The area around the X-point is also a possible major source of line radiation. The main use of the data from the LIDAR system would therefore be used for the analysis of this radiation but no spectroscopic system is planned for this area. As preliminary studies show, in the near UV (200-500 nm), it will be accessible through the slot between the target tiles of the divertor cassettes as used for the X-point LIDAR. Other views from the equatorial port will be necessary to obtain a tomographic analysis as a basis for the measurement of the density of the radiating species.

Not mentioned so far are the pressure gauges, which will be important to assess the pressure in the private flux region in front of the pumping port, determining the pumping speed. This is a strong function of detachment. The system will also comprise several Penning type heads adapted for the observation of He II lines from which the local He concentration can be obtained.

The overall picture is that on the target the planned diagnostics meet most of the specifications for physics evaluation and understanding. In the divertor throat the present design of the impurity measurement system will just meet the parallel resolution but not the perpendicular one. A view into the divertor channel will be needed to achieve this. It remains questionable if the local impurity content can be obtained. Required for this analysis is a high spatial resolution 2D parameter profile. The result of reduced parameter and impurity information is larger uncertainties in the evaluation and understanding of the physics. Therefore the diagnostic design effort should aim to improve the coverage of the divertor region.

6. CONCLUSION AND RECOMMENDATIONS

From the simulated profiles and line integrals it can be concluded that there is adequate capability for kinetic control of the operating point of the divertor. The control requirements of the development stages of the plasma have not been considered and need a separate study. The main control inputs are provided by the Langmuir probes and the line integrals of the spectroscopic arrays. The reflectometers have a supporting role in the control system due to their robustness and they provide an essential confirmation of the density range. The target protection can be achieved by the IR Thermography and the impurity influx from the spectroscopic arrays

For physics understanding to a level that is judged to be necessary, the 2D profile information in both divertor legs and near the X-point will be inadequate in the present situation. 2D n_e and T_e information in the divertor legs would have to be obtained from a Thomson scattering arrangement as suggested and a quantitative interpretation of the spectroscopy is only possible with high resolution primary chords as can be obtained with imaging systems. The configuration suggests that a view from the equatorial port will also be necessary. Near the X-point the LIDAR system should provide a single parameter profile with doubts about the technical realisation. To achieve the desired performance further design effort is necessary. The target measurements have adequate resolution and range for the physics evaluation.

Acknowledgements:

This report is an account of work undertaken within the framework of the ITER EDA Agreement. The views and opinions expressed herein do not necessarily reflect those of the Parties to the ITER Agreement, the IAEA or any agency thereof. Dissemination of the information in this paper is governed by the applicable terms of the ITER EDA Agreement.

References:

1. A. Kukushkin, H.D. Pacher, et al. 12th PSI Conference, St. Raphael, 1996
2. A. Kukushkin, H. D. Pacher, et al. 16th IAEA Fusion Energy Conference, Montreal, 1996
3. G. Janeschitz, H.D.Pacher et al, 16th IAEA Fusion Energy Conference, Montreal, 1996
4. A. Kukushkin, H.D. Pacher, *et al.*, 24th EPS Conference on Controlled Fusion and Plasma Physics, Berchtesgaden, Germany, 9–13 June, 1997. Paper P3.013
5. V.S. Mukhovatov et al., these proceedings
6. A.E. Costley et al, these proceedings
7. R. Reichle et al., these proceedings
8. A. Loarte et al, to be publ. in Nucl. Fusion (1997)
9. L. de Kock et al, these proceedings
10. A. Herrmann, H. Salzmann, these proceedings
11. M. Manso et al., these proceedings
12. G.T. Razdobarin and E.E. Mukhin, these proceedings

INTEGRATION OF VACUUM COUPLED DIAGNOSTICS

P. H. Edmonds, R. Barnsley[1], N. Hawkes[1], A. Kislyakov[2], G. Vayakis[3], C. Walker, L. de Kock, G. Janeschitz, A. E. Costley[3], T. Steinbacher, H. S. Hurzlmeier and the ITER Joint Central and Home Teams.

ITER JCT, Garching JWS, Garching, Germany.
[1] JET, Abingdon, UK,
[2] Ioffe Institute, St. Petersburg, Russia,
[3] ITER JCT, San Diego JWS, USA,

INTRODUCTION

A special class of ITER diagnostics consists of those which extend the primary or tokamak vacuum outside the cryostat wall. This diagnostic set consists of the x-ray crystal spectrometers (XCS) [1], the vacuum ultra-violet spectrometers (VUV) [2], the neutral particle analysers (NPA) [3] and the low-field-side microwave reflectometers [4]. Two XCS systems are included, a high resolution five radial channel array system (XCS-A) designed to view specific spectral lines for temperature and rotation velocity profiles and a survey instrument package (XCS-S) to monitor a broad spectrum for identifying impurity concentrations. The vacuum vessel rough-out line is also located at this port.

The diagnostics described here are combined into an integrated assembly located at the mid plane port 7. This port is located between the neutral beam injector (NBI) port 6 and the Remote Handling (RH) port 8. The NBI secondary enclosure will be extended to include the port 7 diagnostic area. This secondary enclosure, required to provide tritium and contamination containment in the event of a failure of the primary vacuum barrier, is designed to withstand an over-pressure of 30 kPa. The present plan is to also include the port 8 RH access cell inside this extended NBI secondary confinement boundary. Alternatively an isolation door will be installed between ports 7 and 8. In either case, port 7 diagnostic access and maintenance operations will be via the RH cell.

Radiation poses a number of additional problems for those diagnostics with a direct plasma line of sight. The first problem is that the detector background count rates will be increased by the direct streaming neutrons and gammas from the plasma. In addition to this instantaneous effect, a second problem is the increased background count rate caused by accumulation of local material activation from the streaming radiation. And finally there exists the long term problem of neutron induced material damage, this effect is only serious for components close to the plasma edge.. A preliminary study of the radiation expected indicates that long term damage during the life of the machine should be insignificant and that radiation levels at the detectors will be reasonable [5]. This study does not include the reflectometer system or rough-out pump line penetrations, however they are not expected to significantly increase the radiation levels behind the Bio shield.. Typical predicted rates at reference detector locations are 10^5 to 10^6 $cm^{-2}\,s^{-1}$ for both neutrons and gammas.

Diagnostics for Experimental Thermonuclear Fusion Reactors 2
Edited by Stott *et al.*, Plenum Press, New York, 1998

DESIGN DESCRIPTION

An overall perspective view of the port 7 diagnostic assembly is shown in Figure 1, a detailed plan view of the front end is shown in Figure 2 and an elevation detail of the components outside the cryostat is shown in Figure 3.

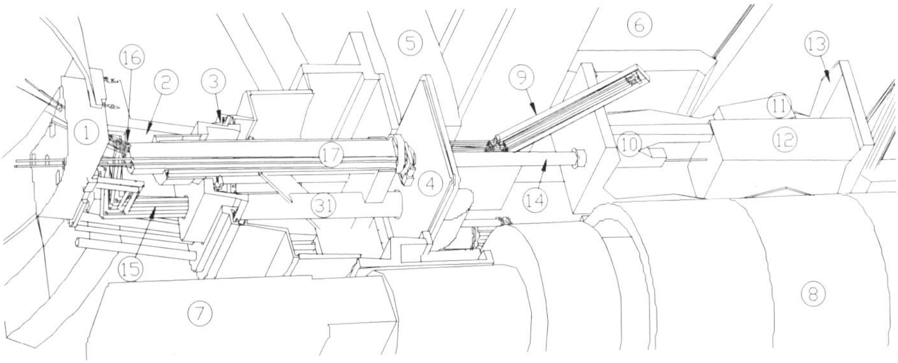

Figure 1. Oblique view of Port 7 and adjacent ports. (1) Port plug, (2) Diagnostic block, (3) Cryostat bellows, (4) Cryostat door, (5) Bio shield, (6) RH cell, (7) NBI duct, (8) Neutral beam injector, (9) XCS-A, (10) XCS-S, (11) VUV Spectrometer, (12) Neutral particle analysers, (13) Diagnostic instrumentation package, (14) Vacuum extension sleeve, (15) Reflectometer waveguides and (31) wave guide bundle.

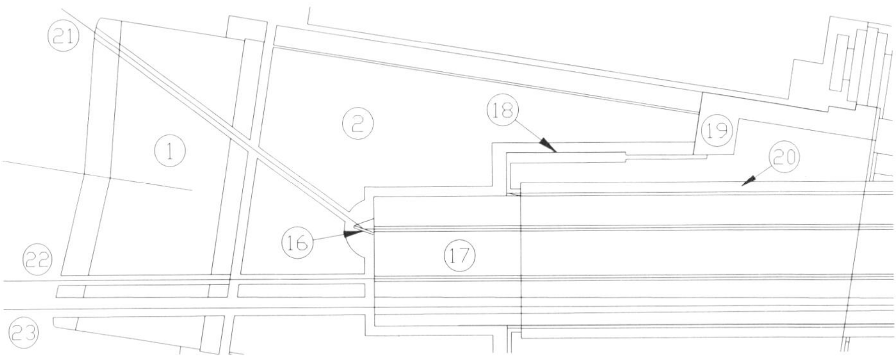

Figure 2. Detail of Port 7 plan view front end. Port plug (1) and diagnostic block (2) with penetrations for the XCS-A array (21), the VUV (22) and the 100mm square NPA (23) sightlines. The graphite Diffractor/reflector array (16) is attached to the end of the shielding plug (17). The shielding plug sleeve (20) is connected to the vessel flange (19) with a large diameter bellows (18).

The diagnostics divide into two very different packages, consisting of those requiring a collimated line of sight of the plasma, the VUV (11) and the NPA (12) or, for the XCS-A system (9), a direct line of sight of a front end graphite mirror (16), and the microwave reflectometry system (15, 31), which uses a series of bends in the waveguide to prevent direct neutron streaming.

The essential design feature is the use of discrete modules, essential to maintain precise sight lines. The first module consists of the front end port plug (1) and the diagnostic block (2), these provide shielding and contain no maintainable components. The oblique sight-lines for the XCS-A (21) and the direct sight-lines (22) for the VUV and XCS-S through the port plug and the diagnostic block are opened up to 50 mm dia. to provide additional tolerance for differential movement, the vacuum vessel movements (thermal and electromagnetic) relative to the cryostat wall are of order 50 mm for normal operation. Similarly the NPA aperture (23) will provide 13 mm of additional clearance. The construction will be for alignment at normal operating conditions, no provision will be made for active alignment.

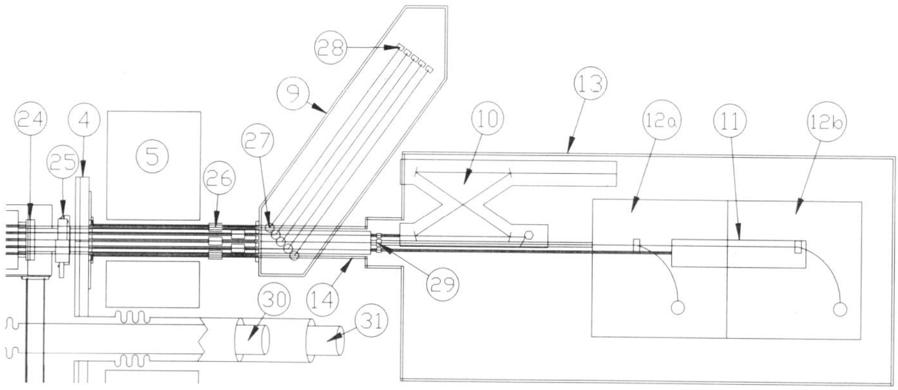

Figure 3. Detail of the outboard end of the Port 7 diagnostic set. (24) XCS-A Beryllium window array, (25) Isolation gate valve for XCS-S, VUV and NPA sightlines, (4) Cryostat flange, (5) Bio shield, (26) Isolation valves for secondary vacuum for XCS-A organ pipe array, (9) XCS-A assembly with crystal diffractors (27) and multistrip gas chamber counters (28), (29) Primary vacuum gate valves for XCS-S, VUV and NPA sightlines, (10) XCS-S scanning and Bragg rotor instrument, (12a) Low energy NPA, (12b) High energy NPA and (15) VUV spectrometer (31) Main vessel rough-out line, (30) Reflectometer waveguide bundle.

Re-entrant into the diagnostic block and establishing the secondary vacuum seal with a bellows (18) connected to the VV flange (19) is the shielding plug sleeve (20), this carries the primary vacuum back to the cryostat flange area and to the isolation valves. The shielding plug is mounted on rails inside the shielding sleeve. The sight-lines for the diagnostics are coaxial penetrations within this plug, the XCS and VUV sight-lines have a bore of 25 mm, the NPA sight-line is a tapered rectangular cross-section hole, 100 mm square at the blanket face and 20 mm by 10 mm at the analyser entrance slit. The viewing chords for the XCS-A profile diagnostic will be established using graphite crystals (with about a 1% bandwidth) mounted on the front surface of the shielding block (16). The orientation will be fixed at assembly, no movable parts or other components requiring maintenance will be installed close to the plasma.

The shielding sleeve is terminated at the cryostat end by an isolation valve (25) for the XCS-S, NPA and VUV sight-lines and by a thin Beryllium window (24) for the XCS-A system.

The XCS-A spectrometer is enclosed in a diagnostic module outside the cryostat (9) and Bio-shield (5) and connects to the cryostat secondary vacuum with an array of five pipes, each with an individual isolation valve on the secondary vacuum (26). The diagnostic module contains the Bragg crystals (27) and the multi-strip gaseous chamber counter array (28). Specific wavelengths matched to the anticipated electron temperatures required to excite the line at the tangency radii will be pre-selected. Extrinsic high Z noble gases (Ar, Kr) will be puffed in as trace impurity sources in addition to the instrinsic impurities (Fe).

The three other diagnostics, the XCS-S (10), the VUV (11) and the NPA (12), are located inside a large secondary vacuum enclosure or diagnostic instrument package (DIP) (13) connected to the cryostat flange by an extension sleeve (14), within this enclosure each instrument is self-contained inside separate individual enclosures, each connected to the primary vacuum through individual isolation valves (29). For maintenance the entire DIP is disconnected from the tokamak and removed to a diagnostic maintenance cell.

The reflectometry system consists of an number of transmit/receive antennas viewing the plasma through an array of apertures in the port plug. The separate W/G runs (15) combine into a bundle behind the port plug (31), the bundle exits through the cryostat door, below the level of the shielding plug sleeve. A two bend labyrinth will prevent direct neutron streaming.

MAINTENANCE AND REMOTE HANDLING

All components inside the cryostat boundary are Class 3 remote maintenance (no maintenance expected) except for the large isolation gate valve (25) which is a Class 2 component. Access to the XCS-A spectrometer and to the DIP is expected to be hands on. Isolation valves (25,29) will allow removal without breaking primary vacuum. Further disassembly procedures require a transfer docking cask which will dock with the special cryostat flange. The standard docking cask will dock to the rear door of this transfer cask. Loads will be supported by rails which will be inserted into the secondary and primary vacuum spaces as disassembly progresses. Space restrictions and the angled access require that the vacuum vessel flange be both rotated and tilted during removal. The secondary bellows may be required to be smaller than the standard cryostat bellows to allow removal, this will be through a series of rotation and translation operations. All procedures inside the cryostat wall will be fully remote handling compatible.

ACKNOWLEDGEMENT

This report has been prepared as an account of work performed under the Agreement among the European Atomic Energy Community, the Government of Japan, the Government of the Russian Federation, and the Government of the United States of America on Co-operation in the Engineering Design Activities for the International Thermonuclear Experimental Reactor ("ITER EDA Agreement") under the auspices of the International Atomic Energy Agency (IAEA).

REFERENCES

1. R. Barnsley et al., X-ray crystal spectrometry, in: *These Proceedings.*
2. N. Hawkes et al., Core VUV impurity monitor, *ibid.*
3. A. Kisliyakov, NPA at ITER: physical basis and design, *ibid.*
4. G. Vayakis et al., Overview of ITER reflectometry systems, *ibid.*
5. V. Khripunov et. al., Radiation conditions along the NPA- and XCS- diagnostic channels in the mid-plane port, *ITER IdoMs No. NA:NAG-47-8-1-97.*
6. R. Barnsley, K. D. Evans, N. J. Peacock, N. C. Hawkes, Bragg rotor spectrometer for tokamak diagnostics, *Rev. Sci. Instrument.* **57**(8) 2159 (1986).
7. R. Barnsley, U. Schumaker et al., Double crystal spectroscopy at JET, *Rev. Sci. Instrument.* **62**(4) 889-898 (1991).

PLASMA CONTROL CONCEPTS FOR ITER

Jonathan B. Lister

Centre de Recherches en Physique des Plasmas,
Association EURATOM-Confédération Suisse,
CRPP-EPFL, CH-1015 Lausanne, Switzerland

INTRODUCTION

The ITER control specifications are in terms of goals and limitations among which are:

- The ignited plasma burn must last for 1000 s;
- The fusion power must be 1500 MW ±20%;
- The shape control must reject specified disturbances, typical of Minor Disruptions and ELMs;
- The 50 mm SOL surface must not touch the first wall for more than 1 s;
- The antenna-separatrix spacing must not vary by more than 50 mm;
- The strike points must be controlled to 100 mm;
- The nominal position must be restored as fast as possible after a disturbance;
- The plasma current must be controlled to within ±0.5 MA or ±0.5%;
- The total site power cannot exceed 650 MW;
- The total PF power cannot exceed 250 MW, with steps less than 60 MW and ramps less than 200 MW/s.

The 21 MA nominal scenario, Figure 1, is the design basis of the ITER Plasma Control System. This scenario has allowed the placing and dimensioning of the PF coils and will also determine the additional heating power. However, other scenarios have subsequently been tested against this PF design to demonstrate adequate flexibility for, among others, low current Ohmic discharges for initial operation, 12 MA reversed shear steady state discharges and 24 MA ignited discharges. The present design handles these within the assumed ranges of the plasma parameters. Due to the cost of ITER, the PF coil current margins will be minimal and operation will be relatively close to these limits.

We could generalise the control problem and design a huge feedback loop in which we have available all the accessible measurements of the system, which we can compare with all our nominal or reference values and then make a decision as to how to react with all of our actuators. This defines the ultimate general feedback controller which would be a daunting prospect to design. However, the overall Plasma Controller can certainly be simplified given our a priori knowledge of the tokamak. Many of the control variables can be treated almost independently with the interactions between them neglected, especially during initial plasma operation. Since the generalisation of the problem leads to an overdose of complexity, the

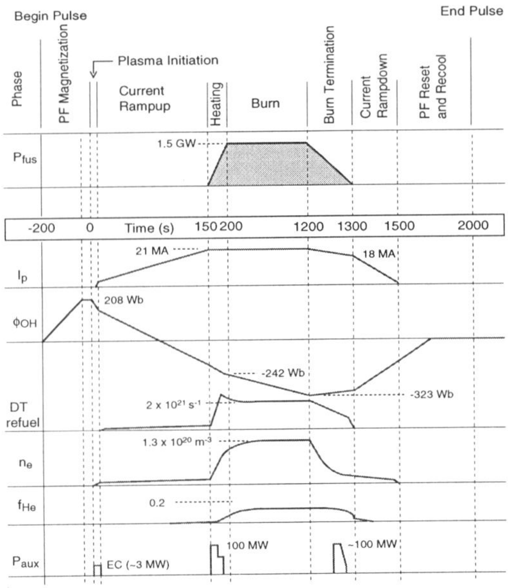

Figure 1. The Nominal 21 MA Scenario

normal procedure is to identify sub-systems which can be treated separately and controlled almost independently, with their interactions treated as perturbations. This is the approach taken, for example, for Magnetic Control and to a lesser extent for Kinetic Control.

However, it should be made clear at the start that this separation is a design approach which will ultimately require integration and testing on the basis of complete knowledge of these interactions. The nature of the tokamak is such that most actuators are linked to most variables, either transiently or statically, examples being:

- Heating, Density control → β_p, l_i → separatrix movement
- Heating, Density control → resistance → plasma current variation
- Current control → safety factor → confinement, stability → density, β_p
- Gap control → q_{95}, divertor → density control, confinement.

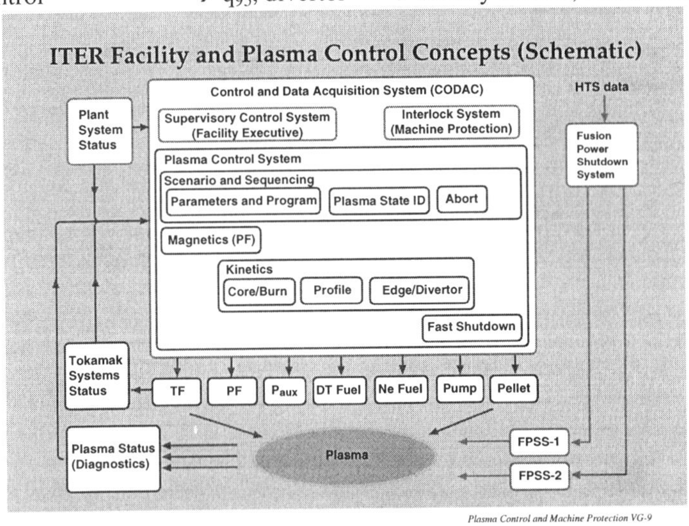

Figure 2. General Overview of the ITER Plasma Control System

At present Magnetic Control of ITER is taken for the purposes of controller design as a separable and linearisable part of the system, which has only minimal physics uncertainties and allows us to demonstrate the existence of a control solution which can be tested on a full non-linear evolutionary code. Due to this fact, Magnetic Control is the most advanced part of the ITER Plasma Control System. Kinetic Control is considered to be a separable and non-linear part of the system with large uncertainties in the model, for which we can only demonstrate the existence of solutions within the assumptions of the models used. At a third level, we consider all the systems as fully non-linear and fully coupled to demonstrate that the interactions between the systems which are neglected in the controller design do not affect the overall performance. Today, the interactions between Kinetic and Magnetic Control are rarely considered in the feedback controllers, with the exception of the pre-programming of the proposed scenario.

We can imagine a mixture of linear controllers, the most frequently encountered in current tokamaks, non-linear controllers such as On-Off control or Neural Networks, "fuzzy-logic" controllers or expert system rule-based controllers. Each tool should be used where it is most suitable and all may find a role to play.

The overall ITER control layout illustrated in Figure 2 shows the planned actuators.

INTEGRATION AND IMPLEMENTATION

The Plasma Control will be fully integrated into an ITER CODAC system. The actual implementation is not yet of concern to us. Some features have already been specified, such as the independence of Plasma Control and Machine Safety and the existence of a Fusion Power Shutdown System. However, a certain number of CODAC features will be have to be specified to satisfy the requirements for Plasma Control.

Since Plasma Control will certainly rely on discrete-time controllers, the CODAC will have to provide data at accurately sampled and regular intervals; delays in transmission must be minimal and constant from sample to sample. For the fastest feedback loops, probably in the Magnetic Control System, a privileged link between the data source and the Magnetic Controller will be required. For the slower feedback loops, this requirement could be relaxed.

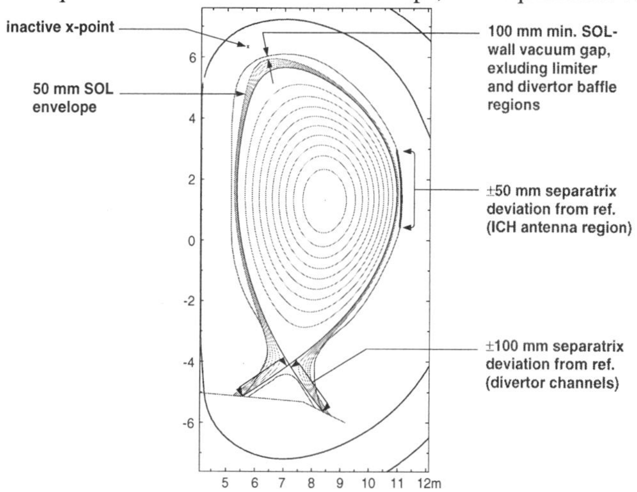

Figure 3. The shape control requirements

Plasma Control will rely for a minimum level of control on a subset of the total diagnostics, for which the diagnostic sensors may be duplicated or even triplicated. However, for some diagnostics this may not be possible. In order not to exclude these diagnostics from the optimisation role of the Plasma Control System, the diagnostic information will have to be validated and their validation assessment transmitted to the different feedback controllers. These controllers will then contain the intelligence to reject or accept the relevant information. This feature could also treat temporarily malfunctioning sensors.

MAGNETIC CONTROL OF ITER PLASMAS

The design requirements of the Magnetic Control have been simply stated. Figure 3 shows the steady state plasma shape control specifications which are the static variation of the separatrix, especially in the divertor region, for different plasma currents, β_p and l_i. Figure 4 shows the PF design and Figure 5 illustrates the evolution of the plasma separatrix during the start-up of the scenario. These fix the current capability of the PF coils. Figure 6 shows how well these goals are achieved by the current PF design.

Having fixed the coil currents, the voltage limits on the PF coils must be chosen with the criterion of minimising cost while guaranteeing controllability. The design methodology of the ITER Poloidal Field Control System is quite advanced. Various linear and non-linear models of the plasma+vessel+coils+power-supplies system have been developed and tested. One of them, the CREATE-L linearised model, has been rigourously tested against TCV experiments in limited and diverted plasmas [1,2]. The results of linearised models also agree well with full TSC simulations.

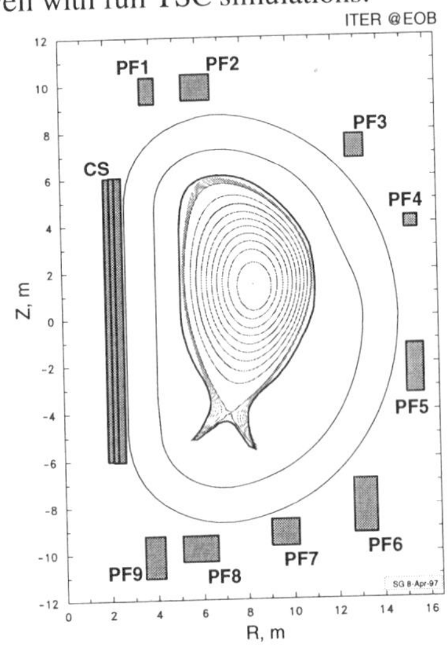

Figure 4. PF Design Configuration

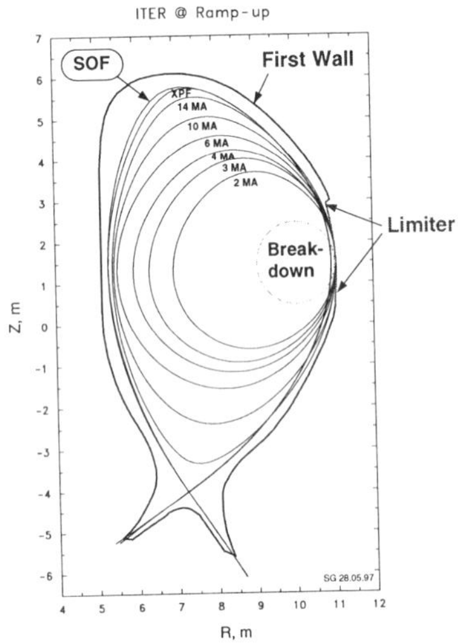

Figure 5. The controlled shape evolution

We assume that the uncertainties in the Magnetics Control models for ITER are small. ITER Magnetic Control does not really present any serious differences with respect to existing tokamaks, expect for four points:

- Minimising costs means minimising the power, voltage and current margins for control;
- Unprecedented power flow to the first wall;
- Superconducting PF magnets;
- Massive passive structure with diagnostics far from the plasma.

On this basis, the design of complete Magnetic Controllers has demonstrated the existence of a solution to the Magnetic Control problem given fixed limits to the coil voltages, coil currents and total PF power. Different approaches to the design of the multivariable Magnetic Controller have been studied, using low order PID controllers typical of current tokamaks as well as higher order LQG controllers and H_∞ controllers. The implementations of all these controllers are identical in structure and expressable as a recursive evolution at discrete time steps of the general form:

$$\underline{X}_{n+1} = \mathbf{A}\,\underline{X}_n + \mathbf{B}\,\underline{\mathrm{Error}}_n \ ;\ \underline{\mathrm{Output}}_n = \mathbf{C}\,\underline{X}_n + \mathbf{D}\,\underline{\mathrm{Error}}_n$$

The different design methods lead to different matrices **A**,**B**,**C**,**D** and different internal state vectors $\underline{X}$. Comparative studies have shown that all methods can provide an adequate

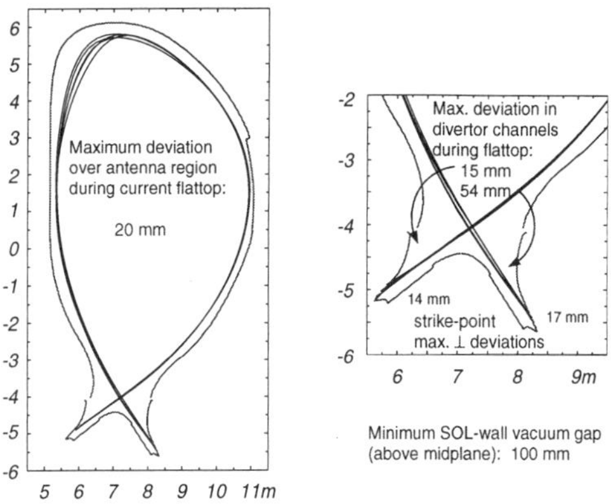

Figure 6. The static separatrix control achievable

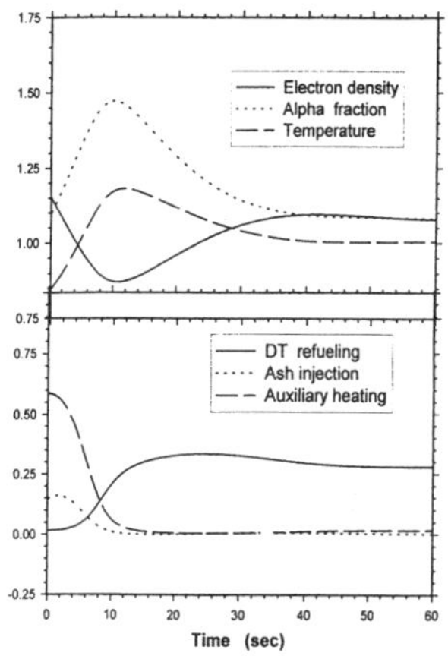

Figure 8. Evolution of burning plasma parameters

performance. What distinguishes them is their "robustness" or tolerance to badly known parts of the model and this is subject to current design work. They have been adapted to minimise the required total PF coil power, the high frequency coil current variations when controlling ELMs as well as minimising the separatrix gap variations. A baseline controller has been designed using the H_∞ approach based on the CREATE-L linearised model of the Plasma+Vessel+Coils system and the controller was tested on the TSC code. Figure 7 shows the response of the separatrix gaps, the coil voltages and currents and the total PF power during the rejection by the controller of a disturbance representative of a minor disruption, defined by a rapid drop of: $\Delta\beta p = -0.2$ and $\Delta li = -0.1$. The nominal requirements discussed above are all met during the rejection of this disturbance.

In this simulation, the input to the controller was the value of the separatrix-wall gap. Although the diagnostics on which Magnetic Control will be based will be dominantly magnetic, this does not mean the exclusive use of magnetic information. Reflectometry is being studied as an additional source of information for the control of the magnetic separatrix. Infra-red imaging of the divertor strike points will give information on the footprint of the divertor on the plates. Integration of information is available from various sources is possible and especially if we have knowledge of the expected errors in the signals, typically as a function of frequency, we can obtain an increase in precision.

Several other questions concerning Magnetic Control should be mentioned. Firstly, there is the requirement of providing a poloidal stray field configuration for plasma breakdown. Although the coil currents can provide the required breakdown structure, obtaining it is a question of pre-programming and feedback control given errors and noise on the measurements. Secondly, There is a requirement that the n>0 poloidal error fields should be $<10^{-5}$

Error field compensation coils are designed to correct the likely error fields, but the control of the currents in these coils is still an open issue. Pre-programming the error currents may not be adequate, in which case a method of determining the amplitude and phase of the error field correction field will have to be developed and tested.

The estimation of the AC losses in the superconductors during control actions and the plasma noise are important issues. If the AC losses are severe, then the extended pulse length for certain scenarios could be limited. If the losses could be reduced, then the cost of the superconductor refrigeration plant could be reduced. However, since the scenario itself provides a significant part of these losses during a cycle, the room for improvement is limited. The variations in the controlled gaps and the feedback power required also depends on the sensor noise which is dominated by plasma noise. Estimating this noise is not simple for operational tokamaks and predicting it for a burning ITER plasma requires courage. This issue will remain open until operation and all we can reasonably do now is to show that the maximum tolerable noise is not too low to be credible.

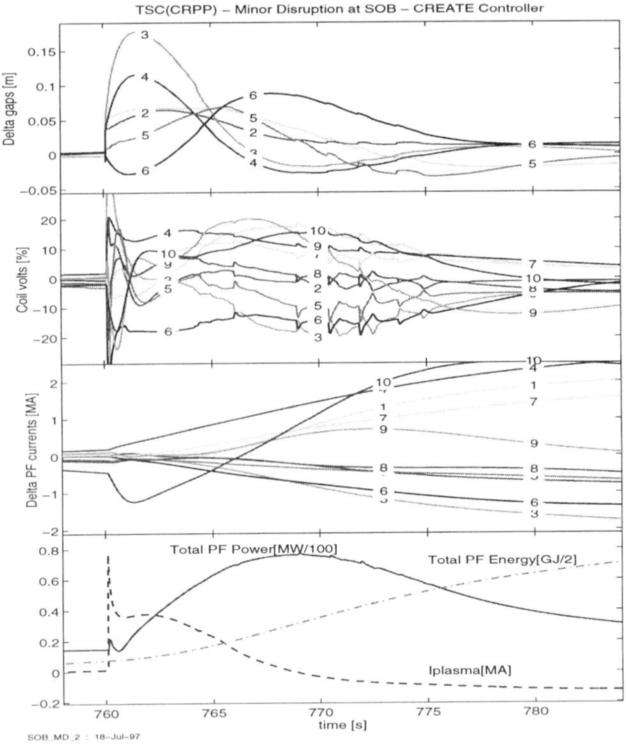

Figure 7. Rejection of a Minor Disruption

KINETIC CONTROL OF ITER PLASMAS

Kinetic control of the plasma is much less advanced than Magnetic Control, since the main EDA design requirement is simply to have adequate heating to achieve ignition and this depends mostly on the assumptions on the confinement time. It is generally assumed that Kinetic Control is of lesser importance to the design. The uncertainties in Kinetic Control are in the properties of the confinement and impurity models rather than in the details of the control dynamics. There are two main new problems of interest for Kinetic Control in ITER.

Firstly we have the problem of burn control, as yet unverified in a tokamak. This problem is interesting from a control point of view since the plasma temperature will be locally unstable at the ignition boundary if the temperature dependence of the confinement time is not sufficiently strong, in which case ignition would be doubtful. The temperature will saturate where $d(n_D n_T<\sigma v>)/dT_i + d(Losses)/dT_i + d(Losses)/dT_e \sim 0$, somewhere midway between the low and high temperature ignition points. This point can be modelled quite simply in 2-D, 1-D or 0-D codes, but the results will always be dependent on the model, since the system is intrinsically non-linear. We return to an example of this later.

Secondly, the control of the divertor flame will be extremely sensitive, requiring a balance between local impurity radiation, local heat conduction, back-flow of the impurities into the main chamber, interaction between the residual heat-flow to the divertor plate with the plasma facing components. This is a highly coupled system for which the modelling is still somewhat uncertain and the object of intensive development.

Control of the plasma density in ITER has been simulated, including the gas-fuelling. There was no problem in obtaining satisfactory density control until the length of the gas-feed piping was included, which is significant on ITER for size and routing reasons. Inserting the delay due to the fuelling pipe removed stable operation of the fuelling feedback loop and to regain control a more advanced controller will be needed which models not only the plasma response, but also the fuelling delay, adding some form of phase advance to the controller response. This is analagous to the effect of Power Supply delay in the magnetic control of small tokamaks.

As mentioned, the burn temperature evolution is unstable once the ignition boundary is crossed and the additional heating is turned off. At this time we lose the "dominant" actuator for controlling the temperature, namely the auxiliary heating. During the subsequent temperature excursion, the fusion power could increase significantly, depending on the ignition temperature for the given confinement time and impurity level.

Recent results [3] have demonstrated the ability of non-linear controllers based on Neural Networks for manipulating the plasma conditions during the burn to displace the working point. The dependence of confinement on the plasma conditions is included and the modelling includes the limited range of three actuators, namely the refuelling rate, the Helium impurity injection rate and the additional power. Given these limits and a 0-D model of the plasma, the non-linear controller uses these actuators to make changes to the plasma which do not simply drift in the direction of the desired change. In order to decrease the density and increase the temperature, for example, the density can be reduced to increase the temperature by a large amount, after which an increase in density can be achieved within the available power range and limiting the fusion power excursion. Figure 8 illustrates this.

The study of kinetic control algorithms will also have to treat the discrete nature of the power supplies themselves and not assume linear "heating/fuelling actuators".

COMBINED CONTROL OF ITER PLASMAS

Although we have treated the Magnetic Control and Kinetic Control as separate, they are clearly linked through β_p and l_i which define the equilibrium. A couple of examples illustrate this.

The antenna-separatrix spacing depends on β_p and defines the coupled power for ICH or LHH. A sudden loss of β_p requires an increase in additional heating to recover it but at the same time provokes a loss of coupled power. This system is thermally unstable and requires feedback, as already performed on existing experiments.

We can demonstrate that the Magnetic Control is possible using the β_p drop as a disturbance. We can demonstrate that the Kinetic Control is possible using the antenna resistance as a control parameter. However, when we combine the Magnetic and Kinetic Control into a single system, we find constraints on the controller dynamics which are not apparent in either separately. If we do not have derivative gain on the magnetic control of the antenna gap, then there is no stable solution to the β_p instability. Studying the Magnetic Control of the antenna gap does not tell us this, nor does studying the Kinetic Control of β_p.

Another practical linking of Kinetic and Magnetic Control is the limitation of the total site power which is not necessarily split between the different controllers in the same way throughout the discharge. The controllers will have to be aware of each others power requirements in order to respect the site limit, to reduce the site requirement and to optimise their control within these limits. We must be careful that a surge in the required additional heating power to regain β_p is not simultaneous with a surge in the PF power to compensate the separatrix movement, for example.

SUPERVISOR ROLES AND CURRENT IMPLEMENTATION

A Supervisor could aid the present shot-to-shot discharge optimisation done by the operational physicists. We will have to build into the control of ITER a set of rules for obtaining and maintaining ignition given varying conditions of the tokamak. Although we will define the scenario of a discharge with as much detail and precision as possible before its execution, the optimisation of the discharge could be assisted by the Supervisor. During initial operation, the supervisor would perform only simple tasks but its role would then expand during operation to solve particular problems encountered or to automate the regular decisions of human operators. Many methods could be envisaged to perform this real-time optimisation and they could be tested against existing models.

An example is the replacement of present operator interventions and shot by shot decisions such as "the XYZ coil saturated in the last discharge – if we change the shape here and here a bit, it will come off saturation and we will get essentially the same shape for our requirements". This intelligence could be embedded in the PFCS Supervisor, since at some points in the discharge, the PF coil currents may be close to their limits. Although simulations have shown that the PF control does not lose its stability during current saturation of one or more PF coils, the loss of precise shape control should not be left to chance. Alternative schemes for switching to a different PF controller once a coil has saturated are being considered.

There have already been significant advances in the direction of autonomous control during tokamak discharges. ASDEX-UG control has achieved an impressive degree of autonomy in some auxiliary heated discharges during which a high-recycling H-mode dropped back into a high-recycling L-mode. The feedback controlled radiated power fraction was immediately decreased by the Supervisor and ramped back to its nominal value once the H-mode was re-established, illustrated in Figure 9 [4].

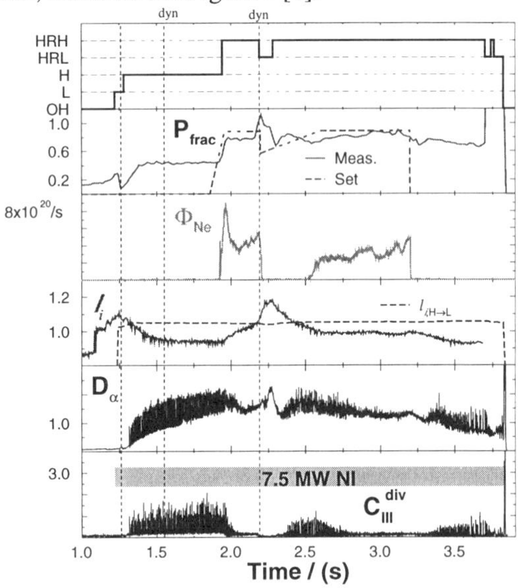

Figure 9. Supervisor autonomy during an ASDEX-UG discharge

The Supervisor could be responsible for maintaining the plasma operation in a safe region, particularly in view of disruptions at high β_N or close to $q_{95} = 3$ or close to the density limit. To do this, the Supervisor needs a model of the dangerous regions and a model of how to "back off" from them safely. An example well known from DIII-D is the high-β, high density, high elongation, low-q operation. The zone with maximum performance is in an acute region of operational space and backing away from any single limit is likely to exacerbate another limit. DIII-D have already started automating this for a single parameter, for which "backing off" is well defined, namely the approach to the high-β limit. In this work [5], a parametrisation of the high-β limit was obtained from a Neural Network description of the disruption boundary based on experimental data. This limit was then imposed on subsequent discharges to avoid crossing it and non-disruptive operation close to the limit was obtained. Interestingly, this Neural Network description of the high-β limit was more precise than the “rule of thumb” modified Troyon limit $\beta_N < 4l_iI_p/aB$. This provides an excellent example of *a posteriori* control demonstrating advantages over *a priori* model control. The DIII-D Supervisor was even able to adapt to Negative Central Shear discharges once they had been included into its database.

An example of control using available observable quantities rather than the quantities which are intended to be controlled is provided by Tore Supra, in which the internal inductance and plasma current were separately controlled in a current profile feedback loop using the parallel wave spectrum and total Lower Hybrid power as actuator inputs [6]. In this way, the safety factor on axis was also indirectly controlled without being measured directly.

In general, to establish the use of advanced Supervisors with intelligent control, we require simultaneous control over all actuators on the basis of more complete diagnostic information and the decisions will have to be based on more accurated models relating the actuators to the controlled variables.

CONTROL FOR OPTIMISATION

The Supervisor will have to execute particularly simple sub-tasks, some of which have already been mentioned, such as avoiding the β_N limit, avoiding current saturation, estimating the accumulated AC losses, interruption of the pre-programmed scenario in particular situations. However, such a level of intelligence is already within sight. A more adventurous challenge for ITER would be the optimisation of the use of the available actuators to obtain the best discharge conditions. At present, this level of intelligence is the realm of the “physicist in charge”, who interprets the available data and reacts to it from discharge to discharge to obtain his optimum. The reduced number of ITER discharges and the length of the nominal discharge suggest that the Supervisor might also contain an element of optimisation. This will have to be done on the basis of the full diagnostic information available, plus an understanding of the underlying trend of the device. Many techniques are available for this type of optimisation, but have not been applied to a tokamak plasma. Although the basic diagnostics will be capable of providing feedback control of the basic plasma parameters, the use of more diagnostic information will be required for optimisation. For this reason, advanced diagnostics are more linked to this Supervisor function than to the lower feedback loops. Maintaining a 10’000 second reversed shear burning plasma discharge does not sound too easy today and it would be negligent to assume that this will be achieved with some nice pre-programming and a few feedback loops.

R & D REQUIREMENTS FOR PLASMA CONTROL

Work on intelligent control with an increasing degree of autonomy has already started on several tokamaks, of which A-UG and DIII-D are the most advanced. Work has progressed on several types of decision making processes and this work will have to continue over the construction period so that the overall framework of ITER Plasma Control becomes clear.

The methodology used for designing the ITER plasma current, position and shape controllers should be demonstrated experimentally in tokamaks with shaped plasma cross-sections and their operational reliability should also be demonstrated.

Methods for diagnosing the error field and correcting them using the plasma response to variations of the error field should be proposed and tested experimentally in large tokamaks, since this represents one of the most serious unknowns for ITER operation.

Finally, a testbed for integrated modelling has to be developed. Each part of the ITER control problem will have its separately identified codes and combining them all approximately will be a hazardous step which could lead to a simulator which does not treat any of the issues correctly. On the other hand, combining them in all their detail could lead to a code so heavy that it can no longer be used for scoping studies. This issue will require care and attention during the coming years to provide the right tools.

An alternative approach is to rely on identified models, derived from the observed behaviour of the plasma rather than the a priori modelled assumptions of the plasma. Work has started on this for Magnetic Control on TCV with encouraging results [7]. This approach gives useable results for an unstable Multiple Input Multiple Output system for Magnetics Control and ought to be extendable to the input-output relationships of all the actuators. This technique also allows on-line identification, within limits, and could be extended to adaptive control during a discharge, due to the very long ITER pulses. Research on present tokamaks could lead to increased confidence in this approach.

CONCLUSIONS

This overview paper has skimmed over a wide range of issues related to the control of ITER plasmas. Although operation of the ITER project will require extensive developmental work to achieve the degree of control required, there is no indication that any of the identified problems will present overwhelming difficulties compared with the operation of present tokamaks. However, the precision of control required and the degree of automation of the final ITER plasma control system will present a challenge which is somewhat greater than for present tokamaks. In order to operate ITER optimally, integrated use of a large amount of diagnostic information will be necessary, evaluated and interpreted automatically. This will challenge both the diagnostics themselves and their supporting interpretation codes. The intervening years will provide us with the opportunity to implement and evaluate most of the new features required for ITER on existing tokamaks, with the exception of the control of an ignited plasma.

ACKNOWLEDGEMENTS

The control concepts for ITER are currently being developed by a large number of people, whose often unpublished work has been extensively plagiarized in this paper, not always with their explicit agreement. My apologies are due to those who feel misrepresented. Special recognition should be given to contributions by particular individuals, namely Lello Albanese, Pepe Ambrosino, Marco Ariola, Dominique Boucher, Dick Bulmer, Alan Costley, Ber de Kock, Sergei Gerasimov, Yuri Gribov, Otto Gruber, Dave Humphreys, Chuck Kessel, Jim Leuer, Brian Lloyd, Vitus Mertens, Yuri Mitrishkin, Pier Luigi Mondino, Alfredo Portone, Gerhard Raupp, Peter Stott, Wolfgang Treutterer, Fabio Villone, Parag Vyas, Mike Walker, David Ward, John Wesley, Izuru Yonekawa. This work was partly funded by the Fonds national suisse de la recherche scientifique.

REFERENCES

1. F. Villone et al., "*Comparison of the CREATE-L plasma response model with TCV limited discharges*", LRP 569/97, accepted for publication in Nuclear Fusion
2. P. Vyas et al., "*The separatrix response of diverted TCV plasmas compared to the CREATE-L model*", LRP 583/97
3. J. Vitela and J. Martinell, "*Stabilization of burn conditions in a thermonuclear reactor using artificial neural networks*", submitted to Plasma Physics and Controlled Fusion
4. P. Franzen et al., "*Online regime identification for the Discharge Control System at ASDEX Upgrade*", 23rd EPS Conf. Contr. Fusion and Plasma Physics, Kiev, Vol 20C (1996) 87
5. D. Wroblewski et al., "*Tokamak disruption alarm based on a neural network model of the high-β limit*", Nuclear Fusion, 37 (1997) 725
6. T. Wijnands et al., "*Feedback control of the current profile in Tore Supra*", Nuclear Fusion, 37 (1997) 777
7. A.C. Coutlis et al., J.B. Lister et al., CDC Conference 1997, LRP 584/97

IGNITION BURN CONTROL AND DIAGNOSTICS IN ITER

Osamu Mitarai[1] and Katsunori Muraoka[2]

[1] Department of EE, Kyushu Tokai University, Kumamoto, Japan
[2] Department of Energy Conversion, Interdisciplinary Graduate School of Engineering Sciences, Kyushu University, Kasuga, Fukuoka, Japan

INTRODUCTION

As design of control algorithms and diagnostic systems for an ignited operation has been a challenging task in the ITER project,[1] it is quite important at the present ITER design stage to show what kind of control laws is applied and how diagnostic system is incorporated in control system for the ignition burn control.

In this study, we analyze the fusion burn rise-up, steady state and fusion burn shutdown phases using zero-dimensional power balance equations with the ITER93HP scaling, and consider the control algorithm including the diagnostic systems to be used in ITER. We have found a unified feedback control law applicable to all burn phases, using the external heating power and the fueling rate which are based on the fusion power and the H-mode power threshold. Discussions are given for areas of further studies.

FEEDBACK CONTROL LAW IN IGNITED OPERATION

In these analyses, we assume the plasma to fill the complete volume of the ITER configuration with the assumed profiles of the temperatures, density and impurities including helium ash, where central values of these quantities are calculated using the zero-dimensional power balance equation. In this situation, the ignited operation is controlled by the external heating power and fueling rate.

The external heating power is determined by the requirement of yielding a specified fusion power as

$$P_{EXT}(P_f) = P_{EXT0}\,(1- P_f/P_{f0})G_{pf} \tag{1}$$

where P_{EXT0} = 100 MW, G_{pf}= 20, P_f is the actual fusion power, and P_{f0} is the target value of the fusion power. The fueling rate is assumed to be given similarly as

$$S_{DT}(P_f) = S_{DT0}\,(1- P_f/P_{f0})G_{SDT} \tag{2}$$

with $S_{DT0} = 1\text{x}10^{19}\ \text{m}^{-3}/\text{s}$ and $G_{SDT} = 25$.

Diagnostics for Experimental Thermonuclear Fusion Reactors 2
Edited by Stott *et al.*, Plenum Press, New York, 1998

In addition, we want the operating regime of the plasma in all three burn phases to be in the H-mode. For the purpose of this criterion, we use an equation for the H-mode power threshold proposed by Takizuka,[2] $M_{HL} = P_{h,net} V_o\, 10^6 / 0.45\, \bar{n}_{20}^{0.75} B_t R^2$. Here M_{HL} is the H-mode indicator ($M_{HL}>1$ for the H-mode), $\bar{n}_{20}$ is the line averaged density in the unit of $10^{20}\ m^{-3}$, B_t is the toroidal field, R is the major radius, $P_{h.net}$ is the net heating power density given by $P_{h,net} = P_{EXT}(HL)/V_o + P_{oh} + P_\alpha - \{P_b + P_s\}$, P_{oh} is the Ohmic heating power, P_α is the alpha heating power, P_b is the bremsstrahlung loss, P_s is the synchrotron radiation loss, and V_o is the plasma volume. No hysteresis effect is assumed in the H-mode power threshold. Rewriting this relation yields the necessary heating power for the H-mode operation as

$$P_{EXT}(HL) = M_{HL}\, 0.45\mathrm{x}10^6\, \bar{n}_{20}^{0.75} B_t R^2 - (P_{oh} + P_\alpha - P_b - P_s) V_o \qquad (3)$$

In the subsequent analyses, we take the larger value in the two external heating powers given by Eqs. (1) and (3),

$$P_{EXT} = \max\{P_{EXT}(HL), P_{EXT}(P_f)\} \qquad (4)$$

in order to satisfy the requirements of yielding a specified fusion power while keeping the operating regime in the H-mode. It is to be noted that the fusion power is also influenced by the fueling rate as given in Eq.(2).

Other constraints for the plasma operations are as follows: the energy confinement time is decide by the ITER93HP scaling with confinement factor of γ_{HH}, the helium ash confinement time to the energy confinement time ratio is $\tau_\alpha^* / \tau_E = 10$, and the impurity is mostly Be and its fraction $n_{Be}/n_e = 2$ %.

CALCULATED RESULTS ON IGNITION ACCESS, STEADY STATE, AND SHUTDOWN PHASE

Ignition Access Phase

The temporal evolution of the plasma parameters and related quantities

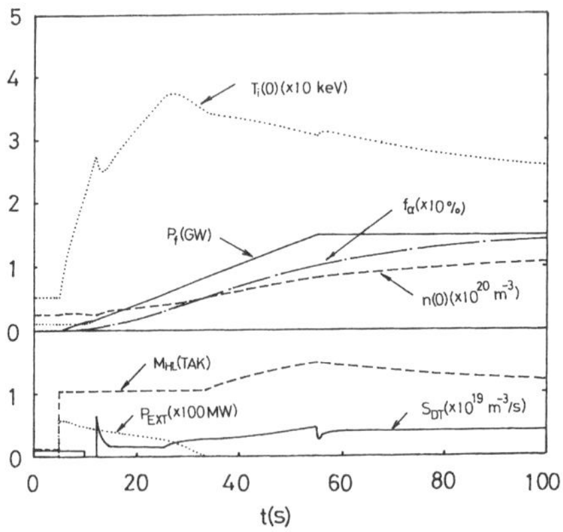

Fig. 1 Ignition access controlled by $P_{EXT}(HL)$ and $S_{DT}(P_f)$

during ignition access is shown in Fig. 1 for the confinement factor of $\gamma_{HH}=1$. Here the Ohmic plasma has been established at t = 0 sec and the heating power is applied at 5 sec, after which the fusion power P_{fo} is specified to increase linearly to 1.5 GW in 50 sec and to be constant after 55 sec. The heating power of ~ 50 MW is gradually reduced to zero, which is adjusted automatically to keep the H-mode with a value above M_{HL}~1.05. As can been seen from the figure, a smooth transition has been realized from the initial Ohmic phase to the fully ignited phase with a fusion power of 1.5 GW.

Effects of the confinement degradation, $\gamma_{HH} \leq 1$, have been studied for attaining the specified fusion power. For $\gamma_{HH} \leq 0.8$, the plasma parameters begin to oscillate after 80 sec, which is suppressed by application of the heating power controlled by $\max\{P_{EXT}(HL),P_{EXT}(P_f)\}$ and $S_{DT}(P_f)$. This control is possible down to $\gamma_{HH} \sim 0.72$.

Steady State Burn Phase

Once a steady state is established, it is maintained unless some changes occur in the plasma confinement. We consider two cases of the sudden change, namely an increase or a decrease in the confinement factor. When the confinement factor suddenly increases from $\gamma_{HH}=1$ to 1.8, the heating power and fueling are automatically switched off during the fusion power surge. After that M_{HL} tends to decrease below 1, then the heating power is increased to keep the H-mode and reaches 100 MW. H-L-H transitions take place at 100 MW, and the H-mode with $\gamma_{HH}=1.0$ is assumed to come back, and then 1.5 GW fusion power is recovered after transition to the H-mode. Thus we have found that the control by $\max\{P_{EXT}(HL),P_{EXT}(P_f)\}$ and $S_{DT}(P_f)$ works properly for the sudden increase in the confinement factor.

When the confinement factor suddenly decreases from $\gamma_{HH}= 1$ to 0.8, the heating power and fueling are both increased to maintain the fusion power at the constant level. The plasma is mainly controlled by $P_{EXT}(P_f)$, but controlled by $P_{EXT}(HL)$ for the short time to maintain the H-mode. Therefore, a set of control law of $\max\{P_{EXT}(HL), P_{EXT}(P_f)\}$ and $S_{DT}(P_f)$ can manage the subignited operation.

Fusion Power Shutdown Phase

It is very difficult to decrease the fusion power linearly in 100 sec without the H-mode. We have applied the same control law of $\max\{P_{EXT}(HL), P_{EXT}(P_f)\}$ and $S_{DT}(P_f)$ in this situation. As the heating power by the H-mode power threshold is always larger than that by the fusion power during the shutdown phase, the shutdown process is determined by the H-mode power threshold. The analysis shows that the fueling is decreased by the feedback control and the heating power is increased to maintain the H-mode and is limited by the available heating power of 100 MW. While the H-mode becomes marginal at this limit in the heating power and H-L-H transitions sometimes take place, the fusion power is linearly decreased to the negligible level as required.

DIAGNOSTIC SYSTEM FOR IGNITION CONTROL

In order to execute the control algorithm proposed here, we have to measure the fusion power P_f to estimate Eqs. (1) and (2), and all the parameters in Eq.(3) and the ITER93HP scaling. These quantities are analyzed, on real time, to yield the necessary feedback values of the external heating power of

$\max\{P_{EXT}(HL), P_{EXT}(P_f)\}$ and fueling rate $S_{DT}(P_f)$. This situation is schematically shown in Fig. 2.

The next stage of the necessary measurements is to demonstrate that plasma conditions are in fact fulfilled, namely measurements of such values as the enhancement factor γ_{HH} over the ITER93HP scaling, the alpha particle density, the H-mode indicator M_{HL} and profiles of various values.

The ultimate goal of this study is to give requirements for each measurement for this kind of control purposes, namely the temporal and spatial resolutions and its accuracy. As the necessary temporal resolutions can be obtained by the response time of the plasma and feedback systems, the numerical model employed here may need more improvements. The necessary accuracy may also be given by the sensitivity analysis in the foregoing calculations. [1]

This ignition burn control should be incorporated into a more general control algorithm for ITER, such as plasma initiation, position and profile controls, and a disruption avoidance.

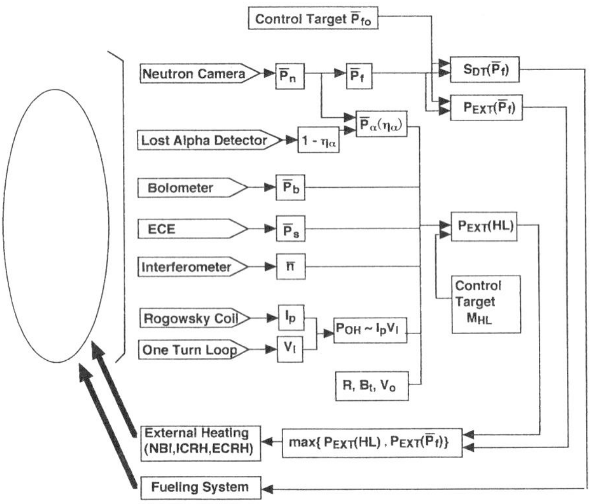

Fig. 2. Schematic diagram of diagnostic and control systems for D-T ignited operations in ITER

CONCLUSION

We have found a unified feedback control law, expressed by a set of $\max\{P_{EXT}(HL), P_{EXT}(P_f)\}$ and $S_{DT}(P_f)$, applicable to all burn phases in ignited or subignited ITER operations, which is based on the H-mode power threshold and the fusion power. Following this law, the diagnostic system supplies feedback values for the external heating power and fueling rate. Diagnostic systems in ITER should be designed based on such algorithm as proposed here.

REFERENCES

1. O. Mitarai and K. Muraoka, Ignition Analyses for Burn Control and Diagnostics Developments in ITER, in press in *Nucl. Fusion* (1997).
2. T. Takizuka et al., Threshold Power and Energy Confinement for ITER, *16th IAEA Fusion Energy Conference* (Montréal Canada 7-11 October 1996) IAEA-CN-64/FP-20

OVERVIEW OF THE ITER REFLECTOMETRY DIAGNOSTIC SYSTEMS

G. Vayakis,[1] T. Ando,[2] N. Bretz,[3] L. deKock,[2] A.J.H. Donné,[4] E.J. Doyle,[5] J. Irby,[6] E. Martin,[2] M. Manso,[7] A. Mase,[8] J. Sanchez,[9] V.A. Vershkov,[10] D. Wagner,[11] C.I. Walker,[2] and the ITER Joint Central Team and Home Teams

[1]ITER Joint Central Team, San Diego JWS, 11025 N. Torrey Pines Road, La Jolla CA 92037, USA

[2]ITER Joint Central Team, Garching JWS, Germany

[3]Princeton Plasma Physics Laboratory, Princeton NJ, USA

[4]FOM, Nieuwegein, The Netherlands

[5]UCLA, Los Angeles, USA

[6]MIT Plasma Fusion Center, Cambridge MA, USA

[7]Associação EURATOM/IST, Lisboa, Portugal

[8]Plasma Research Center, University of Tsukuba, Japan

[9]Association EURATOM/CIEMAT, Madrid, Spain

[10]Kurchatov Institute of Atomic Energy, Moscow, Russia.

[11]Institut für Plasmaforschung, Stuttgart, Germany

INTRODUCTION

Reflectometry exploits the phase shift of electromagnetic waves as they propagate through the plasma. The waves reflect at the cutoff layer defined by the region where the refractive index becomes zero. Reflectometry profile measurements are based on measuring the phase shift or group delay to this layer as a function of frequency. In a simple 1-D picture, the density profile is then recovered by an inversion from the (group delay, frequency) space to the (cutoff frequency, radius) space. In O-mode this involves a simple Abel inversion procedure.

Measurements of density fluctuations can be made by holding the frequency constant, and observing the time evolution of the reflected signal. However, the inversion to obtain the density fluctuation level is not straightforward except for long wavelength, low amplitude modes for which simple 1-D theory can be adequate.

The basic physics of the technique with respect to profile measurement is well established and has been confirmed by many experiments.[1] A wide range of appropriate hardware exists commercially and measurement systems can be readily designed and built. However it is only recently that reflectometry has been deployed extensively, and it is developing into the instrument of choice for the measurement of edge profiles with

Table 1: Measurements for which reflectometry is to play the primary role in the control hierarchy.

Measurement	Category
Plasma Shape and Position (Long pulse)	1b
Divertor Density	1b
Density Profile (Edge)	1b
ELMs (Δn)	1b
Density fluctuations	2

simultaneous high spatial resolution (<~1 cm) and time resolution (<~1 ms). This is due to the recent development and application of strategies to avoid or minimise the effect of density fluctuations on the measurement.

For its implementation on ITER, there are significant problems to overcome: access to the plasma through the plasma facing components, access to the front end for installation and maintenance, calibration to provide immunity to machine movements during the long pulses, long and unwieldy transmission lines and relatively ambitious frequency ranges. It turns out that the most critical problem is that the access to the plasma is restricted to an extreme degree, due to the stringent thermal and mechanical requirements on the first wall and neutron shield structures.

The reflectometry systems required for the ITER device have been identified,[2] and their principal characteristics have been established.[3] There is a reflectometer to measure plasma profiles and transient events in the main chamber, a reflectometer to supplement the magnetics for position control, and a reflectometer system to measure profiles in the divertor legs. Engineering integration of all the systems is in progress. Details on these studies are reported in other papers at this conference.[4-8]

The first part of this paper is a summary of the role that reflectometry is expected to play on ITER, and the generic constraints imposed by a reactor grade device on the implementation of the system designs. A brief report of process of optimisation of the front end is given for each system. Despite the implementation difficulties, it appears that the ITER measurement requirements can be largely met, and that reflectometry will make a substantial contribution to machine operation, plasma control and to the physics understanding of ITER plasmas.

THE ROLE OF REFLECTOMETRY ON ITER

One way of prioritising measurements on ITER is by their likely contribution to the control of the machine. This divides measurements, and by association diagnostics,[9] into three types: basic machine control (1a), that is, measurements essential to control any plasma pulse; advanced control (1b), that is, measurements that are integrated in the control loop for specific programs, or to access particular regions of operating space; finally, measurements for performance evaluation and physics studies (2). Reflectometry is expected to be the main diagnostic for a substantial number of measurements in the (1b) category (Table 1). In addition, it is expected to provide supplementary information for a great number of other measurements, such as plasma-wall gaps (1a), MHD modes (moving and locked, 1b), TAEs (1b), the core density profile (1b), and, possibly divertor plate erosion and flame front location.

The desired measurement capability for each plasma region has been based on the target measurement requirements,[2] and is summarised in reference 3. In general, it appears

Table 2: List of the measurement functions of the main reflectometer system

Measurement	Region of interest	Unique aspects of reflectometry measurement	Minimum required launch set	Category
Density profile	All	continuous sub-ms & sub-cm edge/gradient region density profiles	LFS O	1b
		continuous sub-ms core profiles	HFS X-l	
Locked modes	Core /Gradient	Island size, n_e amplitude	LFS O, HFS X-l	1b
ELMs	Gradient/ Edge	Unambiguous Δn_e	LFS O, X-u	1b
		Inboard-outboard comparisons	HFS O	2
MHD modes	Core /Gradient	Radial structure	LFS O HFS X-l	1b
Sawteeth	Core /Gradient		LFS O HFS X-l	1b
Disruption Precursors	All		LFS O, X-u HFS X-l	1b
Fishbones, TAEs	Core	radial structure	LFS O, HFS X-l	2
Density fluctuations	All		LFS O, X-u HFS X-l	2

possible to meet the measurement requirements for all plasma regions, with the possible exception of the 3 mm density resolution target in the divertor region. This prediction will have to be revisited when more experimental and modeling results from the divertor area become available.

IMPLEMENTATION CONSTRAINTS

The constraints imposed on the reflectometry systems by the machine are generic in nature, although the solutions vary for each system. Starting from the front end they are:

Access through the first wall: Two factors restrict the aperture size that can be provided for antenna access to the plasma: the neutron leakage through the first wall and shield, which must be minimised to ensure hands-on access as close to the machine as possible, and the additional heat loads imposed on the first wall by the presence of the aperture. The latter requirement is particularly important in the divertor area.

Installation and maintenance: Components that require maintenance have to be installed in a way compatible with the standard remote handling requirements and equipment of the area of installation.

Ability to withstand neutron and EM radiation heat loads: The radiative load on parts of the first wall can reach (locally) 0.5 MW/ m^2. For the divertor this figure can be an order of magnitude higher. The neutron first wall loading is ~15 W/cc for stainless steel. Thus, careful analysis of the cooling requirements must be performed for all the front end components. It appears however that reflectometry front ends can be made sufficiently robust, by using suitable materials such as refractory metals and CFCs.[10]

Transmission through the first vacuum boundary: The requirements on the pressure handling capability, temperature range and remote handling of the first window are not very different from the present generation of D-T capable machines. However, the window is usually in a restricted and relatively inaccessible region. Thus, the consequences of window failure on operating time will be severe, so the reliability requirements are commensurate.

Movement compensation: The transitions between operating states, and thermal effects during the plasma pulse, result in relatively large movements (~ cm) between machine components (blanket, vacuum vessel, bioshield). Thermal movements are primarily in the radial and vertical directions. The vacuum vessel, on the other hand moves toroidally during the pulse because of the interaction between the TF and PF coils. These movements have to be taken up in the transmission line, without disturbing the measurement. In general this is accomplished in two ways: by allowing the waveguides to bend in a suitably small deflection, an approach adopted by necessity for the high field side waveguides for ITER, or by providing some sort of swivel arrangement,[3] an approach used in the low field side and divertor systems.

Calibration: One consequence of the machine movements during the pulse is that reference time delays for the reflectometers have to be provided in real time. This is in contrast to the requirements on existing reflectometers, although such schemes are not unknown. For the main reflectometer, a grating/mirror near the front end similar to that presented in ref. 11 may be suitable. For the divertor and high field side systems, a controlled leak or fixed reflector are more suitable. If the calibration is not an integral part of the measurement (for example in the case of a reflective grating), additional instrumentation is necessary to monitor these movements during the pulse.

Location and maintenance of the electronics: Outside the bioshield, the mm-wave electronics are only accessible during maintenance periods, unless extremely long (>100 m) transmission lines are used to bring the signals to the diagnostic hall. This has been found to be impractical, or even impossible in some cases, so that the equipment has to be designed to operate completely unattended for long periods of time, and placed closer to

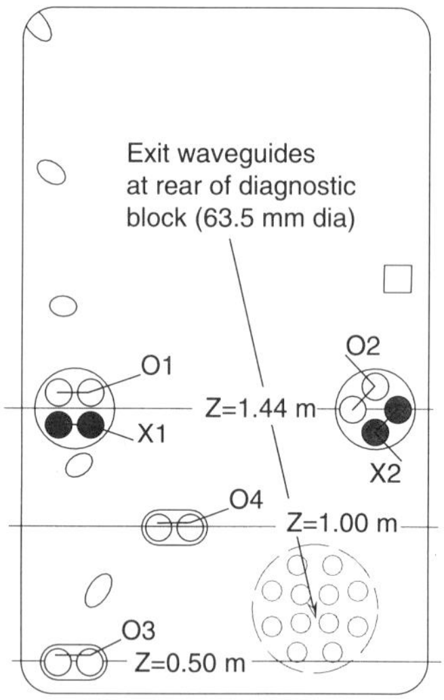

Figure 1: View of the reflectometer low field side port from the plasma. The functions of each antenna cluster are discussed in detail in reference 7. The antennas shown are truncated ends of nominally 90 mm ID waveguide. In practice the antenna pattern will have to be optimised to ensure sufficient overlap for the full frequency range (20-150 GHz in O-mode and 84-225 GHz in X-mode).

the machine. Even so, the transmission line lengths are of order 40 m, near the limits of present operational experience. A program of technology assessment for the waveguides likely to be used on several systems is underway.[12]

DESIGN FEATURES

The main reflectometer

The accessibility of the main plasma cutoff layers for the standard ITER scenarios has been reported elsewhere.[13-15] In brief, determining the position of the plasma edge and measuring the low density edge plasma profile requires the use of the X-mode upper cutoff (X-u). This can also measure beyond the steep edge density gradient region in H–mode plasmas, but will not reach the plasma centre under standard conditions. O-mode reflectometry partially fills this gap for any temperature profile in the gradient region above a minimum density. Access to the core for flat or slightly hollow density profiles requires the use of the X-mode lower cutoff (X-l), launched from the high field side. For these regions (edge/gradient/core) there are a number of measurements required (Table 2).

For the low field side, in order to meet the requirements implied by Table 2 for the majority of ITER operating conditions and plasma shapes, a number of distinct front ends are required. To keep the total number of transmission lines low, the approach is to use relatively few, but broad band, front ends. Based on current operational experience, we are

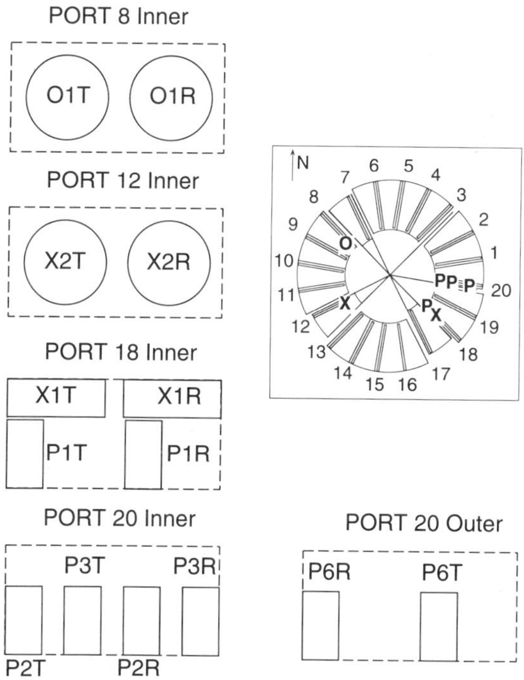

Figure 2: Schematic of the occupation scheme for four nominal diagnostic conduits of ID 70 x 35 mm. P1-P8 refer to waveguide pairs for the plasma position system. O1 refers to a nominal upper band (60-150 GHz) high field side O-mode component of the main reflectometer. X1 and X2 are low (to 40 GHz) and high (to 110 GHz) band waveguide pairs dedicated to the high field side X-l mode system. Note that P4-P5 and P7-P8 are not shown, because they are not expected to use the conduits. Inset: The toroidal distribution shown on a backplate map. The backplate is shown split into assembly sectors. Each conduit is close to the middle of an even (diagnostic) vertical port.

considering bistatic antennas to reduce the possibility of spurious reflections affecting the signal. The arrangement (shown in Figure 1) is optimised for profile performance and MHD measurements for plasmas with Z in the range 0.1-1.75 m. Small scale fluctuation measurements share the same antennas and have affected the design mainly in the grouping of O and X-mode antennas. Details of the design will be presented in accompanying papers.[7,8]

The high field side measurements have to share space with the plasma position system, described below, and also all the other high field side diagnostics using the conduits. Although many measurements are desirable from the physics point of view, it is difficult to accommodate them all within the four conduits presently allocated to reflectometry on the high field side.[7] Nevertheless we expect to be able to perform the key measurements (Table 2), of profiles, MHD and sawteeth in the core region, using the X-mode lower cutoff.

The reflectometer for plasma position

This diagnostic is designed to act as a stand-by position/shape measurement system, in order to correct or supplement the magnetics for plasma position control. It is likely to be essential during pulse operation exceeding 3 x 10^3 s, where the position deduced from

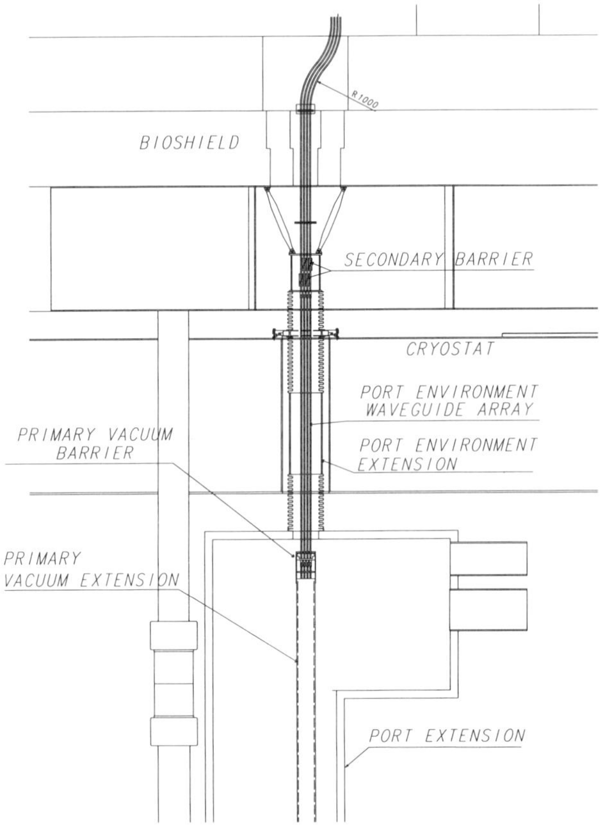

Figure 3: Schematic of the access of microwave guides through a preliminary version of the vertical port for high field side launch (position control and main reflectometer systems). Note the sliding joints beneath the first and second vacuum boundary. The waveguides in the interspace vacuum are allowed individually to flex to take up movements between the vacuum vessel and bioshield.

Table 3: Measurements possible using the divertor waveguide set, and the corresponding waveguide under the assumption that 10% loss per bend is tolerable.

Measurement	Band and Polarisation	Suggested Waveguide size (mm), toroidal x poloidal	Bend type in poloidal plane	Bend type in toroidal plane
Profile reflectometry	mm X	10 x 60	Mitre	None
	low mm O	10 x 10	Optimised non-linear bends (e.g. hyperbolic secant)	
	mm O	10 x 20	Mitre	None
ECA	mm/ sub mm X	10 x 20	Mitre	None
Interferometry, Plasma Dispersion Measurement (including peak density)	sub-mm O	10 x 20 *	Mitre	Mitre

* For these measurements, ohmic losses, (rather than mode conversion at the bends), influence the waveguide choice

the magnetic diagnostics could be subject to substantial error due to drifts. This is an advanced control measurement (category 1b), but can supplement the gap measurement (1a) provided for machine protection by the magnetics in normal pulses.

Clearly, no measurement of a simple plasma quantity can be an exact substitute for the magnetic measurements for the location of the outer flux surfaces. Nevertheless, measurement of the location of contours of a plasma parameter can supply sufficient information to control the plasma shape. For first wall protection, such a measurement has the advantage of providing a more direct estimate of the gap between the plasma and the wall than a magnetic reconstruction.

To keep the system as simple as possible, we have chosen an O-mode system located at several (~8) points around the main plasma chamber. The relevant frequency range[3] is 15-60 GHz. For an O-mode reflectometer, the accuracy of the estimate of the edge gap improves by measuring the group delay for cutoff densities below the chosen control density, because of the initialisation requirement in the inversion process. 15 GHz corresponds to 2.7 x $10^{18}/m^3$ and should ensure more than sufficient accuracy for full performance (and hence dangerous for the first wall) plasmas. At low densities the accuracy degrades, although it is likely to remain reasonable for monotonic edge profiles.[5]

The position control reflectometer shares the implementation problems and features of the high field side main reflectometer, and its engineering implementation is being investigated in parallel with that system. In the present arrangement of diagnostics, four conduits are available for use by the high field side reflectometry systems. Figure 2 illustrates their use. It is seen that in principle there is enough space available if certain features of the high field side main reflectometer and of the plasma position system are consolidated. The antenna distribution, however, is a distinct optimisation problem.[4,5] To illustrate the access path for both systems, Figure 3 shows a schematic of the features of a transmission line taken through a nominal vertical port.

The divertor reflectometer

The divertor reflectometer is mainly a density profile measurement system. For profiles, the resolution target is 3 mm across the divertor leg, and 10 cm along the leg, and the time resolution requirement is 1 ms. However the waveguide set will also have to be used for electron cyclotron absorption and interferometric measurements. Finally, an additional function for this system may be determining the ionisation front location in the semi-detached divertor operating regime. In addition to the difficult access within the cassette, the challenges in designing the divertor system lie in two areas:

1) The divertor operating space (Figure 4). This is so wide, that it forces the use of two distinct types of density measurement. In the mm-wave domain, it is reasonable, based on present technology to plan for continuous swept measurements. In the sub-mm domain, spot measurements at a number of frequencies are planned. By combining transmission and reflection measurements it is expected that the first few moments of the density profile (peak density, width) can be estimated. The frequency operating space requirements also have consequences in the choice of waveguide. For example, near the front end, the toroidal size is fixed at around 10 mm by the exigencies of the first wall, and there are several bends required in the poloidal plane. Table 3 shows suitable types of waveguide given the frequency ranges shown in Figure 4, and a requirement for mode conversion less than 10% per bend.

2) The wide range of possible measurement geometries. For example within reflectometry, one can measure from the private flux or target side, using monostatic or bistatic antennas, in O-mode or X-mode, with a choice of sightlines. In addition ECA, interferometry and plasma dispersion techniques requiring transmission through the plasma share the access space.

The present design choice is for multiple sightline, low transmission line diversity system. In order to estimate the number of possible sightlines in such a system, we have to consider the possibilities for waveguide sharing between measurements. It turns out that, once the limitation on the total number of sightlines is taken into account, it is possible to accommodate 5 sightlines on the outer leg, and 3 sightlines on the inner leg by making full use of two ports (40 waveguides).

More work is necessary to determine the viability of the more difficult schemes of

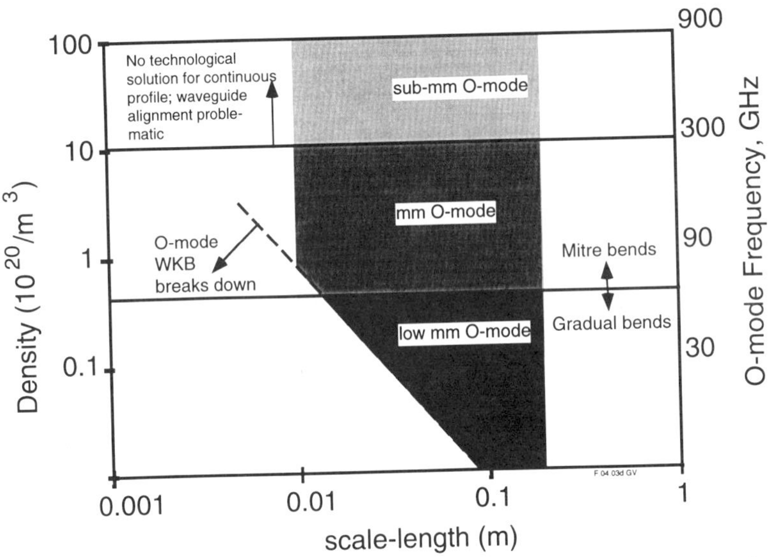

Figure 4: Divertor operating space. The required density measurement range is[1] 10^{19}-10^{22}/m^3. The x-axis represents the local density scale length, deduced from early (no impurities) simulations of the divertor. To cover the density range in O-mode with good accuracy requires frequency coverage from ~9 to 900 GHz. This is to allow for initialisation of the profile reconstruction for the lowest density in this range.[16] In the region where the WKB approximation is violated, early indications[17] are that the resultant errors in the reconstruction are small (~ 1 mm). X-mode can be used to improve accuracy in this area.

Table 3 before determining a final configuration. A crucial decision will be the extent to which X-mode reflectometry need be deployed, because it requires a size of waveguide that is difficult to accommodate, and requires the use of frequencies, particularly in the inner leg, where technological solutions for continuous profiles are not well developed.

In the meantime, since the ITER divertor cassette design is considerably advanced we were able to determine the regions within the divertor cassette where waveguides could be placed with some confidence. A concept of using a semi-independent insert to house and cool the front end waveguides and antennas was developed (Figure 5). The antennas have to view the plasma through the gaps between target plate (10-25 mm). Any other solution produces unacceptable local heat loads on the divertor target.

Providing a reference point for calibration, and resolving plasma reflections (the plasma - antenna distance ranges from a few cm to ~0.5 m, depending on the sightline) is complicated by parasitic reflections in the gap between the cassettes. More details will be presented in a companion paper.[6]

CONCLUSIONS

Reflectometry is expected to be routinely useful in ITER for a wide range of measurements, many of which are required to form part of real-time control loops used to operate the machine. As a result we have planned for thorough coverage of all plasma regions. Nevertheless the redundancy is low, in the sense that for every waveguide there is a clearly identified set of functions. There are great differences in the details of the implementation of each system, but also some common themes (not all of which have been explored in this paper), which will dictate the path of future work on these systems:

One common theme is the requirement for *optimisation of a limited number of antennas for a wide range of functions*. This optimisation depends both on the antenna distribution, which must be sufficient for the coverage of expected plasmas, and the *optimisation of individual front ends for the full frequency and radial range expected* with the minimum disturbance to the first wall or target and neutron shield structures. In principle, solutions have been found for this key problem of access through the first wall for all the systems, but much work remains, including a *comprehensive assessment of the likely effect of fluctuations on front end performance*, which will influence the choice and

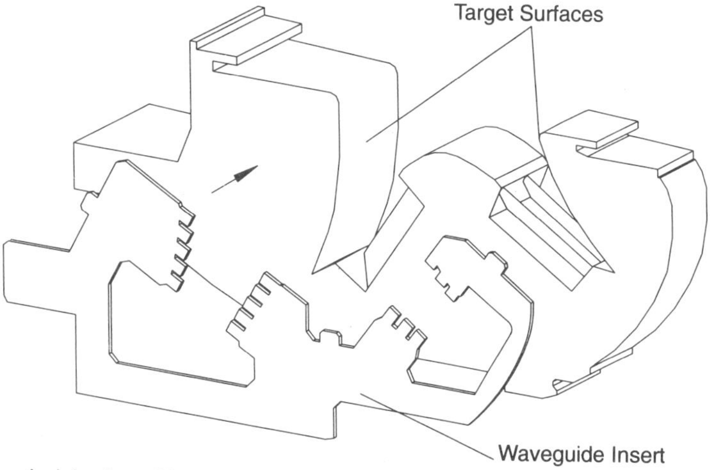

Figure 5: The principle of attaching a pre-assembled waveguide insert onto a divertor cassette. This design allows for considerable freedom within the insert envelope. Outside the cassette the waveguides are coupled to circular corrugated waveguide, then transmitted to the machine pit via a removable set of waveguides that allows maintenance of the cassette and the rest of the divertor.[6]

dimension of the antennas.

Another common element in the designs is the need for rugged calibration schemes that will allow the real time reference of the density profile to a hardware feature of the machine (backplate, vacuum vessel or divertor cassette). Again, this feature is linked to the detailed optimisation of the front end. Finally, for all systems there is a need to mitigate the effects of parasitic reflections that arise from the close fitting apertures through which the antenna views the plasma. This is difficult to arrange, especially in the divertor, where reducing the plasma to antenna distance exposes the antenna to very large (peak of several MW/m^2) radiative heat load.

It is a common observation in complex mm-wave systems that the performance of the whole system depends on interactions between the components, which can produce undesirable and unpredicted results. Given the effort required to place diagnostics within a machine such as ITER, it is expected that prototype transmission lines will have to be thoroughly tested for all the systems before installation.

ACKNOWLEDGMENT

This report has been prepared as an account of work performed under the Agreement among the European Atomic Energy Community, the Government of Japan, the Government of the Russian Federation, and the Government of the United States of America on Cooperation in the Engineering Design Activities for the International Thermonuclear Experimental Reactor ("ITER EDA Agreement") under the auspices of the International Atomic Energy Agency (IAEA).

REFERENCES

1. C. Laviron, A.J.H. Donné, M.E. Manso and J. Sanchez, Reflectometry techniques for density profile measurements on fusion plasmas, *Plasma Phys. Control. Fusion* 38:905 (1996).
2. A. E. Costley *et al.*, Requirements for ITER Diagnostics, in: *Diagnostics for Experimental Thermonuclear Fusion Reactors*, P. E. Stott *et al.*, eds., Plenum Press, New York (1996).
3. G. Vayakis *et al.*, Reflectometry on ITER, *Rev. Sci. Instrum.* 68:435 (1997).
4. N. Bretz *et al.*, ITER position control reflectometry - conceptual design, in: *These Proceedings*
5. E.J. Doyle *et al.*, Design basis for the ITER plasma shape and position control reflectometer system, *ibid.*
6. M. Manso *et al.*, Reflectometry in the ITER divertor, *ibid.*
7. V.A. Vershkov, The ITER reflectometry diagnostic for the main plasma, *ibid.*
8. P. Edmonds *et al.*, Integration of vacuum coupled diagnostics, *ibid.*
9. V. Mukhovatov *et al.*, Role and requirements for plasma measurements on ITER, *ibid.*
10. V.S. Voitsenya *et al.*, On the possibility of using carbon-graphite materials for the inner elements of millimetre and sub-millimetre diagnostics of a fusion plasma, *Fusion Engineering and Design* 34/35:491 (1997)
11. L. Empacher *et al.*, Calculations and experiments on multi-beam transmission for the 140 GHz/10 MW CW ECRH system on the stellarator W7-X, to appear in: *Proceedings 10th Joint Workshop on Electron Cyclotron Emission and Electron Cyclotron Resonance Heating*, Ameland, the Netherlands (1997).
12. D. Wagner *et al.*, Transmission line design for broadband microwave diagnostic systems, in: *These Proceedings.*
13. E.J. Doyle *et al.*, Reflectometry Applications to ITER, in: *Diagnostics for Experimental Thermonuclear Fusion Reactors*, P. E. Stott *et al.*, eds., Plenum Press, New York (1996).
14. M.E. Manso *et al.*, Reflectometry for ITER Density Profiles, *ibid.*
15. V.A. Vershkov, Proposal of Reflectometry System for ITER, *ibid.*
16. P. Varela *et al.*, Initialization of plasma density profiles for reflectometry, *Rev. Sci. Instrum.* 66:4937 (1995).
17. L. Bruskin, A. Mase, T. Tamano and K. Yatso, Application of 1D WKB approximation in Microwave Reflectometry of Plasma Density Profiles, submitted to: *Rev. Sci. Instrum.* (1997).

ITER REFLECTOMETRY DIAGNOSTICS FOR THE MAIN PLASMA

V. Vershkov,[1] M. Manso,[2] G. Vayakis,[3] A. J Sanchez,[4] D. Wagner,[5]
C. Walker,[6] S. Soldatov,[1] L. Kuznetsova,[1] V. Zhuravlev,[1] B. Sestroretskii,[7]
and the ITER Joint Central Team and Rusian and EU Home Teams

[1]Kurchatov Institute of Atomic Energy, Ploshchad'Akademika Kurchatova 1, Moscow 123182, Russia.
[2]Associacao EURATOM/IST, Centro de Fusao Nuclear/ Instituto Superior Tecnico, 1096 Lisboa Codex, Portugal
[3]ITER Joint Central Team, San Diego JWS, 11025 N. Torrey Pines Road, La Jolla CA 92037, USA
[4]Association EURATOM/CIEMAT, 28040 Madrid, Spain
[5]Institut fur Plasmaforschung, Pfaffenwaldring 31, 70569 Stuttgart, Germany
[6]ITER Joint Central Team, Garching JWS, Max-Planck-Institut fur Plasmaphysik, 85748 Garching, Germany
[7]N. P. O. im. S. A. Lavochkina, 141400, Himki, Moscow Region, Russia

INTRODUCTION

Reflectometry is now widely used in tokamak experiments. Reflectometry relies on the fact that, as an electromagnetic wave propagates through the plasma, its phase is shifted due to the departure of the local refractive index from the vacuum value. At a certain critical density, the refractive index goes to zero, and the wave is reflected. This density is the density of the *cutoff layer*. Its application on ITER is especially important because it can provide the density profile (by inverting the phase as a function of frequency) and essential information on density perturbations due to plasma modes, with minimum requirements for the access ports. These measurements are to be used for machine protection, plasma control and for establishing performance characteristics. In addition, reflectometry supplies valuable information on plasma turbulence in all regions of the plasma. It uses well developed mm-wave band equipment and the launched and received waves may be easily transported to and from the tokamak by means of long transmission lines. All in-vessel transmission lines are made from metal and are insensitive to neutron irradiation, while the vacuum windows may be placed away from the high neutron flux zone. Specific problems arise in the reflectometry application to ITER, due to high electron temperature, possible flat density profile, and difficulties with access to high field side of tokamak.

Diagnostics for Experimental Thermonuclear Fusion Reactors 2
Edited by Stott *et al.*, Plenum Press, New York, 1998

As shown in reference 1, ITER reflectometry consists of three systems. This paper presents the project of ITER reflectometry diagnostics for the main plasma, which is the result of common work of JCT team and the Russian and EU home teams. It is based on previous[2,3,4,5] work. It will be shown that reflectometry can meet the ITER requirements for the measurements of the main plasma parameters.

FUNCTIONS AND DESIGN REQUIREMENTS

The target measurements requirements and classifications used in this paper, were formulated in reference 6 and summarised in reference 7. The functions and requirements are also discussed in the reflectometry overview paper.[1] The main plasma reflectometer provides measurements in all three categories: Measurements for machine protection and plasma control (category 1), for performance evaluation and optimisation (category 2), and additional measurements for physics understanding (category 3).

Electron density profile measurements[8,9,10] are the primary function of the diagnostic (category 2). Measurements are required in the core plasma, gradient and SOL regions. For the core plasma the measurements must cover the density range from 0.3 to 3×10^{20} m^3, with accuracy 5%, and provide time resolution 10 ms and spatial resolution 30 cm. For the edge plasma the corresponding requirements are 0.05 to $3\times10^{20}/m^3$, 5%, 10 ms and 0.5 cm.

Locked modes, MHD modes, sawteeth and disruption precursor measurements are category 1. They must be detected in the range of relative poloidal magnetic field fluctuation of 10^{-4}-10^{-2} with 30% accuracy and 1 ms time resolution. The frequency of the MHD modes may be in the range 0 - 3 kHz. Reflectometry[11,12] can provide the radial position, size of the islands and identify n and m numbers for the case of several antenna locations.

Fishbones and TAE mode observations are category 2 and must be carried out in the frequency range of 3 - 30 and 30 - 500 kHz respectively. Reflectometry can resolve the structure and radial positions of the modes.[13]

ELM observation is category 1 and must be performed with an accuracy of 30% and time resolution 0.1 ms. Reflectometry can measure the radial features of the ELMs with high accuracy.[14]

The q profile is a category 2 measurement. Positions of rational q can sometimes be evaluated by means of radial localisation of resonant MHD modes.[11]

Plasma rotation measurements (category 2). Reflectometry has been used to measure the rotation of density fluctuations.[15,16] These measurements do not, in general, give a reliable estimate of a plasma rotation, because the fluctuations move with respect to plasma. In addition reflectometry in principle can't separate toroidal and poloidal rotation.

Density fluctuation measurements[17,16] are of category 3. It is well known that reflectometry is very sensitive to the density fluctuations. However for high fluctuation levels, which may be present in ITER, reflectometry gives only qualitative information with respect to the level and frequency spectrum of small scale turbulence.

Line average density measurements are of category 1 and must cover the density range from 0.01 to 2×10^{20} m^3, with accuracy 1% and time resolution 1 ms. Reflectometry is not the main diagnostic for such measurements, but can be used to check other measurements, for example the calibration of the LIDAR system.

OPERATING FREQUENCY RANGE AND ACCESSIBILITY

The relativistic calculations of cutoff frequencies and absorption coefficients were already published in reference 2. For this paper such calculations were repeated based on reference 18 with essentially the same results. Figure 1 presents the cutoff frequencies, while Fig. 2 shows the plasma absorption for an ITER reference scenario in H-mode. It was found that the relativistic cutoff frequency shift is well described just by the mass correction. When relativistic effects are included, the O-mode cutoff frequency (plasma frequency) profile becomes hollow. Thus O-mode is useful only for density profiles in the gradient region. The O-mode lowest frequency will be in the range of 10-15 GHz, limited by the antenna dimensions. Evaluation of WKB violation effects at low density shows negligible errors.[19]

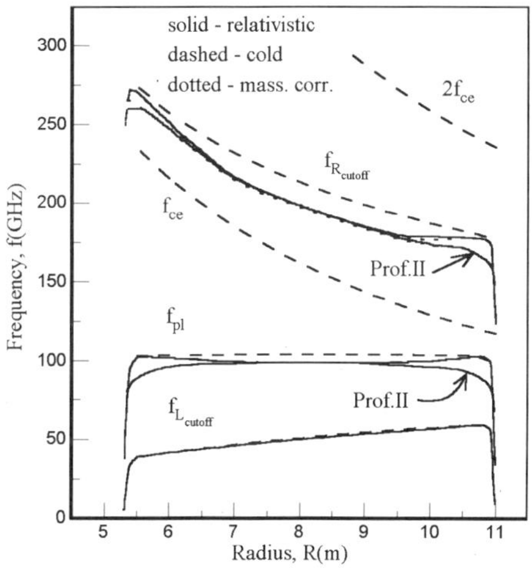

Figure 1. Cutoff frequencies for an ITER reference profile. Profile II - $Ne(r)=(1-(r/a)^2)^{0.15}$.

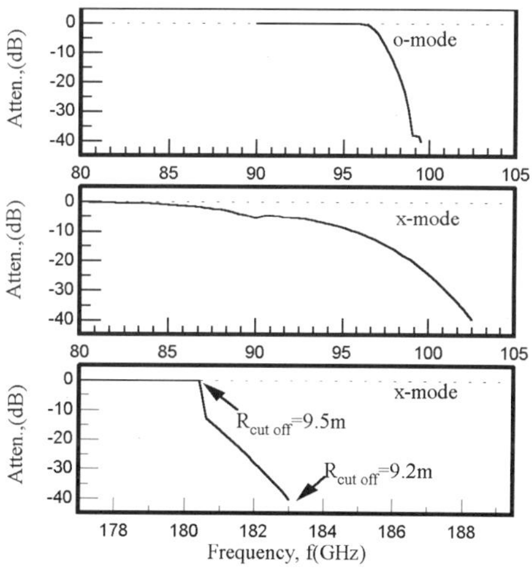

Figure 2. Two pass power absorption.

X-mode launched from the low field side (LFS) and reflected at the upper cutoff layer, (Xu mode), needs a relativistic frequency correction, which creates an inaccessible hollow region near R=10 m. In addition, relativistically down-shifted 2nd harmonic ECA results in prohibitive absorption of the launched signal inside R~9 m. Nevertheless it can measure beyond the steep edge density gradient region in H–mode plasmas in the case of slightly peaked density profile as it is shown in Fig. 1. More importantly, X-u mode can provide the measurements of the plasma edge position and the low density edge plasma profile.

In contrast X radiation launched from the high field side (HFS) and reflected at the lower cutoff (Xl mode) is always able to access the plasma core, with negligible absorption and very small deviations from the non-relativistic case. This makes its use necessary to guarantee core access for flat or even slightly hollow profiles.

Interferometry can be performed in frequency range 105 - 125 GHz for O mode and 65 - 125 GHz for X mode. It uses LFS antennas as the launcher and receiver, with reflection from the HFS wall. The use of O mode is not possible due to the high absorption as it is shown in Fig. 2. But it may be useful for the density build up phase with the frequencies lower 90 GHz. Refraction effects are also likely to be important.

Assuming sufficient overlap in profile coverage to decrease the uncertainties in the initialisation of Xu profiles of the edge (due to errors in the magnetic field estimate), as proposed in reference 20, we conclude that core profile can be measured with Xl mode in frequency range 10 - 110 GHz and with O mode in frequency range 20 - 155 GHz. For the edge profile it is possible to use O mode with frequencies 20 - 155 GHz and Xu mode with 70 - 255 GHz. The lowest frequency for X-u mode includes the interferometric regime.

DESIGN OF THE LOW FIELD SIDE SYSTEM

The LFS system consists of an Xu-mode system for the edge, gradient and interferometry measurements and an O mode system for the gradient region. The LFS antenna configuration is presented in a figure of reference 1, and can also be seen in Fig. 3 together with the rest of the system. Because the Xu mode cutoff is determined by the density and the magnetic field, the return signal at the edge is not very sensitive to changes in the vertical position of the plasma. Thus two antenna systems at one vertical position, separated in the toroidal direction are sufficient. Each antenna consists of a two circular corrugated waveguides, of ~100 mm diameter, terminated at the front edge of the diagnostic block (about 0.7 m from the separatrix). The function of the first antenna pair is to measure SOL / gradient profiles. In addition it performs 1-point MHD measurements and 1-point density fluctuation measurements. The second pair is a back-up of pair 1 and provides, with the first pair, 2-point MHD-scale perturbation measurements.

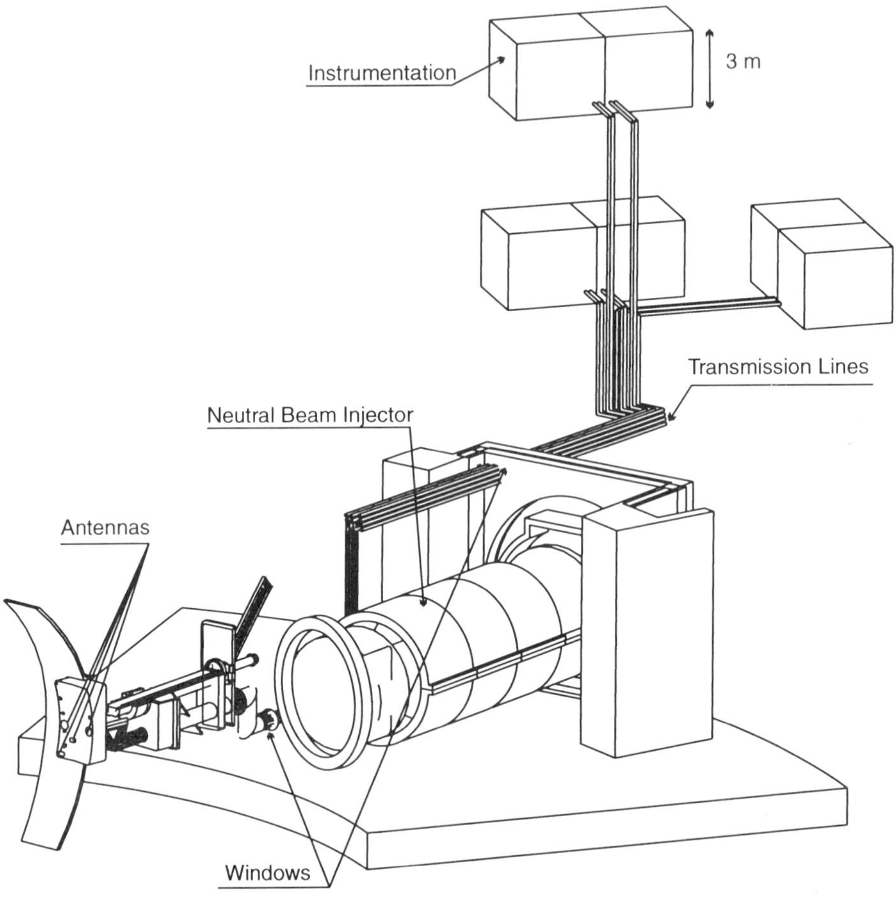

Figure 3. Overview of Low Field Side reflectometry system.

Because the O-mode cutoff is solely determined by the density, the return signal is very sensitive to changes in the vertical position of the plasma, Z. Implicit in the placement of antennas is the assumption that deviations of Z of ±20 cm from the antenna height cause tolerable (< 10 dB) power loss in the signal. In that case four antenna systems are necessary to cover the assumed Z-range of the ITER plasmas. By distributing these toroidally as well as poloidally, multi-point measurements of MHD activity are enabled. Each antenna system is made of the ends of two circular corrugated waveguides, of 100 mm nominal diameter, terminated at the front edge of the diagnostic block. The first antenna pair at nominal height of Z=1.44 m. is used to measure gradient region profiles (full aperture plasmas). In addition it provides gradient region 1-point MHD measurements and 1-point density fluctuation measurements. Antenna Pair 2 is a back-up of pair 1. In addition it provides 2-point MHD measurements (in the toroidal direction, with pair 1). A pair at a nominal height of Z~0.4 m provides gradient region profiles (for start-up plasmas). In addition, it is used for 2-point MHD measurements (in the poloidal direction, with pair 1). A pair at a nominal height of Z~1.0 m provides gradient region profiles (for " advanced scenario " or other plasmas in the Z range 0.8-1.2 m). In addition it is used for 2-point MHD measurements (with any other pair).

All antennas may be used in both polarisations, so they can be integrated to enable 2 and 3-point measurements of density fluctuations, with some degradation in performance due to power attenuation in vacuum windows for the non-specified polarisation. Fig. 4 shows an outline of the transmission line concept. Transmission inside the vacuum vessel is accomplished by means of broad-band circular corrugated waveguide and mitre bends. The diameter of the waveguide is ~90 mm ID, determined by considerations of mode conversion at the mitre bends at low frequencies (< 50 GHz) and due to flexing of the waveguides.

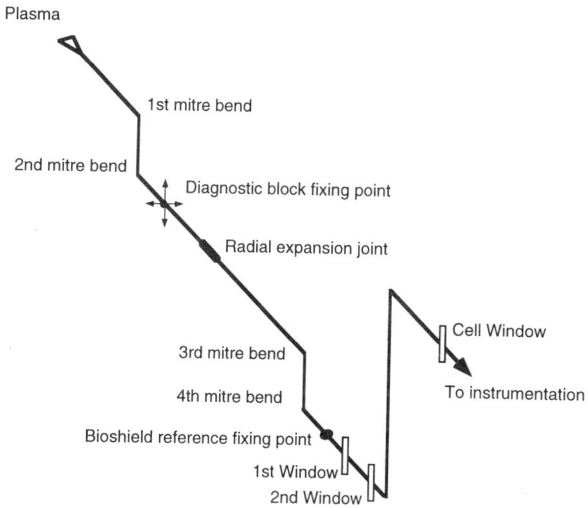

Figure 4. Idealised scheme of the LFS transmission line.

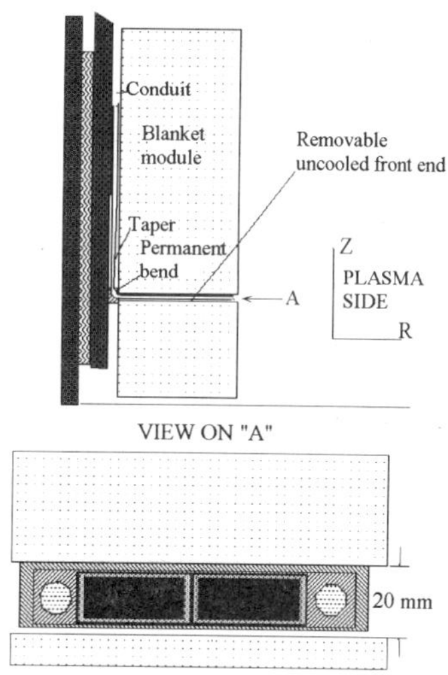

Figure 5. High Field Side antenna concept.

There are 4 mitre bends in the in-vessel transmission line. The first two compose a dog-leg structure through the diagnostic block, designed to reduce neutron streaming. Both bends are specialised designs: The first bend has to cope with direct radiation from the plasma. The second bend incorporates the calibration mechanism. The second two mitre bends form a swivel joint designed to convert the toroidal motion of the vessel into a small vertical bend of the long transmission line to the bioshield. Expansion in the radial direction is accommodated by a waveguide expansion joint. The same joint provides electrical isolation to prevent large currents (eddy currents, currents due to off-normal events) from flowing. Such currents can endanger the integrity of the waveguide structure.

The primary and secondary (cryostat tunnel) vacuum barriers are extended into the neutral beam cell. Millimetre wave radiation is transmitted through double fused quartz windows inclined at the Brewster angle. This pair of windows forms the first confinement barrier. The second confinement barrier is a quartz window inclined at the Brewster angle at the penetration through the NBI cell boundary. In between the confinement barriers there is a remotely operated metal shutter. The primary purpose of this shutter is to isolate maintenance operations from EM radiation from within the torus (ECE emission, ECRH, etc.) In addition it serves as a secondary calibration / line testing point. Finally, in the unlikely event of window failure it can be used to inhibit air flow between the NB cell and atmosphere.

For the purpose of calibration the reflection from the blanket / shield structure on the HFS may be used as the reference. However, it is not a simple mirror and not available during the pulse. Therefore a second reference is provided in the form of a grating structure on the second mitre bend, designed in such a way as to form a well defined out-of-band (> 230 GHz) or in band reflector, without significant effects on the in-band transmission.

In the present design the antenna diagram is narrow (less then 1° at high frequencies for Xu mode), which causes loss of signal due to plasma shift and poor coupling efficiency in the near field zone. Moreover as the spot in plasma illuminated by the antenna becomes close to the turbulence wavelength it is expected to result in strong amplitude modulation of the reflected signal. Therefore it seems reasonable to increase the beam width of the Xu and O mode antennas. It is also desirable to decrease the loss for the second polarisation (now ~3 dB per window) to enable integration for 2- and 3-point correlation measurements and to increase the redundancy.

DESIGN OF THE HIGH FIELD SIDE SYSTEM

The HFS system design is much more difficult task due to the access limitation imposed by the blanket. The key assumptions of the present HFS design were: minimal changes in blanket design, which means that antennas must use only the ~2 cm gaps between the blanket modules, and routing of the waveguides down a standard diagnostic conduit, as described in reference 1. There are 4 sectors which are natural for reflectometry and so the measurements are distributed toroidally.

It may not be possible to design waveguides able to operate adequately over the full frequency range in both polarisations. Thus two antenna pairs, Low Frequency (LF) and High Frequency (HF) are used, both for Xl and O modes, to cover the full frequency range. For Xl mode LF launchers must work in the range 10 - 40 GHz, and HF in the range 40 - 110 GHz. The corresponding frequency bands for O mode are 20 - 60 and 60 - 155 GHz. Thus, the LF O mode launch of the main reflectometer is combined with equatorial launch of the position reflectometry.

All pairs are located near the full aperture plasma current centre height (Z = 1.44 m). An example of HFS LF antenna concept is shown in Fig. 5. The waveguide line may be divided into permanent structures (waveguides in conduit with supports, taper and the rectangular bend) and removable antenna horn. A preliminary design has the antennas for all HFS lines placed with their front end 100 mm behind the surface of the blanket. The internal dimensions of the mouth is 13 x 30 mm. The removable antenna may be made from Carbon Fiber Composite (CFC) and cooled only by radiation, in which case it is heated up to ~700^{0} C.

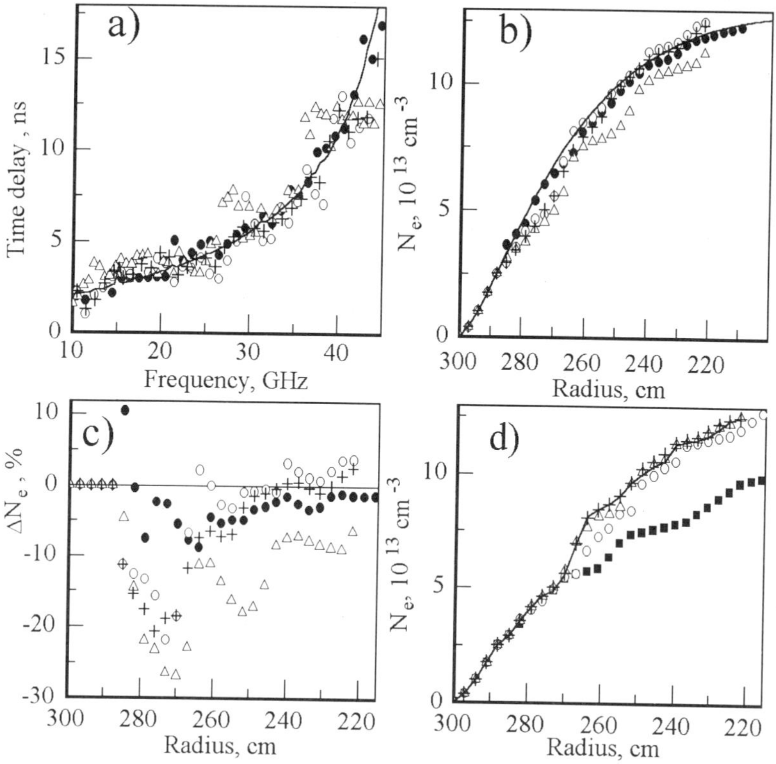

Figure 6. Results of 2D WKB code calculation for Xl mode reflection from the turbulent plasma. a) Group time delay of the launched wave obtained by means of three methods: open circles - broad frequency sweep (FFT filtering), fluctuations do not move; solid circles - broad frequency sweep (FFT filtering), fluctuations move poloidally 130 cm per single sweep ; open triangles - pulse radar; crosses - amplitude modulation; solid line -for non-perturbed density profile. b) Reconstructed density profiles and c) Relative distortions of reconstructed profiles caused by turbulence. The notations are the same. d) Influence of noise on density profile reconstruction in case of broad sweep with FFT filtering: solid line - no noise; solid squares - noise level = 100%; open circles - noise level = 50%; open triangles - noise level = 25%; crosses - noise level = 12.5%

The toroidal map of the HFS launchers is presented in overview paper reference 1. A LF launch for O mode is situated in port 20 and uses ordinary rectangular waveguides with ID 10 x 20 mm. They are tapered to ID 10 x 12 mm before the final bend with curvature radius 70 mm. A LF launch for Xl mode is situated in port 18 and uses also ordinary rectangular waveguides with ID 10 x 30 mm. They are tapered to ID 3 x 30 before the gradual bend. The HF lines for both polarisations use circular corrugated waveguides with 24.5 mm diam. with mitre bends before antenna. HF antenna for O mode is placed in port 8, while for Xl mode in port 12. The waveguides in the conduits are curved, with radii in the range 2-8 m within the conduit, but it was shown[21] that the conversion losses are low for the whole band. Tighter, and thus more problematical, bends, are present within the vertical port.

The first vacuum barrier for each waveguide is mounted on the vacuum vessel seal flange (see reference 1). It is a fused quartz window mounted at a 15° angle. The window is mounted on a gimbaled bellows, designed to accommodate flexing of the waveguide line whilst supporting its weight. Electrical breaks for the waveguide are incorporated on both sides of the window structure. Movement between the blanket backplate and the vacuum vessel is taken up by the elastic bending of the waveguide and /or an expansion joint. The second vacuum barrier is a fused quartz window mounted at the Brewster angle on the top of the cryostat lid. Electrical breaks for the waveguide are incorporated on both sides of the window structure.

Between the first and second vacuum barrier, the waveguides continue with the same cross-section as that used inside the first barrier. An expansion joint is used to take up the large vertical movement range between the cryostat and the vacuum vessel. Toroidal and radial movements are taken up by elastic bending of the waveguide. The waveguides are brought through the bioshield in a removable section incorporating dog-leg like structures to reduce radiation streaming. Another expansion joint in the waveguide is necessary to accommodate cryostat movement with respect to the bioshield. The HFS electronics systems are designed to feed two sets of antennas, one for the low and one for the high frequency band. They are located in special diagnostics room on the bioshield.

For both HFS HF waveguide lines it is possible to employ the same concept of calibration with a grating structure on the last mitre bend. The same concept may be used for LF lines if special reflecting structures will be placed inside waveguides before the last bend.

The four specific antenna systems design for each polarisation and frequency range is not optimal from the point of view of redundancy and MHD mode structure measurements. Full wave simulations show that mitre bends in rectangular waveguide with ID 25 x 30 mm may provide losses less then 2 dB in all frequency range for both polarisations. Thus all four lines may be made identical. The question about optimisation of the windows for two polarisation was already discussed for the LFS systems. The antenna design also needs optimisation. In particular the use of the walls of the blanket as the antenna element will be explored.

REFLECTOMETRY ELECTRONICS SYSTEMS

Reflectometry profile measurements are based on measuring the *group delay* to this critical layer as a function of frequency. In a simple 1-D picture, the density profile is then recovered by an inversion from the (group delay, frequency) space to the (cutoff frequency, radius) space. In O-mode this involves a simple Abel inversion procedure. Measurements of density fluctuations can be made at constant frequency, and observing the time evolution of the phase or group delay.

The technology available for the construction of mm-wave systems is in a state of continuous and rapid improvement and the systems, used on present-day devices, may well be superseded by the time of first operation. It is desirable to employ heterodyne system to have maximum sensitivity and to be compatible with the turbulence measurements. For the LFS there will be 4 O-mode and 2 X-mode systems as it is shown in Fig. 3. Both O and X-mode systems will further divide the signal into appropriate sub-bands. Each sub-band will operate independently, and the full profile reconstructed by merging the information from each sub-band. Quasi-Optical (QO) techniques will be used to perform the split and summation.

In the reference design, each band employs a fast (< 10 ms sweep time) linearly swept source, and a tracking local oscillator to produce a local oscillator (LO) signal. Both fast homodyne and heterodyne detection will be employed in the intermediate frequency (IF). Density profile and fluctuations measurements are supposed to be made simultaneously. The HFS have two similar electronics systems to feed the LF and HF antennas for each polarisation.

PERFORMANCE EVALUATION

The performance of the profile measurements is determined by :

1. Fundamental dependencies of the refractive index on the global plasma parameters, which determine the cutoff frequencies, which where already discussed above.
2. System quality in the absence of turbulence: profile initialisation, calibration, spurious reflections in waveguides and in plasma. Quality also depends on antenna performance in the near field zone and the possibility of standing wave resonances in the case of reflection at small distances from the metal wall.
3. Turbulence, which causes the fluctuations of the reflected wave phase and amplitude and "run away" problems (through plasma rotation effects). Turbulence effects depend on its amplitude and on the slope of refractive index. and do not depend on generator power.
4. Signal to noise ratio, where the signal is determined by generator power, system losses, Electron Cyclotron Absorption (ECA), 3D refraction, the average effect of turbulence (interference and scattering), and the noise is the sum of detector noise, Electron Cyclotron Emission (ECE) and Bremsstrahlung radiation .
5. Correct selection of the optimum bandwidth. In profile time-delay reflectometry measurements that should be used to maximise the reconstruction precision is determined by balancing plasma signal to noise (which should now include the effect of turbulence) with dispersion. A simple calculation for an exponential profile[3], ignoring the change in the speed of propagation in the plasma gives typical values of 1-4 GHz.

Some aspects of these effects are discussed in more detail below.

Initialisation. For O-mode measurements, it is impractical to measure the low density part of the profile because the antenna size becomes difficult to accommodate. In practice this means that a fraction of the profile has to be assumed or extrapolated from the available data, and this introduces an error. Simulations show that, for non-pathological profiles, this error can be neglected for densities ~ 10 times higher then the lowest probed density. For X-mode measurements, the initialisation relies on knowledge of the total magnetic field in the edge. On ITER-scale devices, a 1% error in this measurement results in a 10 cm error in the location of the edge. A combination of O and X-mode data may be used, to correct the X-mode initialisation error as was proposed in reference 20.

Plasma rotation. Depending on the measurement technique, there can be systematic errors in the position due to rapid plasma movement or rotation. The influence of "run away" due to the poloidal rotation of turbulence was investigated with 2D WKB code[22] in reference 23. It was found that misalignment of the antenna of 5^0 results, for the case of FMCW technique[24] in 15% errors in the density or in 5 cm in distance for a 60 GHz frequency scan in 50 μ in the presence of 1.5% turbulence with poloidal wavelength 6 cm and poloidal velocity 15 km/s. (FFT averaging over 1 GHz was used.) It was also shown that pulse radar and AM techniques is not sensitive to this effect. Results of a similar calculations for Xl mode with HFS launch are presented in Fig. 6. A horn antenna with 1.3 cm poloidal width was used. The frequency sweep was performed in the 10-30 GHz range with 20 MHz step and from 30 to 45 GHz with 5 MHz step. The beat frequency for FMCW technique was found with FFT filtering averaging over 1 GHz. It is seen that, for the same turbulence parameters as in ref. 23, the "run away" effect is not visible within the errors of measurements. This may be the advantage of the broad antenna pattern. In addition, reduction of the sweep time to 10 μs will make this error negligible.

Signal to noise ratio. The principle question is to define the frequency band for which the noise is important. The broad band sweep system was taken as reference. The time for 20 GHz sweep was taken 50 μ for Xl central reflection with time delay of 50 ns. Thus the beat frequency is 20 MHz. So the frequency band for the noise was taken 50 MHz for reflectometry and 5 MHz for interferometry regimes. Figure 6 d presents the results of density profile reconstruction with N/S as a parameter. It is clearly seen that a N/S ratio of 1/4 starts to give measurable errors. Thus N/S 1/8 was taken as a target.

The presence of density fluctuations results in the increase of a losses and a scatter in the phase and amplitude of reflected wave. For example, 2D simulations for a turbulence level of 1.5 % gives additional power losses with respect to calm plasma of 14 dB for Xl central HFS reflection with the horns 10 cm wide, but only 6 dB for 1.3 cm horn.

Estimates of the minimum powers for the expected working conditions are presented in Table 1. The EC and Bremsstrahlung powers were reduced to the detector position by taking into account waveguide losses. The antenna transmission losses were found from their apertures using a ray tracing code, and the interference / scattering loss due to the fluctuations by a 2D code. In order to increase the antenna beam width at the LFS, and thus provide sufficient overlap in the plasma, the diameter was reduced to 3 cm. The first column shows the data for the double pass interferometry regime from the LFS. For the LFS Xu and O mode lines the transmission in both polarisations is shown. For the HFS the results for four launchers types are presented. No account is taken of plasma movement. The required mm wave powers are less then are commercially available, except in the LFS Xu case.

Table 1. Estimates of minimum required mm wave power for the different launch systems.

Location:	LFS					HFS			
Waveguide designation:	X mode			O mode		LF band		HF band	
Waveguide used for:	Interf.	Xu	O	O	Xu	Xl	O	O	Xl
Noise power at det. (pW)	0.48	4.8	4.8	4.8	4.8	4.8	4.8	4.8	4.8
ECE power at det. (pW)		7600							
Bremsstrahlung at det. (pW)	0.45	9	1.	4.6	2.3	3.6	3.6	4.6	4.6
Total noise at det.(pW)	0.94	7614	5.9	9.5	7.1	8.4	8.4	9.4	9.4
Loss at summat.\spliter (dB)	6	6	6	6	6	6	6	6	6
Window losses (dB)	6	6	18	6	18	4	4	4	4
Waveguide losses (dB)	10	4	10	10	4	6	6	4	4
Bend losses (dB)						8	8	8	8
Antenna transmission loss (dB)	57.6	25	32	32	25	47.5	29	23	47.5
Interference losses (dB)		9.5	9.5	9.5	9.5	6	6	6	6
Absorption in plasma (dB)		5			5				
Total losses (dB)	79.6	55.5	75.5	63.5	67.5	77.5	59	51	75.5
Required power (mW)	5.4	170.	13.2	1.35	2.5	30.	0.4	0.07	21

Spurious reflections. The possibility of simultaneous reflection at two cutoffs, in the case of Xl mode at HFS, due to the inclination of the total magnetic field must also be mentioned. Although this angle is about 10°, the power of reflected O mode may be equal to the Xl mode for the case of central reflection due to high transmission losses of Xl mode. This effect will not important for the FMCW technique as it can separate two different beat frequencies. For the turbulence measurements at constant frequency it can be eliminated by means of synchronous modulation of the launch power and the receiver efficiency.

Turbulence. The scatter in phase and amplitude is the most important result of turbulence. A 1% level of density fluctuations may give significant errors in profile measurements at r/a <~0.3 for the Xl mode. This effect is worse for the case of flat density profile. Thus 1% density fluctuations will result in reflection 20 cm before the cutoff due to the slow variation of the refractive index with radius. (For the Xu mode, this deviation is 4 cm.) It was found,[23] that a turbulence level of 1.5 % results, for the Xu mode, in radial random errors of 1.5 cm (3% in reconstructed density). Somewhat higher errors of 5 cm and 15% may be seen in Fig. 6 for Xl mode reflection. This error is typical for all simulated techniques. Taking this value for the single measurement in 50 μs, we can conclude that the errors will be reduced to 1.5 cm and 5% with the averaging of 10 sweeps in 1 ms. Thus, the ITER requirements are exceeded in the core region. For a steep gradient zone at the edge the errors of single measurements must be of the order of density decay length or less and averaging over time decreases the errors further to meet the ITER requirements also in this zone. Such estimates are supported by present experiments on tokamaks.

The phase fluctuations, measured with wide aperture antenna, will have a simple relation to the local density fluctuation near cutoff layer only if the distance to cutoff less then the diffraction length,[25] which depends on the turbulence amplitude and wavelength. For ITER the diffraction length is always much smaller then cutoff distance for small scale turbulence with $k \times \rho_i = 1$. In that case imaging systems, which by definition employ a large aperture antenna (real or synthesised) may be necessary to extract quantitative information on small scale turbulence. Apart from the difficulties to accommodate such systems within the blanket/shield structure, this requirement reduces the reliability of profile measurements under shifts of the plasma. In fact special precautions must be made at the LFS to widen the antenna diagram. Thus only qualitative measurements of small scale turbulence may be possible without a special antenna system. Imaging systems are not necessary for the long wave MHD/TAE mode studies, and even qualitative measurements of small scale turbulence can provide vital information.

CONCLUSIONS

The ITER reflectometry diagnostics for the main plasma are able to meet the main requirements of the target task except for quantitative measurements of small scale turbulence. The system performance is likely to be enhanced by the time of ITER operation due to technical progress, further optimisation the system elements, and better understanding of reflectometry physics in the presence of turbulence. In this context, the necessity of experimental work with the HFS Xl mode in existing tokamaks with flat density profile must be pointed out. It is also desirable to develop further the full wave codes to simulate reflection from turbulent plasma in real ITER geometry. Further optimisation of LFS antennas, transmission lines, vacuum windows for two polarisation and waveguides rectangular bends are needed. Although part of work may be done with computer simulation, experimental tests of prototypes in the laboratory are also planned.

ACKNOWLEDGEMENT

This report has been prepared as an account of work performed under the Agreement among the European Atomic Energy Community, the Government of Japan, the Government of the Russian Federation, and the Government of the United States of America on Cooperation in the Engineering Design Activities for the International Thermonuclear Experimental Reactor ("ITER EDA Agreement") under the auspices of the International Atomic Energy Agency (IAEA).

REFERENCES

1. G. Vayakis *et al.*, Overview of the ITER reflectometry diagnostics, this Conference
2. M.E. Manso *et al.*, Reflectometry for ITER Density Profiles, in: *Diagnostics for Experimental Thermonuclear Fusion Reactors*, P. E. Stott *et al.*, eds., Plenum Press, New York (1996).
3. E.J. Doyle *et al.*, Reflectometry Applications to ITER, *ibid.*
4. V.A. Vershkov, Proposal of Reflectometry System for ITER, *ibid*
5. A.J.H. Donne et al, Advantages and limitations of microwave diagnostics in ITER, *ibid*
6. A. E. Costley *et al.*, Requirements for ITER Diagnostics,. *ibid*
7. G. Vayakis *et al.*, Reflectometry on ITER, *Rev. Sci. Instrum.* 68:435 (1997).
8. F. Simomet, Rev. Sci. Instrum., **56**, 664, (1985)
9. A. Silva et al, 22 EPS Conference, Bournemouth (1995), V 19C, part IV, p. 413.
10. E. Doyle et al, Rev. Sci. Instrum.,**66**, 1229 (1992).
11. M.E. Manso et al, 17 EPS Conference, Amsterdam (1990), V14B, part 2, p. 837.
12. R. Nazikian et al, Phys. Plasmas, **3**, (1996) 593.
13. E.D. Frederickson et al, Plasma Phys. and Contr. Nucl. Fus. Res.,IAEA, Vienna, (1995), V. 1, p. 275.
14. K.W. Kim et al, Rev. Sci. Instrum., **68**, (1997), 466.
15. A. Costley et al, Rev. Sci. Instrum. **61** (1990) 3487.
16. V.A. Vershkov et al, Plasma Phys. and Contr. Nucl. Fus. Res.,IAEA, Vienna, (1995), V. 2, p. 65.
17. E. Mazzucatto, Princeton University Plasma Physics Laboratory Report, MATT-1151 (1975).
18. L.K. Kuznetsova, EC Ray Tracing with relativistics effects, to appear in Proceedings of 24 EPS, Berchtehsgarden, 1997, Germany.
19. L. Bruskin, et al, Application of 1D WKB approximation in Microwave Reflectometry of Plasma Density Profiles, submitted to *Rev. Sci. Instrum.* (1997).
20. M.E. Manso *et al.*,Proposal of calibration radially X mode with O mode, this Conference.
21. D. Wagner *et al.*, Transmission line design for broadband microwave diagnostic systems, to appear in: *Proceedings 10th Joint Workshop on Electron Cyclotron Emission and Electron Cyclotron Resonance Heating*, Ameland, the Netherlands (1997).
22. V. Zhuravlev et al, Plasma Phys. Control Fusion **38** (1996) p 2231.
23. T. Estrada, V. Zhuravlev et al, Comparison of reflectometry techniques for density profile measurements in turbulent plasma, toappea r in Proceedings of 24 EPS, Berchtehsgarden, 1997, Germany.
24. C. Laviron, A.J.H. Donne, M.E. Manso and J. Sanchez, Reflectometry techniques for density profile measurements on fusion plasmas, *Plasma Phys. Control. Fusion* 38:905 (1996).
25. E. Mazzucato, R. Nazikian, Rev. Sci. Instr. **66**, 1237, (1995)

DESIGN BASIS FOR THE ITER PLASMA SHAPE AND POSITION CONTROL REFLECTOMETER SYSTEM

E.J. Doyle,[1] N.L. Bretz,[2] K.W. Kim,[1] W.A. Peebles,[1] and T.L. Rhodes[1]

[1]Electrical Engineering Department and Institute of Plasma and Fusion Research, University of California, Los Angeles, CA 90095, USA
[2]Princeton Plasma Physics Laboratory, Princeton University, Princeton, NJ 08543, USA

INTRODUCTION

A fundamentally new approach to plasma position control is required on ITER as a consequence of the anticipated long pulse lengths. The error in conventional magnetic plasma position control systems accumulates with time (due to drifts, etc.), such that magnetic systems are not expected to provide accurate control for discharge lengths in excess of ~1000 s.[1] ITER will have pulse lengths exceeding 1000 s, and potentially up to 10,000 s long, such that an alternative plasma position and shape control system will be essential. The concept that edge density profile measurements via reflectometry at multiple poloidal positions could provide a position control capability emerged through the ITER design process, and was briefly discussed at the time of the 1995 Varenna conference on ITER diagnostics.[2] In this concept a poloidally distributed set of edge gap measurements (specifically the distance from the first wall (blanket face) to the location of a density layer lying at or inside the separatrix) will be used as an estimator of the plasma boundary shape, and utilized for feedback control. In the two years since 1995 the US Home Team has worked to investigate how this concept would work in practice, as well as to provide detailed design calculations. The conclusion of these studies is that a plasma position control system for ITER based on a poloidal set of edge density profile measurements appears to be viable, with performance levels at or close to requirements. As described in the body of this paper, this conclusion is based on: (1) a proven methodology for the design of profile reflectometer systems, (2) demonstrated performance levels of current reflectometer systems, and (3) specific design and modeling for ITER. Recent descriptions of the use of reflectometer systems for density profile measurements are available,[2,3] while a detailed design methodology for optimal system performance has also been developed.[2]

The measurement requirements for the plasma position and shape on ITER are that the inner, outer and top gaps should be measured with 1 cm accuracy and 10 ms time resolution.[4] The US team has studied a poloidal array of eight reflectometers operating in the ordinary mode with frequency coverage from 15 to 60 GHz. This frequency range corresponds to cutoff densities between 0.028 and 0.45x10^{20} m^{-3} [$f_o = 9.0 n_e^{1/2}(m^{-3})\ Hz$], which are estimated to encompass the scrape off layer densities for projected ITER plasma scenarios. Potential poloidal locations for the eight reflectometer systems are shown in Fig. 1. As shown in this paper and an accompanying paper by N.L. Bretz,[5] this system should meet the basic measurement requirements. It is envisioned that the reflectometer position control system will only be used during the steady state (flat top) portion of the discharges, i.e. it is assumed that the magnetic control system will always be used for plasma startup and termination (see Ref. 5 for details of how the reflectometer system could be used to reinitialize the magnetics for control of the termination phase). As already mentioned, the reflectometer system will be essential for plasma control in extended duration discharges (> 1000 s), and can act as a backup/supplement to the magnetics in discharges of shorter duration.

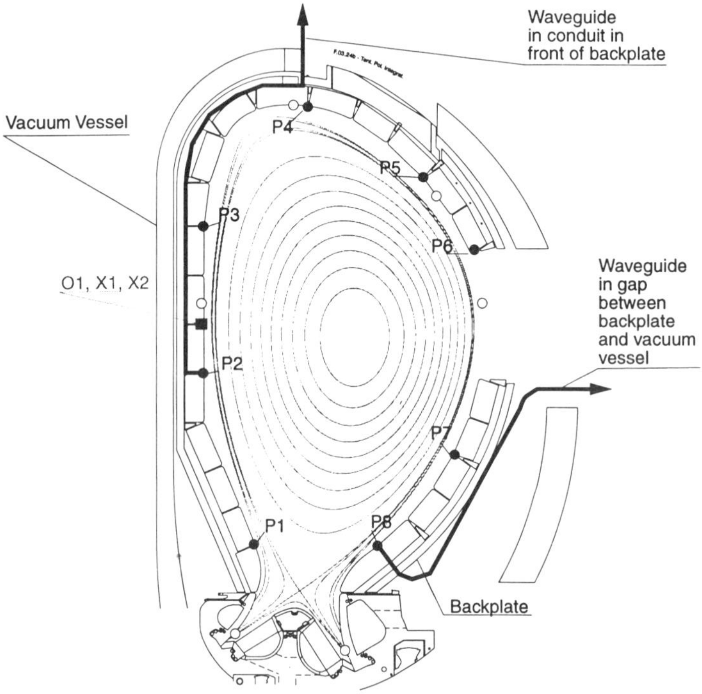

Figure 1. ITER cross section showing a potential poloidal arrangement for eight position control reflectometer antennas (P1-P8,●), and examples of the waveguide routing. The approximate locations of the six magnetic feedback control points (o), and the location of the high field side main reflectometer (O1,X1,X2) are shown for reference.

The purpose of this paper is to describe the general design basis for the position control reflectometer system. Specific questions and issues addressed in the remainder of this paper are as follows: (1) Can edge density profile measurements provide a plasma position/shape control capability? (2) Can reflectometry provide measurements with the required accuracy and reliability needed for ITER? This question is relevant as historically considerable difficulty has been experienced in achieving acceptable reflectometer performance. However, both theory and experiment have evolved to the point that a design for ITER can be undertaken in confidence. With regard to theory, a systematic basis for the optimal design of profile reflectometer systems has been developed at UCLA,[2] which is briefly reviewed. Experimentally, measurements by the current DIII-D edge profile reflectometer system[6-8] exceed ITER accuracy requirements, with demonstrated simultaneous sub-cm spatial and 100 μs time resolution.[2,7,8] (3) Using O-mode propagation the portion of the profile lying below the lowest probing frequency (~15 GHz) will be undetermined and must be modeled, introducing a position error. The magnitude of this position error for a simulated ITER system is examined using measured DIII-D density profiles, and is predicted to lie within acceptable limits.

The objectives of the companion paper[5] are to present details of the actual engineering design for ITER, as well as an investigation of how the choice of the number and location of the reflectometer systems affects the accuracy of the plasma shape control, and an initial consideration of how the reflectometer data could be used to provide a corrected baseline to the magnetics for control of the plasma termination.

PLASMA SHAPE AND POSITION CONTROL USING DENSITY PROFILE MEASUREMENTS

Although closely related, there is a conceptual difference between the functions of plasma shape and position control. Control of the plasma edge gaps (position control) is a vital machine protection function - it is essential that minimum plasma-wall gaps are maintained in order to protect the first wall from plasma damage. Shape control, by contrast, relates to the maintenance of a particular magnetic geometry desired during an experiment. For control purposes, measurements of an edge plasma parameter, such as electron density, are clearly not an identical replacement for magnetic measurements, but can, subject to some limitations, provide similar capabilities. With regard to first wall protection (position control), measurements of the electron density are superior to magnetic measurements, as the edge density profile is a more direct measure of whether the plasma is impinging on the wall. For this application all that is required is a set of measurements of the scrape-off-layer (SOL) density at a number of poloidal locations. For the shape control application, however, the picture is somewhat different. The location of a given density layer in the SOL is *not* in general representative of the plasma shape, i.e. SOL profiles vary from location to location independent of the magnetic geometry of the closed flux surfaces. Instead, what is required is *a set of measurements of the location of a density layer lying on a closed flux surface*, i.e. lying at or inside the separatrix. Subject to the assumption that the density is a constant on a closed flux surface, such a set of measurements would be representative of the plasma shape, and could be used for feedback control. This is the basic premise behind the position and shape control system proposed here. There are of course circumstances, such as the occurrence of MARFEs, when the assumption of constant density on a flux surface is violated. However, the presence of a localized MARFE at a particular reflectometer position should be clear from the measured profiles, allowing for appropriate compensation in the control system.

Having decided to provide feedback control on the position of a density layer lying at or inside the separatrix, how do we determine which densities are in fact inside the separatrix? Fortunately, for any given machine there is in general a scaling relationship between the line average plasma density and the density at the separatrix. An example of such scaling data, which can vary with operating regime, is shown in Fig. 2 for L- and H-mode operation on DIII-D. On DIII-D the density just inside the separatrix, at $\rho \sim 0.95$, is approximately half the line average density. On ITER, line average density measurements will be available in real time from the interferometer system, and an appropriate scaling relationship can be built up as operations progress. Thus, at any given time during extended discharge duration operation on ITER a reasonable estimate of the separatrix density should be available.

Finally, assuming that measurements of an edge plasma parameter such as electron density, temperature or pressure can be utilized to control plasma shape, why opt specifically for reflectometer measurements of edge density profiles? The primary reasons for the choice of reflectometry are: (1) Ease of implementation. The restricted access on ITER, especially to the inboard, high field side of the machine, puts a premium on diagnostic techniques with minimal access demands and the ability to negotiate complex physical paths. Reflectometry meets these requirements: reflectometry can obtain profile measurements from only a single sightline (view) per location, using compact microwave antennas. The transmission system uses small, flexible microwave waveguides, which can incorporate many bends. The use of reflectometry for local measurements in areas such as at RF antenna structures[9] and on the inboard, high field side of tokamaks has already been demonstrated.[10,11] (2) Compatibility with a reactor environment. Reflectometry is one of the few diagnostic systems for which the harsh, reactor-like ITER conditions pose no significant challenge. All the in-vessel components are robust, while other issues such as plasma deposition, etc. are not a problem at microwave wavelengths.

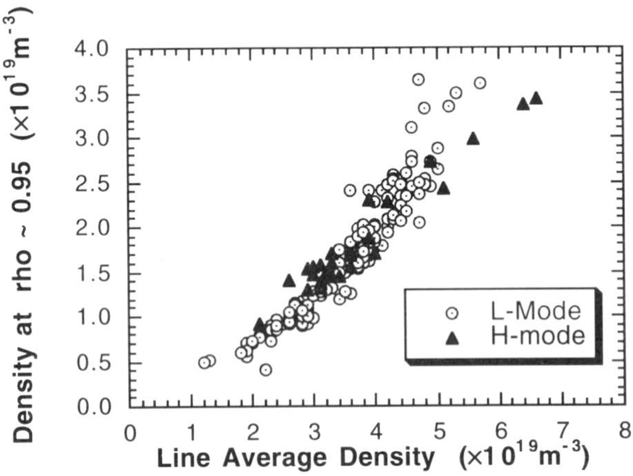

Figure 2. Example of DIII-D data showing the scaling relationship between line average density and the density just inside the separatrix (ρ ~ 0.95) for L- (circles) and H-mode (triangles) operation.

DESIGN BASIS FOR PROFILE REFLECTOMETER SYSTEMS AND DIII-D RESULTS

A central problem in the development of profile reflectometry for many years was the lack of an accepted basis for the optimal design of profile reflectometer systems. Such design criteria were developed by UCLA and are described in detail in Ref. 2. In the same paper the optimal design criteria were used to evaluate the relative advantages and disadvantages of current reflectometer techniques. Experimentally, these criteria were systematically applied in the design of the profile reflectometer systems on DIII-D,[6-8] leading to reliable, accurate profile measurements with simultaneous sub-cm spatial and 100 μs temporal resolution.[2,7,8] Many of the most important of the UCLA design criteria stem from the fact that profile reflectometer systems are in fact specialized close range radar systems, such that suitably modified radar design concepts can be employed. For details of the design criteria and the DIII-D results readers are referred to the original references; only a brief outline is presented here:

Range resolution: range resolution is the ability of a radar system to distinguish multiple targets at varying range; the minimum range at which two targets can be differentiated is $\Delta r = c/2B$, where B is the microwave bandwidth employed. Reflectometer systems have reflections from many "targets;" in addition to the reflection from the desired cutoff layer there are spurious reflections from undesired cutoff layers, single and multiple reflections from system components, waveguide modes, etc. Thus it is vital that the range resolution inherent in reflectometer systems be utilized to measure the range to the desired cutoff layer, as opposed to measuring the range to a weighted average of all the reflections. This implies the use of a minimum bandwidth B sufficient to resolve the desired cutoff layer from the closest source of interference. In many previous systems this inherent capability was not utilized, or else the bandwidth employed was too small to enable the closest sources of interference to be removed.

Precision bandwidth: the spatial precision σ_r of a single radar range measurement depends on the system signal to noise ratio S, and an effective rf bandwidth employed for data smoothing β, viz. $\sigma_r = \frac{c}{2\beta\sqrt{S}}$. Again, the rf bandwidth (signal averaging) in many previous systems was insufficient to obtain accurate results.

Spatial sampling: resolving rapidly varying spatial features in a density profile, such as steep edge H-mode density gradients, implies high spatial sampling. This argues for the use of broadband reflectometer systems with continuous spatial (frequency) coverage, and against multichannel systems with only 4-10 measurement frequencies.

Adaptive data processing system: modern radar systems invariably employ adaptive digital signal processing so as to maximize system performance and flexibility. In reflectometry, adaptive signal processing is important as the optimum range resolution and precision bandwidths are not constant, but vary with the characteristics of the profile being measured, i.e. too large a bandwidth can lead to profile smearing, while too little leads to poor accuracy. Consequently, the filtering and smoothing bandwidths in the UCLA data analysis software are adaptive. Optimal bandwidths can be determined analytically for both O- and X-mode propagation. For O-mode the optimal bandwidth is given simply by $\beta = f\delta r/2L_n$, where f is the microwave frequency, δr the desired radial step size for spatial sampling, and L_n the density scale length. This equation can be used in different ways; for example, β can be fixed, in which case spatial sampling varies with L_n, and f, or δr could be fixed, such that precision varies with β. However, perhaps the most attractive option is to fix a relative sampling rate, i.e. specify $\delta r = L_n/\alpha$, where α is a constant, in which case $\beta = f/2\alpha$ (note, this latter expression is not valid for large L_n).

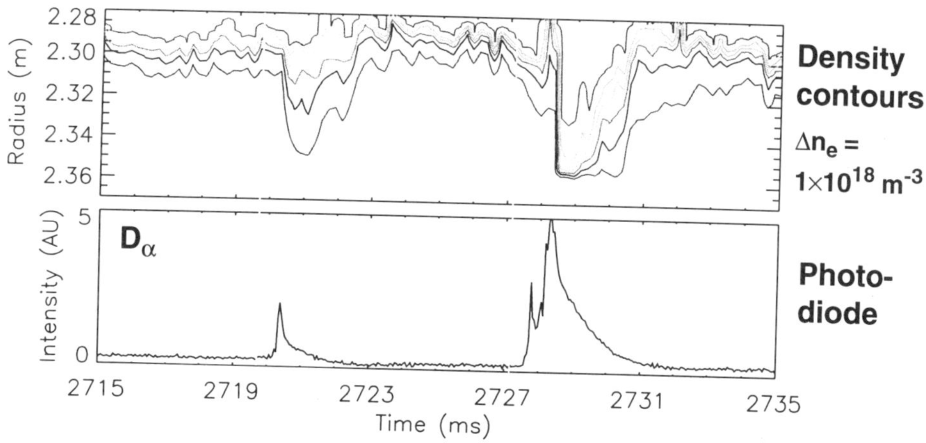

Figure 3. Example of measured DIII-D edge profiles with simultaneous sub-cm spatial and 100 μs temporal resolution. Density contour data are shown for densities in the range of 0-7x10^{18} m^{-3} as a function of time and radius across two ELMs. Note the radial scale: the density contours move outward by ~4-5 cm at the onset of each ELM.

Turbulence immunity: The unexpectedly large effects of plasma turbulence on reflectometer profile measurements compromised many early systems. A better qualitative understanding of the effects of turbulence has lead to the development of several practical approaches which minimize the effect of plasma fluctuations on reflectometer profile measurements.[2,3] While some progress has been made towards a quantitative understanding of turbulence effects,[12-14] an improved theoretical understanding of this issue would be welcome.

Using this set of design criteria a high performance edge profile reflectometer system was implemented on DIII-D.[7,8] The system utilizes X-mode propagation and has a frequency coverage of 33-50 GHz, corresponding to a density range of 0-7x10^{18} m^{-3} at full field. The system has performed well over a number of years, and has coped with a wide range of plasma conditions, including L-mode, H-mode, L-H transitions, ELMs, rf heated plasmas, etc. An example of measured profiles showing simultaneous sub-cm spatial and 100 μs temporal resolution is shown in Fig. 3, where it can be seen that the edge profile *is successfully tracked through ELMs.* Thus, current DIII-D results exceed ITER requirements, and would be sufficient for real time plasma control, giving confidence in profile reflectometer designs for ITER.

OUTLINE DESIGN OF THE ITER POSITION CONTROL REFLECTOMETER

Using the design basis described above and other practical considerations it is possible to outline the design choices for a position and shape control reflectometer system for ITER:

Choice of polarization (O- or X-mode propagation): O-mode is the only practical choice for ITER. X-mode cannot be used to measure the edge profile at the high field side of a tokamak, while use of O-mode polarization means that identical systems covering the same frequency range, and hence density, can be used at all poloidal locations.

Frequency/density range: For this initial design the frequency range chosen was 15 - 60 GHz, corresponding to cutoff densities of 0.028 - 0.45x10^{20} m^{-3}. The minimum frequency should be as low as possible, set by the need to minimize the undetermined portion of the edge profile (described in detail in the next section). This minimum frequency is limited by maximum waveguide and antenna sizes to ~15 GHz, though it may be possible to go as low as 10 - 12 GHz. The upper frequency is set by the need to access to/beyond the separatrix, as described earlier. Using the DIII-D scaling relationship, 0.45x10^{20} m^{-3} at the separatrix would correspond to line densities of ~0.9x10^{20} m^{-3}. This is less than the planned ITER line densities of ~2.0x10^{20} m^{-3}, such that it may be necessary to raise the maximum frequency to ~90 GHz. However, the highest ITER line densities will probably be attained in a high confinement regime, such as H-mode, in which case it may not be essential that the controlling density layer be exactly at or inside the separatrix. This should be investigated experimentally in any future demonstration of the concept.

Antenna geometry: bistatic arrangement (separate launch and receive horns). Bistatic operation has substantial advantages in reducing waveguide mode problems, and monostatic operation (single horn used for both launch and receive) is still possible if one waveguide system should fail.

Choice of reflectometry technique: while a choice of reflectometer measurement technique does not have to be made at this point, use of broadband, continuous sweep, frequency modulated (FMCW) reflectometer systems was adopted as a baseline.

Predicted performance: Ignoring the effect of the need to model the portion of the density profile lying below the minimum probed density layer, which is considered in the next section, predicted performance exceeds ITER requirements. Taking a fixed spatial sampling rate of $L_n/4$ to determine the optimum precision bandwidth, and using the formulae given above, 1 cm accuracy should be obtained at 60 GHz with a system signal to noise ratio as low as 4. As for time resolution, current FMCW reflectometer systems, as on DIII-D, exceed the ITER requirement of 10 ms by two orders of magnitude or more, so the time resolution is not an issue. It should be noted that the radar formulae predict that the best system performance is obtained at the highest frequencies, which are of most relevance to the control system.

EFFECT OF THE UNDETERMINED PORTION OF THE EDGE PROFILE ON MEASUREMENT ACCURACY

Using O-mode propagation, as will be the case on ITER, reflectometer measurements are only available above a minimum frequency f_{min}, and the position of the plasma edge r_0 (position of the zero density layer) is not well determined, leading to position errors in the reconstructed density profiles.[15] It can be demonstrated that in general the size of this error should increase with the magnitude of the density gradient scale length L_n (i.e. larger errors for shallower gradients). That this should be so can be seen conceptually in Fig. 4, which shows an example of three measured DIII-D edge density profiles showing the varying radial extent of the portion of the profile that would be unmeasured by an O-mode system employing a 15 GHz minimum frequency. It should be noted that when using X-mode reflectometry, as on DIII-D, r_0 can be determined directly from the reflectometer data, and measurements can be made from essentially zero density, such that the O-mode problems do not arise. Thus, the additional uncertainty posed by these two edge effects forms the main difference between the proposed edge O-mode profile reflectometer systems for ITER and the current high performance X-mode reflectometer system on DIII-D.

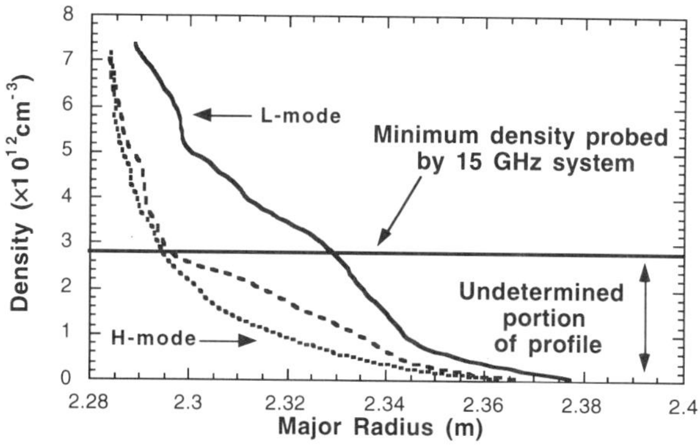

Figure 4. Example of three measured DIII-D edge density profiles showing the varying radial extent of the portion of the profile that would be unmeasured by an O-mode system employing a 15 GHz minimum frequency.

Profile reflectometer systems measure the line integrated phase shift ϕ, or equivalently the time delay, $\tau = \frac{1}{2\pi}\frac{d\phi}{df}$, of an electromagnetic wave reflected from a cutoff layer in a plasma. The phase shift is determined by the same basic equation as for interferometry, viz.

$$\phi\,(f) = \frac{4\pi\, f}{c}\int_{r_0}^{r_c(f)} \mu(r, f)\; dr - \frac{\pi}{2} \qquad (1),$$

where ϕ is the total phase shift in propagating from the plasma edge at r_0 to the cutoff layer at $r_c(f)$, and back, f is the microwave frequency, and μ is the plasma refractive index. By varying the probe beam frequency the cutoff layer position can be moved across the plasma radius, or multiple frequencies can be used to reflect from different positions simultaneously. From a data set of phase (or time delay) as a function of frequency, obtained in either of these ways, Eq. (1) can be inverted to recover the density profile. For O-mode propagation an analytic Abel inversion exists:

$$r(f_c) = r_0 - \frac{c}{2\pi^2} \int_0^{f_c} \frac{d\phi}{df} \frac{df}{\sqrt{f_c^2 - f^2}} \qquad (2),$$

where f_c is the cutoff frequency for reflection at radius r. To apply Eq. 2 it is necessary to know the position of the plasma edge r_0 and have phase measurements for $f > 0$. As already stated, however, using O-mode propagation measurements are only available above a minimum frequency f_{min}, and the position of the plasma edge r_0 is not well determined. In this situation it is necessary to assume or model $\phi(f)$ for $f < f_{min}$, and either assume r_0 or determine the starting position from some other diagnostic.

The O-mode edge modeling problem in addressed in some detail in Ref. 15, but no general treatment of the problem posed by an unknown profile starting position would seem to exist (previous treatments, such as Ref. 15, generally assumed that the start position is known a priori, which will not be the position on ITER). Novel aspects of the work reported here are that measured DIII-D density profiles extending from near zero density are utilized in assessing the modeling induced error, instead of using model data below f_{min} to assess the reliability of the edge modeling, as previously, and the problem of determining r_0 is considered explicitly .

Turning first to the problem of determining r_0, consider a situation in which phase (time delay) data are available for all $f > 0$, and the only uncertainty is in the value of r_0. As reflectometers are in effect radar systems,[2] it might be thought that r_0 could be determined simply from the time delay of the lowest probing frequency. However, this is not so due to the range/gradient ambiguity in reflectometer data (equivalent to the range/Doppler ambiguity in classical radar systems). The time delay to a cutoff layer depends not only on its position, but also on the value of the local density gradient; using Eq. 1, it can be shown that $d\phi/df$ (or the time delay) consists to two terms, one corresponding to the optical path length to the cutoff layer, and a second corresponding to the frequency variation of the path length, which depends on the density gradient. The raw phase or time delay data do, however, impose minimum and maximum distances within which the plasma must lie. In addition, it is easy to demonstrate that for the case under consideration here *Eq. 2 automatically solves the range/gradient ambiguity, such that the correct density profile is recovered independent of the choice of* r_0. Assume that instead of the correct value the actual value used in Eq. 2 is $r' = r_0 + \Delta r$. For this incorrect choice of starting position the phase data inside the integral term are modified to $\phi' = \phi - \Delta\phi$, where $\Delta\phi = 4\pi\Delta r f/c$. The integral in Eq. 2 can be performed analytically for the $\Delta\phi$ term, and results in $-\Delta r$, exactly canceling the original error in the choice of r', and the correct density profile is recovered identically.

In practice the position is different than that just considered, in that phase data have to be modeled/assumed for frequencies lying below f_{min}, in which case Eq. 2 may or may not be self correcting, depending on the form of the model used. Examples of two models which preserve the self correcting feature of Eq. 2 with respect to assumed starting position are simple linear interpolations extending the gradient at f_{min} to zero frequency in either phase or time delay. Using these models the error in the inverted profile is entirely due to the difference between the assumed phase data and the phase corresponding to the actual plasma profile, and does not depend on the assumed plasma starting location. In the modeling presented in this paper a simple self correcting linear phase interpolation model was used. This corresponds to assuming that the undetermined portion of the plasma profile is at the maximum possible distance from the reflectometer system consistent with the data. In addition, a linear phase as a function of frequency corresponds to a mirror reflection, so the assumed model portion of the density profile has a vertical gradient. Obviously this is a conservative choice, more realistic edge profile models may well give better results. The results presented here for the size of the edge modeling error for specific profiles should thus be regarded as upper bounds, reflecting the conservative nature of the actual model used.

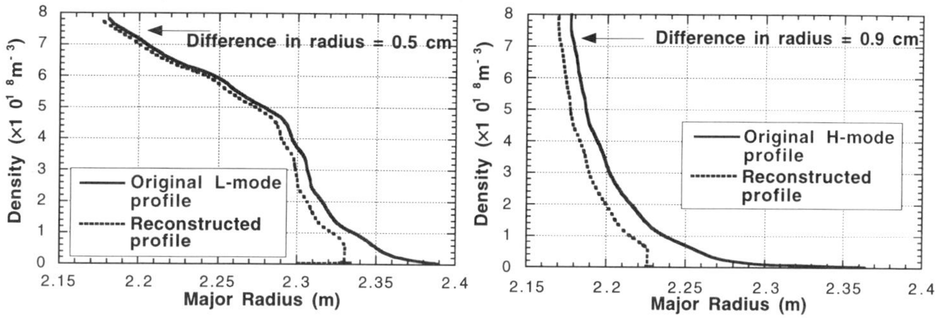

Figure 5. Examples of the position error introduced in reconstructed L- and H-mode density profiles (left and right graphs, respectively) by the edge modeling process for a simulated ITER reflectometer system. The solid curves are the original measured DIII-D edge density profiles, while the dashed curves are the inverted profiles obtained from the simulated ITER system. At the highest density, the position error is 0.5 cm for the L-mode case, and 0-.9 cm for the H-mode.

Using the above edge model the size of the model induced position error for a simulated ITER reflectometer system was investigated using a range of measured DIII-D density profiles. As the measured DIII-D density range was not the same as that projected for ITER, the 15 GHz minimum and 60 GHz maximum ITER frequencies were scaled to DIII-D so as to give the same ratio of measured to unmeasured densities. Examples of the original and reconstructed density profiles are shown for L- and H-mode profiles in Fig. 5. Overall, the scaling of the results is in agreement with expectation, i.e. the error is largest for shallow gradients, which increases the radial extent of the unmeasured portion of the profile. The magnitude of the error, for typical DIII-D L- and H-mode profiles, was less than 1 cm, though up to 1.5 cm was observed for a low density rf heated discharge with shallow edge gradients. Thus, the performance of the simulated ITER O-mode system was within or close to requirements. However, the actual size of the error on ITER will depend critically on the value of L_n in the ITER SOL plasma, which is currently not well defined. If density gradient scale lengths on ITER are comparable to or shorter than on DIII-D (typically 1-2 cm), as in the current “best guess,” then edge modeling induced errors will be comparable to those reported here. If the ITER edge profile is shallower, however, larger errors can be expected. Finally, the results shown here are in general agreement with one of the primary results reported in Ref. 15; the position error induced by the edge modeling process decreases as f_{max} is increased relative to f_{min}, i.e. the most accurate results are obtained at the highest densities, which are the most relevant for the proposed control system.

SUMMARY AND FUTURE RESEARCH AND DEVELOPMENT NEEDS

Design studies indicate that a reflectometer based plasma position/shape control system for ITER is viable, with performance levels at or close to requirements. The application of profile reflectometer systems to ITER can be contemplated with confidence, based on improvements in both the theoretical understanding of optimal reflectometer system design and experimental results from present devices. However, the proposed new plasma position control technique is completely untested and for such a vital machine requirement it is essential to provide a comprehensive experimental demonstration before application to ITER. In order to perform such a demonstration new capabilities must first be developed, such as real time, automatic analysis of reflectometer data. A first step would be to demonstrate real time analysis using a current reflectometer system, after which a full experimental demonstration of plasma shape and position control using only reflectometer measurements could be performed. The LCT-2 tokamak under construction at UCLA would be ideal for such a demonstration since it combines a shaped plasma geometry with unrivaled access to the plasma around the entire poloidal circumference. Other research tasks include the better definition of the expected ITER edge density gradient scale lengths, and a more systematic investigation of the options for modeling the portion of the density profile undetermined by O-mode reflectometry. Finally, an improved quantitative understanding of the impact of plasma turbulence on profile measurements would be useful.

ACKNOWLEDGMENTS

The authors would like to express their thanks to Dr. G. Vayakis (ITER JCT) for many useful discussions and for providing Fig. 1, and to Dr. T.N. Carlstrom (General Atomics) for providing the data shown in Fig. 2. This work was supported by US DOE Grant and Contract Nos. DE-FG03-86ER53225 (UCLA) and DE-AC02-76-CH0-3073 (Princeton). This report has been prepared as an account of work performed under the Agreement among the European Atomic Energy Community, the Government of Japan, the Government of the Russian Federation, and the Government of the United States of America on Cooperation in the Engineering Design Activities for the International Thermonuclear Experimental Reactor ("ITER EDA Agreement") under the auspices of the International Atomic Energy Agency (IAEA).

REFERENCES

1. L. de Kock *et al.*, Magnetic diagnostics for ITER, in: *These Proceedings.*
2. E.J. Doyle *et al.*, Reflectometry applications to ITER, in: *Diagnostics for Experimental Thermonuclear Fusion Reactors*, P. E. Stott *et al.*, eds., Plenum Press, New York (1996).
3. C. Laviron *et al.*, Plasma Phys. Control. Fusion **38**, 905 (1996).
4. V.S. Mukhovatov *et al.*, Role and requirements for plasma measurements on ITER, in: *These Proceedings.*
5. N.L. Bretz *et al.*, ITER position control reflectometry - conceptual design, in: *These Proceedings.*
6. K.W. Kim, *et al.*, Rev. Sci. Instrum. **66**, 1229 (1995).
7. K.W. Kim, Ph.D. Thesis, University of California, Los Angeles (1996).
8. K.W. Kim *et al.*, Rev. Sci. Instrum. **68**, 466 (1997).
9. D.W. Swain *et al.*, Nuc. Fusion **37**, 211 (1997).
10. E.J. Doyle, *et al.*, in: *Plasma Physics and Controlled Nuclear Fusion Research 1992*, IAEA, Vienna (1993).
11. A. Silva, *et al.*, in: *Fusion Technology*, C. Ferro, *et al.*, eds., North-Holland, Amsterdam (1992).
12. E. Mazzucato and R. Nazikian, Rev. Sci. Instrum. **66**, 1237 (1995).
13. G. Conway, Plasma Phys. Control. Fusion **39**, 407 (1997).
14. G. Vayakis, Fusion Engineering and Design **34-35**, 519 (1997).
15. P. Varela *et al.*, Rev. Sci. Instrum. **66**, 4937 (1995).

ITER POSITION CONTROL REFLECTOMETRY - CONCEPTUAL DESIGN

N.L. Bretz, C. Kessel,[1] E.J. Doyle,[2] and G. Vayakis[3]

[1]Princeton Plasma Physics Laboratory, Princeton University, Princeton, NJ 08543, USA
[2]Electrical Engineering Department and Institute of Plasma and Fusion Research, University of California, Los Angeles, CA 90095, USA
[3]ITER Joint Central Team, San Diego, LaJolla, CA 92037, USA

INTRODUCTION

The plasma position reflectometer array will supplement magnetics measurements for plasma position control in ITER by estimating the position and shape of the flux surface near the plasma boundary. Reflectometry can measure the location of a specified density layer near the separatrix and is expected to be especially useful in controlling the location of the plasma boundary during the long equilibrium phase of the discharge when magnetics signals may accumulate offset errors. A poloidal array of eight swept frequency, ordinary mode reflectometers operating between 15 and 60 GHz has been evaluated. This frequency range corresponds to cutoff densities between 0.028 and 0.45×10^{20} m^{-3} [$f_o = 9.0 n_e^{1/2} (m^{-3})\ Hz$] which are estimated to encompass the scrape off density for projected ITER plasma scenarios. The position measurement relies on a time delay technique and, thus, measures the distance from from the plasma surface to a physical reference point near the wall during each frequency sweep. Reflectometry systems for profile measurements using both the ordinary and extraordinary mode cutoffs are used widely in fusion plasmas and are reviewed by Laviron.[1]

For position and shape control ITER requires an accuracy $\Delta R_{in,\ out}$, $\Delta Z_{top} \approx 0.01$ m with an edge radial velocity less than 0.01 m/s and time response less than 1 ms.[2] A swept ordinary mode reflectometer using the frequency range indicated above and frequency sweep rate in the range of 10 to 100 kHz can satisfy these requirements.[3] Reflectometers can be swept at a rate which is high enough to resolve Edge Localized Modes (ELMs) and other fast events. Thus, measurements can be generated much faster than required for position feedback control and not be disturbed by transients. A midplane extraordinary mode reflectometer has been operated on DIII-D over a slightly smaller frequency range which meet ITER requirements.[4] The ordinary mode reflectometer considered here is better suited for an array of poloidal measurements but has slightly different sources of error. In particular, an additional systematic error will result from the unknown density below the lowest cutoff. Doyle[5] shows in an accompanying paper that the error resulting from the ordinary mode inversion procedure is small for typical DIII-D edge profiles. The resulting accuracy is about 0.01 m for profiles with $L_n \leq 0.01$ m and is roughly proportional to $L_n \equiv -n / (\partial n / \partial r)$. In

Diagnostics for Experimental Thermonuclear Fusion Reactors 2
Edited by Stott *et al.*, Plenum Press, New York, 1998

addition the response of ordinary mode reflectometer systems to turbulence is expected to be different and a full evaluation will require more experience on present devices and extrapolation to turbulence expected in ITER. Current experience with ordinary mode systems has been evaluated by Manso[6] and is an area of active investigation. The effect of fluctuations appears to be manageable, and there are a number of techniques to reduce its influence on profile measurements. In all, the accuracy of the technique should satisfy ITER requirements. However, more operational experience with ordinary mode reflectometers on diverted plasmas will be needed to demonstrate reliable profile predictions which are immune from transients.

One significant limitation on the use of a poloidal reflectometer array for position control is that plasmas which are not relatively close to the transmitting and receiving antenna pair do not give consistent return signals. Thus, for plasmas further away from the edge than about 1 m, that is, the start up and shutdown scenarios, magnetics are the appropriate controls. Reflectometry can be used only during the long pulse, equilibrium phase of ITER.

Technical constraints on these measurements imposed by the ITER design are substantial. This paper will concentrate on the position control reflectometer design that impacts the first wall and vacuum boundary. Of particular concern are constraints from the inner wall blanket modules which allow very little room for waveguide and antennas. In this design we assume that antennas can be installed in the 0.02 m gap between modules and that waveguide can run between the inner blanket modules and the backing plate. Away from the inner wall the module gaps are the same but waveguide runs are easier. Thermal, mechanical, electrical, vacuum, and neutronic factors all require a simple, rugged, and reliable design for the antennas, waveguide, and mechanical supports. Minor modifications of the inner blanket modules will be required to accommodate this design. However, the waveguide losses, bends, vacuum feedthroughs, and mechanical stresses can be accommodated with the basic design presented here.

The purpose of this paper is to describe a reflectometer array which meets the control requirements of a plasma position control system, to show how such a system may be implemented on ITER especially those aspects which affect the first wall and vacuum boundary, and to show how reflectometer information can be used for control of the plasma position and shape. A conceptual design will be given of the waveguide route behind the first wall blanket and the antenna location in the gaps between blanket modules. In order to demonstrate what issues are important for position control, a sensitivity study has been done relating the number of reflectometers to the accuracy of the equilibrium reconstruction with and without additional constraints.

REFLECTOMETER INSTRUMENTATION

There are several types of reflectometer which could be employed for this measurement. The technique patterned after Frequency Modulated - Continuous Wave (FM-CW) radar[7] meets the control requirements and is the basis for the discussion given here. However, several other types have been used in the fusion community or are being developed. These include comb reflectometers,[8] short or ultra-short pulse reflectometers,[9] and dual frequency AM or FM reflectometers.[10] A generic CW-FM reflectometer is shown in Fig. 1. The source sweeps linearly and repetitively between the upper and lower frequency limits, 15 to 60 GHz in this case. The output is split into a plasma and a reference leg. The plasma leg reflects from the plasma while the length of the reference leg is adjusted so that, when the reflected and reference leg arms are mixed, there is a low frequency difference or beat frequency, f_B. This beat frequency represents the time delay between the two arms and is a function of the plasma wall distance and the density shape. For sweep rates between 10 and 100 μs and antenna/plasma separations of up to 1 m, one has f_B < 5 and 50 MHz,

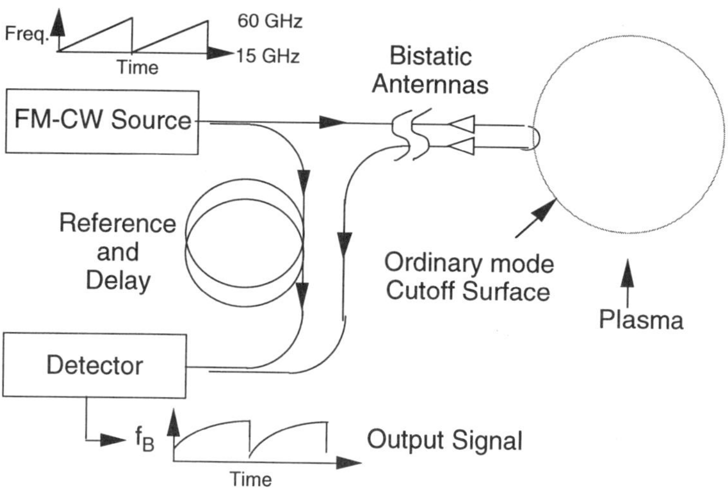

Figure 1. A block diagram of an FM-CW reflectometer is shown. The source sweeps repetitively between the upper and lower frequency limits, 15 and 60 GHz. The detector samples the beat frequency, f_B, between the delayed source and reflected signal from the plasma cutoff.

respectively. A typical receiver noise power in this frequency range is better than $2x10^{-20}$ W/Hz ($T_N \approx$ 1,400 K).

The choice of frequency for the system considered here is determined by the flat-top peak density expected in ITER of 0.1 to $3.0x10^{20}$ m^{-3}. The separatrix density is assumed to be about one quarter of the peak density based on DIII-D experience. The frequency range needed to cover this density (typically 15 - 60 GHz) can nominally be accommodated by sources, receivers, and waveguides consistent with ITER constraints. High power (20 - 100 mW), solid state microwave sources exist which can sweep full waveguide bands in this range in 10 μs. However, the system complexity increases if one needs to multiplex a large number of bands. The range 18 to 60 GHz covers three bands: WR-42 (18 - 26.5 GHz), WR-28 (26.5 - 40 GHz), and WR-19 (40 - 60 GHz). Extensions down to 12.5 GHz or up to 90 GHz would involve two more bands, WR-62 and WR-12, respectively. Fast swept systems in the frequency bands between 18 and 75 GHz have been used on reflectometers on ASDEX-U and DIII-D. The lowest frequency considered here also has the highest waveguide losses in the ITER design. The upper limit is less constrained by waveguide losses and could probably be raised to 90 GHz if there were no constraints on source and receiver complexity.

IMPLEMENTATION CONSTRAINTS

Fig. 2 shows a set of poloidal locations for the reflectometers which has been chosen to fall in blanket gaps and which has been used to analyze position control requirements. These positions have been roughly chosen to control the main sectors of the plasma shape. The six control points used for magnetics feedback control are shown for comparison. Magnetic reconstruction simulation discussed later will be required to determine if this choice is optimal and what, if any, additional constraints might substantially improve the reconstruction. Fig. 2 also shows how the waveguide will be distributed from the top and outer port to the various elements of the reflectometer array.

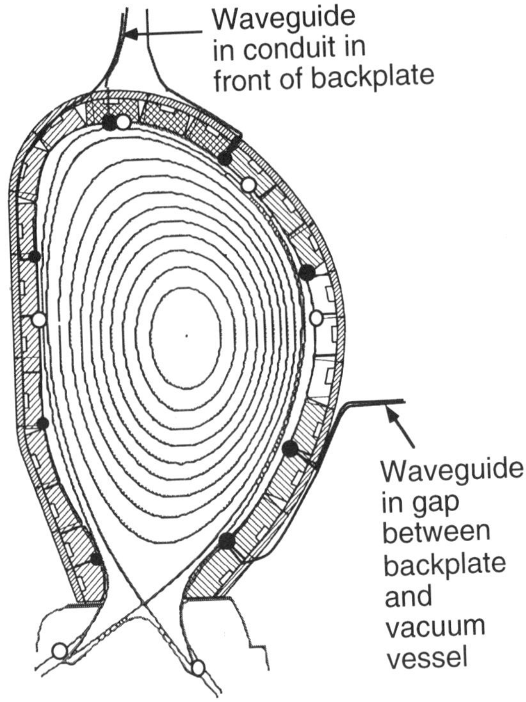

Figure 2. The poloidal distribution of eight position control reflectometer antennas in ITER (●) and the waveguide routing are shown. The six magnetic feedback control points are shown for reference (o).

In order to implement the position reflectometer on ITER several antennas must be installed both on the small and large major radius side of the plasma. We have considered how to install transmission lines and bistatic, that is a pair of transmitting and receiving, antennas in the 15 - 60 GHz range near the blanket modules. Space constraints limit the size of the transmission line to rectangular waveguide of no more than a few centimeters in outside dimensions. The loss for an estimated maximum of 60 m round trip loss between source and receiver for TE_{10} mode in 0.02 m x 0.02 m and 0.02 m x 0.01 cm copper waveguide with E parallel the to short dimension is shown in Table 1. The total tolerable transmission line loss between the source or receiver and antennas should be kept below about 20 dB including bends, vacuum windows, voltage breaks, and transitions.

The waveguide on the inner wall runs in a conduit in the back of the blanket modules between the blanket modules and the backing plate as shown in Fig. 3. The dimensions of the conduit are 0.07 m x 0.035 m and must accommodate the position control reflectometer waveguide as well as that of other systems as described by Vayakis.[11] The smaller waveguide (0.02 x 0.01 m) will be used in the conduit near the inner wall but the larger waveguide will be used everywhere else. Waveguides must be routed around backing plate

Table 1. Attenuation of 60 m of rectangular copper waveguide in the TE_{10} mode with the electric field parallel to the short dimension for the frequency range 15 to 60 GHz.

	15 GHz	60 GHz
0.02 m x 0.02 m ID	6.0 dB/60m	7.0 dB/60m
0.02 m x 0.01 m ID	10.0 dB/60m	14.0 dB/60m

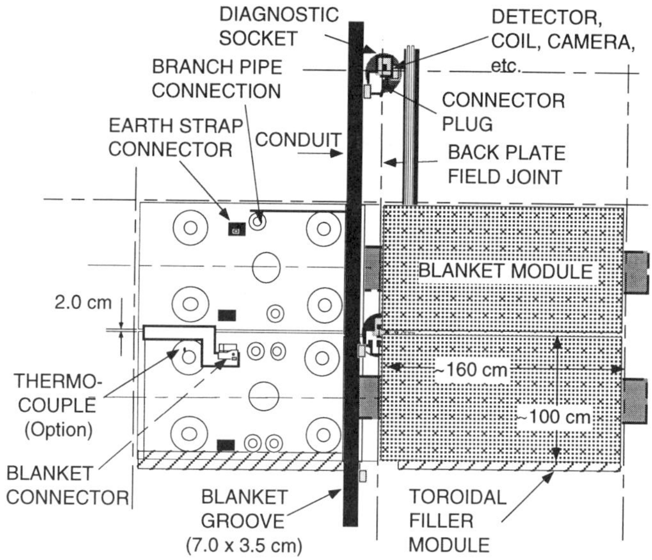

Figure 3. Waveguide route for high field reflectometer antennas (and other mm wave diagnostics) behind the blanket modules. The space between the modules is ~ 0.02 m.

curvatures avoiding water cooling lines and electrical cables. The waveguides from each poloidal section are bundled together and taken out of the vacuum and cryostat boundary through many identical vacuum windows incorporated into the waveguide.

Bends in rectangular guide can be made in a number of different ways depending on the routing, but in general sharp bends cause significant losses and must be minimized.. Waveguides can be bent gently with radius of curvature greater than 2 m with little mode conversion loss. Sharper bends can be made using a hyperbolic secant shape. Right angle hyperbolic secant bends can be made in 0.02 m x 0.02 m copper waveguide in both planes in 0.3 m with losses below 1 dB.[12] Sharp 90 degree bends near the antenna transitions will be more lossy especially at low frequencies. The waveguide to the inner wall is expected to run vertically behind the blanket modules and turn sharply to the antennas located in the space between the blanket modules. Typically, an H-plane bend will be needed and will have about 2 dB loss at 15 GHz and 1 dB loss at 60 GHz. This represents the worst case. The loss can be made less by making a gentler bend than a miter. However, the design is strongly dependent on the extent to which the back of the blanket can be modified and has not been examined in detail. Bend losses can be further reduced near the antenna transition near outer blanket modules where space behind the blanket is less restricted.

The transition between the waveguides and bistatic antennas near the blanket module is shown schematically in Fig. 4. As noted, the poloidal and toroidal spaces between the blanket modules is about 0.02 m, and the antennas must be supported in this space consistent with movement of the blankets. The antennas will be directly exposed to plasma radiation and neutron flux and must withstand flash heating from disruptions as well as the continuous thermal load from plasma emission and neutrons. Preliminary thermal estimates show that thin uncooled antennas made of tungsten or carbon fiber composites can be used between the blanket modules. However, support and installation issues have not been addressed in detail. A fiducial reflection between the antennas and the plasma boundary is needed to provide a reference position for calibration.

Overall, the round trip losses in the waveguide, vacuum windows (3 dB/window), bends, and voltage breaks are not expected to exceed 40 dB over the full frequency range. In addition one expects an antenna coupling loss of 20 to 30 dB depending on the frequency and distance to the reflecting surface. Noise from the plasma will be dominated by electron cylotron emission which is expected to have an equivalent temperature, T_{ECE}, which is similar to that of the plasma at cutoff. This means that the detector noise will be determined primarily by T_{ECE} reduced by $10^{L_{pd}/20}$ where L_{pd} in dB is the antenna and transmission line loss to the detector. Using the most conservative estimate, (1) 10 mW source power, (2) 50 MHz bandwidth (10 μs sweep time), (3) 70 dB total loss and L_{pd} ~ 20 dB, (4) T_{ECE} ~ 2 keV, gives a signal-to-noise ratio of unity. The signal-to-noise will be increased by averaging and using a smaller effective bandwidth consistent with the spatial accuracy required in the profile calculation. Plasma turbulence may contribute to the effective noise but has not been included in this estimate.

There are a number of design issues related to special circumstances when the plasma edge is very close to the antennas and when it is very far away. When the plasma edge is very close to the antenna aperture, the coupling between the launching and receiving antennas can be reduced since the reflected antenna pattern of the launched wave may not overlap well with that of the receiver. This effect is worst at the highest frequency where the antenna pattern is narrowest. However, for the frequencies and antennas discussed here the width of the pattern is never narrower than about $\sin^{-1}(\lambda / D) \approx 15$ deg ($\lambda = c / f$, and D is the waveguide width), which means roughly that if the antennas are separated by the width of the blanket gap, $w \approx 0.02$ m, and if the antennas patterns are normal to the wall, then plasmas closer than about $wD / \lambda \approx 0.08$ m will cause the returned signal to be severely attenuated at the highest frequency. The closest wall-separatrix gaps are expected to be at least 0.10 m in normal ITER scenarios. Antennas can be angled to reduce this effect at the expense of being optimized for a particular wall-plasma spacing. Monostatic antennas do not have this problem but are judged to have other drawbacks. Bistatic antennas could be closer together and incorporated into the blanket module itself, or the antennas could be recessed behind edge of the blanket. Another near plasma effect not accounted for in the usual inversion algorithm is the possibility of a partial reflection from a sharp step in the density whose scale length is less than approximately $[L_n(\lambda / 2\pi)^2]^{1/3} \leq 0.02$ m. but which occurs at a density smaller than the cutoff. In this somewhat pathological case there will be a partial reflection from the step in addition to the reflection from the cutoff layer. In practice this situation has not caused difficulties in interpretation. Finally, MARFEs or physical obstructions from limiter debris

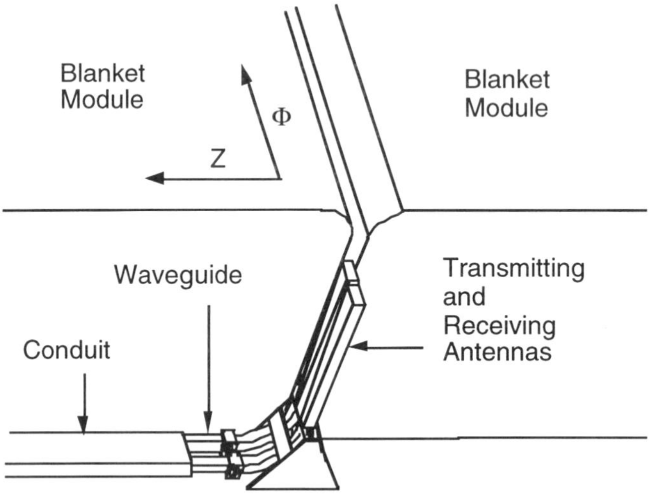

Figure 4. Cutaway view of bistatic position control reflectometer antennas and mounting near blanket modules.

might block the reflection at a particular location. Experience on existing (midplane) systems has shown that such blockages are rare. However, this emphasizes the need for some redundancy in the poloidal array, some judgement in handling transients, and some operational experience in the use of multi-element reflectometers for position control.

Issues associated with plasma surfaces which are far away from the antennas are associated with signal reduction due to the widening antenna pattern for a given frequency which is proportional to $f^2 D^4 / L^2$ where L is the distance between the antenna and the reflection. In addition, if the plasma cutoff surface is not normal to the antenna within an angular range of $\sin^{-1}(\lambda / D)$ deg., then the returned signal will be severely attenuated. All equilibrium scenarios in ITER plasmas are well within the acceptance angle of the antennas shown and will not reduce the signal level significantly. However, outer wall startup and shutdown scenarios can be out of view of all or several of the eight antennas a various stages in the growth and can allow the reflectometer signals to interfere with each other. Magnetics control is clearly better suited for the rapidly changing startup and shutdown. Thus, the role of the position control reflectometer is that of providing redundancy and/or a normalization for the magnetics during the long period of constant conditions so that the equilibrium position of the plasma is well known during the flattop and can be used in conjunction with other information to normalize magnetics before the magnetically controlled shutdown.

PLASMA CONTROL

The question of the placement number of the reflectometer array elements for magnetic reconstruction is addressed in this section. The initial placement of the antenna array was done so that the critical inboard, outboard, top, and bottom gaps were well determined. In addition it was thought prudent to put more elements in the main plasma chamber to be sensitive to shape especially where the boundary is sharpest. However, compared to magnetics, there is a significant lack of information from the reflectometer array (as shown in Fig. 2) in the divertor region. Thus, one may expect that additional constraints on the equilibrium may be needed to accurately determine the strike points from shape information alone.

A Grad-Shafranov equilibrium has used for a standard ITER flattop scenario ($I_p = 21.0$ MA, R = 8.15 m, a = 2.8 m, Z = 1.45 m, κ, = *1.75*, δ = 0.35 where κ and δ are the elongation and triangularity, respectively). From this equilibrium, magnetics loop signals and flux surface shapes can be specified. The magnetic signals are then reproduced with a set of current filaments on a contour in the region $a_f < 2$ m where a_f is the minor radius of the contour of the filament array. Currents in external coils are assumed to be known. It is noted that a set of filament arrays can be found on contours with different values of a_f which satisfy the known magnetic probe signals to an accuracy of less than 0.1%. So far, this is the type of calculation which is done to assess the magnetics feedback control system. In order to make an initial estimate of the sensitivity of the reflectometer reconstruction, a similar type of calculation is carried out. It is assumed that the reflectometers give the position of the boundary flux contour. As before, a set filaments is found for different values of a_f which satisfy the known shape; however, now the values of the boundary flux, φ_b, and I_p are treated as unknown. For this set of filaments one can estimate φ_b, I_p, and the locations of the X and strike points as a function of a_f. Some of these results for 8, 11, 14, and 21 channel reflectometer arrays are shown in Fig. 5. From this figure one can conclude that, if no other information is available, one needs at least 14 elements to obtain both I_p and φ_b to an accuracy of 1%. Variations in the predicted inner and outer strike point of less than 0.02 m are found only for 14 channels and above.

The number of reflectometer channels may be reduced if one imposes additional constraints on the reconstruction. Two such constraints have been considered, (1) the actual plasma current and (2) the actual location of the divertor strike points. When I_p is

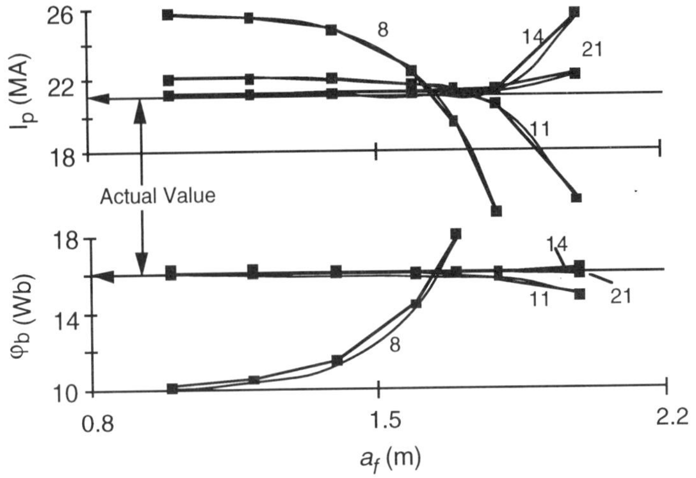

Figure 5. The variation of I_p and φ_b with the filament minor radius is shown for magnetic reconstructions of the flux surface from 8, 11, 14, and 21 reflectometers and no other constraints. The actual values of I_p and φ_b are 21.0 MA and 15.95 Wb, respectively.

constrained, the variation in all reflectometer estimates of φ_b is less than 0.1 % for all reflectometer sets. However, the strike points vary by 0.15 m even for the 14 elements set. When the strike point location alone is constrained, the estimate of I_p and φ_b is not good for any of antenna sets. The total current is apparently necessary to constrain solutions to those with X points. When both the strike point and the total current are constrained to their actual values, the problem still appears to be over constrained, and is substantially improved by eliminating the first singular value. More work needs to be done on this aspect of the control problem. Other constraints may be more apporpriate.

Estimates have also been carried out to determine the significance of errors in each of the reflectometer array elements. A filament reconstruction has been done for random error distributed uniformly on the interval Δ_n = +/- 0.0 m to + 0.05 m normal to the flux surface for each of the n_{max} = 8 and 14 reflectometer elements for a_f. = 1.6 m. The variation of I_p and φ_b, with Δ_n is $\partial I_p / \partial \Delta_n \approx 22 \; MA/m$ and $\partial \varphi_b / \partial \Delta_n \approx \; 25 \; Wb/m$. Furthermore, the errors in predicting the strike point location are about 0.30 m even for Δ_n = +/- 0.01 m. Thus, measurement errors of 0.01 to 0.02 m will dominate reconstruction inaccuracies in I_p, φ_b, and the strike point when the number of reflectometers is greater than about 8.

The conclusion from these calculations is that the gross position and shape of the boundary flux is controlled well from the 8 channel reflectometry array, but the position of the strike points is not well determined failing to meet the ITER specification of 0.10 m even for 14 channels. In addition the possibility of using shape information alone to determine I_p and φ_b requires 14 or more channels. The larger array would be needed to predict magnetic loop signals in order to correct their drift which, in turn, is needed for using magnetics to ramp down the discharge at the end. More work needs to be done to incorporate shape information from the reflectometer array into the overall time dependent feedback control problem to determine what additional constraints may be required to control the strike point and to provide drift corrections for the magnetics

CONCLUSION

An ordinary mode reflectometer operating in the frequency range 15 to 60 GHz can be designed which can measure the location of a plasma density surface near the separatrix with sufficient accuracy to be used to supplement the magnetics for position control. A swept frequency (FM-CW) reflectometer has been analyzed as the most probable candidate for the diagnostic technique. Such a system can be implemented using the same frequency range at each poloidal control location, and it is possible to have many common components. A bistatic antenna arrangement has been proposed in which the antennas are incorporated into edge or core of the blanket modules. Simple overmoded waveguide with several types of low loss bends can bring signals from remote sources and detectors to the antennas, so that complexity near the plasma facing components is minimized. The strength of the position control reflectometer is in the accurate estimate of a specific edge density layer with respect to a physical location near the vessel wall. This estimate relies on a time delay measurement and is independent of the past history of the profile.

Because of a considerable body of experimental work on measuring density profiles with reflectometry, this system can be design with a reasonably high level of confidence. However, there are a number of areas which require addition development. Beside the ITER specific needs to develop windows, evaluate neutron and thermal effects, and evaluate waveguide routing, there are a number of issues that can be addressed on contemporary devices. This includes the development of inversion algorithms for the ordinary mode which demonstrate the required positional accuracy on a tokamak. A poloidal array similar to the ITER design should be used on a contemporary tokamak to obtain operational experience and to directly compare with magnetics.

Finally, the sensitivity of the prediction of I_p and φ_b, to the number of reflectometers in the array has been examined using a filament reconstruction technique. Eight reflectometer elements is sufficient to control the shape of the boundary flux surface provided that the density is a flux quantity and provide a correspondence can be made between a position on the density profile and the density close to the flux boundary. However, the strike point is not well controlled with this system nor is there sufficient information from the shape alone to correct individual magnetic loop drifts. The plasma shape will have to be incorporated into the overall control system to determine if there is sufficient information to accomplish the ramp down in the face of magnetic drifts.

ACKNOWLEDGMENT

This work has been performed under the Agreement among the European Atomic Energy Community, the Government of Japan, the Government of the Russian Federation, and the Government of the United States of America on Cooperation in the Engineering Design Activities for the International Thermonuclear Experimental Reactor under the auspices of the International Atomic Energy Agency. This work was supported by US DoE Contract DE-AC02-76-CH0-3073 (Princeton) and DE-Fg03-86ER53225 (UCLA).

REFERENCES

[1] C. Laviron, A.J.H. Donné, M.E. Manso, and J. Sanchez, Plasma Phys. Control. Fusion **38**, 905 (1996).

[2] A.E. Costley, et al, Proc. Workshop on Diagnostics for ITER, Varenna, Italy, Eds. P.E. Stott, G. Gorini, and E. Sindoni, Plenum Press, NY (1996).

[3] E.J. Doyle, et al, Proc. Workshop on Diagnostics for ITER, Varenna, Italy, Eds. P.E. Stott, G. Gorini, and E. Sindoni, Plenum Press, NY (1996).

[4] K.W. Kim, E.J. Doyle, T.L. Rhodes, W.A. Peebles, C.L. Rettig, and N.C. Luhmann, Jr., Rev. Sci. Instr. **66**, 466 (1997).

[5] E.J. Doyle, et al, Design Basis for the ITER Plasma Shape and Position Control Reflectometer System, These Proceedings

6 Manso, M.E. et al, Proc. Workshop on Diagnostics for ITER, Varenna, Italy, Eds. P.E. Stott, G. Gorini, and E. Sindoni, Plenum Press, NY (1996) and These Proceedings.
7 J. Doane, E. Mazzucato, and G. Schmidt, Rev. Sci. Intrum. **52**, 13 (1981).
8 A.C. Sips and G.J. Kramer, Plas. Phys. Control. Fusion **35**, 743 (1993).
9 S.H. Heijnen, M. de Baar, A.J.H. Donné, M.J. van de Pol, C.A.J. Hugenholtz, and RTP Team, Rev. Sci. Instrum. **66**, 419 (1995).
10 G.R. Hanson, J.B. Wilgen, T.S. Bigelow, I. Collazo, A.C. England, M. Murakami, D.A. Rasmussen, and J.R. Wilson, Rev. Sci. Instrum. **66**, 863 (1995).
11 G. Vayakis, et al, Overview of the ITER Reflectometry Diagnostic Systems, These Proceedings.
12 J.L. Doane, IEEE Trans Mic. Theory Tech. **MTT-32**, 1362 (1984)

REFLECTOMETRY IN THE ITER DIVERTOR

M. Manso[1], L. Cupido[1], G. Leclert[2], D. Wagner[3], G. Vayakis[4] A. Donné[5], L. de Kock[6], C. Laviron[7], J. Sanchez[8], F. Serra[1], A. Silva[1], C. Walker[6], and the ITER Joint Central Team and EU Home Teams.

[1]Associação EURATOM/IST, CFN, Lisboa, Portugal
[2]LPMI, Univ. Henri Poincaré Nancy I, France
[3]Institut für Plasmaforschung, Stuttgart, Germany
[4]ITER JCT, San Diego JWS, USA
[5]FOM, Nieuwegein, The Netherlands
[6]ITER JCT, Garching JWS, Germany
[7]Association EURATOM-CEA, Cadarache, France
[8]Association EURATOM/CIEMAT, 28040 Madrid, Spain

INTRODUCTION

Reflectometry at the ITER divertor will be vital to characterise plasma transport and fluctuations and to optimize machine operation. The main objective is to measure density profiles with high spatial and temporal resolution. Other important measurements are the detection of ELMs and multifaceted axisymmetric radiation from the edge (MARFES), and flame front location. Divertor plate erosion can also, in principle, be measured based on distance measurements.

In the last years reflectometry has experienced great advances[1,2,3] and it is now possible to obtain density profiles with simultaneous high spatial (~1 cm) and temporal resolution (~100 μs). However, although measurements at the main plasma are routine, at the divertor the experience is still limited. High demands are put on reflectometry for the ITER divertor due to the strong refractive effects, wide range of densities to be probed (0.1-100$\times 10^{20}$ m^{-3}) and severe restrictions in access. In spite of the difficulties and through the work by members of the ITER Joint Central Team and EU Home Team, a realistic engineering design was developed[4,5,6] with solutions for the waveguide paths on divertor cassette and transmission line to the outside, compatible with remote handling requirements.

Here, we analyse some aspects of the engineering design in view of their impact on the quality of profile measurement. The propagation conditions in the divertor region are studied with full wave calculations (1D and 2D) and the factors that determine the radial resolution of profile measurements are analysed, in particular initialisation. The extremely high densities in the divertor poses the problem of availability of microwave

Diagnostics for Experimental Thermonuclear Fusion Reactors 2
Edited by Stott *et al.*, Plenum Press, New York, 1998

sources. Several solutions are considered: broadband swept up to 220 GHz, narrow band swept and/or pulse radar reflectometry up to 350 GHz.

The existing engineering design, the present feasibility study and results from experiments on JET[6] and DIII-D[7], show that reflectometry at the ITER divertor is possible and will make an important contribution for the machine operation and to the understanding of the underlying physics.

TYPES OF MEASUREMENTS

The measurements to be performed on the ITER divertor fall into categories: (1b) n_e profiles, Marfes, flame front location; and (2) ELMs, density fluctuations. The resolution target is 0.3 cm across the leg and 10 cm along the leg. The target time resolution is 1 ms and accuracy 10%. The small radial resolution across the leg is imposed by the narrowness of the scrape-off layer. Here reflectometry has the advantage that the radial resolution improves with the steepness of the density gradient and therefore the 3 mm resolution is not a high demand for the system. As we will see later, even the absolute location of the gradient zone within 3 mm can be achieved with reflectometry.

Measurements over the full range of density (ne: 0.1-100$\times 10^{20}$ m^{-3}) require a very large frequency range, 18-900 GHz for O mode propagation, and X-u mode (90-180 GHz) is foreseen to probe the plasma edge. Continuous broadband measurements will be performed up to 220 GHz and for higher frequencies a multichannel system with narrowband channels can be used. All the frequencies can be combined and launched simultaneously into the highest frequency transmission lines and the signal will be detected at different intermediate frequencies IF (heterodyne detection). Due to the strong refractive effects expected at the divertor region, transmission must also be used in order to avoid ambiguities between diffraction of the reflected signal away from the antenna and transmission. Different types of measurements can be performed: density profile, peak density, profile width, delayometry.

IMPLEMENTATION OF THE SYSTEM

A realistic engineering design was developed[4,5,6,7] with solutions for the waveguide paths on divertor cassette and transmission line to the outside, compatible with remote handling requirements. The full profile will be probed from the private flux and target side with O and X mode, with five sightlines on the outer leg and three sightlines in the inner leg, each of them including two bistatic antenna systems for the lower and higher frequency ranges (see Figure 1). Monostatic or bistatic arrangement can be used but bistatic is more convenient for most reflectometry techniques. The distribution of sightlines is a compromise between a detailed description of the divertor profile and the feasibility of the transmission system. With the present number of sightlines the resolution along the leg is only ~20 cm, but as this resolution is most important to resolve the location of the strike zone an additional toroidal location could be used to implement more sightlines.

The waveguides and antennas will be placed in an insert onto the diagnostic cassette and can be shared with Interferometry and Electron Cyclotron Absorption diagnostic. Details about the implementation of the system and optimization of the front end are presented in companion papers[5,6]. Due to restrictions of the available space for the waveguide routing in the divertor, moderately oversized rectangular waveguides are foreseen (see Table 1). These waveguides must be coupled to the oversized corrugated waveguide (D-63.5 mm) outside the divertor used for long distance transmission. To

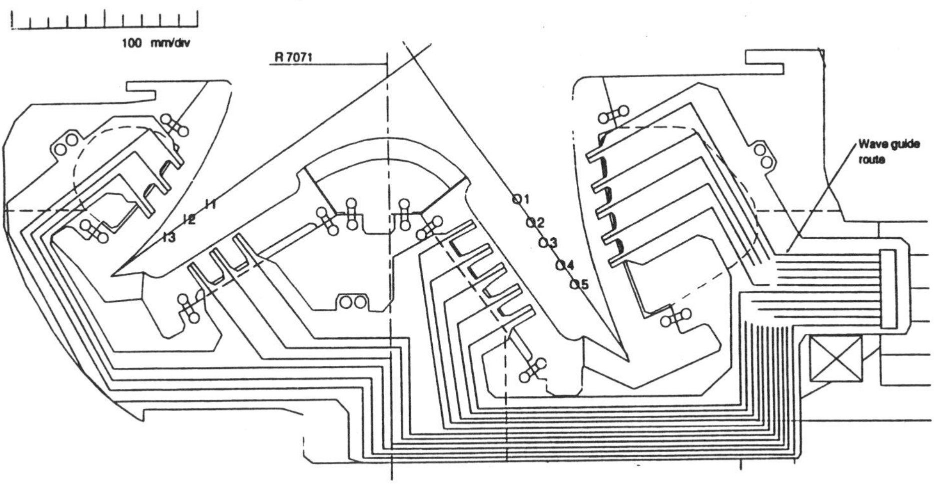

Figure 1 - Central diagnostic cassette port - 8 & 18

Table 1: Waveguide for different frequency bands (O and X modes).

Wave propagation	Frequency range	Waveguides	Bends
O-mode	20-60 GHz	10 mm x 10 mm	gradual bends
O-mode	60-280 GHz	20 mm x 10 mm	quasi-optical mitre bends
X-mode	90-180 GHz	5 mm x 60 mm	quasi-optical mitre bends

make these transitions as compact as possible and broadband at the same time, a quasi-optical transition based on confocal imaging can be used. In order to cover the whole frequency range from 15 GHz to 280 GHz, three different types of rectangular waveguides will be applied. Different types of waveguide bends are necessary to limit mode conversion losses (Table 1).

Gradual waveguide bends offer a better performance in the low frequency range (O-mode, 20-60 GHz). However more detailed design using optimized curvatures is still necessary due to the limited availability of space for the waveguide paths on the divertor cassette. For higher frequencies compact quasi-optical bends are feasible. First experiments[7] for long distance broadband transmission using oversized corrugated circular waveguides with a length of 28 m including three mitre bends demonstrated average losses of only 0.24 dB in the frequency range 45-170 GHz. The measured losses of a quasi-optical transition between the oversized corrugated waveguide and a standard rectangular waveguide were about 1 dB. However in this measurement the available circular horn antenna (length=120 mm, output diameter=16.25 mm) connected of the rectangular waveguide was not optimized for the used mirror setup. For optimized horns, with length and output diameter of 200 mm and 13.2 mm respectively, the expected total loss of the quasi-optical coupler is only 0.5 dB.

In order to avoid the exposure of the antennas to very large heat loads, they must be set back from the plasma facing surface by about 60-70 mm and therefore view the plasma through gaps between the target plates. This will affect the radiation patterns of the rectangular horns. The field radiated from the open waveguide is still guided by the side walls of the divertor cassette which act as a parallel plate waveguide, i.e. there is only beam divergence in one plane. Therefore this could be even an improvement concerning the radiation pattern in the toroidal plane. For the higher frequencies (> 60 GHz) and therefore higher directivity of the milimeter wave beam, radiation from the open waveguide directly into the gap might be possible (Fig. 2a). For lower frequencies and smaller waveguides cross sections (e.g. 10 mm×10 mm rectangular waveguide) such an arrangement would lead to increased side lobes in the pattern radiated out of the gap due to diffraction at the waveguide aperture. This situation can be improved when the waveguide is stuck in the gap; in that way the side wall act like the continuation of the waveguide wall in one plane (Figure 2b). Other improvements are possible if the side walls are inclined and therefore act as a horn antenna in one plane (Figure 2c). The cross talk between neighbouring waveguides, which might be a problem especially for moderately oversized waveguides, must be controlled (Figure 2d). Parasitic reflections should be minimized as much as possible because they will decrease the accuracy of profile measurement. Polarization will be guaranteed by the sections of rectangular waveguides, but it will be necessary to evaluate the power that goes into the wrong polarization due to the plasma. Concerning vacuum windows, a solution could be the use of Brewster angle windows, with high transmission for the wanted polarization.

For calibration purposes it would be convenient to have a hardware reference close to the antenna mouth. For bistatic systems this could be made by a coupling between the transmitting and receiving waveguides but it would originate additional parasitic reflections from the gap between cassettes. The phase and amplitude of the signal recorded in the absence of plasma could also provide a reference for calibration. However, in ITER this is not sufficient because during the plasma pulse relative movements occur and a real time calibration is needed. Another possibility is to use two frequencies (narrow band swept) and to perform a differential calibration.

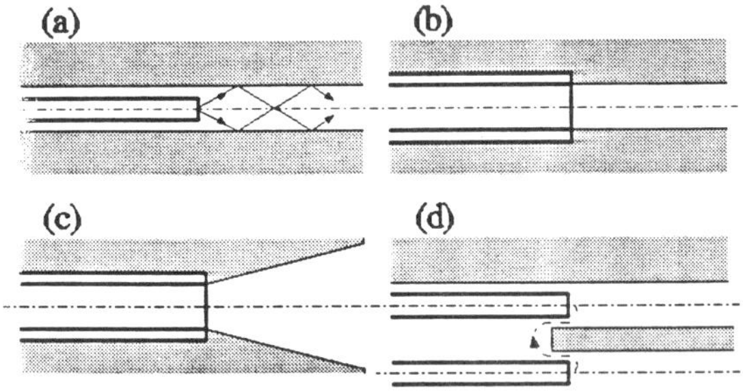

Figure 2 - Different waveguide implementations in the gap between target plates

PROFILE RECONSTRUCTION FROM REFLECTOMETRY

Wave propagation studies with 1D and 2D full wave codes[8]

The possibility of reconstructing the density profile in the nonmonotonic case (detached case), has been tested with the 1D code. A 1D density profile along the line of sight of an antenna (i.e. perpendicular to the separatrix) has been obtained from the 2D density data of the ITER data base. The 1D full wave code has been used with this density profile to give the phase shift versus the frequency; the density profile is then reconstructed by Abel inversion. The profile can be recovered almost everywhere even in the nonmonotonic case as in the example Figure 3. Since the phase velocity increases in the density well, at first the main peak is seen further away than it is in reality. However, as the frequency is increased, the true profile is quickly recovered.

Due to the large amount of time needed for 2D calculations, only specific frequency cases could so far be studied, and a density profile cannot be obtained. In order to speed up the calculations, the waveguide has been located at the plasma edge, even when there is a vacuum layer between the plasma and the wall. It is also assumed that the waveguide excites only the fundamental mode.

Two different cases are shown on Figure 4. The field distribution for the attached case is given in 4a, for the waveguide at the location 5 (in Fig. 1). The incident frequency is rather low (50 GHz). (Note that the grid spacing of the density data introduces some coherent fluctuations that still do not perturb the reflected signal very much). Figure 4b has been obtained in the same case as the 1D calculation of Figure 3, for a frequency of 80 GHz (cutoff layer before the density well). The 2D evaluation of the field along the waveguide shows that the electric field remains fairly regular despite the assumed density irregularities.

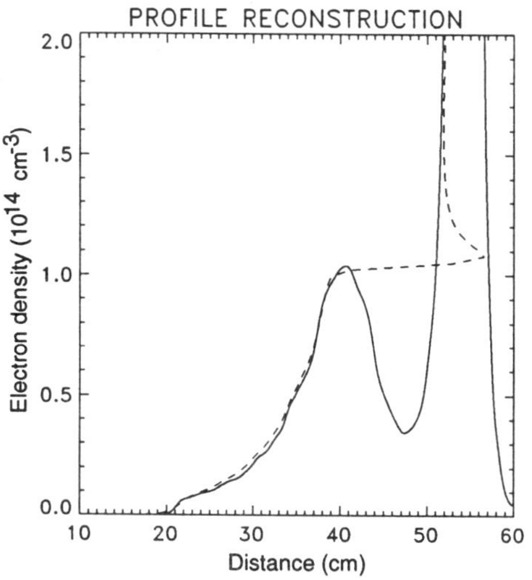

Figure 3 - Profile reconstruction from the 1D full wave code (- - -) for the detached case, nonmonotonic profile (___). Antenna 2.

Reflected power

The evolution of the main reflected power with frequency has been analysed, in the range 80-120 GHz. For the incident frequency 80 GHz, the reflected field is about 20 dB below the incident field. As expected the power decreases when the frequency grows, but only very slowly. The influence of the density clip (around 100 GHz, in Figure 3) is seen as an increase of the slope. Above the 110 GHz the decrease is even less important, since the density profile is very steep. The average slope is about 3 dB / 50 GHz. Hence one should be able to make most of the profile measurements (up to 300 GHz) with losses less than 30 to 40 dB. This will easily be achieved with heterodyne detection.

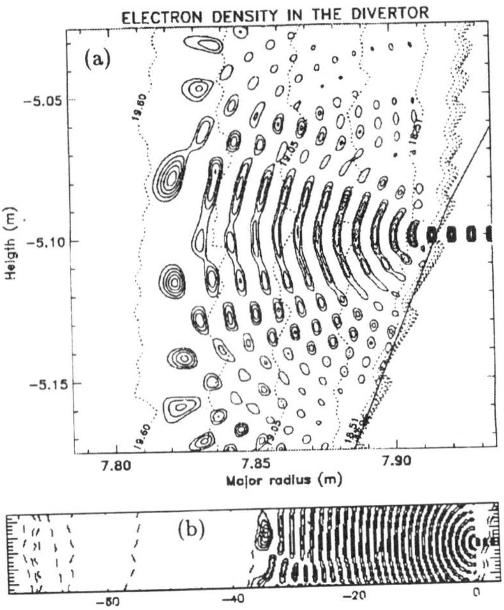

Figure 4(a) - Field distribution for the attached case, antenna 5, f = 50 GHz (2D full wave code); (b) Detached case (nonmonotonic profile of Fig., 3), f=80 GHz, distances normalized to the vacuum wavelength. Solid lines: contour plot of the positive real part of the electric field; dashed lines: electron density.

Initialisation

In order to optimize the absolute location of the profile, detailed studies have been performed about the initialisation process[9,10,11].

With O-mode the distance d_α from the antenna to any plasma layer is obtained from:

$$d_\alpha(n_{e\alpha}) = \frac{c}{\pi}\int_0^{F_1} \frac{\tau_g(f)}{\sqrt{F_\alpha^2 - f^2}} df \quad + \quad \frac{c}{\pi}\int_{F1}^{F_\alpha} \frac{\tau_g(f)}{\sqrt{F_\alpha^2 - f^2}} df \quad = \quad I_1(F_\alpha) + I_2(F_\alpha)$$

The first integral (I_1) represents the contribution of the non measured profile (below F_1) and the second integral (I_2) is the contribution from reflectometry.

The contribution from the non measured profile at $F = F_\alpha$ can be written as:

$$I_1(F_\alpha) \quad \leq \quad \frac{c}{\pi}\Big(\int_0^{F_1} \tau_g^2 df\Big)^{1/2} \Big[\int_0^{F_1} \Big(\frac{1}{\sqrt{F^2 - F_\alpha^2}}\Big)^2 df\Big]^{1/2}$$

The second term in the above expression gives the decaying of I_1 with frequency. The first term represents the contribution of the non measured profile to the localisation of plasma layer $n_{e\alpha}(F_\alpha)$. This term depends only on the quantity A_1'

$\frac{c}{\pi}\Big(\int\limits_0^{F_1} \tau_g^2 df\Big)^{1/2}$, that is related to the area A_1. Therefore the detailed shape of the profile (including the localisation of the first plasma layer) may not be relevant for the initialization, as long as the estimation of A_1 does not differ significantly from the correct value. The errors due to initialisation can be estimated as follows[10]:

$$\delta r_\alpha(F_\alpha) \quad \leq \quad \frac{c}{\pi}\Big(\int\limits_0^{F_1} \delta\tau_g^2 df\Big)^{1/2} \Big[\frac{1}{2F_\alpha} \cdot ln\frac{F_\alpha + F_1}{F_\alpha - F_1}\Big]^{1/2}$$

where $\delta\tau_g(F)$ is the diference between the correct value of τ_g below F_1 and the estimated value, τ_g' at each frequency. The above expression can also be used to quantify the errors due to any uncorrect group delay measurements over some frequency region $[F_\beta - F_\gamma]$ just by replacing the frequencies zero, F_1 by F_β, F_γ.

Reconstruction of ITER divertor density profiles

We analysed two cases[10]:

Case (a): attached case (BL/EIRENE, 100 MW, no impurities)
The density profile along sightline 1 and the corresponding O-mode group delay characteristic are depicted in Figure 6a and Figure 6b respectively. The profile was reconstructed from the τ_g curve above F_1=18 GHz. The initialisation was based on a model[9] that uses the measured τ_g curve close and above F_1 to obtain an estimate of the non measured curve below F_1. It is assumed that this curve can be modelled as $\tau_g = A\big(\frac{f}{F_1}\big)^{2/s}$ where the parameter A is determined from $\tau_g(F_1^+) = \tau_g(F_1^-)$ and the parameter s is obtained by matching the slope of the experimental and estimated curves τ_g at $F = F_1$. The errors (Figure 6c) of the reconstructed profile (dashed line in Figure 6a), shows that it should be possible to measure the high density with an accuracy of 3-5 mm.

Case (b): detached case (B2/EIRENE, 40 MW, no impurities).
In this case the profile (Figure 6a) has a non monotonic shape at low density including a hollow part. The group delay due to this profile (Figure 6b) shows a clear jump due to the hollow region. This is "seen" by O-mode reflectometry as a density plateau (dashed line in Figure 6a). In spite of the inability of reflectometry to recover the hollow part of the profile, the high density part of the profile beyond the hollow region is reconstructed with only small radial errors (a few millimeters), as seen in Figure 6c.

Sources of errors

The above results show that the errors in the absolute localisation of density profiles are within the target parameters (3 mm). However, those studies did not account for effects that may decrease the accuracy of the experimental τ_g curve, such as plasma fluctuations and modes and parasitic reflections (due to transmission line, antennas, target structure). As a result a decrease of the reflected power and some phase mixing is normally observed. This problem can be circumvented to a great extent with the improvement of signal to noise ratio with heterodyne detection and the effect of fluctuations can be further minimized by averaging in time[11].

But the parasitic reflections can be an important source of errors due to the fact that the antennas view the plasma through the gaps between target plates. Both the

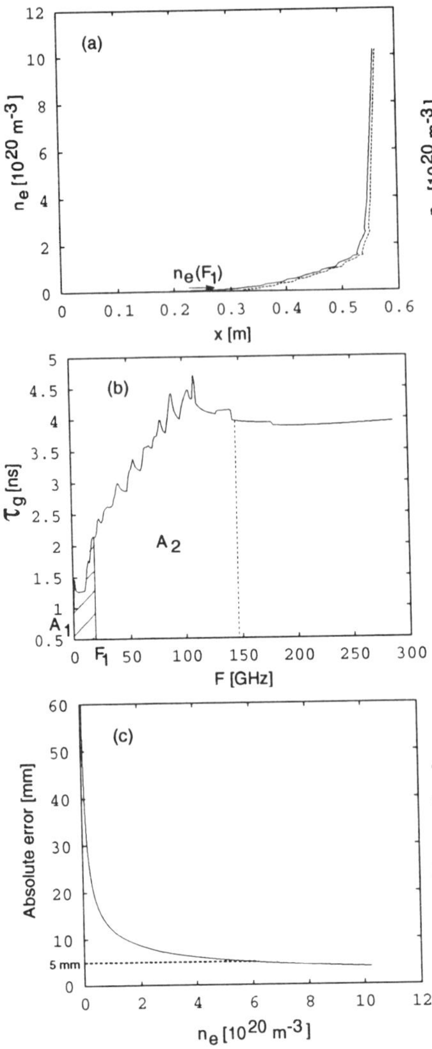

Figure 5 - Attached case

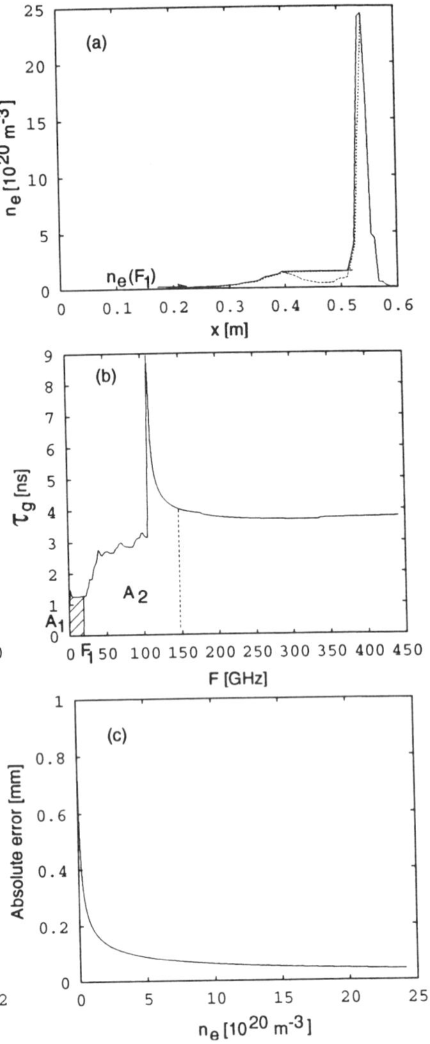

Figure 6- Detached case

modification of the design of the antennas and the target structure around the antennas is being studied in order to minimise this problem, as discussed above.

In order to measure the plasma with 3 mm radial resolution it is important to use X mode to probe the plasma edge which will contribute to the better estimate of the group delay below F_1. Combined information from O/X reflectometry[11] will improve significantly the radial resolution of the profile measurement at the outer edge.

TECHNOLOGICAL ISSUES

Due to the extremely high densities to be probed the problem of availability of microwave sources was considered[12]. In the range of 30-300 GHz solid state oscillators as VTOs and HTOs followed by broadband active and passive multipliers offer the best choice for fast swept sources able to operate in full fundamental waveguide bandwidths.

Fundamental mode oscillators for the lower frequencies will be available very likely in the coming years. On the other hand, the most probable scenario for the higher frequencies will be that multipliers will still be required for broadband operation. In addition, the use of heterodyne detection techniques in swept operation also requires the master oscillator (before multiplication) to operate in at frequencies where PLLs or frequency synthetiser can be used, that is up to 26 GHz.

The power which could be generated with these sources is summarised on Table II. In this table, which shows the state of the art of source performance, we have considered broad band active multipliers up to 60 GHz (which are since 1996 comercially available) followed by high efficiency diode multipliers for the range above 60 GHz.

The near future prediction is based on active multipliers up to 90 GHz and passive ones above that. (According to some manufacturers non-offical information these products are already under development and will be commercially available in less than 10 years time).

Receiver sensitivity figures assume a heterodyne receiver configuration, being in most of the cases the plasma noise the main limiting factor.

The bands above 220 GHz are difficult to implement in broadband with the present solid state technology although it is possible to use modest bandwidths of 10% to 20%.

Another problem is related to the sensitivity of the receivers, which is limited not only by the plasma noise (if detectable) but essentially by the IF bandwidths employed. These must be larger than the coherence bandwidth of the plasma and local oscillator signals. This fact suggests again that rather than using fundamental gunn oscilators it might be advantageous to use cleaner sources at low frequency followed by multipliers which might give the same power output with a much better phase noise performance. These sources with low phase noise could be phase locked gunn or inpatt oscillators or even lower frequency multiplied sources.

Optical sources down converted to the sub-millimeter region would be a solution for large to extremely large bandwidth sources, but the phase noise of the resulted signal would be extremely poor for broadband swept operation. Therefore, the bandwidth of the receiver must be extremely large and the noise from the plasma (or just termal noise) would represent the main limitation. Although some recent developments in tuned laser seem promissing we are still a little far from the desirable figures that would be required for a good sub-millimetre swept heterodyne receiver.

Table 2: Performance of signal generators for the range 30-300 GHz.

Wave guide band:	State of Art (1996) Power dBm		Near future Power dBm		Receiver NF Harm. mixer dB	Receiver NF Fund. mixer dB
Ka - 26.5/40 GHz	23		27			7
Q - 33/50 GHz	13	(1)	20			8
U - 40/60 GHz	13		17			8
V - 50/75 GHz	5	(1)	7	(1)	12	8
E - 60/90 GHz	3	(1)	10		16	9
W - 75/110 GHz	1	(2)	5	(1)	22	9
F - 90/140 GHz	0	(2)	3	(1)	24	10
D - 110/170 GHz	-7	(2)	0	(2)	30	10
G - 140/220 GHz	-10	(3)	-5	(2)	34	12
Q.O. - >220 GHz	-20?	(4)	-10?	(4)	> 35	>14

1) - Double diode doubler; 2) - Varactor tripler; 3) - Double diode quadrupler; 4) - Corner cube structure.

On the sub-millimetre bands it is possible to predict the availability of series of fixed frequency oscillators (modulated or not) multiplied (by diodes in corner cube or quasy optical structures) in comb arrangements, or BWO tubes, now going up to 1000 GHz. These type of schemes shall be used if it is possible to construct reception devices operating up to 1000 GHz achieving good sensitivity at room temperature.

Alternative methods

Due to the specific requirements and scenarios in the ITER divertor, some new measurement proposals have been raised: the first is the use of correlation measurements between neighbouring radial positions to obtain the time delay curve[13]. The method requires a longer measuring time and a more complex hardware than the standard ones but offers the possibility to measure in scenarios with complex waveguides and strong turbulence. The second proposal is the use of the whole set of crossed sightlines between the five front ends to establish a mesh of transmission reflection measurement. This would lead to a 2-dimensional comb-like system which could be used to determine the position of the ionization front[14]. The implementation of this solution would require a detailed study of the beam cross propagation with antennas which in principle should be optimised for direct propagation.

EXISTING EXPERIENCE

DIII-D divertor reflectometer system

The reflectometer for the DIII-D divertor was designed to be a modular, low loss, broadband system, supporting frequencies in the range 50 - 110 GHz[15]. The use of circular waveguide allows a variety of polarization (O-, X-mode and combination of each). The system currently views directly upward towards the X-point of the tokamak; an offset double mitre bend mirror system has been designed to view either the inner or outer divertor leg.

Density profiles were obtained in the private flux region using a swept source (BWO, 50 - 75 GHz) with a broadband FM technique (Figure 7). These profiles, with spatial resolution of ~1 cm, complement the Thomson scattering. They will add to the needed data base and provide feedback to divertor modeling efforts.

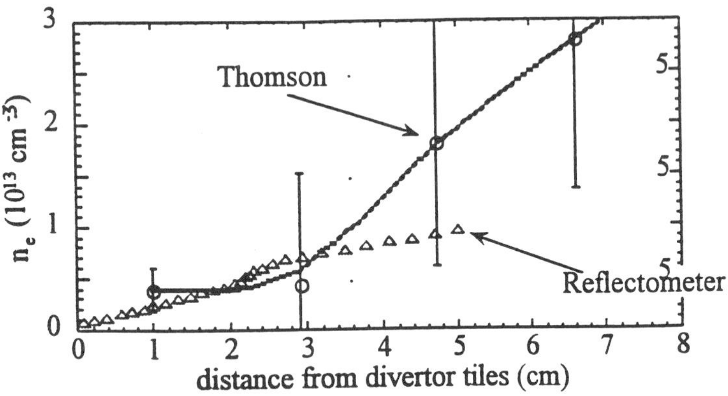

Figure 7 - Private flux region density profiles at DIII-D obtained with FM CW reflectometry [from ref. 15].

Reflectometer for the JET divertor

The "comb" reflectometer for the JET divertor uses up to eight fixed frequency sources (free running Gunn oscillators), launching simultaneously into the plasma O-mode probing signals in the frequency range 50 - 98.5 GHz[16]. Amplitude modulation of the radiation source and the use of a narrowband heterodyne receivers centred on the modulation frequency, provide a reliable way of determining if a given frequency is in reflection or transmission. The maximum frequency of the signals reflected and the minimum frequency of the signals transmitted will give lower and upper limits to the peak density on the sightline through the plasma.

Figure 8 shows an example of the influence of ELMs on the divertor plasma. The large ELM at ~13.7 s produces a large signal on all reflection channels and cuts off all of the transmission signals for times between 50 ms and 100 ms; during the main spike of the ELM the peak density on the sightline is at least 1.2×10^{20} m^{-3}. Subsequent ELMs produce only very small reflection signals most probably due the non-optimized reception. In fact, the antenna originally foreseen could not be used due to the large parasitic reflection from the transmission waveguide and an antenna located away from the main reflected beam was employed. Nevertheless, transmission signals provides reliable measurements in many divertor plasma scenarios.

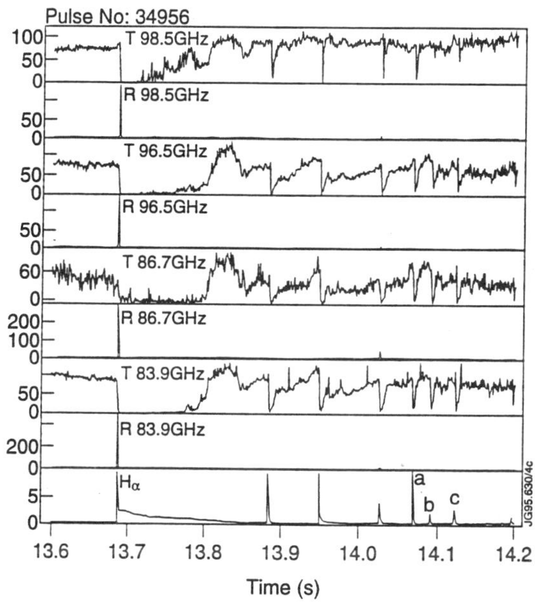

Figure 8 - The effects of large ELMs on both the transmission (T) and reflection (R) signals, for emitting frequencies: 83.9 GHz-98.5 GHz, obtained with the JET divertor comb system [from ref. 16].

DISCUSSION AND FURTHER WORK

Overall we have shown that the extremely challenging problem of reflectometry in the divertor can be overcome: the expected profiles for the divertor can be handled with reflectometry and the required resolution achieved. Also we are able to design a feasible waveguide and antenna system to have access to the divertor and have technological solutions for the difficult geometry and wide density range.

The main problem is the design and test of the suitable waveguide system. The access is very complex and issues like alignement are still a problem for the higher frequencies. Also the analysis of overall losses is important: line plus bends plus plasma and comparison with what can be obtained from technology of emission/detection.

Existing reflectometers (or new ones) can be used to address the critical issues of divertor reflectometry on ITER time scale.

ACKNOWLEDGEMENT

This report has been prepared as an account of work performed under the Agreement among the European Atomic Energy Community, the Government of Japan, the Government of the Russian Federation, and the Government of the United States of America on Cooperation in the Engineering Design Activities for the International Thermonuclear Experimental Reactor ("ITER EDA Agreement") under the auspices of the International Atomic Energy Agency (IAEA).

REFERENCES

1. E.J. Doyle et al., *Diagnostics for Experimental Thermonuclear Fusion Reactors*, edited by P.E. Stott et al, Plenum Press, New York, 1996, p.117
2. C. Laviron et al., *Plasma Phys. Control. Fusion* 38, 905, (1996).
3. M. Manso, *to appear in Proc. "8th Intern. Symp. on Laser-Aided Diagnostic* (1997).
4. G. Vayakis et al., *Rev. Sci. Instrum.* 68 (1) (1997).
5. G. Vayakis et al., These Proceedings.
6. C. Walker et al., These Proceedings.
7. D. Wagner et al., These Proceedings.
8. G. Leclert et al., *Proc. III Work. on Micro. Reflect. for Fusion Diagn.* Madrid, 1997.
9. P. Varela et al., *Rev. Sci. Instrum.* 66 (10) (1995).
10. M. Manso, presentation at *III Work. on Micro. Reflect. for Fus. Diagn.* Madrid, 1997.
11. M. Manso et al., These Proceedings.
12. L. Cupido et al., *Proc. III Work. on Micro. Reflect. for Fus. Diagn.* Madrid, 1997.
13. T. Estrada et al., *Proc. III Work. on Micro. Reflect. for Fusion Diagn.* Madrid, 1997.
14. A. Donné, presentation at *III Work. on Micro. Reflect. for Fus. Diagn.* Madrid, 1997.
15. T.L. Rhodes et al., *Rev. Sci. Instrum.* 68 (1) (1997).
16. L. Cupido et al., *Plasma Phys. Control. Fusion* 39 (1997).

REFLECTOMETRY DENSITY PROFILE MEASUREMENTS IN TURBULENT PLASMAS

J. Sánchez, T. Estrada, E. de la Luna, V. Zhuravlev[1], M. Francés, B. Brañas

Asociación EURATOM-CIEMAT, 28040 Madrid Spain
[1] Institute Kurchatov, Moscow Russia

INTRODUCTION

One of the limiting factors in reflectometers when trying to measure the density profile in fusion plasmas is the existence of a fraction of experiments where the measurement becomes impossible: very noisy signals leading to nonsense density profiles. Those so called "bad shots" are usually linked with strongly turbulent plasmas and the remaining problem is to be able to assess "a priori" if the data coming from a given experiment will correspond to a normal profile or if they will be meaningless.

The application of reflectometry for ITER requires maximum reliability and therefore the analysis of the limitations caused by reflectometry becomes highly important.

The effect of turbulence in the reflectometer signal is twofold: temporal and spatial.

The temporal effect is seen in swept frequency systems which use consecutive measurements to determine the time delay ($\partial\phi/\partial f$) function: the profile changes locally during the measurement (false fringes in frequency swept homodyne systems). These effects are usually overcome in the experiments by sweeping fast (frequency sweep rates in the order of 1 GHz/μs have been achieved, leading to clean results).

The spatial effects come from the fact that the reflecting layer is a corrugated surface: both phase and amplitude of the reflected beam become affected by interference effects and the perturbation cannot be avoided by sweeping fast.

Under strong turbulence the reflectometer will aim to determine the average density profile but, due to the several nonlinear magnitudes involved (phase, time delay), the achievement of this goal is not guaranteed. The main question here is whether or not this average profile can be determined and how the different techniques are affected.

SIMULATION PROCESS

In order to compare the ability of the different techniques (AM, FM, Pulse) to provide an acceptable average profile we have developed a simulation procedure:

A steady state profile is assumed and a 2-dimensional distribution of turbulence is added. With a propagation code we obtain the transfer function for each relevant cutoff frequency: amplitude A(f) and phase $\phi(f)$. Once we have this functions, we can synthesize the different reflectometers, obtain the averaged time delay function $<\tau>(f)$, perform the profile inversion (by a simple 1 d reconstruction algorithm) and compare with the original unperturbed profile.

The wave propagation is simulated with a 2-dimensional WKB code [1] (multi-ray tracing): many rays are launched for each relevant frequency and those collected at the

Diagnostics for Experimental Thermonuclear Fusion Reactors 2
Edited by Stott *et al.*, Plenum Press, New York, 1998

receiving antenna interfere to give a resulting phase and amplitude. The use of the WKB approximation is based on the fact that we have a short wavelength beam with a diameter similar to the size of the dominant turbulent structures.

As it has been shown in a previous work [1], most techniques are able to produce reasonable profiles if the proper average is taken. The clue to understand how we can get good profile reconstructions even from noisy phase data is the concept of "vector average" which we explain below. Either during the frequency sweep or during a fixed frequency observation of a evolving plasma the signal has some instants with destructive interference (as a result the amplitude of the received signal is low and the phase random) and some others with large amplitude and meaningful phase, due to constructive or absent interference. If we first obtain the phase of the reflected field and then take its time average, the result includes errors because we are averaging "good" with "bad" phase values. On the other hand, if we take the average of the electric field at the receiver (phase & amplitude) then the random values corresponding to strong interference are less weighted (because of their smaller amplitude) and the final average becomes free of error. The vector average is done by hardware at detection in the AM and Pulse radar technique (the pulse radar naturally averages over many frequencies whereas the AM system performs the average at the last filter). FM systems perform the average by post detection processing (with the FFT technique, FM systems do the average in the same way as pulse radar does).

In the present work we have analyzed three aspects of the influence of turbulence: effects of plasma rotation, turbulence level and antenna beam size.

EFFECTS OF PLASMA ROTATION

Plasma rotation in addition to slight antenna tilting leads to phase-runaway phenomena. This is both observed in the experiments (unidirectional phase drift which depends on tilting angle and rotation speed) and also predicted by the theoretical models [2].

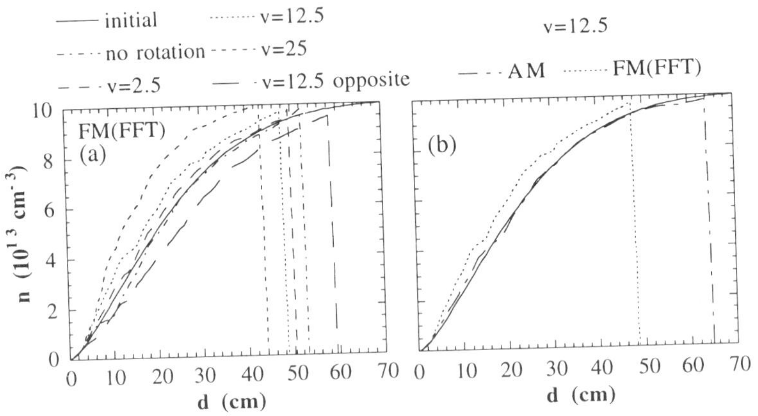

Fig.1. Reconstructed density profiles for different rotation speeds (ITER geometry).
a) Profiles reconstructed from a FM reflectometer for different values of the rotation parameter v. (5° antenna misalignment)
b) Comparison of the reconstructions from AM and FM profiles.

Using the 2D WKB model, the performance of the different reflectometers (AM, FM, Pulse) is analyzed for a rotating ITER like plasma. We have chosen as relative speed parameter "v": the number of turbulence wavelengths passing in front of the antenna during the (linear) sweep time. For example, v=2.5 corresponds to a sweep rate of 4 GHz/μs with 200 kHz turbulence. In the simulations, the antennae were tilted 5°. The results of the simulations for the FM technique with FFT analysis are shown on fig. 1a. For small values of the parameter v the effect of the rotation is small in the reconstructed density profile. With

increasing v the systematic error increases, reaching unacceptable levels. Opposite direction of rotation leads to the inversion of the error. The same calculations for the AM reflectometer with v=12.5 have shown no such errors due to the plasma rotation (fig. 1b). This result could be expected being the AM measurement an instantaneous measurement of the time delay (with the three spectral components reflecting simultaneously in the plasma). The same results would be obtained for the Pulse technique, which is based on the same principle of simultaneous reflection of several frequencies.
The error introduced in the FM system is minimized by fast sweep. Today's available technology on swept oscillators gives a high confidence that phase runaway phenomena could be overcome by fast sweeping. Also a possibility would be the use of hybrid techniques: dual frequency FM... etc. If a fast sweep is required to avoid rotation effects, turbulence average could be achieved by averaging consecutive sweeps.

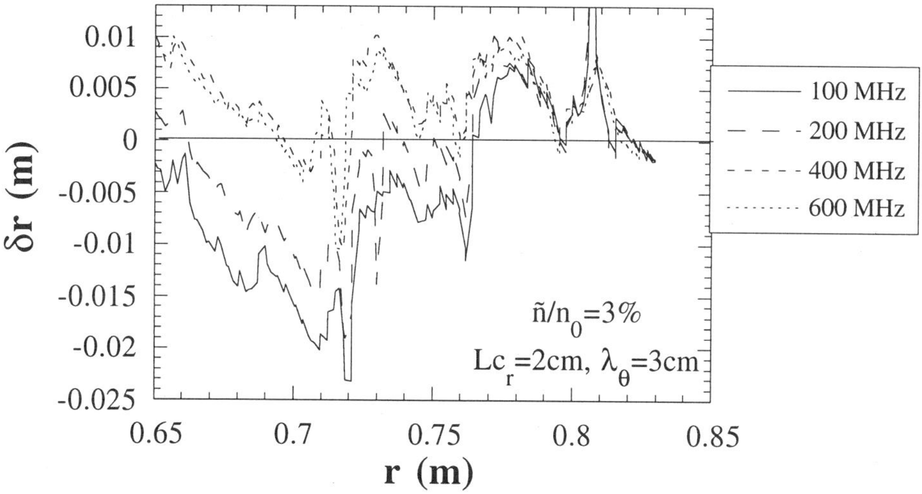

Fig.2. Radial error in the reconstructed density profiles for different level of average. AM technique. The average extension is performed along the frequency intervals of 100 to 600 MHz during the frequency sweep.

EFFECT OF TURBULENCE LEVEL AND ANTENNA BEAM SIZE

One of the key questions related to the effects of turbulence in profile reflectometry is to which point can the average method improve the profile reconstruction in strong turbulent plasmas. Fig.2 shows the radial error in the density profile measured with the AM technique for a turbulence level of 3% of the central density value. The error improves by more extensive average (in this example by extending the interval of mm-wave frequency over which the reflected E-field is averaged), but the improvement shows some trend to saturation which requires further investigation: different average techniques and more extensive statistics. For the presented case, with relative high turbulence level (which means up to 30% for r/a=0.8) and short poloidal wavelength (λ_θ= 3cm) we still have errors below 1cm, which is close to ITER specifications.

In order to assess the effect of antenna beam size, the analysis was also performed for a turbulence level $\tilde{n}/n_0$=2% using antennas of different divergence (given by the 3dB angle). Fig.3 shows the radial error of the reconstruction (AM technique with average along 400 MHz) for the different radial positions at the plasma edge: the error becomes smaller for the narrower beams: the narrower the beam the smaller the effect of corrugations. This is in contradiction with some other work [3] and suggest the strong dependence of the turbulence effects on the poloidal wavelength of the density perturbation for a given antenna pattern. The subject will also require a deeper analysis with a larger statistics, comparing the beam size with the dominant poloidal wavelength of the plasma turbulence and for different turbulence levels. The amplitude of the reflected signal (not shown in the figure for clarity reasons) varies strongly for the different frequencies but in average it does not show a clear dependence with the antenna beam size (as it does in the absence of turbulence: the signal is stronger for the narrower beam size).

The calculations for the effects of average level and beam size were performed for the AM system as a reference case. Very similar results are expected for the FM and pulse systems because plasma rotation was not included in the study.

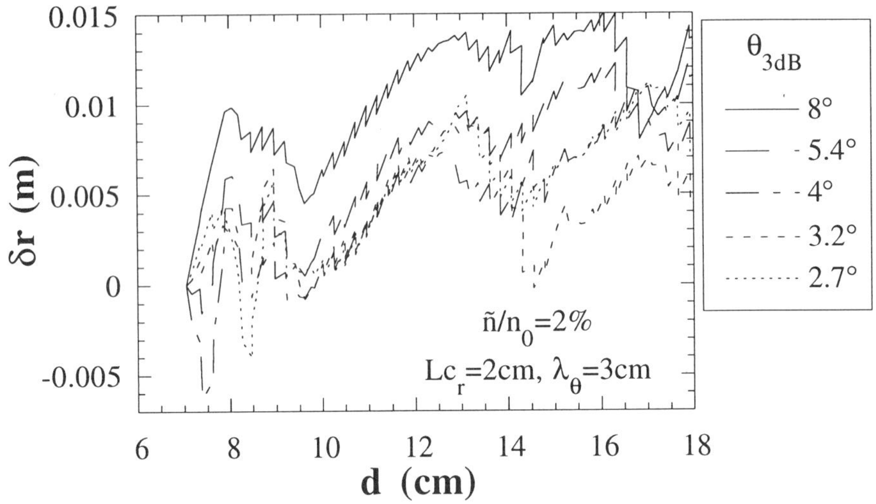

Fig.3. Radial error in the reconstructed of the edge density profile for different antenna beam sizes. AM technique

CONCLUSIONS

The 2-dim WKB code, able to compute interference effects, was used for qualitative investigation of the perturbations in reflectometry measurements induced by turbulence. Different reflectometry techniques: AM, Pulse Radar, FM (FFT method), are able to measure the density profile in a turbulent ITER size plasma with acceptable accuracy. Plasma rotation together with non-perpendicular reflection that cause no additional errors in the results of the AM reflectometry measurements, lead to systematic errors for the FM reflectometer. Those errors could be eliminated by fast frequency sweep (frequency sweep times shorter than the characteristic rotation times).

A preliminary study of the effects of increased turbulence level and antenna beam size on the final performance of the reflectometer has been performed for the reference case of an AM reflectometer. Profile reconstruction improves by more extensive average as well as for narrower beam size. Effects are more clear at the edge, where the level is higher and the beam size is more governed by the original geometry. Further studies are required

REFERENCES

[1] V. Zhuravlev, J. Sánchez, E. de la Luna. Plasma Phys and Contrl. Fusion 38 , 2331 (1966).

[2] B. Brañas , M. Hirsch, V. Zhuravlev, J. Baldzuhn et al. " Phase Runaway Effects on W7AS" Proceedings.of the III Reflectometry Workshop for Fusion Plasmas. Madrid 5-7 May 1997.

[3] G. Conway, submitted for publication to Plasma Phys and Contrl. Fusion

PROFILE INITIALIZATION FROM O-MODE REFLECTOMETRY

M. Manso, F. Silva, I. Nunes, J. Santos

Associação EURATOM/IST, CFN
Instituto Superior Técnico
1096 Lisboa Codex

INTRODUCTION

Reflectometry will be used on ITER to study plasma fluctuations and to measure the density profile in the main and divertor plasmas. The basic information for profile evaluation is the group delay τ_g due to the propagation and reflection in the plasma of ordinary (O) or extraordinary (X) waves, with variable frequencies F. With O-mode the profile inversion is simpler because the cutoff depends only on the plasma density. However, as frequencies below F_1 (~15 GHz) cannot be used, plasma densities $n_e \leq 0.28 \cdot 10^{19}$ m^{-3} are not probed in reflection and the distance information about the lowest density profile cannot be obtained. X mode waves can be reflected from zero density but the cutoff depends both on n_e and on the total magnetic field B_{tot}. Accurate information about B_{tot}, necessary to invert the profile, may be difficult to obtain namely in long plasma ITER pulses (exceeding 3000s). It is therefore important to study the reconstruction of profiles from O-mode only and to assess the errors due to the initialization model used to complement the data below the first probing frequency.

RECONSTRUCTION OF DENSITY PROFILES

Initialization process from O-mode

The density $n_{e\alpha}$ at the cutoff is determined by the incident wave frequency ($n_{e\alpha} \propto F_\alpha^2$). The distances ($d_A$) from the antenna to the different cutoff layers [$n_{e\alpha}(F_\alpha)$] are obtained from the group delay history between 0 and F_α. As τ_g is not measured below $F_1 \leq 15$ GHz, each distance d_α is obtained part from reflectometry, $I_2(F_\alpha)$, and part from the initialization model, $I_1(F_\alpha)$, according to:

$$d_\alpha(n_{e\alpha}) = \frac{c}{2\pi}\int_0^{F_1}\frac{\tau_g(f)}{\sqrt{F_\alpha^2}}df + \frac{c}{2\pi}\int_{F_1}^{F_\alpha}\frac{\tau_g(f)}{\sqrt{F_\alpha^2 - f^2}}df = I_1(F_\alpha) + I_2(F_\alpha)$$

In Figure 1 the different steps of the profile reconstruction are shown. The assumed density profile $n_e(r)$ and the corresponding (O-mode) group delay are depicted in Figure 1a and Figure 1b, respectively. As illustrated in Figure 1c the contribution from reflectometry $I_2(F)$ increases with frequency whereas $I_1(F)$ decreases, showing that the importance (and the errors) of the

Diagnostics for Experimental Thermonuclear Fusion Reactors 2
Edited by Stott *et al.*, Plenum Press, New York, 1998

initialisation decreases to the plasma interior. In a second step we assume that the τ_g curve below F_1 was incorrectly estimated (τ'_g in Figure 1b), resulting the contribution $I'_1(F)$ represented in Figure 1c. The profile $n'_e(r)$ reconstructed from $I'_1(F)+I_2(F)$ is represented in Figure 1a. The errors in the radial position, δr, can be evaluated either from the exact value or from the approximate quantity[1], $\delta' r$, as follows:

$$\delta r_\alpha(F_\alpha)=\frac{c}{2\pi}\int_0^{F_1}\frac{\delta\tau_g(f)}{\sqrt{F_\alpha^2-f^2}}df\ ;\ \delta' r_\alpha(F_\alpha)\leq\frac{c}{\pi}\left[\int_0^{F_1}\delta\tau_g^2\,df\right]^{1/2}\cdot\left[\frac{1}{2F}\ln\frac{F_\alpha+F_1}{F_\alpha-F_1}\right]^{1/2}$$

where $\delta\tau_g=\tau_g-\tau'_g$. The second term of $\delta' r$ is the decaying factor with frequency. The first term shows that the errors of initialization depends essentially on a quantity $\delta' A_1=\left[\int_0^{F_1}\delta\tau_g^2\,df\right]^{1/2}$ related to the deviation between the value of the area A_1 (under the τ_g curve below F_1, as seen in Figure 1d) and the estimated value A'_1. Therefore, in order to initialize the profile the shape of the plasma below $n_{e1}(F_1)$, does not need to be known but it is crucial to have a good estimation of A_1.

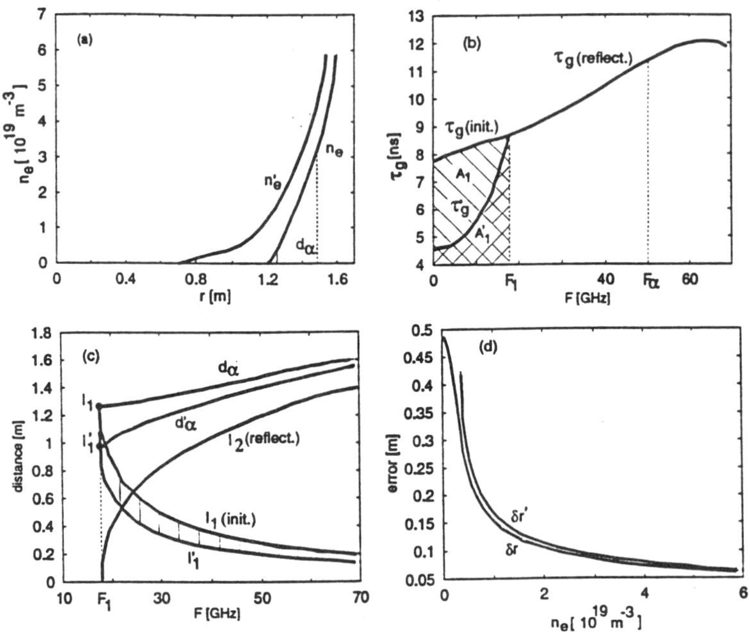

Figure 1. Profile reconstruction from O-mode reflectometry: (a) assumed profile $n_e(r)$ and reconstructed profile $n'_e(r)$; (b) group delay curve τ_g and incorrect τ'_g curve simulating the errors due to initialisation; (c) contributions to the distances d_α from the initialisation I_1 and from reflectometry I_2; (d) exact value δr of radial errors and approximate value $\delta' r$.

Initialization model A model has been developed[2] to estimate A_1 based on the measurements performed at the lowest frequencies. The τ_g curve obtained from reflectometry is extrapolated down to zero frequency by assuming a shape of the curve below the first probing frequency F_1 and the continuity at $F = F_1$ of the complete curve resulting from the estimated one (below F_1) and the curve measured by reflectometry (above F_1). A basic requirement for the successful application of the model is the reliable reflectometry measurement of the initial group delay curve.

Reconstruction of ITER density profiles A study was performed to simulate the reconstruction of ITER profiles from O-mode reflectometry. Reference cases of main and divertor ITER plasmas were analysed. The study for the divertor plasma is shown in a companion paper[3]. The assumed ITER main profile $n_e(r)$, and the corresponding group delay curve are depicted in Figure 2a and Figure 2b, respectively. The profile reconstructed using the initialisation model is represented in Figure 2a but cannot be seen because the errors are negligible (in the sub-millimetre range). This results from two facts: the area A_1 is small and it is well estimated with the model (as seen in Figure 2c). For the divertor plasma it was also found[3] that the errors are also very small (typically 1-3 mm). The numerical studies, however, did not include errors due to perturbations of τ_g that are present in the experiments caused by parasitic reflections and plasma turbulence. This is illustrated in the following sections with FMCW reflectometry data from ASDEX Upgrade.

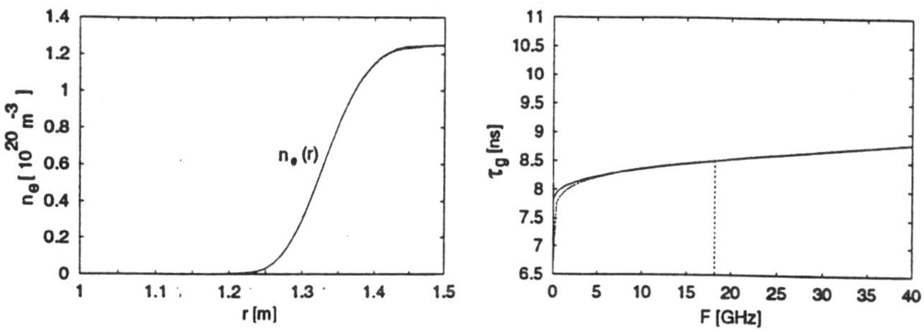

Figure 2. Simulation study for ITER main plasma: (a) assumed $n_e(r)$ and reconstructed profile (this cannot be seen because it coincides with $n_e(r)$); (b) τ_g curves corresponding to the assumed and reconstructed profiles.

Profile reconstruction from O-mode FMCW reflectometry in ASDEX Upgrade

The results presented here were obtained in shot #9352 with swept measurements performed in 100 μs with an interval of 20 μs. The group delay curves in Figure 3a were measured before an

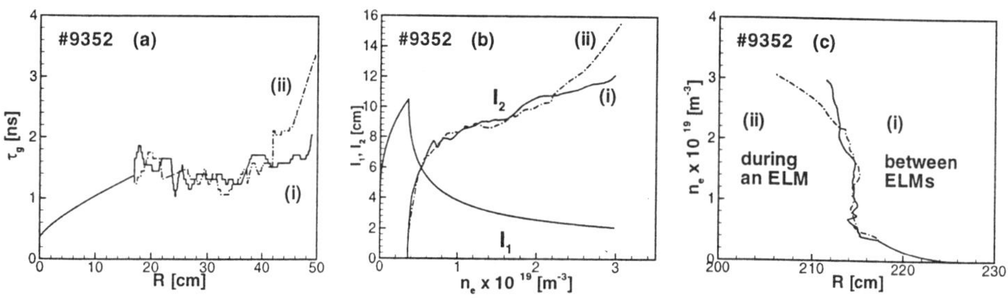

Figure 3. Reflectometry measurements performed in ASDEX Upgrade shot #9352: (i) before an ELM and (ii) during the ELM: (a) group delay curves; (b) contributions I_1 from the initialisation model and I_2 from reflectometry; (c) reconstructed density profiles showing that the peaked profile observed before an ELM flattens during the ELM.

ELM (i) and during the ELM (ii). The group delays were evaluated from the time-frequency distribution of the power of the reflected signals using a sliding Fast Fourier Transform[4]. The contributions to the distance from the initialization model, I_1, and from reflectometry, I_2, are depicted in Figure 3b. The $I_2(F)$ curves indicate that the profiles should differ only for $n_e \geq 0.28 \cdot 10^{19}\ \mathrm{m}^{-3}$ which is confirmed by the reconstructed profiles (Figure 3c). It also seen that the peaked profile observed before an ELM flattens close and inside the separatrix (R_α ~216 cm) during the ELM.

Effect of plasma perturbations It is very important not only for initialization but also for profile evaluation to average the perturbations of the τ_g curve caused by plasma fluctuations. One possibility is to use several consecutive measurements and average them in time. This is equivalent to increase the signal to noise ratio and therefore it improves the precision of the inverted profiles. We present two examples: the first one refers to the L phase of #9352 where the level of fluctuations at the edge (Low field side) is very high (δn/n: 50 – 100 %). Eight group delay curves were obtained from consecutive single sweeps (Figure 4a). Although some of the curves are strongly disturbed a simple average of the eight curves gives a smooth characteristic and an average profile (over 960 μs) is obtained (Figure 4b). The second example refers to the ELMy phase of the H mode regime in the same shot. In this case the level of perturbations is much lower, as seen in the two τ_g curves (Figure 4b) obtained from consecutive sweeps between ELMs. (Note that the vertical scales in (a) and (b) are different).

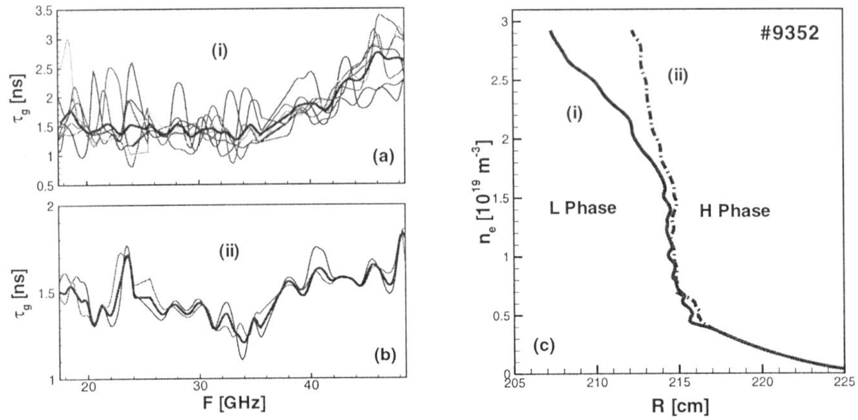

Figure 4. Reflectometry measurements performed in ASDEX Upgrade shot #9352: (i) at the L phase and (ii) at the H-mode phase. (a) group delay curves from eight consecutive sweeps and time average curve; (b) group delay curves from two consecutive sweeps between ELMs and τ_g curve resulting from the cross power spectra at each window of analysis; (c) reconstructed profiles showing the peaking from L to H-mode phase.

Here we evaluate the cross power spectra in each window of analysis (we use a sliding Fast Fourier Transform to extract τ_g[4]) which permit to enhance the common features of both spectra and reduce the group delay perturbations. In Figure 4c it can be seen the corresponding density profile and the peaking of the profile from L to H- mode phase.

Combined O and X-mode probing

X-mode is very useful to complement the O-mode data at the lowest densities. Profile inversion from X mode requires accurate information about the magnetic field as mentioned before. Another problem is some uncertainty about the first reflected signal (at n_e = 0) and the corresponding probing frequency F_0. This can be overcome by combining the O and X-mode data, as shown in the example of Figure 5 obtained at shot # 9291. In this case two values of F_0 (44.3 and 44.7 GHz) were considered and two profiles were reconstructed from X-mode between 0 and

$0.4 \cdot 10^{19}\ m^{-3}$. We calculate then the τ_g curves that would result if O-waves were propagated in those profiles. The curve for $F_0 = 44.3$ GHz (dashed-dot line in Figure 5a) was selected because it matched better to the τ_g curve obtained from O-mode measurements (solid line in Figure 5a). The profiles evaluated from O plus X-mode and from O-mode only, using the initialization model, are depicted in Figure 5b. They present only very small deviations for $n_{e1} \geq 0.4 \cdot 10^{19}\ m^{-3}$ because the area A_1 under the group delay curve below $F_1 \approx 17$ GHz was estimated with a good approximation (Figure 5a), although the exact curve could not be obtained from the model and significant profile deviations are observed for $n_e \leq n_{e1}$.

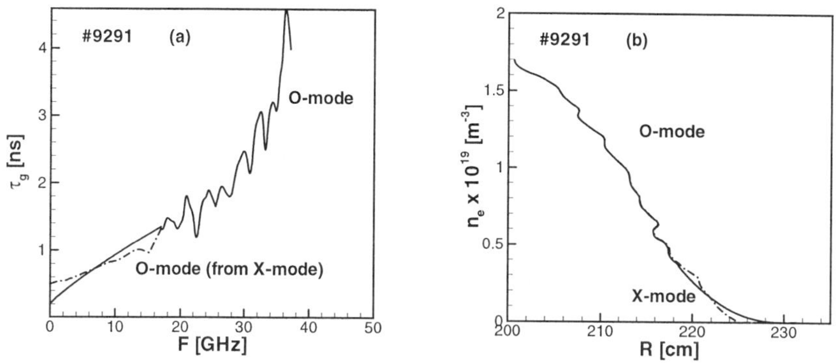

Figure 5. Reflectometry measurements performed in ASDEX Upgrade shot # 9291 with O and X-mode probing. (a) group delay curves from O-mode and from calculating the group delay of O-mode waves (dashed-dot line) reflected in the X-mode edge profile; (b) reconstructed profiles from O-mode and combined O and X-modes.

DISCUSSION

Density profiles can be reconstructed from O-mode only using an initialization model based on the group delay curve measured at the lowest frequencies. The errors of the evaluated profiles are due to deviations of the area A_1 (under the group delay curve below the first probing frequency) obtained from the initialization model from its correct value. For ITER reference profiles it is shown that the errors should be very low (sub-millimeter to millimeters). However, the numerical study did not take into account errors due to perturbations of the group delay curve (mainly caused by plasma fluctuations). Examples from FMCW reflectometry on ASDEX Upgrade are presented to illustrate some of the techniques that can be used to improve the accuracy of profile reconstruction. The errors due to initialization should be further investigated in present reflectometry diagnostics using combined O and X mode information and data from other edge density diagnostics.

1 M. Manso, et al. to be submitted to Plasma Physics and Controlled Nuclear Fusion
2 P. Varela, et al. Review of Scientific Instruments, 66 (10) October 1995
3 M. Manso et al. in Proceedings of this conference
4 J. Santos et al. III Work. Microwave Reflect. For Fusion Plasma Diagnostics Madrid (1997)

TRANSMISSION LINE DESIGN FOR BROADBAND MICROWAVE DIAGNOSTIC SYSTEMS

D.Wagner,[1] W.Kasparek,[1] M.Thumm,[2] G.Gantenbein,[1] and H.Zohm[1]

[1]Universität Stuttgart, Institut für Plasmaforschung, Pfaffenwaldring 31, D-70569 Stuttgart, Germany
[2]Forschungszentrum Karlsruhe, Association EURATOM-FZK, ITP and Universität Karlsruhe, Institut für Höchstfrequenztechnik und Elektronik, P.O.Box 3640, D-76021 Karlsruhe, Germany

INTRODUCTION

Up to now mostly smooth-wall waveguides are used for low power applications in fusion plasma diagnostic systems. However ohmic losses and mode conversion in mitre bends are a serious problem for precise amplitude and phase measurements e.g. for ECE and reflectometry. For microwave diagnostic systems on future fusion experiments like ITER with very long waveguide runs (typically several tens of meters), oversized corrugated waveguides propagating the HE_{11} mode which exhibits extremely low losses are a good candidate. Such type of transmission line is widely used for high-power applications, especially for electron cyclotron heating of plasmas, however only at single frequencies.[1] The theory is presented in several publications, e.g. Ref.[2,3] and measurements at single frequencies have confirmed the low attenuation of the HE_{11} mode.[4,5] This paper assesses the possibilities and problems which arise if these transmission lines are used in broadband (low-power) diagnostic systems. Besides the low ohmic attenuation such waveguides offer several additional advantages in comparison to smooth-wall oversized waveguides like the possibility to build compact quasi-optical mitre bends. Radiated from the open-ended waveguide, the HE_{11} mode couples almost perfectly to the Gaussian free-space mode and therefore simplifies the antenna design. In such waveguides high order spurious modes can be suppressed by including gaps into the transmission line which cause only very low losses for the main HE_{11} mode. At the same time these gaps act as DC breaks and movement compensators. Due to the almost linear polarization of the HE_{11} mode, broadband Brewster angle windows can be applied. Two combined Brewster angle windows can be used as an effective beam splitter.

The theoretical bandwidth of the corrugated HE_{11} waveguide, depending on the waveguide diameter, corrugation depth and period, is evaluated numerically. The frequency dependence of the losses in quasi-optical mitre bends is calculated and compared with measurements.

To validate the theoretical results a test transmission line with a diameter of 87mm and a length of 13.9m, including 3 mitre bends was used. An effective overall length of 27.8m was obtained by inserting a plane mirror at the end of the line and the measurement of the reflected signal.

BROADBAND EXCITATION OF THE HE_{11} MODE IN A CORRUGATED WAVEGUIDE

In diagnostic systems, low-power sources and receivers usually operate in fundamental rectangular waveguides carrying the TE_{10} mode, which has to be matched to the HE_{11} mode in the oversized circular corrugated waveguide. Several solutions exist for this task. Waveguide mode converters,[7] based on a variation of the wall impedance or on the generation of mode mixtures,[8] reach very high efficiencies, but are rather narrow-band. Other adiabatic transitions use dielectric rod guides, but no broad data base on the performance is available. A versatile solution is a confocal two-mirror system which is inherently broadband.[6] The source or receiver to be matched is fed into a circular horn with a small flare angle or a waveguide taper to achieve a sufficiently plane phase front at the aperture. If a scalar horn is used, the bandwidth of the arrangement is usually determined by the horn. In case of simple conical horns or tapers, the lower frequency limit is determined by the beam truncation due to the finite size of the mirrors, the upper limit is determined by the onset of mode conversion in the horn (i.e. deviation of the plane phase front at the horn aperture $\geq \lambda/10$, λ being the wavelength of the radiation). However, some additional loss due to the non-perfect matching between the TE_{11} mode of the transmitter and receiver antennas and the HE_{11} mode of the corrugated waveguide (about 0.5 dB, varying due to the fact, that the power distribution in the waveguide is a function of the frequency for a given corrugation[3,9]) must be taken into account.

By combining two or more such arrangements, the summation of several sources or detectors to one broadband line, polarization diplexers or directional couplers can be built.[6] For the experiments described here, a bidirectional coupler with an alumina beam splitter (see Figure 1 for a principle sketch) was used to measure reflections from the waveguide under test and to measure the transmission properties in case of no reflections by terminating the line with a mirror. For pure transmission measurements, only the input channel was used for the coupling to the waveguide. The device allowed the broadband excitation over the frequency range 45 GHz $\leq f \leq$ 175 GHz with losses in fair agreement with theory.

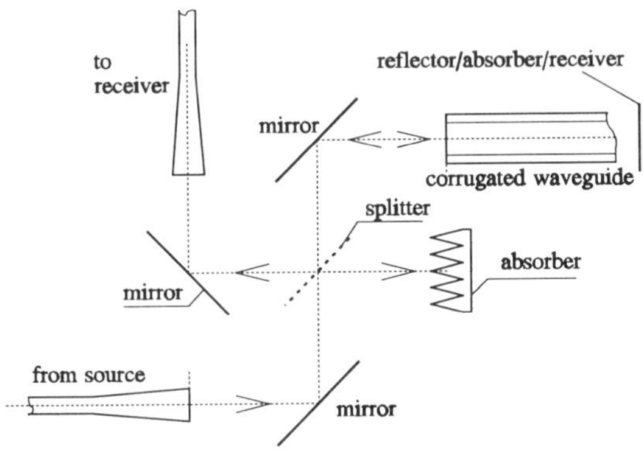

Figure 1. Sketch of the setup for transmission and reflection measurements.

TRANSMISSION CHARACTERISTICS OF THE CORRUGATED WAVEGUIDE

First experiments were performed with a highly oversized corrugated waveguide system. The inner diameter (D) of the corrugated waveguide is 87mm. The corrugation profile is almost rectangular (with rounded corners) with a depth of d=1.0mm and a period of p=0.89mm. The eigenmode of the corrugated waveguide can be expressed as a sum of eigenmodes of the smooth wall waveguide.[9] The low loss HE_{11} mode corresponds to a mixture of mainly 84.6% TE_{11} and 14.5% TM_{11} with a phase difference of 180° (balanced hybrid mode). Figure 2 shows the calculated eigenmode mixture for the above described waveguide as a function of frequency. The calculated frequency dependence of the ohmic losses is plotted in Figure 3. A comparison between Figures 2 and 3 shows that the ohmic losses increase where this balanced hybrid mode condition between TE_{11} and TM_{11} is not fulfilled.

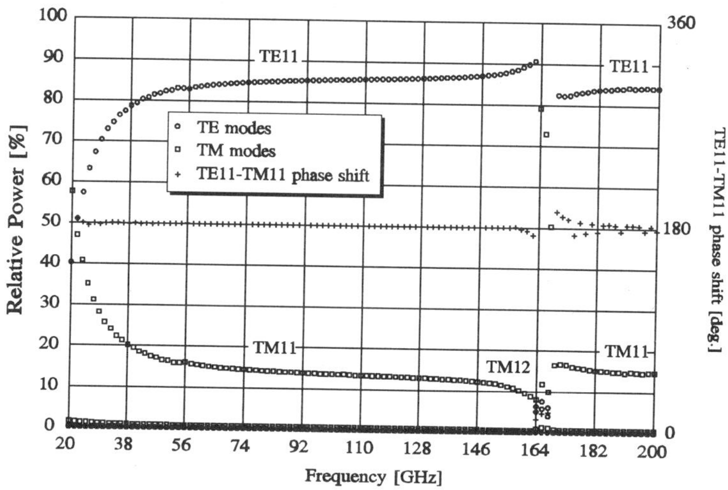

Figure 2. Eigenmode mixture for a corrugated waveguide with D=87mm, d=1.0mm, p=0.89mm.

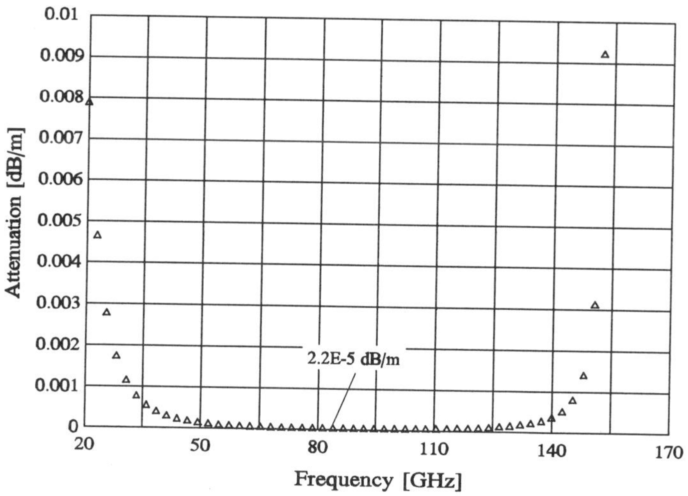

Figure 3. Ohmic attenuation for a corrugated waveguide with D=87mm, d=1.0mm, p=0.89mm.

The calculated mode conversion losses in a quasi optical mitre bend propagating the HE_{11} mode with a diameter of 87mm are plotted in Figure 4.

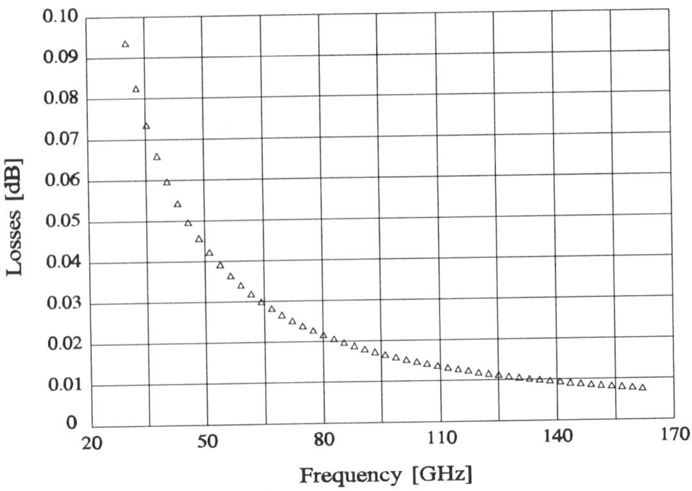

Figure 4. Mode conversion losses in a quasi-optical HE_{11} mitre bend with D=87mm.

The results of transmission measurements are shown in Fig. 5. In the frequency range 45 to 160 GHz , a negligible loss was measured in average by comparing the reflections from the end of the line with signals from a reflector inserted near the input. Standing wave effects (the reflected power from the receiver horn was about 1 - 3 %) were eliminated by evaluating Fourier transforms of frequency sweeps. The variation of the transmitted power is probably not due to absorption, but is determined by the phasing of additional spurious modes excited at the waveguide input as was also seen in other experiments.[5] The main part of the losses is due to mode conversion in the 3 mitre bends.

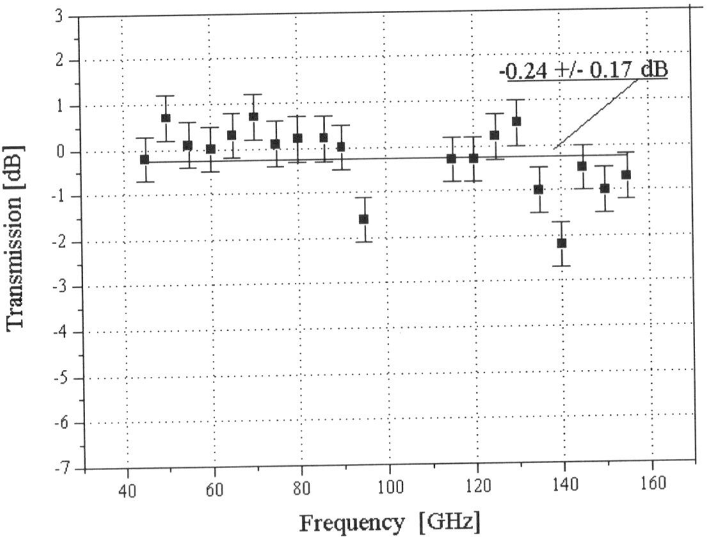

Figure 5. Measured transmitted power through 27m of corrugated waveguide including 3 mitre bends as a function of frequency.

When the wavelength λ becomes identical to twice the period of the corrugated waveguide, Bragg reflections from the waveguide are expected according to the Bragg condition

$$k_{Bragg} = \frac{2\pi}{p} = k_{HE11} + k_{1n} \quad , \tag{1}$$

where k_{HE11} and k_{1n} are the propagation constants of the forward travelling HE_{11} mode and reflected TE_{1n} or TM_{1n} modes, respectively. The onset for these reflections is expected at a frequency of $f \geq c/2p = 168$ GHz. To investigate the behaviour near the Bragg limit, transmission and reflection measurements with high frequency resolution were performed. In the transmission characteristic, which is shown in Figure 6, a series of sharp drops caused by Bragg reflections from the incoming HE_{11} mode to higher-order TE_{1n} and TM_{1n} can be seen as expected, which sets an upper limit for the use of the waveguide. Additionally, strong attenuation is seen at frequencies slightly below 168 GHz which result from the excitation of slow surface waves when the effective groove width is about $\lambda/2$.[10] This effect is also seen in the calculated eigenmode mixtures in Figure 2 where the main surface wave EH_{11} is characterized by a high TM_{11} mode content and 0° phase difference between TE_{11} and TM_{11}. Besides the stronger ohmic losses of such waves they are mainly filtered in the quasi-optical coupling section. This is due to the fact that most part of the power radiated from the open ended waveguide is concentrated in sidelobes.

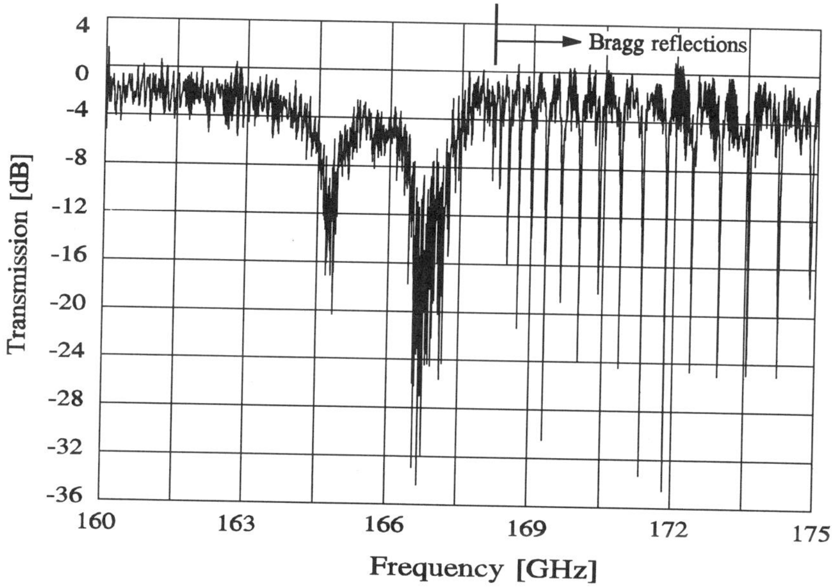

Figure 6. Measured transmission behaviour of the corrugated waveguide near the Bragg limit.

Calculations to study this behaviour in detail have been performed. The result from a scattering matrix calculation taking into account a straight corrugated waveguide section of 28 m length is shown in Figure 7. In this case the HE_{11} is again represented by a mixture of TE_{11} (84.6%) and TM_{11} (14.5%). Although the frequency resolution is lower than in the experiment, a qualitative agreement is already achieved.

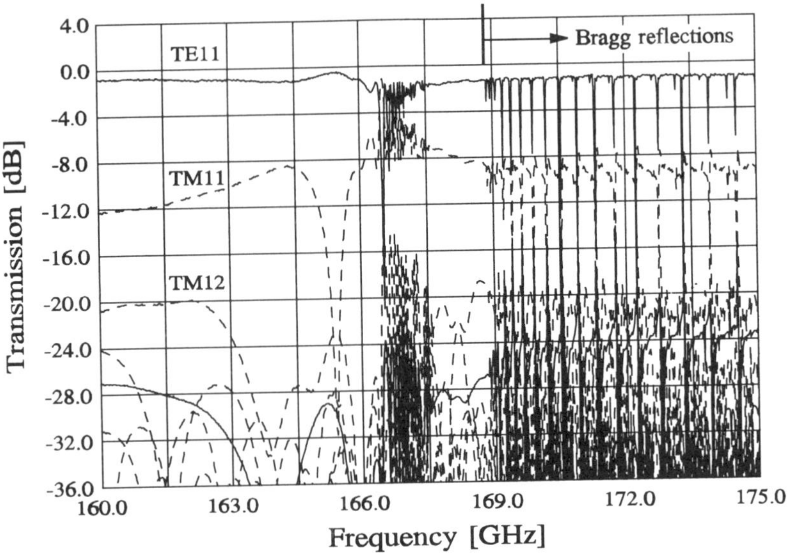

Figure 7. Calculated transmission through 28m corrugated waveguide in the Bragg region.

CHARACTERISTICS OF COMPLEX TRANSMISSION SYSTEMS

In long and complex transmission lines, which include several mitre bends, the excitation of spurious modes at the bends as well as misalignments cause a rippled frequency dependence. This holds especially for moderately oversized waveguides as the expression for mode conversion losses in mitre bends shows:[11]

$$P_L = 2.4 \cdot \left(\frac{\lambda}{D}\right)^{\frac{3}{2}} \ [dB] \qquad (2)$$

(λ is the wavelength and D is the inner waveguide diameter), where 75% of the power is transmitted and 25% is reflected. The reflected power consists of a spectrum of high-order modes [12] which are characterized by high ohmic losses and can lead to standing waves and therefore resonances in the transmission line. To estimate the effects calculations have been performed using a scattering matrix code where the mitre bends are represented by large diameter steps, using the fact that the diffraction losses of mitre elbows are very similar to the losses of waveguide gaps. The broadband performance can be improved by including waveguide gaps at each bend. Here one takes advantage of the strong dependence of the gap losses on the Bessel zero of each mode.[13] This means that fundamental modes are almost not affected by the gaps, whereas higher-order modes suffer strong damping. Figure 8 compares the calculated transmission characteristics of a corrugated line with D=63.5mm (100m, 10 miter bends) without and with mode filtering gaps with length L=D.

To demonstrate the mode filtering by gaps, corresponding measurements with moderately oversized smooth waveguides (e.g. D=27.79 mm) propagating the TE_{11} mode were carried out. In this experiment the TE_{11} mode was excited in a monomode waveguide followed by a nonlinear uptaper. The same arrangement was used at the end of the section to measure the transmitted signal. The length of the transmission line was 2 m including two miter bends with plane mirrors. Gaps with a length of 28 mm were introduced in front of each mitre bend. Figure 9 shows a comparison between the power of the transmitted signals without and with gaps relative to a straight waveguide section with the same length. The introduction of mode filtering gaps leads to a smoothening of the frequency characteristics as predicted by the calculations, however at the expense of a higher total loss. Due to space restrictions on future experiments like ITER such moderately oversized waveguides and antennas will have to be used. In this case additional waveguide transitions like the one described in Figure 1 and gradual bends must be applied.[19] However, for a highly oversized corrugated waveguide the gap losses are almost negligible. Further theoretical and experimental work to get rules to optimize the behaviour of such complex transmission lines is planned.

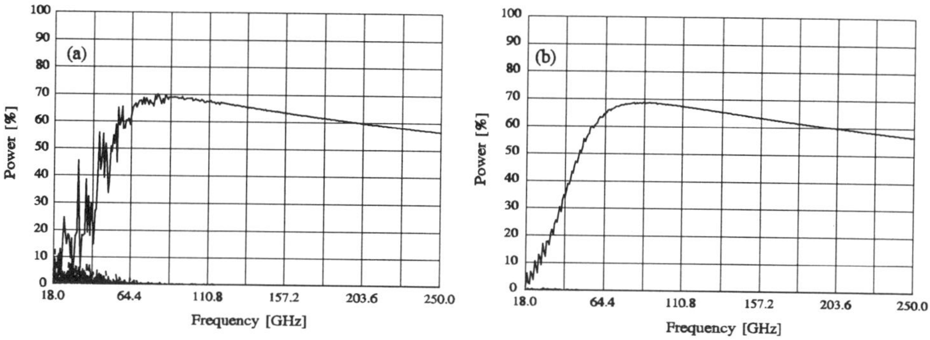

Figure 8. Estimated transmitted power through a corrugated transmission line of 100m length (D=63.5mm) including 10 mitre bends without (a) and with (b) gaps.

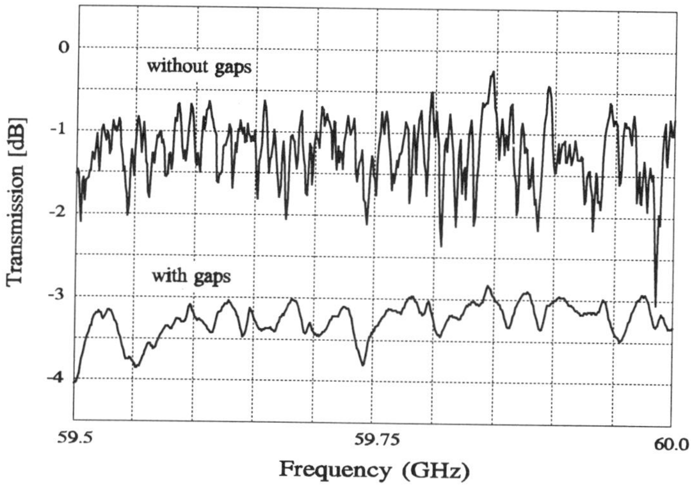

Figure 9. Comparison of transmission characteristics of a TE_{11} line without and with gaps in the waveguide.

CHARACTERISTICS OF VACUUM BARRIER WINDOWS IN THE WAVEGUIDE

In oversized transmission lines, mostly vacuum barrier windows which consist of one or two discs oriented perpendicular to the waveguide axis are used. This leads to periodically varying transmission and reflection, due to multiple interference of the waves reflected from the window surfaces. Especially for the case of two windows in the line (which might be necessary for fusion reactors like ITER), the resulting transmission characteristics are strongly modulated.[14] Several techniques to reduce the reflection and thus to provide broadband transmission are in use,[15,16,17] but these are mostly limited to less than one octave. Alternatively or additionally it is recommendable to incline the discs against the waveguide axis by an angle of at least $\alpha \geq \lambda/D$. In this case, the discs are decoupled and the transmission characteristics is determined by the sum of the characteristics of the individual discs; reflections that occur in higher-order modes easily can be filtered out by gaps and thus do not influence the transmission characteristics any more. Additional losses due to the walk-off effect in the discs remain negligible for sufficiently oversized waveguides.[18]

For the HE_{11} mode with a single linear polarization, the window inclined at the Brewster angle given by $\tan\alpha_B = n$, where n is the refractive index, is often the best solution. For a non-dispersive window material, there is in principle no bandwidth limitation. The mode conversion loss should be similar to the loss of a gap in the waveguide corresponding to the window thickness and therefore tolerable, if the window is designed properly, i.e. including an offset of the connecting waveguides according to the propagation path in the material:

$$\Delta x = d(\sin\alpha - \cos\alpha \cdot \tan\beta) \tag{3}$$

where d is the thickness of the plate, α the angle of incidence and the angle $\beta = \arcsin(\sin\alpha/n)$ is given by the law of refraction (see also Figure 10).

To check the behaviour of Brewster angle windows in corrugated waveguides, two samples with 10 mm thick PVC-plates were built and installed in the waveguide with D=87mm. (Note: PVC is an inexpensive plastic material with a refractive index similar to quartz and was readily available for these tests, however, it has an absorption of about 1 dB/cm at 60 GHz, i.e. absolute values can be extrapolated to quartz windows only by a correction for this absorption). Reflection measurements were performed with the bi-directional coupler as described above; and referenced to a mirror at the end of the line. For the transmission measurements, a receiver horn at the waveguide output was used, and the results were related to the situation without window. Note, that the measurements comprehend mainly the contribution of the HE_{11} mode and reflect the situation that the coupling to the oversized waveguide is performed via a quasi-optical section or that the waveguide includes absorbing gaps to filter higher-order modes.

The measured frequency characteristics of two Brewster windows are shown in Figure 11. In the correct polarisation, the transmission is flat over a broad frequency range. Because the length of the waveguide behind the windows was equal to several beat wavelengths (about 3 λ_B at 60 GHz) between HE_{11} and TM_{02}/HE_{21} which mainly could be excited by the window, frequency scans would react sensitively to the presence of these modes. Neither the frequency spectrum and its Fourier transformation nor measurements of the cross-polarization (peak-value -19 dB) show a significant level of wrong modes. The reflections in the HE_{11} mode are down by at least -45 dB.

The same windows also were tested with a cross-polarized HE_{11} mode. The transmission of the single window shows the typical shape given by the Airy function for an effective thickness of $d \cdot \cos\beta$.[14] In case of two windows, the transmission is approximately given by the product of two single window transmissions, as the inclined discs are fairly decoupled, as shown in Figure 12. The remaining coupling of the discs which occurs via high-order modes can be strongly reduced by a gap between the windows. The average reflections are of course high, but occur in very high-order modes, which are filtered out in the quasi-optical coupler. In the experiments, which are sensitive mainly for the HE_{11} mode, only a level below -30 dB is measured for one or two windows.

Such Brewster windows can be used to design polarization splitting arrangements, where the parallel component is transmitted, the cross-polarized component is reflected from the window surface forming a mitre bend. This offers the possibility of polarization diplexing with compact and rugged devices and a saving of waveguides, and is therefore in discussion for ITER diagnostic lines. To get high polarization discrimination over a broad band, a wire grid on the window surface is necessary.

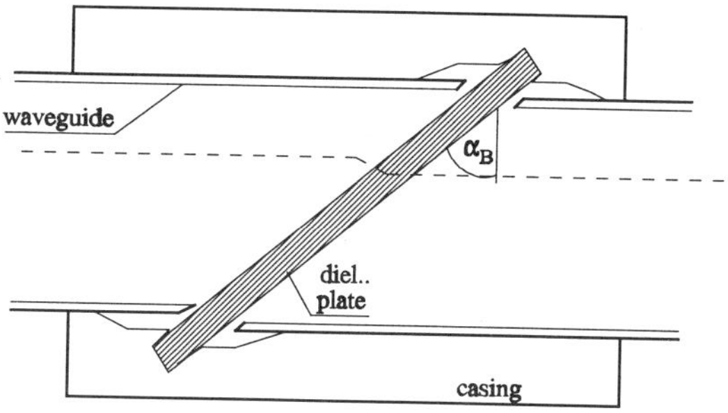

Figure 10. Schematic view of an in-waveguide Brewster angle window.

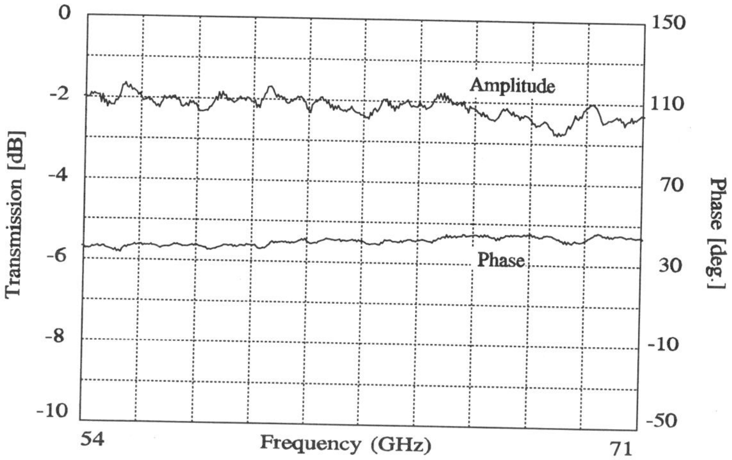

Figure 11. Transmission characteristics of two Brewster angle windows built into the corrugated waveguide with D=87mm (Note, that the average damping comes from the absorption of the material used for the tests).

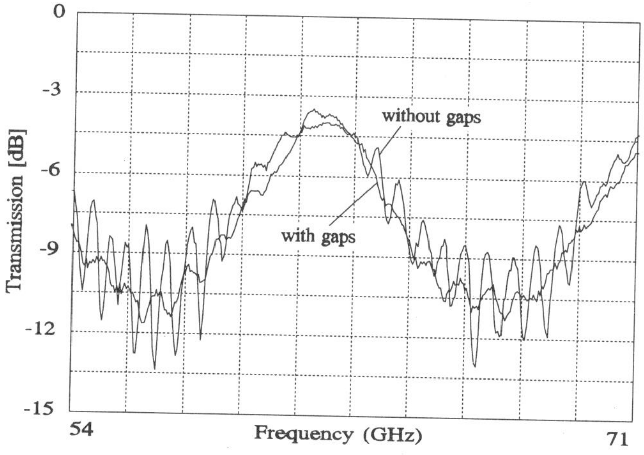

Figure 12. Transmission of two Brewster angle windows with polarisation perpendicular to the plane of incidence.

CONCLUSIONS

The calculations and experiments presented here illuminate the characteristics of corrugated waveguide transmission lines. Broadband excitation from fundamental waveguide and low-loss transmission over two octaves has been shown and the behaviour at the Bragg-limit has been investigated. The possibility to optimize transmission characteristics by mode-filtering gaps is demonstrated. For vacuum barriers, Brewster windows are a good choice giving a smooth broadband transmission with low loss. However, where the available space is not sufficient for highly oversized HE_{11} waveguides (e.g. divertor and high-field-side antennas on ITER), transitions to moderately oversized waveguides might be necessary. Such complex transmission lines with several changes in cross section and direction still need to be investigated in more detail. Since a pure theoretical description is very difficult in this case, experiments on a future prototype transmission line would be very helpful.

REFERENCES

1. M. Thumm and W. Kasparek, *Fusion Eng. and Design* 26:291(1995).
2. J.L. Doane, 1985, in: *Infrared and Millimeter Waves* 13, ed. K. Button, Academic Press, New York (1985).
3. J.P. Crenn and C. Charollais, *Int. J. of Infrared and Millimeter Waves* 17:1475 (1996).
4. A. Cavallo et al., Conf. Digest 18th Int. Conf. on Infrared and Millimeter Waves, Honolulu, Hawaii, in: *SPIE Proc.* 1039:113 (1988).
5. K. Ohkubo et al., *Proc. 18th Symp. on Fusion Technology 1994 (SOFT94)*, Elsevier Science Publishers B.V., Karlsruhe, 513 (1995).
6. J.C.G. Lesurf, *Millimetre Wave Optics, Devices and Systems*, Adam Hilger, ed., Bristol (1990).
7. M. Thumm, A. Jacobs and M. Sorolla Ayza, *IEEE Trans. Microwave Theory Tech.* 39:301 (1991).
8. G.G. Denisov et al., to appear in: *Proceedings of the EC-10 Workshop*, Netherlands (1997).
9. D. Wagner, J. Pretterebner, M. Thumm, *Proceedings of the EC-8 Workshop*, IPP-Report III/186, Garching, 575 (1992).
10. J.L. Doane, *Proc. 19th Int. Conf. of Infrared and Millimeter Waves*, Sendai, 232 (1994).
11. J.L. Doane and C.P. Moeller, *Int. J. Electronics* 77:489 (1994).
12. D. Wagner et al., *Proc. 15th Int. Conf. on Infrared and Millimeter Waves*, Orlando, SPIE Proc. 1514:437 (1990).
13. D. Wagner et al., *Int. J. of Infrared and Millimeter Waves* 17:1071 (1996).
14. M. Born and E. Wolf, *Fundamentals of Optics*, Pergamon Press, Oxford (1980).
15. J.Y.L. Ma and L.C. Robinson, *Optica Acta* 30:1685 (1983).
16. H.U. Nickel, H. Massler, and M. Thumm, *Digest of the 18th Int. Conf. on Infrared and Millimeter Waves*, Essex, SPIE Proc., 2104:172 (1993).
17. T.P. Hughes, S.D. Richards, and D.V. Bartlett, *Int. J. of Infrared and Millimeter Waves* 15:61 (1994).
18. H.G. Danielmeyer, *IEEE J. Quant. Electronics* 6:101 (1970).
19. G.Vayakis et al., Overview of the ITER Reflectometry diagnostic systems, this conference.

PHYSICS ASPECTS OF ECE T_e MEASUREMENTS IN ITER

D V Bartlett & H Bindslev

JET Joint Undertaking, Abingdon Oxon, OX14 3EA, UK
and members of the European Home Team for Microwave Diagnostics *

ABSTRACT

This paper summarises the study which has been made of the physics of ECE for ITER. Two ECE simulation codes have been used to obtain quantitative information about the spatial localisation and accessibility limits of T_e measurements. The calculations include the effects of spatial averaging due to the antenna pattern. Results are presented for three different scenarios: the core of full performance plasmas, the edge region and the startup plasma. It is shown that for thermal plasmas it will generally be possible to make core T_e measurements with spatial resolution at or close to the ITER target (~0.1 m) using the second harmonic extraordinary mode, although access to the plasma centre will be limited at high T_e. Access is better for the first harmonic ordinary mode, but the resolution is poor. Edge T_e measurements using ECE will have a spatial resolution >50 mm, much worse than the ITER target (5 mm), and it will generally not be possible to measure low values of T_e (<1 keV) close to the separatrix. It is shown that it will be possible to make core T_e measurements during much of the discharge startup if there is a suitable sightline.

INTRODUCTION

This paper is concerned with the physics aspects of obtaining T_e from ECE measurements on Maxwellian plasmas in ITER. Companion papers at this workshop and the previous Varenna workshop on ITER diagnostics discuss the instrumentation [1], and the design of the antennas [2]. There is close interaction between the physics issues and the antenna design, since the spatial resolution is determined by the combined contributions of the antenna pattern and spectral broadening. The antenna pattern should not significantly degrade the resolution limits set by spectral broadening, but trying to achieve too narrow an antenna pattern would give rise to unnecessary practical problems, since large apertures in the port plug would be needed [2]. The physics issues also determine the optimum choice of harmonic and polarisation in any measurement scenario, and this also influences the

* see list in Acknowledgements

antenna design. There is less interaction between the physics issues and the instrument design, since the spectral resolution requirements will not be difficult to meet for the instrumentation which is foreseen.

There were papers on ECE physics at the previous Varenna workshop on ITER diagnostics [3,4]. In [3], a typical simulated ITER ECE spectrum was shown and the physical origin of the various characteristic features of that spectrum were explained. An analysis of the spatial localisation of the emission over the whole optically thick region of the spectrum was shown. The parts of the spectrum suitable for T_e measurements were identified and attention was drawn to the effects of the relativistically enhanced harmonic overlap. The conclusions of that paper were that the ITER target spatial resolution for core T_e measurements could be met using the second harmonic extraordinary mode, but that access to the core would not be possible at high T_e. A preliminary analysis of the resolution of edge T_e measurements suggested that ECE would fall short of the ITER target (5 mm) by an order of magnitude.

This paper extends these earlier results by presenting a detailed analysis for three measurement scenarios: the core of high performance plasmas, the plasma edge, and the startup plasma. In each case, the importance of all the various physical effects has been examined and taken into account where necessary. Parameter scans have been made to determine the sensitivity of the results to changes in T_e and n_e.

THE ANALYSIS TECHNIQUES

We describe the limitations on the localisation of ECE T_e measurements under three separate but related headings. The first two of these are fundamental limits, while the third can be treated as a systematic effect when measurements are analysed:

1) **Access.** A region of plasma which does not give rise to a unique contribution to some part of the ECE spectrum is inaccessible for T_e measurements. Lack of access is usually due to harmonic overlap, "shine through" of core plasma emission preventing T_e measurements close to the separatrix, or low optical depth in the startup plasma.
2) **Spatial Resolution.** The measured ECE at any frequency is an average over a volume of plasma whose size and shape is determined by spectral broadening (a function of plasma parameters, frequency and direction of view) and the antenna pattern characteristics.
3) **Location.** The "centre-of-gravity" of the averaging volume of the emission at a given radiation frequency is shifted from its nominal location on the antenna axis and at the resonance radius by asymmetric spectral broadening and refraction. Provided this shift is reasonably small, it can be estimated and included in the analysis of data.

The results shown in the following sections are presented in terms of the above quantities for several different measurement scenarios and a wide range of plasma conditions. The relative importance of various factors in determining the limitations is first investigated, then detailed calculations taking into account only the significant factors are made. This avoids substantial computational overhead as well as allowing the underlying causes of the behaviour to be more easily seen. The various analysis tools which have been used for this purpose are described below.

The primary tool needed to quantify the above effects is an ECE simulation code which solves the radiation transport equation. Such codes usually integrate backwards along a ray path from the antenna, using an equilibrium which specifies profiles of T_e, n_e and the magnetic field to calculate the emission and absorption and the accumulated radiation intensity. The ray path may be straight or may include bending due to refraction. The model used to calculate emission and absorption may include or exclude various physical factors, and may assume certain approximations (such as $kT_e/mc^2 << 1$). Two different codes have been used for the simulations reported here.

For simulations which will be accurate under all the circumstances considered here, a comprehensive code with a variety of options from cold plasma to fully relativistic ray tracing has been used [5]. The weakly relativistic approximation for the refractive index and emission and absorption is adequate for the simulations reported here [5]. Because of the computational overhead of this code, it has been mainly used to determine the importance of beam bending by refraction, and to benchmark a simpler and faster code which has been run for the parameter scans.

The simpler code ("ECELS") was described at the previous Varenna workshop on ITER diagnostics [3]. It solves the radiation transport equation for a Maxwellian plasma, using a formulation of the absorption coefficient due to Bornatici [6] which is valid for perpendicular propagation, T_e up to about 25 keV, and includes the "finite density" effects. Deviation of the rays due to refraction is not included.

As explained in [3], the codes solve the radiation transport equation at each frequency by integration of the function $G(R) = T_e(R).\alpha(R).e^{-\tau(R)}$ backwards along R (the ray path from the antenna). The absorption coefficient is $\alpha(R)$, and $\tau(R)$ is the integral of α from the antenna to R. The final value of $\int G(R)$ (which occurs where $e^{-\tau(R)}$ falls to zero for the optically thick cases of interest here) is the radiation temperature, T_{RAD}, which is the measured quantity. T_{RAD} is a weighted average of T_e over the region where G(R) is non-zero. It is re-absorption within the resonance layer (the $e^{-\tau(R)}$ term) which is responsible for good localisation at high T_e, even though $\alpha(R)$ is non-zero over a wide radius range.

We characterise the localisation of the emission due to spectral broadening at each frequency by the radii at which the cumulative integral of G(R) reaches 2.5%, 50% and 97.5% of its final value. These are denoted $R_{2.5}$, R_{50} and $R_{97.5}$, and are used to define:

1) The **major radius resolution,** $\Delta R_{res} \equiv R_{2.5} - R_{97.5}$.
   This is the radius range from which 95% of the observed intensity originates.
2) The **mean emission radius,** $R_{mean} \equiv R_{50}$.
   Since the function G(R) is generally very close to being symmetric, this is a good approximation to the radial centre-of-gravity of the emission volume.
3) The **major radius shift,** $\Delta R_{shift} \equiv R_{NR} - R_{mean}$.
   R_{NR} is the non-relativistic resonance radius, ie. the solution to $\omega = m.e.B(R)/m_o$ when the rest mass, m_o, is used in the resonance condition which gives the frequency, ω, as a function of the harmonic number, m, and the magnetic field, B(R).

In addition to spectral broadening, the effect of the antenna pattern on localisation needs to be included in the calculations. The antennas described in [2] are Gaussian telescopes. In the absence of refraction, the antenna pattern is fully defined by the major radius variation of the 1/e radius, w(R), of the transverse intensity distribution, which is a function only of the wavelength, λ, and the location, R_{b0}, and size, w_0, of the beam waist in the plasma:

$$w(R) = w_0 \times (1 + (\lambda(R - R_{b0})/(\pi\, w_0^2))^2)^{1/2}.$$

We use a "composite" Gaussian beam in which λ varies with R according to the magnetic field profile and the harmonic being considered. For this composite beam, we define $D_{eff} = 2.448 \times w(R)$ as the **effective beam diameter** (since this contains 95% of the total beam power, by analogy with ΔR_{res} above).

The spatial resolution is defined by the intersection of the antenna pattern and G(R). To obtain a single number to represent the resolution we combine the transverse resolution (ie the beam diameter) and the longitudinal resolution (ie the radial resolution) in the following way, which is applicable to horizontal and near-horizontal beams. The averaging volume is represented by a vertical cross in the poloidal cross-section, centred on R_{mean} and z_{beam} of the beam axis. The height of the cross is D_{eff} and its width is ΔR_{res}. From the values of the normalised flux at the four points of the cross, the corresponding minor radii in the plasma midplane can be determined by translating along flux surfaces. Thus the transverse and longitudinal resolutions can each be expressed as equivalent minor radius resolutions, and the extrema of the four points used to give an **effective minor radius resolution**. This is illustrated in the sketches of Figure 1 which show two cases, one near the plasma centre and one near the edge. Even though ΔR_{res} is shown as decreasing towards the edge, the relationship to the flux surfaces causes D_{eff} to dominate near the centre and ΔR_{res} near the edge. This analysis is suited only to poloidally symmetric cases. To study the resolution of, for example, high order MHD modes a 2-dimensional integration would be required.

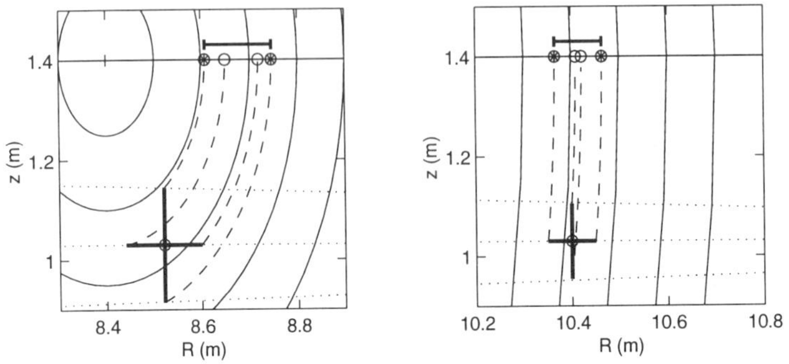

Figure 1: Sketches illustrating the technique used to characterise spatial resolution. The radiation beam is shown as dashed lines, and the height of the cross is equal to its diameter, D_{eff}. The width of the cross, which is centred on (R_{mean}, z_{beam}), is equal to ΔR_{res}. Translating the four points of the cross to the plasma midplane allows the effective minor radius resolution (horizontal bars) to be determined.

SIMULATION SCENARIOS

Equilibria and profiles derived from transport code simulations of a sawtooth free discharge have been used for the core plasma calculations [7]. The T_e profile is peaked, with $T_{eo} \approx 28$ keV, while n_e is flat at 1.3×10^{20} m^{-3}. The plasma centre is at $R \approx 8.4$ m and $z \approx 1.4$ m, with a mid-plane minor radius ≈ 3 m and elongation ≈ 1.5. The on-axis magnetic field strength is about 5.9 T. The equilibrium and profiles are shown in Figure 2.

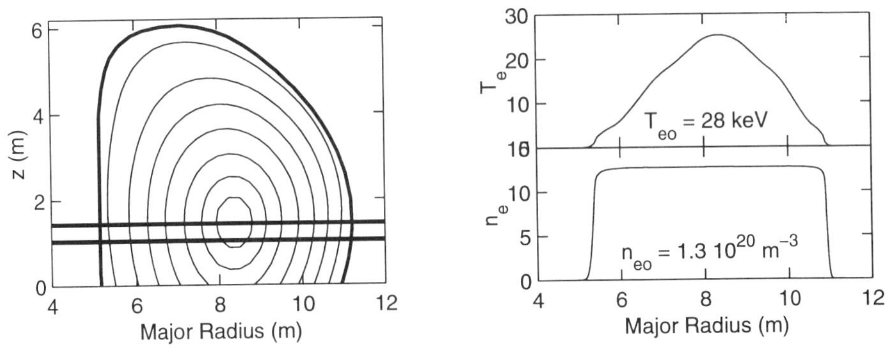

Figure 2: Equilibrium of a high performance plasma. Flux surfaces and the upper two antenna sightlines are shown on the left, and the T_e and n_e profiles on the right.

Simulations of edge T_e and n_e profiles made for ITER [8] show gradients which are modest compared to those sometimes observed in present experiments such as JET [9]. Because of this, a simple model which allows a wide range of edge gradients to be investigated has been created for use here. It represents the T_e and n_e profiles as three linear segments: a flat central plateau, an intermediate region of moderate slope, and a high gradient edge region. The T_e and n_e gradients in this outermost region have been scanned over a wide range of values. Examples of these profiles are shown in Figure 8. As the right side of Figure 1 shows, the antenna pattern does not influence spatial resolution in the edge for a sightline near the midplane, and need not be included in the simulations. Similarly, refraction is not significant in the edge and can be ignored.

Measurements of the startup plasma are not explicit in the ITER requirements, but have been investigated in this study because they pose different measurement problems to the high performance plasmas. The plasma size and location, as well as T_e and n_e will be quite different to the high performance phase. Simple equilibria (concentric elliptical flux surfaces, vacuum magnetic field) have been constructed from sketches provided by ITER [8] and are illustrated in Figure 10. The poloidal cross-section shows the beam of the antenna at z=0.562 m and the separatrix for I_p = 0.5, 1, 2, 4 and 8 MA. A set of T_e and n_e values for the different plasma currents have been assumed: increasing from 1 keV and 1×10^{19} m^{-3} at 0.5 MA to 4 keV and 4×10^{19} m^{-3} at 8 MA, with moderately peaked profiles.

The design of the collection optics is described in [2]. The current design consists of three antennas which view the plasma horizontally along major radii, two at heights near the expected centre of high performance plasmas, and one at a lower height to observe the startup plasma. The antenna parameters are listed in Table 1. The antennas are described by their height (z_{beam}), major radius of the beam waist (R_{b0}) and waist radius (w_0).

Table 1: Parameters of the ECE antennas in the current design.

Ant #	z_{beam}	R_{b0}	w_0
A	1.440 m	11.71 m	36.0 mm
B	1.040 m	11.71 m	36.0 mm
C	0.562 m	11.94 m	33.8 mm

COMPARISON OF THE TWO CODES AND CORE PLASMA RESULTS

The two simulation codes have been compared for high performance plasma scenarios for which any shortcomings of the ECELS code will be greatest, either because of high T_e effects or due to neglecting refraction at high n_e. The plasma profile shapes and the equilibrium illustrated in Figure 2 have been used. To make scans of T_e and n_e, the profile shapes are maintained and the profiles scaled from 10% to 100% of the "reference" value.

Antenna Pattern and Refraction

To investigate the effects of refraction, the ray tracing code has been run for both 2nd/E-mode and 1st/O-mode at the frequencies resonant at the plasma centre, using antennas A and B. Five rays are used to represent the antenna pattern: one on axis and four distributed symmetrically up/down and left/right of the axis. In vacuum, the four outer rays cross the beam axis at the waist radius (R_{b0}) and pass through the 95% intensity points of the beam at the resonance radius. This give the correct spot size at the resonance (in vacuum) but is a conservative representation of the beam, since the rays chosen have a larger divergence than does the Gaussian beam at the resonance radius (0.95° as opposed to 0.4° for the E-mode). It therefore overestimates both the beam deflection and increased spectral broadening due to the Doppler effect.

It is found that the deflection is only weakly dependent on T_e, so the scaling with n_e has been investigated at fixed T_e values. For the 2nd/E-mode the T_e scaling used is 50% (T_{eo} = 14 keV) since this is the highest T_e which allows access to the plasma centre. A 100% scaling (T_{eo} = 28 keV) has been used for the 1st/O-mode. The results for antenna A (sightline near the midplane) are summarised in Figure 3 which shows, as a function of n_{eo}, the displacement of the beam centre from the antenna axis, and the increase in vertical and horizontal width. For antenna B (~0.4 m below the plasma midplane) the deflection and widening are about 1.5 times larger in the E-mode and twice as large in the O-mode.

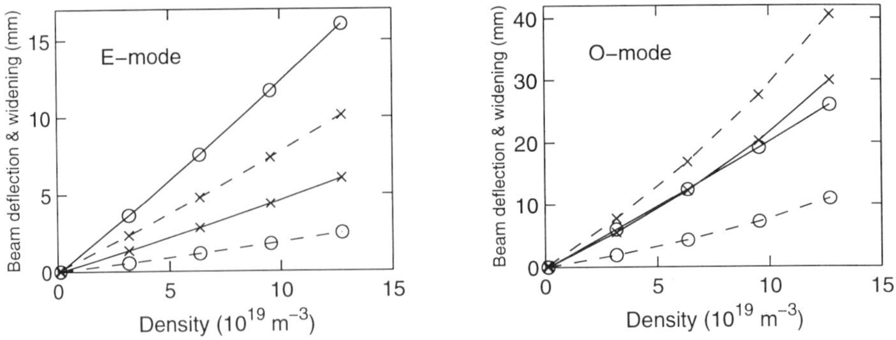

Figure 3: The effect of refraction in both E- and O-modes as a function of n_e for antenna A. The solid lines show the deflection of the beam axis (crosses for vertical deflection, circles for toroidal) and the dashed lines the increase in beam width (crosses for vertical widening, circles for toroidal).

We conclude from these results that for an antenna sightline near the plasma midplane, the deflection has a very small effect on location (which we can neglect here) and the increase in beam width is a secondary effect in the determination of resolution. For an accurate calculation of resolution in antenna B refraction may need to be taken into account, but it can be ignored for antenna A. These conclusions might need to be revised for n_e profiles which are less flat.

Code Comparisons

To check for inaccuracy at high T_e in the expressions for the absorption coefficient used in the ECELS code, and for errors due to ignoring Doppler broadening, the localisation information produced by the two codes has been compared. Figure 4 compares the function G(R) calculated by the two codes for the central ray of antenna A, at frequencies resonant at the plasma centre. The T_e profile has been scaled to 50% (T_{eo} = 14 keV) for the E-mode because of the access problem, and the O-mode has been calculated for T_{eo} of both 14 and 28 keV. In both cases the density has its full value. The difference in the O-mode at 28 keV is probably due to deteriorating accuracy of the ECELS code at high T_e, while the difference at 14 keV appears to be due to the effects of refraction. Nevertheless, we find that the widths of the curves from the two codes are nearly identical, and we conclude that the ECELS code is sufficiently accurate to use in the resolution study for the antenna sightline near the plasma midplane. The results will suffer some inaccuracy above 20 keV (only relevant for the O-mode), and for the off-midplane antennas at high n_e.

The results of a combined analysis of spectral resolution (ECELS code) and Gaussian beam antenna pattern are shown in Figure 5. For part (a), the full performance plasma conditions and 2nd/E-mode have been assumed. The beam diameter is generally larger than the major radius resolution, but as part (b) shows, when these are converted into effective minor radius resolution using the technique of Figure 1, the spectral broadening is entirely dominant. This behaviour is typical for a midplane antenna under high performance conditions (ie high T_e). However, when an off-midplane antenna (B) is considered and both T_e and n_e are reduced by a factor 2, as shown in (c) and (d), the results are somewhat different . In this case there is access in as far as R = 8 m (since T_e is lower) and the antenna pattern dominates the resolution at around R = 8.3 m. For part (d) the magnetic field has also been reduced by a factor 2 and the effect of the larger antenna pattern at the lower frequency can be clearly seen. This reduction in field has the same effect on antenna pattern as operating in the 1st harmonic.

In all these cases, the effective minor radius resolution is within the 0.1 m target, except near R = 8.3 m at low toroidal field where it may be up to 0.12 m.

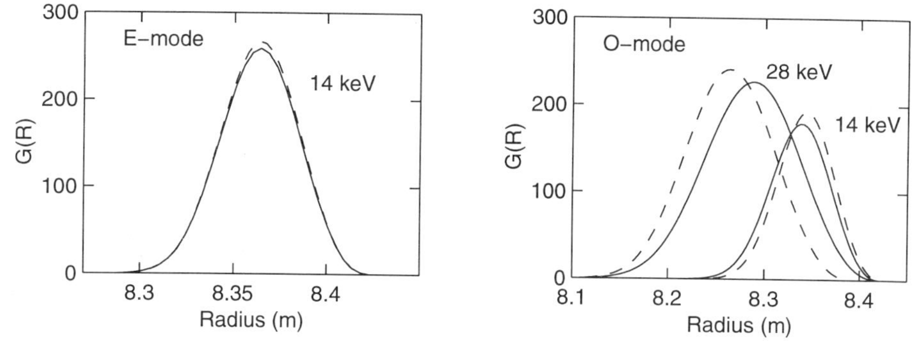

Figure 4: Comparison of G(R) calculated by the ECELS code (solid lines) and the weakly relativistic code (dashed lines). There is some difference at the highest T_e, but the widths of the curves (ie. ΔR_{res}) are very similar in all cases.

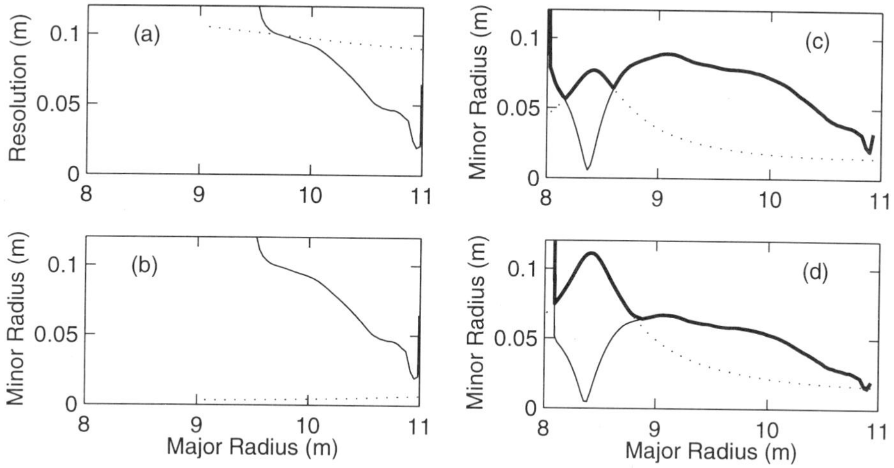

Figure 5: Combined effects of spectral resolution (E-mode) and antenna pattern. Parts (a) and (b) are for antenna A and a full performance plasma. Part (a) shows that the major radius resolution (solid line) is generally smaller than the beam diameter (dashed line), while in part (b) it can be seen that the spectral resolution dominates the effective minor radius resolution. Parts (c) and (d) show the minor radius resolution for antenna B and reduced performance: T_e and n_e reduced to 50% in part (c), and the magnetic field is also reduced to 50% in part (d).

Parameter Scans

In figure 6, T_e profiles (scanned from 20% to 100% of the full value) are plotted, with the accessible region shown by the bold lines. The lack of access on the inboard side is due to relativistically enhanced harmonic overlap which becomes more important as T_e increases. The curves are all plotted for the highest density. The results are not strongly n_e dependent for the E-mode. The inboard O-mode results may be slightly inaccurate at high T_e because of a slight overestimate of the second harmonic optical depth in the O-mode.

Since the resolution is generally determined by the spectral broadening, we have made scans of T_e and n_e (at full field), and calculated the major radius resolution for antenna A. Figure 7 shows contour plots of the major radius resolution at frequencies resonant at the plasma centre in both 2nd/E-mode and 1st.O-mode. Up to the 15 keV limit set by access, the 2nd/E-mode resolution is generally better than the 0.1 m target, whereas the 1st/O-mode is generally worse: at high T_e (>12 keV) the target is never reached, and at 28 keV the resolution is between 0.17 and 0.27 m depending on n_e.

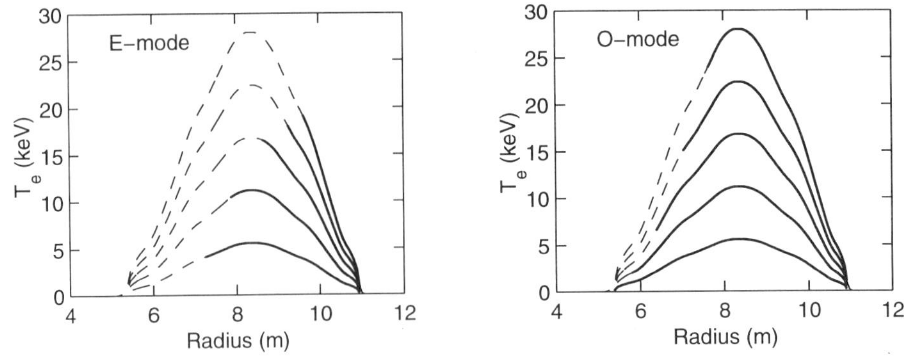

Figure 6: Core access. Each curve is a T_e profile (scaled to 20, 40, 60, 80 & 100% of the full value). The part of the profile in bold is accessible for T_e measurements, while the inboard regions are obscured by relativistically enhanced harmonic overlap. The density used is the full value.

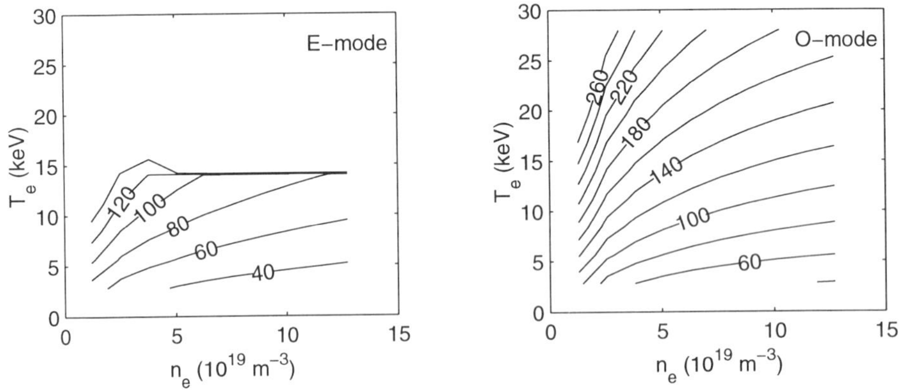

Figure 7: Contours of major radius resolution for both E- and O-modes, as a function of T_e and n_e. A frequency resonant at the plasma centre has been used, and the profiles scaled from 10% to 100% of the high performance values. The contour interval is 20 mm. Above 15 keV there is no access to the plasma centre in the E-mode.

EDGE PLASMA RESULTS

The families of T_e and n_e profiles which have been used (left side of Figure 8) have five values of edge ∇T_e from 25 to 85 keV m^{-1} and five values of edge ∇n_e from 2.5×10^{20} to 8.5×10^{20} m^{-4}. The T_{RAD} spectra on the right side of Figure 8 (calculated with the ECELS code) correspond to all five ∇T_e values for the lowest ∇n_e (upper part) and the highest ∇n_e (lower part). The low frequency emission (f < 232 GHz) is due to relativistically downshifted radiation from further into the plasma (where T_e is larger) which can pass through the low optical depth edge region ("shine through"). This shine through is an effect which does not occur in the core, and it sets a limit on how close to the edge T_e measurements can be made. The value of the low T_e limit increases with ∇T_e and decreases with ∇n_e. Examination of these spectra also shows that above this lower limit of T_e the slope of the T_e profile is well recovered, but there is some smoothing due to the resolution.

Figure 9 is a quantitative presentation of the lower limit of measurable T_e and the resolution. Resolution contours are plotted over a range of edge gradients. The left side shows contours for the emission frequencies for which T_{RAD} = 1.5 keV, and on the right for T_{RAD} = 2 keV. The bold lines on the upper left corners of these plots mark the boundary of the inaccessible region (ie above these lines the emitted intensity never falls to the target value). Where the edge T_e can be measured, the resolution is generally between 40 and 70 mm. The O-mode has not been investigated in the edge since it has lower optical depth.

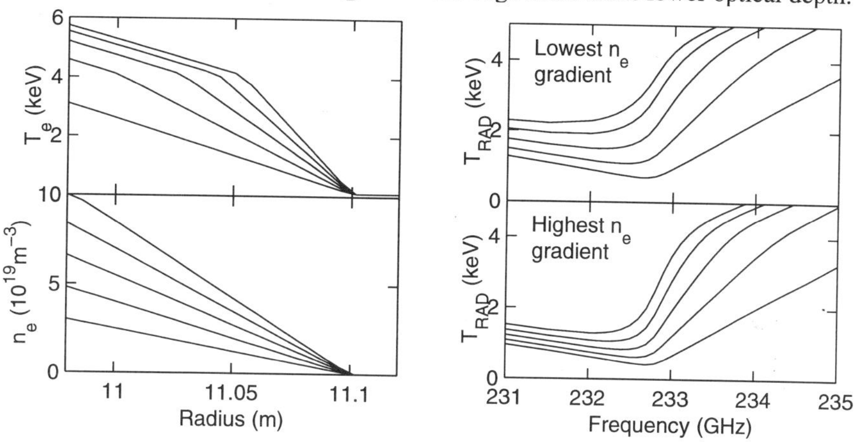

Figure 8: On the left side, the families of edge T_e and n_e profiles used in the edge simulations. On the right are spectra calculated for all five values of ∇T_e, at the lowest ∇n_e (top) and highest ∇n_e (bottom). The features of the high gradient region are rounded in the spectra by broadening, and the intensity does not fall to zero, because of the "shine through" effect.

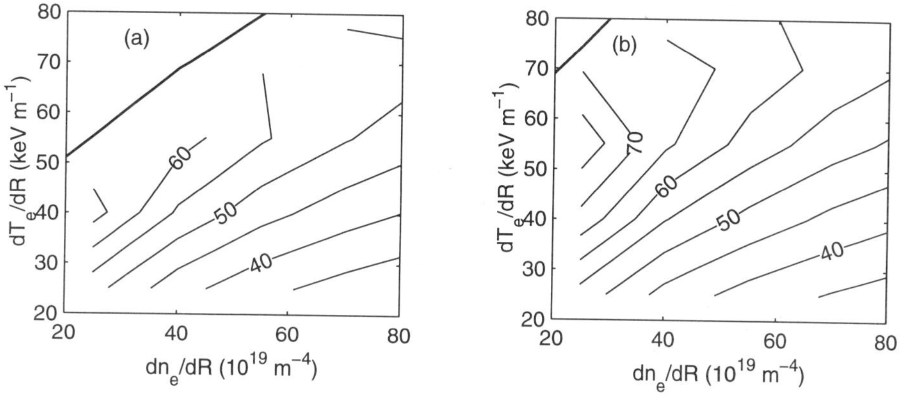

Figure 9: Contour plots of major radius resolution in ∇ne, ∇Te space for the profiles of Figure 8. Part (a) is for $T_{RAD} = 1.5$ keV, and part (b) for $T_{RAD} = 2.0$ keV. The bold diagonal lines mark the maximum gradients for which the given T_{RAD} is accessible. The contour interval is 5 mm.

STARTUP PLASMA RESULTS

The ECELS code has been used for the startup plasma because it should be accurate at the low T_e and n_e values which have been assumed. Calculations have been made for antenna C, and the 2nd/E-mode has been used since it has the larger optical depth. The limit on access in these cases is due to low optical depth: we have taken a strict criterion ($\tau > 4$) for the validity of T_e measurements. The plots on the right of Figure 10 are (from top to bottom): the T_e profiles (solid for $\tau > 4$ ($e^{-\tau} < 2\%$), dashed for $\tau < 4$), the transmission, and the minor radius resolution in the optically thick region. For the conditions assumed, τ is too low for a reliable T_e measurement at 0.5 MA, and only the central part of the profile is accessible at 1 MA and 2 MA, for which the transverse resolution dominates. The 4 MA and 8 MA cases show how the transverse resolution has a significant effect in a narrow region around the plasma centre. Nevertheless, in all cases it is again found that a limit of about 0.12 m in minor radius resolution is not exceeded.

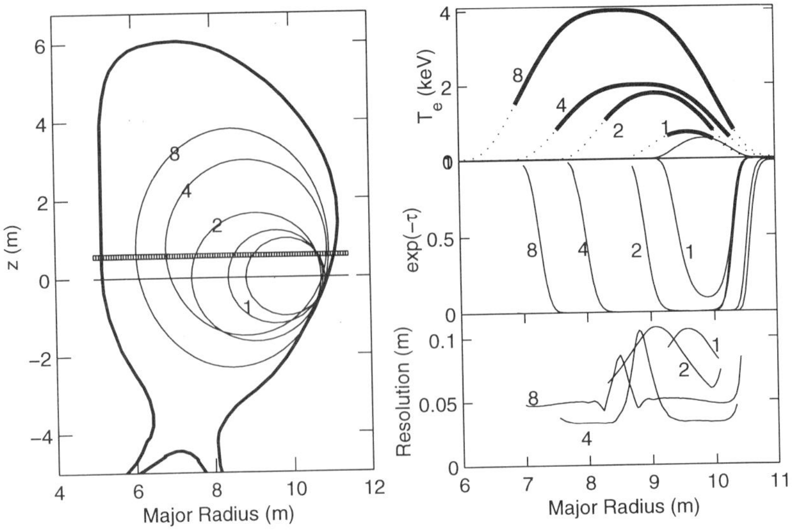

Figure 10: Resolution analysis for the startup plasma. On the left, the separatrix is labelled for I_p = 8, 4, 2 & 1 MA (0.5 MA not labelled) and the beam of antenna C is shown. On the right, from top to bottom, are the T_e profiles (in bold where good T_e measurements are possible), the transmission ($e^{-\tau}$) and the minor radius resolution.

CONCLUSIONS

The ITER requirements for spatial resolution of T_e measurements (~0.1 m) can be met in and near the centre of high performance plasmas using the second harmonic extraordinary mode at T_e up to about 15 keV. At higher T_e the first harmonic ordinary mode can be used, but the resolution is poorer (up to 0.27 m at 28 keV). At all temperatures, harmonic overlap restricts the access of the 2nd/E-mode to radii outboard of the plasma centre. The first harmonic ordinary mode has access to radii inboard of the plasma centre at all temperatures, and almost all of the profile for $T_e < 15$ keV.

In the edge region the spatial resolution is typically 40 mm to 70 mm, much larger than the ITER target. It will not be possible to make measurements close to the separatrix, at T_e below about 1 keV. The exact values of all these figures depend on the steepness of the edge T_e and n_e gradients.

The limitation in the startup plasma is likely to be low optical depth. However, if an antenna viewing through the plasma centre is installed, good T_e measurements should be possible during much of the startup phase.

ACKNOWLEDGEMENTS

Members of the European Home Team for Microwave Diagnostics who have participated in this work are: A Airoldi, P Buratti, L Cupido, T Donné, G Giruzzi, H Hartfuss, E Joffrin, M E Manso, S Nowak, R-L Meyer, G Ramponi, J Sanchez, B Schokker, A Silva, P Stott, V Tribaldos, G Waidmann. The authors wish to thank Drs G Vayakis and D Boucher for providing valuable information required for the calculations presented here.

REFERENCES

[1] H Hartfuss et al, Proc. of the Intl. Workshop on Diagn. for ITER (Varenna, 1995)
[2] G Vayakis et al, this workshop.
[3] D V Bartlett et al, Proc. of the Intl Workshop on Diagn. for ITER (Varenna, 1995)
[4] M Bornatici, Proc. of the Intl. Workshop on Diagn. for ITER (Varenna, 1995)
[5] H Bindslev, Proc 9th Joint Workshop on ECE & ECRH (Borrego Springs, 1995)
[6] M Bornatici, F Engelmann & U Ruffina, Sov J Quantum Elec, 13, 68 (1983)
[7] D Boucher, private communication.
[8] G Vayakis, private communication.
[9] L Porte et al, Proc 10th Joint Workshop on ECE & ECRH (Ameland, 1997)

ECE DIAGNOSTIC ON ITER IN THE PRESENCE OF SUPERTHERMALS

Marino Bornatici,[1,2] and Umberto Ruffina[2]

[1]Physics Department, University of Ferrara, Ferrara, Italy
[2]INFM, Physics Department "A. Volta", University of Pavia, Pavia, Italy

ABSTRACT

The effects of superthermals on ECE T_e–measurements in the core plasma of ITER are investigated on the basis of a model distribution function for which a generalized Kirchhoff's law is valid. On varying both the spatial profiles and the position of the superthermals, the feasibility of T_e–measurements using the first harmonic ordinary–mode as well as the second harmonic extraordinary–mode is assessed in respect to the accuracy requirement set by ITER.

INTRODUCTION

The potential of electron cyclotron emission (ECE) for electron temperature (T_e) measurements in ITER requires a careful assessment since the projected ITER plasma parameters, in particular, the very high T_e and dimensions, differ significantly from those of existing tokamaks. Preliminary studies of ECE diagnostic on ITER have indicated that, for a Maxwellian electron distribution, and $T_e \gtrsim 20keV$, the accuracy of 90% and the spatial resolution of $10cm$ set by ITER for core T_e– measurements can be met by the 1st harmonic ordinary (O) mode over a significant portion of the plasma cross–section.[1,2] The 2nd harmonic extraordinary (X) mode, instead, which is more strongly affected by harmonic overlap, does not appear to be suitable for T_e–measurements in the core plasma for $T_e \gtrsim 20keV$, its T_e–diagnostic potential being restricted to the outboard plasma region.[3,4]

Here the analysis of ECE T_e–measurements in the core plasma of ITER is extended to include the presence of superthermals. A convenient model distribution to study the effects of a low–density population of high energy electrons (hereafter referred to as superthermals (s)), including a motion along the magnetic field with respect to the bulk (b) electrons, consists of a (relativistic) bi–Maxwellian distribution for which Kirchhoff's law is valid for each electron population separately. For the specific case of a plane plasma slab of thickness $2a$ and observation on the outboard side of the plasma, the radiation temperature of the ith mode ($i = X, O$) obtained on solving the equation of radiative transfer, is[5]

$$T_{rad}{}^{(i)}(\omega, N_{\|}) = \frac{1 + R_o e^{-\tau^{(i)}}}{1 - R_w R_o e^{-2\tau^{(i)}}} \int_{-a}^{a} dx G^{(i)}(\omega, N_{\|}, x), \tag{1a}$$

$$G^{(i)}(\omega, N_{\|}, x) \equiv T_b(x)\tilde{\alpha}_b^{(i)}(\omega, N_{\|}, x) + \frac{T_s(x)\tilde{\alpha}_s^{(i)}(\omega, N_{\|}, x)}{\gamma_s(x)[1 - N_{\|}\overline{v}_s(x)]} \equiv G_b^{(i)} + G_s^{(i)} \tag{1b}$$

and

$$\tilde{\alpha}_k^{(i)}(\omega, N_{\|}, x) \equiv \alpha_k^{(i)}(\omega, N_{\|}, x) e^{-\int_x^a dx'(\alpha_b^{(i)} + \alpha_s^{(i)})(\omega, N_{\|}, x')}, \quad k = b, s \tag{1c}$$

Diagnostics for Experimental Thermonuclear Fusion Reactors 2
Edited by Stott *et al.*, Plenum Press, New York, 1998

The first factor on the right–hand–side of (1a), for which $\tau^{(i)}(\omega, N_\parallel, x_0) \equiv \int_{x_0}^{a} dx(\alpha_b^{(i)} + \alpha_s^{(i)})(\omega, N_\parallel, x)$ is the effective optical thickness, accounts for the reflections from both the outboard wall at $x = a$, with reflection coefficient R_w, and a reflecting layer at $x_0(-a \le x_0 < a)$, with reflection coefficient R_0. In particular, $R_0 = 1$ and $R_0 = R_w$ for the case in which the reflecting layer at x_0 is connected with, respectively, a cut–off and the inboard ($x = -a$) wall. The $G_k^{(i)}$–function, defined in (1b,c), yields the spatial distribution of the emission from the kth electron population ($\approx T_k \alpha_k^{(i)}$, with T_k the temperature) on account of the re–absorption by both the electron populations, as described by the exponential factor in (1c).

NON–THERMAL ECE SPECTRA AND T_e–MEASUREMENTS ON ITER

Let us refer specifically to the ITER projected parameters of the Interim Design, namely, minor radius $a = 2.8m$ and inverse aspect ratio $\epsilon = 0.34$; central toroidal field $B(0) = 5.68T$; for the bulk, central density $n_e(0)) = 1.2 \times 10^{20} m^{-3}$ and central temperature $T_e(0) = 20keV$. With reference to the superthermals, the peak values of density and temperature are, respectively, $n_s(a_0) = 1.8 \times 10^{17} m^{-3}$ ($= 1.5 \times 10^{-3} n_e(0)$) and $T_s(a_0) = 80keV$ ($= 4T_e(0)$), a_0 denoting the central position of the superthermals, such that $0 \le a_0 \le a$; $\gamma_s(a_0) = 1.01$ is taken as central value of the streaming parameter of the superthermals $\gamma_s \equiv [1 - (v_s/c)^2]^{-1/2}$. As for the spatial profiles used as input to the calculation of the ECE spectra, along with the vacuum toroidal field profile, the generic profile shape is of the form $[1 - (|x - a_0|/a_k)^\delta]^\beta$, $a_0 - a_k \le x \le a_0 + a_k$, with, in particular, $a_0 = 0$ and $a_k = a$ for the bulk, and three types of profiles are considered, namely, flat profiles with $\delta = 5$ and $\beta = 0.25$, to be referred to as trapezoidal–like profiles, parabolic profiles for which $\delta = 2$ and $\beta = 1$, and (3/2)–profiles with $\delta = 2$ and $\beta = 3/2$. With respect to the bulk, the spatial distribution of the superthermals is taken to be steeper (Case I), equally flat (Case II), equally steep (Case III). Also, we consider observation along a major radius ($N_\parallel = 0$), and assume a wall reflection coefficient $R_w = 0.6$.

In Fig.Ia the radiation temperature of the **O–mode** is shown for the frequency range extending up to the 5th harmonic, for trapezoidal–like profiles for the bulk and parabolic and (3/2)–profiles, respectively, for the density and temperature of the su-

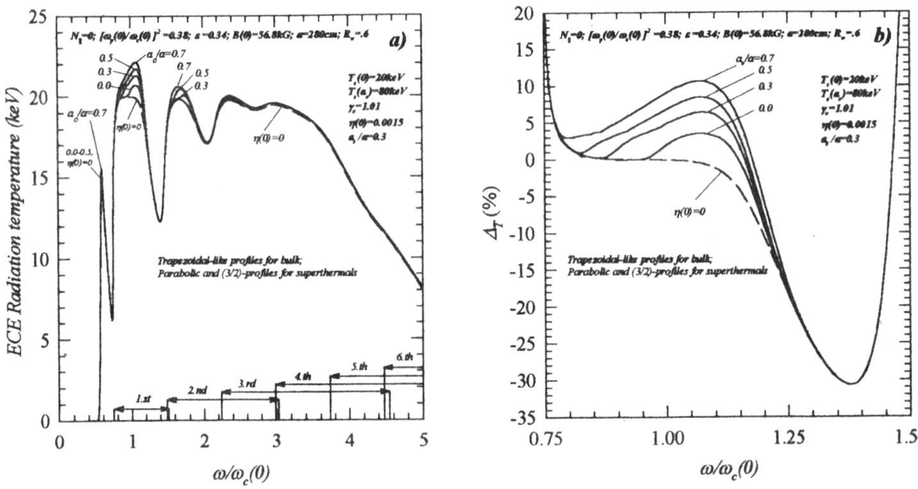

Fig. I: a) Non–thermal ECE spectra of the perpendicular O–mode. The location of the peak of the spatial distribution of the superthermals is varied from the center of the bulk, $a_0 = 0$, to near the outboard edge, $a_0 = 0.7a$; the width of the profile is $a_s = 0.3a$. The superthermals carry a $720kA$ current. The $\eta = 0$–curve refers to the spectrum without superthermals.

b) The uncertainty of the T_e–measurements, $\Delta_T = (T_{rad} - T_e)/T_e$, for the 1st harmonic O–mode on account of the downshifted 2nd harmonic emission from superthermals.

perthermals (**Case I**), for different central locations of the superthermals. With respect to the thermal spectrum, the $\eta = 0$-curve, it appears that the overall effect of the superthermals consists in an enhancement of the radiaton temperature, the effect being the more significant the nearer to the outboard plasma–edge are the superthermals.

Note, however, that the downshifted 1st harmonic feature is largely unaffected by the superthermals. In particular, the effect of the downshifted 2nd harmonic emission from the superthermals on the low–frequency part of the 1st harmonic frequency range of the thermal spectrum could spoil the potential of the 1st harmonic O–mode for T_e–measurements. For a quantitative assessment of this effect, the uncertainty of the T_e–measurements, $\Delta_T \equiv (T_{rad} - T_e)/T_e$, is evaluated for the 1st harmonic O–mode, for the same parameters of Fig.Ia, and shown in Fig.Ib. It appears that the uncertainty $|\Delta_T| \lesssim 10\%$, i.e., an accuracy $(1 - |\Delta_T|) \gtrsim 90\%$, set by ITER can be attained over most of the 1st harmonic frequency range, namely, $0.76 \lesssim (\omega/\omega_c(0)) \lesssim 1.20$, provided that the superthermals are located not too near the outboard plasma–edge.

Let us consider now **Case II** for which trapezoidal–like profiles are assumed for the bulk as well as for the superthermals. A sample of non– thermal ECE spectra of the O–mode is shown in Fig. IIa, and the corresponding uncertainty of the T_e– measurements is shown in Fig. IIb for the 1st harmonic frequency range. The salient points to be noticed are: i) the downshifted 1st harmonic feature can be significantly

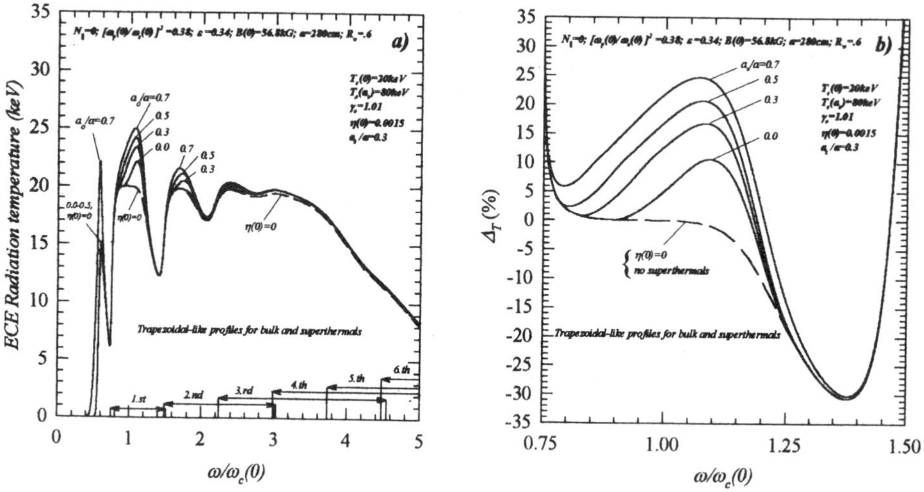

Fig. II: a) Same as Fig.Ia; b) Same as Fig.Ib. Trapezoidal–like profiles for the bulk and the superthermals are assumed.

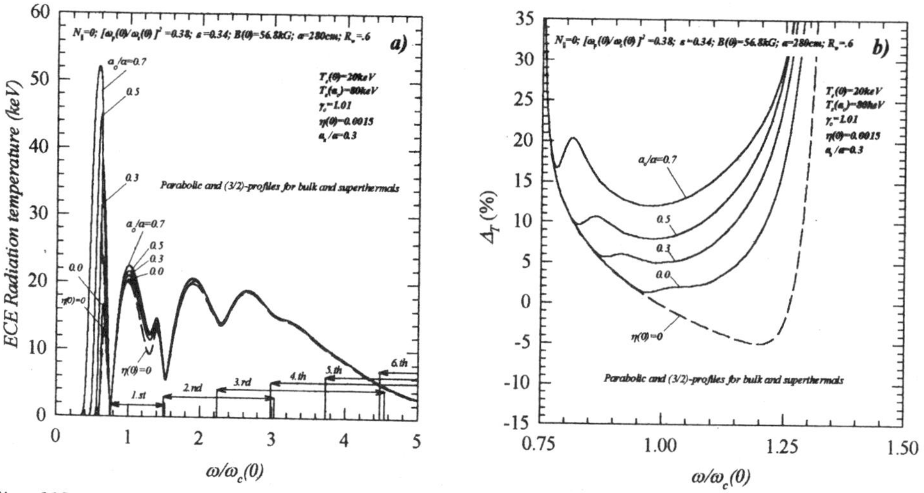

Fig. III: a) Same as Fig.Ia; b) Same as Fig.Ib. The density profile is parabolic and the temperature profile is of (3/2)–type, for both the bulk and the superthermals enhanced by the emission from superthermals located near the outboard plasma edge, to be compared with the corresponding downshifted feature of Fig.Ia; ii) the downshifted 2nd harmonic emission from the superthermals can significantly enhance the 1st harmonic radiation temperature with respect to its value in the absence of superthermals, with the result that the accuracy of T_e–measurements tends to be worse than that for the case for which the profiles of the superthermals are steeper than the bulk profiles (compare Fig.IIb with Fig.Ib).

The downshifted 1st harmonic emission from the superthermals can be dominant with respect to that from the bulk for the case for which the profiles of density and temperature are, respectively, parabolic and of (3/2)–type for both the superthermals and the bulk (**Case III**), as it appears from the non–thermal ECE spectra shown in Fig.IIIa. Because of the downshifted 2nd harmonic emission from the superthermals, the uncertainty of the 1st harmonic T_e–measurements, shown in Fig.IIIb, can be significantly higher than its value with no superthermals, the requirement set by ITER ($|\Delta_T| \lesssim 10\%$) being not satisfied for the case of superthermals near the outboard plasma–edge, e.g., $\Delta_T \gtrsim 12\%$ for $a_0 = 0.7a$.

A comprehensive analysis of the non–thermal ECE spectra has been made also for the **X**–mode. In view of the accuracy of 2nd harmonic T_e– measurements in a thermal plasma,[3,4] the most favourable case corresponds to the (3/2)–profile for the bulk temperature (the shape of the density profile does not appear to be so important), which is shown in Fig.IIIc along with the effect of the emission from superthermals with the same peaked profiles as for the bulk. On comparing with Fig.IIIb, the 1st harmonic O–mode appears to be the best choice for T_e–measurements in the core plasma of ITER, notwithstanding the issue of the spatial resolution.

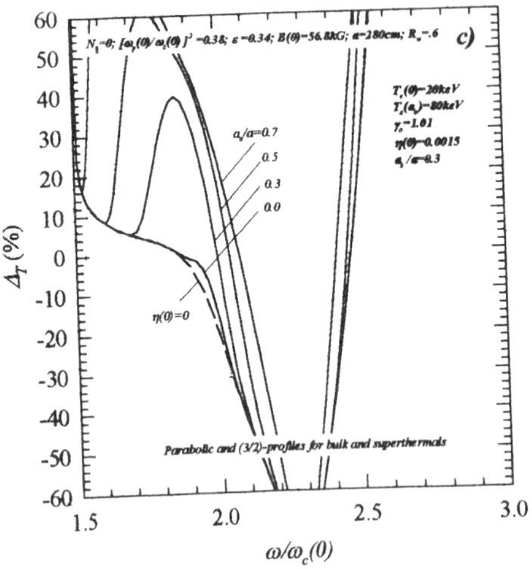

Fig. IIIc The uncertainty of the T_e–measurements for the 2nd harmonic X–mode, for the same parameters as Fig.IIIb.

REFERENCES

1. M. Bornatici, and U. Ruffina, in **Diagnostics for Experimental Thermonuclear Fusion Reactors**, P.E. Stott, G. Gorini and E. Sindoni (Eds.), Plenum Press, New York, 1996, pp. 193–196.
2. M. Bornatici, and U. Ruffina, in Proc. 10th Joint Workshop on ECE and ECRH (Ameland, 1997).
3. D.V. Bartlett, in **Diagnostics for Experimental Thermonuclear Fusion Reactors**, P.E. Stott, G. Gorini and E. Sindoni (Eds.), Plenum Press, New York, 1996, pp. 183–192.
4. D.V. Bartlett and H. Bindslev, this workshop.
5. M. Bornatici, and U. Ruffina, Plasma Phys. Contr. Fusion **38**, 1561(1996).

THE ITER ECE DIAGNOSTIC FRONT END DESIGN

G. Vayakis,[1] D. Bartlett,[2] P. Edmonds,[3] H. Hartfuß,[4] D. Wagner,[5]
C.I. Walker,[3] and the ITER Joint Central Team and Home Teams

[1]ITER Joint Central Team, San Diego JWS, 11025 N. Torrey Pines Road, La Jolla CA 92037, USA
[2]JET Joint Undertaking, Abingdon, United Kingdom
[3]ITER Joint Central Team, Garching JWS, Germany
[4]Max-Planck-Institut für Plasmaphysik, Garching, Germany
[5]Institut für Plasmaforschung, Stuttgart, Germany

INTRODUCTION

The ECE diagnostic is one of the two primary systems for the measurement of electron temperature in ITER. It is capable of viewing the plasma for the whole pulse, including the startup and termination phases. In addition to the measurement of electron temperature profiles and related quantities,[1] it will also be used to provide measurements from which the total ECE power loss can be deduced. In this paper we present the outline design of the front end of the ECE diagnostic, which has been shaped by the following considerations: (i) The choice of an appropriate front end, (ii) the selection of an adequate sightline distribution, (iii) the optimisation of the front end to minimise the perturbation to the first wall and shield, and (iv) the incorporation of a suitable calibration mechanism.

CHOICE AND LOCATION OF FRONT END

In order to reach the spatial resolution limits imposed by the physics of the emission process,[2] and thus achieve the target resolution in the core (0.1 m), the front end optics has been designed as a focusing system. Well defined antenna patterns over a very broad band (100 GHz-1 THz) are a requirement to avoid potential degradation of the resolution due to the antenna side lobes.

The present design employs a Quasi-Optical (QO) Gaussian beam telescope arrangement (figure 1). An alternative design using Flux Concentrators (FC) is also being considered.[3] Both types of system can be made to provide good neutron shielding through the diagnostic block: The QO system exploits the narrowing of the beam near the confocal point, and the FC system has natural tapers and a "neck point".

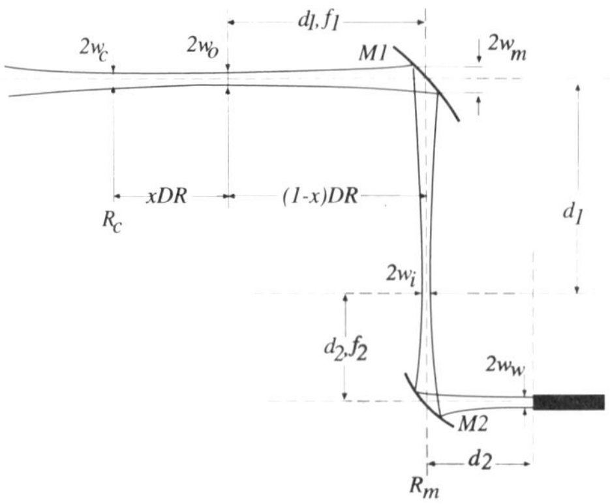

Figure 1: Schematic of a Gaussian telescope. The minimum waist w_0 and the waveguide waist w_w are frequency independent. The intermediate waist w_i is maximum for the lowest frequency, and is about 1/2 the size of the waist at the mirror for the ITER antennas.

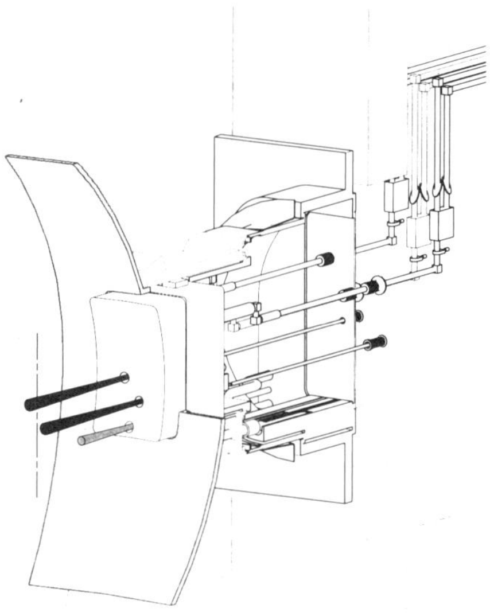

Figure 2: View of the ECE system from the plasma. The beams shown trace out the spot size of all three lines at 110 GHz; the vertical straight line is at the major radius of the plasma centre.

The first mirror of the QO system has been placed some distance behind the first wall (~1 m). This reduces the EM radiation heat flux on these components by about a factor of 10 compared to the first wall values, or ~ 50 kW/m^2. This load is low enough that, combined with the need to operate only at long wavelength (>300 μm), means that simple stainless steel mirrors can be used, and cooled without much difficulty.

NUMBER AND DISTRIBUTION OF SIGHTLINES

In order to measure the required parameters for a wide range of plasma shapes, and in order to accommodate the necessary instrumentation, more than one sightline is necessary. Figure 2 shows the sightline distribution. The nominal system instrumentation[6] consists of a pair of heterodyne radiometers to perform the profile measurement (one each in X and O-mode) and a pair of Michelson interferometers for spectrum overview, again one for each polarisation. Thus at least two lines are necessary to be able to couple the signal to all instruments with reasonable loss. At the same time, if properly located, these two sightlines provide access to the core with some flexibility, and a measure of redundancy.With the present plasma scenario, a third sightline is required to cover the startup and termination phases. This sightline is intended for lower temperature plasmas at large major radius, so that its mirror can be placed further out, thus easing congestion in the port. A way to switch instruments between sightlines in real time is envisaged.

OPTIMISATION OF THE REQUIRED APERTURE

For the QO front end, the size for the first mirror has been iteratively decreased, subject to the resolution requirement.[4,5] Two considerations are pertinent here: Minimising the first wall side wall cooling requirements, and minimising the total neutron throughput

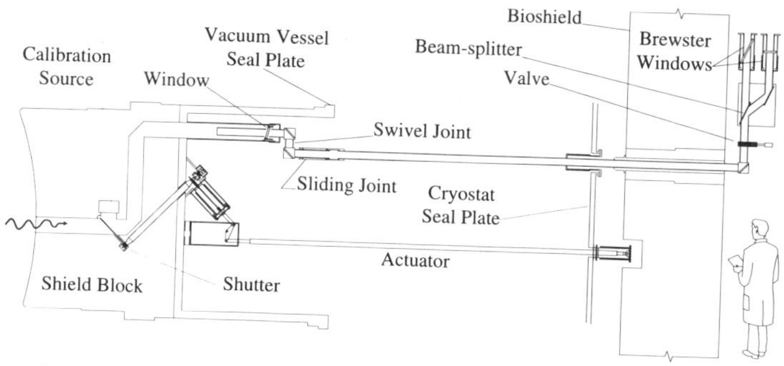

Figure 3: Schematic of one of the ECE sightlines to the second vacuum boundary, showing key components. Note the swivel/sliding joint assembly to compensate for vessel movement.

which is linked to the tunnel diameter. If the overriding constraint (or iterative parameter) is made the first wall aperture, rather than the mirror size, then slightly smaller first wall apertures result, but with much larger mirrors, which would greatly increase the neutron throughput given the extended neutron source shape. Thus, the mirror size was used.

For each iteration, an optimisation procedure is followed, which consists of calculating the mirror focal length that results in the minimum spot size at the innermost radius of interest for an appropriate wavelength, subject to the beam truncation requirement: This demands that the QO condition (beam diameter = 5 x waist size) be met at the lowest frequency of intrest. The resolution of the arrangement is then calculated.[2] It is found that a mirror size of 200 mm gives acceptable performance for all sightlines.

The optimisation is done for O-mode, so that for the main sightline the two frequencies used were 150 GHz (near the core) and 110 GHz (for no beam truncation). The same calculation in X-mode at full field would result in much smaller mirrors, but extreme performance degradation at lower than full fields and in O-mode, which would translate to a serious lack of flexibility. With the present setup, good performance in X-mode is available down to 1/2 field (2.85 T).

CALIBRATION

In order to guarantee the accuracy of the measurements, an end-to-end calibration mechanism has been incorporated in the two main sightlines.[6] Unlike the situation on present devices, on ITER it will be difficult, and expensive in operation time, to introduce an in-vessel calibration source at frequent intervals. At the same time, there is no other comprehensive way of checking the full system performance.

Two types of calibration scheme have been considered. In the first type, the first mirror is made to rotate and view a calibration source in the diagnostic block.[6] In the second type, a mirror is introduced at a 45° angle to the sightline. Although both could be implemented, in the first method the mirror has to satisfy both the mechanical and alignment requirements of the shutter during calibration, and the thermal and mechanical stability requirements of the first mirror during plasma operation. Thus we have explored in more detail the second case (figure 4). Here the shutter is rotated using an axle driven by a connecting rod, attached to a guided bellows on the first vacuum boundary.

The shutter is made out of a thin sheet of refractory material attached to a frame, and is polished on the side facing the antenna. The shutter and frame are designed to withstand indefinite exposure to the plasma during operation so they can be used to protect the antenna, if required. However in normal machine operation they are shielded from the plasma in a recess within the diagnostic block. In both extreme positions, the frame of the mirror is wedged in well defined slots, so that the axle bearing tolerance is not critical.

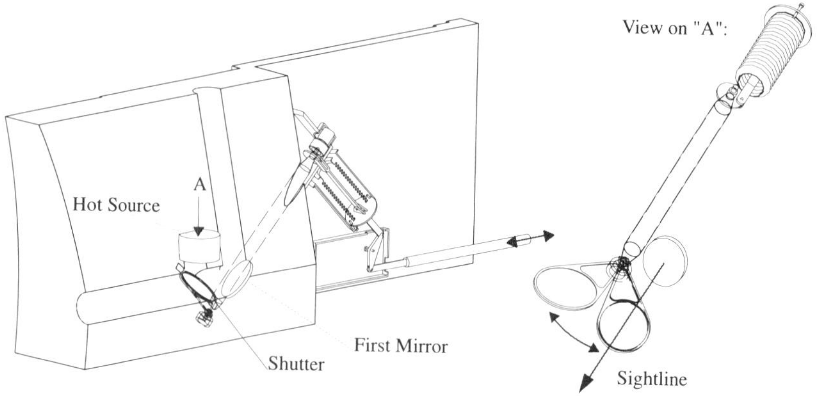

Figure 4: Cutaway view of the calibration setup. All moving parts inside the primary vacuum are cooled by radiation. Thus, the mirror, and the thin hollow cylinder rotating the shutter, are designed to form effective radiation heat sinks to the surrounding shielding (cooled to ~140 °C) for the frame assembly and the bearing, respectively. It appears to be possible to meet the required lifetime, without maintenance, of a few x 10^5 operations for the components inside the primary vacuum.

CONCLUSIONS

The ECE front end, from the first mirror to the second vacuum boundary has been designed in outline, and is now in a form likely to be compatible with the assembly, maintenance and lifetime requirements for installation in ITER. Two good alternatives for the antenna are being explored. A number of calculations (neutronics, electromagnetic) and much detailed design remains. In particular a long-lived hot source designed to operate in the hostile environment of the ITER diagnostic block remains to be designed.

ACKNOWLEDGEMENT

This report has been prepared as an account of work performed under the Agreement among the European Atomic Energy Community, the Government of Japan, the Government of the Russian Federation, and the Government of the United States of America on Cooperation in the Engineering Design Activities for the International Thermonuclear Experimental Reactor ("ITER EDA Agreement") under the auspices of the International Atomic Energy Agency (IAEA).

REFERENCES

1. A. E. Costley et al., Requirements for ITER diagnostics, in: *Diagnostics for Experimental Thermonuclear Fusion Reactors*, P. E. Stott et al., eds., Plenum Press, New York (1996).
2. D.V. Bartlett et al, Physics Aspects of the ITER ECE System Design, in: *These Proceedings*
3. H. Hartfuss et al, The non-imaging flux collector as a compact ECE antenna for ITER, in: *These Proceedings*
4. D. Bartlett et al, Physics aspects of the ITER ECE system design, to appear in: *Proceedings of the EC-10 Workshop*, Netherlands (1997)
5. G. Vayakis et al, Technical aspects of the ITER ECE front end design, to appear in: *Proceedings of the EC-10 Workshop*, Netherlands (1997)
6. H.J. Hartfuss, Instrumentation of ECE for ITER, in: *Diagnostics for Experimental Thermonuclear Fusion Reactors*, P. E. Stott et al., eds., Plenum Press, New York (1996).

THE NON-IMAGING FLUX COLLECTOR AS A COMPACT ECE ANTENNA FOR ITER

H. Hartfuss[1], M. Zerbini[2], P. Buratti[2], P. de Bernardis[3] and M. De Petris[3]

[1]Max-Plank-Institut fur Plasmaphysik, EURATOM-IPP Ass., Boltzmannstr. 2, 85748 Garching, Germany
[2]Associaz. EURATOM-ENEA sulla Fusione, CR Frascati, CP 65, 00044 Italy
[3]Dip. di Fisica, Universita' di Roma, Piazza A. Moro 2, 00185 Rome, Italy

ABSTRACT

The characteristics of flux collectors based on nonimaging optical elements (Winston cones with circular aperture) have been tested in the laboratory in order to check the possibility of employing these elements for Electron Cyclotron Emission (ECE) diagnostics on ITER. In particular the effects of diffraction on the achievable spatial resolution and on the concentrator-lightpipe coupling have been verified by antenna pattern measurements at microwave frequencies.

While the directivity properties turned out compatible with the spatial resoulution required for the ECE spectral survey, cross-polarisation measurements pointed out the need to develop Winston cones with square or rectangular aperture.

INTRODUCTION

Light collection and transport systems based on combinations of light pipes and optical elements have been successfully used for ECE diagnostics[1]. The application of this scheme in a hostile environment could be difficult because absorbing foam is needed to realise the stop and to baffle stray light. For this reason the possibility to replace the lens-stop combination by a nonimaging flux concentrator has been considered. The use of flux collectors based on nonimaging optical elements[2,3] allows to design robust self-baffling systems which define the field of view at the input aperture and do not require any in-situ alignment; this is particularly advantageous when the diagnostic front-end has to be placed in a hostile environment such as the one expected in ITER. In this paper the effect of diffraction on spatial resolution, light grasp and concentrator-light pipe coupling at the ITER relevant wavelength range will be discussed on the basis of laboratory tests.

DESIGN PRINCIPLES

The non-imaging concentrator (fig. 1) is a hybrid component which allows to collect the radiation entering one of the apertures. In the compound parabolic concentrator (CPC), all rays

Diagnostics for Experimental Thermonuclear Fusion Reactors 2
Edited by Stott *et al.*, Plenum Press, New York, 1998

entering with inclination $\theta < \theta_e$ emerge through the exit aperture, while rays with larger inclination are reflected back. The ratio of the exit and entry apertures is:

$$a'/a = \sin\theta_e$$

and the overall length L is:

$$L = (a/2) \bullet (1 + \sin\theta_e) / \tan\theta_e$$

The 3D concentrator is obtained by revolution about the optical axis.The CPC exit aperture is quite small if a small cutoff angle is required, furthermore the emerging rays are spreaded over a solid angle of 2π. For these reasons a multiple CPC "throat -to-throat" configuration is investigated, where a second concentrator is mounted in reversed direction.

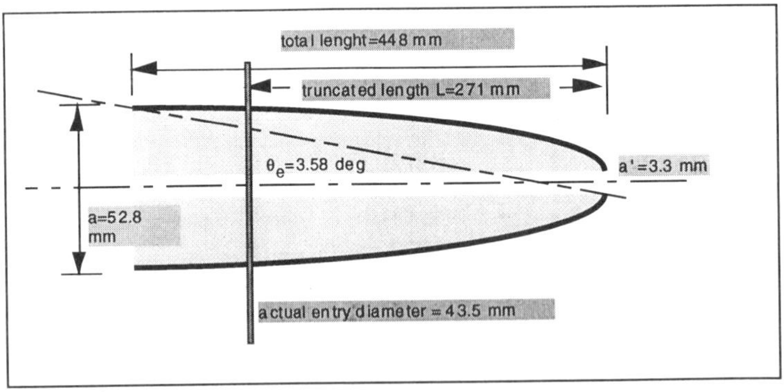

Figure 1:Meridian section of the CPC profile; The component used in this work is a truncated section of a CPC having 3.58 deg cutoff angle. Both the basic design parameters and the ones of the truncated section are indicated.

ITER REQUIREMENTS

The spatial localization required for ECE temperature measurements on ITER gives rise to two main constraints for the light collection system: the antenna pattern width has to be smaller than $\Delta z = 15$ cm, and in order to avoid excessive Doppler broadening, very little power has to be accepted outside the angular range θ_e with $\theta_e < 5$ deg. in toroidal direction. The required antenna pattern can be obtained using Gaussian beam collection optics with limited etendue. For the Fourier transform instruments covering an extreme wide spectral band, multimode collection and transportation is preferable because of the higher etendue. For ITER $\theta_e < 5$ deg, $\Delta z < 15$ cm, $\lambda < 3$ mm is assumed. With a typical value of the detector etendue of an InSb detector, 0.4 sr cm^2, and a distance between the low field side antenna and the plasma center of 3 m, it is found that the optimum acceptance angle including diffraction effects for ITER is $\theta_e = 2.30$ deg with an antenna diameter in the range $10 < a < 30$ cm.

MULTIMODE AND MONOMODE DETECTION MEASUREMENTS

CPC's were tested in different configurations. In the multimode configuration[4], a small source has been scanned along a circular path centered at the CPC input aperture. A pyroelectric detector has been placed at the CPC output. The sources were a high pressure mercury lamp to check the features of the concentrator in the visible range and a 150 GHz IMPATT microwave source, to evaluate the effects of diffraction at ECE relevant frequencies. The measurements have been repeated for the combined CPC where the input CPC is matched with a light pipe by a second concentrator and the radiation exciting the lightpipe is concentrated onto the detector by a

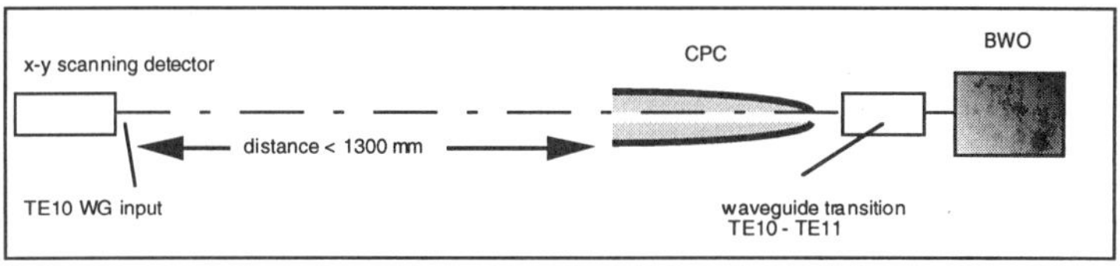

Figure 2: Schematic view of the experimental set-up used for 2D measurements. The CPC is fed with a TE11 mod from a BWO oscillator at 156 GHz. Intensity measurements are made with a scanning detector in a plane at various distances.

third CPC. The measured acceptance angle (HWHM) in the optical case is 3.2 deg. and the profile reasonably approaches the ideal shape which is rectangular with a half width of 3.58 deg. At 150 GHz the HWHM of the intensity Airy pattern for diffraction by a circular aperture is 1.47 deg. The expected final width of the transmission curve can be roughly estimated with 3.9 deg. in reasonable agreement with the measured 4.4 deg.

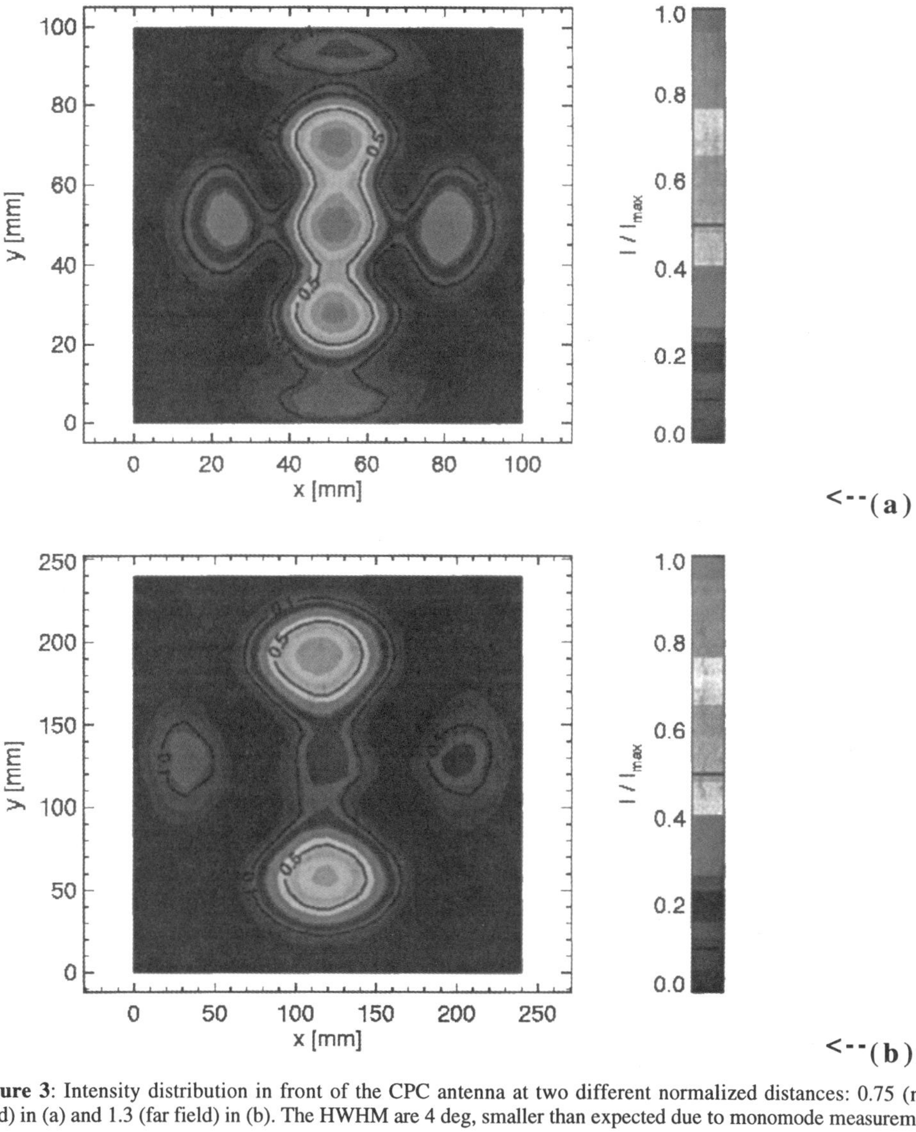

Figure 3: Intensity distribution in front of the CPC antenna at two different normalized distances: 0.75 (near field) in (a) and 1.3 (far field) in (b). The HWHM are 4 deg, smaller than expected due to monomode measurement.

The measurements have been repeated in monomode configuration, using CPC's with the same apertures as before but without the straight untapered section at the input side. Pattern measurements were performed with the 2D setup shown in fig. 2, exciting the CPC with a TE11 single mode at 156 GHz from a BWO source. The expected acceptance angle is about 4.5 deg. The detection is made monomode as well with a slightly tapered D-band piece of waveguide and a detector diode operated in the power proportional regime. The detector scans the plane at constant distance to the CPC aperture. As shown in fig. 3, measurements are conducted at various distances in the near and far field range between $0.4 < a^2 / \lambda < 1.3$.

High order modes are generated by the CPC and strong polarization scrambling occurs due to the multiple reflections in the concentrator. The figures give contour plots from which the HWHM angles can be determined. It is 4 deg. and due to the single mode measurement smaller than expected. The cross polarization component shows a power distribution with four maxima instead of two in the direction of the exciting wave. The total power in the two polarization directions is identical demonstrating the polarization scrambling of the CPC. Averaged over the power pattern, the degree of polarization is zero. The HWHM angle measured in the cross polarization direction is 4.8 deg. With a polarization insensitive detector the superposition then would lead to a measured pattern in good agreement with expectations.

Measurements in the far field under the same excitation and detection conditions with the multiple CPC system were carried out. The patterns differ only very little from those obtained with a single CPC. The power in both polarization directions is about the same.

CONCLUSIONS

The measurements demonstrate that the non-imaging flux collectors can be used in cyclotron radiation collection systems with narrow antenna pattern and large etendue, as required for Fourier transform spectrometers. Total radiation scrambling occurs, then for polarization sensitive measurements polarizers at the entrance are necessary. An alternative way is to use squared section CPC's, a technique which is still under investigation.

For overview measurements of the total emission in O- and X-mode with polarization insensitive detection, the circular CPC's seem to be well suited.

REFERENCES

1. P. Buratti, O. Tudisco, and M. Zerbini, A broadband light collection system for ECE diagnostics on the FTU tokamak, *Infrared Phys.*, **34**, 533 (1993).

2. W.T. Welford and R. Winston, *The Optics of Nonimaging Concentrators*, Academic Press, New York (1978).

3. P. Buratti and M. Zerbini, Multimode Light Collection Systems For ECE Diagnostics, in Diagnostics for Experimental Thermonuclear Fusion Reactors, Plenum Press, New York, London (1996), p. 211.

4. P. Buratti, M. Zerbini, P. de Bernardis, M. de Petris, *Flux Collectors For ECE Diagnostics On ITER*. Proceedings of Tenth Joint Workshop on ECE and ECRH Ameland (The Netherlands), April 1997.

BASELINE DESIGN OF A MULTI-CHANNEL INTERFEROMETER AND POLARIMETER SYSTEM FOR DENSITY MEASUREMENTS ON ITER

T.N. Carlstrom, R.T. Snider, C.B. Baxi, C.L. Rettig,* and W.A. Peebles*

General Atomics
P.O. Box 85608
San Diego, CA 92186-5608

*University of California, Los Angeles
Los Angeles, CA

ABSTRACT

A design of a multi-channel system to measure line average plasma density for ITER has been developed using vibration compensated interferometry and Faraday rotation measurements. The purpose of this study is to establish a consistent design based on current technology, that can be used to examine performance and reliability issues and to evaluate the effect of future design improvements. The system uses a vibration compensated interferometer, with a CO_2 laser at 10.6 μm for the probe beam and a CO laser at 5.2 μm for the vibration compensation. The CO_2 laser probe beam also allows Faraday rotation measurements to be made along the tangential beam path. This improves the reliability of the system because density measurements made from the Faraday rotation, although less accurate than interferometry, do not require cumulative knowledge of the phase shift over the long time scales expected in ITER. Plasma effects on both types of measurements and ITER specific operational considerations, such as plasma start-up, pulse length, and vibration effects will be discussed. Survivability and placement of the plasma facing optics in the ITER structure are critical issues and a discussion of the material selection, and integration into the ITER design are given.

1. INTRODUCTION

The electron density measurement is expected to play a critical role in the operation of ITER and in the basic understanding of the ITER plasmas. The avoidance of density limit disruptions, burn control and fueling all require accurate active density feedback control.[1] While active density feedback control is routinely used in present day tokamaks, the size, environment and pulse length of ITER places new constraints and requirements on a density control diagnostic. The severity and possible consequences of disruptions are of particular concern when considering the reliability of a density control diagnostic. The critical nature of the density measurement has lead us to a design using laser interferometry techniques with adequate density accuracy and that have proven to be robust and reliable for density control in other tokamaks. In the ITER baseline design presented here, a short wavelength probing laser beam is used both as an interferometer to measure line average

Diagnostics for Experimental Thermonuclear Fusion Reactors 2
Edited by Stott *et al.*, Plenum Press, New York, 1998

electron density and as a polarimeter to measure the plasma induced Faraday rotation which is proportional to $\int n_{e(z)} B(z)dz$.[2] The addition of the Faraday rotation measurement improves the reliability of the overall system, particularly during long pulse operation, since the phase rotation is less then 2 π and hence does not have to rely on keeping track of a cumulative fringe count during the entire pulse. The line averaged density measured with Faraday rotation can be used to check and correct any counting errors that may occur in the higher resolution interferometer measurement.

Previous studies have shown that a vibration compensated interferometer system based on a CO_2 laser is near optimum for ITER in terms of refraction, density resolution, plasma facing optics survival and access to ITER.[3,4] Here we describe the ITER baseline design of the density measurement system based on a vibration compensated CO_2 interferometer and polarimeter using existing technology that we believe meets the needs of ITER. This baseline design which self consistently fits all of the major components together, can then be used to benchmark and test future design improvements or new concepts.

2. ITER INTERFEROMETER/POLARIMETRY BASELINE DESIGN

System Layout

A plan view of ITER showing five tangential probe beam paths through the ITER plasma is shown in Fig. 1. This geometry was chosen over radial or vertical viewing systems because of the severe access constraints on ITER.[4] A radial system for example requires optical surfaces on the center, inside wall of the torus where neutron streaming to the inner TF coils prevent adequate penetrations through the inner shield modules for mirrors. The five beams shown in Fig. 1 provide coverage for the various stages during the plasma growth as shown in Fig. 2 and give some crude real-time density profile information. Work by Bower has shown that these few interferometer channels combined with an edge reflectometer can provide density profiles with good spatial resolution throughout the plasma.[5] The multiple beam paths also improve the reliability of the entire diagnostic by giving some redundancy during the flattop portion of the plasma pulse.

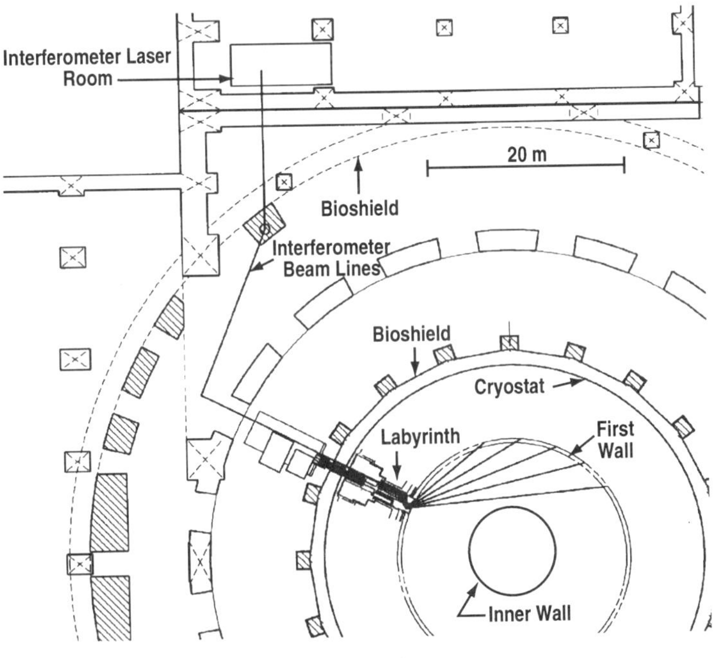

Figure 1. Plan view of ITER and proposed beam paths and laser room.

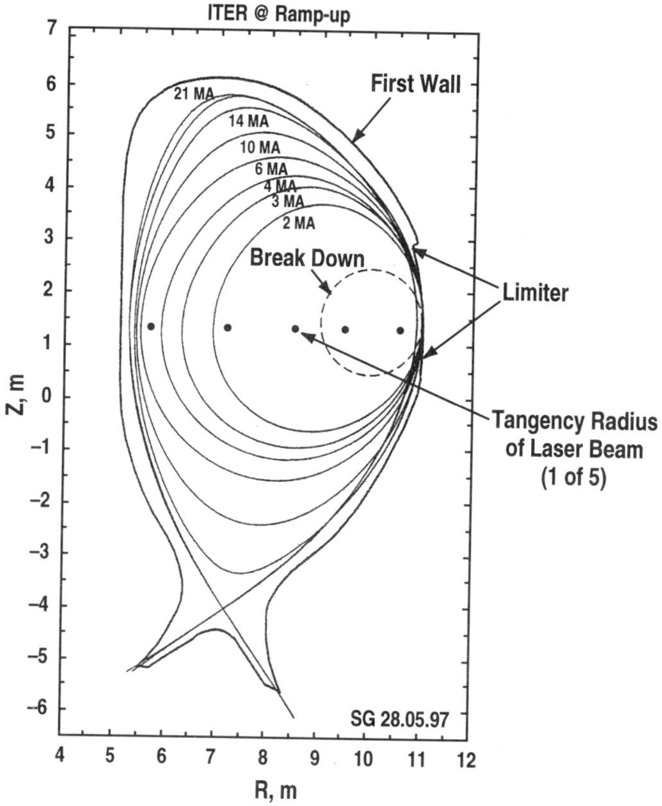

Figure 2. Cross section of ITER and the current plasma startup scheme along with the tangency radius of the five laser beam paths.

All five of the laser beam paths enter and exit the plasma chamber from a single port and are reflected back down the same beam path (with an offset) by retroreflectors embedded behind the ITER shield modules. A plan view of the beam paths through the port shield plug are shown in Fig. 3. By requiring only a single port with no large vibration damping structure the total penetration area through the shield modules is minimized and the neutron leakage both to the bioshield and to the TF coils is kept small thus reducing the total external shielding required.

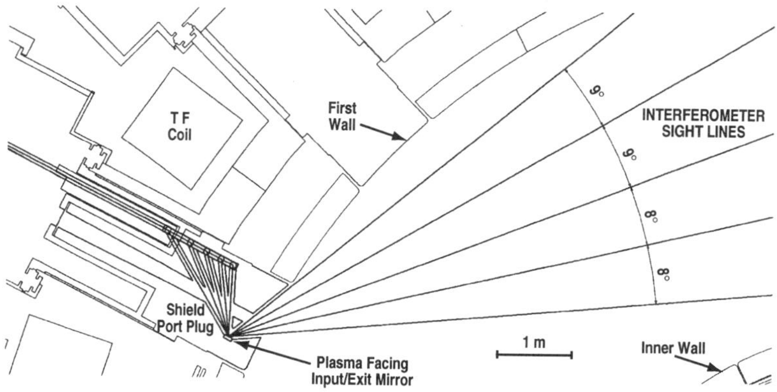

Figure 3. Plan view of the input/exit port plug and laser beam paths.

The optical path through the shield plug on the entrance/exit port forms a labyrinth in order to reduce the escaping neutrons to acceptable levels. The retroreflectors are recessed in the shield wall to minimize their exposure to the plasma and neutrons as shown in Fig. 4. All of the optical components within the bioshield of ITER are reflective except for the vacuum windows which are well behind the port plug and in a relatively benign radiation environment. Reflective optics were chosen to avoid neutron damage issues associated with transmission optics.

Each beam path has collinear 10.6 μm (CO_2) and 5.3 μm (CO) laser probe beams. This choice of laser wavelengths has acceptable refraction and phase resolution[3,4] and commercial laser detector combinations with good power/sensitivity are available. The tokamak community has experience with both CO_2 and CO laser interferometer systems.[8–11]

The interferometer portion of the diagnostic uses standard two color heterodyne interferometry using acousto-optic modulators (Bragg cells) to generate the intermediate frequency (IF). The Bragg cells shown in Fig. 5, split the incoming laser beam into two beams separated by a small angle with one beam at the initial frequency (ω) and the other at the initial frequency plus the modulator frequency (ω + 42 MHz for example). Figure 5 shows a schematic layout of both arms of the two colors (CO_2 and CO laser wavelengths). To make the Faraday rotation measurement right and left circularly polarized beams at different frequencies are generated from a single CO_2 laser beam and recombined to produce an elliptically polarized beam which rotates at half the difference frequency. The rotating elliptical beam is passed through the plasma and allowed to interfere with a reference beam at a single detector as shown in Fig. 5(a). The output from the detector is then split, filtered and the phase detected (relative to the IF frequencies) as shown in Fig. 5(b). This method is currently used with FIR laser systems in TEXTOR[12] and RTP[13] and can be extended for use with a CO_2 laser system. The signals from the phase detectors shown in Fig. 5(b) can be used to extract the line average electron density using both the CO and CO_2 interferometer signals to measure and subtract the phase shift induced by vibrations or motion of the mirrors ($\phi_{vib}=2\pi\Delta L/\lambda$ where ΔL is the motion of the mirrors in the direction of the beam) from the phase shift induced by the plasma density ($\phi_{nel}=2.8\times10^{-15}\lambda \int n(z)dz$). A signal from the same CO_2 detector can be used to measure the Faraday rotation angle, $\alpha_F=2.6\times10^{-13}\lambda^2 \int n_{el}(z)B(z)dz$.

This method for making the interferometer and Faraday rotation measurements requires a minimum of detectors (one CO_2 and one CO detector for each spatial channel), uses separate CO_2 and CO lasers for each spatial channel, and has no mechanically moving parts such as rotating grating or polarization rotator. In addition, all of the required components are commercially available. This combination of attributes helps assure that the system will be robust and reliable. The high modulation frequency (4 MHz in Fig. 5) permits the system to follow rapid changes in the density without loss of accuracy.

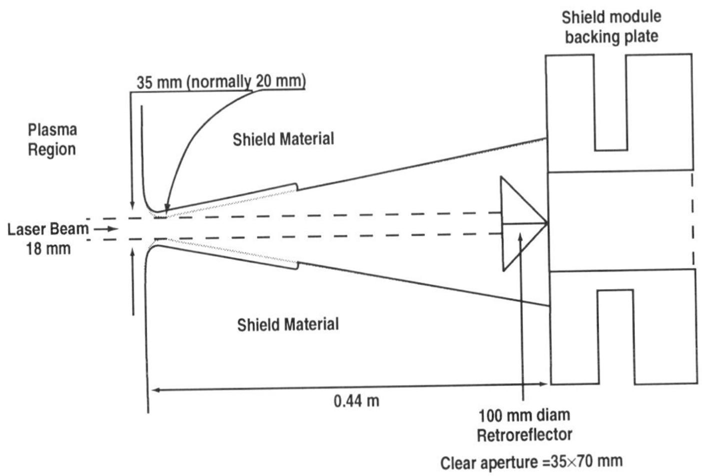

Figure 4. Elevation view of a retroreflector behind the shield modules in the space between modules.

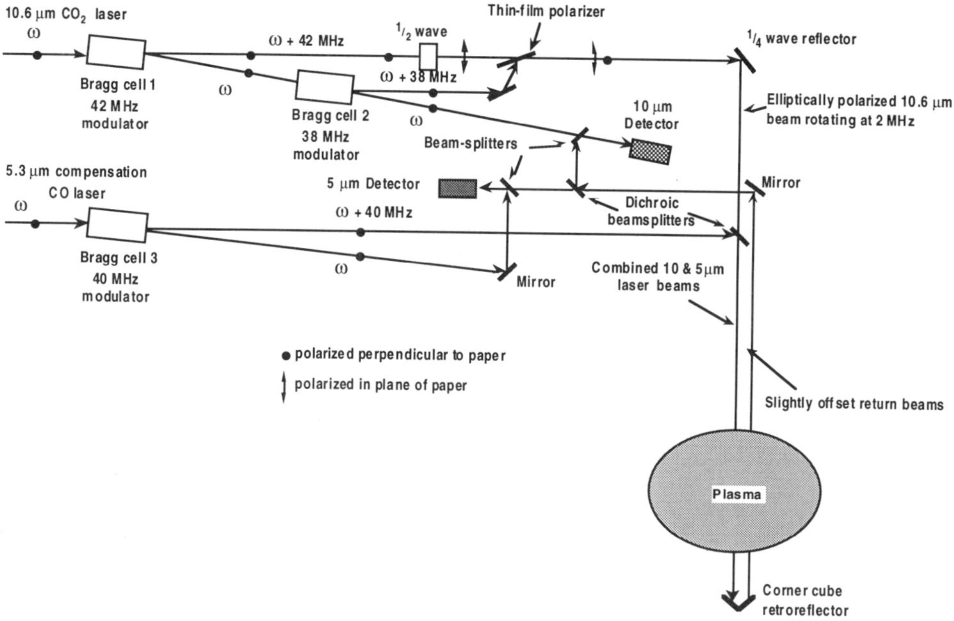

Figure 5(a). Schematic layout of the vibration-compensated interferometer (CO and CO_2 lasers) and the additional Bragg cell and polarizers for making the elliptically polarized CO_2 laser beam for the Faraday rotation measurement.

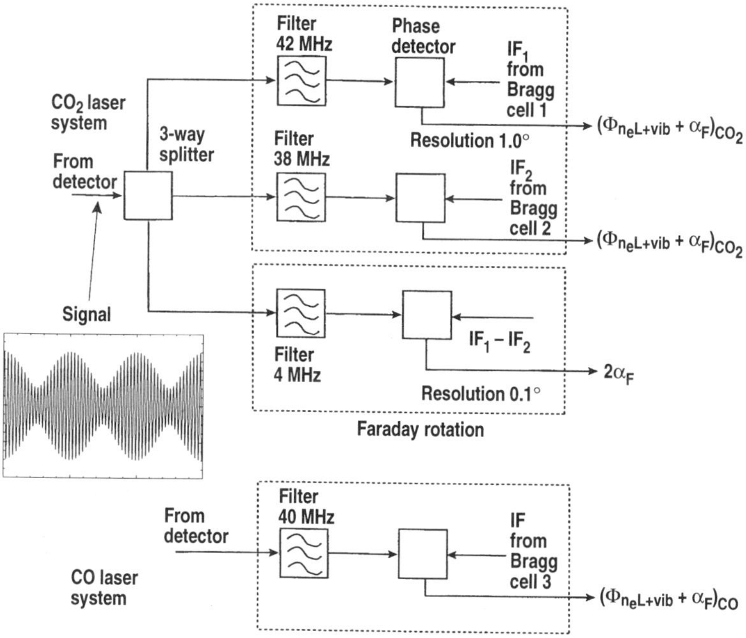

Figure 5(b). The signal from a single CO_2 detector is split and filtered to give three signals that can be used to calculate $\int n_e dz$ and $\int n_{e(z)}B(z)dz$ The signal from the CO detector is used to subtract the phase shift due to vibrations of the mirrors.

Beam Path Layout

We have done an approximate Gaussian beam optics calculation to estimate the laser beam size along a typical beam path as shown in Fig. 6. We start with a near diffraction limited CO_2 laser beam (DEOS[14] LC-40, 40 W, 1.54 mm output beam diameter, full angle divergence θ=9.7 mrad; diffraction limit $\theta = 4\lambda/\pi d_0 = 8.8$ mrad), which is then 3× expanded and focused to a beam waist 2 m away. This keeps the beam diameter small, between 4.62 mm and 6.46 mm, while on the optics table where it passes through two Bragg cells, several retardation reflectors, mirrors, and beam combiners. A typical Bragg cell (IntraAction[15] AGM-406) has an aperture of 6 mm and power density capability of 320 W/cm^2. These specifications are easily met if the Bragg cells are placed near the first beam expander. The resulting beam is further expanded by 3.6× before it is combined with the CO laser beam and leaves the optics table to start its path to the plasma. At this point, the beam diameter is 23.2 mm and the divergence is 0.9 mrad. The CO laser beam is handled in a similar manner so that it has approximately the same beam diameter and divergence. An 18.0 mm beam waist is formed about 20 m from the beam expander and this beam waist is relayed to the position of the corner cube by a focusing mirror located midway between these two positions. The return beam is offset horizontally 35 mm at the corner cube (clear aperture 35×70 mm), and returns through the same optics as the incident beam. This offset is required to prevent the return beam from re-entering the laser and affecting its stability. Due to the relay optics, the return beam has the same diameter as the incident beam. At the optics table the beam is combined with the reference CO_2 and CO laser beams and focused on the detectors.

There are several possibilities for feedback alignment control. In Fig. 6 we show a CO_2 beam position sensor before the second alignment mirror, close to the vacuum window. This sensor is used to control the first alignment mirror so the beam is properly located at the vacuum window. The second CO_2 beam position sensor is located at the optics table, just before the beam is combined with the reference beams. This sensor is used to control the second alignment mirror, and keeps the beam focused on the detector. The CO beam is made collinear with the CO_2 beam using a similar method (not shown in Fig. 6). To

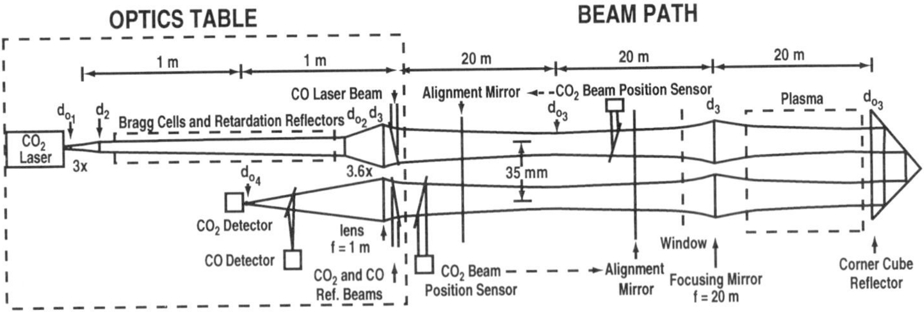

d_{o_1} = 1.54 mm, θ_1 = 9.7 mrad

d_{o_2} = 6.46 mm, θ_2 = 3.2 mrad

d_{o_3} = 18.0 mm, θ_3 = 0.9 mrad

d_{o_4} = 0.94 mm

d_2 = 4.62 mm

d_3 = 23.2 mm

$d_o = {}^1/_{e^2}$ beam waist

θ = full angle diverence

$d = {}^1/_{e^2}$ beam diameter

$d_o \approx f\theta$

$d = \sqrt{2}d_o$ (confocal resonator)

beam expander: $\frac{d_{out}}{d_{in}} = M$, $\frac{\theta_{in}}{\theta_{out}} = M$

Figure 6. An estimate of the beam diameter along the entire beam path using Gaussian beam optics. Note that the vertical and horizontal scales are different and not to scale.

maintain a good interference signal, the beam splitter that combines the plasma and reference beams can be dithered in a feedback alignment scheme to give the maximum interference signal (38–42 MHz).

The plasma facing mirrors (one steering mirror and three corner cube mirrors) are made from tungsten, whose reflectivity is about 70% at these wavelengths. The five reflections from these mirrors reduce the laser power to 17%. Assuming the transmission of 20% for the rest of the optics (30 reflections at 95%), a 30 W CO_2 laser would produce about 1 W at the detector. This is adequate for room temperature detectors. Cooled detectors have about 100 times improved detectabilty, so signal levels should not be an issue.

Estimate Of Signals And Errors

An estimate of the phase shift and Faraday rotation angle from a central viewing channel for a normal ITER plasma discharge[6] is shown in Fig. 7. Small variations of the laser wavelength will result in a vibration compensation error that is proportional the change in path length due to mirror motion. This error is anticipated to be larger than the phase measurement error (assumed to be 1°) because of the large (2 cm) motion of the vacuum vessel of ITER during the discharge. Experience from DIII–D[4] suggests that the amount of vibration will be proportional to the OH coil flux, shown in Fig. 7(b). The phase shift due to the expected mirror motion is shown in Fig. 7(e) and Fig. 7(f) shows the resulting error in the density measurement for a 250 MHz variation in the laser frequency (typical of commercial lasers). The last two plots in Fig. 7 display the Faraday rotation angle for the ITER discharge and the uncertainty based on a phase detection error of 0.1°. The expected errors throughout the discharge are well within the ITER requirements.[6] The error in the Faraday rotation measurement during the flattop is less then 1%. This very low error, if realized, suggests that the Faraday rotation measurement from this system has the accuracy to make a measurement of the internal diamagnetism of the plasma as described in Ref. 3.

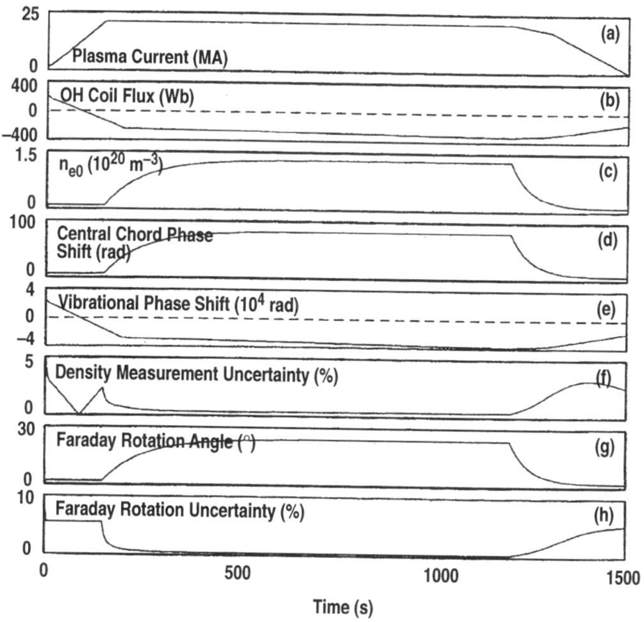

Figure 7. Time traces of plasma parameters and density measurements signals.

3. PLASMA FACING MIRRORS

A major concern for any optical diagnostic on ITER is the survivability of the plasma facing optical components. Surface roughness and distortion of the flat mirror should be kept below 1.2 μm (λ/4 for 5 μm) to maintain the phase fronts of the probing beams. A further requirement for the retroreflectors is that the perpendicularity of the three flat surfaces that make up the retroreflector must be maintained to less then 10 arc sec corresponding to a 1 mm shift in the return beam at the input/exit port.

A finite element thermal analysis code was used to calculate the distortion of water-cooled tungsten mirrors for a full power steady-state ITER discharge. Thermal distortion of the mirrors due to non-uniform heating from plasma radiation and neutrons is less then 0.9 μm as shown in Table 1. A refined water cooling design and minor change in the geometry of the mirrors reduced the thermal distortion of these mirrors from that reported in Ref. [3]. The change in perpendicularity of the retroreflectors due to thermal distortion from plasma and neutron heating is calculated to be less then 4 arc sec.

Sputtering from charge exchange neutrals does not appear to be an issue for the tungsten mirrors with the geometry shown in Fig. 1. An estimate of the CX sputtering rates for the input/exit mirrors is shown in Table 2. The calculations and assumptions from which this estimate were made follow the work done by Mayer and de Kock.[16] We assume a CX-flux (Γ_{tot}) of 2×10^{19} atoms m^{-2} s^{-1} hitting the walls of ITER and that the flux incident on the mirrors is $\Gamma_{mirror} = \Omega/\pi\ \Gamma_{tot}$ where Ω is the solid angle of the plasma as viewed by the mirror. The flux consists of 50% D and 50% T and the energy distribution of the neutrals (both D and T) were taken from Fig. 2 in Ref. 16. The mirrors for the interferometer are not normal to the plasma surface and an increase in the sputtering erosion of up to $\cos^{-f}\Theta$ can result, where $1 < f < 2$ and Θ is the angle of incidence.[17] Because of this effect, the erosion rates have been increased by a factor of two over the values calculated from the normal incident sputtering yields. As shown in Table 2, the estimated erosion from sputtering during the lifetime of ITER[7] is well below the 1.2 μm required for operation of the diagnostic for tungsten, rhodium, and molybdenum mirrors.

The plasma facing optics in the present design have anticipated thermal distortion and sputtering levels that are acceptable. The key to reducing both the sputtering and the thermal distortion is the choice of materials (with tungsten as the clear favorite), the cooling geometry, and the minimization of the solid angle view of the plasma by the optics.

Table 1. Finite element thermal/stress analysis results for the plasma facing mirrors. Distortion is below the required 1.2 μm. The water flow velocity in the cooling channels was adjusted to give reasonable temperature and distortion.

	Maximum Distortion	Plasma Radiation Heat Load	Neutron Heating	Water Flow Velocity	Temperature Variation Across Surface
Flat input/exit mirror	0.9 μm	4.95 W/cm^2	4.2 W/cm^3	10 m/s	8°C
Retroreflector	0.36 μm	0.09 W/cm^2	0.36 W/cm^3	1 m/s	5°C

Table 2. Calculated maximum sputtering removal rates for the flat input/exit mirror. Tungsten, Rhodium and Molybdenum have removal rates that will allow the mirror to survive (less than 1.2 μm of materials removed) the life of ITER.

Mirror Material	Sputtering Removal Rate (μm/s)	Material Removal per ITER Shot	Material Removal During Expected ITER Life
Tungsten	1.3×10^{-9}	1.3×10^{-6} μm	0.13 μm
Rhodium	2.2×10^{-9}	2.2×10^{-6} μm	0.22 μm
Molybdenum	5.3×10^{-9}	5.3×10^{-6} μm	0.53 μm
Copper	4.4×10^{-8}	4.4×10^{-5} μm	4.4 μm

While these results have reasonable safety margins, significant increases in the plasma radiated power (factors of two or more) over the estimates used in the calculations could result in thermal distortion problems. A remaining unaddressed issue is coating of the optics from the plasma. Anecdotal information from existing tokamaks suggest that if the optical component is recessed from the first wall by more then one or two diameters of the penetration through the wall then coating of optics is minimal. While the present design satisfies that criteria the very long pulse length of ITER may require other ways to reduce the plasma coating. Systematic studies in present day tokamak should be able to better address this issue.

4. DISCUSSION

Knowledge of the electron density is important for the control of the plasma and safety of the tokamak. Therefore, the interferometer system must be robust and have suitable redundancy and reliability. In this regard, we recommend that the five chords of the proposed interferometer be independent systems, each with their own lasers, optics, detectors and electronics. Only a few common elements of the beam path, such as the final plasma facing turning mirror should be shared. (Even the plasma facing mirror could be made separate for each spatial channel, however there is a significant penalty to be paid in terms of port space and neutron streaming). This would offer significant redundancy, insuring that at least one interferometer is always operational. All electronic and delicate components should be located in an area where personnel access is permitted and an inventory of spare parts maintained.

We have chosen to avoid the use of mechanical polarization rotators or frequency shifters in favor of solid state acousto-optic modulators. We feel the solid state technology offers much higher reliably. For this same reason, we have chosen not to pursue efficient designs that use a rotating grating to produce multiple chords from a single laser beam.[18,19]

In summary, we have developed a baseline design for an interferometer and polarimetry density measurement for ITER that meets the ITER requirements both in terms of performance of the diagnostic and integration into the ITER device. The design uses well established techniques and components and should be viewed as a measure for future improvements or new concepts for density measurements. Future work includes a demonstration of a complete system on an existing tokamak, detailed neutronics calculations to verify the rough calculations to date, systematic studies on window coatings by tokamak plasmas and a detailed design of the optical system including the alignment system.

ACKNOWLEDGMENTS

We wish to thank C. Walker and N. Kobayashi of the ITER Joint Central Team for the machine drawings of ITER. This work supported by U.S. Department of Energy under Contract No. DE-AC03-94SF20282 under Raytheon Engineers & Constructors, Inc., Subcontract No. ITER-GA-4002.

REFERENCES

1. ITER Report TAC-95-15; TAC-JCT informal technical reviews.
2. F.C. Jobes and D.K. Mansfield, Rev. Sci. Instrum. **63**, 5154 (1992).
3. R.T. Snider, et al., Proceedings of the International Workshop on Diagnostics for ITER. Plenum Press, 225 (1996).
4. R.T. Snider, et al., Rev. Sci. Instrum. **68**, 728 (1997).
5. D. Brower, et al., submitted to Plasma Physics and Controlled Fusion.
6. A.E. Costley, et al., Proceedings of the International Workshop on Diagnostics for ITER. Plenum Press, 23 (1996).
7. Dv. Orlinski, Ibid p 51.
8. T.N. Carlstrom, D.R. Ahlgren, and J. Crosbie, Rev. Sci. Instrum. **59**, 1063 (1988).
9. Y. Kawano, et al., Rev. Sci. Instrum. **63**, 4971 (1992).

10. J.H. Irby, et al., Rev. Sci. Instrum. **59**, 1568 (1988).
11. P. Innocente, S. Martini, Rev. Sci. Instrum. **63**, 4996 (1992).
12. H. Soltwisch, "Combined interferometric and polarimetric diagnostics for TEXTOR," Jülich Report Jul-1638 (1980).
13. J. Rommers, Ph.d. Thesis, FOM-Instituut voor Plasmafysica, Nieuwegiun.
14. DEOS, 1280 Blue Hills Avenue, Bloom Field, Connecticut
15. IntraAction Corporation, 3719 Warren Avenue, Bellwood, Illinois.
16. M. Mayer and L. deKock, CX Erosion of Mirrors, IPP Report in progress.
17. E.W. Thomas, Atomic Data for Controlled Fusion Research, Vol. III, "Particle Interactions with Surfaces," ORNL-6088/V3 (1985).
18. J. Howard, Rev. Sci. Instrum. **61**, 1086 (1990).
19. F.C. Jobes, Rev. Sci. Instrum. **66**, 386 (1995).

POLARIMETRY FOR POLOIDAL FIELD MEASUREMENTS

A.J.H. Donné[1], T. Edlington[2], E. Joffrin[3], H.R. Koslowski[4],
Ch. Nieswand[5], S. Segre[6], P. Stott[7], C. Warrick[2] and V. Zanza[6]

[1]FOM-Inst. voor Plasmafysica Rijnhuizen, 3430 BE Nieuwegein, NL
[2]UKAEA Culham Laboratory, Abingdon OX14 3DB, UK
[3]CEA Cadarache, 13108 Saint-Paul-lez-Durance, F
[4]Forschungszentrum Jülich GmbH, 52425 Jülich, D
[5]CRPP/EPFL, 1015 Lausanne, CH
[6]ENEA, CR Frascati, 00044 Frascati, I
[7]JET Joint Undertaking, Abingdon, Oxon, OX14 3EA, UK

I. INTRODUCTION

To operate the International Experimental Thermonuclear Reactor (ITER) in a safe and reliable way, it is of vital importance to have a proper knowledge of the current density distribution in the plasma. Moreover, the fact that very promising results have been achieved with the socalled advanced tokamak scenarios indicates that active feedback on the current density profile would be very desirable.

A standard diagnostic to measure the current density distribution is multichord polarimetry.[1] This diagnostic is based on measuring the Faraday rotation of the plane of polarization of a far-infrared beam that is sent through the plasma. Although the application of this technique to ITER is straightforward from the physical point of view, major technical difficulties are encountered due to the (multi-chord) access that is required to measure the full current density profile. The set-up that is used nowadays on most tokamaks is the one with a set of discrete vertical probing beams or, alternatively, with a slab-like beam covering a substantial part of the poloidal plasma cross section.[2] On ITER this set-up cannot be employed due to the difficult access via the top ports and the presence of the divertor at the bottom. A polarimeter system with lateral probing chords, using 'standard' penetrations through the diagnostic plugs was for a long time thought to be impossible due to necessity to have retroreflectors embedded in the blanket modules at the high field side, thus decreasing the shielding of the superconducting toroidal field coils.

Despite all these difficulties the Microwave Diagnostics Group of the European Home Team has carried out a feasibility study for a poloidal polarimeter system for ITER. The emphasis has been put on the following issues:

- what is the optimum probing wavelength and chord geometry for the polarimeter system?

Diagnostics for Experimental Thermonuclear Fusion Reactors 2
Edited by Stott *et al.*, Plenum Press, New York, 1998

- how large are the effects of beam refraction and the absorption at very high harmonics of the electron cyclotron frequency?
- is it possible to employ the Cotton-Mouton effect to measure the electron density simultaneously by ellipsometry along the same probing chords?

This paper is organized as follows: The results of the feasibility calculations will be presented in Section II. In the first part of this Section initial calculations for some ITER equilibria will be briefly presented. More detailed calculations for a range of density profiles and equilibria are worked out in Section II.B. Whether or not the Cotton-Mouton effect can be used as a novel diagnostic for measuring the density is discussed in Section II.C. The effects of calculations on refraction and electron cyclotron absorption are then briefly presented, after which the wavelength choice is discussed in Section II.E. Since the technical implementation of the poloidal polarimeter system has not yet been elaborated extensively, Section III on the assessment of technical solutions will be rather brief.

II. CHOICE OF WAVELENGTH AND VIEWING GEOMETRY

II.A. Initial calculations and possible chord geometries

To find the optimum chord geometry for the polarimeter system initial calculations have been done for five different geometries and two sets of ITER plasma parameters.[3] All calculations have been done for three different wavelengths (10, 50 and 100 μm). The Faraday rotation as well as the Cotton-Mouton effect have been calculated for polarimeter beams injected in the O-mode. This implies that the orientation of the Stokes vector $\mathbf{S} = (\cos2\chi\cos2\psi, \cos2\chi\sin2\psi, \sin2\chi) = (1,0,0)$, with ψ the orientation of the polarization ellipse and χ related to the ellipticity ε via $\tan\chi = \varepsilon$. The ITER parameters used in the calculations have been derived from transport code simulations for ITER reference ELMy H-mode conditions. One parameter set simulates a sawtoothing discharge with a flat region at the centre of the T_e profile, while the other set with a peaked temperature profile simulates a sawtooth free discharge. The density profiles in both parameter sets are rather similar: very flat with a central density of 1.3×10^{20} m^{-3}. Because of the similarity of the density profiles no big differences were observed between the calculations for the two cases. The full magnetic equilibrium of two discharges have been used in the calculations.

The five viewing geometries included three options with a fan of polarimeter beams launched from the low field side towards the high field side. These three options mainly differed by the fact that beams were launched from points below, on and above the midplane. In all three cases the beams covered almost the complete poloidal plasma cross section. In the fourth geometry a fan of beams was launched from the top of the ITER plasma towards the bottom part of the plasma at the low field side. The fifth geometry that has been incorporated consisted of a set of parallel horizontal viewing chords penetrating through the diagnostic plugs (see Fig. 1). This geometry is the one that is most straightforward to implement from the technical point of view.

The calculations were done by cutting the plasma in 1 cm thick slabs, calculating the transformation matrices for an electromagnetic wave travelling through this slab and applying these matrices to the electric field vector of the probing beams.[4]

The result of the initial calculations is that polarimetry is physically feasible at all three wavelengths studied. The typical rotation angles (at single passage through the plasma) are 1°, 20° and 80° for 10, 50 and 100 μm. The ellipticity is negligible at 10 μm. Care should be taken to correct for the ellipticity at 50 μm. At 100 μm the ellipticity can become very large and should be explicitly taken into account in the analysis. Because of the special conditions at ITER it becomes for the first time possible to derive the density profile from the ellipticity measured by an ellipsometer.[5] The sensitivity of the polarimeter system to plasma position changes in a direction perpendicular to the viewing chord has been determined in the initial calculations. For the polarimeter system with a horizontal set of probing chords the sensitivity for plasma position changes is 0.43, 0.12 and 0.005 degrees/cm at 100, 50 and 10 μm respectively.

In general no large differences were observed between the five viewing geometries in terms of absolute rotation angles and ellipticity. Only the vertical geometry resulted in somewhat larger values because of the longer integration path through the plasma. Because it was thought that the technical implementation of a polarimeter system featuring a set of horizontal probing chords is the simplests it was decided to concentrate further work on this special geometry. Furthermore, to get an idea of the sensitivity of the polarimeter system it was decided to do also calculations for plasmas exhibiting completely different n_e profiles.

II.B. Sensitivity

A set of seven ITER equilibria generated by the CHEASE code[6] was used to investigate the dependence of the Faraday rotation angle and ellipticity on the central safety factor q_0. The same current profile shape was used in all equilibria. The different values of q_0 were achieved by scaling the total plasma current (see Table 1). The typical equilibrium used for the calculations is shown in Fig.1.

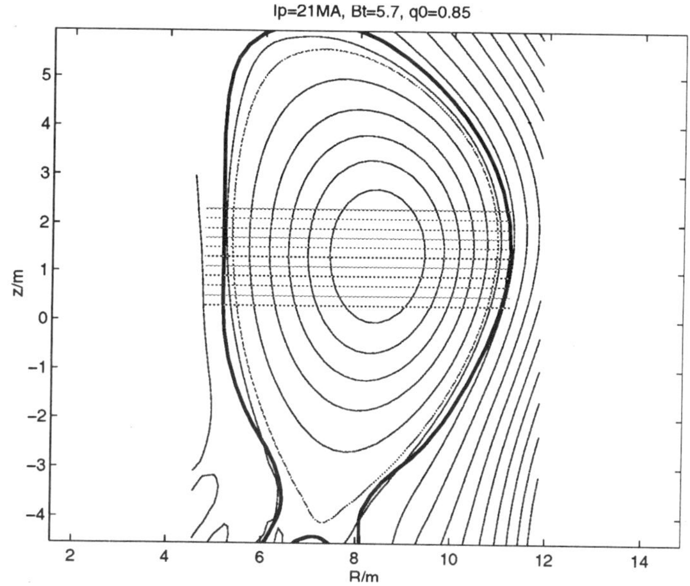

Figure 1. Typical ITER equilibrium used for the calculations. The viewing chords used in the calculation of the horizontal polarimeter system are also indicated.

Table 1. Set of equilibria with equal current profile shape.

q_0	I_p [MA]	$p_{tot}(0)$ [MPa]
0.81	*21.0*	*1.13*
0.85	*20.0*	*1.02*
0.90	*18.8*	*0.90*
0.95	*17.7*	*0.80*
1.00	*16.8*	*0.72*
1.05	*15.9*	*0.65*
1.10	*15.2*	*0.59*

Also a pair of equilibria was generated with the same plasma current (21 MA) but with a different current density profile. Finally, also a calculation was done for a reversed shear scenario. The plasma in this scenario has a smaller shape and carries only 12 MA of plasma current.

A set of density profiles of the following shape was generated:

$$n_e(r) = \{n_e(0) - n_e(a)\}\{1 - (\psi_{norm})^{\alpha}\}^2 + n_e(a);\ n_e(a) = 10^{19}\ m^{-3}$$

To separate the effect of changes in the magnetic field from changes in the density profile the parameters were chosen such that they are consistent with a line-averaged density exceeding the Greenwald limit by a factor of two (for the case with the highest current 21 MA), knowing that this leads to an inconsistency in the low current case. Nevertheless, any scaling factor in the density appears linearly in both the Faraday rotation and Cotton-Mouton effect and can be corrected for afterwards. In Table 2 the density profile parameters and the line averaged densities are listed. The set of density profiles includes relatively flat profiles (α = 10) as well as peaked profiles (α = 2), see Fig. 2.

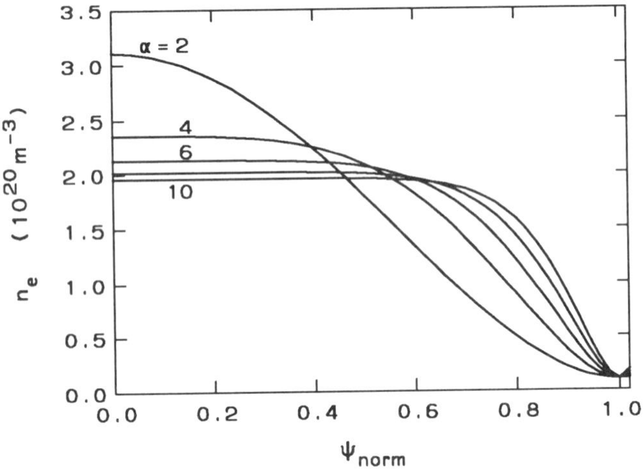

Figure 2. Density profiles used for the calculations.

Table 2. Density profile parameters and line averaged densities.

α	$n_e(0)$ $10^{19}\ m^{-3}$	$\bar{n}_e/10^{20}\ m^{-3}$ $q_0 = 0.81$ 21 MA	$\bar{n}_e/10^{20}\ m^{-3}$ $q_0 = 1.10$ 15.2 MA	$\bar{n}_e/10^{20}\ m^{-3}$ $q_0 = 0.94$ 21MA	$\bar{n}_e/10^{20}\ m^{-3}$ $q_0 = 5$ (rev. sh.) 12 MA
2	31.0	1.98	1.92	1.93	1.84
4	23.5	1.81	1.78	1.79	1.61
6	21.2	1.76	1.73	1.74	1.53
8	20.1	1.74	1.70	1.71	1.49
10	19.5	1.72	1.69	1.70	1.47

Because of the length restrictions it is not possible here to present the details of all calculations of the Faraday rotation and Cotton Mouton effect. These will be published elsewhere.[7]

For a peaked density profile a change of 0.05 in the central safety factor can be observed (see Fig. 3), provided that a change of 4°/m can be detected in the slope of the Faraday rotation angle versus $1/q_0$. At 100 μm this is easy to achieve. Around 10 μm this seems to be more difficult, but it is certainly not impossible. A flat density profile results in a somewhat smaller slope, but changes in the central safety factor of 0.05 can still be observed at 100 μm provided the accuracy is 2.5°/m. The slope of the Faraday rotation profiles is shown in Fig. 4.

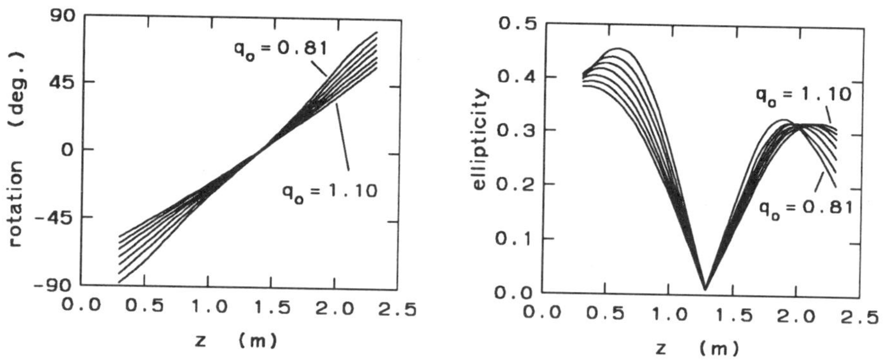

Figure 3. Faraday rotation angle and ellipticity for a peaked density profile. The values are for a single passage of the beam through the plasma only.

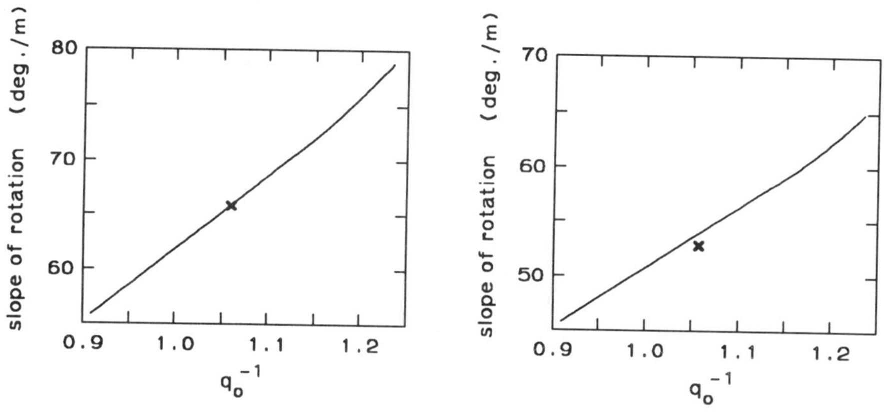

Figure 4. Slope of the Faraday rotation profiles for a peaked (left) and a flat (right) density profile at the zero-crossing of the rotation profile. The cross denotes the case where the current is 21 MA but q_0 is increased from 0.81 to 0.94.

Calculations in which the total plasma current was kept constant (21 MA), but the shape of the current profile was changed such that q_0 increased from 0.81 to 0.94 led to the conclusion that the slope of the q-profile depends stronger on the value of q_0 than on the plasma current (see Fig. 4).

Changes in the central density value, keeping the line-averaged density constant, have a similar effect as variations in q_0. In particular a decrease in the central electron density from 3 to $2 \times 10^{20}\,m^{-3}$ has the same effect on the slope of the Faraday rotation profile as an increase of q_0 from 0.81 to 0.95 in the case of a peaked density profile. The same statement can be made for the absolute value of the Faraday rotation angle. The Cotton-Mouton effect, however, is more strongly affected by changes in q_0 than by changes in the density profile.

Calculations for the reversed shear case, with densities scaled down by a factor of 0.57 (being the ratio of the current in the reversed shear case to the maximum current) resulted in very small Faraday rotation angles (-5° to +5°) at the highest wavelength of 100 μm. It is questionable whether good measurements can be carried out at a wavelength of 10 μm in this particular case. Also the Cotton-Mouton effect is small in the reversed shear scenario. A set of vertical chords could be applied here to increase the sensitivity, albeit that this is at the cost of a much more complicated port interfacing with the blanket and vacuum vessel.

Considering the quoted values for the sensitivity one should realize that all calculations have been done for densities exceeding the Greenwald limit excessively. In practice, however, the densities in ITER might turn out to be smaller, thus resulting in smaller rotation angles. If one also takes the small rotation angles in the reversed shear scenario in mind, then one should conclude that the optimum wavelength for a poloidal polarimeter is in the range from 50 to 100 μm. Longer wavelengths do not make much sense, since one then quickly runs into adverse effects like refraction and electron cyclotron absorption (see Section II. D)

II.C. Density determination from the Cotton-Mouton effect

An explicit study was made to find out whether the Cotton-Mouton effect can be explored as an independent density measurement. In this study density profiles of the form $n_e(x) = n_e(0)(1-x^2)^\alpha$ were used with $x = \rho/\rho_{LCMS}$, and current profiles of the form $j(x) = j_0(1-x^2)^\beta + j_1(1-x^2)^\gamma$, with $j_1/j_0 < 0$ in cases of shear inversion. The value of $n_e(0)$ was fixed by assigning a value for the line-averaged density on a vertical chord through the magnetic axis, whereas the value of j_0 was chosen to obtain a certain value for the total plasma current.

Calculations were done for beams with input polarization **S** = (1,0,0) as well as for **S** = (0,1,0), corresponding to $\psi = \pi/4$ and $\chi = 0$ and amongst others for a polarimeter featuring a set of parallel horizontal probing beams. The other values were taken to: λ = 100 μm, $\alpha = 2$, $I_p = 21$ MA, $B_TR = 46.2$ Tm, $j_1 = 0$, $\beta = 1.7$, corresponding to $q_0 = 0.81$ and $q_{LCMS} = 5.49$.

The calculations for O-mode injection are very similar to those presented in Section II.B. The values of $\Delta\psi$ and $\Delta\chi$ for the chord probing the plasma at z = 0.52 m are listed in Table 3.

Table 3. Values of $\Delta\psi$ and $\Delta\chi$ for **S** = (1,0,0) and for three different values of the line-averaged density

$\bar{n}_e/10^{20}\,m^{-3}$	$\Delta\psi$	$\Delta\chi$
0.16	*5.96°*	*0.06°*
0.5	*18.46°*	*1.81°*
1.6	*57.90°*	*17.37°*

The value of $\Delta\psi$ is approximately linearly proportional to $\bar{n}_e$, whereas $\Delta\chi$ scales roughly as $\bar{n}_e^2$. This is not surprising since for O-mode polarization, ellipticity is only a second order effect. Calculations with the polarization vector oriented under an angle of 45° with the toroidal direction, e.g. **S** = (0, 1, 0), are shown in Fig. 5. The values of $\Delta\psi$ and $\Delta\chi$ for the chord probing the plasma at z = 1.4 m are listed in Table 4. The Faraday rotation angle is not largely effected by the input polarization and still shows the same scaling with $\bar{n}_e$, but now also $\Delta\chi$ scales in the same way.

Table 4. Values of $\Delta\psi$ and $\Delta\chi$ for **S** = (0,1,0) and for three different values of the line-averaged density

$\bar{n}_e/10^{20}m^{-3}$	$\Delta\psi$	$\Delta\chi$
0.16	*5.96°*	*2.22°*
0.5	*18.80°*	*6.99°*
1.6	*63.63°*	*22.53°*

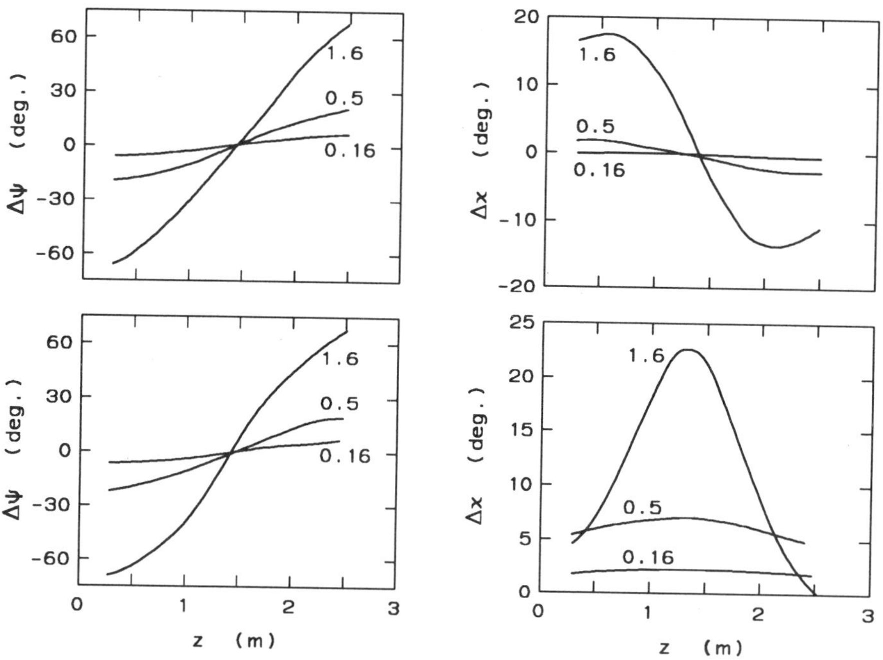

Figure 5. $\Delta\psi$ (left graphs) and $\Delta\chi$ (right graphs) for **S** = (1,0,0) (top graphs) and **S** = (0,1,0) (bottom graphs). The values are for a single passage of the beam through the plasma only.

Calculations for a set a q-profiles and both a low and high value of the line-averaged density led to the conclusion that at low density ($1.6 \times 10^{19} m^{-3}$) $\Delta\chi$ depends only weakly on the q-profile and mainly on the density profile, whereas $\Delta\psi$ depends on both the q- and n_e-profile. At high density ($1.6 \times 10^{20} m^{-3}$) $\Delta\psi$ and $\Delta\chi$ both depend on the q- and n_e-profiles, albeit that for the central chord at z = 1.4 m, $\Delta\chi$ only depends on the n_e-profile. This implies that a measurement of $\Delta\chi$ at 1.4 m provides a direct monitor for the line-averaged density.

A drawback of a poloidal polarimeter system with only a set a horizontal probing chords is the fact the areas around the maximum of the $\Delta\psi$-profile are not diagnosed. This is a serious shortcoming since the positions of these regions and their value are important input values in the determination of the plasma profiles or in the reconstruction of MHD equilibria. Moreover, in advanced tokamak scenarios one would like to have active feedback on the value of q_{min}. This minimum value of q is found close to the position where $\Delta\psi$ is maximum.

For this purpose it is required to have additional to a polarimeter with a horizontal set of viewing chords a limited number of oblique viewing chords in the poloidal plane.

II.D. Effects of refraction and electron cyclotron absorption

Extensive calculations have been done with the BANDIT 3D and TORAY ray-tracing codes to investigate the effect of refraction and electron cyclotron absorption. These studies have been done for the same geometries and density profiles as used for the calculations in Section II.B. Only the longest wavelength of 100 μm was considered, since here the effects are the largest.

The results of the ray-tracing simulations are very encouraging and can be summarized in only a few words. Neither refraction nor electron cyclotron absorption will hamper the operation of a poloidal polarimeter system at 100 μm. To illustrate the smallness of refraction, the end point of two identical vertically launched rays, but for a peaked and a flat density profile, respectively, differ by only 16 mm in the vertical direction and about 2 mm in the radial direction.

Absorption of the rays by the 12th - 19th harmonics of the electron cyclotron frequency is also negligible at 100 μm. Typically a fraction of 10^{-7} to 10^{-6} of the injected power is absorbed.

No calculations have been performed for wavelengths of 10 and 50 μm since refraction and electron cyclotron absorption will be even more negligible.

II.E. Choice of wavelength

Summarizing all calculations presented above there are strong arguments to choose a wavelength of about 100 μm (e.g. there is a strong infrared line at 118 μm) for the following reasons:

- the Faraday rotation angles are very large and can easily be determined. A polarimeter with a modest accuracy of 2.5°/m can still determine changes in the central safety factor of about 0.05.
- the Cotton-Mouton effect is relatively large and should certainly be incorporated in the analysis of the Faraday rotation angle. However, this should not be regarded as a disadvantage since it becomes for the first time possible to measure the line-integrated density by means of ellipsometry. In this way a robust measurement is achieved, which is insensitive to fringe jumps.
- although effects of refraction and absorption by high harmonics of the electron cyclotron frequency increase with wavelength, they are still very small at 100 μm.
- the tolerances of optical components, like the flatness and positioning of mirrors, as well as their degradation due to the vulnerable environment in which they are placed is much less severe for the operation of a system at 100 μm then for a 10 μm system.

At wavelengths much longer than 100 μm one quickly runs into problems with refraction and electron cyclotron absorption.

III. ASSESSMENT OF POSSIBLE TECHNICAL SOLUTIONS

In Section II it has been demonstrated that a poloidal polarimeter system featuring a set of horizontal probing chords along with a limited number of oblique probing chords, and operating at a wavelength of approximately 100 μm, is an ideal instrument to diagnose the q- and n_e-profiles in ITER from the physical point of view. The only big issue that needs to be resolved is how such a polarimeter system can be incorporated in the ITER design.

Because the probing beams are launched from the low field side (LFS) towards the high field side (HFS) special care should be taken that sufficient power is reflected back towards the LFS. Two possible solutions for the reflector at the HFS have been proposed: Firstly, one could employ direct reflection from the blanket tiles. Although it seems that the smoothness of the surface of these tiles on the design table is good enough to ensure that enough power is reflected, it is difficult to assess what will happen with the quality of the surface once ITER comes into full operation. Secondly, one can mount narrow retroreflectors in the space between adjacent blanket tiles at the HFS and slightly indented with respect to the front surface of the tiles. By focussing the far-infrared beam to a waist of approximately 10 cm diameter one can thus ensure that several tens of percent of the injected beam power are reflected back to the LFS. Although reflection of the far-infrared beam at the HFS is certainly a difficult issue, it is thought possible to find a proper solution.

At the LFS penetrations for the probing beams are needed. Most of the beams can be brought into the machine via so-called dog-legs through penetrations of approximately 25 - 30 cm diameter. To determine the optimum number of probing chords extensive simulations are needed, which also incorporate information from other diagnostics on the magnetic equilibrium (e.g. Motional Stark effect, magnetics). The tentative idea is to have five to six horizontal probing chords with four oblique probing chords (two with a positive and two with a negative angle with respect to the equatorial). To enable the use of oblique viewing lines it might be necessary to use some oversized waveguides in slots between the blanket modules[8] instead of the quasi-optical beam penetrations. Whether or not oversized waveguides can be applied depends very much on the available laser power and the fraction of the probing beams that is received back after reflection from the back wall. Any polarization changes and depolarization of the probing beams in the waveguides or in the quasi-optical penetrations can be corrected for by making measurements without any plasma present. In this way the optical transmission matrix of the beam transfer system can be determined. The same matrix is then used to correct the plasma discharges.

VI. CONCLUSIONS

It has been demonstrated that a poloidal polarimeter system can be employed in ITER to measure the current density profile as well as the electron density profile (via the Cotton-Mouton effect). The system does not rely on fringe counting but on measuring the polarization vector of the probing beam after passage through the plasma. This makes the system robust (e.g. not sensitive to fringe jumps). The Faraday rotation angles are large (tens of degrees) and also the ellipticity is relatively straightforward to measure with good accuracy, making it possible to achieve a high accuracy. Relative changes in the value of q_0 of 0.05 can be easily measured. Also the position of zero-crossing of the Faraday rotation angle profile can be measured with an accuracy in the range of a few mm. Because of the elongation and triangularity of the ITER plasma, it is not straightforward to calculate the current density profile from the measured rotation profiles. To do this in a proper way, all information that is available on the magnetic topology of the ITER plasma (e.g from Motional Stark Effect, magnetics, etc.) should be plugged into an equilibrium solver.

Although the outcome of the feasibility calculations is very positive, a warning should be made. Up till now no experimental demonstration of an electron density measurement by the Cotton-Mouton effect has been reported. This is mainly due to the fact that in the present generation of magnetic confinement devices the Cotton-Mouton effect is too small. ITER will be the first device on which this type of measurement can be explored. Furthermore, it still needs to be studied in detail how the current density and electron density profiles can be extracted in a self-consistent way from combined ellipso-metric/polarimetric measurements.

Reversing the last argument: to get an idea of the total number of probing chords of a poloidal polarimeter system for ITER one needs to have an idea what information can be obtained from other diagnostics. One then can insert simulated measurements for a known magnetic topology into an equilibrium solver and compare the equilibrium thus found with the input equilibrium. Although first generic ideas exist how the penetrations of the probing beams through the blanket should look like as well as how the beam can be reflected from the HFS, one needs first to determine the optimum number of probing chords before the penetrations can be worked out in detail.

It has been demonstrated that refraction and electron cyclotron absorption are negligible at wavelengths of 100 μm or smaller. It should still be investigated what will happen to the refraction in case MARFEs are present near the HFS.

Acknowledgements

The contents of this paper is based on discussions in the Microwave Diagnostics Group of the European ITER Home Team and we would like to thank the members of this group, especially R. Behn, S. Barry, C.A.J. Hugenholtz, F.A. Karelse, R.W.M. Polman and J.H. Rommers for their active participation at the various meetings. The work of AJHD was performed as part of the research programme of the association agreement of EURATOM and the 'Stichting voor Fundamenteel Onderzoek der Materie' (FOM) with financial support from the 'Nederlandse Organisatie voor Wetenschappelijk Onderzoek' (NWO) and EURATOM.

References

1. F. De Marco and S.E. Segre, The polarization of an e.m. wave propagating in a plasma with magnetic shear: the measurement of poloidal magnetic field in tokamak,*Plasma Phys.* 14:245 (1972).
2. A.J.H. Donné, High spatial resolution interferometry and polarimetry in hot plasmas, *Rev. Sci. Instrum.* 66:3407 (1995).
3. D. Boucher, private communication.
4. H. Soltwisch, Application of Jones matrices to the analysis of far-infrared wave propagation in tokamak plasmas, *Plasma Phys. Contr. Fusion* 35:1777 (1993)
5. S.E. Segre, On the simultaneous measurement by polarimetry of both electron line density and poloidal magnetic field in tokamaks, *Plasma Phys. Control. Fusion* 35:1261 (1993).
6. O. Sauter, private communication.
7. C. Nieswand, Faraday Rotation Calculations for a FIR Polarimeter on ITER, *these Proceedings;* C. Nieswand, to be published.
8. W.R.Geck, C.W. Domier and N.C. Luhmann Jr., Multiple view fan beam polarimetry on tokamak devices, *Rev. Sci. Instrum.* 68:416 (1997).

FARADAY ROTATION CALCULATIONS FOR A FIR POLARIMETER ON ITER

Christoph Nieswand

Centre de Recherches en Physique des Plasmas
Ecole Polytechnique Fédérale de Lausanne
Association EURATOM - Confédération Suisse
CH-1015 Lausanne, Switzerland

INTRODUCTION

The measurement of the safety factor profile has been considered as an essential diagnostics for ITER.[1] Without the presence of a neutral beam, the only reliable diagnostics which can fulfill the requirements for the q-profile determination is at present the polarimetry.

This paper presents the results of calculations of the Faraday rotation and the Cotton-Mouton effect for various plasma configurations (considered as typical) and various beam geometries which can eventually be realized in spite of the restricted access.

The calculations should help to find a decision for the wavelength and the number and the position of the observation chords of a possible polarimeter system on ITER.

The paper does not deal with technical questions concerning the implementation of such a system on ITER. The potential use of internal retro-reflectors or waveguides for the beams is not discussed here but elsewhere in this volume.[2]

CALCULATION METHOD

Plasma equilibria

All calculations are based on a set of 10 ITER plasma equilibria generated by O.Sauter using the CHEASE code[3] (see figure 1a). The toroidal magnetic field of all investigated equilibria has been 5.7 T on the magnetic axis.

In order to determine the dependence of the Faraday rotation on the central safety factor q_0 the current profile shape was kept constant in 7 equilibria while the total current was scaled such that q_0 varies from 0.81 to 1.1 corresponding to a total plasma current from 21.0MA to 15.2MA, respectively (see figure 1b).

Another pair of equilibria was generated keeping the plasma current constant at 21.0MA but changing the current profile. As a consequence the q-profile and q_0 are different.

Finally a reversed shear case was investigated which was significantly different from the other equilibria in current (12MA) and shape.

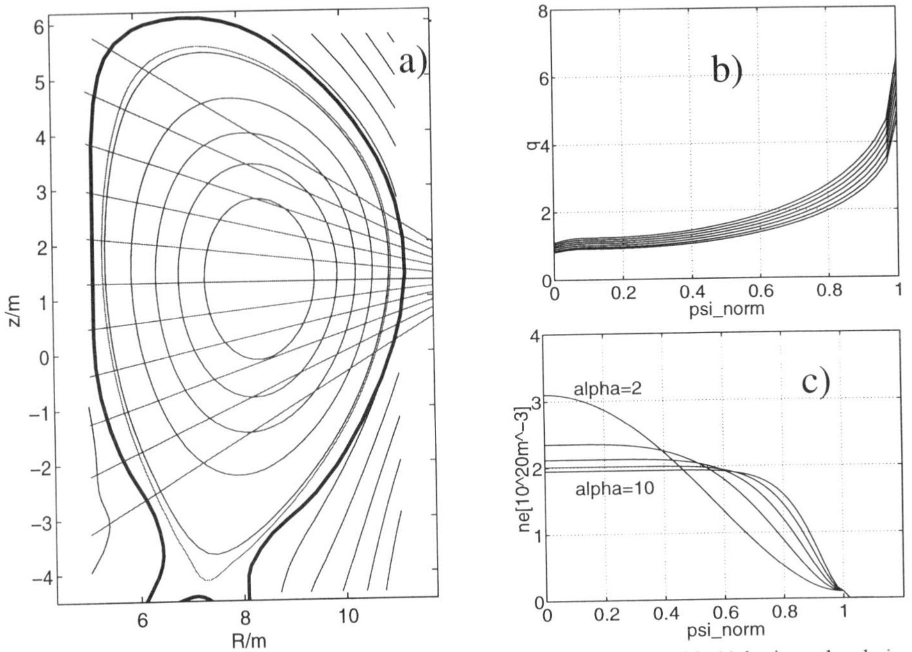

Figure 1. a) Typical ITER equilibrium used for the calculations together with 11 horizontal polarimeter chords; b) q-profiles of a set of 7 equilibria used for the calculations; c) Density profiles used to the calculations

Density profiles

Density profiles have been assumed to be a function of the normalized poloidal flux.

$$n_e(\psi_{norm}) = \{n_e(0) - 10^{19}m^{-3}\}\{1 - (\psi_{norm})^{\alpha}\}^2 + 10^{19}m^{-3}$$

α has been varied from 2 (peaked) to 10 (flat profile) while the central density $n_e(0)$ has been adjusted such that the line averaged density is about twice the Greenwald limit for the 21 MA case ($3.1 \cdot 10^{20} m^{-3}$ to $1.95 \cdot 10^{20} m^{-3}$, figure 1c). Consequently at lower currents the Greenwald limit has been exceeded excessively. However in order to separate magnetic field effects from density profile effects, it is justified to apply these profiles to all equilibria.

Beam parameters and geometries

The calculations have been performed for a wavelength of 100μm. The polarization of the beams has been chosen to be O-mode for most of the calculations.

The main effort has been focused on a scenario with 11 chords distributed equidistantly over a diagnostics port in the vessel and slightly inclined (figure 1). Additional vertical and skew chords have been considered as well.

Beam refraction has been neglected which is discussed elsewhere in this volume.[2]

Method

The plasma has been cut into slabs of 1cm thickness along the observation chords. The transformation matrices for an electromagnetic wave travelling through these slabs have been calculated based on the Appleton-Hartree formula and have been applied successively to the electric field vector of the probing beam. Finally the polarization state of each beam after passing through the plasma once has been evaluated.

RESULTS AND DISCUSSION

A typical profile of Faraday rotation and Cotton-Mouton effect for a flat density profile and 11 horizontal chords is shown in figure 2. The total plasma current was modified resulting in a change of q_0 and in the slope of the Faraday rotation profile (figure 3).

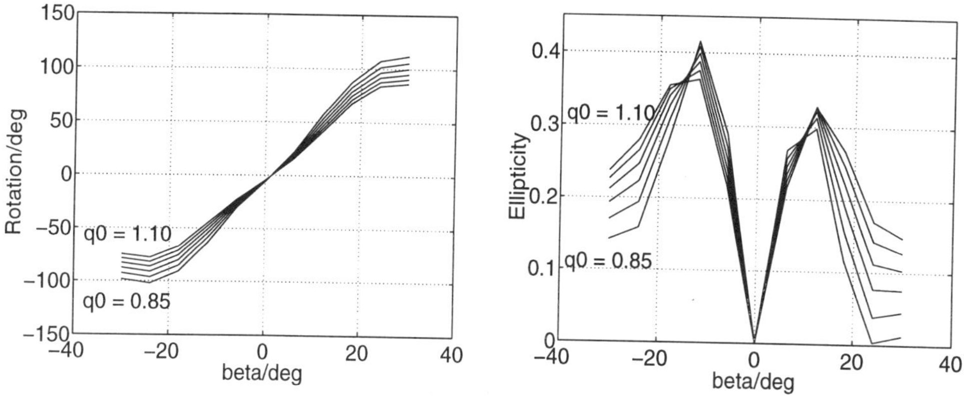

Figure 2. Faraday rotation and ellipticity for a flat density profile, wavelength : 100μm

The Faraday rotation angles for a wavelength of 100μm for a single path through the plasma are in the order of some tens of degrees. Nowadays polarimeters are able to resolve Faraday rotation angles of 0.1° or less. From the slope of the rotation profile (figure 3) at the zero crossing of about 50°/m it can be assumed that a vertical movement of the plasma in the order of 1 centimeter can easily be detected.

In this geometry the chords cover most of the plasma cross section. Specially in reversed shear cases it might be important to have chords passing the region of the minimum of the q-profile, which can be far from the plasma center.

The Cotton-Mouton effect adds significant rotation and ellipticity to the polarization of the beams so that it cannot be neglected at 100μm wavelength. Eventually it can be used to determine the line integrated density simultaneously along the polarimeter chords and provide so informations about the density profile.[2]

At longer wavelengths refraction of the beam becomes important. At shorter wavelengths the Cotton-Mouton effect can completely be neglected. But it has to be taken into account that the density of the calculated cases is at the high density limit and that in the low current cases the density even excessively overestimated. A wavelength of about 100μm (e.g.

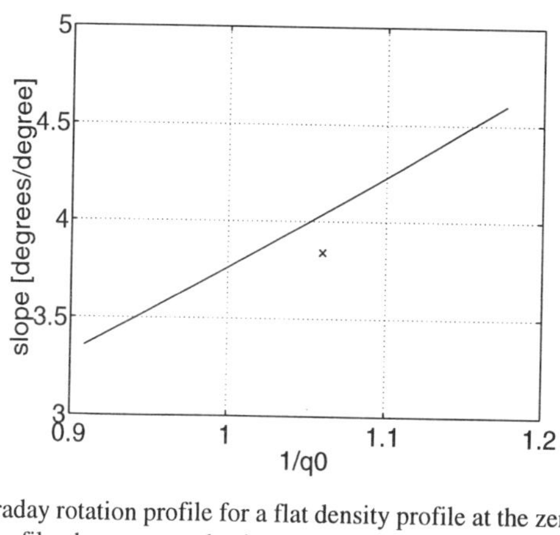

Figure 3. Slope of the Faraday rotation profile for a flat density profile at the zero crossing of the rotation profile, the cross marks the case with 21MA but q_0=0.94.

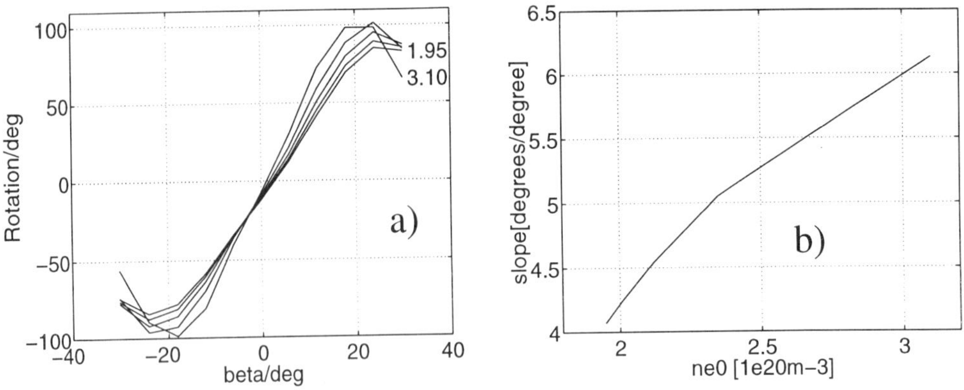

Figure 4. a) Faraday rotation for low q_0. Numbers in the plot denote the central density in $10^{20}m^{-3}$; b) Slope of the Faraday rotation profile for a flat density profile at the zero crossing of the rotation profile

the 118µm CH_3OH laser line) is therefore the optimum choice for a polarimeter on ITER.

In case of equal q_0 but different total current and current profile the slope of the Faraday rotation profile near the zero crossing was found to be equal (figure 3).

However modifying the density profile for a given equilibrium changes the slope of the Faraday rotation profile significantly (figure 4).

CONCLUSIONS

A wavelength of 100µm is the best choice for a FIR polarimeter on ITER.

Slightly inclined chords launched from the diagnostics port allow for measurements across the whole plasma.

It was shown that the slope of the Faraday rotation is linearly dependent on q_0 but also on the density profile. An absolute value of q_0 can only be derived having more informations about the magnetic configuration (symmetry).

The capabilities of a polarimeter on ITER can therefore only be determined completely by equilibrium reconstruction simulations including all relevant diagnostics.

ACKNOWLEDGMENTS

This work was partly supported by the Swiss National Science Foundation. I would like to acknowledge the helpful discussions of this work with the members of the Microwave Diagnostics Group of the European ITER Home Team, namely A.J.H.Donné, T.Edlington, E.Joffrin, H.Koslowski, S.Segre, P.Stott, R.Behn, S.Barry, C.A.J.Hugenholtz, F.A.Karelese, R.W.M.Polman and J.H.Rommers.

REFERENCES

1. K.M.Young et al., The ITER diagnostic programme, in : *Diagnostics for Experimental Thermonuclear Fusion Reactors*, P.E.Stott, G.Gorini, E.Sindoni, ed.,Plenum Press, New York, 1996
2. A.J.H Donné, Polarimetry for poloidal field measurements, this volume
3. H.Lütjens et al., The CHEASE code for toroidal MHD equilibria, *Comput.Phys.Commun.* 95 : (1996) 47-57.

LIDAR THOMSON SCATTERING FOR THE ITER CORE PLASMA

P. Nielsen,[1] L. de Kock,[2] C. Gowers,[1]
F. Orsitto,[3] H. Salzmann,[4] and C. Walker[2]

[1] JET Joint Undertaking, Abingdon, OX14 3EA, UK
[2] ITER JCT Garching, 85748 Garching, Germany
[3] ENEA C.E. Frascati, 00044 Frascati, Italy
[4] MPI für Plasmaphysik, EURATOM Association, 85748 Garching, Germany

INTRODUCTION

For physics interpretations as well as for control of advanced operational scenarios the knowledge of electron temperature and density profiles of the ITER plasma is absolutely necessary. The specifications for the core measurements are 10% accuracy for the electron temperature at a spatial resolution of 10 cm over the temperature range 0.5 - 30 keV, 5% accuracy for the density at a spatial resolution of 30 cm over the range $0.1 - 3.0 * 10^{20}$ m^{-3}. The repetition rate of the measurements should be 100 Hz.

Comparing Thomson scattering with competing techniques, it offers the advantages that it measures both quantities indisputably at the same position and time and that the interpretation of the signals is independent of the plasma scenario. Possible drawbacks of Thomson scattering are that the front end in the ITER environment is vulnerable, that permanent alignment between the laser beam and the collection optics is required and that a large solid angle of collection is needed to achieve the required SNR. Thus the main effort of the design study on Thomson scattering[1] was related to these points of view. Here the system will be outlined only briefly and the emphasis is put on the critical material issues.

LAYOUT OF OPTICS

The problem of maintaining the alignment is much simpler in a backscatter time-of-flight or LIDAR (Light Detection And Ranging) scheme as it is realised on the JET tokamak[2] than in a conventional 90° scattering set-up. This is due to the fact that both the laser beam and the beam of collected light pass the same front optics.

For LIDAR, the spatial resolution ΔL is given by $\Delta L = (\tau^2_{Laser} + \tau^2_{Detection})^{1/2}$, where c is the speed of light, τ_{Laser} is the laser pulse duration and $\tau_{Detection}$ is the response time of the detection system. Thus, in a LIDAR scheme, good spatial resolution requires a short laser pulse and a high-speed detection system. As can be concluded from the experiences with the JET LIDAR diagnostic, the specifications given above can be realised. In addition, the scheme offers the advantages that i) the background radiation added to the signal is small due to the high-speed detection, thus improving the SNR, ii) it only requires one access route/window to the plasma, iii) all spatial channels are transmitted via one line of optics. This means that simpler relay optics (and much less space close to the machine) are required than for the wide field-of-view collection optics needed in a conventional system, and iv) all spatial channels are measured with a single detection system. This has the advantage that the absolute calibration of the system is not essential, but that comparison to a density line integral is sufficient. These arguments, especially those pertaining to the access problem, lead to the conclusion that a LIDAR system is to be preferred for the Thomson scattering diagnostic for the ITER core plasma[3].

The requirement for a vacuum window as small as possible and the requirement for a solid angle of the collection optics as large as possible are opposing. It is overcome by introducing an imaging optics into the vacuum vessel.

Considering that the ITER magnetic coils and all transparent optics such as windows and lenses have to be shielded against neutrons there has to be a shielding labyrinth inside the port and the front optics has to be reflective. The first mirror should be as far away as possible from the blanket penetration to avoid damage by sputtering and to reduce nuclear heating.

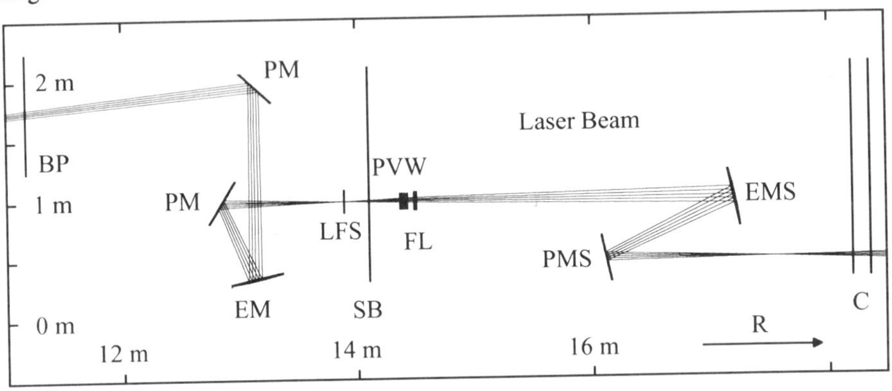

Figure 1. The optical labyrinth in the diagnostic shielding block and the transmission optics in the cryostat. Shown is only the incoming laser beam. This beam is enclosed by the beam of collected light. BP: Blanket Penetration; PM: Plane Mirror; EM: Ellipsoidal Mirror; LFS: Laser Focal Spot; SB: End of Shielding Block; PVW: Primary Vacuum Window; FL: Field Lens; PMS: Plane Mirror (Steerable); EMS: Ellipsoidal Mirror (Steerable); C: Cryostat.

To optimize the shielding efficiency of the labyrinth the bend angles should be around 90°. On the other hand, imaging wide-field reflective optics require small off-angle incidence because otherwise the imaging quality deteriorates too much. These two requirements can be met by a design as shown in Fig. 1. The labyrinth has been designed for a line of sight in a poloidal plane passing through the plasma centre. A LIDAR system has to consider the problrm of vignetting[1,3]. This has led us to image the blanket penetration onto the primary vacuum window. To allow the window to be as small as possible, the front mirror is plane and the second mirror is ellipsoidal.

The laser beam is focused at the plasma centre. A focal spot of the laser also occurs between the imaging mirror and the window, well away from the window, maintaining an energy density on the window well below damage threshold.

Inaccessible optics within the vessel immediately raise the question of alignment of the collection optics with respect to the laser beam. This problem is solved by using the same optics for laser input and collection. Thus relative alignment with respect to each other is guaranteed once the laser and the collected light beam are aligned colinearly outside the vessel (in an accessible area), no matter how the mirrors inside the vessel move. However, there exists the problem that the vessel, and with it the shielding block, moves relative to the building. This movement is not only caused by thermal expansion but also by magnetic forces and therefore it has even a time-varying toroidal component.

Due to these changing conditions the alignment of the scattering diagnostic outside the primary vacuum has got to be feedback controlled: Inside the cryostat a pair of adjustable mirrors (an off-axis ellipsoidal mirror and a plane mirror) is introduced. Approximately 1:1 imaging optics are required inside the cryostat anyway since the cryostat window should also be kept small. The first mirror deflects the optical axis of the front optics so that it intersects the fixed optical axis of the external optics on the second mirror. The second mirror then adjusts the direction of the axis of the front end optics to be collinear with the axis of the external optics.

The laser beam, which outside the cryostat has its own transmission line separated from that for the collected light must be made collinear with the collected light beam. Among the possible methods the easiest and most reliable is to use a hole in one of the mirrors outside the vessel. Vignetting problems are avoided by choosing a mirror located in an image plane of the elliptical mirror[1].

The biological shield outside the cryostat is passed via a labyrinth. Lasers and the detection optics are located in areas permitting 'hands-on' operation. These areas are reached via two optical transmission lines.

LASER, TRANSMISSION OPTICS AND DETECTION EQUIPMENT

The chosen laser wavelength is about 790 nm. With this laser wavelength the 'blue' wing of the scattered spectrum is covered for the full envisaged temperature range by the sensitivity range of photocathodes. A 10 ps / 0.6 J / 100 Hz Ti:sapphire laser system (λ = 800 nm) is outlined in [4]. This system would produce 500 ps / 1 J pulses without the final grating pulse compressor, parameters which would be sufficient for the planned LIDAR system with 30 cm spatial resolution for both the density and the temperature for measurements over the outer minor radius. We conclude from this that the required laser technology is available already now.

An estimate of the optical transmission @ 800 nm of the laser transmission line including the front optics yields a value of about 0.5, whereas the transmission of the collection optics amounts to about 0.3[1].

The spectrometer must offer a high étendue. This requirement is met by interference filters. Since the complete temperature profile is to be measured with a single instrument, the spectrometer is to provide a dynamic range of 0.5 to 30 keV. Approximately 10 spectral channels are required. Further, the spectrometer has to cope with the problem that the scattering volume moves along the line of sight. Consequently the intensity distribution in an arbitrarily chosen, fixed image plane could change from a small, well-defined image to a larger, blurred spot. This in turn would lead to changes in the spectral transmission due to filter inhomogeneities. This effect can be minimised by placing the filters at image planes of the ellipsoidal (second) mirror[1]. The described solution is the same as the one adopted for the JET LIDAR system[2,5] and thus essentially the spectrometer used at JET will be copied.

As detectors we envisage using proximity focused MCP photomultipliers as on JET. These detectors show an excellent gating behaviour[6,7]. Gating the photomultipliers allows efficient discrimination against stray light pulses occurring before the scattered signal. Such pulses are created when the laser pulse passes optical components common to the laser input path and the collected light path, e.g. the vacuum window and the mirrors in the labyrinth. Without gating the detectors would be charge depleted before the measurement.

Digitising of the output signals is done with state-of-the-art transient digitisers which are capable of the short integration time needed for 10 cm spatial resolution. Processing of the data with a neural network will allow real time density and temperature profiles with about 10 ms delay.

In the radial set-up the laser pulse hits the inner wall at the end of the measurement and produces a stray light (and microplasma light) pulse which depletes the charge stored in the MCPs of the detectors. In the low repetition-rate JET device these signal pulses of about 300 V amplitude are clipped electronically so that they do not destroy the input amplifiers of the digitisers[7]. In the ITER set-up fast gating-off of the detectors before the laser pulse hits the inner wall should be foreseen in order to facilitate the high repetition rate and to increase the lifetime of the detectors.

SIGNAL-TO-NOISE RATIO

The expected error of the temperature fit[1] is shown in Fig. 2. The error curves plotted in this figure are derived by calculating the expected signal in 20 equally sized spectral channels over a fixed wavelength range. The numbers for the error bars in Fig. 2 are absolute values. All the curves are calculated using fixed geometry (f/16), fixed laser energy of 1 J, fixed transmission of 20%, 0.15 m scattering length, electron density of 1.0×10^{19} m^{-3} and including 1% bit noise. Including bit noise of this level brings the identical estimate done for the JET LIDAR system in line with the experimentally observed accuracy.

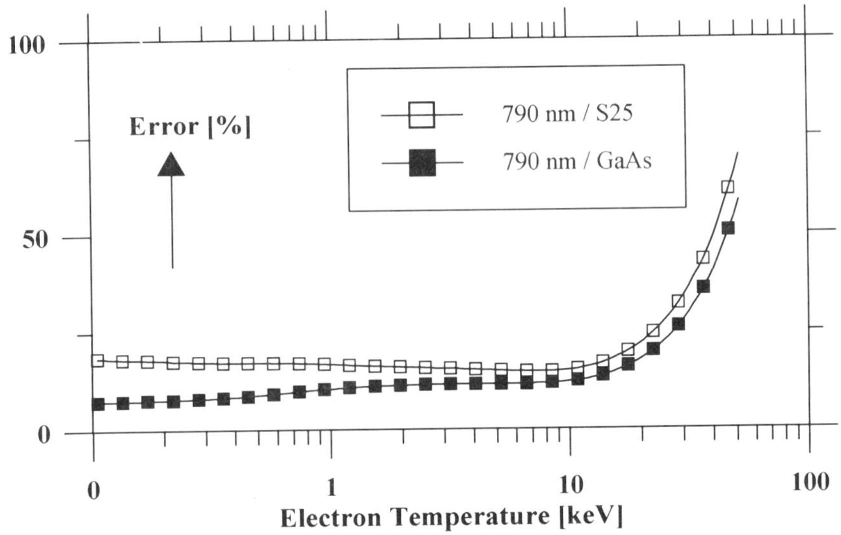

Figure 2. Dynamic range of the electron temperature measurements. i) (790 nm,S25) - 790 nm laser, detector η= 0.2 @ 400 nm dropping to 0.01@ 800 nm. ii) (790 nm,GaAs) - 790 nm laser, detector quantum efficieny η = 0.2 @ 400 nm dropping to 0.1 @ 900 nm.

MATERIAL PROBLEMS

Material problems are envisaged for the front mirror and the primary vacuum window. In the case of the mirror sputtering by CX neutrals, blistering due to ion implantation, laser damage and mirror deformation due to nuclear heating could all be a problem, whereas in the case of the window radiation induced optical absorption, radiation induced luminescence and laser damage have to be considered. With respect to radiation induced damage the shielding properties of the proposed labyrinth have been investigated using approximate methods[8] as well as 3D Monte Carlo calculations[9].

The total shielding efficiency of the LIDAR penetration on its own is 3 times better than the uniform shielding of the blanket and vacuum vessel. Outside the shielding block, behind the first mirror, a higher total neutron flux is found than at the exit hole indicating that locally the wall behind the first mirror is too thin. A 15 cm shift of the labyrinth towards the plasma or some additional shielding material will be sufficient to reduce this flux to an acceptable level.

Primary Vacuum Window

The primary vacuum window will be a double window of less than 15 cm diameter. At its position a total neutron flux of $6.4 \times 10^9\ cm^{-2}\ s^{-1}$ is estimated[9].

The influence of the background luminescence radiation of the window on the SNR of the scattering measurements can be estimated from a comparison of the numbers of photoelectrons in a spectral channel due to Thomson scattering and due to luminescence, respectively. For the Thomson scattering signal we expect about 150 signal photoelectrons / spectral channel / spatial point. The assumed parameters were: 1 J laser energy @ 800 nm, 10 cm scattering length, $n_e = 3 \times 10^{19}\ m^{-3}$, f/16 at the plasma centre, optical transmission 0.1, quantum efficiency 0.1, blue wing of spectrum divided into 10 spectral channels, each 40 nm wide.

The luminescence signal is estimated as follows: The window is positioned at the image plane of the blanket penetration which is to be imaged onto the detectors by the relay optics. Given the optical throughput of the system (20 mm detector with f#1 optics) and the size of the vacuum window (120 mm dia.), the solid angle of collection for luminescence from the window is inferred to be $d\Omega = 0.022$ sr (f/6). With the radioluminescence of the KU-1 fused silica glass[10] of 5×10^8 photons $cm^{-3}\ nm^{-1}\ sr^{-1}\ Gy^{-1}$, a window thickness of 4 cm (a double window) and the integration time of the scattering experiment of 0.3 ns (corresponding to 10 cm scattering length), we get 0.6 background photoelectrons / spectral channel / spatial point /(Gy/s). Independent measurements[11] at neutron fluxes of $4.3 * 10^{16}\ cm^{-2}\ s^{-1}$ confirm the used coefficient for radioluminescence.

Taking the total neutron flux at the window position[8] of $6.4 \times 10^9\ cm^{-2}\ s^{-1}$ and assuming that this neutron flux corresponds to ionising radiation of about 0.07 Gy/s, we arrive at $n_e^L = 0.05$ photoelectrons/channel/spatial point which does not effect the SNR of the measurements in the least.

To investigate radiation induced darkening KU-1 samples were irradiated at room temperature up to a neutron fluence of $10^{18}\ cm^{-2}$ and a dose of 45 MGy[12]. Measurements up to fluence of $6\times10^{19}\ cm^{-2}$ were performed with the same material using a pulsed fission reactor[11]. Subtracting reflection losses, the radiation induced optical density amounts to D $\approx 0.08\ cm^{-1}$ @ 600 nm which for a total optical path length in the double primary vacuum window of 4 cm would correspond to intolerable absorption losses of 52%. Allowing 0.1% absorption losses corresponding to an optical density of $1.085*10^{-4}\ cm^{-1}$ (caused by absorption) a fluence of $8*10^{16}\ cm^{-2}$ could be tolerated. For the neutron flux at the window location as calculated with Monte Carlo methods[9] such a fluence would be reached after

47,800 discharges of 1000 s pulse length. Thus it is not necessary to keep the window at elevated temperature in order to anneal the created colour centres.

To conclude, radiation induced absorption is not a problem with the outlined shielding labyrinth.

Front Mirror

The design criteria for the front mirror are: i) High optical reflectivity over a large spectral range (400 nm - 800 nm); ii) Since the laser beam has to pass the mirror, the laser damage threshold of the reflective layer must be high enough; iii) Low sputtering yield; iv) The thermal deformations of the mirror due to nuclear and laser heating must be compatible with the required optical imaging quality.

Our approach is based on a solution with a reflective layer deposited upon a mirror substrate. This should help to untangle problems of optical reflectivity from those of mechanical stability, thermal conductivity and nuclear heating effects. Considering the requirements for broadband optical reflectivity and low sputtering yield we favour metallic reflective coatings.

Among the metals offering high reflectivity gold and copper are excluded since their reflection coefficient shows a roll-off at 550 nm. In addition gold transmutes to mercury and forms a dark amalgam. This leaves us with the choice among Ag ($R \approx 0.98$ @ 40° incidence, s-pol), Al ($R \approx 0.92$), and Rh ($R \approx 0.84$).

First laser damage measurements using the 300 ps ruby laser of the JET LIDAR system yield laser damage thresholds for normal incidence of 620 mJ/cm^2 for Ag, 350 mJ/cm^2 for Al and 120 mJ/cm^2 for Rh. The single pulse measurements indicate an increase of the damage threshold with the reflection coefficient. Thus, slightly higher threshold values are to be expected for 40° incidence and s-polarisation.

The dependence of the damage threshold on the reflection coefficient indicates thermo-mechanical effects depending on the absorbed laser energy. These effects have been studied in literature for a number of materials and wavelengths[13,14,15,16,17,18,19]. Two main effects leading to mirror damage have been identified:

i) At high laser intensity the laser energy absorbed within a thin surface layer and spread over the penetration depth of the heat wave during the laser pulse can be sufficient to cause melting at the surface. The theoretical single shot threshold laser fluences for damage by melting for a 1 ns laser pulse and an initial mirror temperature of 150 C are 3.58 J cm^{-2} for Ag, 0.29 J cm^{-2} for Al and 0.73 J cm^{-2} for Rh. These threshold data would allow to pass huge laser energies across metallic mirrors. There is, however, a second effect:

ii) The absorbed laser energy heats up a thin layer of material at the surface. The heated matter expands and can do so freely normal to the surface. However, at the radial boundary of the laser spot the dilatation is clamped by the surrounding cold material. This leads to stresses which, if strong enough, can lead to slips. The effect obviously depends upon the radial distribution of the laser energy across the mirror. We do not believe this effect to be limiting for a practical beam. However, we should make sure that the mirrors are not exposed to "hot spots", as produced by diffraction or interference from refractive surfaces.

For multiple-pulse laser irradiation this damage effect in addition shows a reduction of the damage threshold fluence which is cumulative with the number of laser pulses (independent from the laser interpulse time): At each laser pulse the material is stressed. If the stresses exceed the elastic limit repeated plastic deformation occurs which can lead to damage by material fatigue[16].

The models for laser damage on metallic reflectors indicate that all three materials in question could be used. Experiments with large laser spot sizes are planned.

A comparison of the sputtering efficiency of Ag[20], Al[20] and Rh[21] for normal incidence of Deuterium CX neutrals is shown in Fig. 3. The asterisks show experimentally determined values at 100 eV and 2 keV deuterium energy. Note that the experimental error of the measurement at 100 eV is large (the sputtering coefficient is determined from the weight difference of the sample which was only 10 μg in this case).

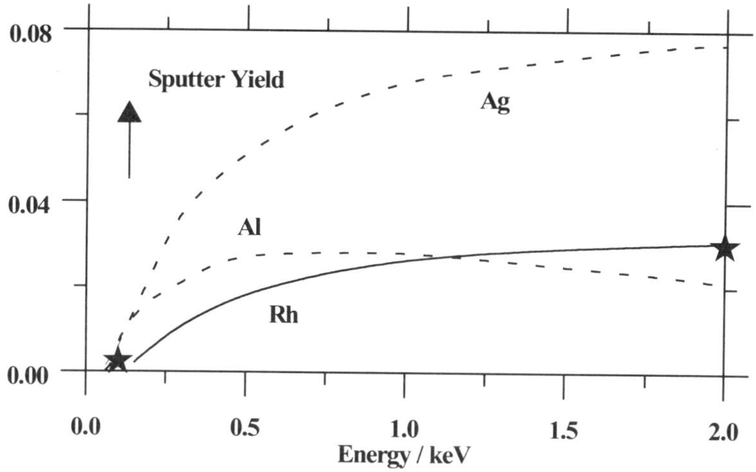

Figure 3. The sputtering yield of aluminum, silver and rhodium for incident deuterium. The asterisks show experimental values for rhodium.

The foreseen lifetime of the mirrors as limited by CX sputtering from a metal surface located at the bottom of a 20 cm dia., 2 m long duct by charge-exchange neutrals with a composition of 50% D and 50% T is inferred[21] using these data. The incident flux onto the wall is estimated as $2x10^{19}$ atoms m^{-2} s^{-1} and is assumed to be isotropic. For the assumed energy spectrum see[21]. With these assumptions one gets an average erosion of 0.031 monolayers of Ag for an ITER discharge of 1000 s duration, and of 0.006 for Rh, respectively. Thus, taking 500 discharges per year, after 4 years of operation about 12 monolayers of Rh would be eroded, corresponding to about 5 nm.

However, the erosion is not homogeneous across the surface but will roughen up the polished mirror, which will lead to an increase in scattering losses. In a review article on metallic mirrors for optical telescopes[22] roughness values attainable by polishing different materials and the related values of the diffuse reflectance are given. Typically, a roughness about 5 nm leads to a diffuse reflectance of less than 1% at 630 nm. Thus we can conclude that after 4 years of operation the maximum surface roughness of the first mirror due to sputtering will only be of the order of its original roughness.

These estimates were checked experimentally: Solid Rh samples were polished to optical quality with a roughness of less than 1.5 nm rms. They were then irradiated with Deuterium ions of 2 keV energy with a fluence which is expected for the above given numbers for 4 years of ITER operation ($5x10^{18}$ atoms cm^{-2}). No measurable deterioration of both the surface roughness and the specular reflection coefficient was found. A 10 fold increase of the fluence was required to enable the measurements of the sputtering yield shown in Fig. 3. Under these conditions the specular reflectivity decreased significantly to a value of about 20%.

It should be noted that only sputtering by CX neutrals from the plasma is taken into account. The problem of secondary particles sputtered from the wall of the duct and the question whether baffling could be a suitable remedy has not been treated yet.

Considering the choice of the substrate the main argument is the stability against thermal deformations. For simplicity cooling channels inside the rather large substrate should be avoided and the cooling should be accomplished by thermal contact to the (water cooled) shielding block. Two possible solutions have been investigated: A sandwich type of substrate with low thermal conductivity between the front and the rear plate and a solid plate of high thermal conductivity. For the first case the worst situation was assumed, namely that the only the front plate is heated. If the substrate can expand freely it will approximately shape into a sphere of radius $R = t\,\alpha^{-1}\,\Delta T^{-1}$ where t is the thickness of the sandwich, α is the linear thermal expansion coefficient and ΔT is the temperature rise of the front plate. Assuming that R = 100 m is tolerable for the nominally plane front mirror, a permissible temperature rise of 135° is inferred for SiC, of 30° for copper and stainless steel and of 20° for aluminum. Nuclear heating at the position of the first mirror however only leads to an adiabatic temperature rise of 2.7° for SiC, 2.9° for Al, 4.9° for stainless steel and 6.2° for Cu after the end of the discharge. These numbers are based on the Monte Carlo calculations[9] which indicate a nuclear heating of 20 $mWcm^{-3}$ for solid stainless steel at this position. However, if Rh is used as reflecting layer, heating by the laser beam leads to much higher temperatures and only SiC (ΔT =74°) would still fulfill the requirements.

For the solid substrate an estimate shows that the thermal equilibration time is much shorter than the ITER discharge time for all the considered materials and it can be concluded that thermal deformations do not pose a serious problem.

CALIBRATION

The relative spectral sensitivity of the detection system which must be known to infer the electron temperature from the scattering signals is measured by illuminating with a calibrated, tunable, monochromatic light source. This light source is usually placed outside the vessel and thus the transmission of the window and of optical in-vessel components cannot be re-measured after the startup of the plasma experiments. However, the presence of inaccessible optical components, whose spectral calibration may change over time (e.g. vacuum windows, mirrors inside the vacuum system), poses a problem in a Thomson scattering diagnostic[23,24,25]. As possible means for updating the spectral calibration, a method is proposed[26,1] which consists in doing Thomson scattering with two lasers at different wavelengths on the same plasma within a short time interval.

The absolute calibration has two objectives. One is to provide a number to allow absolute determination of the plasma density. A more important factor in a LIDAR system is determining of the variation of the solid angle of collection as function of the scattering position. This latter quantity allows us, even in the absence of an absolute measurement, to determine the relative variation of the density profile. By fixing either one point or the line integral, through comparison with other diagnostics, the density profile is inferred.

Density calibration of a Thomson scattering arrangement can be done by performing either Rayleigh or Raman scattering from a rather high pressure (up to 1 bar) molecular gas filling of the discharge vessel. However, Rayleigh scattering may in practice be ruled out due to stray light problems. Raman scattering using nitrogen is a common calibration technique on a number of tokamaks. If the spectral filter characteristics are known in detail this can also provide a complete absolute calibration. However, in the case of a LIDAR system using an interference filter polychromator, the application of the Raman scattering method is difficult: The rotational anti-Stokes Raman spectrum of nitrogen is spectrally very close to the laser line, that is to an intense source of stray light. Therefore interference filters with a very steep slope towards the laser wavelength are needed. The difficulty arises from the fact that the scattered light from different positions along the line of sight intersects the interference filters at different angles of incidence and that the spectral transmission slope of the filter shifts with the angle of incidence. On JET, where a ruby laser is used, this difficulty is overcome by applying a strongly attenuating absorption filter which is insensitive against changes of the angle of incidence. For the 3-level ruby laser system such a specific filter exists in the form of highly Cr^{3+} doped, unpumped ruby[5]. In the case of a laser at 790 nm such a filter would have to be developed.

Fortunately with the geometry chosen for this design the variation of the solid angle of collection is easily calculated. We therefore only need an absolute measurement at one location, e.g. the plasma centre. There scattering is from a point source and it is possible to place a filter at a location such that a parallel beam is incident for this scattering position, avoiding the above mentioned problem.

The unknown ratio to the absolute density, which is constant for all spatial points, can be inferred either from comparison to i) a line integral of the electron density n_e obtained from an interferometer, ii) a line integral of $n_e * B_{par}$ obtained from a polarimeter, where B_{par} is the component of the magnetic field strength parallel to the line of sight, or iii) a comparison with the density at a single point measured by another diagnostic, such as reflectometry. Method i) is well-established on JET and will work on ITER, too.

SUMMARY

A Thomson scattering scheme for measuring electron temperature and density profiles on a central chord of the ITER core plasma has been designed. A backscattering time-of-flight (LIDAR) scheme is used. A spatial resolution of 10 cm along the chord and repetition rates of the measurement of up to 100 Hz are possible but we suggest reducing the resolution to 30 cm to allow for safer operation. The required dynamic range with respect to the high electron temperatures foreseen is achieved by using a short pulse laser at 800 nm together with fast detection covering the whole visible spectrum.

To accomplish a large solid angle of collection for the scattered light together with sufficient shielding of the primary vacuum window (and the machine structure) against the streaming neutrons, an imaging reflective optics is located in a labyrinth penetrating the shielding block which fills the radial port. Neutron scattering calculations yield that the attenuation is sufficient with regard to the machine components. In addition, considering the radioluminescence and the radiation induced absorption measured for high-purity quartz glass, the radiation level at the window location is low enough as to avoid problems.

Thus, the main concern is about the survivability of the front mirror which directly views the plasma. Metallic coatings on different substrate materials have been considered and the effects of nuclear heating, impinging CX neutrals, heating by the laser beam as well as laser damage threshold problems have been considered. A solution is suggested which should survive a reasonable experimental period of the order of 4 years.

Solutions are found to the problems of calibration, both relative (spectral) and absolute. In the case of a slowly changing optical transmission (deposition on the front mirror) a novel method is suggested, where by performing Thomson scattering on the same plasma with a second laser at different wavelength the relative spectral transmission of the complete optical system can be calibrated. Absolute calibration for density measurements is based on cross-calibration against either reflectometer or interferometer/polarimeter measurements.

REFERENCES

1. ITER DDD 5.5.C.01, Thomson scattering (core), (1997)
2. H. Salzmann, J. Bundgaard, A. Gadd, C. Gowers, H.B. Hansen, K. Hirsch, P.Nielsen, K. Reed, C. Schrödter, K. Weisberg, The LIDAR Thomson scattering diagnostic on JET, *Rev. Sci. Instrum.* 59:1451 (1988)
3. C. Gowers, P. Nielsen, F. Orsitto, F.J. Pijper, H. Salzmann, B. Schunke, A Thomson scattering scheme for obtaining Te and ne profiles of the ITER core plasma, in: *Proc. of the Workshop on Diagnostics for ITER*, Varenna, Italy, September 1995 and Report JET-P(95)44
4. A.W. Molvik, R.A. Lerche, D.G. Nilson, M.D. Perry, E.B. Hooper, LIDAR Thomson scattering for advanced tokamaks - final report, *Report UCRL-ID-123434* (1996)
5. C. Gowers, K. Hirsch, P. Nielsen, H. Salzmann, A high rejection ruby filter for laser light scattering experiments, *Appl. Optics* 27:3625 (1988)
6. K. Hirsch, M. Köchel, H. Salzmann, Shutter ratio of a gated ITT F4128 microchannel-plate photomultiplier, *Rev. Sci. Instrum.* 58:2339 (1987)
7. J. Bundgaard, K.B. Hansen, K.-V. Weisberg, Electronics for microchannel photomultipliers in the LIDAR Thomson scattering diagnostic on JET, *Rev. Sci. Instrum.* 60:3265 (1989)
8. H. Iida, R.T. Santoro, S. Yamamoto, Preliminary estimates of radiation streaming through LIDAR penetrations, *ITER Report No. G 73 RI 5 96-09-09 W 1.1* (1996)
9. ITER Report of RF Diagnostic DDG for the third Quarter 1996, Part1, Neutronic Analysis for Diagnostics, Sec. 1, LIDAR system (1996)
10. E.R. Hodgson, A. Morono: 'Sample EC W-T28-01' , ITER R&D Task T28 'Irradiation effects on diagnostic components', *presented at the ITER Technical Meeting of Irradiation Tests on Diagnostic Components based on the T246 Task Agreement, Garching,* 20 June 1996
11. A. Gorshkov, D. Orlinsky, V. Sannikov, K. Vukolov, S. Goncharov, Yu. Sadovnikov, A. Kirillov, Measurement of radioluminescence spectral intensity of the quartz glass under neutron and gamma pulse irradiation at Pulse Fission Reactor "BARS", *Paper given at ITER Diagnostics Progress Meeting, Garching,* 15.-17. Jan. 1997
12. A. B. Berlizov, O.V. Kachalov, A.R. Matesov, Sample RF W-T28-01, *ITER R&D Task T28 'Irradiation Effects on Diagnostic Components'*
13. J.O. Porteus, D.L. Decker, S.C. Seitel, M.J. Soileau, Dependence of metal mirror damage thresholds on wavelength, material, pulse length and preparation method, in *Laser Induced Damage in Optical Materials: 1980*, NBS Special Publication 620, p. 201, Boulder (1980)
14. H.M. Musal, Thermomechanical stress degradation of metal mirror surfaces under pulsed laser irradiation, in *Laser Induced Damage in Optical Materials: 1979*, NBS Special Publication 568, p. 159, Boulder (1979)
15. H.M. Musal, Pulsed laser initiation of surface plasma on metal mirrors, in *Laser Induced Damage in Optical Materials: 1980*, NBS Special Publication 620, p. 227, Boulder (1980)
16. N. Koumvakalis, C.S. Lee, M. Bass, Single and multiple pulse catastrophic damage in diamond-turned Cu and Ag mirrors at 10.6, 1.06 and 0.532 μm, *Optical Engineering* 22:419 (1983)
17. C.S. Lee, N. Koumvakalis, M. Bass, A theoretical model for multiple-pulse laser-induced damage to metal mirrors, *J. Appl. Phys.* 54:5727 (1983)
18. Y. Jee, M.F. Becker, R.M. Walser, Laser-induced damage on single-crystal metal surfaces, *J. Opt. Soc. Am. B* 5:648 (1988)
19. M.F. Becker, Chunchi Ma, R.M. Walser, Predicting multi-pulse laser-induced failure for molybdenum metal mirrors, *Appl. Phys.* 30:5239 (1991)
20. W. Eckstein, C. Garcia-Rosales, J. Roth, W. Ottenberger, Sputtering data, *Report IPP 9/82* (1993)
21. M. Mayer, L. de Kock, CX erosion of mirrors, *presented at the 4th Meeting of ITER Expert Group on Diagnostics, Moscow,* 1 March 1996
22. M.M. Miroshnikov, S.V. Lyubarskii, Yu.P. Khimich, Optical telescope mirrors, *Sov. J. Opt. Technol.* 57:523 (1990)
23. B. W. Brown, C. Gowers, P. Nielsen, B. Schunke, Window transmission monitoring and cleaning schemes used with the light detection and ranging Thomson scattering diagnostic on the JET tokamak', *Rev. Sci. Instrum.* 66:3077 (1995)
24. D. H. McNeill, Effect of changes in viewing window transmission on high-temperature Thomson scattering data, *Rev. Sci. Instrum.* 61:1263 (1990)
25. H. Yoshida, O. Naito, T. Hatae, A. Nagashima, Approach to a window coating problem by *in situ* transmission monitoring and laser blow-off cleaning developed in the JT-60U Thomson scattering system, *Rev. Sci. Instrum.* 68:256 (1997)
26. O.R.P. Smith, C. Gowers, P. Nielsen, H. Salzmann, A self-calibration technique for a Thomson scattering system, *Rev. Sci. Instrum* 68:725 (1997).

ITER EDGE THOMSON SCATTERING

D. W. Johnson and B. Grek

Princeton Plasma Physics Laboratory
P.O. Box 451
Princeton, NJ 08540 USA

INTRODUCTION

Many of ITER's performance limiting mechanisms are expected to exist and interact at the plasma edge. The H-mode edge pedestal parameters are high leverage variables of models used for predicting ITER's success, in measures ranging from core reactivity to divertor erosion. For example, the pedestal temperature is critical in determining the L-H transition. According to many transport models, once H-mode is reached, the H-factor and Q are strongly related to the pedestal pressure. Several classes of ELM instabilities are sensitive to the edge pressure profile shape and also affect performance. The ITER divertor scenario counts on impurity radiation from the plasma edge to reduce power flow to the divertor plates, a mechanism that also affects the edge pedestal region. It is therefore reasonable to expect that, during ITER operation, the edge T_e and n_e profiles are likely to become key operational indicators, and ideally, they would be available for use in feedback control for discharge optimization.

One of the tools identified to probe this important area is edge Thomson scattering, providing local measurements of electron temperature and density. Distinct from other Thomson scattering systems planned to probe the ITER core and divertor regions, the edge Thomson system was conceived to provide profiles of the edge pedestal region with high spatial resolution. In this report, we will first review the current expectations for the ITER edge parameters, and compare these with the measurement goals currently defined for edge T_e and n_e. We will then summarize the evolution of thinking in this area, which is currently at a Phase 2 (of 4) level of design. The design consists of a conventional imaging system making use of a top port on ITER for both the laser and the collection optics. There are several advantages of the top (versus midplane) geometry which will be highlighted. We will then describe a preliminary design which will serve as a reference for further discussion concerning access/performance tradeoffs. While we are aware of no fatal problems at present, there are challenging issues which will strongly affect the design. It appears that target measurement goals can be achieved with adequate access, using available laser/detector technology. However, the important assessment of radiation streaming, and its effect on the survivability of the collection optics and relay fibers await further study. Another key issue is contention for space with other systems in the top port, such as the glow discharge system and the vessel inspection system.

EDGE PROFILE EXPECTATIONS AND TARGET MEASUREMENT GOALS

An excellent summary of the expectations for the ITER edge pedestal characteristics was given by Janeschitz.[1] Based on measurements of edge parameters on CMOD, ASDEX-U, DIII-D and JET (many of which are edge Thomson measurements), and on accepted models describing operational limits and edge transport barrier mechanisms, projections are made to predict the ITER edge operational space in terms of n_e and T_e. Of particular importance for the ITER edge Thomson design is the expected pedestal width. This will dictate the required spatial resolution, with strong impacts on all aspects of the design. For an H-mode scenario with Type I ELMS, the case with the steepest gradients, reference 1 projects that the ITER pedestal width at the midplane will be in the range of 9-19 cm. This compares to an experimental value of 6 cm on JET. For an H-mode pedestal density limit of $6 \times 10^{19} m^{-3}$, assuming pedestal pressure is limited by ideal ballooning, this range in widths corresponds to pedestal temperature projections in the range of 4 - 8 keV, increasing at lower densities.[1]

Gradient scale lengths for the electron temperature and density in the ITER scrape off layer are expected to be in the range of 2-4 cm at the midplane, projecting linearly with machine size from existing devices. The gradient scale lengths in both the edge pedestal and the SOL are found to shorten as the H-factor improves on existing devices.

Table 1. Measurement specifications as recommended by ITER Expert Group for edge electron temperature and density.

parameter	param. range	spatial res. (referred to midplane)	time res.	precision
T_e	0.05-10 keV	0.5 cm	10 ms	10%
n_e	$(0.05\text{-}3)10^{20} m^{-3}$	0.5 cm	10 ms	5%

Target measurement goals for edge T_e and n_e are given in Table 1. It appears that the target spatial resolution of 0.5 cm is more than adequate to resolve the expected edge pedestal gradient and also adequate for the SOL gradients. This is, however, a technically challenging goal when coupled with achievment of the desired precision at the lowest densities. We have chosen to interpret the spatial resolution goal as appropriate for the image quality of the optical design. However, we also recognize that access constraints on the optical throughput may result in the need to average spatially or temporally to achieve the desired precision at low density. Given the expectations discussed above, spatial averaging for 1 cm resolution would likely be acceptable. With an ELM period expected to be >0.5 s, temporal averaging may also prove useful.

SUMMARY OF PREVIOUS DESIGN EFFORT

At the 1995 Varenna ITER Diagnostics Workshop, several design concepts for edge Thomson systems were presented.[2] It was concluded that considerable laser development was needed to demonstrate the feasibility of a 10-20 ps laser needed for edge LIDAR measurements with the resolution and sensitivity called for in Table 1. Such a laser would need to be capable of high repetition rate, continuous operation in the blue or green with ~10 joule/pulse output energy. Propagation of such a high power density laser beam without component damage or beam filamentation in the vacuum window would also be very challenging.

Several conventional imaging options were also considered in the previous study. In anticipation of blanket penetration restrictions we considered what could be achieved with f/10 viewing optics. As an additional baseline constraint, we also considered readily available laser technology, with $E_{laser} \approx$ 1 joule at 100 Hz. We looked at systems with the laser beam both near-normal and also near-tangent to the outer flux surfaces. Systems with beams nearly tangential to the outer flux surface benefit from long scattering lengths and hence improved sensitivity for the same effective radial resolution. With these constraints, only the geometry with the laser tangent to the flux surfaces had the sensitivity called for in Table 1. However, a single laser pulse would probe only a single spatial position in this arrangement, and scanning the beam between shots or using multiple lasers at different positions would be required to obtain an full edge profile. With the beam at a slight angle to the flux surfaces, it would also be possible to obtain a full profile per shot, and to achieve the require precision at low densities, many shots could be averaged together.

However, geometries with the beam tangent or at a slight angle to the flux surface are problematic, since they rely on redirecting the laser beam with a mirror located within a blanket module. We now doubt the alignment stability and survivablility of the optical quality of such a mirror. It would be subject to possible damage from the laser and the many hazards of close proximity to the plasma. These include eddy current forces on the mounting structure and possibly the mirror itself, nuclear heating and radiation induced damage of the optical coating, as well as surface degradation due to charge exchange bombardment and contamination from plasma impurities and debris. The focusing optics and final aiming mirror for the laser should be located in a well shielded area, preferably outside the primary vacuum. This means that we must concentrate on geometries with the beam near-normal to the flux surfaces.

In the near normal geometry with a small laser beam, the scattering length is ~ 0.5 cm, the specified spatial resolution. With such a small scattering length, it impossible to achieve the required precision under the baseline constraints mentioned above (1 joule laser and f/10 viewing optics). For example, at the midplane with f/10 optics, this geometry would require $E_{laser} \approx$ 17 joule, or with $E_{laser} \approx$ 1 joule, a viewing system with f/1.9 optics would be required. Thus to achieve the required precision in this geometry, we will thus need to design for both higher throughput viewing optics and higher laser energy.

TOP PORT VS. MIDPLANE PORT ACCESS FOR EDGE THOMSON SYSTEM

In principle, access for a near-normal system exists both at the midplane and at the top of ITER. A drawing of a top viewing geometry is shown in Fig. 1. There are a number of advantages in using the top port. When top access was first considered for this diagnostic, there was encouragement from ITER planners since midplane port access was fully allocated. This is likely to be a temporary advantage, since all ITER port space will ultimately be in demand.

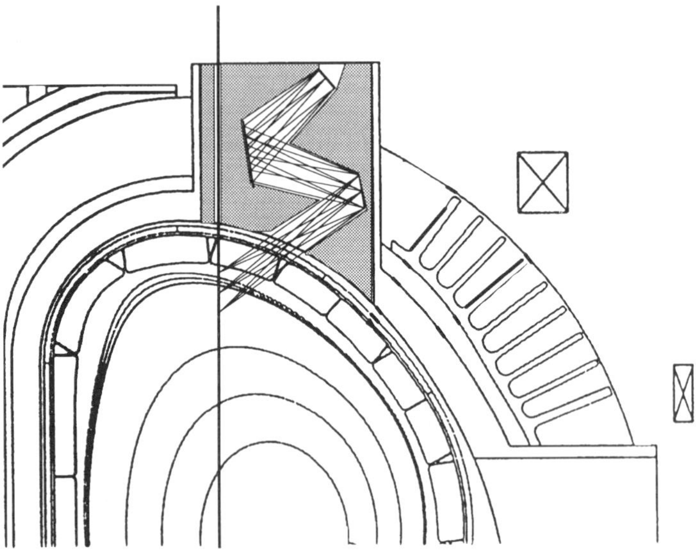

Figure 1. Viewing geometry using dedicated top port.

With its high elongation, there is an expansion in the spacing of the edge flux surfaces by roughly a factor of 3 at the top relative to the midplane on ITER. The most important technical advantage of the top vs. midplane access for edge Thomson is to take advantage of this factor. Other features being the same, the scattering length, and therefore the scattered signal, can be 3 times larger for the same effective spatial resolution. The flux expansion has advantages and disadvantages from the viewpoint of the optical design. The image quality of the viewing optics can be less stringent in absolute terms in the top port geometry. However, the optical field size is increased by this same factor to cover the desired range of positions. This also results in a longer slot in the blanket module.

Another important advantage for the top geometry as shown in Fig. 1 is that the viewing sightlines have a relatively short path length in the plasma. Thus the bremsstrahlung contribution to the plasma light background is reduced. Similar sightlines in a midplane system would have much longer path lengths in the plasma. This benefit could be offset at the extreme edge positions where edge line emission could be significant, although a reduced spectral bandwidth in this low temperature region will aid in rejecting this background light.

Compared to a geometry with a similar viewing angle to a horizontal beam at the midplane, in the top port geometry, the sightlines are inclined farther from the hot core, and there is more shielding between the first mirror and the plasma. This will reduce the radiation exposure of the first mirror and minimize the streaming of neutrons and gammas through the port plug, where they might endanger the window and relay fibers, as well as other components.

Notice that all three of the factors noted above: 1) the flux expansion, 2) the reduction in viewing path length through the plasma, and 3) the shielding from the plasma core; are of maximum benefit if the laser beam can be located as far to the inside of the top port as possible. As we will see later, this configuration also has advantages from the viewpoint of the optical design.

For an imaging system, it is always an advantage to have a narrow laser beam through the viewing volume. If the focusing optics for the beam are far from the plasma, as they should be for ITER, the beam will be small for some distance. The beam used in a midplane geometry would be radial and would need to be dumped on the inner wall. It would be difficult to devise a dump in this environment that would avoid laser ablation of the dump material. The vertical beam from a top system would have more distance to expand and could be dumped in a convenient location in the divertor region.

An important disadvantage of the top port is that the port space available in the toroidal dimension is significantly smaller than for the midplane port, due to shielding requirements for the toroidal field coils. As we will see below, this necessitates the use of elongated optical elements, with adverse effects on the imaging. In the balance, however, we feel that the top geometry is the most favorable for ITER edge Thomson scattering. The existing Thomson system which most closely meets the ITER edge measurement goals probes the top of the DIII-D device.[3]

REFERENCE OPTICAL DESIGN

The ray trace shown in Fig. 1 is the result of an initial optical design effort. The problem posed was to image a 60 cm long section of the vertical laser beam to an array of fiber bundles located outside a primary vacuum window at the port closure plate. We assumed a 20 cm slot width (toroidal) in a dedicated vertical port plug available for the elongated imaging mirrors. The design was done to optimize the optical throughput. The effective collection f/# of this design is f/7.6, although the half angle of rays in the poloidal plane corresponds to f/5.5. With conic section mirrors, spot sizes in the range of 3-5 mm are obtained in the plasma, the low values for cases in which there is a refractive corrector plate at the window location. While far from a final design, we feel this illustrates the optical feasibility of this approach.

PERFORMANCE EXPECTATIONS

Using techniques described in reference 4 and the parameters listed in Table 2, we have predicted the performance of this reference design. We are basing these calculations on the use of a fundamental Nd:YAG laser with 3 joule output energy. Such lasers are becoming commercially available with 10 Hz repetition rate. Diode pumped, higher repetition rate versions will be available in the near future. Multiple lasers have also been used effectively to increase the time resolution of Thomson systems.[3] We assume conventional avalanche photodiode detector technology with filter polychromator analyzers. Noise due to background plasma light can be a significant factor degrading the precision, particularly in the measurement of the low density region of a steep edge pedestal. In this calculation we assume a uniform background plasma with the parameters shown in Table 2.

Table 2. Parameters used in performance analysis of preliminary design of ITER top port Thomson system

Laser	
Laser energy (1064 nm)	3.0 joules
Optics	
Collection f/#	7.6
Scattering length	1.5 cm
Scattering angle	124°
Transmission	10%
Detector	
APD excess noise factor	1.5
Preamplifier noise	15 electrons
Spectrally dependent quantum efficiency	
Spectrometer	
Long wavelength edge	1060nm
Channel widths	5,10,20,40,80,160,250 nm
Background light (assume uniform background plasma)	
Integration fime	50 nsec
T_{brem}	4.0 keV
n_{brem}	$6.0x10^{19}m^{-3}$
Z_{eff}	1.5
Bremsstrahlung multiplier	3.0

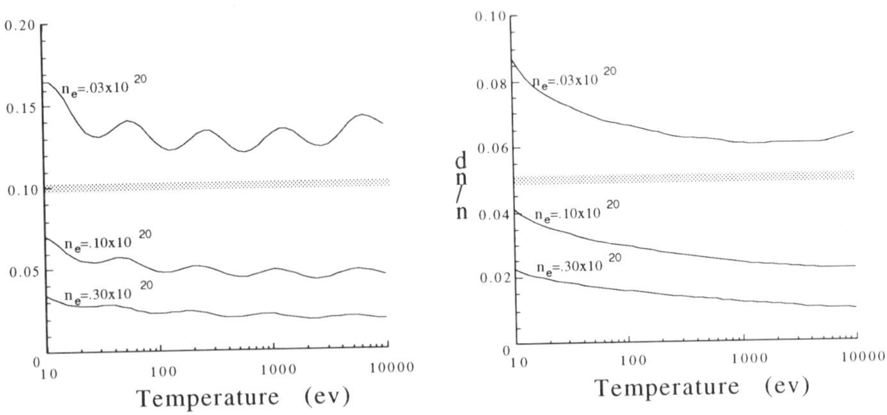

Figure 2. Calculated fractional uncertainty in electron temperature and density measurements for parameters listed in Table 2. The shaded region indicates the specified target measurement precision.

Fig. 2 shows the predicted measurement precision as a function of temperature and density. The oscillations in the plot of dT/T result from the discreteness of the choice of spectrometer channels. These calculations demonstrate that the parameters of Table 2 represent a reasonable design basis for meeting the specified measurement goals.

TOP PORT EDGE THOMSON DESIGN CONSTRAINTS AND ISSUES

The basic features of the conceptual design for a top edge Thomson system are shown schematically in Fig. 3. Below we examine in detail the design constraints used in the optical design and in the layout of components, and highlight issues that need to be addressed to bring this design further.

Field length. The system is to probe along a vertical line 60 cm long extending approximately 50 cm below the separatrix for a fully developed ITER plasma. Note that with a flux expansion of ~3 and an anticipated pedestal width ranging up to 19 cm, this field length may be too small. Also, changes in the vertical location of the separatrix for different scenarios will affect the utility of this diagnostic. Increasing the field length challenges the optical design and also increases the length of the slot required in the blanket and the port plug to accommodate the view, compromising the shielding. ISSUE: Is the 60 cm field length an appropriate balance between physics requirements and technical difficulty?

Use of fiber optics. We have assumed that it is possible to use fiber optics located behind a window at the top of the port plug, outside the primary vacuum boundary. These fibers would be used primarily in the red part of the visible spectrum and in the near IR, since we would be viewing the short wavelength side of the scattered spectrum from a Nd:YAG laser. In this design the fiber optic bundles would be continuous through secondary vacuum seals until they emerge from the biological shield. Although it is possible to couple light between separate fibers through a vacuum seal, there is typically a significant loss of light involved. It is likely that the fibers would have to be heated, to anneal radiation damage which would otherwise cause a loss of transmission.[5] ISSUES: What is the usable lifetime of heated quartz fiber optics at this location? What seal designs can be used to permit continuous quartz fiber to penetrate secondary vacuum seals?

Slot width available for mirrors. We have assumed that there is room within the slot in the port plug for two mirrors 20 cm wide in the toroidal direction. In order to maximize the collection efficiency and image the full 60 cm field, these mirrors are long. The lower mirror is 50 cm long and the upper is 90 cm long. In our preliminary design shown in the figure, both mirrors are curved. It would be possible to make the first mirror a flat mirror. However, under the constraint of a fixed slot width, the collected light for fixed slot length scales roughly as the inverse of the distance to the first curved surface. ISSUES Is a vertical port plug slot width of 20 cm compatible with shielding requirements? Can the optical stability and lifetime of the mirrors at these locations be made adequate?

Window Design. Although the loss in optical throughput would be severe, one benefit of a planar first mirror would be more demagnification, resulting in a smaller window and smaller fiber bundles. In the our preliminary design with curved mirrors, the window is ~ 40 cm long and ~ 8 cm wide. We have also assumed that the window can be angled with respect to

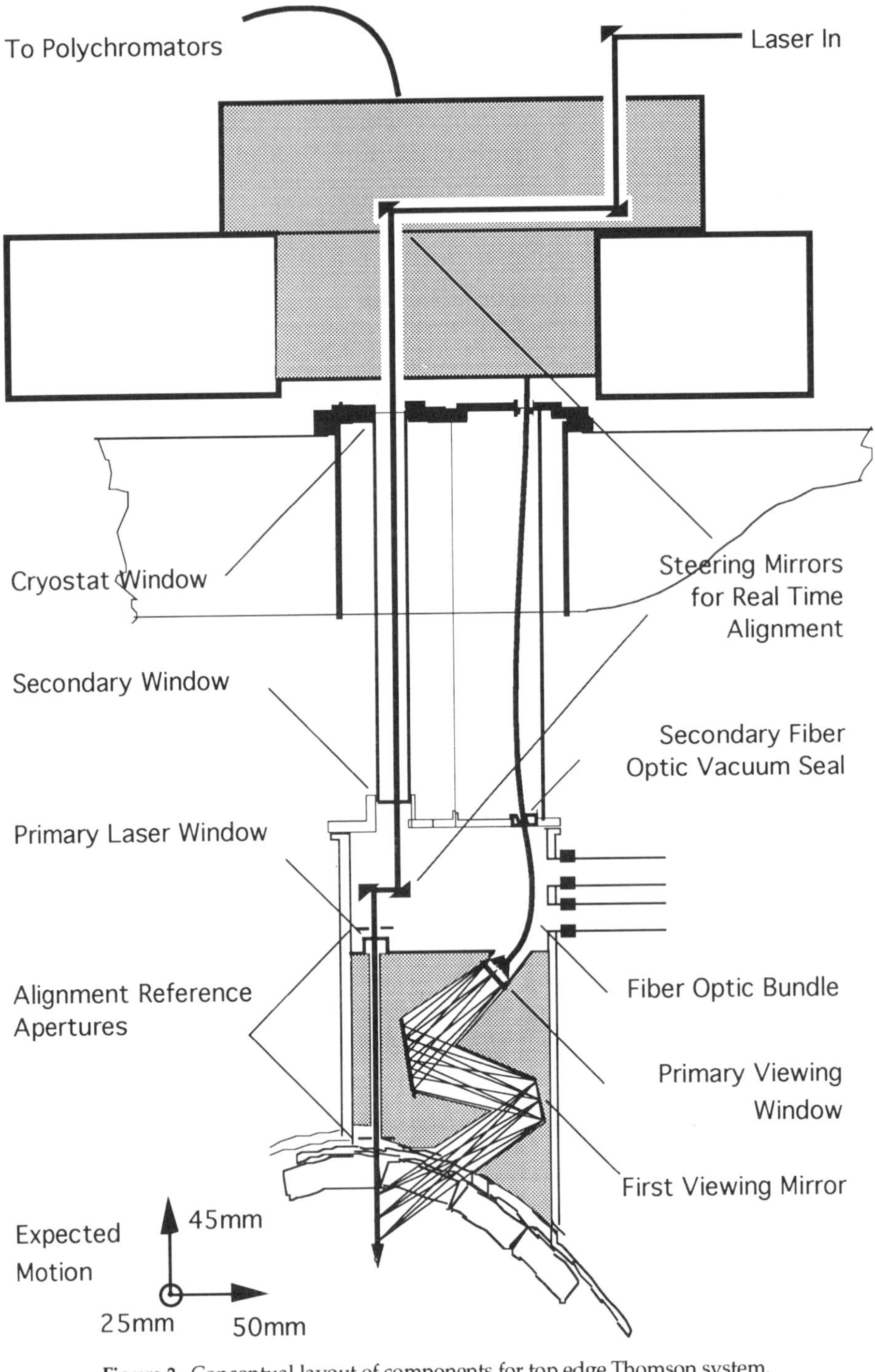

Figure 3. Conceptual layout of components for top edge Thomson system.

the horizontal plane to accommodate the optical design. ISSUE: What are the requirements for window design as a primary vacuum seal at the top of the vertical port plug?

Alignment. In the course of thermal cycling, the ITER vessel moves roughly as shown in Fig. 3. Remote steering capability is required to establish alignment between the laser and the collection optics. There may be smaller motions during a plasma discharge requiring real time compensation. Alignment apertures have been used for real time alignment control.[6] Dedicated alignment bundles straddling the main collection fibers can also be used to monitor misalignment using Thomson scattered light. ISSUE: What are the expected relative motions of the optical components during an ITER discharge? Is real time alignment compensation practical?

Blanket Penetration. In the reference design, a viewing slot would be necessary in a top blanket module. This slot would be ~ 70 cm long with a toroidal width on the plasma facing surface tapering from ~ 7 cm to ~ 3 cm. This slot would broaden toroidally moving away from the plasma, reaching widths of ~ 10 cm and ~ 6 cm, respectively, at the back plate. A hole or slot of ~ 2 cm would be needed for the laser beam. ISSUE: Is the required viewing slot compatible with blanket cooling and mounting requirements?

Use of dedicated top port access. The most questionable assumption in the reference design is the dedicated use of a top ITER port. Recently, we have become aware of the needs for vessel inspection and glow discharge probe hardware requiring roughly the inner half of the top port. We have considered the impact of restricting the access for the edge Thomson diagnostic to the outer half of the port plug. There are several obvious effects. First, the flux expansion factor decreases from 3 to 2, reducing the scattering length and the field length Second, the sightline path lengths increase by ~ x2, increasing uncertainties due to noise from the plasma light. Also, the sightlines are more inclined toward the core, reducing the shielding efficiency of the viewing labyrinth, possibly precluding the use of relay fibers. Finally, our initial attempts to obtain good imaging at comparable f/# failed, due to the large angles of incidence. It appears likely that a workable solution will involve a significant reduction in the viewing aperture. This can be explored further in the future. Even at the same viewing aperture, the laser energy required to meet the required precision at the lowest density increases from 2.2 joules in the reference case to 3.8 joules in the radially compressed case. Increasing the f/# to f/10 results in a requirement of 5.5 joules. High repetition rate lasers with such high output energies are not commercially available. ISSUE: How much of the top port space should be considered available in future edge Thomson design efforts? If access is restricted to the outer half of the top port, can a relaxation of measurement requirements be considered to allow spatial or temporal averaging to recover the desired precision with commercially available lasers?

CONCLUSION

In light of current expectations of ITER parameters for the edge pedestal and SOL, the measurement goals defined for ITER edge measurements appear to be appropriate. Given the interface constraints, it is a difficult technical challenge for an edge Thomson scattering system to meet these goals. There are significant advantages in locating such a system in a top port, particularly if the entire radial extent of the port is available. Assuming a dedicated top port, a reference design concept has been presented which meets the goals with modest extensions in available technology. However, a number of interface issues have been identified which will strongly impact this design. The Phase 3 design effort should address these issues. A key question is the availability of space in the top plug. Radial compression of the reference geometry to accommodate other systems in the vertical port plug will seriously degrade the measurement capability. Significant spatial and/or temporal averaging will probably be needed, in this case, to recover the desired measurement precision.

ACKNOWLEDGEMENTS

The authors wish to acknowledge useful discussions with S. Wolfe, A. Hubbard, and C. S. Pitcher regarding expectations for ITER edge pedestal and SOL characteristics, and kind encouragement from K. M. Young. This work is supported under U. S. Department of Energy Contract No. DE-AC02-76-CH03073.

REFERENCES

1. G. Janeschitz, A. Hubbard, Y. Igitkhanov, J. Lingertat, T. Osborne, H. D. Fischer, O.P.Pogutse, D. E. Post, M. Shimada, M. Sigihara, W. Suttrop, ITER operational space in terms of T_e and N_e at the plasma edge. *Proceedings of the 24th EPS Conference on Controlled Fusion and Plasma Physics*, Berchtesgaden, Germany, June 9 - 13, 1997.
2. D. W. Johnson, T. Carlstrom, B. Grek, and R. Snider, Thomson scattering at the edge of ITER, *Diagnostics for Experimental Thermonuclear Fusion Reactors*. P. Stott, G. Gorini and E. Sindoni, ed.,Plenum Press, New York. (1996).
3. T.N. Carlstrom, G. L. Campbell, J.C. DeBoo, R. Evanko, J. Evans, C.M. Greenfield, J. Haskovec, C.L. Hsieh, E. McKee, R.T. Snider, R. Stockdale, P.K. Trost, and M.P. Thomas, Design and operation of the multipulse Thomson scattering diagnostic on DIII-D, *Rev. Sci. Instrum.* 63:4901.
4. J. R. Wolberg, *Prediction Analysis*, D. Van Noistrand, New York, (1967)
5. A. T. Ramsey, Radiation effects on TFTR diagnostics, *Rev. Sci. Instrum.* 66:871.
6. D. Dimock, B. Grek, D. Johnson, B. Labombard, B. Lipshultz, and G. McCracken, A compact Thomson scattering system, *Rev. Sci. Instrum.* 68:700.

STATUS OF THOMSON SCATTERING DIAGNOSTIC DESIGN FOR ITER X-POINT AND DIVERTOR PLASMAS,

G.T.Razdobarin and E.E.Mukhin

IOFFE Physical-Technical Institute,
Russian Academy of Sciences,
St.Petersburg 194021, RUSSIA

ABSTRACT

For a few years now the Thomson scattering diagnostic design for ITER X-point vicinity is constantly developing and improving. The beneficial use of LIDAR technique is the basis for the design guidelines. The penetration and safety severe constraints force to compromise a number of features like the lower density limit and viewing chord length. Nevertheless, the compromised electron density $10^{19}m^{-3}$ and viewing chord length of no less than 1m basically accommodate the target measurement specifications. The conventional Thomson scattering coupled to LIDAR diagnostic arrangement is considered as well to penetrate a divertor leg. It makes the provisions to T_e,n_e profile recording along the vertical chord with pretty good spatial resolution of 1cm within the anticipated range of plasma parameters. The background and details of the design are reviewed in the paper.

1. INTRODUCTION

The ITER Thomson scattering diagnostic design confined to the X-point and the divertor outer leg regions is conceptually outlined in the paper regarding the reference "vertical target" divertor concept. The strong constraint must be placed on the design to avoid any significant cut-out through the target plate. This one meets the objections for the parallel heat flux is enormous and directly hits the cut-out hole edges. The convoluted beam path inside the plasma supported by high power laser mirrors is also contradicted by safety requirements. These constraints refuse to favor of scattered emission continuous recording from the X-point to the divertor bottom. With a given access restrictions the two kinds of straight line penetrations shown schematically in Fig.1 are considered.

The main beam coming radially nearby the X-point is compatible with LIDAR Thomson scattering. The LIDAR technique enables the scrape-off parameters (T_e,n_e) measurement in a region of significant flux expansion. The plasma in X-point vicinity and

Diagnostics for Experimental Thermonuclear Fusion Reactors 2
Edited by Stott *et al.*, Plenum Press, New York, 1998

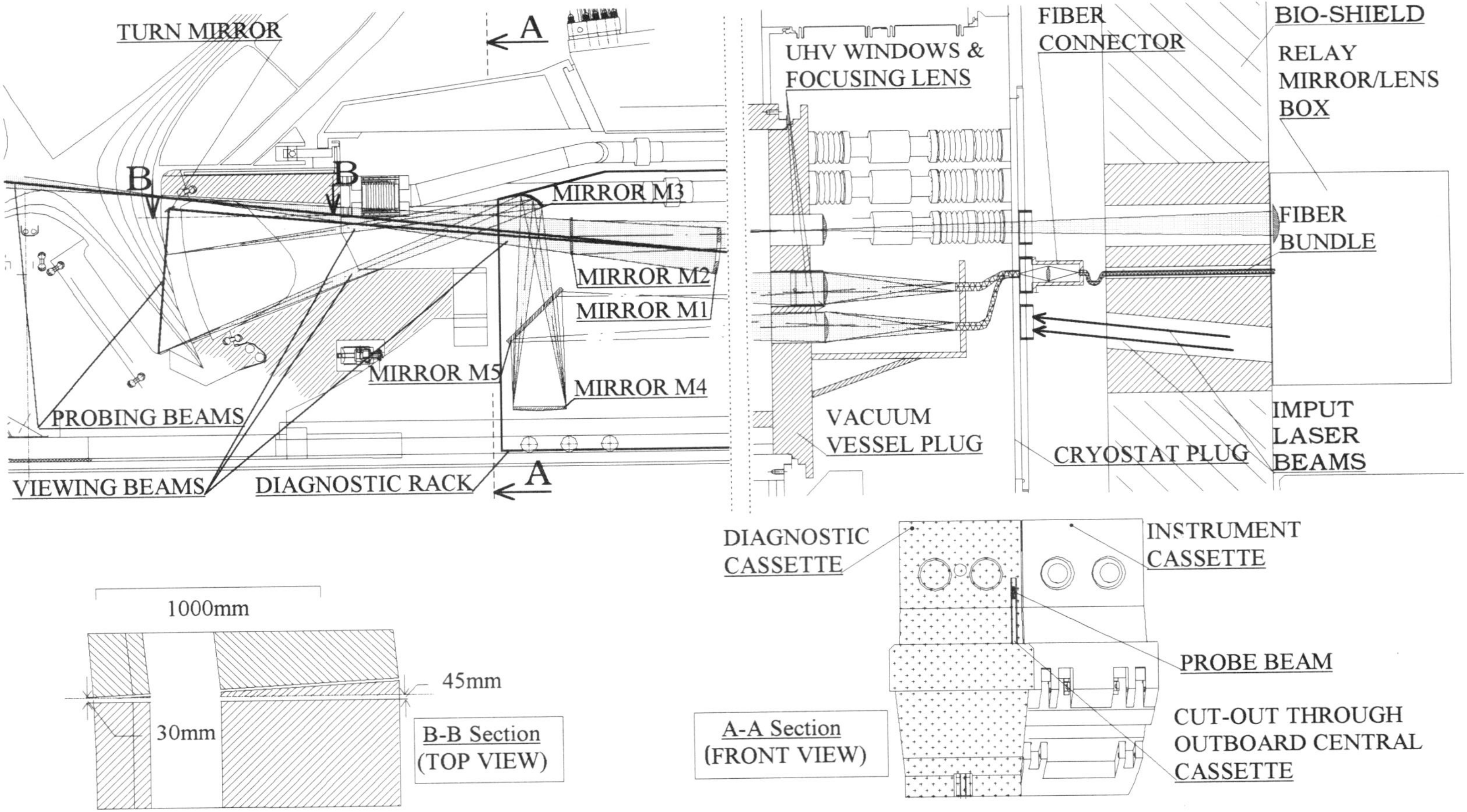

Figure 1. Conceptual drawing of remote maintenance port and optical penetration system

mostly ahead of X-point turns out accessible with spatial resolution of ~ 5cm. Rather high temperatures of hundreds eV and electron density in the range 10^{19}-10^{20} m^{-3} are recognized for this region implying the edge steep gradient zones. The main probe beam is splitted at a cassette margin to produce the auxiliary route pointed to the divertor outer leg. The conventional Thomson scattering with use of crossed probing and viewing beams is suggested to overview poloidally the divertor scrape-off parameters with cutting the ionization front. This particular plasma is of the most sophisticated formation. The existing model calculations predict the divertor recycling will tend to rise the density in the scrape-off well above the average bulk plasma density. The electron temperature drastic rolling-off is predicted as well. The target measurement specifications[1] have indicated the plasma parameter range n_e=10^{19}-$10^{22}$$m^{-3}$, T_e=1-200eV. Of course it is an extremely demanding design mission that guarantees the achievement of the needed divertor parameters. Thus, the greater creditable is the diagnostics part in study of the divertor plasma properties. These are required for the model validation but, more importantly, to execute the tokamak and divertor operations in accord with the expected regimes and margins. The divertor operational framework is evidenced by a number of features. Among them the plasma-of-target detachment and T_e, n_e profiles along the field lines are mostly indicative. Thus, the suggested probe beam routing downward from the outer to inner side of the separatrix looks favorable physically as well.

The single composed laser system optimized for the both X-point LIDAR and divertor conventional Thomson scattering is to be designed. This one is illustrated in the paper instrumentation part. As for the divertor leg Thomson scattering, the provisions are made of selecting the probe wavelength matched to the plasma radiation line-free spectrum windows. The probe pulse duration continuos controlling from 0.1 to a few ns is considered as well for matching to the detector response time. The laser pulse nanosecond range is to fit the avalanche photodetectors. The gained encouraging experience on a large format chip and photodiode array designing is referred herein. The avalanche diode matrix is given the first priority in project to ensure the essential sensitivity gain within a wavelength region adjacent to 900nm. The laser pulse subnanosecond range is alternatively reserved for photocathode multichannel devices at an emergency of strongly enhanced plasma background radiation. The trends of LIDAR diagnostics detection system development are outlined as well in the paper instrumentation part. The starting point of the design is obviously the admitted access to plasma that is the dominant to define the entire system configuration.

2. ACCESS CONSIDERATION AND SENSITIVITY ESTIMATES.

2.1 X-Point Access with LIDAR Thomson Scattering.

Two basic principles of access routing may be considered in LIDAR experiment. The first one, which is beneficial for a core plasma[2] provides the plasma-viewing mirror simultaneous use for scattered light collection and probe beam pointing. The routing of probe beam through the same optical components as for the scattered light collection is advantageous to solve the complicated alignment problem.

The alternative approach is based on the separate paths for the probe and backscattered beams provided the independent alignment of laser and the collection optics. The use of a narrow straight penetration along the entire beam path from entrance window to X-point is strongly advantageous to avoid the mutual action of the hostile plasma environment and the laser beam radiological loads on the front mirror surface. This our solution is recognized the basic one for the design guidelines.

The access to the divertor central cassette adapted to special diagnostic requirements is provided through remote maintenance port inside the cryostat. Each of the plugs in outer port extension is supplied with a pair of input (∅20cm) and output (∅25cm) vacuum optical windows made of fused quartz. To secure against tritium leaks all windows are double assemblies with separate pumping of the interspace. The optical system copes with looking through the 3cm vertical gap between the diagnostic and instrumented divertor cassettes. All optical units except of viewing and periscope mirrors are carried away the cassette at a distance more than 7m. The collected light cone waist is band shaped in accordance with laser beam transverse dimensions 1.5x10cm^2. It is positioned at outboard target plate margin to reduce the vignetting losses. The plasma viewing slab-like spherical mirror M1 of 5.3m curvature radius is of 10cm in horizontal and 45cm in vertical planes. The mirror is installed at 4.5m opposite the gap between the diagnostic and the instrumented cassettes to direct the line of sight to the X-point. The supplementary narrow slot cut-out through outboard central cassette like in Fig.1 (see top and front view sections) is provided. The spherical mirror M1 is supplied with a central slot opening (2.5x13cm^2) for the laser beam emerging. The periscope plane slab mirror M2 of 10x40cm^2 cross section is installed at 1.3m away the curved mirror with a rear side facing the plasma. The clearance between the mirror center points is 18cm in horizontal plane. Both mirrors are immersed in the main vacuum inside the diagnostic shielding block. They are required to maintain own focusing and optical properties for the full life of the tokamak. As is known, the pure polished metallic mirrors are appropriate for this purpose[2]. The front mirror is imaging the waist of the collected light cone under 1.45 magnification onto a field lens F2.3m located outside the vacuum vessel plug. The field lens useful area is limited by image stop of 2.2x14.5cm^2 cross section. The image stop aperture defines the viewing chord length to be ~1m on allowance of ~30% edge vignetting losses. The next step imaging is produced behind the bio-shield.

The overall conjugate ratio is kept 5:1 to obtain the final image band of 0.3x2cm^2 cross section. This one is converted to 0.6x1cm^2 at a spectrometer entrance slit using a dual-prism assembly. The auxiliary twofold image reduction is achieved by the spectrometer itself using a fast F/1 camera lens. The resulting 0.3x0.5cm^2 resolved element makes a good fit to the anticipated streak camera detector. The detection system is located at a cell in the Pit area behind the bio-shield. The magnetic flux local shielding starting from nearly 1kG to below 50G at a detector location is provided with use of 10cm soft iron shell.

For the reason of manual handling the laser set is preferred to be installed outside a cell in the upper elevation diagnostics area. The probe beam is transported in the cell area towards the machine using a sealed beam guides filled with nitrogen of 0.01atm overpressure.

2.2 The Capabilities of X-point LIDAR Thomson Scattering.

The LIDAR diagnostics detection capabilities are actually limited for a demanding time-of-flight spatial resolution of 3-5cm and essentially narrow opening for beam penetration. A fast respond streak camera is stipulated to comply with high resolution time-of-flight detection principle. The indicated in the paper part3 photocathode dimensions provide an overall throughput restriction to certain extent that enables no more than F/10 collection optics under 10cm probe beam height. Secondly, the probe beam narrowed to 1.5cm by penetration opening implies F/66 collection optics to ensure the unvignetting region of approximately 1m length. Thus, on going with sensitivity estimate we are limited to an F/10 height and F/66 width collection system which provides the light collection solid angle 1.5x10^{-3}sr. The following assumptions could be made in addition: overall transmission and photocathode quantum efficiency are of 10% each; resolved spectrum band is 0.2 of the scattered spectrum profile halfwidth. If so, a 3J laser output is needed to yield nearly one hundred photoelectrons per spectrum channel at the lower limit of the electron density $10^{19}m^{-3}$. This provides nearly 10% accuracy of temperature measurement respectively.

The given above assessment is correct on the neglection of plasma background continuous radiation. To our estimate[3], the bremsstrahlung and recombination contributions are tolerable in a condition of very short subnanosecond laser pulse duration. More dangerous are the multiplets of low-Z impurities overlapping the spectral diagnostic channels. To make any predictions it is normal to multiply the bremsstrahlung contribution at worst by 100. On allowances of this pessimistic prediction a 3J laser output behaves like a lower limit estimate.

2.3 Divertor Leg Penetration.

The sharing of the common beam line and the laser itself for penetration to either X-point or divertor leg plasmas is highly recommended for methodological and economical reasons. The two angular deviated probe beams (the 'green' for X-point and 'red' one for the divertor) are cut by one another at the mirror M1 center slot opening and transmitted to a cassette margin with some vertical displacement. Here the "red" beam is intercepted by turn mirror and directed to the divertor outer leg for multiposition scattering from the vertical chord of 0.8m height. The assembly of M3,M4,M5 mirrors inside the diagnostic block is used for probe beam imaging. The slab $10x15cm^2$ aspheric mirror collector M3 of 16° viewing angle is located 3m away the probe beam to collect the scattered radiation through a vertical slot between the cassettes. It provides the probe beam imaging under twofold demagnification in front of spherical field mirror M4 of $10x44cm^2$ cross section. The precision curvature profile making is required for aspheric mirror to secure a good imaging quality. The spherical mirror M4 of 2.5m curvature radius together with plane mirror M5 are to convey the beam outside the machine about 7m away the outboard cassette boundary. An M3 collector pupil is conjugated to the composed aperture of a dual-lens assembly behind UHV windows at vacuum vessel plug. The admitted window/lens diameter of 25cm accommodates the total collection solid angle $0.7x10^{-3}sr$. Each of two F/3.4 focusing lens is properly matched to the coherently packed quartz fiber guide of $0.17x7.3cm^2$ input window. The bundle is secured at the plug board to absorb the movement and vibrations of viewing optical axis. The two separate fiber guides are composed coherently to a single one of twice as large aperture width thus matching to the polychromator slit opening. The array of 4 grating polychromator (like in table1) of 2cm slit height is taken for light dispersing. The fiber arrangement along each of the slits is matched to the 20cm vertical chord section in the plasma. The instrument dispersions are optimized to fit a respective temperature range spread over the vertical chord. The array of polychromators is placed in diagnostic room for the maintenance and safety reasons. An optical fiber of more than 10m length is required for scattered light convey. But only a short guide piece of 2m length is tracing inside the bio-shield which facilitates the fiber survival during a full lifetime of ITER[4].

2.4 Alignment Problem.

The optical units painstaking adjustment has always been a problem for the conventional Thomson scattering. The hard structural frame comprising tightly fixed probing and viewing apertures supplied with adjustable matching optics usually accommodates the operations at small and moderate size tokamaks. The ITER divertor cassette itself with a fixed vertical slot for probe and viewing beams emerging suggests the structural frame functions. The engineering provisions should be additionally made to remote handle the mirror assembly inside the diagnostic block. The major constraint follows from the remote handling of the front mirror located at a cassette margin which is scarcely possible. The reasonable alternative is the use of metallic grating as a beam deflector instead of the adjustable front mirror. The metal blank (of gold or copper) is furnished for the heat sink and keeping the acceptable erosion. The laser beam angular adjustment is carried out by wavelength tuning. The conceptual drawing in Fig.2 illustrates the admitted operation principle.

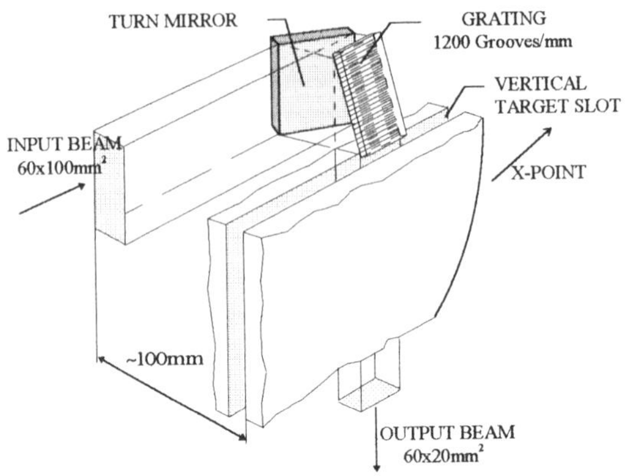

Figure 2. Conceptual drawing of beam penetration arrangement for divertor leg Thomsin scattering.

In a condition of wavelength-to-groove spacing ratio kept over unity the grating first order efficiency for S-polarized emission is known[5] to approach 100% (apart from metallic layer absorption) for a broadband diffraction angular range like in Fig.3.

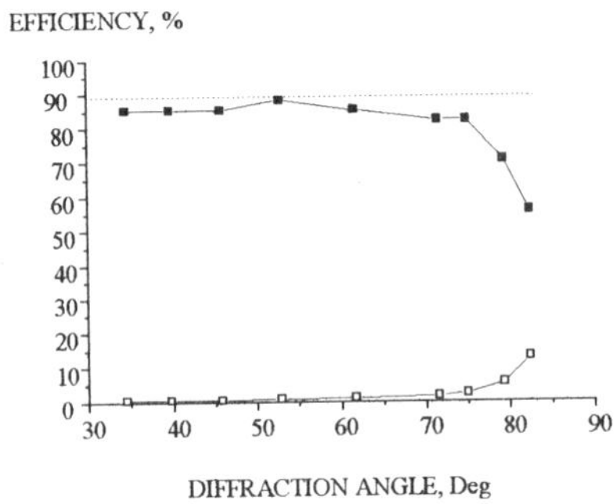

Figure 3. Experimental curve for grating S-plane first order efficiency as a function of diffraction angle. 1800grooves/mm aluminum coated grating of 36.5° blaze angle under HeNe laser illumination is sampled herein. Hollow points show the zero order specular reflection. Dotted line corresponds to Al film reflectance.

For the limiting diffraction angle as large as 80° the beam deflection sensitivity would be nearly 1cm/nm at a divertor bottom. It means that wavelength tuning within 5nm range fits the laser beam scanning across a slot opening between the cassettes.

The engineering provisions should be made to isolate the high frequency mount vibrations, below which there is no concern. The protective covers must be supplied as well to hold the grooves etched-off layer in a tolerable limit of approximately 10 monolayers (3nm or so) for an extended operation period[6]. This could be done provided CX neutrals channeling via vertical slot from underneath the divertor within the reduced solid angle.

The critical design issue for a divertor leg Thomson scattering is the anticipated number of laser probe pulses that warrants the grating/mirror safely operation. This number n is given sometimes by the empirical relation $n=[P_{threshold}/P]^{\gamma}$ where γ is much above unity[7]. The magnitude n is strongly dependent on safety factor $P_{threshold}/P$ where P is a probe flux density impinging on the surface and $P_{threshold}$ is a respective damage threshold value. The laser beam is to fill up the grating ruled area as much as possible being collimated in a space of plasma location. It is the grating anamorphic magnification that meets these contradictory requirements. Nearly 5 times beam narrowing from 10 to 2cm is apparent at the diffraction angle approaching 80°. The grating acceptance clear aperture of $10x6cm^2$ is likely compatible with 1J laser driving under 10Hz repetition rate to accommodate approximately a thousand of 10^3s ITER discharges. This rather a lower limit estimate is related to the extreme fast degradation rates γ for some metallic blanks[7] provided the incident flux density is 1% the damage threshold. The dedicated experiments are required, however, to proceed with more accurate estimations.

2.5 The Capabilities of Divertor Leg Thomson Scattering.

The limit spatial resolution over the probe beam is obviously attributed to the both aberration image blur at polychromator sagittal plane and the matrix detector finesse. So far, there is no seen hard objections to predict 0.1cm resolution over slit height and 1cm resolved length in plasma respectively. The more realistic prediction follows from the sensitivity considerations. The two basic magnitudes are defined more or less accurately on this occasion. These are the probe energy of nearly 1J and the collection solid angle limited to $0.7x10^{-3}$sr. This would deliver over 2000 photons per 1cm resolved length at the density $n_e=10^{20}m^{-3}$ on assumption of 10% overall transmission. Even after the spreading over the several spectral channels the signal still remains well above the detection limit of avalanche photodetector. Thus, the resolved scattering length which is reciprocal of electron density scales as 1cm at $n_e=10^{20}m^{-3}$. The resolved depth in viewing axis direction determined by probe beam transverse extent is illustrated in Fig.2 to be 6cm. In so far as it was necessary[1], the depth resolution below 1cm could be satisfied sacrifying the probe energy. The laser output matching is actually required to protect the turn mirror/grating against damage. Thus the depth and scattering length resolutions are compromised each other. The lower recorded temperature limit according to the reference data in table1 turns out nearly 1eV under a resolved spectrum band down to 0.6nm provided a reasonable measurement accuracy perhaps of no worse than 10%.

3. INSTRUMENTATION (DESIGNS AND FEASIBILITY ASSESSMENTS)

3.1 Probe Laser.

The considered laser system (see Fig.4) is based on the Nd:YLF master oscillator-amplifier chain. The pulse length is controlled by means of a pulse slicer using the transverse Pokkels effect in $LNbO_3$ or alike crystal. The pulse slicer output is amplified to 5J.The beginning of a chain enclosed by frame has been manufactured to proceed with testing experiments. Pulse duration continuos controlling in the range from 4 to 0.15ns has been demonstrated experimentally.

After amplification the beam is converted into the second-523.5nm and third-349nm harmonics of nearly 3J and 2J output respectively. Under 5J driving and ~100ps pulse duration the required clear aperture of the amplifier is defined by selffocusing and damage thresholds. Thus, the intensity of the beam is to be below 10^{10}W/cm^2 corresponding to energy density less than 0.5-1J/cm^2. This requires the Nd:YLF crystal aperture of 5-10cm^2. The high quality crystals of such dimension are not available today but a respective slab amplifier could be manufactured in a short future. A strongly desirable 100Hz driving is looked for on the way of diode pumping owing to at least 10-20 times higher efficiency compared to conventional flash lamp operation. YLF material is more preferable to diode pumping because of nearly twice as large inversion lifetime and higher level of stored energy compared to YAG. The 3J output at second harmonic is taken somewhat arbitrary to fit the given detection limit. The optimistic forecast could be made regarding the experiments on Nd:YAG diode pumping. The 5J output at second harmonic has been demonstrated under 33Hz repetition rate and a few ns pulse duration[8]. As far as the accurate wavelength tuning is considered for a divertor leg Thomson scattering, the second Nd:YLF beamline is used for Ti:Sapphire (Ti:Sa.) laser pumping. The 4J pumping energy at the second harmonic 532.5nm is provided under the "long"-15ns pulse duration. The YLF rod of 20mm in diameter or disk amplifier will operate at 2J/cm^2 which is below the damage threshold. The Ti:Sa. master oscillator-amplifier chain is basically the same as for a main Nd laser. It consists of Ti:Sa. master oscillator followed by synchronized pulse slicer and power amplifier. The energy conversion efficiency from Nd second harmonic to Ti:Sa. is about 35% at 810nm. The estimated output energies for Ti:Sa. laser via tuned wavelength are shown in Fig.4.

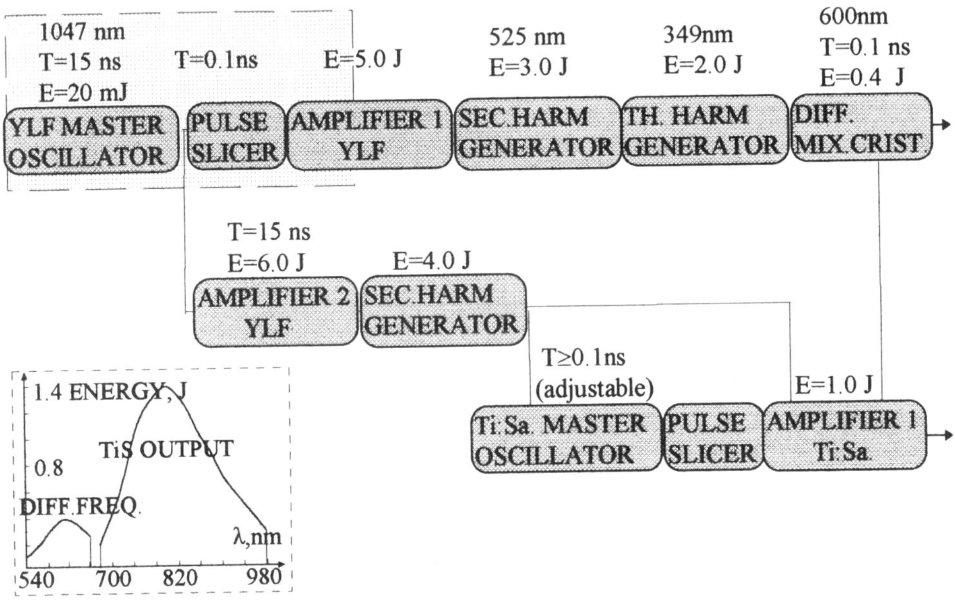

Figure 4. Laser system schematic diagram and performance specifications.

The auxiliary extension of probe wavelength range into 540-700nm band could be obtained if necessary using the differential frequency mixing. Thus the extremely wide range of probe beam wavelengths became available:

- 700-1000nm -Ti:Sa. fundamental frequency,
- 360-490nm -Ti:Sa. second harmonic (not shown in Fig.4),
- 540-660nm -Nd third harmonic with Ti:Sa. fundamental frequency differential mixing.

The frequency mixing provides the extremely high energy contrast ratio of probe beam (about 10^{10}). The pulse length continuous control starting from 0.1ns is available owing to the electrooptical pulse slicing.

3.2 Grating Polychromator.

The polychromator of high spectral resolution and flux luminosity adapted to visible and near-IR range has been designed in a frame of the project.

The special mount based on a double diffraction principle is suggested using the large diffraction angles to above 80°. Besides of enhanced flux luminosity, a wide range adjustable dispersion is available. The dispersion controlling without change of grating/collimator makes the instrument a valuable tool for measurements in a wide temperature range from 1eV up to hundreds eV. The quick reference data on two polychromators models are illustrated in the table1.

Table 1. Grating polychromator quick reference data.

Wavelength scanning, nm	Focal length, mm	Aperture size ratio (at 75° diffraction angle)	Grating ruled width, mm	Groove number lines/mm	Reciprocal linear dispersion, (adjustable) nm/mm	Limit resolution, (down to) nm
above 400 (LIDAR)	140	F/3	130	1800	1.0-0.25	$5x10^{-3}$
above 600 (conventional)	300	F/4	290	1200	0.7-0.17	$3.5x10^{-3}$

The array of instruments of different focal lengths has been elaborated and verified in a test facility concentrating on grating efficiency behavior at large diffraction angles. The absolute grating S-plane efficiency expressed as a percentage of the incident intensity for single diffraction is presented in Fig.3 as a function of diffraction angle. The experimental curve demonstrates the admitted diffraction angular range capable of good efficiency performance. The enhanced angular dispersion owing to large diffraction angles provides a rather small influence of aberration image blur. This offers a simplicity of design utilizing a customary imaging optics of a reduced focal length. The outlined design compares favorably with commonly used instruments developed with Czerny-Turner scheme.

3.3 Streak Camera.

Special purpose streak cameras of increased photocathode dimensions are needed for LIDAR diagnostics to improve an overall throughput performance. Unfortunately, this is contradicted by the requirement to secure a large number of independent measurements along the streak trace and viewing chord respectively. The streak camera comprising the new type streak-tube has recently been developed. The device is distinguished by the large format photocathode of $40x4mm^2$ area and a strong twofold electron-optical demagnification factor. These properties are of major importance to overcome the outlined shortcomings of serial devices. The dedicated technology is developed to proceed with the large format photocathode coating inside a vacuum boundary. This guarantees a uniform photocathode sensitivity and accelerating mesh protection from atomic alkali pollution, thus reducing a background noise. The designing of fourfold demagnifying streak tube is now underway to preserve the number of independent measurements as great as 40 along the 40mm streak length. The matter of special concern is a camera locking capability to discriminate against the spurious backreflected light pulses on a streak trace. The recommended instantaneous pulse shuttering faces the obvious difficulties owing to a shutter recovery time. A point highlight saturation of the supplied microchannel plate intensifier is relied upon to discriminate between the useful and extremely large spurious pulses on avoid of significant blurring. To demonstrate a high blurring resistance capability the test experiments were carried out using a LIDAR Rayleigh scattering in atmosphere[3]. The experiments make the evidence to the detection of an extremely weak scattered signal composed of the discrete photoelectron counts at the streak trace in a presence of very strong spurious pulse. The time interval between the both signals was a fraction of ns.

4. SUMMARY.

The reliable T_e, n_e profile measurements in ITER divertor address the highly promising Thomson scattering diagnostics. The limitations imposed by ITER environment and poor access have lead to the design involving the particular penetrations to X-point and divertor leg plasmas under two different diagnostic concepts. The LIDAR concept for the X-point and the conventional Thomson scattering for the divertor leg are fitted well to a vertical target divertor configuration with minimal impact on ITER design. The given analysis gains a credit to diagnostics ability to accommodate the target objectives and specifications. The provisions are made to use the unified composed diagnostic facility for both approaches whenever it is possible. The diagnostic design is basically oriented to the performance optimization functions. However, the measured quantities may provisionally be included in a feed back control system on the occasion the divertor kinetics becomes a crucial issue.

REFERENCES

1. A.E.Costley, R.Bartiromo, L.de Kock et al., Requirements for ITER diagnostics, in: *Diagnostics for Experimental Thermonuclear Fusion Reactors*, Plenum Press, NY, p.23 (1996).
2. C.Gowers, P.Nielsen, F.Orsitto et al., A Thomson scattering scheme for obtaining T_e and N_e profiles of the ITER core plasma, in: *Diagnostics for Experimental Thermonuclear Fusion Reactors*, Plenum Press, NY, p.249 (1996).
3. G.T.Razdobarin, A.Daavittila, V.K.Gusev et al., The concept of ITER divertor plasma diagnostics by means of LIDAR technique, in: *Diagnostics for Experimental Thermonuclear Fusion Reactors*, Plenum Press, NY, p.259 (1996).
4. Divertor impurity monitoring system, *Design Description Document* , WBS 5.5.E.04.
5. E.G.Loewen, M.Neviere, D.Maystre, Grating efficiency theory as it applies to blazed and holographic gratings, *Applied Optics* 16:2711 (1977).
6. Thomson scattering (core), *Design Description Document,* WBS 5.5.C.01.
7. V.Voitsenya, V.Konovalov, O.Motojima et al., Evaluations of different metals for manufacturing mirrors of Thomson scattering system for the LHD divertor plasma, in: *Research Report NIFS Series,* NIFS-463 (1966).
8. M.Yamanake, Recent topics of laser developments, in: *Proc.7 Int. Symp. On Laser-Aided Plasma Diagnostics,* Fukuoka, Japan, p.217 (1995).

TI:SAPPHIRE LASER FOR ITER LIDAR THOMSON SCATTERING SYSTEM

F. Orsitto[1], E. Nava[2], P. Nielsen[3], H. Salzmann[4]

[1]Associazione EURATOM-ENEA sulla Fusione
[2]CISE S.p.A.-C.P.12081.I-20134Milano(Italy)
[3]JET Joint Undertaking, Abingdon, Oxon, OX14 3EA, UK
[4]MPI fur Plasmaphysik, 85748 Garching, D

INTRODUCTION

The choice of the laser wavelength for ITER LIDAR THOMSON SCATTERING SYSTEM for the core plasma is linked to the extended range of plasma temperatures[1] to be measured, i.e. 0.5-30keV. The additional requirements[1] are: temperature relative error $\Delta T_e/T_e$=0.1 with spatial resolution of 0.1m, plasma density relative error $\Delta n_e/n_e$=0.05 and spatial resolution 0.3m, at a temporal resolution of 10ms.These last parameters determine the laser energy, the pulsewidth and the repetition rate. It has been found[2] that a **single** laser working at a wavelength of 790nm solves the problem of measuring the ITER temperature with a relative error less than or equal to 20%. A Titanium-Sapphire laser ($Ti:Al_2O_3$) is the natural candidate for such a source since the fluorescence peak is at 800nm, with a very large tunability range between 700 and 1000nm . Other important characteristics of the material are the large gain cross section, and relatively short fluorescence lifetime, its hardness and good thermal conductivity properties inherent to the sapphire host, and the availability in large size, good quality crystals. High power short pulse Ti:Sapphire lasers can be built. The paper is organized as follows: the required laser figures are summarized, the main Ti:Sapphire(TiS) material parameters are listed in comparison to other laser materials, the scheme of a 10Hz TiS laser is presented, the possibilities of improving the previous scheme to 100Hz source are discussed and a comparison with other proposed schemes[4] is cited, finally the conclusions are drawn about the possibility of building these sources with present day technology .

LASER FIGURES

The main required laser figures can be inferred from the requirements cited on the measurements: the spatial resolution range of 0.1m - 0.3m gives a laser time pulsewith of τ_L=0.6-2ns (the detector rise time is assumed τ_D=0.3ns), while the temperature and density relative error gives the laser energy about 1-2J for 0.1m and 0.3m spatial resolution respectively, finally the time resolution gives the repetition rate that must be 100Hz.
As starting point of the present work, a basic set of laser figures, shown in Table 1, is assumed.

Table 1. Reference TiS Laser Figures

wavelength	750-800nm
energy/pulse	1.5J
pulsewidth(FWHM)	0.6/1.5ns
divergence(full angle)	<10*diffraction limit
beam pointing instability(90% of pulses)	~ 1/10 divergence
repetition rate	10Hz

The main limitation, in this basic set of parameters, is the laser frequency(10Hz), which reduce strongly the complexity of the source lay-out, being still useful for most of the diagnostic necessities of ITER. The reference source can give the required relative errors in electron temperature and density, with spatial resolution of 0.3m(τ_L=1.5ns), and temporal resolution of 100ms, for most of the plasma density profile at the lowest density $n_e = 3.\ 10^{19} m^{-3}$; while for a spatial resolution of 0.1m(τ_L=0.6ns) the required accuracy is achieved marginally only at the plasma center, for the same density.

TiS MATERIAL PARAMETERS

Chemically the TiS consists of sapphire(Al_2O_3) in which a small amount (0.1%) of the ion Al^{3+} has been replaced by Ti^{3+} ion. It shares the host material with Ruby, where the Cr^{3+} ion replace the Al^{3+} ion. So the main material characteristics are similar for both. In Table 2 some important parameters[3] of Ruby, Alexandrite(Be:Al_2O_4), Nd:YAG and TiS are reported.

From Table 2 it can be noted that TiS has the same parameters of Ruby in terms of thermal shock strength , damage threshold, and non-linear index of refraction, while higher cross section (same as Nd:YAG) and lower fluorescence lifetime. The high thermal shock parameter and excitation cross section of TiS respect to the Alexandrite as well as the laser gain parameters, reported in Table 3, give an explanation of suitability of TiS for short pulse high power applications, respect to the Alexandrite laser.

Because of the low τ_{fl} fast flash lamps or (better) Nd:YAG frequency doubled (@532nm) are used as pump sources.

Table 2. Comparison between laser material parameters.

	σ $10^{-20}cm^2$ @300K	n2 10^{-13} esu	τ_{fl} 10^{-6} s	R W/cm	D.Thr. GW/cm^2 τ_p<1ns
Alexandrite (@755nm)	0.7-1	0.8	260	35-74	1-2
Nd:YAG @1064nm	50	2.7	230	7.9	3
TiS (@785nm)	30	1.3	3.9	100	1-2
Ruby @694.3nm	2.5	1.4	3103	100	1-2

The meaning of symbols in Table 2 is:

σ excitation cross section

τ_{fl} fluorescence lifetime

R thermal shock parameter

D.Thr. damage power threshold

Table 3. Laser gain parameters

	Gain coefficient for 1J/cm^3 stored energy(cm^{-1})	Saturation fluence (J/cm^2)
ruby	0.18	5.7
alexandrite	0.04	26
TiS	1.2	0.85
Nd:YAG	2.7	0.4

TiS 10Hz laser scheme

The scheme for the reference source (see Table 1) can be made by an oscillator , a preamplifier with double pass amplification , and a final amplifier.
Each amplification section is preceded by a magnifying telescope with vacuum spatial filter.
The laser pump can be a fast flash lamp or a 10Hz Nd:YAG laser frequency doubled, pumping the TiS rods in longitudinal configuration, so reaching a 25% of efficiency.
A 11J/10Hz Nd:YAG(@1064nm) laser is needed, for the reference TiS source pumping necessities: since from this source a laser beam of 6J at 532nm can be obtained, just enough to give the 1.5J TiS laser.

Oscillator

A TiS oscillator pumped by flash-lamps or doubled Nd:YAG can be realized.
This scheme is suitable for the short(τ_L=0.6ns) as well as long pulse (τ_L=1.5ns) laser:
In the first case an intracavity mode-locking system(acusto-optic modulator) with an etalon must be inserted, and flash lamps or Nd:YAG doubled can be used as pump sources.
In the second case a very short TiS cavity plus a preamplifier can be used to give about 15mJ, with a pump energy of 200mJ from a doubled Nd:YAG.

First amplifier

A first amplifier can be made by a double pass section and realized by a 8mm/diameter rod: on the first pass the small signal energy gain G_0=7-8, and on the second the gain G_0=6, giving at the output 400-450mJ at 10Hz.

Final amplifier

This amplifier can be realized using a rod approximately doubled in diameter (i.e.15mm).A small signal energy gain G_0=3 seems feasible giving raise to the 1.5J projected output.

TiS 100Hz laser

The scheme for the 100Hz source can be similar to that proposed for 10Hz TiS source: i.e. one oscillator, a preamplifier in double pass, a final amplifier.
For the high frequency TiS source the flash lamp pump scheme is no longer suitable, so 100Hz Nd:YAG laser must be used.

The TiS pump efficiency is a strong function of the laser beam quality so Nd:YAG laser accurately designed for this purpose must be used.

The Nd:YAG laser can still be pumped by flash lamps (while a diode pumped source could be preferred because of reduced thermal stress on the laser rods); to compensatie for the thermal lens effect and distorsion induced by the pumping procedure, phase conjugate mirrors are used, allowing to obtain good wavefront quality.

This type of compensation must be applied for TiS laser itself, thermal lens effect in the sapphire can be important at high frequency.

A scheme for this source was also proposed in ref.4.

CONCLUSIONS

The present paper presented a detailed discussion of the feasibility of a TiS laser for the ITER LIDAR Thomson Scattering System.

A 10Hz TiS source with basic figures presented in Table 1 seems feasible with present day technology, at relatively low cost of order of 1M/US$/1997.

The 100Hz source can also be built but the technological complexity is quite high, and this raises the cost.

REFERENCES

1. A.E.Costley , R.Bartiromo, L.DeKoch, E.Marmar, T.Matoba, V.Mukhovatov, K.Muraoka,A.Nagashima,D.V.Orlinsky,M.Petrov,P.E.Stott,V.Strelkov, S.Yamamoto, and K.M.Young, Requirements for ITER Diagnostics , in : *Diagnostic For Experimental Fusion Reactors* , Edited by P.E.Stott, G.Gorini and E.Sindoni, Plenum Press,NY(1996),p.23.
2. C.Gowers,P.Nielsen,F.Orsitto,F.J.Pijper,H.Salzmann, and B.Schunke, A Thomson scattering scheme for obtaining Te and ne profiles of the ITER core plasma, therein p.249.
3. W.Koechner,*Solid State Laser Engineering*, Springer-Verlag, Berlin 1988.
4. A.W.Molvik, R.A.Lerche, D.G.Nilson, M.D.Perry, E.B.Hooper, LIDAR Thomson scattering for advanced tokamaks-final report, UCRL-ID-123434(march 1996).

MIRROR CONSTRUCTION AND CHARACTERIZATION FOR ITER LIDAR THOMSON SCATTERING SYSTEM

F. Orsitto[1], A. Brusadin[1], M. Di Fino[2], C.Gowers[3], P.Nielsen[3], H.Salzmann[4]

[1]Associazione EURATOM-ENEA, C.R. Frascati, Frascati (ITALY)
[2]ENEA-Dipartimento Innovazione, Divisione Laser, C.R. Frascati.
[3]JET Joint Undertaking, Abingdon, Oxon, OX14 3EA, UK
[4]MPI fur Plasmaphysik, 85748 Garching, D

INTRODUCTION

The requirements for the plasma facing mirror for the ITER LIDAR THOMSON SCATTERING (ILTS) SYSTEM[1] are: high and as uniform as possible reflectivity in the spectral range 400-800 nm, low sputtering yield, and high laser damage threshold. Thermal deformation due to nuclear heating is also one of the effects to be taken into account and it must be compatible with acceptable optical imaging quality. Rhodium coated mirrors with a copper substrate are considered as possible candidates meeting these requirements. The paper describes the techniques used for building prototype mirrors with thick (30-80 nm) Rhodium coatings on copper substrates and their optical characterization.

MIRROR CONSTRUCTION

The mirror construction starts preparing the copper substrate, after that a thin nickel coating is deposited on the substrate, finally the rhodium coating is applied.

Copper substrate preparation

Copper substrates (two or three inches diameter) were obtained machining pure Copper (Cu-OFHC) by means of a diamond fly cutting machine. The surfaces were hand polished and chemically etched for smoothing out the machine grooves. In this way plane mirrors with a good optical quality were obtained with $\lambda/10$ ($\lambda = 652.8$ nm) surface deviation from planarity on (at least) 75% of the field. Typical roughness of these mirrors is 10-30 nm.

Diagnostics for Experimental Thermonuclear Fusion Reactors 2
Edited by Stott *et al.*, Plenum Press, New York, 1998

Coating preparation

Two techniques are used : evaporation by means of an electron gun and electrodeposition using a galvanic bath. To realize the mirrors a thin coating (8-15 nm thick) of nickel is deposited on the copper substrate. The Ni layer is crucial to obtain a uniform deposition with both techniques since i) the rhodium direct deposition results in microholes on the copper surface while the Ni intermediate layer smears out uniformly the rhodium, when evaporation is used; ii) the Cu substrate is directly attacked by Rh ions in the bath, when electrodeposition is used, so Ni serves as a protection for the Cu substrate.

Rhodium coated mirrors

Eight mirrors of various diameter with Cu substrate were built.

The evaporation technique was used for obtaining Rh coatings with a maximum thickness of 35 nm on Ni layers 10 nm thick.

A Rh mirror of 3" diameter was built by electrodeposition with 1.5 μm thick Ni layer resulting in a 0.8 μm thick Rh coating.

All these mirrors exibit very good reflectivity in the visible.

CONCLUSION

The electrodeposition technique gave the best result in terms of Rh coating thickness, so far.

1.H.Salzmann, L.deKock, C.Gowers, P.Nielsen, F.Orsitto, C.Walker, LIDAR Thomson Scattering for the ITER Core Plasma, these proceedings.

COLLECTIVE THOMSON SCATTERING(CTS) SYSTEM ON ITER

F.Orsitto[1], G.Giruzzi[2] and S.Nowak[3]

[1]Associazione EURATOM-ENEA sulla Fusione
C.R.Frascati-C.P.65-00044Frascati(Italy)
[2]Association EURATOM-CEA sur la Fusion
Centre d'Etude de Cadarache-13108 St.Paul-lez-Durance(France)
[3]Istituto di Fisica del Plasma, Associazione EURATOM-ENEA-CNR
v.R.Cozzi 53- 20125Milano(Italy)

INTRODUCTION

In a previous work the analysis[1] on a possible CTS system on ITER has led to the following preliminary conclusions: two options can be taken into account seriously, i.e. i) CO_2 (λ=10.6μm) extreme foreward scattering(scattering angle θ<0.3°) without spatial resolution, ii)nearly back scattering(θ=170°) configuration using two-colour gyrotron scattering, at frequencies f_0=65 and 72GHz, which can give also an acceptable spatial resolution of the order of 1/10 minor radius[2], launching the principal beam from an equatorial port, just below the equatorial plane, and collecting the scattered radiation in the same port, placing the array of collecting antennas above the equatorial plane. The first option brings the tangential injection of the beam in order to take advantage of null refraction for an easy beam separation on long paths, furthermore the signal to noise ratio(SNR) is limited only by the noise equivalent power of the receiver, being negligible the ECE noise of the plasma. The second option, while having the possibility of a spatial profile measurements with good SNR, identify the scattering volume by means of the ray-tracing code evaluation of the antenna beam position inside the plasma[2], and the SNR is influenced by the ECE emission. In the present paper the proposal of two-color X-mode gyrotron CTS below the electron cyclotron frequency is reviewed and some points already mentioned in that paper are specifically discussed: i) the theoretical model used in the calculations of ref.2 is commented; ii) a more refined calculation of ray propagation has been performed using realistic ITER equilibria and gaussian beam ray-tracing code.

REVIEW OF TWO COLOR GYROTRON X-MODE CTS ON ITER

The two color proposal for a CTS on ITER was derived since the allowed propagation window for the incident and scattered wave shrinks and the middle frequency of this interval

Diagnostics for Experimental Thermonuclear Fusion Reactors 2
Edited by Stott *et al.*, Plenum Press, New York, 1998

increases as the plasma density is raised. The figure 1 shows the minimum(maximum) frequency f_{min} (f_{max}) versus central plasma density for n_α/n_e=0.3%: f_{min} is the lowest frequency at which a ray launched at $\phi=15^o$ (ϕ is the angle between the scattering plane and the direction normal to the magnetic field) reach the plasma centre, while f_{max} is the highest frequency for which the SNR is greater than 10. The dotted line describes the wave upper frequency deduced from the condition that $R_{\alpha e}=S_\alpha/S_e>2$ (the ratio between the alpha signal and the electron signal). The f_{min} is regulated by the refraction effects , while f_{max} is regulated by the ECE maximum radiation temperature.

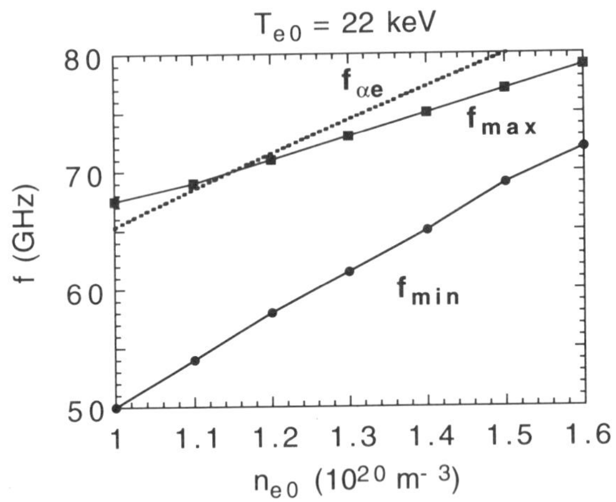

Figure 1. Wave accessibility diagram.

In this way, it is of some advantage using two frequency launching(and receiving) systems which can be used in two intervals of plasma density. In particular[2] i) for densities $n_{e0,20} < 1.2$, (n_{e020} is the plasma density measured in unities of $10^{20}m^{-3}$) a gyrotron frequency at 65GHz can be used and the scattered spectrum is measured in the range 59-65GHz while in the interval $1.2 < n_{e0,20} < 1.4$ the upper side of the scattered spectrum is measured; ii) for higher densities a gyrotron frequency of 72GHz can be used and the possibility of measuring the scattered wave goes in the same way as 65GHz.

Taking $\theta=170^o$ scattering angle, the scattering length, defined as the maximum length across the volume where the incident and scattered beam overlap, can be estimated as L= $2w_0/\sin\theta$, where w_0 is the transverse size of the wave beams, i.e. typically the gaussian beam waist. Assuming $w_0=25.\ 10^{-3}$m, and $\theta=170^o$, it is found L=0.28m, namely L~a/10 (a= ITER minor radius), which is still a good spatial localisation for the measurement. This localisation obviously improves decreasing θ , but because the signal itself decreases with θ the ratio $R_{\alpha e}=S_\alpha/S_e$, i.e. the ratio between the signal due to the alphas and that due to the free electrons, decreases. The signal-to-noise ratio can be restored increasing the integration time of the measurement, moving from $\theta=170^o$ to 90°, from τ=1ms to τ=30 ms, so spatially resolved measurements appear feasible. In this way the possibility of measuring the radial profile of the alpha particles on ITER mainly depends on the availability of access for diagnostics. However a possible scheme could be the following: the incident beam is launched from below the plasma equatorial plane, at elevation z=-0.5m aiming to the plasma centre. The scattered waves are received above that plane, around z=+0.5m at various poloidal angles. This configuration helps in minimisation of refraction effects. as shown in fig.2, where several rays are traced(see the following sections for a discussion). This arrangement can be used for the radially resolved ion population velocity distribution, leading also to the measurements of fast ions and important as well of bulk ion temperature.

DISCUSSION OF THE THEORETICAL MODEL.

The theoretical model used in ref.2 takes the electron density correlation as the leading term in the CTS dynamic form factor.

Formulas were obtained using a model for the source current j^{σ} of the scattered wave field E^{σ} derived on the basis of plasma kinetic theory[3] in the low temperature plasma approximation. The source current j^{σ} has contributions from density ,electric field, current and magnetic field fluctuations associated to the wave propagating into the plasma which couples with the incident wave generating the scattered wave. The CTS cross section is proportional the autocorrelation of the source current $<j^{\sigma}j^{\sigma}>$, giving raise to various correlation terms (<nn>, <EE>, <jj>, <BB>) and crossed correlations like <nB> etc. Now it was shown[3] that the terms different from <nn> are important in geometries where $\phi \sim 0^{o}$, where the resonances due to the lower hybrid begin to be important. This subject was also treated[4] using a fluctuation-dissipation theorem for non-equilibrium plasmas.

GAUSSIAN RAY-TRACING ANALYSIS

A ray tracing analysis was performed using realistic ITER equilibria and gaussian beam propagation[4]. The results are shown in fig.2 : the poloidal(fig.2a) and toroidal (fig.2b) projections confirm the results already presented in ref.2.

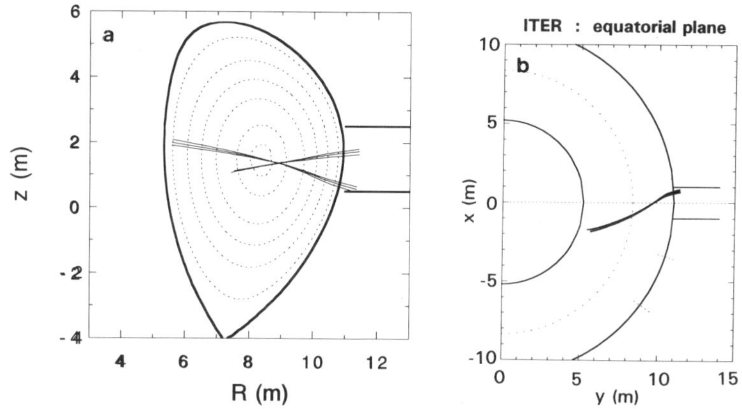

Figure 2 Gaussian raytracing code analysis results: 2a poloidal projection, 2b toroidal projection.

CONSIDERATIONS ON THE EXPERIMENTAL APPARATUS

The ITER equatorial ports (3 x 1.8 m^2, HxW) allow for an integrated launching-collecting system arranged for the backscattering geometry. The main geometrical constraints on the launching-collecting systems are illustrated in figures 3a and 3b. The toroidal launching angle ϕ, which in a backscattering geometry, is also the angle between the scattering wavevector **k=Ki-Ks(Ki incident wavevector,Ks scattered wavevector)** and the magnetic field, must be $\phi=15^{o}$. The emerging scattering geometry is one with the scattering plane (identified by **Ki** and **Ks**) tilted by 15° (or more) respect to the normal to the magnetic field. The fig.3a shows the launching system (as projected on the equatorial plane), its cohordinate is z=-0.5m from that plane, and the orientation is chosen for focalize into the plasma center at a beam waist of w0=0.05m. Two quasi-optical systems are arranged inside the cryostat and in the plug: the beam after emerging from a circular corrugated waveguide pass through a first window and, entering the cryostat vacuum, it is imaged on the second window with magnification one, in the plug the second optical system matches the beam waist formed onto the first window into a waist w0=0.05m in the plasma center.

The collecting system is shown in fig.3b, this is a projection on a vertical plane. A mirror looks at a line of sight passing through the plasma center, at 10° from the equatorial plane (corresponding to 170° scattering angle), with toroidal angle $\phi=15^{o}$. It can be put approximately 0.3m into the blanket, facing the plasma through a hole 0.3m diameter. The imaging optics is behind this mirror. These optics exhibit a variable (versus the torus major radius) vignetting, thus the overlapping efficiency between the input and antenna beam varies through the object space. This could be tolerated because of the margin on the signal-to-noise ratio. The imaging optics, again made of two telescopes, map each volume of the object space into an array of Gaussian antennas. This is similar to the LIDAR Thomson scattering system proposed for ITER.

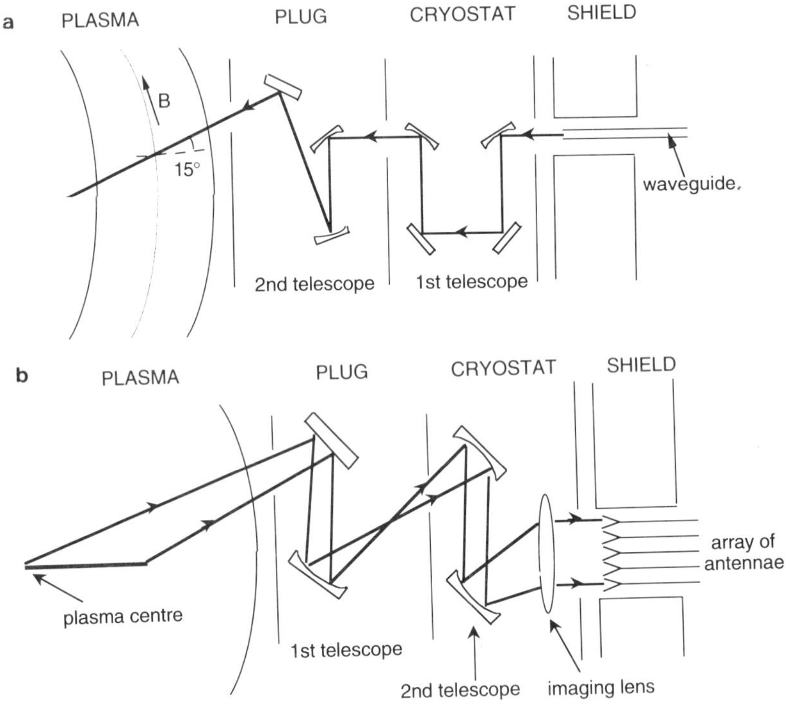

Figure 3. Launching (3a) and collecting optical system (3b) lay-out.

ACKNOWLEDGEMENTS

The authors are pleased to acknowledge helpful discussions with G.Vayakis, C.I.Walker and E.Doyle.

REFERENCES

1.F.Orsitto and G.Giruzzi, Rev.Sci.Instr. 68(1997)626.
2.F.Orsitto, G.Giruzzi, Nucl.Fus. 37(1997)915.
3.H.Bindslev, Journal of Atmospheric and Terrestrial Physics 58(1996)983.
4.S.C.Chiu, Phys.Fluids B3(1991)1374.
5.S.Nowak, E.Lazzaro and G.Ramponi, Phys.Plasmas 3(1996)4140.

DEVELOPMENT OF NEW VACUUM WINDOW SEAL FOR ITER OPTICAL DIAGNOSTICS

Akira Nagashima, Tsuneshi Fujisawa, Tatsuo Sugie and Satoshi Kasai

Department of Fusion Plasma Research,
Naka Fusion Research Establishment,
Japan Atomic Energy Research Institute,
Naka-machi, Naka-gun, Ibaraki-ken, 311-01, Japan

INTRODUCTION

Many windows will be required on ITER for optical, infrared, and microwave diagnostics. The diagnostic transmission systems will be incorporate labyrinths in port shielding blocks and so the windows will not be located in high levels of neutron and gamma radiation. Nevertheless it is necessary for the window seals to be able to withstand pressure transients, temperature excursions and mechanical shocks which are significantly different from those experienced on present-day tokamaks. The objective of the R&D is to develop suitable vacuum seal for ITER diagnostic windows concerning to mechanical capability. The main required features of the window specification for ITER diagnostics are the following. (1) Vacuum Seal:He Leak < 1 x 10^{-10} Pa m^3/s, (2) Capacity to Resist Inner Pressure Rise:> 5 atm, (3) Temperature:> 200 °C, (4) Temperature ramp rate:> 20 °C/hr, (5) Temperature gradients across windows:3~4 points for each surface of window, (6) Mechanical acceleration of structure:> 15g, duration 10ms, 1000cycles., (7) Optical property

At present, no commercial window can guarantee all the specification. Under the R&D, a reliable window seal will be developed to guarantee the specification with materials of Fused Quartz, Sapphire and ZnSe. At the beginning stage, a prototype window seal has been examined in successfully with 120mm$^{\phi}$ x 20mm^{t} fused quartz. The paper describes the performances of the new window seal method.

SCHEMATIC OF THE WINDOW SEAL

New vacuum seal for diagnostic window is being developed. The schematic of the window seal is shown in Figure 1. The important objectives we are considering are the capacity to resist 5 atm. inner pressure rise and 200°C temperature survivability. To develop a suitable seal, the important component is the seal ring with V-shape cross section (V-ring) which has a role of vacuum tight, and no-welding contact with cylindrical edge surface of the window material and elastic feature of V-ring eliminates to happen a shearing stress usually grown at the welding contact boundary originated by the difference of thermal expansion of materials.

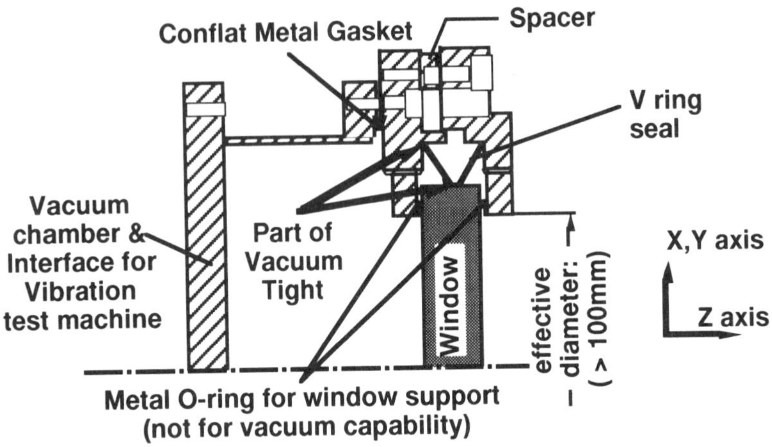

Figure 1. Schematic of newly developing vacuum optical window (JAERI Prototype). The vacuum seal is kept by the two part of V-ring contact. One is at the cylindrical edge surface of the window another is at the corner of vacuum side flange. The capability of inner pressure rise resistance is maintained by the window support including metal O-ring.

The remarkable advantageous properties of the window seal using a no welded V-ring are summarized in Table 1. The table shows that the new seal structure will resolve many critical problems of conventional window seal method. The important feature of freedom to select material which will not become radioactive is accomplished only by the no-welding structure.

Table 1 . The features and advantages of the newly developing window seal

Features of the new window seal	Advantages
no welding method on window material is used for seal.	• freedom to select material which will not become radioactive • no effect on optical property of window • no restriction of using weld capable material for window • re-usable of window flange • no weld contact between V ring and window edge surface makes it better for thermal expansion problem
elastic feature of V ring	• better for thermal expansion problem • better for mechanical shock
necessary notes	• polishing of window's cylindrical edge surface and contact part of V ring for leak tight • precise manufacturing of diameter of V-ring and window

PRELIMINARY RESULTS

The prototype examination has been curried out in successfully, concerning the specification as follows. For (1)vacuum seal, no He leak was detected under the sensitivity limit 1 x 10^{-11} Pa m^3/s; (2) capacity to resist inner pressure rise: no He leak was detected after 5 atm. pressure supply; (3) temperature, 3 heat cycles were performed up to 200°C and after each cycle, and no He leak was detected at each heat cycle; (4) temperature ramp rate of ~ 50 °C/hr was done by manual control; (5) temperature gradients across windows is about 10°C; (6) mechanical acceleration of structure of 15g, duration 10ms, 1000cycles was supplied for each axis direction ±Z and X and after each 1000cycles no He leak was detected; (7) Optical property was examined for cracks and twinning before and after the capability examination, and no change was observed. In these condition, the typical stress at the contact surface caused by 5 atm. pressure, 15g acceleration and thermal heat cycle are estimated as 0.75kg•f/mm^2, 0.01kg•f/mm^2 and ~10kg•f/mm^2, respectively. The values show the thermal heat cycle is the most important property to be cared for reliability.

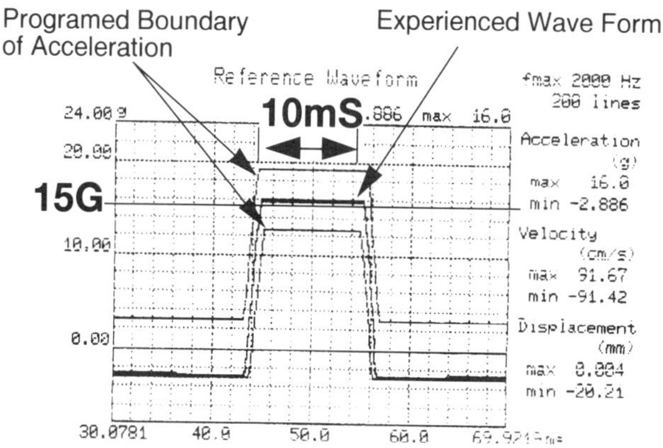

Figure 2. Mechanical acceleration wave form to examine window reliability. 1000cycles survivability has been examined to every axis of the window assembly under the acceleration condition of 15G and 10mS duration.

STRESS ANALYSIS FOR THERMAL CONDITION

A linear stress analysis was curried out by a finite element method assist system (FEMAS). The stress property at V-ring contact was examined respect to the difference between room temperature and 200°C higher condition. Figure 3 shows the element structure and the result by the FEMAS analysis. Materials are assumed as fused quartz, copper and stainless steel for window, V-ring and flange, respectively. In the figure, no serious change is observed between the two different temperatures at the V-ring contact. The contour line in Fig.3 (b) and (c) shows the equi-stress lines with unit of kg•f/mm^2. The positive/negative sign indicate tension/pressure stress. The damage stress threshold of Fused Quartz is recognized as 115 for pressure and 5 for tension. At the contact surface, only the pressure damage is of concern, and the estimation is suitably under the threshold. At the higher temperature, slightly higher stress is observed for higher temperature case, but the stress is suitably under the damage threshold.

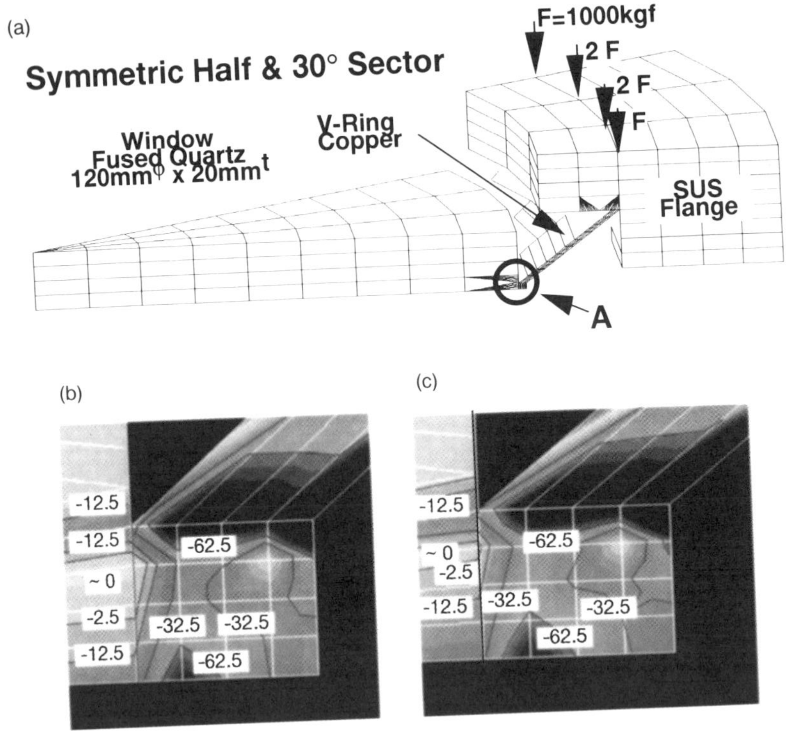

Figure 3. A linear stress analysis was curried out by a finite element method assist system (FEMAS) for the window seal structure and applied force condition of(a). The detail of V-ring contact part which is indicated "A" are shown in (b) and (c). No serious change is observed between (b) the stress map at room temperature and (c) at 200°C higher temperature. A slightly higher stress can be seen for higher temperature case, but the stress is suitably under the damage threshold.

DISCUSSION AND FUTURE PLAN

The seal method described in the paper was examined under the conditions which are considered to happen in ITER operation and has been accomplished by the prototype. The key structure of the window seal is the elastic feature of V-ring which enables tight vacuum seal without welding between window material and flange. The structure permits free selection of material free from weldable condition and makes it possible to eliminate the problem of using radio-active material as cobalt which usually be used in welding material. The examination for remained materials Sapphire and ZnSe will be performed under the similar structure. It will be reported the examined result for those.

ACKNOWLEDGMENTS

This report has been prepared as an account of work assigned to the Japanese Home Team under Task Agreement number S 55 TT 02 96-08-09 FJ within the Agreement among the European Atomic Energy Community, the Government of Japan, the Government of the Russian Federation, and the Government of the United States of America on Cooperation in the Engineering Design Activities for the International Thermonuclear Experimental Reactor ("ITER EDA Agreement") under the auspices of the International Atomic Energy Agency (IAEA).

RADIATION PROBLEMS AND TESTING OF ITER DIAGNOSTIC COMPONENTS

E. R. Hodgson

Euratom/CIEMAT Fusion Association, Madrid, Spain.

INTRODUCTION

It is envisaged that by the beginning of the next century ITER will come into operation. It is hoped that this intermediate next-step "technology" machine will bridge the gap between the present day large "physics" machines and the pre-commercial DEMO reactor. Although ITER will help to solve many of the problems which still remain in the field of plasma physics, it will also present additional operational and experimental problems due to radiation damage effects as a result of the intense radiation field from the "burning" plasma. This will give rise to a high energy neutron and gamma flux extending well beyond the first wall. In other words one foresees a serious materials problem which has to be solved. In the initial physics phase of ITER it is the radiation flux which will be of concern, whereas in the later technology phase both flux and fluence will play important roles as the radiation damage builds up in the materials. For structural metallic materials the problem of radiation damage is expected to be severe only near to the first wall, however the problem facing the numerous insulating components is far more serious due to the necessity to maintain not only the mechanical, but also the more sensitive physical properties intact.

Following a short introduction on the problem of radiation damage in both metals and insulators, the paper will concentrate on the problem facing the use of insulators. Apart from their use in the heating and current drive systems, insulators will be required for many diagnostic components including those which directly affect the control and stability of the fusion reactor plasma. For all these uses, in addition to mechanical strength and volume stability, the physical properties of interest will be electrical conductivity, dielectric loss and permittivity, thermal conductivity, and optical properties (transmission and luminescence, coloration). The effect of radiation on these properties will be discussed, and in particular electrical and optical aspects will be highlighted.

RADIATION DAMAGE

Radiation damage is generally divided into two components; displacement damage and ionization effects. Displacement damage results from the direct knock-on of atoms/ions from their lattice sites giving rise to vacancies and interstitials. These point defects may either recombine in which case no net damage is produced, or they may stabilise and even aggregate producing more stable extended defects. Displacement damage affects both metals and insulators, although the situation for insulators is in general more complex due to the polyatomic nature of these materials. The initial damage is measured in "dpa" (displacements

Diagnostics for Experimental Thermonuclear Fusion Reactors 2
Edited by Stott *et al.*, Plenum Press, New York, 1998

per atom) where 1 dpa is equivalent to displacing all the atoms once from their lattice sites. At the first wall of ITER the displacement dose rate will be of the order of 10^{-6} dpa s^{-1}.

Ionizing radiation although absorbed by both metals and insulators, in general produces only heating in metals. The effects of ionization on insulators is however quite marked due to the excitation of electrons from the valence to the conduction band giving rise to charge transfer effects. Ionizing radiation is measured in absorbed dose Gy (Gray) where 1 Gy = 1 J kg^{-1}. At the first wall of ITER the dose rate will be of the order of 10^4 Gy s^{-1}.

One other aspect of radiation damage that is worth mentioning is nuclear transmutation. The high energy neutrons will produce nuclear reactions in all the materials giving rise to transmutation products. These will build up with time and represent impurities in the materials which may modify their properties. However with the possible exception of He this is not expected to present a serious problem for ITER.

The result of these radiation damage processes will be a change in the physical and mechanical properties of the materials. Unfortunately, in general, these modifications do not improve the materials. The properties of concern which suffer modification are the mechanical strength and volume, thermal and electrical conductivity, dielectric loss and permittivity, and optical properties. The effects of such changes are that the materials become more brittle, they may swell, insulators may suffer Joule heating, and windows become opaque.

Radiation damage affects all materials, but the polyatomic nature of insulators and their susceptibility to ionizing radiation makes these materials far more sensitive to radiation damage than metals. While stainless steel for example can withstand several dpa and GGy with no problem some properties of insulating materials can be rendered useless by as little as 10^{-5} dpa or a few kGy. Because of this the present on-going programmes of radiation testing for diagnostics are concentrating mainly on the insulating component of the systems

Displacement damage and ionizing radiation degrade both the mechanical and physical properties of insulating materials. Of the numerous insulating materials it is the refractory oxides and nitrides which in general show the highest radiation resistance. Of these the ones which have received specific attention include MgO, Al_2O_3, $MgAl_2O_4$, BeO, AlN, and Si_3N_4. In addition different forms of SiO_2, and materials such as diamond for ECE have been examined.

SIMULATION EXPERIMENTS

At the present time no entirely suitable irradiation testing facility exists and as a consequence experiments are being performed in nuclear reactors and particle accelerators in an attempt to simulate the real operating conditions of the insulating materials and components. The experiments required must simulate the neutron and gamma radiation field i.e. the displacement and ionization damage rates, the radiation environment i.e. vacuum and temperature, and also the operating conditions such as applied voltage. It is furthermore essential that in-situ testing is carried out to determine whether or not the required physical properties of the material or component are maintained during irradiation.

Experimental nuclear fission reactors clearly have the advantage of producing a radiation field of neutrons and gammas, although in most cases the actual neutron energy spectrum and the displacement to ionization ratio are not those which will be experienced in a fusion reactor. The main difficulties with in-reactor experiments come from the inaccessibility of the radiation volume and are concerned with the problem of carrying out in-situ measurements and achieving the correct irradiation environment. While considerable success has been attained in the in-situ measurement requirement, the problem of irradiating in vacuum still remains, with most experiments being performed in a controlled He environment. Furthermore many experiments rely on nuclear and gamma heating to reach the required temperature and hence have difficulty with maintaining a controlled temperature.

Particle accelerators on the other hand are ideal for carrying out in-situ experiments in vacuum and at well controlled temperatures due to the easy access and very localised radiation field. It is however in the radiation field where their disadvantage is evident, and great care has to be taken to ensure that appropriate displacement rates are deduced. A further disadvantage is due to the limited irradiation volume and particle penetration depth. This in general means that only small material samples or components can be tested.

The present day situation of materials and component radiation testing for diagnostic systems takes full advantage of both fission reactors and particle accelerators. In-situ measurements can now be made during irradiation of the important electrical, dielectric, and optical properties. In addition other aspects such as mechanical strength and tritium diffusion are being assessed during irradiation.

DEGRADATION OF INSULATOR ELECTRICAL RESISTANCE

Diagnostic systems with components which rely on the insulating material maintaining a high stable electrical resistance will have to contend with radiation induced conductivity (RIC), radiation induced electrical degradation (RIED), and surface effects. Such components include electrical feedthroughs and standoffs, and mineral insulated cables and their terminations. In the case of MI cables the additional problem of radiation induced electro-motive force (RIEMF) must also be considered.

RIC has been studied for many years and an extensive data base exists together with a sound theoretical understanding[1-3]. This marked increase in the electrical conductivity during ionizing irradiation is due to the excitation of electrons from the valence to the conduction band and their subsequent trapping in levels near to the conduction band from where they are thermally excited once again into the conduction band. It has been shown that;

$$\sigma_{RIC} = \sigma_o + KR^{\delta}$$

where σ_o is the conductivity in the absence of radiation, R the dose rate, K a constant, and δ is approximately 1. From the many materials which have been examined two general conclusions / indications may be drawn;

RIC (single crystal) > RIC (polycrystal) and RIC (pure) > RIC (impure)

The following table illustrates this and gives conductivity data[4];

<u>RIC values at 450C 700Gy/s</u>

Material		RIC S/m
MgAl2O4 U.C.	sc (170)	9x10-7
MgO Spicers 4N	sc (500)	6x10-7
Al2O3 U.C.	sc (230)	5x10-7
Al2O3 Deranox	pc 999	3x10-7
Al2O3 Vitox	pc 999	1x10-7
Al2O3 Wesgo	pc 995	6x10-8
BeO G'fellows	pc 995	3x10-8

NOTE; solid material (numbers in brackets approximate impurity content). σ_o for these materials is in the range 10^{-11} to 10^{-10} S/m at 450 °C. One expects powders (i.e. MI cables) to show less conductivity, most probably being determined by the gas conduction. The data in this table shows that even at first wall dose rates of about 10^4 Gy s^{-1} insulating materials exist with an electrical conductivity of $\leq 10^{-6}$ S m^{-1}. However it is important to notice that RIC is a flux dependent effect and will be present from the on-set of operation of ITER. Hence devices which are sensitive to impedance changes, which will occur for example in MI cables, must take RIC into account.

RIED is more controversial and at the present time there is no general agreement as to the possible danger this type of electrical degradation poses for future devices such as ITER. RIED is due to a permanent increase in the volume conductivity σ_o and is fluence dependent. It has been observed to occur in various materials irradiated with an electric field applied and at temperatures of between about 250 and 500 °C. RIED like behaviour has been observed during electron[5-7], proton[8,9], alpha[10,11], and neutron[12-14] irradiation. However in contrast some experiments have not observed RIED or have observed enhanced surface degradation[15-17]. See also reference[18] for an extended discussion.

From the wide range of experimental results it is extremely difficult to come to a firm and reliable conclusion as to the importance of RIED for fusion. However it is important to note that essential parameters such as material type / grade, irradiation environment (vacuum or air / He), and irradiation temperature are not being taken fully into consideration. It is now clear that high purity Vitox alumina is susceptible to RIED at relatively low voltages (< 100 V/mm), whereas Wesgo 995 alumina is quite resistant to RIED up to about 1 kV/mm. It would also appear that Deranox alumina which is nominally the same as Vitox is also RIED resistant[19]. However recent results show that even with one nominally uniform grade large differences exist, as may be seen in figure 1 below for six different Deranox 999 samples[20].

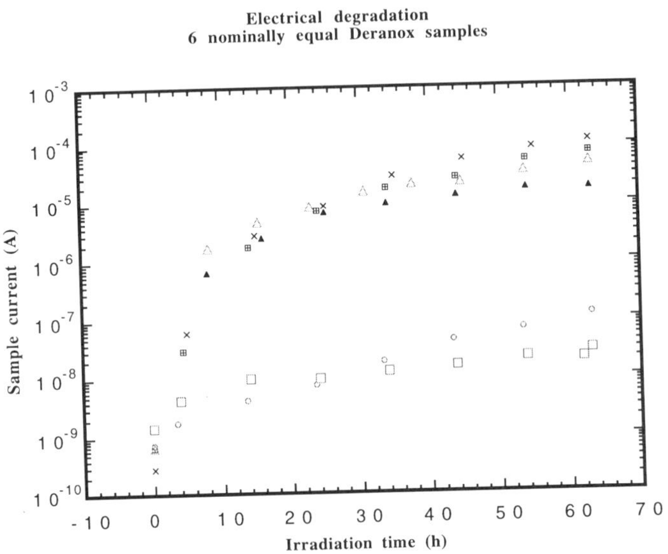

Figure 1. Electrical degradation of Deranox disc samples due to electron irradiation.

As may be seen the results for the samples fall into two groups, although they are from the same material and were all irradiated under identical conditions.

One recently completed in-reactor irradiation[21] in addition to highlighting the technical difficulties with such experiments brings to the forefront the problem of the irradiation environment. The experiment like several other in-reactor experiments was performed in He and significant sample degradation was only observed for one of the 13 samples. This contrasts with the two in-reactor observations of RIED carried out in vacuum[12,13] where marked degradation was observed. That the effect of irradiating in vacuum or air / He can play an important role on the electrical degradation is also evident from the differences observed in RIED during electron irradiation[6,7]. Both the experiments observe RIED but the one carried out in air[7] observed the on-set of degradation at a higher fluence, but more important observed a saturation in the degradation at a level which would make it difficult to detect in a reactor type experiment (10^{-7} S m^{-1} at 500 ^{o}C).

Surface degradation can also play an important part in insulator failure. Such surface effects can be large and may be due to contamination[15,16], but they may also occur in the material itself possibly due to impurity segregation or vacuum reduction of the surface[22]. It is moreover very important to carry out irradiation testing in the correct environment as it has been observed that Wesgo 995 alumina shows no surface degradation when irradiated in air or He, but severe effects are observed for irradiation in high vacuum[23].

RIC, RIED, and surface degradation can lead to unacceptably high losses at the insulator component of any diagnostic system. These possibilities must be carefully assessed to take into account both material and environmental effects.

DEGRADATION OF OPTICAL PROPERTIES

Optical transmission systems will suffer from surface degradation of front mirrors, together with optical absorption and radioluminescence in lenses / windows and optical fibres. In this section attention will be concentrated on the results for materials suitable for transmission in the UV and visible wavelengths. Two materials have been considered, these are high purity sapphire (Al_2O_3) and silica (SiO_2) including silica glass.

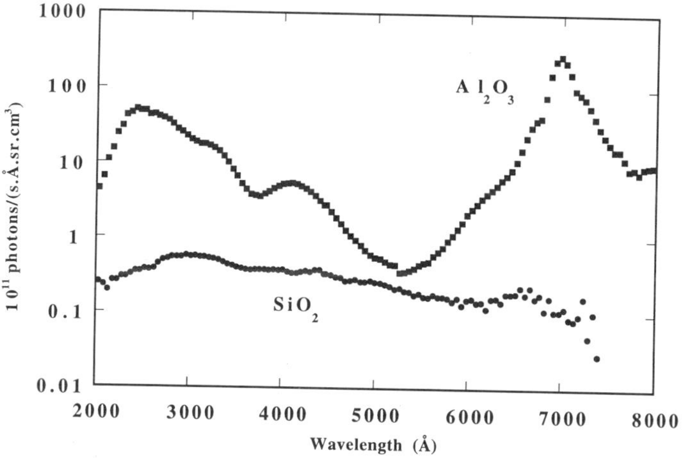

Figure 2. Radioluminescence for sapphire and KU1 quartz glass at 700 Gy/s, 20 ^{o}C [24,25].

It became clear at an early stage that one of the serious problems facing optical components in the UV-visible range would come from the radiation induced light emission, the so-called radioluminescence[24,25]. The intensity of this light for sapphire and some silicas has been shown to be comparable with the intensity of the collected plasma emission. However it has been shown that certain grades of SiO_2 show virtually no radioluminescence in the UV-visible region, the emission being limited almost to the Cherenkov background. Radioluminescence for Union Carbide UV grade sapphire and RF KU1 quartz glass are shown above in figure 2. Radioluminescence like RIC is dependent on the ionizing flux and hence will be a problem from the onset of operation of ITER. However it would appear that a material is available with low emission. It is necessary to check the radioluminescence as a function of fluence to ensure that the radiation induced defects do not begin to emit light.

The other major problem with optical transmission components is the radiation induced absorption. Although ionizing radiation can cause increased optical absorption it is in general the displacement damage mechanism which induces severe absorption. This is a fluence effect which will mean that with time the optical signal received will be reduced and distorted as a function of wavelength. Figure 3 gives the initial and final absorption spectra for KU1 quartz glass following 500 hours of gamma irradiation at 6 Gy s^{-1}, 20 ^{o}C. As may be seen the induced absorption is essentially confined to below 400 nm. This UV absorption as a function of irradiation temperature is given in more detail in figure 4. It is clear that for low radiation levels and for temperatures at or above about 100 ^{o}C very little absorption is induced in this material.

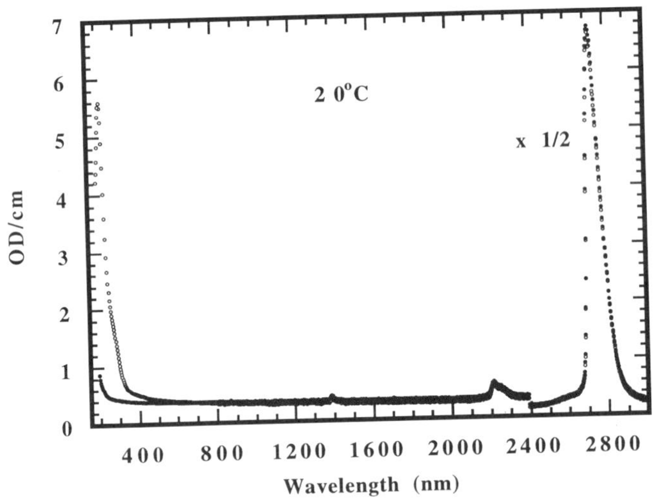

Figure 3. Radiation induced absorption in KU1 quartz glass.

As the radiation damage builds up however, the oxygen vacancy related absorption bands shown in figure 4 increase and their absorption tails begin to overlap into the visible region[25]. Sapphire shows more radiation resistance than SiO_2 in terms of optical absorption, but as mentioned above it gives intense radioluminescence. However as may be seen in figure 2 in the region near 500 nm the luminescence is almost as low as in the KU1 material. So sapphire may provide a useful window in this region. Furthermore sapphire is highly radiation resistant in the IR region up to about 5000 nm.

Radiation induced optical absorption and light emission are problems for optical window and lens components, but in the case of optical fibres the situation is far worse due to the length of the optical path. Furthermore due to the manufacturing techniques, fibres with characteristics as good as those observed for the KU1 quartz glass have so far not been obtained, however rapid progress is being made[25].

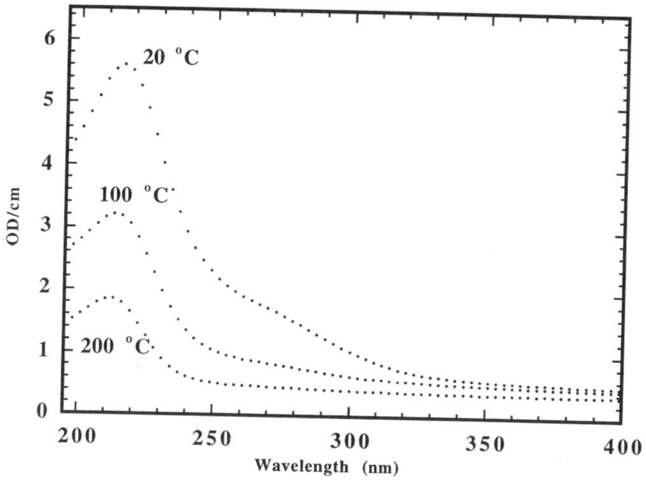

Figure 4. Detail of absorption in the UV for figure 3 together with the effect of irradiation temperature.

At the present time two particularly difficult problems need attention. The first is concerned with the windows for the LIDAR system, and the second is concerned with the requirement to make observations in the UV region. The LIDAR windows will be required to transmit high power laser pulses in the 500 to 1000 nm region, with only a few percent absorption in the window. Using standard spectrometer techniques it is extremely difficult to reliably determine such a small change in the absorption, this is particularly true for samples which have to be measured post-irradiation. To determine such small changes required in-situ measurement. In-situ measurement is also required in order to determine possible radiation enhanced absorption which can easily reach such small values. The possibility of radiation enhanced dielectric breakdown due to the intense laser pulse and the ionizing radiation is also being considered. However such a determination requires an elaborate in-situ experiment.

As may be seen in figures 3 and 4 even low levels of radiation introduce marked optical absorption in the 200 to 400 nm UV region. This a common feature of all the possible radiation resistant materials; MgO, Al_2O_3, $MgAl_2O_4$, and SiO_2, and is due to oxygen vacancy type point defects produced by displacement damage. However another common feature of these materials is that the complementary defect, the oxygen interstitial, becomes unstable on increasing temperature. As a result, as may be seen for example in figure 4, irradiation at high temperatures results in lower absorption. If a window of sapphire or SiO_2 could be employed and maintained at about 400 ^{o}C most of the radiation induced defects would be continually annealed out. As the window would be used for the UV region any possible interference from the black body emission would be negligible.

CONCLUSIONS

Without careful design it will prove extremely difficult to separate out the required plasma information from the radiation induced perturbations. At the present time a wide data base exists for several candidate insulator materials for fusion applications, and work is now underway or planned for specific diagnostic component tests. However we are at the beginning and much still remains to be done.

REFERENCES

1. R.W.Klaffky, B.H.Rose, A.N.Goland, and G.J.Dienes, Phys. Rev. B 21 (1980) 3610.
2. E.R.Hodgson and S.Clement, J. Nucl. Mat., 155 - 157 (1988) 357.
3. G.P.Pells, Rad. Effects, 97 (1986) 39.
4. E.R.Hodgson, ITER Technical Meeting of Irradiation Tests on Diagnostic Components Garching, June 1996.
5. E.R.Hodgson, Cryst. Latt. Def. and Amorph. Mater., 18 (1989) 169.
6. E.R.Hodgson, J. Nucl. Mat., 212 - 215 (1994) 1123.
7. X.F.Zong, C.F.Shen, S.Liu, Z.C.Wu, Y.Chen, B.D.Evans, R.Gonzalez, and C.H.Sellers, Phys. Rev. B 49 (1994) 15514.
8. G.P.Pells, UKAEA Fusion Report, AEA FUS 86, 1990.
9. G.P.Pells, J. Nucl. Mat., 184 (1991) 177.
10. A.Möslang, E.Daum, and R.Lindau, Fusion Technology 1994 (Proceedings 18th. Symposium on Fusion Technology, Karlsruhe Aug.1994) 1313.
11. J.D.Hunn, R.E.Stoller, and S.J.Zinkle, J. Nucl. Mat., 219 (1995) 169.
12. T.Shikama, M.Narui, Y.Endo, T.Sagawa, and H.Kayano, J. Nucl. Mat., 191 - 194 (1992) 575.
13. T.Shikama, M.Narui, H.Kayano, and T.Sagawa, J. Nucl. Mat., 212 - 215 (1994) 1133.
14. E.H.Farnum, T.Shikama, M.Narui, T.Sagawa, and K.Scarborough, J. Nucl. Mat., 228 (1996) 117.
15. W.Kesternich, F.Scheuermann, and S.J.Zinkle, J. Nucl. Mat., 206 (1993) 68.
16. P.Jung, Z.Zhu, and H.Klein, J. Nucl. Mat., 206 (1993) 72.
17. E.H.Farnum, F.W.Clinard Jr., W.F.Sommer, J.C.Kennedy III, and T.Shikama, J. Nucl. Mat., 212 - 215 (1994) 1123.
18. C.Kinoshita and S.J.Zinkle, J. Nucl. Mat., 233 - 237 (1996) 100.
19. A.Möslang, private communication.
20. E.R.Hodgson and A.Moroño, to be published.
21. S.J.Zinkle, IEA Ceramics Workshop, Cincinnati, May 1997.
22. A.Moroño and E.R.Hodgson, J. Nucl. Mat., 233 - 237 (1996) 1299.
23. A.Moroño and E.R.Hodgson, to be presented at ICFRM 8, Sendai Japan, Oct.1997.
24. A.Moroño and E.R.Hodgson, J. Nucl. Mat., 224 (1995) 216.
25. E.R.Hodgson, T.Nishitani, D.Orlinski, and S.Yamamoto, ITER Technical Meeting of Irradiation Tests on Diagnostic Components Garching, July 1997.

IRRADIATION TESTS ON ITER DIAGNOSTIC COMPONENTS

S. Yamamoto[1], L. deKock[1], V. Belyakov[2], D. Orlinski[2], F. Engelmann[3], E. Hodgson[3], S. Kasai[4], T. Nishitani[4], E. Farnum[5], K. Young[5], S. Bender[2], V. Chernov[2], D. Griscom[5], R. Heidinger[3], T. Iida[4], E. Ishitsuka[4], T. Kakuta[4], H. Kawamura[4], A. Krasilnikov[2], V. Lopatin[2], R. Macaulay-Newcombe[3], K. Noda[4], T. Matoba[4], Y. Oyama[4], P. Pells[3], A. Ramsey[5], Y. Revyakin[2], T. Shikama[4], R. Snider[5], P. Stott[3], V. Stepanov[2], U. Tarabrin[2], I. Valeri[2], K. Vasili[2], V. Voitsenya[2], K. Vukolov[2], V. Zaveriaev[2], S. Zinkle[5], A. Costley[6], T. Ando[1], K. Ebisawa[6], P. Edmonds[1], H. Iida[1], L. Johnson[6], R. Santoro[1], S. Tanaka[1], V. Mukhovatov[6], G. Vayakis[6], C. Walker[1], G. Janeschitz[1]

1 ITER Joint Central Team, Garching, Germany
2 RF Party
3 EU Party,
4 JA Party
5 US Party
6 ITER Joint Central Team, San Diego, USA

ABSTRACT

Radiation effects on key components of diagnostic systems expected to be subjected to high neutron and gamma fluxes and fluences are being examined in irradiation tests to evaluate and establish an ITER-relevant database to support the design. A comprehensive irradiation database has been accumulated and permits conclusions to be drawn on the application of these components in ITER. The design studies on prototypical assemblies of diagnostic components are continuing based on the irradiation data bases, neutronics calculations for evaluating irradiation environment of diagnostic components and required specification of diagnostic systems. These studies will aid the recognition of detailed requirements of diagnostic systems leading to more specific irradiation tests on diagnostic components.

Diagnostics for Experimental Thermonuclear Fusion Reactors 2
Edited by Stott *et al.*, Plenum Press, New York, 1998

INTRODUCTION

The main objective of irradiation tests on diagnostic components is to establish a database of irradiation effects on properties of diagnostic components and to specify appropriately the working conditions for developing diagnostic systems which can be operated reliably in the ITER irradiation environment.

The first meeting of irradiation effects on ITER diagnostic components was held at St. Petersburg, Russia in 1991 in order to provide an opportunity for radiation experts and diagnosticians to develop a common base for understanding radiation effects on diagnostic components [1]. The second meeting was held at the Garching ITER Joint Work Site in 1993 [2]. Based on the discussions of those meetings, an task was initiated. Irradiation tests on ceramics, windows, optical fibers, mirrors/reflectors and wires/cables were started under the task because they are key components of diagnostic systems and are expected to be used under relatively high irradiation environments.

For the irradiation tests on diagnostic components, in-situ measurements should be emphasised. For ceramics, radiation induced conductivity (RIC), radiation induced electrical degradation (RIED, which is a permanent degradation observed in ceramic insulators under applied voltage), dielectric, thermal and mechanical properties, and tritium diffusion have been studied. For potential window materials, the optical properties including radioluminescence and optical transmission have been studied. Similar studies have been carried out for potential optical fibres. For potential mirror materials, reflectivity is measured at relevant wavelengths. For wires/cables, RIC, RIED, radiation induced electromotive force (RIEMF) between sheath and centre conductor of MI cables and conductor resistance are measured.

Round robin experiments on Wesgo 995 polycrystalline alumina (the IEA reference ceramic material), fused silica glass KU-1 (distributed by Russia) and Anhydro-guide (low-OH) fiber with 200/220/280 μm core/cladding/jacket diameters (distributed by USA) were also performed by the four ITER Home Teams using various kind of irradiation sources.

As the design of diagnostic systems progresses, the irradiation tests will focus on the solution of well defined problems and irradiation of prototypical assemblies including prototype sensors (magnetic coils, bolometers, pressure gauges etc.) and transmission components (vacuum seals of windows, vacuum electric feedthroughs, remote handled electrical connectors, vacuum seals for optical fibers etc.).

In this paper we will give an overview of the results of the irradiation tests on diagnostic components and assess the present diagnostic integration design in the light of these results.

IRRADIATION TEST FACILITIES

Displacements and ionisation effects (namely, electronic excitation effects) are very important irradiation effects on materials. The ratio of absorbed doses used for electronic excitation to those for atomic displacements (REA) under 1.8 MeV electrons, 20 MeV protons, fission neutron and 14 MeV neutron irradiations are ~ 10000, ~4000, ~10-100 and ~30, respectively. The REA value near the ITER first wall is around 10 and varies by many orders of magnitude depending on the distance from the plasma. Fission neutron irradiation sources seem to be the best available irradiation facilities for testing ITER components except for transmutation-related issues. However, all irradiation facilities, i.e. fission reactors, 14 MeV neutron sources, charged particle accelerator facilities and gamma ray sources should be utilised to provide timely and cost-effective data for ITER. A balance between low fluence and high fluence experiments and between in-situ and post-irradiation measurements must be found in order to make use of the resources in the most efficient way.

SUMMARY OF PRINCIPAL IRRADIATION TEST RESULTS

Recommended irradiation test conditions for clarification of the irradiation effects on material properties and important factors affecting these properties are summarised in Table 2 [3]. Important factors influencing the properties, including dose rate, fluence, environment i.e. temperature and vacuum (or gas), impurities and defects in the material and sputtering, these are designated by X in the table. Some comments on the necessity for in-situ measurements and additional needs are also added to the table. Measurements designated U will help to characterise the processes involved in the response to irradiation. A detailed description of the important influencing factors will be given in another paper [4].

The principal irradiation test results are as follows [5, 6, 7].

Ceramics:

• In the case of an alumina ceramic Wesgo 995, RIED was not observed under electron beam or neutron irradiations.

• No catastrophic RIED was observed up to neutron fluence of 3 dpa (displacements per atom) in helium.

• The maximum RIC at room temperature in ceramics for ionising dose rates relevant for ITER first wall conditions (~ 10^4 Gy/s) is about 1×10^{-6} S/m.

• The value of RIC due to 14 MeV neutrons is one of the reference values obtained under other sources of the fission neutron, electron beam, ion beam or gamma irradiations.

• The swelling behaviour of ceramics is strongly dependent on the irradiation temperature and the particular ceramic. Typical swelling values of 0.1 to 5% are observed after neutron irradiation to a damage level of about 10 dpa.

• Ceramics generally experience a slight degradation in strength and fracture toughness after fission neutron irradiation to ~ 1 dpa.

• Post-irradiation measurements performed on neutron irradiated ceramics have found that the thermal conductivity typically decreases to about 1/3 of the unirradiated value, although significantly larger and smaller changes have been observed in numerous studies.

• Results for sapphire and alumina indicate an enhancement of the tritium diffusivity at room temperature by a factor of about 10^{20} at 10^6Gy/s, with electronic excitation effects playing an important role. Initial results for BeO indicate that the radiation enhanced diffusion is related to displacement damage rather than electronic excitation.

Table 2. Recommended irradiation test conditions and important factors

Factor / Property	Important Factor: Dose rate	Fluence	Env.: Temp.	Env.: Vac. Gas	Imps De-fects	Spu-tter	Comments: In-Situ	Comments: Need to be clarified	Conditions: Temperature (°C), Dose rate (Gy/s), Flux ($n/cm^2/s$), Fluence, E-field
RIC (Mag. probe)	X:U	X	X:U	X	X		X:U	Transmu-tation	150-650 °C up to 10^4 Gy/s, Target flux $4x10^{14}$ $n/cm^2/s$ up to 20 dpa Low voltage
RIC (Ceramics)	X:U	X	X:U	X	X		X:U	Transmu-tation	Same as for magnetic probe
RIED (Ceramics)	X	X	X	X	X		X		150-600 °C E-field >100V/mm Flux 10^{11}-10^{14} $n/cm^2/s$
Mechnical (Ceramics)	? X							Swelling Shock SCCG	
Thermal (Ceramics)								Shock	
Tritium Diffusion	X		X				X		Windows/feedthroughs: < 350°C Low pressure Low flux < $10^{10}$$n/cm^2/s$ Max
Optical absorption	X	X	X		X		X		25-400 °C up to 10^3 Gy/s
Radio Luminescence	X	X	X		X		X		25-400 °C up to 10^3 Gy/s
Reflectivity		X	X			X			25-400 °C flux higher for first mirror up to 10^4 Gy/s, Target flux $4x10^{14}$ $n/cm^2/s$ up to 20 dpa

X: Important influence
U: Understanding

Windows:

• Significant irradiation effects on optical absorption of fused silica glass KU-1 for visible regions (from 350 nm to 850 nm) were not observed up to $10^{22}n/m^2$. Transparency in the wavelength from 450 to 650 nm decreased by 5 % to 15 % in the range of the fluence from 10^{22} to $5x10^{23}n/m^2$.
• No irradiation effect on optical absorption of sapphire windows for IR regions (from 850 nm to 5000 nm) was observed up to $4x10^{24}n/m^2$.
• Radioluminescence in sapphire is very large in the region 200 nm to 750 nm and two or three orders of magnitude greater than in quartz. Those values at 410 nm are 10^{10} photons/Gy/Å/st./cm^3 for sapphire and 10^7 photons/Gy/Å/st./cm^3 at 410 nm for fused silica glass or quartz.
• The transient optical absorption in silica glass KU-1 under gamma or electron irradiations was measured. It was shown that the absorption in the UV spectral region varies very rapidly during the first few minutes after the start and after the termination of irradiations. In the middle and red part of the visible spectrum the sample transparency changes very slowly or does not change at all.
• Annealing effects with the temperature of up to 500 °C on optical absorption in sapphire and quartz windows were very small. For UV regions of KU-1, at least 700 °C is required for significant annealing.

Fiber optics:

• Pure-silica core fibers with polymer jacket can be radiation hardened against gamma-ray-induced visible absorption (400-700 nm) by high-dose gamma-ray pre-irradiation ($8.3x10^6$Gy) in the presence of propagating visible light (~1.4mWh). This phenomena were not observed in metal-jacketed fibers.
• Noticeable part of the radioluminescence of optical fibers is a Cherenkov radiation.
• Operating of silica core fibers at >350°C (or periodic heating) appears a promisimg way to substantiallly reduce the absorption in a relatively low fluence region.
• Pure silica core fibers with low OH content show better radiation resistance than other fibers for the visible regions.
• Fibers with extremely low OH (0.5 ppm) and Cl (<20ppm) contents in their cores demonstrated a rather flat induced loss spectrum in the visible region at a level of 2-2.5 dB/m for gamma-doses of about 10^7Gy.
• Pure silica core fibers with low OH content have an induced loss in the wavelength region $\lambda > 400$ nm less than 10dB/m up to neutron fluence of $1.6x10^{20}n/m^2$ and in the wavelength region $\lambda > 800$ nm less than 10dB/m up to neutron fluence of $10^{25}n/m^2$ (1dpa).
• Neutrons causes displacement damage associated with electronic excitation. And the electronic excitation is introduced densely in a very localised zone. On the other hand, gamma rays cause mainly electronic excitation. Comparison between neutron irradiated and gamma-ray irradiated optical fibers showed that neutrons might have larger effects in causing in-situ absorption than gamma rays.

Mirrors/Reflectors:

• Metal mirrors made of copper can be used as first mirrors up to 40 dpa.

Molybdenum Corner Cube Reflectors (CCR) were irradiated up to $1.4\times10^{24}n/m^2$. The surface condition of the Mo CCR was examined by an interferometric technique and SEM and no irradiation effects was observed.

- Post irradiation measurements of layered synthetic microstructures (LSMs) for soft x-ray reflectors were completed. Mo/Si, W/B_4C and W/C LSMs were irradiated up to $1.1\times10^{23}n/m^2$ (0.01 dpa). A shift of peak reflectivity to shorter wavelength regions was observed and the reflectivity of the shifted peak decreased by 20% compared with the original peak reflectivity of an unirradiated reflector.
- Preliminary reflectivity measurements were completed on dielectric-coated mirrors for Thomson scattering laser beams. (a) HfO_2/SiO_2 - 37 layers, (b) ZrO_2/SiO_2 - 121 layers, (c) ZrO_2/SiO_2 - 29 layers, (d) TiO_2/SiO_2 - 39 layers and (e) TiO2/SiO_2 - 39 layers (plate polariser) were irradiated up to $1.1\times10^{23}n/m^2$ with temperature cycled between 60°C to 270°C. Of the 5 types of mirrors irradiated, two (HfO_2/SiO_2 - 37 layers and TiO_2/SiO_2 - 39 layers) showed little mechanical damage. But, other three types of mirrors showed flaking or blistering of the coating. The reflectivity of the undamaged mirrors was measured and to within ±5% reflectivity, no degradation was observed. Zirconia coatings were physically damaged but surviving parts retained high reflectivity.
- Some organic x-ray crystals change their characteristics at a relatively low fluence level. However, inorganic crystals including Ge, Si, SiO_2 and graphite do not have a change in their reflectivity, interplane distance and width and form of diffraction lines up to $10^{23}n/m^2$. Mica monocrystals preserve their characteristics up to $10^{22}n/m^2$.

Wires/Cables and Magnetic coils:

- RIC values of coaxial cables with 8 different types of powder mineral insulation (Al_2O_3, Al_2O_3 doped with Cr, MgO, diamond, mullite and cordierite) were measured. The lowest levels of RIC were observed in the fine mesh high-purity alumina powder (~0.3 μm particle size). The extrapolated conductivities of these cables are $<10^{-8}$ S/m at temperatures up to ~450°C for ionising dose rates as high as 10^4 Gy/s.
- No catastrophic RIED was observed up to 1.8 dpa ($\sim1.8\times10^{25}$ n/m^2 (E>0.1 MeV)) in MI cables irradiated with temperature of below 440°C and with applied electric field of 80 to 490 V/mm .
- No major change of MgO insulation resistance of MI cables under neutron irradiation up to fluence $3\times10^{25}n/m^2$ (3 dpa) and then gradual decrease by a factor of 3-4 to a final fluence of $10^{26}n/m^2$ were observed.
- No inductance change of a magnetic coil made of MI cables was observed up to neutron fluence of $6\times10^{24}n/cm$.
- The observed values of RIEMF in about 10m long MI cables under fission reactor irradiations are about 3.0V. The RIEMF may be current driven.

The recommended materials and maximum allowable dose or dose rates to these materials are summarised in Table 3. A set of one-page-summaries for all irradiation tests and the summary of irradiation test results under this ITER programme will be available as an ITER memo.

EVALUATION OF LIFE TIME OF DIAGNOSTIC COMPONENTS

Diagnostic components will be located at several locations in the ITER device. In order to simplify the maintenance scheme of the diagnostic systems, in-situ replacement of diagnostic components due to irradiation damage should be avoided. If the diagnostic components installed in in-vessel components, for example, diagnostic divertor cassettes, diagnostic plugs in horizontal diagnostic ports or in vertical diagnostic ports, survive longer than the lifetime of these in-vessel components, the diagnostic component can be repaired when the related in-vessel component is replaced. According to the ITER maintenance classifications, divertor cassettes and diagnostic plugs in the horizontal or vertical ports are categorised as Class 1 components. Blanket modules and back-plates are categorised as Class 2 and Class 3 components, respectively. The required numbers of full power discharges with 1000 seconds for each class and the maximum dpa in ceramics before the replacements of the in-vessel components are shown in Table 4. The total flux (>0.1 MeV) and ionisation dose rate at the first wall of ITER are $3x10^{18}/m^2/s$ and $3x10^3 Gy/s$ during operation with a wall load of $1MW/m^2$, respectively. A total fluence of $10^{25} n/m^2$ makes about 1 dpa damage in ceramics, 0.5dpa in stainless steel and 0.7dpa in copper. Therefore, a total damage of 5dpa in ceramics and 2-3dpa in metals and a total ionisation dose of 10^{10}Gy in ceramics are expected at the first wall at the end of Basic Performance Phase (BPP) with a total fluence of $0.3MWa/m^2$. Gamma-ray radiation produced by the residual ^{16}N in cooling water resulting from the $^{16}O(n,p)^{16}N$ reaction ($T_{1/2}$ = 7.13 s) is about 0.06 Gy/s at 1 m from the coolant channels. This effect cannot be ignored especially near the cryostat. Each location of the component and the radiation shield configuration for the component must be designed taking into account the life time and required signal to noise ratio of the diagnostic system.

SUMMARY AND DISCUSSIONS

The quality of electrical insulation by ceramics is one of the key issues for ITER diagnostic systems. Ceramic insulators will be used in electric feedthroughs, connectors, mechanical supports and general standoffs, MI cables and their seals, substrates (bolometers), and other sensor devices such as pressure gauges.

The important factors of RIC are the operating temperature and irradiation dose rate. The RIC was quantified and should not a problem in diagnostic design. For example, the magnetic coils made of MI cables require an insulating conductivity of less than 10^{-6} S/m (preferably much less i.e. 10^{-7} S/m). MI cable resistivity on the back plates of the blanket structure meets these specifications.

The RIED is still only partly understood [8]. The important factors of RIED are an applied voltage, operating temperature and irradiation dose. The RIED only occurs if the following two conditions are satisfied at the same time: Experiments have shown that if an electrical field greater than 50 kV/m is applied during displacive irradiation and the working temperature is between 175±25°C and 600±50°C. If we avoid the above mentioned RIED expected conditions or select particular materials (for example, Wesgo AL995), the RIED will not be a serious issue for the use of ceramic insulators. Some early RIED measurements are now believed to be due to surface contamination through surrounding environments and this is serious concern for installation. We must take possibilities of surface contamination and degradation into account and consider

Table 3 Recommended materials and maximum allowable dose or dose rates

Diagnostic components	Recommended materials	Accumulated effects	Dynamic effects
Ceramics (electrical insulators)	Single crystal sapphire and polycrystal alumina (Al_2O_3)	3dpa in helium (RIED: No catastrophic degradation)	10^4Gy/s (RIC: <10^{-6}S/m)
Wires /Cables	MI-cables: SUS, Inconel (sheath)/MgO, Al_2O_3 (insulator)/ Cu, Ni (centre conductor)	1.8dpa (RIED: No catastrophic degradation)	10^4 Gy/s (RIC: <10^{-6}S/m) 10^3Gy/s (RIEMF: <10V)
Windows	Fused Silica/Quartz (400-1200 nm)	10^{-3}dpa (Transmission; 5% degradation)	Radioluminescence: 10^7photons/Gy.A.sr.cm^3
	Sapphire (800-5000nm)	0.4dpa (Transmission; No degradation)	Radioluminescence: 10^{10}photons/Gy.A.sr.cm^3 at 410nm
Optical fibers	Pure silica (core)/F doped (clad)/Al jacket (Visible region)	10^7Gy (Transmission: 2-2.5 dB/m) $6x10^{-6}$dpa (Transmission: 10dB/m)	Radioluminescence:
	Pure silica (core)/F doped (clad)/Al jacket (IR region)	1dpa (Transmission: 10dB/m)	Radioluminescence:
Mirrors /Reflectors	First mirrors: Metal (Cu, W, Mo, SS, Al)	40dpa (Cu) (Reflectivity: No degradation)	
	First mirrors for LIDAR: Single coated (Rh/V)?		
	Dielectric mirrors: (HfO_2/SiO_2, TiO_2/SiO_2) ?	< 10^{-2}dpa (Flaking, Blistering)	
	LSMs: (Mo/Si, W/B_4C and W/C) ?	< 10^{-2}dpa (the shift of the peak reflectivity to shorter wavelength)	
	X- ray crystals: (Ge, Si, SiO_2, Graphite)	10^{-2}dpa	

Table 4 The Maximum Flux and Fluence to Surfaces of the Plasma Side of each In-Vessel Component

Maintenance Classification	Example	Required lifetime Numbers of discharges with 1000 sec	Flux [n/m^2/s] (>0.1 MeV)	Fluence [n/m^2] (>0.1 MeV)	dpa in ceramics
Class 3	Back plate	10000-30000	$2x10^{16}$	$(2-6)x10^{23}$	$(2-6)x10^{-2}$
Class 2	Blanket module	3000-5000	$3x10^{18}$	$(9-15)x10^{24}$	0.9-1.5
Class 1	Divertor cassettes Diagnostic plug (Horizontal and Vertical)	1000-3000	$3x10^{18}$	$(3-9)x10^{24}$	$(3-9)x10^{-1}$

counter measures to mitigate such effects.

Fused silica glass or quartz windows should be used in spectroscopic systems for visible regions and sapphire windows should be used for IR regions.

During operation optical fibers for the IR regions can be used inside the cryostat but optical fibers for visible regions can only be used outside the bioshield.

Windows, electric feedthroughs and optical feedthroughs constitute very important vacuum and tritium barriers in diagnostic systems. A window seal design is required to accommodate the high bakeout temperature of 250°C. Windows as a first vacuum barrier must withstand a 5 bar accident. Window seal configurations will be accepted only after they are subjected to at least 5 bakeout and pressure cycles, a full expected dose of neutron and gamma radiation and mechanical acceleration tests to 15g.

Metal mirrors (Copper and Molybdenum) are preferable for first mirrors in the ITER irradiation environment. There is a possibility that it will not be necessary to change the mirrors close to the first wall during the ITER life time if other adverse effects of sputtering, evaporation or coating are not too serious. Single coated metal mirrors of suitable combinations have to be used as a LIDAR first mirror. The design of mirror assemblies to reduce mirror distortion due to radiation heating is very important. Especially, dielectric mirrors and LSMs should be used in well-shielded locations and under temperature control.

The experiments show that magnetic coils with MI cables on the back plates will survive the life time of ITER. Observed radiation induced drift of an integrator/coil combination may limit an observable pulse duration to ≤3000s.

The following irradiation issues have still to be resolved for ITER diagnostic systems in the near future.

- Irradiation effects on seals of windows [9], electric feedthroughs and optical feedthroughs.
- Irradiation effects on remote handled electrical connectors.
- Laser power absorption and thermal shock in lenses and windows; in order to evaluate the life time of the high power laser window, irradiation effects on transient and permanent absorption should be investigated with 1 % accuracy.
- Damage threshold of dielectric mirrors and LSMs under irradiation.
- Laser damage threshold in irradiated dielectric mirrors or single coated mirrors.
- Reliable cable termination techniques with low leakage currents for MI cables.
- RIEMF effects on total magnetic measuring systems; origin of the RIEMF has not been understood yet. Recently, an integrator concept for much longer times than 10^3 sec has been demonstrated Measurements of the radiation induced noise including the RIEMF and any radiation induced drift in the long pulse integrator are under way.
- Development of radiation hardened bolometers; the JT-60-type bolometer made of a stainless-steel mounting frame, gold absorber and polyimide sheet was irradiated up to 220 MGy under gamma irradiations. No significant change in the bolometer sensitivity was observed. A JT-60-type bolometer made of AlN substrates was completed and will be irradiated under fission neutron fields. The JET-type bolometer made of mica substrates, usable at elevated temperatures of 300°C is under neutron irradiation. A few candidates of substrate materials, Si_3N_4, AlN, Al_2O_3 or $MgAl_2O_4$ for the JET-type bolometer were proposed for further irradiation tests [10].
- Development of radiation hardened pressure gauges; based on the ASDEX-type pressure gauge, the ITER-relevant pressure gauge will be developed [11].

Irradiation effects on the mechanical strength of hot cathodes should be addressed. Parasitic signals due to RIC, surface leakage current due to surface contamination, degradation or gas ionisation due to tritium or gamma irradiation should be evaluated.

ACKNOWLEDGEMENTS

This paper was prepared as an account of work performed under the Agreement among the European Atomic Energy Community, the Government of Japan, the Government of the Russian Federation, and the Government of the United State of America on Co-operation in the Engineering Design Activities for the International Thermonuclear Experimental Reactors (ITER EDA Agreement) under the auspices of the International Atomic Energy Agency (IAEA).

REFERENCES

[1] Summary Report, ITER Workshop on "Radiation Effects on Diagnostic Components" (October 14-17, 1991, St. Petersburg, USSR) compiled by K.M.Young, V.S.Mukhovatov, S.Yamamoto, and F.Engelmann.
[2] Summary Report, ITER Technical Meeting on "Radiation Effects on In Vessel Components" (15-19 November 1993, ITER Garching JWS, Germany), S.Yamamto, S.E.Bender, G.Bosia L.deKock, E.H.Farnum et al.
[3] Summary Report, ITER Technical Meeting on "Irradiation Testing of In Vessel Structural Materials and Components" (12-16 December 1994, ITER Garching JWS, Germany), S.Yamamoto, E.H.Farnum, E.R.Hodgson et al.
[4] E.R.Hodgson, "Radiation Problems and Testing of ITER Diagnostic Components", these Proceedings.
[5] Reports of four Home Teams for ITER Technology R&D Tasks of T28 and T246, " Irradiation Tests on Diagnostic Components".
[6] D.V. Orlinski, "Radiation Hardening of Diagnostic Components", Diagnostics for Experimental Thermonuclear Fusion Reactors, p51-p60, Plenum Press, New York, 1996.
[7] V.S. Voitsenya, A.F.Bardamid, V.L.Berezhnyi et al., Imitation of Fusion Reactor Environment Effects on the Inner Elements of Spectroscopical, mm and Sub-mm Diagnostics", Diagnostics for Experimental Thermonuclear Fusion Reactors, p61-p70, Plenum Press, New York, 1996.
[8] Summary Report of IEA Workshop on "Radiation Effects in Ceramic Insulators", (May 7-8, 1997, Cincinnati, Ohio, USA), S.J.Zinkle, E.R.Hodgson, T.Shikama.
[9] A. Nagashima, T. Fujisawa, T.Sugie, S.Kasai, "Development of New Window Seal for ITER Optical Diagnostics", these Proceedings.
[10] R.Reichle, M.Di.Maio, L.C.Ingesson, K.F.Mast, "Progress of the reference design for ITER bolometers and development of a high performance alternative", these Proceedings.
[11] G.Haas, H.-S.Bosch, D.Coster, L.deKock, J.Neuhauser, R.Schneider, J.Schweinzer, "Pressure Gauges and Neutral Pressure Measurement in ITER", these Proceedings.

CHANGE OF THE OPTICAL REFLECTIVITY OF MIRROR SURFACES EXPOSED TO JET PLASMAS

M. Mayer, [1] R. Behrisch, [1] C. Gowers, [2] P. Andrew, [2] and A.T. Peacock [2]

[1]Max Planck Institut für Plasmaphysik, EURATOM Association, Boltzmannstr. 2, D-85748 Garching, Germany
[2]JET Joint Undertaking, Abingdon, Oxfordshire OX14 3EA, United Kingdom

INTRODUCTION

For passive and active optical plasma diagnostics mirrors are needed at the vessel walls. However, all surface layers of plasma exposed surfaces have been found to be modified by erosion, redeposition and hydrogen implantation from the plasma [1]. These modifications will generally decrease the reflectivity of the surfaces [2] and frequent recalibration of the diagnostics are necessary. They may even result in damage of the mirrors, for example in the case of high power laser pulses.

Diagnostic mirrors are typically far away from the plasma. They are, however, bombarded with neutral hydrogen and impurity atoms created in charge-exchange (CX) collisions [3, 4, 5], resulting in surface layer modifications. Neutral CX-hydrogen atoms have an energy distribution with a high energy tail which reaches into the several ten keV range, resulting in erosion of the exposed surfaces and hydrogen implantation [6, 7, 8]. There is only little knowledge about the change of the optical reflectivity of different materials due to CX-bombardment. In order to get more insight into the mechanisms resulting in changes of the reflectivity we have examined the surface modifications and the reflectivity change of carbon, aluminium and nickel samples exposed to JET plasmas. From these results lifetime predictions for mirrors used in ITER can be obtained.

EXPERIMENTAL

About 30 polished carbon long term samples (LTS), partly covered with 530 nm aluminium were mounted at the JET vessel walls at distances between 150 mm and

Diagnostics for Experimental Thermonuclear Fusion Reactors 2
Edited by Stott *et al.*, Plenum Press, New York, 1998

500 mm from the plasma surface in the shadow of the protection limiters in March 1996 and replaced in October 1996 after about 2000 discharges (about 50000 plasma seconds). The optical reflectivity of the LTS was measured at a wavelength of 670 nm before installation and after removal. The change of the surface topography of the LTS was investigated with scanning electron microscopy. The surface roughness was measured with a mechanical profiler with a vertical resolution of about 5 nm. The compositions of the LTS surface layers were determined with MeV ion beam techniques (RBS, NRA, PIXE) [7].

The CX-flux to the inner vessel wall of JET, as determined by the measured erosion of the LTS, is of the order of 5×10^{19} D-atoms/m^{2}s [7], giving a total CX-flux hitting the LTS of about 2.5×10^{24} D-atoms/m^2 during the whole discharge period. This value is an upper bound due to the additional erosion by He glow discharge cleaning [7].

RESULTS

Change of Reflectivity

The LTS mounted at the outer torus wall showed thick deposits mainly of Be and a large decrease of reflectivity. This is mainly due to the regular beryllium evaporations applied for wall conditioning in JET.

LTS mounted at the inner torus wall showed no or only minor Be deposits. This is due to the lower Be evaporation rate at the inner wall and a probably higher erosion rate. The Al and Ni layers of these LTS were eroded.

Table 1 summarises the measured erosion or deposition and the change of reflectivity of some samples. The reflectivity of eroded samples after exposure has decreased by a factor of roughly 2–3. For the Al samples 240–380 nm of initially 530 nm Al were eroded. The samples with the highest erosion showed also the highest decrease of reflectivity. As has been shown by Bohmeyer *et. al.* [2] for plasma exposed mirror surfaces eroded at the Berlin plasma generator the decrease of the direct optical reflectivity is associated with an increase of the diffuse reflectivity. The decrease of the direct optical reflectivity is even more pronounced for shorter wavelengths [2].

For samples with deposited layers on the surface the decrease of reflectivity was large. For deposited layer thicknesses around 2 μm the reflectivity decreased to nearly zero.

The decrease of the reflectivity of eroded samples is mainly due to two effects:

1. Surface roughening due to erosion
2. Change of the chemical composition of the near surface region.

Additionally recrystallisation of the surface may occur [2].

Surface Roughening

Fig. 1 shows the surface roughness of an Al layer before and after exposure. The total erosion of this Al-layer was 240 nm. Due to the erosion the roughness of the surface has increased. For a quantitative description of surface roughness we introduce the mean roughness y_{mean}:

$$y_{\text{mean}} = \frac{1}{N} \sum_{i=1}^{N} |y_i| \qquad (1)$$

with y_i the vertical deviation and N the number of data points.

Fig. 2 shows the measured surface roughness of the Al-layers as a function of the

Table 1: Reflectivity of some LTS at a wavelength of 670 nm before and after exposure.

Material	Reflectivity before exposure	Reflectivity after exposure	Total erosion	Total deposition	Position
C	1–3%	≈ 0	?	–	inner wall
Al	19%	11%	236 nm	–	inner wall
Al	19%	7%	383 nm	–	inner wall
Ni	65%	32%	–	100 nm	outer wall
Al	22%	≈ 0	–	2000 nm	outer wall

amount of eroded aluminium. The mean roughness of the non-eroded samples was about 20 nm. As shown in fig. 2 the surface roughness of the eroded samples increases about linearly with increasing erosion. The mean roughness is about equal to the total erosion. There is also some experimental evidence from laboratory experiments that for low bombarding fluences the surface roughness increases about linearly with incident fluence. This was demonstrated for the bombardment of pyrolytic graphite by 2 keV deuterium ions [9]. At large fluences (> 10^{19} D-atoms/cm^2) the surface roughness saturates and does not increase further.

Compositional Change of the Surface Layer

Fig. 3 shows the depth profiles of Al, C, O, D and Ni of an Al-layer before and after exposure. Initially the evaporated Al-layer contained about 5% O and 8% C due to the vacuum conditions during evaporation. After exposure about 300 nm Al are eroded. The surface is oxidised within a depth of about 50 nm. Ni, Fe and Cr, but also other impurities such as P, S, Cl, K and Ca, are present in the whole layer with a total concentration of about 0.5–1%. Deuterium has been implanted up to a mean concentration of about 4%. Because of the poor depth resolution for D it cannot be determined if the D is present mainly at the surface or is distributed throughout the whole layer. Additionally some per cent Be may be present in the layer. The reflectivity of this sample has decreased from 19% to 10%.

On carbon samples the main impurities observed at the surface are implanted hydrogen isotopes and oxygen, the amount of other impurities is generally much smaller than for aluminium layers (of the order of 10^{15} atoms/cm^2).

PREDICTIONS FOR ITER

Energy Distribution of the CX-Flux

The energy distribution of the CX-flux is measured at ASDEX-Upgrade in detail in the outer torus midplane with the **L**ow **E**nergy **N**eutral Particle **A**nalyser (LENA) in the energy range 30–1000 eV [10, 11] and with the **N**eutral **P**article **A**nalyser (NPA) for energies > 500 eV [11, 12]. The flux below 30 eV is based on simulations with the EIRENE code [13, 14, 15]. The CX energy spectrum measured at ASDEX-Upgrade during a standard shot (# 6535, $n_e = 3 \times 10^{19}$ m^{-3}) is shown in fig. 4 together with a preliminary calculation for the flat-top phase of ITER [16]. Both spectra are normalised

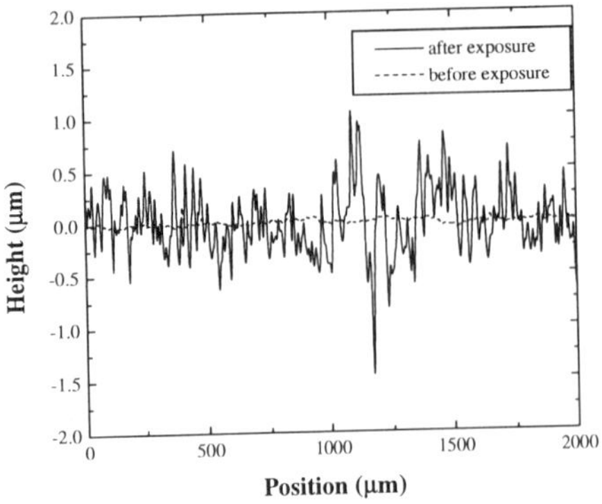

Figure 1: Surface roughness of an Al layer before and after exposure. The mean roughness before exposure was about 20 nm and increased to about 240 nm for the eroded sample.

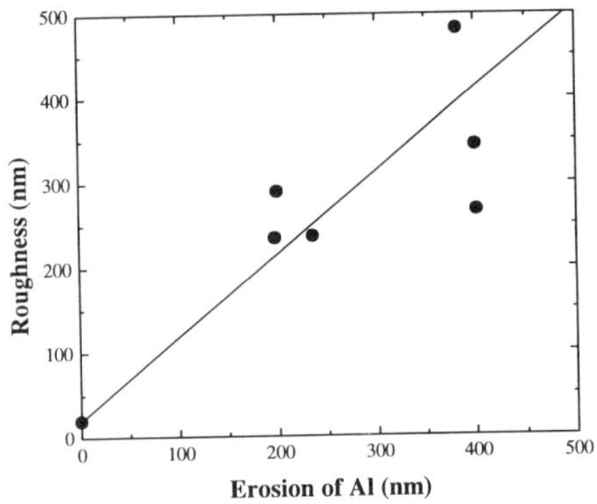

Figure 2: Mean surface roughness of different Al-layers as a function of Al-erosion.

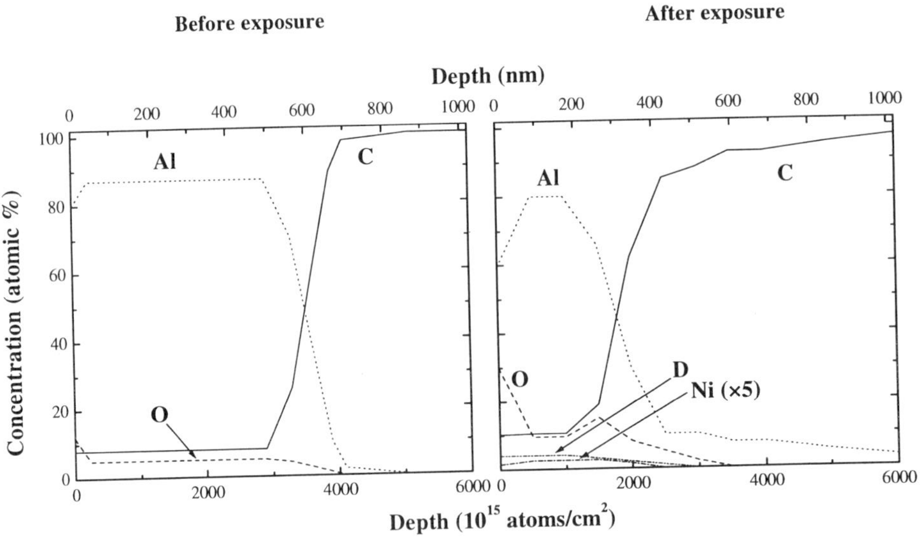

Figure 3: Depth profiles of Al, C, O, Ni and D of an Al-layer before and after exposure. The depth given in the top scale uses the theoretical Al density.

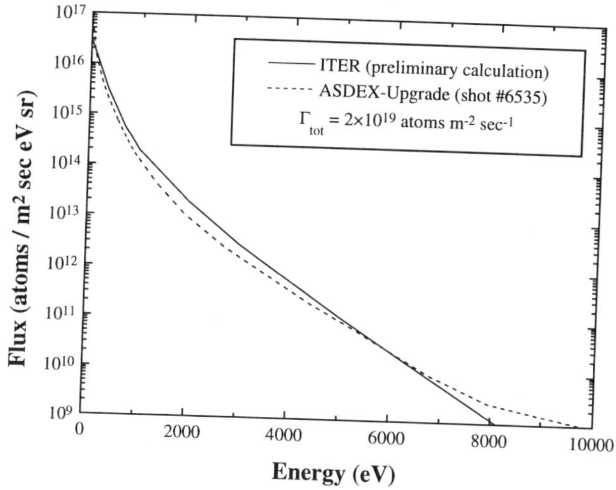

Figure 4: Energy distribution of the CX-flux measured at ASDEX-Upgrade and a preliminary calculation for the flat-top phase of ITER. The total flux Γ_{tot} is 2×10^{19} atoms/m^{2}s in both cases.

to the same total flux of 2×10^{19} atoms/m^{2}s (see below). The high energy component which originates from the core plasma is small, and the CX-flux decreases by several orders of magnitude with increasing energy. The differences between the two spectra are low.

For an incident flux with an energy distribution $\Gamma(E)$ an effective sputtering yield Y_{eff} can be defined by

$$Y_{eff} = \frac{1}{\Gamma_{tot}} \int Y(E)\,\Gamma(E)\,dE \tag{2}$$

where $Y(E)$ is the energy dependent sputtering yield and Γ_{tot} the total incident flux, $\Gamma_{tot} = \int \Gamma(E)\,dE$. The effective sputtering yields Y_{eff} for D, T and He bombardment at normal incidence are tabulated in table 2 for different materials. The sputtering data from [17] were used for the calculation of Y_{eff}. The effective sputtering by T atoms is typically 3–4 times higher than the sputtering by D atoms due to the lower sputtering threshold energy of T. The sputtering by He is about 5–10 times higher than the sputtering by D.

Total CX-Flux to the First Walls of the Main Chamber

The mean CX-flux at the inner torus wall of JET, as determined from the measured erosion of long term samples, is of the order of 5×10^{19} D-atoms/m^{2}s [7]. At the outer torus midplane of ASDEX-Upgrade the measured CX-flux is 10^{19}–10^{20} D-atoms/m^{2}s, depending on the discharge conditions. Preliminary calculations for ITER indicate a smaller CX-flux than in present days machines due to the smaller recycling of neutral hydrogen in ITER. For an estimate of the expected CX-erosion in the torus midplane of ITER we use a conservative value for the total CX-flux of 2×10^{19} Atoms/m^{2}s.

However, it should be noted that the charge-exchange flux is strongly dependent on the position in the torus. It has been demonstrated at ASDEX-Upgrade that the total flux varies by about 3 orders of magnitude depending on the position in the torus [8]: The highest fluxes are observed at the divertor entrance and exceed the given values by about two orders of magnitude. At the top of the torus the smallest CX-fluxes are

Table 2: Effective sputtering yields Y_{eff} for the bombardment of different materials with D, T and He at normal incidence. Energy distribution of the incident particles as in ASDEX-Upgrade shot #6535, fig. 4.

Material	Bombarding species		
	D	T	He
Be	1.3×10^{-2}	1.5×10^{-2}	2.5×10^{-2}
Al	5.9×10^{-3}	2.3×10^{-2}	4.9×10^{-2}
Fe	5.7×10^{-3}	1.3×10^{-2}	2.7×10^{-2}
Ni	7.8×10^{-3}	1.2×10^{-2}	3.9×10^{-2}
Cu	1.4×10^{-2}	1.6×10^{-2}	6.3×10^{-2}
Mo	5.2×10^{-4}	4.2×10^{-3}	5.7×10^{-3}
Rh	2.4×10^{-3}	5.0×10^{-3}	1.1×10^{-2}
Ag	8.7×10^{-3}	1.3×10^{-2}	4.4×10^{-2}
W	1.2×10^{-4}	7.4×10^{-4}	1.4×10^{-3}
Au	1.9×10^{-3}	4.4×10^{-3}	1.2×10^{-2}

observed. At these positions the CX-flux is about one order of magnitude lower. The energy distribution of the charge-exchange flux is also position dependent [8]. Also near sources of neutral gas such as limiters much larger fluxes may be expected. These positions should be avoided for mirrors.

Erosion and Reflectivity Change of Mirror Surfaces in the Main Chamber

We assume a total CX-flux of 2×10^{19} atoms/m^{2}s hitting the walls. The flux consists of 50% D and 50% T. Although the concentration of He may be 10% in the core plasma of ITER, the charge-exchange cross-section for the creation of neutral He is much lower than the cross-section for creation of neutral D or T. The concentration of neutral He in the CX-flux hitting the walls is considered to be below 1% and is neglected.

The mirror is located in a duct with diameter 20 cm and is 50 cm away from the plasma surface. The solid angle $\Delta\Omega$ of visible plasma is 0.126 sr. If we assume a cosine angular distribution of the incident CX-flux then the number N of removed atoms/m^2 from the mirror surface during a discharge time T is given by

$$N = \frac{\Gamma_{tot}}{\pi} Y_{eff} \Delta\Omega \, T. \tag{3}$$

Secondary effects, such as reflection of particles at the walls of the duct and sputtering at the duct walls, are not considered.

The results obtained at JET and predictions for ITER are summarised in table 3. The Al layers exposed at JET see an effective plasma solid angle of nearly 2π. The measured erosion of the exposed Al layers at JET is about one monolayer Al during a shot. For ITER about 500 shots per year are expected. We assume a linear relation between surface roughening and decrease of reflectivity and normal incidence. As can be seen from eq. 3 the erosion of mirror surfaces may be largely reduced if the solid angle is decreased, i.e. by placing the mirror further away from the plasma. From the viewpoint of surface roughening due to CX-sputtering tungsten is the most promising material due to the smallest sputtering yield of all materials. Gold is excluded as mirror material due to transmutation to mercury by neutron bombardment and is listed in table 3 only for comparison.

Table 3: Measured erosion and reflectivity decrease of the Al layers exposed at JET and predictions for ITER. The CX-flux is given for the torus midplane. The solid angle assumed for ITER is for a mirror with 20 cm diameter which is 50 cm away from the plasma surface. The erosion is given in monolayers [ML] per shot, the reflectivity is at 670 nm.

	JET	ITER
CX-flux [Atoms/m^2s]	$< 6 \times 10^{19}$	2×10^{19}
Composition	D	50% D, 50% T
Solid angle [sr]	2π	0.126
Shot length [s]	25	1000
Erosion [ML per shot]		
Al	1	1
Mo	-	0.1
Rh	-	0.2
Ag	-	0.6
W	-	0.02
Au	-	0.2
Reflectivity decrease after 2000 shots		
Al	40–60%	50%
Mo	-	5%
Rh	-	10%
Ag	-	30%
W	-	1%
Au	-	10%

For tilted mirrors the sputtering yields are higher by a factor of about 5 in the worst case, which is 70–80° particle incidence angle relative to the surface normal. However, the erosion given in table 3 may be exceeded by a factor of about 100 if the mirrors are placed close to the plasma or in regions with large CX-flux. For mirrors located at the top of the torus the erosion may be ten times smaller than given in table 3 due to the smaller CX-flux.

Reflectivity Change of Mirror Surfaces in the Divertor

In the JET divertor thick deposits with thicknesses up to 100 μm, consisting mainly of carbon with some beryllium and oxygen, are observed besides the strike points. For mirrors in the ITER divertor therefore not erosion, but deposition may be the main mechanism for changes of the optical reflectivity. As can be seen from table 1 the optical reflectivity of deposition dominated samples is largely reduced and decreases to nearly zero for layer thicknesses of 2 μm. However, currently no predictions about the change of reflectivity of mirrors in the ITER divertor can be made.

CONCLUSIONS

The optical reflectivity of erosion dominated carbon and aluminium surfaces from the inner torus wall decreases by a factor 2–3 after exposure to about 2000 JET plasma discharges. The erosion is due to sputtering by charge-exchange hydrogen and He glow discharge cleaning. The decrease of the surface reflectivity is mainly due to two effects:

1. Surface roughening of eroded surfaces
2. Change of the chemical composition of the surface.

The roughness of eroded surfaces increases with increasing total erosion. To minimise the effect of erosion induced surface roughening materials with low CX-erosion yields should be used, such as high Z materials. The change of the chemical composition of the surface layers at the JET conditions is material dependent. The change of the chemical composition due to bombardment with hydrogen, carbon, oxygen and beryllium is hard to predict and should be investigated for the mirror materials anticipated for use in ITER.

Samples located at deposition dominated areas show thick deposits. The reflectivity of these samples decreases to nearly zero.

REFERENCES

[1] R. Behrisch, M. Mayer, and C. García-Rosales, J. Nucl. Mater. 233-237 (1996) 673.

[2] W. Bohmeyer, G. Fussmann, P. Kornejew, H. Reiner, and H. Grote, in: 23rd EPS Conference on Controlled Fusion and Plasma Physics, europhysics conference abstracts Vol. 20 (1996) p. 1128.

[3] H. Verbeek and the ASDEX-team, J. Nucl. Mater. 145-147 (1987) 523.

[4] R. Behrisch, J. Roth, G. Staudenmaier, and H. Verbeek, Nucl. Instr. Meth. B18 (1987) 629.

[5] G. Staudenmaier and W. Wampler, J. Nucl. Mater. 162-164 (1989) 414.

[6] M. Mayer, R. Behrisch, V. Prozesky, P. Andrew, and A. Peacock, in: 22nd EPS Conference on Controlled Fusion and Plasma Physics, europhysics conference abstracts Vol. 19C (1995) p. 301.

[7] M. Mayer, R. Behrisch, P. Andrew, and A. Peacock, J. Nucl. Mater. 241-243 (1997) 469.

[8] H. Verbeek, J. Stober, D. Coster, and R. Schneider, in: 24th EPS Conference on Controlled Fusion and Plasma Physics, europhysics conference abstracts Vol. 21A (1997) p. 1457.

[9] M. Küstner, Zerstäubung rauher Oberflächen, PhD thesis, Universität Bayreuth, 1997, (in german).

[10] H. Verbeek, J. Phys. E 19 (1986) 964.

[11] H. Verbeek, O. Heinrich, R. Schneider, H. Fahrbach, W. Herrmann, J. Neuhauser, U. Stroth, the ASDEX-team, and D. Reiter, J. Nucl. Mater. 196-198 (1992) 1027.

[12] H. Fahrbach, W. Herrmann, and H. Mayer, in: 16nd EPS Conference on Controlled Fusion and Plasma Physics, europhysics conference abstracts Vol. 13B (1989) p. 1537.

[13] D. Reiter, Technical Report Jül-1947, Forschungszentrum Jülich, 1984.

[14] D. Reiter, The EIRENE Code, Users Manual, Technical Report Jül-2599, Forschungszentrum Jülich, 1992.

[15] J. Stober, H. Verbeek, D. Coster, H.-U. Fahrbach, O. Heinrich, W. Herrmann, G. Haas, O. Kardaun, D. Reiter, R. Schneider, J. Schweinzer, W. Suttrop, ASDEX Upgrade Team, NI Team, ICRH Team, and ECRH Team, in: 22nd EPS Conference on Controlled Fusion and Plasma Physics, europhysics conference abstracts Vol. 19C (1995) p. 249.

[16] A. Kukushkin, private communication, 1995.

[17] W. Eckstein, C. García-Rosales, J. Roth, and W. Ottenberger, Sputtering Data, Technical Report IPP 9/82, Max-Planck-Institut für Plasmaphysik, Garching, 1993.

INVESTIGATION OF THE FIBER OPTIC RADIATING RESISTANCE AND RADIOLUMINESCENCE UNDER IR-8 REACTOR CONDITION

Andrei A. Ivanov, Sergei N. Tugarinov, Igor N. Rastyagaev, Vladimir N. Amosov, Yuri A. Kaschuck, Anatoli V. Krasilnikov, Sergei E. Bender[1]

TRINITI, Troitsk, Moscow region 142092, Russia
[1]NIIEFA, St. Petersburg, 189631, Russia

INTRODUCTION

Optical fibers have a promising application in diagnostic systems of nuclear sites. In particular, it is revealed that fiber optics has become a common technique to transmit analog light signals from the limited-access areas near tokamak fusion reactors to remote diagnostic instruments. Typical examples of fiber-based diagnostic are visible bremsstrahlung diagnostic (λ=535 nm) and beam emission spectroscopy system using D_α fluorescence (λ=656 nm). However, the problems of the fiber optic radiating resistance connected with sharp degradation of their optical characteristics and the induced luminescence under neutron and γ flux are appeared.[1-5] Previous experiments have shown that the optical fibers with pure quartz core are more stable to irradiation, but radiation resistance such kind of fibers produced using different technology is also different.[6]

In this paper three fiber types simultaneously exposed to reactor irradiation are compared from the view point of both the radiation induced transmission loss and the radioluminescence, observed in green at 535 nm and in red at 656 nm, vs neutron flux/fluence. The total fast neutron fluence and γ dose were $\approx 8\times10^{14}$ n/cm^2 and $\approx$17 Mrad, respectively.

SAMPLES, IRRADIATION CONDITIONS AND SETUP

The three different types of fibers used in the study are described in Table 1. The core material of fibers was SiO_2, and SiO_2 claddings were fluorine dropped. All fibers have the same cladding diameter, numerical aperture and polymer coating but differed in OH-group contents, except for fibers #2 and #3. A KU-1 silica core material (fiber #3) is a commercial product of the Russian industry analogous to Heraeus F100 silica (fiber #2). All fibers were produced in the Steko, St. Petersburg, Russia, by plasma outside deposition technique.

Diagnostics for Experimental Thermonuclear Fusion Reactors 2
Edited by Stott *et al.*, Plenum Press, New York, 1998

Table 1. Specifications of optical fiber studied.

Fiber No.	Name and Core/Clad type	Core/Clad Diameter, (microm.)	Jacket type	NA	OH-group contents, (ppm)	Producer
1	DSF-3 SiO_2/SiO_2:F	200/240	epoxyacrylate	0.2±0.02	25	Steko S.-Pet., Russia
2	Heraeus F100 SiO_2/SiO_2:F	220/240	epoxyacrylate	0.2±0.02	>900	Steko S.-Pet., Russia
3	KU-1 SiO_2/SiO_2:F	200/240	epoxyacrylate	0.2±0.02	>900	Steko S.-Pet., Russia

All three fiber types were simultaneously exposed to reactor irradiation at reactor powers of 50, 100 and 200 kW (see Table 2) and temperature of 55 C. A neutron/gamma reactor output was changed due to changing of reactor power.

Table 2. Neutron/gamma reactor output,[1] and total fluence/dose at the moment of spectral measurement.

Reactor power, (kW)	Fast neutrons		Thermal neutrons		Gamma radiation	
	Flux, ($n/cm^2/s$)	Fluence, (n/cm^2)	Flux, ($n/cm^2/s$)	Fluence, (n/cm^2)	Dose rate, (rad/s)	Dose, (rad)
0	-	-	-	-	147	$8.1*10^5$
50	$1.1*10^{11}$	$8.5*10^{13}$	$1.5*10^{12}$	$1.2*10^{15}$	$2.2*10^3$	$1.9*10^6$
100	$2.2*10^{11}$	$3.7*10^{14}$	$2.9*10^{12}$	$5.0*10^{15}$	$4.4*10^3$	$7.8*10^6$
200	$4.4*10^{11}$	$8.3*10^{14}$	$5.9*10^{12}$	$1.1*10^{16}$	$8.7*10^3$	$1.7*10^7$

[1]Research reactor IR-8, FSC "Kurchatov Institute", Moscow, Russia, wet channel no. 6-3.

The experimental setup[7] involved 51 m total length of both sample and reference fiber, coiled inside the reactor zone (58 cm height) into double and single loops, respectively, of each fiber type studied. The length difference between the homogeneously irradiated parts of the sample and reference fibers was 56 cm. These fibers were illuminated with a spectrometric tungsten lamp that has been calibrated for absolute measurements in terms of photons/s/cm^2/Å/sr. To avoid photobleaching effects the distance between W-lamp and fibers input was adapted so that input light power was below than 0.03 μW.

Both the radioluminescent and transmission spectra were detected by charge-coupled-device (CCD) detector (1024 pixels, peltier cooled, has time resolution from 4 ms up to 1 min) with image intensifier providing the brightness intensification coefficient of 5×10^4 and the dynamic range of 4×10^3 counts. It was placed downstream of the output slit of a polychromator with the spectral window from 440 to 740 nm and the resolution of 1 nm. The two camera objectives (f/#) were used to conform the fibers light output to the polychromator light input. A personal computer was used for detector control and data storage.

RESULTS AND DISCUSSION

To investigate both the induced transmission loss and radioluminescence, step by step *in-situ* spectral measurements have been carried out according to reactor power (see Table 2). The results are summarized in Figures 1 and 2.

The incremental transmission loss was evaluated in terms of dB using the relation $10\times\log(I_1/I_2)$, where I_1 is the signal intensity before and I_2 is the intensity after the irradiation. In Figure 1, the incremental fiber attenuation, observed in green (λ=535 nm) and in red (λ=656nm), as a function of fast neutron fluence is given for fibers studied. This shows

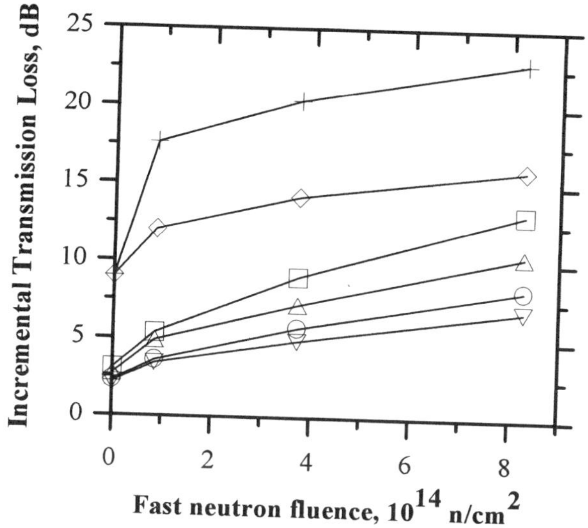

Figure 1. Incremental fiber attenuation in green and red as a function of fast neutron fluence for the fibers studied. Fiber 1 - 535 nm (✚) and 656 nm (◆), Fiber 2 - 535 nm (■) and 656 nm (●), Fiber 3 - 535 nm (▲) and 656 nm (▼).

a linear loss with neutron fluence for all fibers. Comparing three fibers with one another one can see clearly that the loss level in the fiber #3 is noticeable lower than that of the fibers #1 and #2. On reactor power of 200 kW (with $8.3*10^{14}$ n/cm^2 at the fibers loop) the fiber #3 lose 6.9 dB in the red while the fibers #1 and #2 loses 16 dB and 8.3 dB, respectively. It is similar to behavior observed in the green. It is to be noted that the fiber transmission loss at 656 nm associated with non-bridging oxygen hole centers (NBHOC), which form an absorption band at 630 nm. Therefore, to fabricate a radiation-resistant fiber, it is necessary to reduce the concentration of hydroxyl.

Just as for fiber transmission, we measured the radioluminescence as a function of fast neutron fluence at 535 and 656 nm. The luminescent spectra were corrected for fiber transmission. As can be see in Figure 2, the linear dependence on neutron fluence/flux was retained for all fibers. Note that in high purity silica fibers, radioluminescence appears to be dominated by Cerenkov radiation from electrons of 200 keV or higher energy in the core.

CONCLUSION

It follows from presented results that KU-1 based optical fibers with fluorine dropped cladding are more stable to reactor irradiation, but radiation resistance such kind of fibers produced using different technology is also different.

Our *in-situ* measurement technique allows to determine most promising technologies for producing radiation-resistant optical fibers.

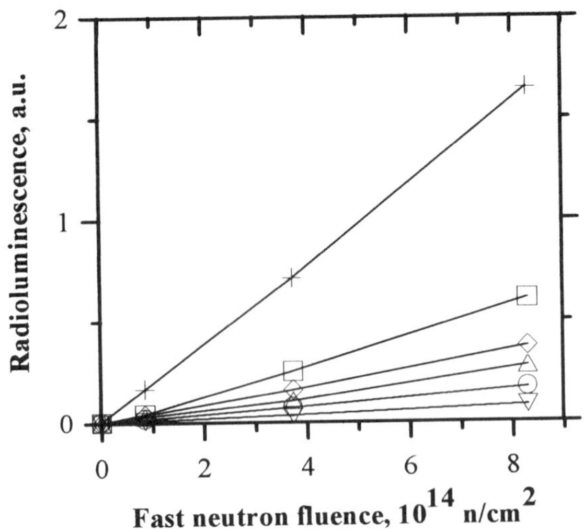

Figure 2. Plot of measured luminescence at 535 nm and 656 nm vs fast neutron fluence. It is corrected for fiber transmission. For curves note see Figure 1.

Finally, it is important to note that pure quartz optical fibers manufactured with different radiation-hardening technologies could be successfully used in ITER conditions.

ACKNOWLEDGMENTS

We would like to thank V.A. Belyakov and E.A. Azizov for their support and helpful discussions. This work has been supported by the ITER Russian Home Team (ITER tasks T28 and T246).

REFERENCES

1. M. Decreton, V. Massaut, and P. Borgermans, Potential benefit of fibre optics in nuclear applications: the case of the decommissioning and the waste storage activities, in: *Proc. SPIE,* 2425:2 (1994)
2. A.T. Ramsey, D-T radiation effects on TFTR diagnostics, *Rev. Sci. Instrum.* 66:871 (1995).
3. H.G. Adler, K.W. Hill, A.T. Ramsey, and W. Tighe, Luminescence and transmission measurements on fibers exposed to high neutron fluxes, *Rev. Sci. Instrum.* 66:904 (1995).
4. W. Tighe, P. Morgan, H. Adler, D. Cylinder, D. Griscom, D. Johnson, D. Palladino, and A. Ramsey, Proposed experiment to investigate use of heated optical fibers for tokamak diagnostics during D-T discharges, *Rev. Sci. Instrum.* 66:907 (1995).
5. S.F. Paul, J.L. Goldstein, R.D. Durst, and R.J. Fonck, Effect of high-energy neutron flux on fiber optics in an active diagnostic on TFTR, *Rev. Sci. Instrum.* 66:1252 (1995).
6. D.L. Griscom, K.M. Golant, A.L. Tomashuk, D.V. Pavlov, Yu.A. Tarabrin, Gamma-radiation resistance of aluminum-coated all-silica optical fibers fabricated using different types of silica in the core, to be published in *Appl. Phys. Lett.*
7. A.V. Krasilnikov, A.A. Ivanov, S.N. Tugarinov, I.N. Rasyagaev, A.Yu. Tsutskih, V.N. Amosov, Yu.A. Kaschuck, S.E. Bender, Transmission and radioluminescence of all-silica optical fibers exposed to gamma-source and reactor radiation, in: *Proc. 9th National Topical Conference on High-Temperature Plasma Diagnostics,* Saint-Petersburg, 2-4 June, 1997.

RAD-HARD OPTICAL FIBRES FOR DIAGNOSTICS OF EXPERIMENTAL FUSION REACTORS

O. Deparis [a], P. Mégret [a], M. Decréton [b], M. Blondel [a], K.M. Golant [c], A.L. Tomashuk [c]

[a] Faculté Polytechnique de Mons (Belgium)
[b] SCK•CEN, Belgian Nuclear Research Centre, Mol (Belgium)
[c] Fibre-Optics Research Centre at the General Physics Institute of the Russian Academy of Sciences, Moscow (Russia)

INTRODUCTION

Because they can be used for monitoring and controlling the plasma conditions, radiation-resistant optical fibres are among the key components for diagnostics on fusion reactors. Ideally, the optical fibre spectrum should cover the wavelength range from 200 nm to 2000 nm, namely from the ultraviolet to the visible and infrared ranges. Whereas a lot of data exist on the radiation tolerance of optical fibres at telecommunication wavelengths (0.85, 1.3 and 1.55 μm), much less is available below, in the visible (400-700 nm) and especially in the ultraviolet (< 400 nm) range. Moreover, the spectrometry of the radiation-induced loss in the ultraviolet requires measurement devices such as light sources and detectors that are quite different from the standard equipments used for visible-near infrared spectrometry (e.g. tungsten-halogen light sources, semiconductor detectors). Key information however can be gained from recent investigations around radiation hardened fibroscopy, performed as part of the ITER remote handling studies.

FIBRE OPTIC IMAGE GUIDE

As image guides, optical fibres offer a very flexible possibility of visual control during in-vessel maintenance tasks. For such application, optical fibres ought to maintain good transmission only in the visible range and over relatively short lengths (about 10 to 20 meters). Rad-hard fibres with pure silica core and F-doped silica cladding have proven their capability to withstand very high gamma doses (up to 10^6-10^7 Gy) due the saturation phenomenon of the radiation-induced loss with dose. In the absence of data in the ultraviolet range, one can nevertheless infer some useful features from what has been measured in the visible range.

RADIATION TOLERANCE OF PURE SILICA CONVENTIONAL FIBRES

It is for instance known that radiation-induced loss between 400 and 700 nm mainly results from the superposition of an attenuation band near 600 nm attributed to nonbridging oxygen hole centres (NBOHCs), together with band tails extending from the ultraviolet due to intrinsic or extrinsic defect centres [1][2][3]. The concentration of precursors of radiation-induced colour centres depends primarily on the specific fabrication technology of the silica used as the fibre core material. In conventional fibres with a core fabricated from $SiCl_4$, the intensity of a near-UV band tail is related to the chlorine impurity concentration in the core. The radiolytic atomic chlorine species are believed to absorb light around 340 nm and therefore their concentration should be minimised for use in the ultraviolet and the visible ranges. On the other hand, it is also desirable to minimise another kind of impurity : the hydroxyl (OH) groups, which are

Diagnostics for Experimental Thermonuclear Fusion Reactors 2
Edited by Stott *et al.*, Plenum Press, New York, 1998

NBOHC precursors. The simultaneous reduction, however, of Cl and OH concentrations is a contradictory goal in the conventional technology. With conventional pure silica core fibres, reduction of radiation-induced loss by a factor of 100 has been achieved by in-situ thermal annealing (at 400°C) during deuterium-tritium operation of the TFTR fusion test reactor at Princeton University [4].

LOW HYDROXYL LOW CHLORINE FIBRES FABRICATED USING KS-4V SILICA

A new glass fabrication approach overcomes this dilemma and is able to produce silica with very low concentrations of both OH and Cl [5]. This is KS-4V silica developed in the Silicate Chemistry Institute of the Russian Academy of Sciences and brought into Russian industry. Fibres with a KS-4V core and an F-doped silica cladding demonstrate a rather flat induced absorption spectrum in the visible spectral region at the level of ~ 2 dB/m in a wide dose range from 1 to 10 MGy [1]. A shortcoming of KS-4V fibres is a strong transient absorption band at 670 nm, arising at low doses (~ 100 Gy) and vanishing with increasing dose. This transient absorption band is believed to be inherent in Cl- and OH-free silicas [6]. Recently it has been found that this band can be totally suppressed by loading fibres with molecular hydrogen [7].

IRRADIATION RESULTS

a. Comparison of conventional (Suprasil F-100) and KS-4V fibres

For fibroscopy, a conventional polyimide-coated high-OH low-Cl fibre (Suprasil F-100) and an aluminium-coated low-OH low-Cl (KS-4V) fibre have been tested with γ rays (5 kGy/h) up to very high doses (Figs. 1 and 2). Both fibres have an undoped silica core and a F-doped silica cladding. Their characteristics are shown in Table 1.

Table 1

Fiber	Suprasil F-100	KS-4V
Technology	$SiCl_4$ based	KS-4V glass
Diam. (μm)	100/110/125	100/120/150
Coating	polyimide	aluminium
[OH] (ppm)	600-800	< 0.2
[Cl] (ppm)	100-125	< 20

Typical radiation-induced attenuation spectra are shown on Figs. 1 and 2 for the conventional and the new fibre respectively. For both fibres, the level of radiation-induced attenuation reached in the MGy range is lower than ≈2.5 dB/m (maximum measurable level). Due to its induced absorption band at 600 nm, the Suprasil F-100 fibre shows a dose or time-variant induced attenuation spectrum during both irradiation and recovery cycles. The UV band tail associated with chlorine is here not very intense. By contrast, in the KS-4V fibre, the 600 nm band is absent. On the other hand, the increasing but stable induced attenuation at shorter

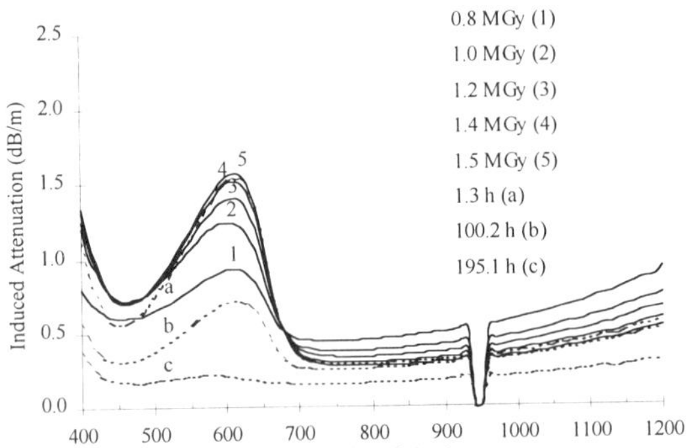

Fig. 1 : Visible-infrared radiation-induced attenuation spectra in a conventional rad-hard fibre. Legends (1-5) give the cumulated dose in MGy (SiO_2). Legends (a-c) give the recovery time after irradiation (hours) [2].

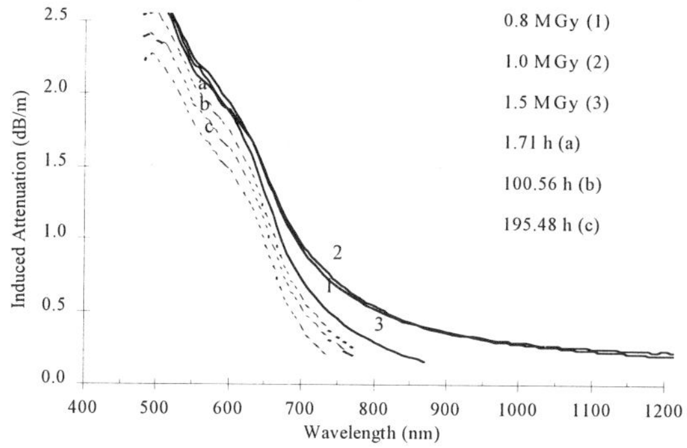

Fig. 2 : Visible-infrared radiation-induced attenuation spectra in a low OH low Cl rad-hard fibre. Legends (1-5) give the cumulated dose in MGy (SiO_2). Legends (a-c) give the recovery time after irradiation (hours) [2].

wavelengths suggests the existence of intense but stable intrinsic defect centres in the ultraviolet (Fig. 2). The comparison of the radiation tolerance of conventional and new rad-hard fibres can be made on the basis of data recorded in the visible range. Some trends can be stated for the optimisation of the fibre fabrication (e.g. low concentrations of OH and Cl impurities should be desirable). However, for the use of fibres in the widest wavelength range (200-2000 nm), one still needs of course to investigate the radiation-induced defect centres and their associated absorption bands in the ultraviolet range.

b. Hydrogen loading as an efficient technique for radiation hardening

In order to study the influence of H_2 loading on the radiation tolerance of silica fibres, pieces of KS-4V and high-OH KU polymer-coated fibres (KU silica is similar to Suprasil F-100 glass) were kept in a pressurized H_2 atmosphere (240 MPa, room temperature) for 11 days. For good comparison purpose, the fibres were chosen with the same geometry (core and cladding diameters of 100 and 120 microns respectively). All samples were then gamma irradiated in two steps with a ^{60}Co source (dose rate of 6.6 Gy/s and T~40°C), up to 297 Gy (step 1) and 1.7 MGy (step 2). During step 1, the transient 700 nm absorption band in KS-4V

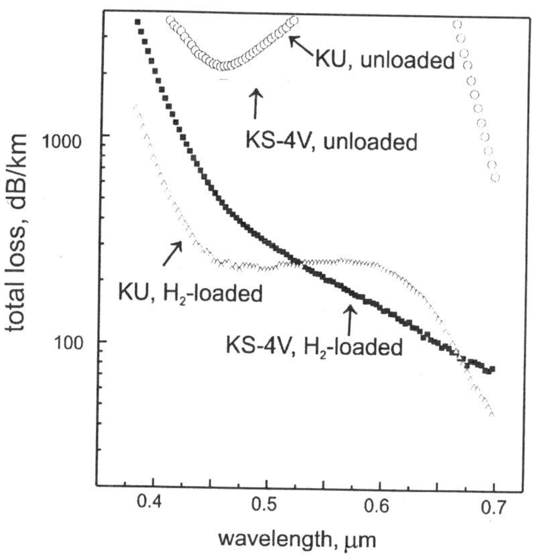

Fig. 3 : Loss spectra in H_2-loaded and unloaded fibres with KU and KS-4V cores measured within several hours after gamma-irradiation to 1.7 MGy (Si)

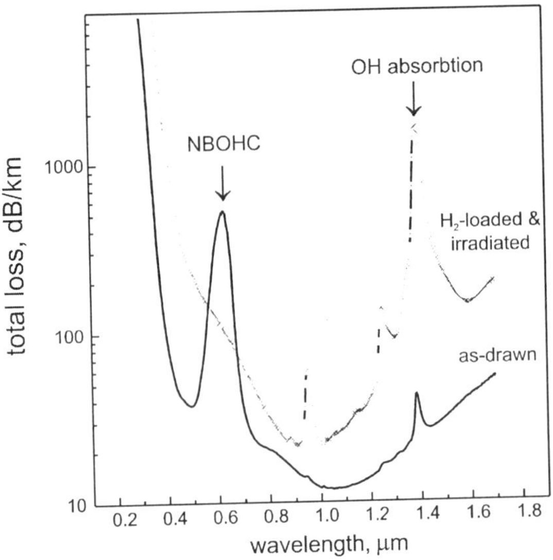

Fig. 4 : Evolution of the loss spectrum of the KS-4V fibre as the result of hydrogen loading and gamma-irradiation to 1.7 MGy (Si)

fibres was totally suppressed as the result of the H_2 loading. After step 2, the induced absorption in the visible spectral region for both types of H_2-loaded fibres, turned out to be more than an order of magnitude lower than in the as-drawn fibres (Fig. 3). The absorption levels in both H_2 loaded fibre types were similar. It is also worth noting that during the three days of irradiation, a major part of the H_2 molecules entered into the glass matrix through radiolytic reactions. This was for instance manifested by a tremendous rise of the OH-group absorption peak at 1380 nm in the KS-4V fibre, from 45 to 1600 dB/km (Fig. 4). The test assessed hydrogen loading as an efficient technique for radiation hardening of fibres, to be used for diagnostics purposes.

CONCLUSION

Experimental results were presented for the radiation tolerance of pure silica fibres in the visible region of the light spectrum. Different fibre types were compared, showing the favourable influence of low Cl concentration and of H2 loading. Although some trends can be extrapolated towards the UV region, further tests should be needed on these fibres to assess their characteristics fro UV plasma diagnostics.

ACKNOWLEDGEMENTS

Financial support is gratefully acknowledged by A. Tomashuk to the Russian Basic Research Foundation (Grant 97-02-16552), and by M. Decréton to the European Commission (Fusion Underlying Technology Programme, Contract ERB 5004 CT 97 0058).

REFERENCES:

[1] D.L.Griscom, K.M.Golant, A.L.Tomashuk, D.V.Pavlov, Yu.A.Tarabrin, Gamma Radiation Resistance of Aluminum-Coated All-Silica Optical Fibers Fabricated Using Different Types of Silica in the Core, Appl. Phys. Lett., vol. 69, pp. 322-324, 1996.

[2] O. Deparis, P. Mégret, M. Decréton, M. Blondel, Gamma Radiation Tests of Potential Optical Fibre Candidates for Fibroscopy, IEEE Trans. on Nucl. Sci., vol. 43, n°6, pp. 3027-3031, 1996.

[3] D. L. Griscom, Gamma and fission reactor radiation effects on the visible-range transparency of aluminum-jacketed, all-silica optical fibers, J. Appl. Phys., vol. 80, n°4, pp. 2142-2155, 1996.

[4] A. T. Ramsey, W. Tighe, J. Bartolick, P. D. Morgan, Radiation effects on heated optical fibers, Rev. Sci. Instrum., vol. 68, n°1, pp. 632-635, 1997.

[5] V.A. Bogatyrjov, I. I. Cheremisin, E. M. Dianov, K. M. Golant, A. L. Tomashuk, "Super High Strength Metal Coated Low Hydroxyl Low Chlorine All Silica Optical Fibres", IEEE Trans. on Nucl. Sci., vol. 43, n°3, Part I, pp. 1057-1060, 1996.
[6] D.L. Griscom, Visible/Infra-Red Absorption Study in Fiber Geometry of Metastable Defect States in High-Purity Fused Silicas, Proc. 13th International Conf. On Defects in Insulating Materials (Wake Forest University, Winston-Salem, NC), July 1996.
[7] A.L.Tomashuk, E.M.Dianov, K.M.Golant, A.O.Rybaltovskii, Gamma Radiation-Induced Absorption in Pure-Silica-Core Fibres in the Visible Spectral Region: the Effect of H2-Loading, RADECS'97 Conference, Cannes, September 15-19, 1997.

XUV AND VUV SPECTROSCOPY OF ITER

N. C. Hawkes, R. DeAngelis[1], S. Dillon[2], P. Edmonds[3], J. Edwards, R. M. Giannella[4], P. Gray[5], G. Janeschitz[3], K. D. Lawson, M. O'Mullane[2], N. J. Peacock, M. Valisa[6], C. I. Walker[3]

UKAEA/Euratom Association, Culham, Abingdon, Oxon, OX14 3DB, UK.
[1] ENEA–C. R. E. Frascati, Rome, Italy.
[2] JET Joint Undertaking, Abingdon, Oxon, OX13 3EA, UK.
[3] ITER JCT, Garching JWS, Garching, Germany.
[4] CEN Cadarache, 13108 Saint-Paul-lez-Durance, FRANCE
[5] Rutherford Appleton Laboratory, Chilton, Didcot, Oxon, OX11 0QX, UK.
[6] Istituto Gas Ionizzati del CNR, I-35020 Padova, Italy.

ABSTRACT

The vacuum ultraviolet (VUV) and extreme ultraviolet (XUV) spectral regions, together comprising the wavelength range from about 1-100 nm, contain important information about the state of a tokamak plasma. The short wavelength end of this band, in particular, contains much useful information about light impurity distributions yet is one of the most difficult spectral regions to diagnose. This paper will describe the uses to which information from this spectral region will be put on ITER, the difficulties to be encountered and the solutions envisaged.

INTRODUCTION

The XUV spectral region (1–10 nm) encompasses the Lyman emission lines of the light impurities found in present tokamaks (C, O, Be) while the VUV region (10–100 nm) includes the n=2–2 emission lines of metallic impurities (Fe, Ni) and strong emission features of likely plasma dopants for radiative edge operations. The monitoring of these impurity species will be important for ITER for the diagnosis of overall machine conditions, for diagnosis of impending fault conditions and for the real time control of the level of impurity dopants. Measurements of the spatial distributions of impurity ion emission features and their time evolution following transient events (the deliberate introduction of trace impurities, for example) can be used to determine the profiles of the transport coefficients. Differences in the transport properties of metals and light impurities are expected due to Z-dependencies within neoclassical theory. As well as their direct relevance to impurity and ash control in a steady-state reactor, the measured transport profiles yield important clues as to the nature and form of the background ion and energy transport. It has therefore always been a goal of experimentalists to be able to measure the radial profiles of XUV and VUV emission from plasma devices. VUV emission has also been used to measure ion temperatures and plasma flows,[1] although care is needed in the interpretation of measurements from multichannel grazing incidence spectrometers where the instrument function may play a large role in determining the lineshapes.[2]

In this paper we consider the specific needs of ITER and the constraints imposed by its radiation environment. Two schemes are presented, one vertically viewing giving space resolution of the plasma edge and wavelength coverage to below 2 nm, the other a radially viewing VUV survey spectrometer.

REQUIREMENTS OF VUV AND XUV SPECTROSCOPY

The broad requirements for this diagnostic are outlined in the specification of the expert group,[3]

Impurity	Content range	Spatial res.	Time res.	Accuracy
$Z \leq 8$	0.5–20%	30 cm	100 ms	20%
$Z > 8$	0.01–0.3%	30 cm	100 ms	20%

For a more detailed specification of the desired performance of the diagnostic we must examine the predictions of impurity transport simulations of the ITER plasma. Simulations have therefore been run using the radial profiles of temperature and electron density given in ref. 4 with transport coefficients appropriate to both L and H-mode conditions, figure 1. For intermediate Z impurity ions (iron), the radial positions of the emission shells lie inside the region of the transport barrier and thus are not oversensitive to the details of the model. For the low Z impurities (carbon), the ratio of emission from different ion stages is very sensitive to the details of the H-mode transport model, in particular to the location of the transport barrier.[5] The edge T_e profile determines whether the C VI shell lies within the last closed flux surface (LCFS) or in the scrape-off layer (SOL). Therefore, the calculations of the positions and intensities of the light impurity emission has been based on an L-mode transport model. An important function of this diagnostic will be elucidate the details of the H-mode transport at the edge of ITER and the role of ELMs in controlling impurity levels (of particular relevance where edge radiating impurities are deliberately introduced to reduce the divertor heat load, but where occurence of these impurities in the bulk plasma is to be avoided).

The results of the transport modelling show that the space resolution of the diagnostic must be of the order 5 cm at the plasma edge, reducing towards the plasma centre where the profiles are expected to be flatter and the relevant ionisation stages exist over a larger temperature range. However, it should be borne in mind that in present reversed-shear experiments a sharp barrier to transport is found at about half radius and the diagnostic should ideally be capable of resolving details such as this.

As a design goal we therefore aim to have a space resolution of 5 cm over the outer 50 cm of the plasma becoming rapidly coarser towards the plasma centre.

ITER SOLUTIONS

From the foregoing remarks it is clear that the edge plasma is the most important in terms of space resolution. It is the important Lyman lines of the light impurities that radiate near the edge. These lines lie in the shortest wavelength region (1–10 nm) to be covered by the diagnostic. In this region the optical system must be designed with

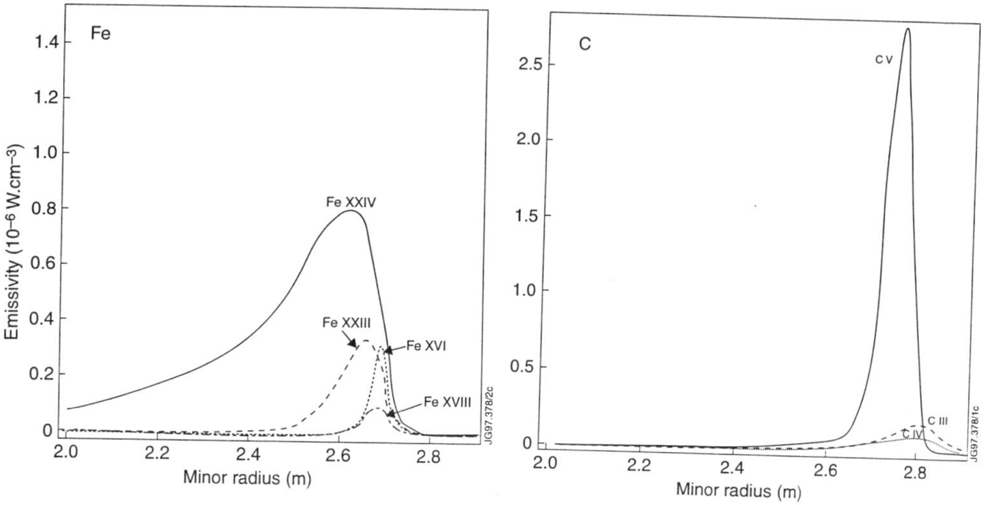

Figure 1. The radial emission profiles of an intermediate Z impurity (iron) and a light impurity (carbon) using the plasma T_e and n_e profiles given in ref. 4. The transport model used for the simulation is L-mode like for the C simulation (since the details of the edge transport barrier in H-mode have a profound effect on the C emission profiles) while the Fe simulation is based on an H-mode transport model.

grazing incidence (5° or less) reflections throughout. These twin constraints severely restrict the range of options available for the layout of this diagnostic. The only scope for tangential viewing of the edge plasma with a grazing incidence flightline is from the vertical ports, figure 2. Access to the plasma edge from the radial ports entails reflection angles of the order of 45° restricting the operating wavelength to above 30 nm.

Where longer wavelength coverage only is required, or where no space resolution is needed, considerable choice exists on the location of the spectrometer. It is proposed to include such a longer wavelength spectrometer as part of the diagnostic package in radial port 7 as a relatively simple survey diagnostic.[6]

Vertical Viewing XUV System

The general layout of the vertical viewing design is of a duct containing two mirrors, figure 2. A first (plane) mirror deviates the light providing shielding against the direct neutron flux while a second (curved) mirror focusses the light into the spectrometer. Each spatial chord has a separate spectrometer and beamline, the optical components would be assembled and aligned as a single entity before installation into the vacuum assembly and shielding structure in the port.

The space restrictions within the vertical port force the use of 5° incidence deviating mirrors for the edge channels. The more central channels are less restricted and could therefore use shallower incidence mirrors, however the direct neutron flux will be higher for these central chords and the higher deviation angle will offer better radiation attenuation. Additionally, radiation below 5 nm is not expected for these chords and so extreme grazing incidence optics are not required.

In order to maintain the angle of incidence at 5° for the deviating mirror the viewing chords make use of the gap between blanket tile segments near the top of the machine.

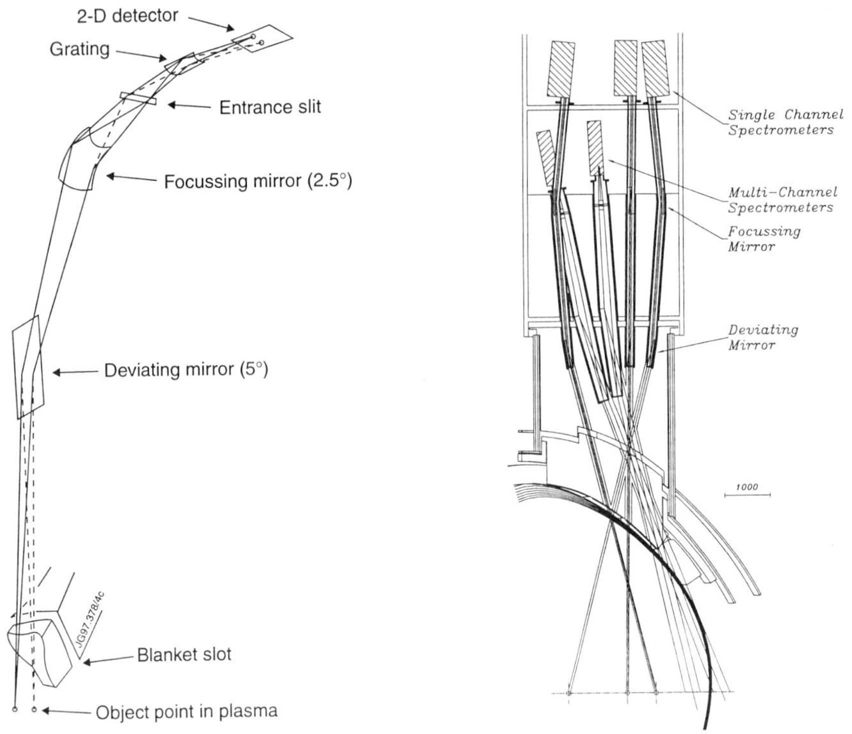

Figure 2. Sketch of optical arrangement and engineering layout of the complete vertical port system.

Normally this gap is about 2 cm but this would have to be enlarged to 5 cm on this port to match the light collection aperture of the diagnostic and ensure maximum signal levels.

The central channels of the diagnostic each collect light from a single chord and use a focussing mirror near the spectrometer to match its acceptance cone, keeping the shield duct diameter as small as possible. The density of chords required at the plasma edge would involve many (ten or more) individual spectrometer and mirror assemblies. Such an arrangement would be both expensive and extremely difficult to accomodate in the restricted space of the vertical port extension, while siting the instruments above the biological shield is discouraged. Instead, an imaging arrangement has been used, splitting the cluster of edge chords into two groups. The focussing mirror of the central chords is replaced by a toroidal mirror that forms an image of the plasma, seen through the blanket slot, along the length of the spectrometer entrance slit. Relatively wide ducts are needed in the shield to accomodate the full image size, however these need only be 5 cm thick slots, and will not receive radiation directly from the centre of the plasma.

The central chords are arranged to avoid viewing the divertor volume. This region is a strong emitter of soft X-ray radiation which can swamp the relatively weak emission from the plasma centre, figure 3.

Optical Design The optical design for each of the edge imaging systems is identical and is based on a 2 metre Rowland circle grazing incidence spectrometer. A toroidal mirror 600 mm in front of the spectrometer images the plasma view (seen though the blanket module slot) onto the spectrometer entrance slit. In the raytrace plots, figure 4,

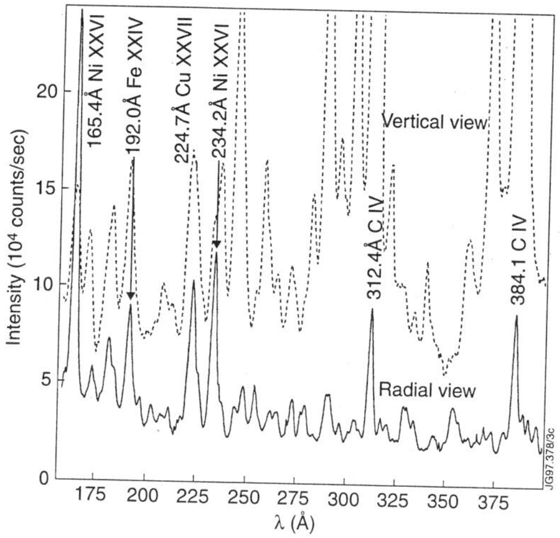

Figure 3. Comparison of spectra from a radial viewing and a vertical viewing spectrometer (SPRED[7]) on JET. The vertical spectrometer looks into the divertor volume through the plasma centre and demonstrates how the spectra are completely dominated by the divertor emission. For measurements of the plasma centre it is therefore important to avoid lines of sight that intercept strong divertor radiation.

a single point in the plasma forms an image of 3–6 mm in extent (at various wavelengths). The magnification of the system is 12 which therefore results in a resolution in the plasma of 35–70 mm with the poorest resolution occurring at long wavelengths and at the extreme ends of the imaging region. It is likely that the design can be further improved to give better resolution in these regions.

The highest space resolution is seen at the entrance slit position. It is possible to place a pinhole at this position and allow only light from a single spatial position into the spectrometer thereby obtaining higher resolution (17–30 mm) than is possible with the multichannel system (the spectral resolution would also be improved). This technique has been used in high resolution VUV solar imaging spectrometers.[8] However, this scheme requires the pinhole to be moved across the slit to obtain a spatial scan and hence would not allow simultaneous measurements across the whole profile which would be a significant disadvantage.

Radiation Shielding The radiation shielding has to be sufficently effective to reduce the neutron (and gamma) induced noise in the detector to significantly below the signal countrate. In a multichannel detector neutron induced signal constitutes a pedestal on which the spectral information sits.[4] In principle this pedestal can be subtracted away leaving the spectral information alone, however the shot noise associated with the neutron signal remains and degrades the signal to noise ratio of the measurements.

The transport calculation of figure 1 can be used to estimate photon fluxes in the spectrometers, including the reflection losses at the two mirrors. Data on the reflectivies of gold for the deviating mirror at 5° and the focussing mirror at 2.5° incidence are taken from.[9] For a line of sight viewing the plasma tangentially to flux surfaces and a concentration of Fe in the plasma of 0.007% a countrate of 4×10^4 cts.s^{-1} (Fe XXIV) is anticipated. For the L_α line of carbon the intensity in H-mode is sensitively dependent on the transport assumtions. However, for L-mode transport a countrate in the spectro-

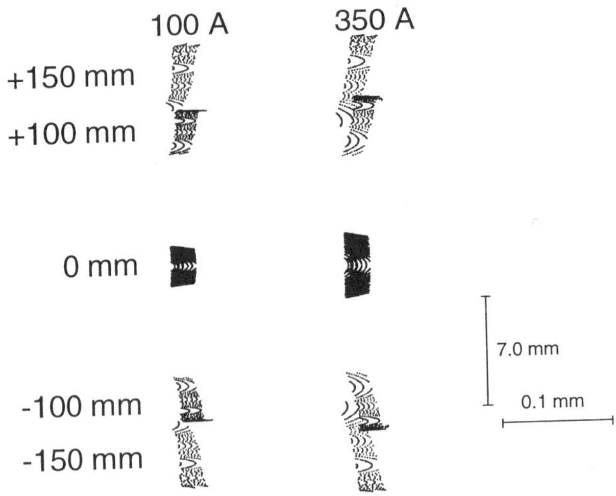

Figure 4. Raytrace calculations of the images in the spectrometer at two wavelengths (10 and 35 nm). The spot patterns shown are from points in the plasma at 0, ±10 and ±15 cm from the optical axis, *ie* a coverage of 30 cm. The sizes of the spot patterns yield a resolution between 30 and 70 mm in the plasma.

meter of 5×10^5 cts.s^{-1} is predicted at a carbon concentration of 1% and $n_e = 1 \times 10^{20}$ m^{-3}

Neutron and gamma radiation levels for this diagnostic are estimated using the 'HANDC' code.[10] The calculations are based on two shields in series, the first between the plasma and the deviating mirror, the second between the mirror and the spectrometer. No account has been taken of the extra reduction in neutron flux obtainable by a moderate amount of tight baffling near the spectrometer entrance slit, where only a very small optical aperture is needed. An effective shield can be constructed here which could also lead to weight savings elsewhere.

In the calculations it is assumed that neutron radiation reaching the deviating mirror is collimated by the penetration in the first shield and therefore does not enter directly into the duct of the second shield. The effectiveness of this shielding arrangement depends critically on this assumption for without it the neutron flux entering the spectrometer is higher by a factor of 10^3. Preliminary MCNP calculations agree with the results of the HANDC code with this assumption.

Care should be taken to keep the exposed mass of the deviating mirror low to avoid forward scatter of radiation towards the second shield. (The mirror could be constructed of lightweight material supported on a stable structure to either side of the path of the streaming neutrons.)

The calculations give a neutron rate of 3×10^{11} n.m^{-2}.s^{-1} at the spectrometer position. The radiation sensitivity figures for microchannelplate detectors[11,12] indicate that these devices are more sensitive to gamma rays than to neutrons at 2.2×10^{-2} c.γ^{-1}, so assuming that the neutron flux is accompanied by an equal gamma flux a noise signal is induced in the spectrometer of 4×10^2 c.pixel^{-1}.s^{-1}. Even over a typical spectral linewidth of 3–5 pixels this countrate is much lower (at least a factor of 100) than the estimated signal levels and indicates that the shield is effective. However, the spectrometers are within the biological shield and so subject to the ambient radiation flux of the cryostat thus there would probably have to be extra shielding installed around the sides of the vertical port to give adequate attenuation of this radiation.

Instrumentation The calculations of the imaging system have assumed grazing incidence spectrometers as are required for coverage down to 1–2 nm for the strong emission lines of light impurities. The maximum wavelength of such instruments is typically 35 nm, wavelengths above 10 nm being the domain of higher incidence angle devices. As the incidence angle is increased instrumental aberrations become less and imaging becomes simpler; above 50 nm normal incidence spectrometers can be used to good effect,[13] however the high f-number of the relay optics will impose a limit on the optical thoughput of such a system.

Since a large part of the complication and cost of installing this diagnostic on ITER will involve the shield and transmission optics it is important to exploit these components to the full by making the spectrometer as flexible as possible. Options for extending the wavelength range of the system with a second spectrometer on each chord include using the zero order light from the XUV system[4] or tilting the spectrometer so that light from the entrance slit illuminates a second grating. The coverage of the grazing incidence instrument itself can also be extended (to ~100 nm) with a longer Rowland circle.

Detectors In the radiation calculations above a microchannelplate based detector was assumed although other possibilities are available. A critical issue for the detector choice, apart from those of sensitivity to XUV photons and neutrons, is that of sensitivity to tritium poisoning. The spectrometers and detectors lie within the ITER primary vacuum boundary and as such will be exposed to the tritium present in that environment. Microchannelplates are likely to be poor candidates with regard to tritium poisoning since they present a large surface area for absorption and the decay βs would each trigger an electron avalanche in the detector (a photon counting energy discriminating readout system could be an advantage in this situation). A better approach could be to use a very thin photon converter (scintillator) and relay the visible light to a CCD (or CID) outside the vacuum chamber, but at present data is not available to make the comparisons. In the event that tritium poisoning remains an issue, steps would have to be taken to exclude it from the spectrometer chamber (perhaps by a low pressure purge of hydrogen) or to clean the detectors in-situ (outgassing of channelplate detectors with a UV radiation source).

Certain detectors, notably photomultiplier tube and image intensifiers, but also microchannelplates (to a lesser extent) are sensitive to the magnetic environment. For the location of the spectrometers within the cryostat the magnetic fields can be of the order of 0.5 T (ref. 14). These types of detectors would all require shielding against such fields.

First Mirror Lifetime The radiation levels calculated at the location of the deviating mirror are low, $7 \times 10^{14}\,\mathrm{n.m^{-2}.s^{-1}}$, while the damage threshold for a steel substrate mirror is probably in the region of $10^{23}\,\mathrm{n.m^{-2}.s^{-1}}$.[15] Thus a lifetime in excess of 10^6 ITER discharges is expected for this component. The neutron heating rate is less than 23 W which again is well within safe limits.

While the deviation mirror is well protected from radiation effects there remains the issue of contamination of the mirror surface. The indications from JET[4] are pessimistic, giving a lifetime for contamination of only 50 ITER shots (to 50% transmission at 10 nm). However, the JET mirror contamination data translate into rather high impurity fluxes which may have occurred outside of plasma pulsing (for instance during cleaning).

Radial Impurity Survey

It is planned to include a broadband survey spectrometer as part of the integrated diagnostic package in the radial port 7 (see ref. 6) operating above 10 nm, figure 5. The

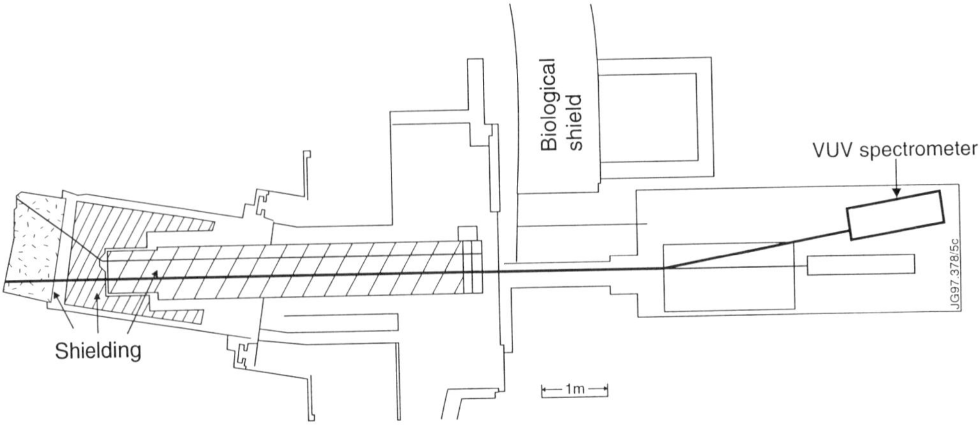

Figure 5. Plan view of radial port 7 showing the location of a VUV survey spectrometer.

instrument would incorporate a focussing mirror to image the aperture in the port shield onto the spectrometer entrance slit and match the spectrometer acceptance aperture. A plane deviating mirror at 5° incidence is included within the access cell to allow the sharing of the port with other diagnostics.

The optical and radiation design considerations are broadly similar to the vertical XUV spectrometers considered above, but with some reduction in neutron and gamma radiation flux due to greater depth of shielding.

Calibration

The calibration of XUV and VUV spectrometers is always an area of great difficulty; there are very few sources in this spectral region that can be used as either a transfer or absolute standard. The sensitive components of the spectromter, its reflectors and detectors, are all part of the tokamak primary vacuum and thus vulnerable to surface contamination that can alter their characteristics over time. These problems mean that the most successful techniques for the calibration of spectrometers in this spectral region would either use an in-situ calibrated source or make use of the plasma itself as a transfer standard. The latter technique utilises branching ratios from the visible,[16] referenced back to an absolutely calibrated visible instrument. A combination of these two techniques seems the most promising for ITER, although VUV spectrometers have been successfully calibrated using synchrotron sources[17] and X-ray tubes. However, the 20% accuracy specified by the expert group is probably unattainable; the typical best accuracies obtained in present day XUV/VUV spectrometers are 50%.

CONCLUSIONS

In this paper we have presented a scheme for the design of a spatially resolving XUV/VUV spectrometer system for ITER. Areas of uncertaintly are higlighted, in particular the neutron shielding efficiency of shallow bend angles in the shield ducts. Nevertheless, the design appears to hold promise and fulfills the requirements specified for its design.

ACKNOWLEDGEMENTS

We wish to acknowledge useful discussions with Dr. P. Van Belle and the support of the UK DTI and Euratom.

References

1 M. Mattioli, J. Ramette, B. Sautic, B. Denne, E. Källne, F. Bombarda and R. Giannella, *Journal of Applied Physics* **64**, (7), 3345

2 K. D. Lawson and N. J. Peacock, *Optics Communication*, **68**, (1988) 121. (see also JET-R(88)12)

3 *Minutes of 4th Meeting of the ITER Physics Expert Group on Diagnostics*, S CX MI 9 96-01-03 F1

4 N. J. Peacock, R. Barnsley, N. C. Hawkes, K. D. Lawson, M. G. O'Mullane, *Diagnostics for Experimental Thermonuclear Fusion Reactors*, P. E. Stott et al., (editors), Plenum Press, 1996, page 291

5 D. Pasini et al. *Nuclear Fusion* 30:2049 (1990)

6 P. H. Edmonds, R. Barnsley, N. Hawkes, A. Kislyakov, G. Vayakis, C. Walker, L. de Kock, G. Janeschitz, A. E. Costley, T. Steinbacher, 1997, this conference.

7 R. J. Fonck, A. T. Ramsey, and R. V. Yelle, Applied Optics 21:2115 (1982)

8 K. Lidiard et al. to be published in *Optical Engineering*, August 1997

9 G. P. Williams and M. R. Howells, *Photon Flux Calculations at the NSLS UV Facility*, Brookhaven National Laboratory Report BNL 26121

10 H. Iida, D. Valenza and R. T. Santoro, *A Handy Method for Estimating Radiation Streaming Through Holes in Shield Assemblies*, ITER Report G 73 CC 1 96-07-18 W 1.1.

11 J. G. Timothy and R. L. Bybee, *Review of Scientific Instruments*, 50:743 (1979)

12 S. S. Medley and R. Persing, *Review of Scientific Instruments*, 52:1463 (1981)

13 N. C. Hawkes and N. J. Peacock, *Nuc. Fusion*, 25:971 (1985)

14 ITER Diagnostic Group, 1995, Report S 55 RE 1 95-07-11 F 1, *ITER Information: Updated Background Information relating to Diagnostic Task Agreement.*

15 E. H. Farnum, F. W. Clinard, Jr., S. P. Regan and B. Schunke, *Neutron Damage to Diagnostic Mirrors*, JET Internal Report, JET-IR(94)03

16 B. Denne and E. Hinnov, *Journal of the Optical Society of America B: Optical Physics*, 1:699 (1984)

17 B. C. Stratton, R. J. Fonck, K. Ida, K. P. Jaehnig and A. T. Ramsey, *Review of Scientific Instruments*, 57:2043 (1986)

X-RAY SPECTROMETRY SYSTEMS FOR ITER

R Barnsley[1], P H Edmonds[2], R M Giannella[3], M G O'Mullane.

JET Joint Undertaking, Abingdon, OX14 3EA, UK.
[1]Leicester University, LE1 7RH, UK.
[2]ITER JCT, Garching JWS, Garching, Germany.
[3]CEA, Cadarache, France.

INTRODUCTION

This paper continues the progress, reported[1] at the 1995 ITER diagnostics workshop in Varenna, toward the design of an integrated soft x-ray spectroscopy system for ITER. Instrumentation for the soft x-ray region has several advantages; absolute sensitivity calibration is possible, surface effects are relatively unimportant, vacuum isolation windows can be used, and radiation-hard detectors are available whose photon-counting nature allows additional background rejection. Details of the proposed installation on Port 7 of ITER are contained in Edmonds et al[2], these proceedings.

The integrity and condition of the machine will be monitored in the soft x-ray band between about 1 Å and 100 Å, where all impurities have strong lines characteristic of their radiated power losses. This function is fulfilled by a broadband soft x-ray survey spectrometer (XCS-S) which is at present unchanged from the 1995 design, and uses a two-reflection instrument to achieve excellent shielding.

High resolution crystal spectroscopy in the band below about 10 Å is an established technique for measuring tokamak plasma T_i and rotation profiles[3]. Due to access restrictions, the ITER crystal spectrometer array (XCS-A) uses graphite prereflectors in the blanket to relay relatively narrow spectral regions to the spectrometers outside the cryostat. Because the viewing angles will be fixed in the blanket, the corresponding wavelengths and chordal views cannot be varied and must be chosen carefully. Most of the work presented here relates to the feasibility study for this scheme.

The radial distribution of impurity line radiation must be predicted, so that suitable transitions can be selected, firstly for the (central-chord only) impurity monitor, and secondly for radial profiles of T_i. Line emission from relevant ionisation stages of the expected intrinsic impurities has been modelled with the SANCO impurity transport code, using transport profiles shown in figure 1. The T_e and n_e profiles used here[4] are for an ignited plasma, and are relatively broad and flat, whereas ITER can be expected to operate with a wide range of profiles before it achieves reactor conditions. Therefore these simulations probably represent the outer limit of r/a for each emission shell - this being the most important condition to model, particularly when choosing transitions suitable for measurement of T_i profiles.

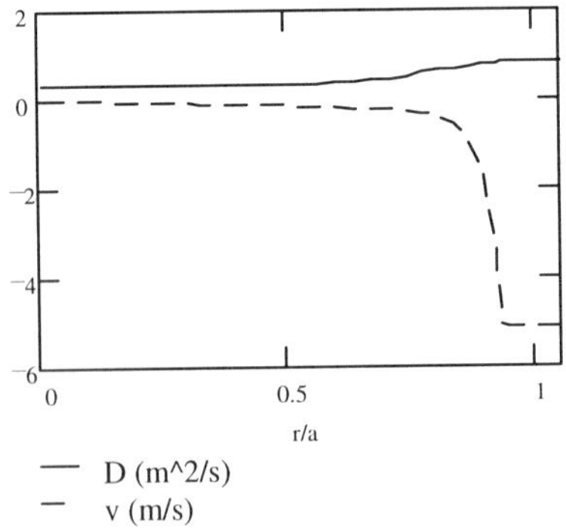

Figure 1. Profiles of transport parameters D and v, used in the SANCO impurity transport code.

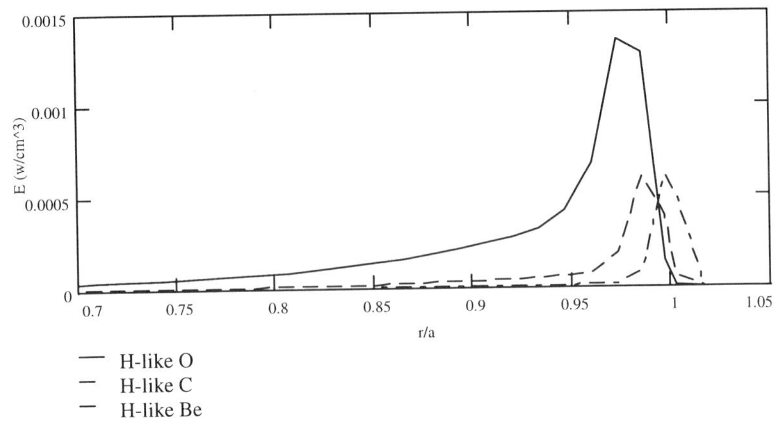

Figure 2. Emissivity profiles of light impurity ions, from the SANCO code.

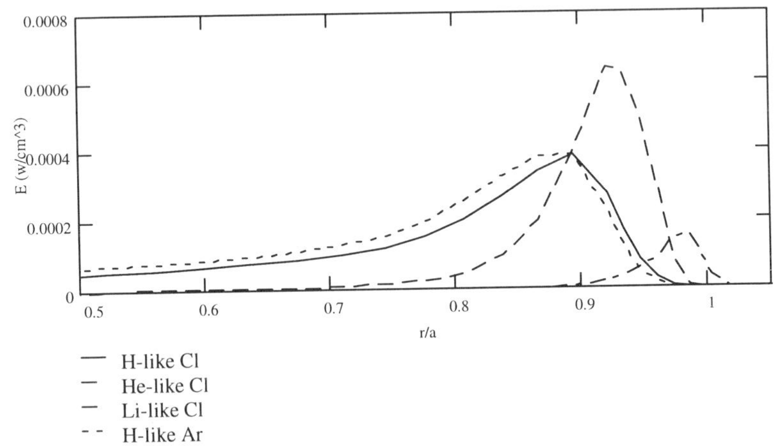

Figure 3. Emissivity profiles of Cl and Ar ions, from the SANCO code.

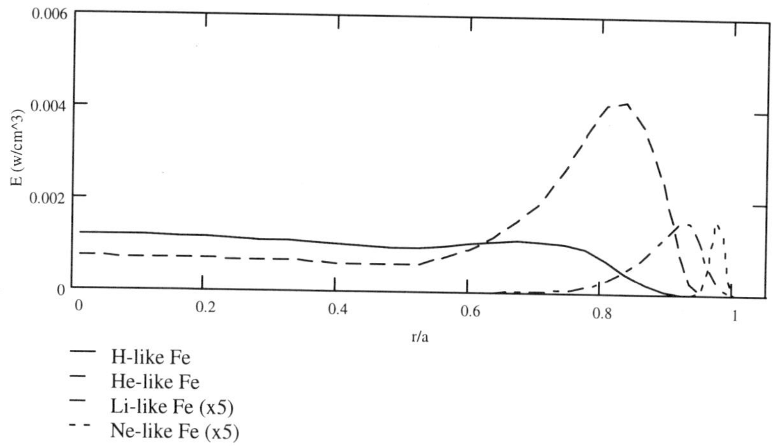

Figure 4. Emissivity profiles of Fe ions, from the SANCO code.

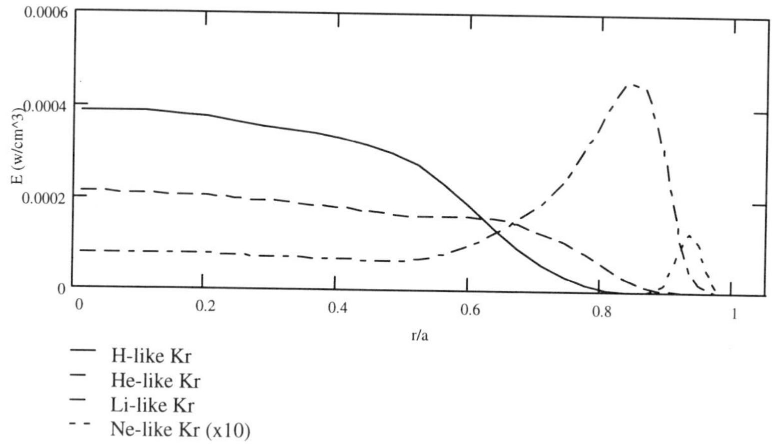

Figure 5. Emissivity profiles of Kr ions, from the SANCO code.

SURVEY X-RAY CRYSTAL SPECTROMETER (XCS-S)

The SANCO modelling shows that the innermost (H-like) line radiation from the light impurities Be, C and O, is localised close to, or overlapping the separatrix (fig.2). There are suitable lines from all impurities in the region outside r/a~0.8 (figs.2-5), making this a good first-choice for the sight-line of a single-chord soft x-ray impurity monitor. Using such an outer chord would reduce the bremsstrahlung component of the signal, and to some extent reduce the neutron flux at the spectrometer, factors which would improve detection thresholds for the line spectrum. However, with the access available for the present design, the broadband impurity monitor uses a central chord.

The survey spectrometer (XCS-S) is based on DT experience with the single and double reflection instruments at JET, and uses a combination of crystals and multilayer mirrors to survey the spectrum between about 1 Å and 100 Å, with a resolving power $\lambda/\delta\lambda$ of ~500 below 25 Å and ~50 above 25 Å, the chosen diffractors and wavebands being unchanged from ref. 1. It requires a vacuum beamline with a direct view of the plasma, but has no optical elements within the cryostat.

The present design, detailed in ref. 2, includes a two-reflection instrument similar to that installed at JET[5] since that has been shown to have excellent shielding properties[6]. However, even the single-reflection monitor spectrometer[7] showed good signal-to-noise performance during recent DT discharges at JET. Figure 6 shows the raw data from the JET single reflection x-ray monitor spectrometer during about 2 MW of DT power, where the noise in the background channel is about 20 times less than the continuum level in the line-scan. Due to the small crystal aperture relative to the beamline aperture, the neutron/gamma etendue of the instrument is 20 times greater than the x-ray etendue, so there is considerable scope for a signal-to-noise improvement if the sight-line shielding can be matched to the x-ray aperture. Given that further improvements in detector shielding and pulse-processing are also possible, these results indicate that a single-reflection survey instrument should be feasible for ITER, in which case other optical schemes become possible. Figure 7 shows the two main types of broadband crystal spectrometer, which, for a given resolving power and Bragg angle range, have equivalent sensitivity[8]. However, the polychromator has a better signal-to-noise ratio because the neutron/gamma etendue of the input slit can match the x-ray etendue, and there is scope for some additional shielding between the crystal and detector, whereas a large-area collimator tends to be transparent to neutrons and gammas. The polychromator also has the advantage of needing no moving parts. Suitable position-sensitive detectors are available.

Further to the spectra and radiated power components presented in ref. 1, figures 8 and 9 illustrate additional features of the x-ray spectroscopic data obtained at JET. The ability of a soft x-ray spectrometer to monitor the bremsstrahlung continuum is shown in figure 8, where the measure spectrum from Ge(111) is compared with the calculated sensitivity function. The presence of the Ar-K edge at the calculated ratio confirms that the continuum signal is due to the plasma free-free/free-bound continuum. Figure 9 shows an example of long-term soft x-ray spectroscopic data, where the trend in the Cl XVI 4.444 Å line is plotted over a series of several thousand discharges.

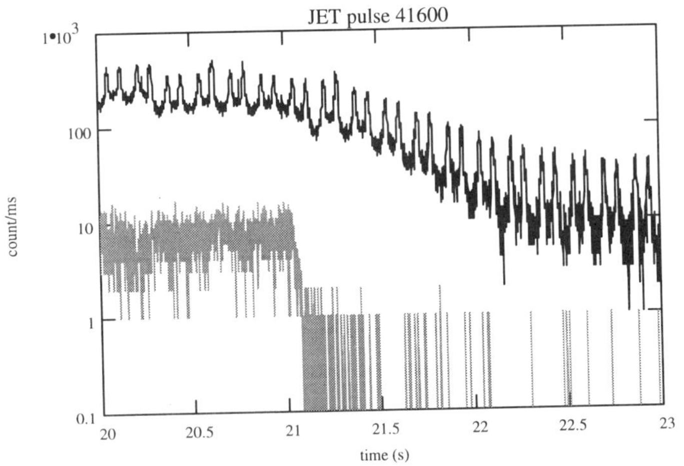

Figure 6. Raw data from the JET single-reflection x-ray monitor spectrometer during about 2MW of DT power terminating at 21s. (Above) line-scan of C VI Lyα . (Below) Noise in the background channel. Due to the small crystal aperture relative to the beamline aperture, the neutron/gamma etendue of the instrument is 20 times greater than the x-ray etendue.

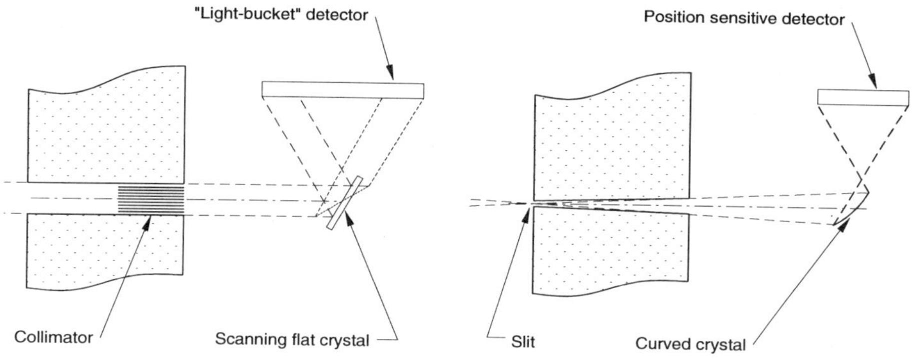

Figure 7. The two main types of broadband crystal spectrometer.
(Left) Scanning monochromator, with input collimator, flat crystal and "light bucket" detector.
(Right) Polychromator, with input slit, fixed crystal and position sensitive detector.

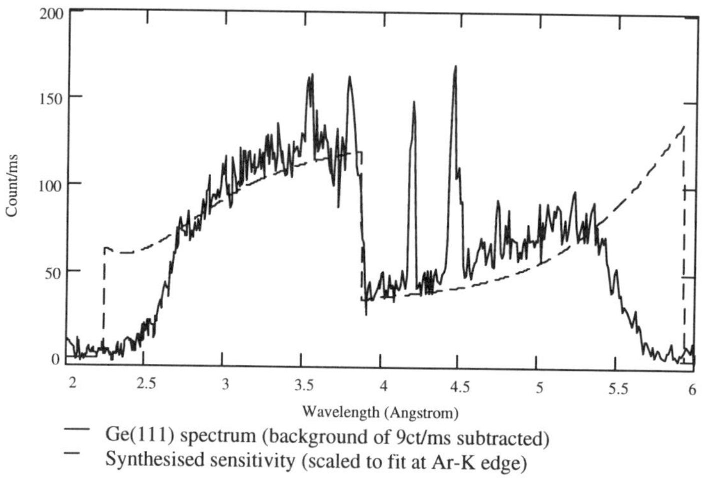

Figure 8. Ge(111) spectrum from the JET broadband monitor. Main features are the detector Ar-K absorption edge at 3.87Å, and the (unresolved) principal lines of H- and He-like Cl at 4.2Å and 4.45Å respectively. The presence of the Ar-K edge at the calculated ratio confirms that the continuum signal is due to the plasma free-free/free-bound continuum.

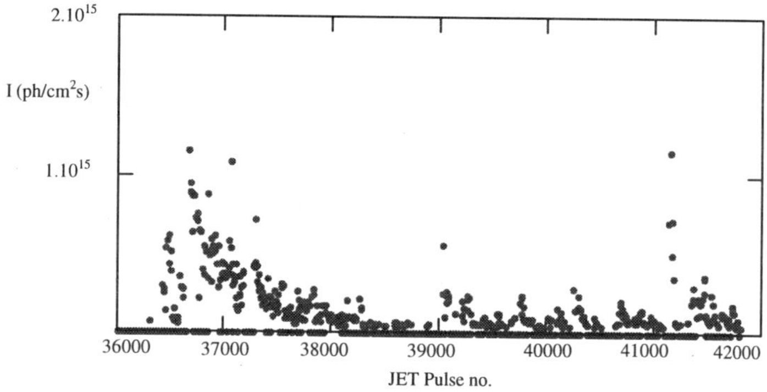

Figure 9. Intensity of He-like Cl at 4.44A, measured at 5.5 s into the discharge, for a long series of JET discharges. Similar data are available for all the major intrinsic impurities.

HIGH RESOLUTION X-RAY CRYSTAL SPECTROMETER ARRAY (XCS-A)

Within the constraints of the allocated Port 7, it is not possible to view a useful range of minor radii with a fan-array of direct-viewing chords, which would be the preferred option. A practical solution is to use an array of graphite prereflectors to view a wide range of minor radii, each with a toroidal component (fig.10). The relay sight-tubes are parallel, are contained within a shielding plug, and are isolated from the spectrometers by beryllium windows. Advantages of this scheme are that moving parts inside the cryostat are unnecessary, and that a large reduction in the neutron flux at the spectrometers can be expected, since there is no direct line-of-sight to the plasma.

The lowest measurable T_i is limited to about 200 eV by the resolving power of available crystals, and by the excitation of suitable ions (KAP(002), Ne IX). The use of graphite as a prereflector raises the lower limit to about 500 eV (Cl XVI) by limiting the longest observable wavelength to about 5Å. SANCO modelling shows that lines suitable for high-resolution spectroscopy can be observed across almost the entire minor radius, out to an r/a of at least 0.95 (figs. 3-5). Depending on impurity transport, medium-Z intrinsic impurities such as Fe are likely to be fully ionised on axis, so to be certain of obtaining good measurements within r/a~0.5, this study has been based on added Krypton (Z=36). Kr is almost ideal for the ITER plasma since the H-and He-like ionisation stages will predominate on-axis, while the H-like stage will not burn out at the expected central T_e. If tungsten (W, Z=74) is included in the divertor, it will be a rich source of line radiation suitable for Doppler measurements.

Measurements will be avoided at minor radii within any emission shell, because there will be insufficient viewing chords to make a satisfactory Abel inversion of a hollow emission profile. Therefore one aspect of the modelling is to enable the selection of ionisation stages whose line emission is unlikely to peak at an r/a greater than that of a given chordal view.

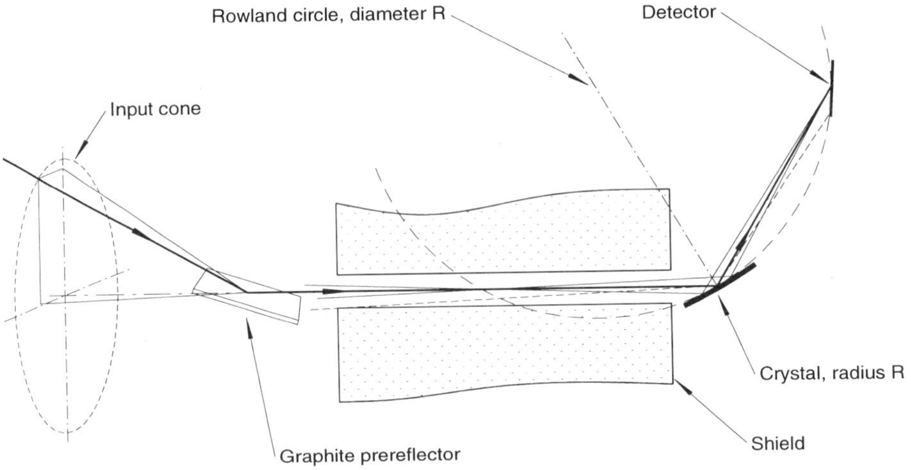

Figure 10. Schematic of a Johann spectrometer with graphite prereflector to provide shielding and allow poloidal and toroidal components of the line of sight. The wavelength range $\Delta\lambda$, and the crystal filling factor κ, are both determined by the sight-tube dimensions.

Graphite(001) 2d=6.708Å has a peak reflectivity of typically 10%, and a mosaic spread between 0.3° and 1.5° depending on sample preparation[9]. A mosaic spread of 0.8° has been used for this study. In addition to the signal loss due to the graphite peak reflectivity, the main disadvantage is the severe band-width constraint imposed by the requirement to have the graphite input and output angles fixed within the blanket module. Therefore a major purpose of this study was to find a working set of Bragg angles for the prereflectors. The requirement is that the Bragg angle should access a spectral region that contains suitable spectral lines, preferably from more than one ionisation stage of more than one impurity. The input sight-line then lies on a cone, and must be orientated so that the plasma can be viewed along a chord where the relevant spectral lines are emitted, preferably on the outside of any emission shell.

Sensitivities S_λ have been synthesised (fig.11) from the energy-dependent transmission, absorption and reflectivity of each component, using Henke's[10] atomic scattering factors and crystal reflection integrals (except for Ge(440) which was calculated at ENEA Frascati[11]). The sight-line geometry, with its constraints on the free spectral range, has been included, as shown in the plots of overall sensitivity and band-pass. The count-rate N'_λ (count/s) from a spectral line with intensity I_λ (photon/cm^2.s), is given by

$$N'_\lambda = I_\lambda \cdot S_\lambda$$

where, for a Johann spectrometer with graphite prereflector, the sensitivity function S_λ (cm^2) is given by

$$S_\lambda = Pgr_\lambda \cdot \frac{\kappa \cdot \psi \cdot Rc_\lambda}{4 \cdot \pi} \cdot hx \cdot hy \cdot \eta_\lambda$$

The terms are: graphite peak reflectivity Pgr_λ, vertical divergence ψ (rad), crystal reflection integral Rc_λ (rad), projected crystal width hx (cm), crystal height hy (cm), and the combined detector and window efficiencies η_λ. The fraction κ of the crystal aperture filled at a given wavelength depends on the source and beamline geometry.

In this application, where owing to space restrictions the source-crystal distance will be two to three times the crystal-detector distance, κ may be kept close to unity by using an asymmetric crystal cut[12] to move the virtual entrance slit into the plasma.

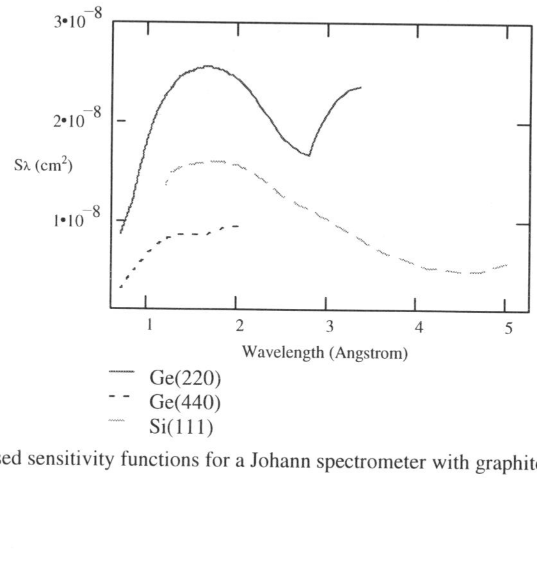

Figure 11. Synthesised sensitivity functions for a Johann spectrometer with graphite prereflector

To mitigate the sensitivity losses imposed by the prereflectors and by the relatively narrow (25 mm dia.) sight-tubes, it is important to maximise the throughput of the spectrometers, mainly by using a crystal with the highest possible reflection integral consistent with sufficient resolving power. A near-optimum crystal is Ge(220) 2d=4.00 Å, which has a high reflection integral and adequate resolving power (Rc~0.05 mr and $\lambda/\delta\lambda$~7000 respectively). Other factors being equal, the use of doubly-curved diffractors[13] will improve the signal-to-noise ratio, by allowing for a smaller input aperture and smaller detector. As a complement to calculated calibrations, the complete instruments should be calibrated with a purpose-built large-area source[14]. Any *in situ* calibration will be limited to the ex-vessel spectrometers, while the attenuation due to the graphite will be determined by using the direct-viewing monitor sight-line as a reference.

The derived sensitivities were superimposed on synthetic spectra of the main H- and He-like lines of Cl, Ar, Fe, Ni and Kr, and Li- and Be-like lines of Kr. Representative results are shown in figures 12 to 15, and if the 2nd order of graphite is also used (about a factor two less intense than 1st order) there are several useful bands. Figure 13 shows that the overall sensitivity is comparable to that of the existing JET instrument. A summary of the calculations for a 5-chord array is given in table 1, with the preferred options highlighted. A set of viewing-chords compatible with the Port 7 installation is shown in table 2. Based on the SANCO simulations and the calculated instrument sensitivities, estimates of count-rates in impurity Lyα lines have been derived, and are shown in table 3, together with the estimated radiated power loss for seeded Kr.

The spatial resolution of this system depends on several factors, including the beam-spread of each sight-line, and the errors associated with deconvolving the chord-integrals. However, if the beam-width is minimised, and measurement inside hollow emission profiles is avoided, then the main limitation on spatial resolution will be the number of chords. The present design is based on 5 chords, but an increase to at least 10 chords would greatly improve the quality of the derived T_i and ω profiles. The accessible chordal views give components of toroidal rotation velocity which are similar to those observed at JET (sin15° to sin20°) and the lowest measurable rotation velocity should be about 10 km/s.

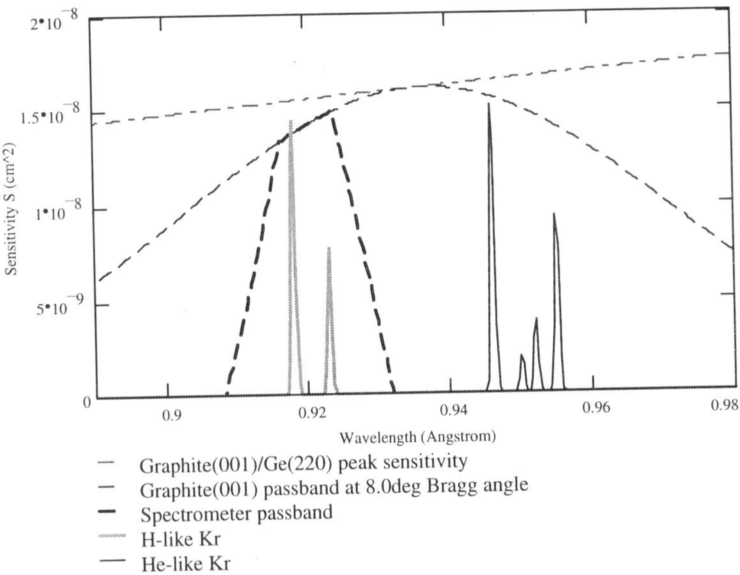

Figure 12. Accessible spectrum and spectrometer passband, in the region of the principal lines of H- and He-like Kr, for a Johann spectrometer with graphite prereflector.

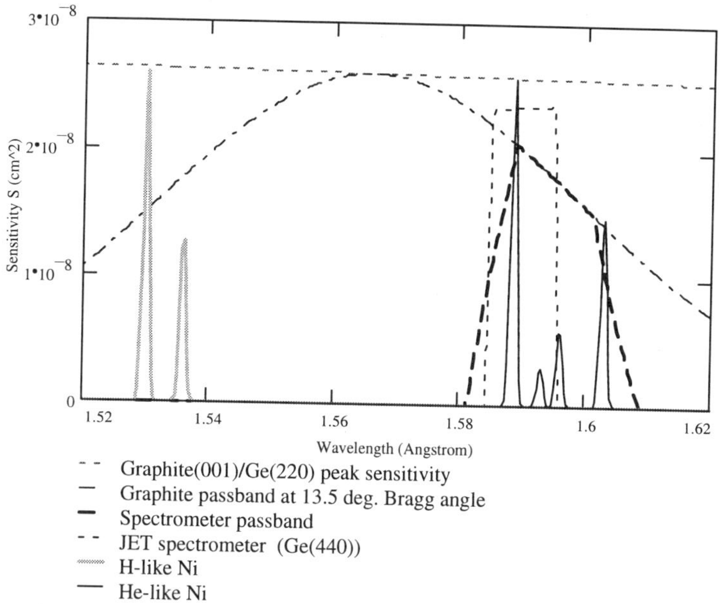

Figure 13. Accessible spectrum and spectrometer passband, in the region of the principal lines of H- and He-like Ni, for a Johann spectrometer with graphite prereflector. The sensitivity and passband of the existing JET instrument are shown for comparison.

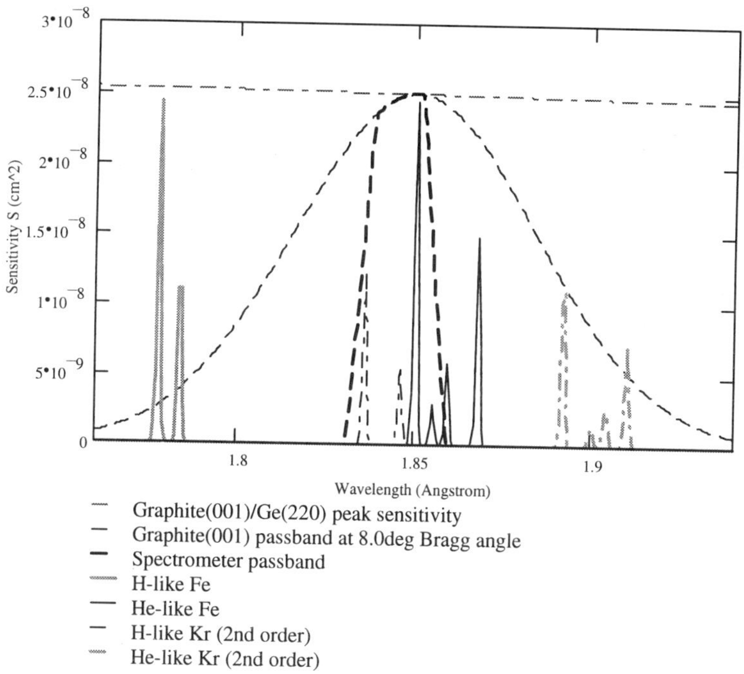

Figure 14. Accessible spectrum and spectrometer passband, in the region of the principal lines of H- and He-like Fe and Kr, for a Johann spectrometer with graphite prereflector.

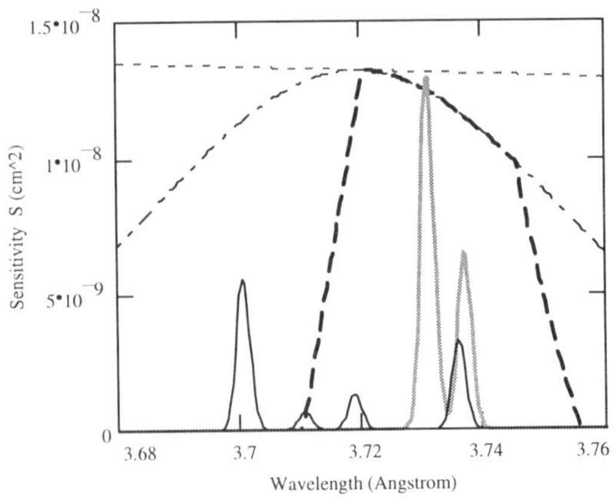

Figure 15. Accessible spectrum and spectrometer passband, in the region of the principal lines of H-like Ar and He-like Fe, for a Johann spectrometer with graphite prereflector.

Table 1. Observable impurity ions and T_i ranges at five chords, using graphite prereflectors.

ϑ_B gr. (deg.)	Chord r/a (1)	Ion	T_i min (keV)	T_i max (keV)	r/a min	r/a max	Comments
8.0	**0, 0.25**	'H'-Kr	10 (2)	**50**	0	0.75	
	0.5	'He'-Kr	**4**	20	0.3	0.85	
16.0	0, 0.25	'H'-Kr	10	**50**	0	0.75	Kr in 2nd order
	0.5, 0.75	'He'-Kr	4	20	0.3	0.85	
		'H'-Fe	3	20	0	0.7	
		'He'-Fe	**1.5**	10	0.3	0.8	
33.7	0.75, 1.0	'He'-Fe	1.5	**10**	0.3	0.8	Fe in 2nd order
		'H'-Ar	**1**	5	0.85	1.0	
36.3	1.0	'Li'-Kr	3	**8**	0.7	0.95	Kr lines weak,
		'He'-Ar	**0.5**	3	0.9	1	Lowest measurable T_i
							No intrinsics
38.5	**1.0**	'Li'-Kr	3	**8**	0.7	0.95	Needs only Kr added
		'Be'-Kr	2.5	6	0.8	1.0	$2\vartheta = 77°$ may be too great
		'H'-Cl	**1**	4	0.85	1.0	

(1) First choice for each of the five chords is shown in bold.
(2) Indicative max and min T_i at each chord is shown in bold. Since the measurable T_i range is related to T_e and impurity transport, the values given can only be approximate.

Table 2. Chords and graphite angles accessible with the port 7 design.

Major radius (mm)	Minor radius (mm)	r/a	Graphite angle (deg)
8317	0	0	direct
9121	804	0.25	16
9700	1383	0.5	16
10321	2004	0.75	16
10690	2373	0.875	16
11085	2768	1	16

Table 3. Count-rate and radiated power estimates.

Transition	Count-rate for $f_{imp}=10^{-6}$ (count/s)	P_{rad} for $f_{imp}=10^{-6}$ (kW)
Ar Lyα	6 000	
Fe Lyα	30 000	
Ni Lyα	30 000	
Kr Lyα	20 000	260

CONCLUSION

The technique of broadband soft x-ray spectroscopy is fully demonstrated at JET and provides unique data during DT discharges. Results from the JET single-reflection spectrometer indicate that a single-reflection survey polychromator could be developed for ITER. It would have greater sensitivity and be mechanically much simpler than a two-reflection system, but until an acceptable signal-to-noise ratio is demonstrated we include a two-reflection system in the design. The need for real-time display and analysis of spectroscopic data is being addressed at JET, where a real-time data acquisition system for the survey spectrometer will be commissioned.

The study for the high-resolution array shows that the sensitivity and band-width losses imposed by graphite prereflectors can be accommodated, and that a compatible set of sight-lines, wavelengths and spectral lines can be found. A further iteration of the design is required, because several features require clarification or improvement. The most significant improvement will be to increase the number of chords from the present five, to the largest number practical. Energy dispersive ray tracing of the full system is then required, partly to improve the sensitivity calculation, but mainly to minimise the beam-width in the plasma by optimising the free parameters in the optics (graphite/crystal radii, crystal-cut asymmetry etc.). These results, combined with modelling of the plasma emissivity and flux surfaces for the expected range of plasma parameters, will facilitate optimum chord placement, and allow the accuracy of the deconvolved profiles to be assessed.

Also required are detailed calculations of the neutron and gamma fluxes throughout the sytem, and tests of neutron damage to graphite. A study of the tungsten spectrum should extend existing work[15] to as near ITER-relevant conditions as possible. Although krypton has been used for this design study, if tungsten is present in ITER there will be an abundance of suitable spectral lines throughout both the spectrum and the plasma.

REFERENCES

[1] R Barnsley et al, Applications of x-ray spectroscopy to ITER, publ. in: *Diagnostics for experimental thermonuclear fusion reactors,* Ed. P E Stott et al, Plenum Press, New York, (1996) 353-364.

[2] P H Edmonds et al, Integration of vacuum coupled diagnostics. *These proceedings.*

[3] M Bitter et al, *Rev. Sci. Instrum.* **59**(10) (1988) 2131.

[4] ITER Diagnostic Group report, S 55 RE 95-07-11 F 1.

[5] R Barnsley, U Schumacher et al, Double crystal spectroscopy at JET, *Rev. Sci. Instrum.* **62**(4) (1991) 889-898.

[6] R Barnsley et al, Bragg spectroscopy during the JET preliminary tritium experiment, *Rev. Sci. Instrum.* **63**(10) (1992) 5023.

[7] R Barnsley et al, Bragg rotor spectrometer for tokamak diagnostics, *Rev. Sci. Instrum.* **57**(8) (1986) 2159.

[8] R Barnsley et al, JET Diagnostic Design Study 12-6, Euratom (1982).

[9] K D Evans et al, Calibration data for the Ariel 5 Bragg spectrometer, *Space Sci. Instrum.* **2** 313-323 (1976).

[10] B L Henke et al, X-ray interactions: Photoabsorption, scattering, transmission and reflection, *Atomic and Nuclear Data Tables* **54**(2) (1993) 181-342.

[11] F Bombarda, private communication (1989).

[12] R Bartiromo et al, The asymmetric Johann spectrometer and its application to plasma diagnostics in tokamaks, *Nucl. Inst. Meth.* **221** (1984) 453-459.

[13] V.A.Bryzgunov and A.B.Gil'varg, X-ray crystal spectrometer - proposal for ITER *These proceedings.*

[14] U Schumacher et al, Investigations on calibration sources for soft x-ray plasma spectroscopy and impurity monitors. *Rev. Sci. Instrum.* **67**(8) (1996) 2826-2830.

[15] R Neu et al, Observations of x-ray spectra from highly charged tungsten ions in tokamak plasmas, IPP Garching report 10/7 (June 1997).

PHOTOELECTRON PLASMA DIAGNOSTICS

Yu.V.Gott, V.A. Shurygin
RSC "Kurchatov Institute", Moscow, Russia

INTRODUCTION

The development of plasma diagnostics has been associated historically with the borrowing of methods and means of detection from other areas of physics: astrophysics, nuclear and atomic physics, ets. The need for such borrowing increased significantly when it became apparent that even in the planning stage for diagnostic array in a thermonuclear reactor there must be modernization of a significant fraction of the measurement techniques and development of new diagnostics. In this connection special attention should be directed at present time to certain unconventional measurement techniques which have been used successfully outside the area of thermonuclear studies. One of these is the *photoelectron technique* for analyzing X-ray spectra and detection of X-ray fluxes.

This development may be said to have begun with A.Einstein, who explained the photoelectric effect (1905), and the work of H.Robinson (1914) who formulated the fundamental ideas of X-ray photoelectron spectroscopy. At present photoelectron spectrometers (PES) are used to determine the energy of γ rays (up to 3 *MeV*) in nuclear physics, in order to measure the electron binding energy in atoms (with a resolution of up to 10^{-5}) in electron spectroscopy for chemical analysis (ESCA), and to study secondary electron emission produced by effect of photons in irradiate a solid objects.

The study of plasmas with help of the photoelectron spectra exited by X rays from them began with the work of V.Afrosimov et al.

Historically, measurements of the soft X-ray spectrum from a hot plasma have been performed over a wide energy range using puls-height analysis (PHA) systems, operating in the pulsed detection mode and accumulating the analysed spectra. In practice these measurements involve a spectrum that dies away rapidly (usually exponentially). In the high-energy region such a spectrum may be represented by a dozen pulses accumulated over a period of 5-50 *ms*, which clearly shows the "value" of detecting every photon. In addition, these detectors usually have a detection efficiency close to unity. In this connection an erroneous idea regarding the small X-ray emission flux from the plasma is sometimes encountered. It arises because it seems at first glance contradictory that PES that have a low quantum efficiency (10^{-4} - 10^{-5} *electrons/photon*) can operate in small devices in the current mode of detection with a time resolution less than 1 ms , and the measurements themselves consequently appear inefficient.

Diagnostics for Experimental Thermonuclear Fusion Reactors 2
Edited by Stott *et al.*, Plenum Press, New York, 1998

This misunderstanding is easily explained: on the one hand, the actual number of photons from the "warm" part of the plasma X-ray spectrum that can reach the spectrometer, depending on the experimental parameters, is usually equal to 10^8 - 10^{14} *photon/(s.cm²);* on the other hand , it is necessary to limit the load of a PHA system (using filters and diaphragms) to a level of less than 10^4 - 10^5 *pulses/s* over the whole measured spectrum. This imposes a significant limitation on the sensitivity of the measurement system, regardless of the plasma parameters and the size of the device. This drawback manifests itself in a bad time resolution of PHA. There is no such limitation in photoelectron spectrometry; its advantages in sensitivity are therefore especially noticeable in small devices and PES signal increases as a function of the quantity $n_e^- \cdot a$, where n_e is the plasma density and a is the characteristic plasma dimension along a sight line.

If we take into account the ITER conditions, we can state that semiconductor detectors are completely inappropriate for D-T experiments both because of sensitivity to and damage by neutron and gamma irradiation. Design of photoelectron spectrometer gives as possibility to shift its detector from the plasma line of sight and so to shield it from neutron and hard X-ray radiation effectively.

The purpose of the present paper is to demonstrate the diversity of the ways in which this unconventional method can solve the majority of conventional problems in X-ray plasma diagnostics.

A lot of References on this problem one can found in our Review article [1].

THEORY OF THE PHOTOELECTRON METHOD

To analyse the X radiation from a plasma by the photoelectron technique we need information about the structure of photoelectron spectra and the sources of the component of the spectrum: how the bremsstrahlung and recombination continua and the line spectrum are transformed into the corresponding photoelectron components.

1. FORMATION OF PHOTOELECTRON SPECTRA

The main process that occurs when X radiation with energy less than some E_{crit} interacts with material is the *photoelectric effect*. The value of E_{crit} depends on material atomic number and, for example, is about 50 *keV* for Al, 150 *keV* for Cu and 500 *keV* for Pb. The photoelectron energy E_e emitted by a free atom due to the photoelectric effect is related to the photon energy E_γ by Einstain relation

$$E_e = E_\gamma - E_{nl} \tag{1}$$

where E_{nl} is the electron binding energy in the nl suborbital (here n is the principal and l is the orbital quantum number).

The general appearance of the electron spectrum (Fig.1) emitted from the surface of a material subjected to X radiation is determined by a number of processes associated with the photoelectric effect, the passage of the resulting photoelectrons through the layer of material, and the production of secondary emission fluxes: slow secondary electrons with energies less than 50 *eV* , Auger electrons and fluorescence (See Fig.1.)

From Eq.(1) it follows that each X-ray line is represented in the photoelectron spectrum by a combination of photoelectron lines resulting from the photoelectric effect in those sublevels for which $E_{nl} < E_\gamma$.

From Fig.1 one can see that spectrum of Auger and photoelectrons consist from narrow line produced by photoelectrons which have not lost their energy and from the low energy part that is produced by the photoelectrons that are lost part of their energy in leaving the irradiated target.

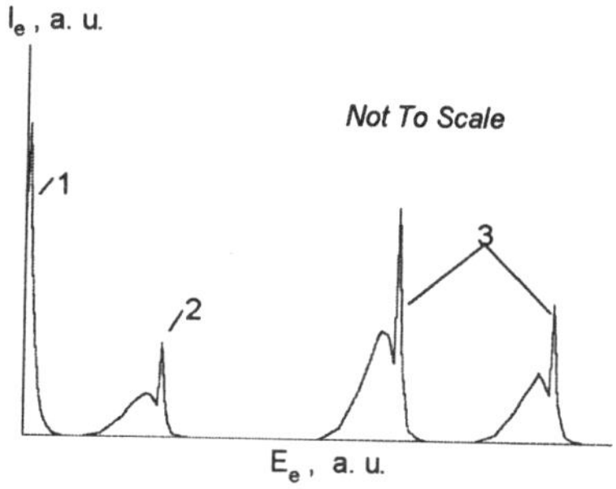

Fig.1. Electron emission spectrum of an irradiated solid target: 1) slow (less than 50 eV) secondary electrons; 2) Auger electrons; 3) photoelectrons from K and L shells of target atoms

The straggling of the narrow photoelectron line is very small: the full width at half-maximum (FWHM) of this distribution is less than 3 eV and to detect it we need the spectrometer with very good resolution - 10^{-4} - 10^{-5} . The FWHM of low energy part of electron distribution depends on the target thickness. If the target thickness greater than the photoelectron range in the target we have so called equilibrium target and the relative value of FWHM is about 45%.

Consider the conversion of the continuous X-ray spectrum, i.e., the continuum from a Maxwellian plasma, into the photoelectron spectrum. Initially we will neglect the slowing down of photoelectrons when they leave the target material. The photoelectron spectrum thus produced can be viewed as the result of adding up the corresponding partial spectra of all the nl sublevels. The partial spectra are displaced relative to one another and with respect to the X-ray spectrum in energy in accordance with Eq. (1).

A hot plasma whose electron component is heated up to temperature T_e is a source of bremsstrahlung . In this case the bremsstrahlung X-ray spectrum has the form

$$I_\gamma \sim exp\left(-\frac{E\gamma}{T_e}\right) \tag{2}$$

Substituting Eq. (1) in (2) we find that photoemission from all the nl sublevels of target atoms converts it into photoelectron spectrum of the form

$$I_e = \sum_{nl} I_e^{nl} \sim exp\left(-\frac{E_e}{T_e}\right) \cdot \sum_{nl} f_{nl} \cdot exp\left(-\frac{E_{nl}}{T_e}\right) \tag{3}$$

where f_{nl} is the efficiency with which a photon is converted into photoelectrons. If we assume that f_{nl} varies as a function of energy more slowly than I_γ , than, comparing the original X-ray spectrum (2) and photoelectron spectrum (3) , we note that the exponential shape has been preserved by this transformation. The discontinuities that characterise the corresponding dependence of the total photoabsorption cross section on the radiation energy are absent from the photoelectron spectrum because each partial photoelectron spectrum is shifted with respect to the X-ray spectrum by an amount E_{nl} .

The calculations fulfilled in Ref. 1 show that

$$f_{nl} \sim \mu_{nl} R \quad \text{for} \quad d > R$$
$$f_{nl} \sim \mu_{nl} d \quad \text{for} \quad d < R \tag{4}$$

here d is the target thickness, μ_{nl} is the linear photoabsorption coefficient in the nl subshell and R is the effective distance required for photoelectrons to escape from the target.

For the R value we have

$$R = 900 E_e^{1.3} / \rho^{0.8} \quad \text{for } 0.1 \leq E_e \leq 4\, keV$$
$$R = 560 E_e^{1.7} / \rho^{0.9} \quad \text{for } 4\, keV \leq E_e \tag{5}$$

here R is in Å, E_e is in keV, ρ is in g/cm^3 (ρ is the target density). The values of μ_{nl} one can find in Ref. 1.

To calculate the shape of the photoelectron continuum in real situation we need to take into account the target thickness and material, the contributions of different subshells, the slowing down of photoelectrons when they leave the target, and the photoelectron spectrometric instrument function.

In this case the shape of the photoelectron continuum in the region $E_e > (2-3)T_e$ can be represented in the form

$$\frac{dN_e}{dE_e} = const \cdot n_e^2 \cdot \varsigma \cdot T_e^{-1/2} \cdot \Phi(E_e, T_e) \cdot exp\left(-\frac{E_e}{T_e}\right) \tag{6}$$

where ζ is an enhancement factor, i.e., the enhancement of the radiation intensity in the X-ray continuum above the radiation from a purely hydrogenic plasma due to an impurity presence, the function $\Phi(E_e, T_e)$ describes all the elements of conversion of X-ray photons into the photoelectrons. The efficiency of the PES target (Φ) not only increases with increasing atomic number of the target material , but also it becomes more weaker function of the energy in a wide energy range.

The plasma electron temperature may be found from Eq. (6) as follows

$$\frac{1}{T_{exp}} = -\frac{d\ln\left(\frac{dN_e}{dE_e}\right)}{dE_e} = \frac{1}{T_e} - \frac{d\ln\Phi}{dE_e} \tag{7}$$

where T_{exp} is the temperature found experimentally. From Eq. (7) one can see that $T_{exp} \approx T_e$ if $1/T_e >> \left| d\ln\Phi / dE_e \right|$. In the reality the difference between T_{exp} and T_e usually is enough small.

In this paper we mainly discuss the measurements of X-ray continuum. The measurements of line radiation are described in Refs. [1-3].

3. EXPERIMENTAL RESULTS.

The description of several types of spectrometers one can find in Ref. [1]. Here in Fig. 2 one can see the tokamak T-10 photoelectron spectrometer.

The photoelectrons emitted from the target 1 are analysed in energy in longitudinal magnetic field produced by the coils 4 and then registered by the microchannel electron multiplier 3. The plasma X-rays pass through the carbon filter (with thickness of about 500 Å), which is placed before the PES enter and intended to protect the detector 3 from ultraviolet radiation and neutrals coming from the plasma.

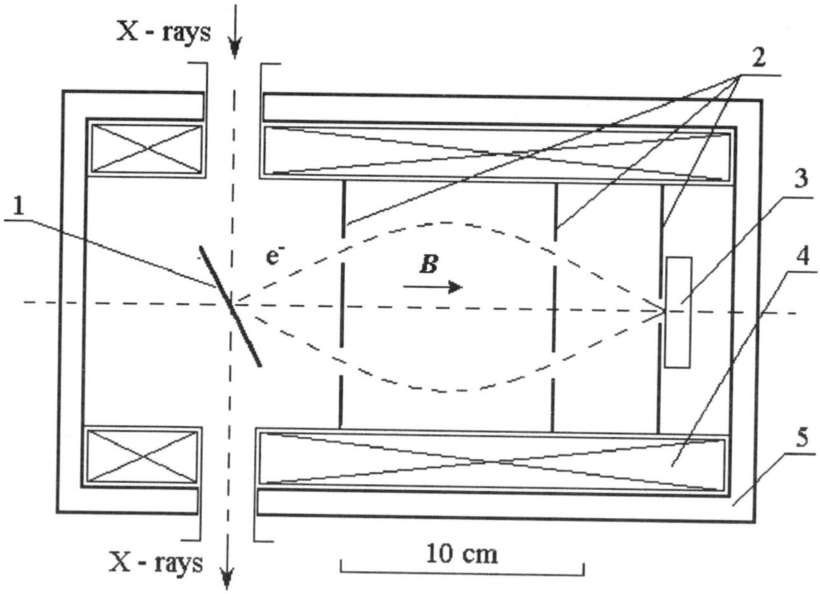

Fig.2. PES with magnetic analyzer: 1-target, 2-diaphragms, 3-detector, 4coils, 5-magnetic shielding.

In Fig. 3 one can see the photoelectron spectrum obtained on small tokamak T-13 (minor radius a = 6.5 *cm*, major radius R = 41 *cm*, plasma density $n_e = 10^{13}$ cm^{-3}, plasma current I = 30 *kA*, PES target was from carbon 380Å thick). This spectrum was obtained on the ramp up stage of discharge. The plasma electron temperature was 110 *eV* at that moment.

The photoelectron spectrum and an example of T_e determination on tokamak T-10 (minor radius a = 36 *cm*, major radius R = 150 *cm*, plasma density $n_e = 5{\cdot}10^{13}$ cm^{-3}, plasma current I = 300 *kA*, PES target was from silver with ~ 400Å) are shown in Fig.4 [4]. The spectrum was recorded in the time interval of about 200 *ms*. This time interval was limited due to low intensity of the X-rays incident on the PES target from the plasma. Unfortunately, in these first experiments the X-ray flux coming to the PES target was strongly restricted by the small size collimator with the spatial resolution in the plasma centre of about 1.5 *cm*. It is evident that the increase of this size up to 5-8 *cm* leads to significant (squared) increasing of the measured signal. One can see that electron temperature in this discharge was about 1 *keV*

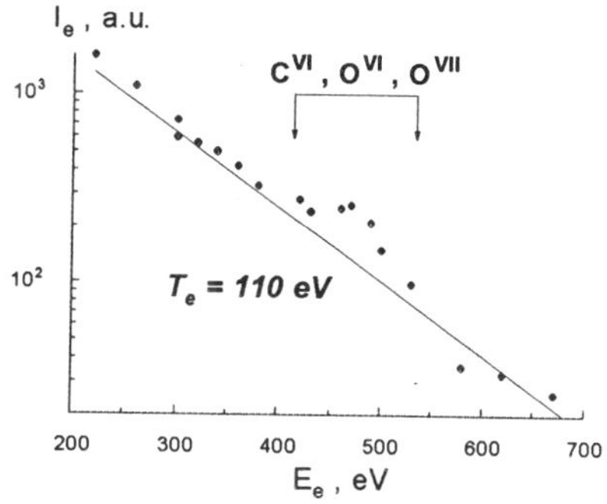

Fig. 3. Photoelectron spectrum on tokamak T-13

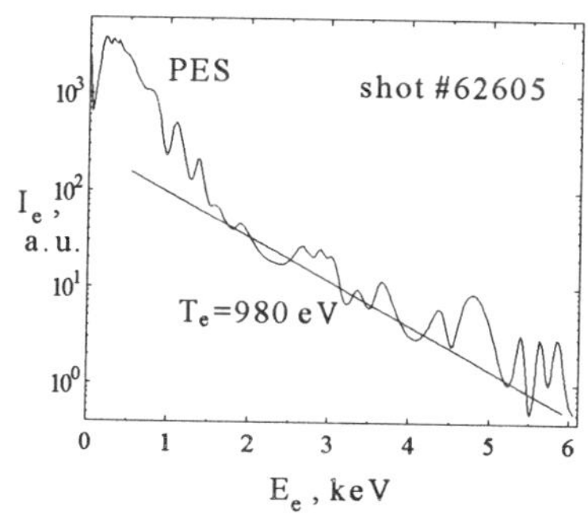

Fig. 4. Photoelectron spectrum on tokamak T-10

The recording of the time traces for the spectral signals may have of special interest, since the correlation of these traces for example with electron cyclotron emission signals (ECE) reveals the valuable information about interaction processes between the plasma electrons and impurities. The correlation between the PES signal ($E_\gamma \approx 1$ *keV*) and the ECE signal recorded just after the disruption in T-10 is shown in Fig. 5. From Fig. 5 one can seen that the PES signals are shifted in time compared with the ECE signals and the delay time is about 1.9 *ms*. This shift may be depend on some peculiarities in process of impurity ionisation. It is clear that the high temporal resolution of the PES spectral signals and their scanning in a wide energy range give the new approach to study the nonstationary processes, particularly, for the edge plasma study with proper analysis [4].

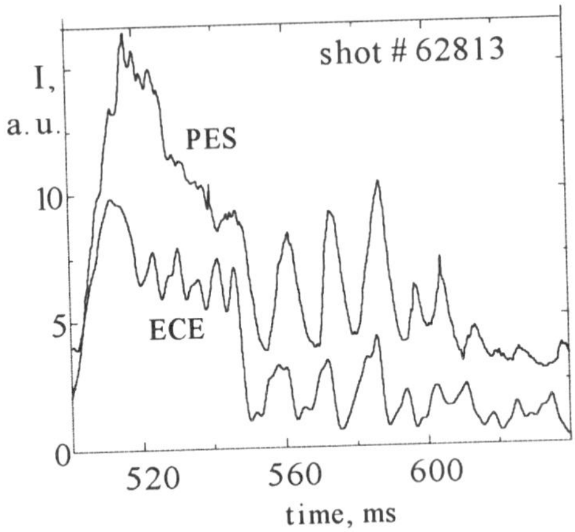

Fig. 5. Correlation between PES and ECE signals just after disruption

VACUUM-PHOTODIODE DETECTOR.

For the measurements of full X-ray flux from plasma (without energy resolution) and for plasma imaging it is convenient to use vacuum-photodiode detectors (VPD) [1], [5]. The design of such detector proposed in [5] one can see in Fig.6. The radiation flux is incident on the conical cathode, the photoelectrons from which are directed toward the conical or cylindrically-symmetric anode. Between the cathode and anode various potentials up to 1-2 *kV* are applied. In order to distinguish flux of neutrals from the X-ray flux or to obtain a spectral resolution $\Delta E_\gamma / E_\gamma \sim 1$, filters must be placed in front of the entrance to the device.

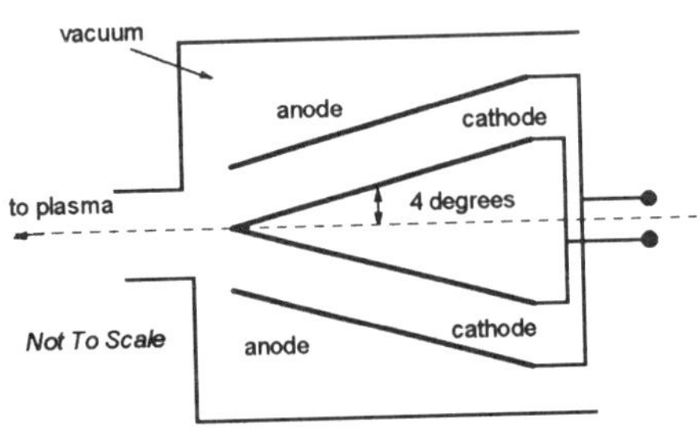

Fig. 6. Schematic of vacuum-photodiode detector

A VPD has a short response time - it can operate in the picosecond range, and the linear dependence between the radiation dose and output signal is maintained as the dose varies over 15 orders of magnitude.

In additional to ultraviolet and X-ray radiation from plasma a VPD can also register charge-exchange neutral fluxes if it is used without filters. The contribution of neutrals to the total VPD signal depends on the specific experimental conditions.

ITER APPLICATION

It is well known that the semiconductor detectors are completely inappropriate for ITER, both because of sensitivity to and damage by neutron and gamma irradiation.

The proposed PES techniques for investigation of plasma X-ray radiation can be successfully utilised in the ITER - experiments with deuterium-tritium plasma. Indeed, the dimensions of the instrument are sufficiently small that considerably alleviates their shielding against hard X-ray and neutron radiation. Detectors in these instruments (microchannel plate multipliers) have a relatively weak sensitivity to such radiation and they do not face the plasma directly that significantly reduce the background signal.

For ITER parameters it will be possible to obtain the signal from PES in analogue mode of operation with amplitudes 0.01- 1 *V* in the range of X-ray energies 0.1-200 *keV*.

The time resolution may be not worse than 1-10 *ms*, and the spatial resolution may be not worse than 10 *cm*.

The fact that the PES detector may be shifted from the diagnostic penetration axe on any desirable distance gives us the possibility of effective shielding of instrument against hard X-ray and neutron radiation.

Estimation of PES background signal due to neutron radiation near the diagnostic penetration with diameter of 3 *cm* shows that the value of background signal after proper shielding may be 10^2 - 10^3 *pulse per second*. In the same time the useful signal must be 10^6 - 10^8 *p/s*.

So we believe that the PES diagnostic will meet no difficulties from the hard radiation.

The more serious problem may be the influence of tritium on the detector. Now we have no enough information to discuss this problem.

Vacuum-photodiode detectors will be placed near ITER walls and so its shielding from the effect of γ and neutron radiation will be impossible. Let us to estimate the effect of hard radiation on VPD signal.

The bremsstrahlung flux on the chamber wall is

$$W = 1.7 \cdot 10^{-21} \cdot \frac{a \cdot n_e^2}{\sqrt{T_e}} \quad photon/m^2 \cdot s \qquad (8)$$

For ITER we have that $a = 2.8\ m$, $T_e = 20\ keV$, $n_e = 5 \cdot 10^{19}\ m^{-3}$, and $W \approx 3 \cdot 10^{18}$ $photon/m^2 \cdot s$. The mean energy of the gamma-rays is in the range of 0.5-1 *MeV* and gamma flux near the first wall is $2 \cdot 10^{18}\ \gamma / m^2 \cdot s$ [6]. The full flux of photoelectrons is proportional to the value $\mu R \sim E_e^{-1.3}$, where μ is the full linear photoabsorption coefficient. So one can see that the background signal from γ flux is about two orders of magnitude less than signal from X-ray flux.

To our great regret we did not find any data concerning the secondary electron emission coefficient under effect of neutrons but our estimation show that signal from neutron flux must be one-two orders of magnitude less than signal from γ flux.

CONCLUSION

PES and VPD are extremely useful and versatile due to their simplicity, compactness and high sensitivity.

The PES give us possibility to obtain a great amount of information in a wide range of plasma electron temperatures with high time and space resolution.

The VPD is very convenient for plasma imaging, for MHD oscillation investigations and for charge-exchange flux measurements.

The above-done consideration allows one to state that the proposed techniques for X-ray spectrum analysis and plasma imaging can be successfully utilised in the ITER - experiments with deuterium-tritium plasma.

REFERENCES

1. Gott Yu., V., Shurygin V.A., Plasma Phys.Rept., 1993, **19**, 559
2. Shurygin V.A., Plasma Phys.Rept., 1996, **22**, 975
3. Rantsev-Kartinov V.A., Shurygin V.A., Plasma Phys.Rept., 1997, **23**, 1
4. Gott Yu., V., Shurygin V.A., 22nd EPS Conf. on Contr. Fus. and Plasma Phys., Bournemouth, GB, 1995, **19C**, pt.IV, 221
5. Séguin F.H., Petrasso R.D., Li Ch..K., Radiation-Hardened X-Ray Imaging for Burning-Plasma Tokamaks, Accepted for publication in Reviews of Scientific Instruments
6. ITER Diagnostics, ITER Documentation series, #33, IAEA, Vienna, 1990

DIVERTOR IMPURITY MONITOR FOR ITER

Tatsuo Sugie, Hiroaki Ogawa, Jun Katsunuma[1], Mitsumasa Maruo[1], Katsuyuki Ebisawa[2], Toshiro Ando[3], Yoshio Kita[4], and Satoshi Kasai

Naka Fusion Research Establishment, Japan Atomic Energy Research Institute, 801-1 Mukoyama, Naka-machi, Naka-gun, Ibaraki-ken, Japan
[1]Nikon Corporation, Shinagawa-ku, Tokyo, Japan
[2]ITER - EDA Joint Central Team, San Diego, CA, USA
[3]ITER - EDA Joint Central Team, Garching, Germany
[4]Toshiba Corporation, Fuchu-shi, Tokyo, Japan

INTRODUCTION

The design of the divertor impurity monitoring system for ITER is now well advanced. The main functions of this system are to identify impurity species and to measure the two-dimensional distributions of the particle influxes in the divertor plasmas. The expected impurities are carbon, tungsten, beryllium and copper originating from the divertor target plate and from the surface of the first wall in the main chamber. Neon and other impurity gases injected into the plasma for radiation cooling in the divertor will also be observed. The wavelength range is 200 nm to 1000 nm. This system, which is one of the most important diagnostic systems for plasma control, is included in the start-up set of ITER diagnostics[1].

The temperature of the divertor plasma is lower than that of the main plasma. Many spectral lines originating from neutral and ionized atoms and molecules are emitted in the ultraviolet and visible region as well as in the vacuum ultraviolet region. These spectral lines have information of plasma-wall interaction. In existing tokamaks, visible spectroscopy is used extensively to study divertor plasmas and also edge plasmas, because the apparatus of visible spectroscopy is relatively simple compared to that of the vacuum ultraviolet spectroscopy which needs a vacuum extension and a pumping device. For example, impurity species identification, particle influx measurements and studies of the impurity generation and particle recycling mechanism[2,3,4,5,6,7] have been carried out in various tokamaks. Electron temperature and density measurements[8] and ion temperature measurements have been also attempted. These techniques will be able to extrapolate to ITER divertor diagnostics except in the very high electron density and low temperature region where the recombination process is dominant.

In this paper, the conceptual design and the detailed optical design of the divertor impurity monitoring system are described. In addition, the measurable limit, the neutron and

Diagnostics for Experimental Thermonuclear Fusion Reactors 2
Edited by Stott *et al.*, Plenum Press, New York, 1998

γ-ray irradiation effect on windows, a calibration method and an alignment method are considered.

REQUIREMENT

The identification and monitoring of the impurity species, and the two-dimensional measurement of particle influxes in the divertor plasma are very important for ITER plasma control. As shown in the section 5.5.E.04 of the ITER design description document[9], the objective of the divertor impurity monitoring system is to obtain spatially resolved measurements from the plasma in the divertor channel to identify and quantify the impurity species. The wavelength range of 200-1000 nm and two-dimensional measurement in the poloidal plane are required.

The more detailed requirements for parameters to be measured, parameter ranges, spatial resolutions, time resolutions and accuracies are summarized in Table 1.

Table 1. Target requirements for impurity monitoring system.

• Impurity Species Monitor

Parameter	Parameter range	Spatial res.	Time res.	Accuracy
Be influx	TBD	Several point	10 ms	10 % (rel.)
C influx	TBD	TBD	10 ms	10 % (rel.)
Cu influx	TBD	Several point	10 ms	10 % (rel.)
Ne influx	TBD	TBD	10 ms	10 % (rel.)

• Key Divertor Parameter

Parameter	Parameter range	Spatial res.	Time res.	Accuracy
"Ionization front" position	0 - 2 m	~ 10 cm	1 ms	10 %

• Impurity and D, T Influx in Divertor

Parameter	Parameter range	Spatial res.	Time res.	Accuracy
Γ_{Be}, Γ_C, Γ_W	10^{17} - 10^{22} at/sec	10 mm	1 ms	30 %
Γ_D, Γ_T	10^{19} - 10^{25} at/sec	10 mm	1 ms	30 %

• Divertor Helium Density

Parameter	Parameter range	Spatial res.	Time res.	Accuracy
n_{He}	10^{17} - 10^{20} m^{-3}	-	1 ms	20 %

• n_T/n_D, n_H/n_D in Divertor

Parameter	Parameter range	Spatial res.	Time res.	Accuracy
n_T/n_D	0.1 - 10	-	100 ms	20 %
n_H/n_D	0.01 - 0.1	-	100 ms	20 %

Table 1 (continued). Target requirements for impurity monitoring system

• Ion Temperature in Divertor

Parameter	Parameter range	Spatial res.	Time res.	Accuracy
Ti	1-200 eV	10 cm (along legs) 0.3 cm (across legs)	1 ms	20 %

CONCEPTUAL COMPOSITION AND ARRANGEMENT

In order to realize the required measurements, the divertor impurity monitoring system has three different types of spectrometers; one for each function.

The first are visible survey spectrometers for impurity species monitoring and particle influx measurements with a time resolution of 10 ms. These spectrometers have more than 12 lines of sight in the divertor legs. The spectral lines emitted from 200 nm to 1000 nm will be measured simultaneously.

The second are filter spectrometers for two-dimensional measurements of particle influxes with the spatial resolution of 10 - 15 mm and the time resolution of 1 ms. These spectrometers have almost 500 lines of sight and will be able to measure over 12 spectral lines for every line of sight simultaneously. The ionization front and helium density will also be measured by these spectrometers. The measurement of electron density and the electron temperature will be attempted.

The third are high dispersion spectrometers for measuring the ion temperature, the particle energy distribution and the ratios of tritium density to deuterium density (n_T/n_D) and hydrogen density to deuterium density (n_H/n_D) in the divertor plasma. The ion temperature will be derived from the Doppler broadening of impurity lines. The ratios of n_T/n_D and n_H/n_D will be estimated from the intensity ratios of tritium Tα to deuterium Dα and hydrogen Hα to deuterium Dα respectively.

The functions and outline specifications of each spectrometer are summarized in Table 2.

Table 2. Functions and outline specifications of proposed spectrometers.

• Visible survey spectrometer	
Parameters to be measured:	Impurity species, Impurity and D/T influx
Wavelength range:	200 nm - 1000 nm (simultaneously)
Wavelength resolution:	~ 0.1 nm
Time resolution:	10 ms
Spatial resolution:	~ 12 sight lines for divertor legs
• Filter spectrometer	
Parameters to be measured:	2-dimensional impurity and D/T influx, Ionization front, Helium density, (n_e, T_e)
Wavelength range:	200 nm - 1000 nm (~ 12 lines)
Wavelength resolution:	~ 1 nm
Time resolution:	1 ms
Spatial resolution:	~ 10 mm
• High dispersion spectrometer	
Parameters to be measured:	n_T/n_D and n_H/n_D ratio, Ion temperature
Wavelength range:	200 nm - 1000 nm (simultaneously)
Wavelength resolution:	< 0.01 nm
Time resolution:	10 ms
Spatial resolution:	~ 12 sight lines for divertor legs

First mirror

Because of intense nuclear radiation the first mirror should be metallic[10]. The reflectivity of molybdenum (Mo), tungsten (W), copper (Cu) and aluminum (Al) are shown in Figure 1 as a function of the wavelength[11]. In the region of 200 - 500 nm, the reflectivity of Mo is better than that of Cu and W. In the region of 500 - 1000 nm, Cu is better than Mo and W. The reflectivity of Al is high from 200 nm to 1000 nm. The sputtering yield of Mo with deuterium is almost 1/40 of that of Cu and Al at the deuterium energy of 200 eV[12]. Therefore, we selected molybdenum as the material of the first mirror in the divertor cassette. On the other hand, it is better to use aluminum for the mirror located behind the biological shield.

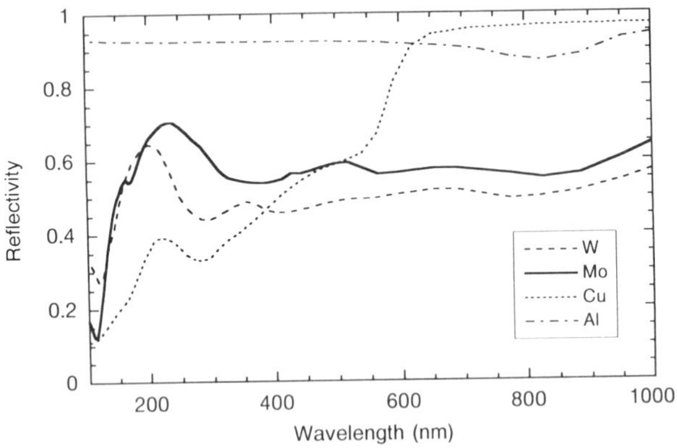

Figure 1. Normal incidence reflectivity as a function of wavelength for tungsten (W), molybdenum (Mo), copper (Cu) and aluminum (Al)[11].

Transmission loss of optical fiber

The transmission loss of the optical fiber is large at short wavelengths even if it is optimized for the ultraviolet region. For example, the transmission loss is 1 dB/m at 200 nm, 0.02 dB/m at 450 nm and 0.005 dB/m at 800 nm. It is difficult, therefore, to measure spectral lines below 450 nm through long optical fibers. Therefore, spectrometers for the wavelength range of 200-450 nm should be located just behind the biological shield.

Conceptual arrangement

From the considerations above, the conceptual arrangement shown in Figure 2 has been chosen. The light from the divertor region passes through the quartz windows on the divertor port plug and the cryostat, and goes through the dog-leg optics in the biological shield. The light is then focused on the ends of the fiber bundle by collecting and focusing optics. The fiber bundle guides the light to the spectrometers. The spectrometers with the wavelength region below 450 nm are installed just behind the biological shield to minimize the transmission loss in fiber. On the other hand, the light with $\lambda > 450$ nm is guided by long optical fibers to the spectrometers which are located remotely in the diagnostic room in order to have good accessibility. We will be able to change the spectrometers easily corresponding to the change of the experiment.

The spectrometers for 200 - 450 nm and the local controller are installed on a movable trolley so that they can be removed in a short time before the divertor maintenance.

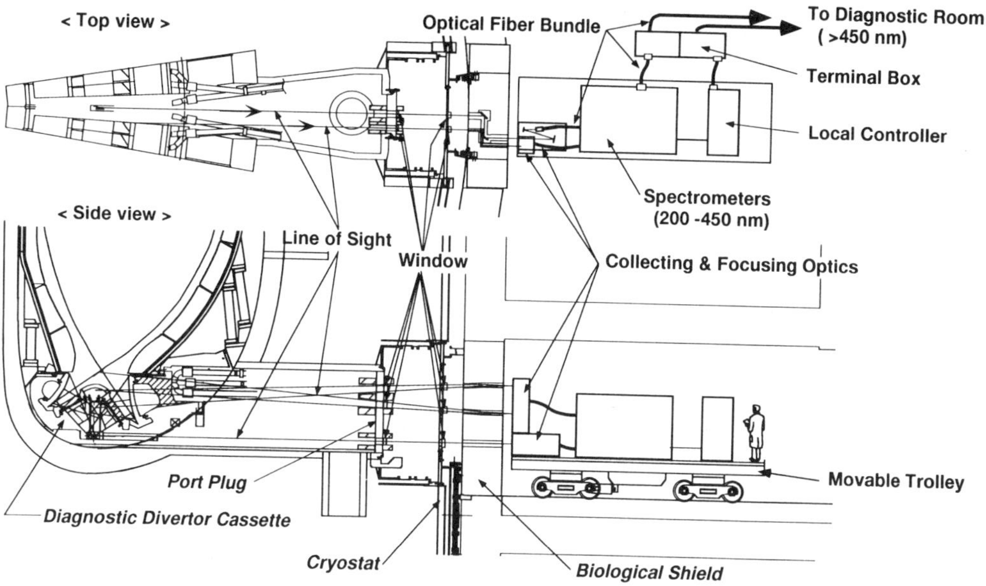

Figure 2. Conceptual arrangement of impurity monitoring system.

LINE OF SIGHT AND RAY TRACE ANALYSIS

Line of sight

The two-dimensional measurement in the poloidal plane is performed with two viewing fans, which intersect, namely OV and OH for the outer divertor region, and IV and IH for the inner region. These viewing fans are realized by metallic mirrors (made of molybdenum) located in the divertor cassette as shown in Figure 3. The number of lines of sight for each viewing fan is given in Table 3. The 10 mm spatial resolution will be realized by filter spectrometers. The region from approximately up the divertor leg to the x-point will be observed with the additional viewing fans named XL and XU through the gap between the divertor cassettes.

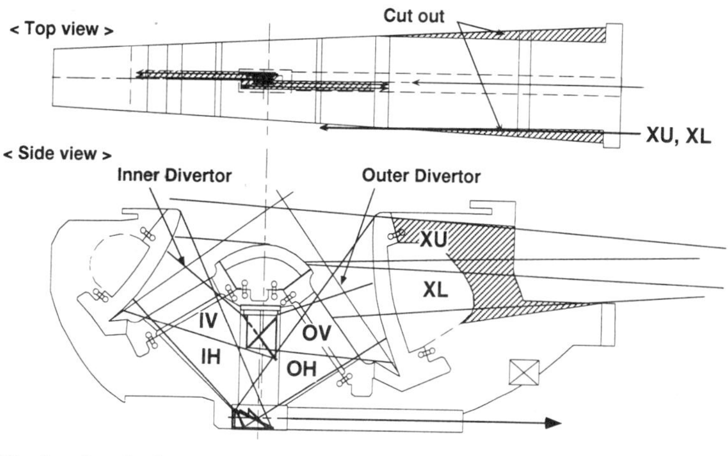

Figure 3. Viewing fans in the divertor cassette. The additional viewing fans XL and XU observe the region near the top of the divertor leg to the x-point through the gap between the divertor cassette[13].

Table 3. Number of lines of sight for spectrometers in each viewing fan.

Viewing fan	Visible survey spectrometer	Filter spectrometer	High dispersion spectrometer
OV	4	100	4
OH	-	100	-
IV	4	100	4
IH	-	100	-
XL	2	50	2
XU	2	50	2

Ray trace analysis

The optimum optical arrangement is determined for each viewing fan by a ray trace analysis. For each viewing fan, rays are emitted from the several points of the fiber bundle and go to the divertor plate through the collecting and focusing optics and the penetration optics. The results confirm that 10-15 mm resolution will be satisfied for all regions of the divertor plate. A typical result for the viewing fan IV, which observes the inner divertor from the dome, is shown in Figure 4. The traced rays, the spot diagrams on the divertor plate and the contours of irradiance for the images of the fiber cores are shown in this figure. Here, it is assumed that the core emits the light uniformly.

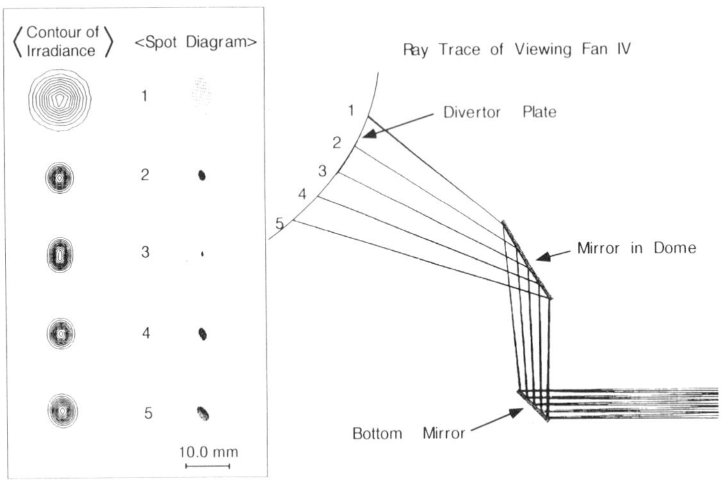

Figure 4. The traced rays, the spot diagram and the contours of irradiance for the viewing fan IV. Rays are emitted from the five points on the end of the fiber bundle and go to the divertor plate through the penetration optics.

SPECTROMETERS

Spectrometers have been designed in accordance with the required functions as shown in Table 2.

Visible survey spectrometer

The spectral lines in the wavelength range of 200 - 1000 nm are observed simultaneously by sixteen grating spectrometers. Each spatial line of sight has sixteen fibers and each fiber guides the light to the spectrometer. The light from over twelve spatial lines of sight are observed by each spectrometer simultaneously. Here, it is assumed that spectral lines are detected by an ICCD detector with 1024 x 512 pixels (imaging area: 2.5 cm x 1.25 cm). If we use a larger detector, the number of spectrometers will be decreased.

Filter spectrometer

There are two sets of filter spectrometers. One is the set for the viewing fans OV, OH, IV and IH. Another is the set for the viewing fans XL and XU. Each set has four spectrometers and each spectrometer can observe three different spectral lines simultaneously. The spectrometer for the wavelength region of 200 - 450 nm is set on the movable trolley just behind the biological shield. The other spectrometers for the wavelength range 450 - 1000 nm are installed in the diagnostic room. The number of lines of sight and fibers are shown in Table 3. The set for the viewing fans of OV, OH, IV and IH has 400 lines of sight and each line of sight has four fibers in order to guide the light to each spectrometer.

The optical design has been carried out for twelve selected lines: 1) Hα+Dα+Tα, 2) HeI 667.8 nm, 3) HeI 706.5 and 706.6 nm, 4) HeI 728.1 nm, 5) HeII 468.6 nm, 6) BeII 313.0 and 313.1 nm, 7) BeIII 372.0, 372.1 and 372.3 nm, 8) BeIV 465.9 nm, 9) CII 657.8 nm, 10) CV 227.1 nm, 11) CuI 521.8 and 522.0 nm, and 12) NeI 640.2 and 640.1 nm. As an example, the schematic view of the filter spectrometer for 200 - 450 nm is shown in Figure 5. The light emitted from the optical fiber bundle, which is composed of 400 fibers, passes through a collimator. After that, the light of a required wavelength region is reflected by a dichroic mirror. It passes through a band-pass filter (full width of half maximum: ~1 nm) corresponding to the selected spectral line and is focused on a 2-dimensional detector by a camera. The light penetrating the dichroic mirror goes to the next dichroic mirror. The spot diagrams on the detector at 300 nm are also shown in Figure 5. The spot size even at the position of 3 mm away from the axis is about 50 μm. This size is small compared with the diameter of the fiber core of 200 μm so that each image of the 400 fiber cores will be resolved on the detector.

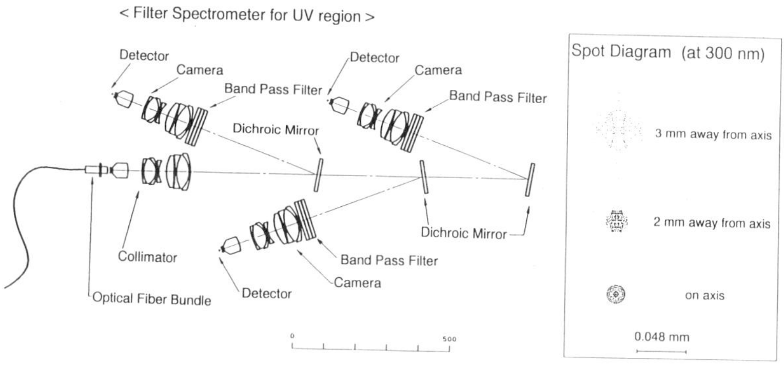

Figure 5. Schematic view of filter spectrometer for UV region and the spot diagrams at 300 nm.

High dispersion spectrometer

The high dispersion spectrometer is a Littrow type spectrometer with an echelle grating.

EVALUATION OF OPTICS

Number of photons coming into the detector

The numbers of photons N_d (photons·s^{-1}) incident on the detector is

$$N_d = I_{div} \cdot T \cdot \Omega_f \cdot S_f , \tag{1}$$

where I_{div} is the intensity of the spectral line emitted in the divertor, T is the transmissivity of the optics from the divertor to the detector, Ω_f is the effective solid angle on to the fiber and S_f is the area of the fiber core.

Ω_f , S_f and T for the filter spectrometers with the viewing fans of OV, OH, IV and IH are 3.8 x 10^{-3} sr, 1.26 x10^{-7} m^2 and 0.06 - 0.15 respectively. On the other hand, Ω_f and T for the filter spectrometers with the viewing fans of XL and XU are 5.1 x 10^{-2} sr and 0.12 - 0.24 while S_f remains 1.26 x10^{-7} m^2.

Intensity of spectral lines in the divertor

The particle influx Γ is given by

$$\Gamma = 4\pi \cdot K \cdot I, \tag{2}$$

where K is the number of ionization events per photon for the observed line with the intensity of I.

The required range to be measured for the deuterium total influx is 10^{19} - 10^{25} at·s^{-1} as shown in Table 1. Since the power deposition area on the divertor targets is about 10 m^2, the measurable range of the deuterium influx density Γ to the divertor is 10^{18} - 10^{24} at·m^{-2}.s^{-1}. In a similar way, the influx density to be measured for carbon and beryllium are 10^{16} - 10^{21} at·m^{-2}.s^{-1}.

We estimated the intensities of Dα, CII(3s-3p, 657.8 nm) and BeII(2s-2p, 313.06 nm) lines for each required influx density region using equation (2) under the conditions of electron density of 1x10^{19} - 1x10^{22} m^{-3} and electron temperature of 1 - 50 eV. These values are expected for ITER divertor plasmas. The values of K are calculated in references[14,15,16] for wide ranges of electron density and temperature.

The calculated results are shown in Figure 6.

Measurable limit

If we assume the lower limit of the number of photons is 1000 photon·ms^{-1} at the detector, the measurable intensity limit in the divertor derived from equation (1) is about 1.4x10^{16} - 3x10^{16} photon·m^{-2}·sr^{-1}·s^{-1} for the optics of viewing fans of OV, OH, IV and IH. From Figure 6, it is expected that the measurable limits for influxes will be different depending on the electron density and temperature. For example, it is difficult to measure the carbon influx density of 1x10^{16} m^{-2}·s^{-1} for T_e=5eV. Also, it is difficult to distinguish between line emission and bremsstrahlung emission by filter spectrometers. The calculated line intensities of the bremsstrahlung emission (Z_{eff}=1.5, T_e= 5-50 eV) are also shown in Figure 6. Here, we have assumed that the integrated length is 5 cm and the wavelength band width is 1 nm which is the same as that of the band pass filter of the spectrometer. In the high density region, it is necessary to confirm the bremsstrahlung component by survey spectrometers.

In the low temperature and high density region, more detailed consideration of atomic and molecule process is necessary.

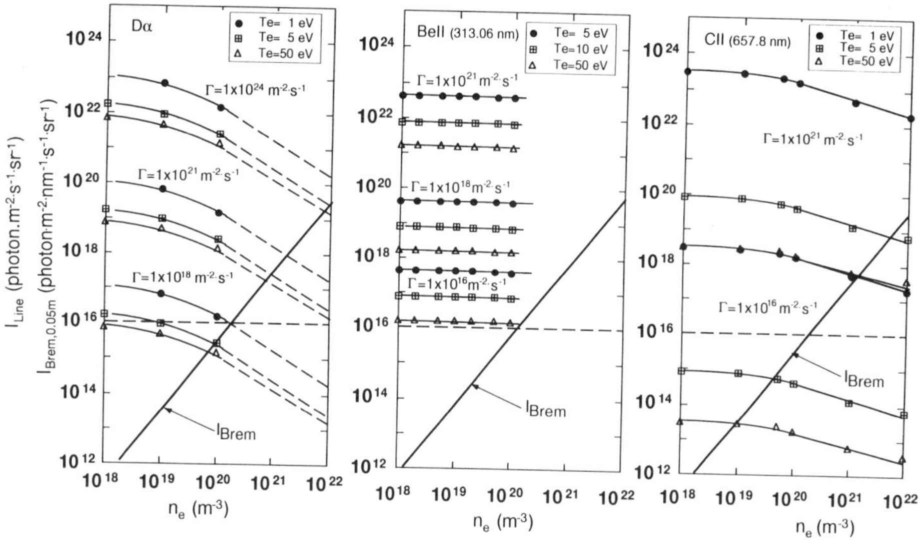

Figure 6. Intensities of Dα, BeII (313.06 nm) and CII (657.8 nm) lines as a function of electron density for various influx densities and electron temperatures, and bremsstrahlung emission I_{Brem}.

ALIGNMENT

The alignment of the optic axis will be carried out with a laser, a diffuser, a plane mirror and a corner cube reflector (CCR) mounted between the line of sight in the divertor cassette. The laser is shone through the optical axis of the system and the position of light spots created by the diffuser, the plane mirror and the CCR are monitored with a CCD camera.

CALIBRATION

Before the installation of the diagnostic divertor cassette and the penetration optics, the sensitivity calibration will be carried out by arranging the real components in the same arrangement as the real optics with standard light sources set in the divertor. After the installation, the standard light will be set behind the biological shield. The light will be applied to the molybdenum CCRs through the same optics as used for the measurements. The CCR will be installed in the pocket which is free from neutral particle bombardment and from dust deposition. The spectral intensity of the reflected light from the CCR will be monitored by a spectrometer in order to monitor the degradation of the penetration optics. Only a few areas of the optics will be calibrated by this method. The details of the alignment and calibration are still under development.

NEUTRON AND γ-RAY IRRADIATION EFFECT ON WINDOW

The neutron flux and γ-ray dose rate at the port plug are calculated to be 2×10^{13} $m^{-2} \cdot s^{-1}$ and 200 Gy/h[17] respectively. The windows located at the port plug are shielded by water cooled stainless steel tubes as shown in Figure 2. The neutron flux and γ-ray dose rate at the windows are expected to be 6×10^{11} $m^{-2} \cdot s^{-1}$ and 3 Gy/h. The degradation of trasmissivity (λ>350 nm) of fused silica KU is less than 5 % at the neutron fluence of 10^{21} m^{-2} and the γ-ray dose of 10^7 Gy[10,18]. At 200 nm, the degradation is 17 % at the γ-ray dose of 10^5 Gy.

From these consideration, the windows will survive during ITER operation (50,000 shots x 1000 sec) for the wavelength region of λ>350 nm.

For the wavelength region of 200 - 350 nm, the windows will survive against the γ-ray irradiation. For the neutron irradiation, more careful consideration is necessary

ACKNOWLEDGMENT

This paper has been prepared as an account of work assigned to the Japanese Home Team under Task Agreement numbers S 91 TD21 95-01-20 FJ and S 91 TD31 95-08-04 FJ within the Agreement among the European Atomic Energy Community, the Government of Japan, the Government of the Russian Federation, and the Government of the United States of America on Cooperation in the Engineering Design Activities for the International Thermonuclear Experimental Reactor ("ITER EDA Agreement") under the auspices of the International Atomic Energy Agency (IAEA).

The authors are grateful to Drs. A. Costley, S. Yamamoto, L. de Kock, C. Walker, V. Mukhovatov, H. Iida, H. Nakamura and the members of ITER EDA Joint Central Team for their fruitful discussions and cooperations. We also greatly appreciate fruitful comments by Dr. H. Kubo. We would like to express our gratitude to Drs. T. Tsunematsu and S. Matsuda for their continuous encouragement of this work.

REFERENCES

1. A. E. Costley, K. Ebisawa, P. Edmond, et al., Overview of the ITER diagnostic system, These proceedings (1997)
2. K. H. Behringer, Spectroscopic studies of plasma-wall interaction and impurity behavior in tokamaks, *J. Nucl. Mater,* 145-147: 145 (1987).
3. H. Kubo, M. Shimada, T. Sugie, et al., Impurity generation mechanism and remote radiative cooling in JT-60U divertor discharges, *J. Nucl. Mater,* 196-198: 71 (1992).
4. P. Bogen, D. Rusbüldt, Velocity distribution of carbon and oxygen atoms in front of a tokamak limiter, *J. Nucl. Mater,* 196-198: 179 (1992).
5. D. Reiter, P. Bogen, U. Samm, Measurement and monte carlo computations of Hα profiles in front of a TEXTOR limiter, *J. Nucl. Mater,* 196-198: 1059 (1992).
6. B. Unterberg, H. Knauf, P. Bogen, et al., *J. Nucl. Mater,* 220-222: 462 (1992).
7. H. Kubo, T. Sugie, H. Takenaga, et al., High resolution visible spectrometer for divertor study in JT-60U, *Fusion Eng. Design,* 34-35: 277 (1997).
8. B. Schweer, G. Mank, A. Pospieszczyk, Electron temperature and electron density profiles measured with a thermal He-beam in the plasma boundary of TEXTOR, *J. Nucl. Mater,* 196-198: 174 (1992).
9. A. E. Costley, et al., ITER Diagnostic System, in: *ITER Design Description Document,* (1996).
10. D. V. Orlinski, Radiation hardening of diagnostic components, in: *Diagnostics for Experimental Thermonuclear Fusion Reactors,* P. E. Stoott, G. Gorini and E. Sindoni, ed., Plenum Press, New York (1996).
11. J. H. Weaver and H. P. R. Frederikse, Optical properties of metals and semiconductors, in: *CRC Handbook of Chemistry and Physics,* D. R. Lide, ed., CRC Press, Boca Raton (1994).
12. Y. Yamamura, H. Tawara, Energy dependence of ion-induced sputtering yields from monatomic solids at normal incidence, *Report NIFS-DATA*-23 (1995).
13. K. Ebisawa, T. Ando, A. Costley, et al., VUV divertor impurity monitor for ITER, These proceedings (1997)
14. L. C. Johnson and E. Hinnov, Ionization, recombination, and population of excited levels in hydrogen plasma, *J. Quant. Spectrosc. Radiat. Transfer,* 13: 333 (1973).
15. K. Behringer, H. P. Summers, B. Denne, et al., Spectroscopic determination of impurity influx from localized surfaces, *Plasma Phys. Cont. Fusion,* 14: 2059 (1989).
16. H. P. Summers, W. J. Dickson, A. Boileau, et al., Spectral emission from beryllium in plasma, *Plasma Phys. Contr. Fusion,* 34: 325 (1992).
17. R. T. Santro, H. Iida, V. Khripunov, Neutron and gamma-ray flux and dose rate levels in the cryostat and gallery cells. Impact on electronic instrumentation, *ITER Internal Report NA:NAG*-31-16-05-97 (1997).
18. S. Yamamoto, The present status and future plan of task T246, *in Technical Meeting on Irradiation Tests on Diagnostic Components based on the T246 task agreement,* Garching (1997).

VUV DIVERTOR IMPURITY MONITOR FOR ITER

K. Ebisawa[1], T. Ando[2], A. Costley[1], G. Janeschitz[2], E. Martin[2] and T. Sugie[3]

[1]ITER Joint Central Team, San Diego JWS, 11025 N. Torrey Pines Road, La Jolla CA 92037, USA
[2]ITER Joint Central Team, Garching JWS, Germany
[3]Japan Atomic Energy Research Institute, Naka, Ibaraki-ken, Japan

ABSTRACT

The function of measurements of VUV radiation emanating from the plasma in the divertor region are briefly summarized. The engineering constraints that would have to be satisfied in the implementation of VUV measuring systems for this region are identified. Two proposed configurations for the installation of a VUV spectrometer are considered and the further work that is necessary to develop these design options is identified.

INTRODUCTION

The ITER Divertor is designed to cope with a radiated power loading of order 100 MW, and to control the location of the radiation region by confining the recycling of neutral impurities such as Ne and Ar, and DT fuel to the divertor channels. The electron temperature in the divertor plasma ranges from 1-20 eV near the divertor plate to 200-500 eV near the separatrix at the mid plane, hence Neon and higher Z impurities can radiate 50-70 MW from the divertor and the scrape-off layer (SOL) plasma for a separatrix density of 10^{20} m^{-3}. H, He, Be and C radiation and charge exchange losses supplement total power losses.

Measurements of spatially resolved radiation emissivities of the impurity radiation by spectroscopic means are very important to ensure the proper operation of the divertor[1,2,3]. Since the vacuum ultraviolet (VUV) region is expected to contain useful emission lines arising from recombination, charge exchange collisions between light and medium Z atoms as well as Lyα emissions, the possibility of installing a VUV spectrometer at the divertor level is being considered.

Diagnostics for Experimental Thermonuclear Fusion Reactors 2
Edited by Stott *et al.*, Plenum Press, New York, 1998

FUNCTION REQUIREMENT FOR IMPURITY MONITOR

It is necessary for the plasma control and divertor optimization to measure line emission over the wide range of wavelength (10 nm>λ>1000 nm) from the plasma near the divertor plate and/or the X-point. Analyzing the measured line spectrum gives information on impurity species and transitions. Impurity fluxes, Γ, can be obtained from the line intensities together with the excitation and ionization rates. The functions requirement for VUV and visible spectrometers and parameters to be measures are listed in Table 1. Here measurements are divided into three categories for the priority purpose[4,5]; basic machine protection and plasma control (1a), advanced plasma control (1b), evaluation and physics studies (2).

Table 1: Target Measurement Requirements for Divertor Impurity Monitors

Measurement (Category)	Parameters of interest	Parameter range	Spatial resolution	Time resolution
Impurity species monitoring (1a)	Be	TBD	Several points	10 ms
	C	TBD	Several points	10 ms
	Cu	TBD	Several points	10 ms
	W	TBD	Several points	10 ms
	Extrinsic gas (Ne, Ar)	TBD	Several points	10 ms
Impurity and D,T influx (1b)	Γ_{Be}, Γ_C, Γ_W	10^{17} - 10^{22} at/sec	3 mm	1 ms
	Γ_D, Γ_T	10^{19} - 10^{25} at/sec	3 mm	1 ms
Key Divertor parameter (1a)	Ionization front	0 - 2 m	10 cm	1 ms
Electron density and Temperature (1b)	n_e	10^{19} - 10^{22} /m^3	10 cm along legs 0.3 cm across legs	1 ms
	T_e	1 - 200 eV	10 cm along legs 0.3 cm across legs	1 ms
Ion temperature (2)	T_i	1 - 200 eV	10 cm along legs 0.3 cm across legs	1 ms

IMPLEMENTATION CONSTRAINTS

There are four Divertor remote handling ports and 16 Cryo-pump ports at the divertor level where potentially VUV monitors could be installed. Installation in the pit area outside a remote handling port has the disadvantage that it would delay the changing of the divertor cassettes since it would have to be removed before this activity could commence. For the initial work, therefore, we have examined the possibility of installing the spectrometers at a pumping port. Direct view of VUV spectrum into the ITER divertor channel is very restricted at the divertor level. A narrow gap of order of 10 mm between divertor cassettes could possibly be used as an opening for a line of sight. This opening would permit a viewing cord from X-point to the upper half of the divertor leg. Due to the internal components such as a cryogenic pump and divertor cooling pipes, only a small solid angle would be possible.

Two possibilities are being considered for the installation of the spectrometers and detectors ; namely installation in the pit area outside the Bio-shield, or installation in the pumping port inside the cryostat. Several key engineering factors have to be considered in each case including Tritium barrier boundary, neutron and gamma-ray flux, and movement of torus components. Many of the factors are common to other diagnostic systems but for spectroscopic systems which employ photomultiplier devices the magnetic filed is also an important consideration. The factors and the corresponding design requirements are shown in Table 2. Additional factors not shown in Table 2 are operation temperature, cooling, vacuum pressure, out gassing and maintenance for which requirements will be addressed more specifically at a later stage.

Table 2: Engineering Aspects and Design Requirements for Installation of the VUV Spectrometers and Detectors

Engineering Aspect	Conditions	Design Requirement
Outside the Bio-shield		
Tritium barrier	In an atmosphere, surface contamination is restricted & controlled.	Minimum extension of double wall of vacuum boundary.
Neutron flux	2×10^2 /cm^2 sec (E_n>0.1 MeV)	(Fluence< 10^{14} /cm^2)
Dose rate	37 Gy/h at 1 m from the cooling pipes.	Total Dose < 10^4 Gy
Magnetic field	0.2 T at R=21 m, z=-5 m	B< 100 Gauss
Port/Cryostat, cassette movement due to Baking, Vertical Displacement Event.	Max. ΔR=50 mm, ΔTroidal= 25 mm, Δz= 45 mm for port movement	Design must account for movement
Interference with other components	Remote handling machine at the head of divertor cooling pipes.	Removed evry 1000 pulses.
Inside the cryostat		
Tritium barrier	In the secondary vacuum	Extension of torus vacuum boundary
Neutron flux	1×10^8 /cm^2 sec (E_n>0.1 MeV)	labyrinth and shield
Dose rate	37 Gy/h at 1 m from the cooling pipes.	Total Dose < 10^4 Gy
Magnetic field	0.6 T at R=16.5 m, z=-5 m	Error field at plasma< 10^{-5}
Cassette movement due to Baking, VDE	TBD	Optical path adjustment
Loads on the port extension	7.5 ton	Support against disruption

DESIGN FEATURES

Installation Outside the Bio-shield: Vacuum Extension

For installation outside the bioshield, a line of sight to the divertor channel is only possible in between the divertor cooling pipes because a large removable closure plate on the cryostat occupies a wide area in the port entrance (Figure 1). A primary vacuum duct directly coupled with the torus vacuum is extended into the pumping pit for about 3 m outside the Bio-shield. At the end of the vacuum extension, a detector box is fixed on a support flange from the pit floor with a small auxiliary cryopump. The vacuum duct, detector box, auxiliary cryopump and valves for isolation are 5 bar rated and covered by the secondary vacuum boundary which prevents tritium from leaking at any accidental failure.

Bellows mounted both on the primary and the secondary boundary absorb relative displacements due to the port and cryostat movements. Since it is very difficult to mount any focusing or reflecting mirror inside the flexible double wall duct, a straight viewing line from the detector to divertor channel without a mirror is employed.

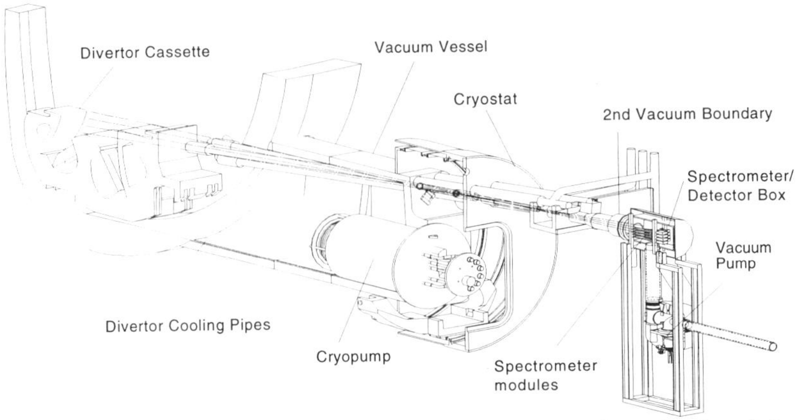

Fig. 1. Isometric view of the VUV spectrometer looking at divertor leg region through the gap. For clarity, Bio-shield and divertor cassette of the other side are not shown. The auxiliary vacuum pump is just for an example.

Even though the duct size is kept to a minimum diameter of 165 mm inside the Bio-shield, an additional shielding around the extension duct is necessary to reduce neutron streaming to an acceptable level. Gamma-ray flux from the divertor cooling pipes which run parallel with the vacuum extension is so strong that any reflecting mirrors in front of the detector with a grazing angle would not decrease the streaming effectively. Hence the gamma-ray shielding is achieved by shielding blocks inside the detector box with a normal incident mirror.

Alignment for this configuration is difficult. Differential movement of the long, narrow slot between the divertor cassettes (1.4 m x 10 mm) and the apertures in the cryostat could restrict the direct line of sight The position of the detector box would have to be automatically adjusted to compensate for this movement. Another disadvantage of this configuration is that the detector box will interfere with a special machine which is mounted on the cooling pipe header to cut pipes when divertor cassettes are removed for repair.

The spectromter/detector could be similar to the Off Roland, normal incidence spectrometer employed in the VUV spectrometer of JT-60U[6]. Figure 2 shows the general layout of a defining slit, a deviating mirror, a grating mirror and a detector mounted on a solid mono-block base firmly as a spectrometer module. The detector consists of an open micro channel plate (MCP) image intensifier, a phosphor plate, fiber optics and a photo diode array. The concave grating is made by a holographically recorded grooves covering 60 - 160 nm spectral range over the 60 mm focal plane with resolving power $\lambda/\Delta\lambda$ about 300. Both deviating and grating mirrors are coated with $Al+MgF_2$ for the normal incident angle of 7° - 14°.

The dimensions of the module are width 30 cm, length 98 cm length and height 5 cm. It is also possible to cover the 160 - 200 nm spectral range by changing the relative position of the deviating mirror and employing a larger incident angle on the grating mirror. Similar spectrometers are used elsewhere[7], and the technology of these components, particularly the semi-conductor devises, are making rapid progress. A wider spectral range with a compact detector size and high resolution is anticipated.

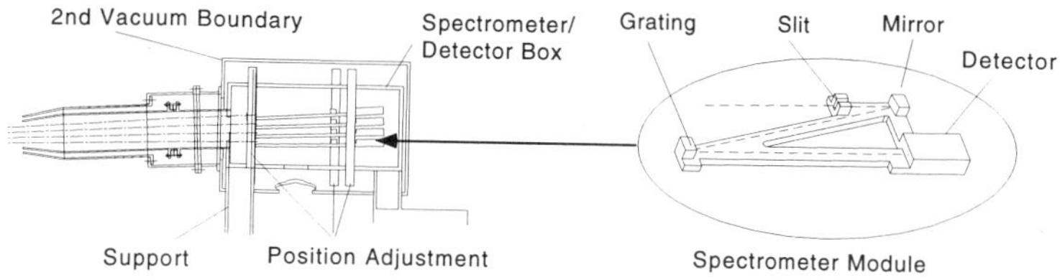

Fig. 2. Four spectrometer modules are stacked inside the doubly confined box. Open space in the box is filled with Lead blocks for gamma-ray shielding. Feed through for cabling, cooling are not shown.

With this flexible module, many important spectral lines of D I Lyα, C II-IV, O III-VI could be measured and used in studies of the radiation power loss. Impurity lines of Ne I-II, Ar I-II are expected but more precise analysis is needed to predict their intensities. Intensity calibration using the branching ratio method is possible by comparison of a line pair, for example, C III 78.44 nm in VUV and 298.21 nm in the visible. Typically high count rates are expected: for example, a count rate of $2x10^4$ cts/sec is anticipated for C IV 154.8 nm line at $n_e = 1x10^{20}$ m^{-3}. This should be an order of magnitude higher than the counts originating from the background gamma rays. Studies of the performance of the spectrometer - signal/noise, spatial resolution etc - are in progress and will be reported seperately.

Inside the spectrometer box, four modules are supported by the position adjustment mechanism so that four spatial chords will cover the area from the X-point to the upper half of the divertor channel with an average separation 20 cm (Fig. 3). Since a grating with a long focal length is employed, there is enough room to install gamma-ray shield around the detector and viewing line inside the box. Lead shielding blocks with 10 cm thickness reduce the gamma-ray flux to the level of 10^7 γ/cm^2sec so that the gamma induced background noise will be much less than the expected signal levels.

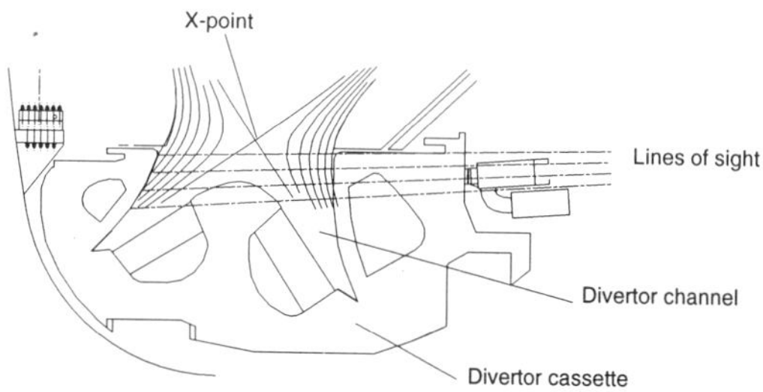

Fig. 3. Four lines of sight cover the area from the X-point to the upper half of the divertor channel. The seperation of the lines can be adjusted.

Spectrometer/Detector mounted inside the cryostat

For the mounting inside the cryostat, the relative displacement between the divertor cassette and the detector is expected to be relatively small and the requirement for optical alignment is less demanding. Figure 4 shows the isometric view of a proposed installation inside the cryostat. Viewing lines are deviated by a common mirror about 105° at the end of vacuum port into a labyrinth which reduces the direct neutron flux. At the end of the viewing duct, spectrometer modules similar to those described above are installed for measuring in part of the wavelength range 60 - 200 nm.

An important practical aspect for this configuration is the magnetic field at the proposed detector location (about 0.6T). Substantial shielding would be necessary to reduce it to below 0.01T. Without the mirror used to deviate the radiation, the spectrometer/detector module can be reduced in size to about width 20 cm, and length 40 cm and four modules can be mounted in a cavity of 32 cm diameter. A prolate shape iron shield of 1.06 m diameter by 1.5 m length and 37 cm thickness reduces the magnetic field to below 100 Gauss at the detector as required. A simple dipole model approximation indicates that the relative error field due to the magnetic shield at the plasma center may reach 3×10^{-5}, three times larger than permissible. A more careful multipole analysis is needed to determine more accurately the magnitude of the associated error field.

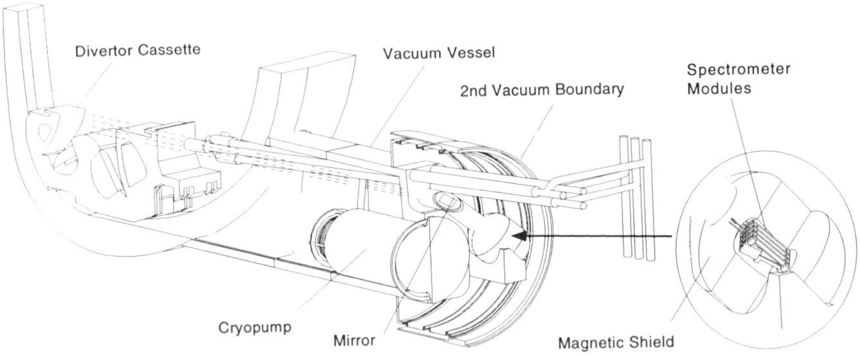

Fig. 4. Isometric view of the VUV spectrometer located inside the cryostat. The mirror is revealed just for explanation. The auxiliary vacuum pump is to be connected to the cavity in the magnetic shield.

The dominant source of radiation would still be the water in the cooling pipes. The iron shell reduces the radiation flux by two order of magnitudes to an acceptable level of 10^8 γ/cm^2sec. The modules inside the magnetic shield are confined in the primary vacuum boundary. A small cryopump, not seen in the figure, exhausts gas into high vacuum side of the pumping port.

The maintenance of the spectrometer is also a critical issue for this configuration. If a substantial part of the magnetic shield has to be removed along with a malfunctioned spectrometer module, the entire cyopump assembly will need to be taken out beforehand. The disconnection of the foreline and removal of the cryogenic valve box will have to be carried out carefully to prevent tritium contamination in the pit area, and such a procedure would be possible a few times in the operational life of ITER.

The magnetic shield will have a mass of about 7.5 ton. Analysis of the dynamic and electromagnetic force caused by a sudden motion or plasma disruption is necessary and assessment of dynamic stress on the supporting structure will be important if this arrangement is pursued.

SUMMARY AND FUTURE WORKS

Two configurations for installing a VUV spectrometer at the divertor level are under consideration: in one case the spectrometer/detector is mounted outside the cryostat in the pit while in the other it is mounted inside the croystat. The key engineering aspects have been identified in each case and solutions to some of the key interface problems have been found. Both cases have the merit of a direct view through a narrow gap between the divertor cassettes which removes the need for a first mirror. The reflectivity of such a mirror would most probably be seriously and rapidly degraded by material deposition and would have to be frequently replaced. Employing a compact spectrometer/detector with a concave grating in normal incident it is possible to monitor the 60 - 200 nm spectral range. The detector must be shielded against the strong gamma flux from the water in the nearby cooling pipes.

For the case of installation outside the Bio-shield, the method of coping with the differential motion of the port and cryostat remains to be devised. Interference with the remote handling machine for the divertor cooling pipes is also a key issue. For the case of installation inside cryostat, the error field due to the large amount of magnetic shielding a concern but the maintenance scheme is probably the most critical issue in this configuration.

Further feasibility studies are necessary, including detailed optical design for the long, narrow line of sight with the compact spectrometer/detector modules, and evaluation of the error field due to the magnetic shield for the detector. Remote maintenance, local vacuum pumping, detector cooling and methods of alignment remain to be developed. The alternative location for the spectrometer, at a remote handling port, should also be considered.

ACKNOWLEDGMENT

This report has been prepared as an account of work performed under the Agreement among the European Atomic Energy Community, the Government of Japan, the Government of the Russian Federation, and the Government of the United States of America on Cooperation in the Engineering Design Activities for the International Thermonuclear Experimental Reactor ("ITER EDA Agreement") under the auspices of the International Atomic Energy Agency (IAEA).

REFERENCES

1. N.J. Peacock et al., Spectroscopy for Impurity Control in ITER, in: *Diagnostics for Experimental Thermonuclear Fusion Reactors*, P. E. Stott et al., eds., Plenum Press, New York (1996).
2. R.C. Isler, Spectroscopy of Divertor Plasma, in: *Diagnostics for Experimental Thermonuclear Fusion Reactors*, P. E. Stott et al., eds., Plenum Press, New York (1996).
3. A.T. Ramsey, Tokamak Spectroscopy in the UV - Is This the Golden Age?, in: *UV and X-ray Spectroscopy of Astrophysical and Laboratory Plasma,* Cambridge Univ. Press (1993)
4. V.S. Mukhovatov et al., Role and Requirements for Plasma Measurements on ITER, *These Proceedings* (1997)
5. A.E. Costley et al., Overview of the Diagnostic System, *These Proceedings* (1997)
6. H. Kubo et al., Study of Impurity and Radiative Loss in Dovertor Plasma with Absolute Calibrated VUV Spectrometers in JT-60U, *Nuclear Fusion,* **33** (1993) 1427
7. R.J. Fonck, A.T. Ramsey and R.V. Yelle, Multichannel grazing-incidence spectrometer for plasma impurity diagnosis: SPRED, *Applied Optics*, **21** (1982) 2115

LUMINESCENT EFFICIENCY OF THIN PHOSPHOR SCREENS

A. Baciero, K. J. McCarthy, L. R. Barquero, A. Ibarra and B. Zurro

Asociación EURATOM/CIEMAT para Fusión. E-28040 Madrid

INTRODUCTION

In a previous work, we presented a feasibility study for a broadband X-ray/VUV radiation detector, based on the luminescence of a phosphor, for ITER-like plasmas[1]. Its immunity to electromagnetic interference, its radiation hardness plus its capability of operation when access to a machine is limited, make it attractive as a global radiation detector for harsh plasma environments, such as ITER.

In this work we extend the work to estimate the efficiency of phosphor screens as a function of screen thickness and incident photon energy. Two models are compared using efficiencies measured with an X-ray tube source to provide fitting parameters for extending the models over a broader range of photon energies. In this way screen thicknesses can be tailored to energy ranges of interest whilst reducing background noise signal.

DESCRIPTION OF MODELS

Two models are considered here. In the first model the phosphor screen is assumed to be a uniform infinite slab having a mass thickness W (mg/cm^2). When VUV or X-ray radiation is incident on the screen (see Fig. 1), the ratio between the forward luminescent light output and the incident radiation, η_u (efficiency in transmission mode) is given by[2,3]

$$\eta_u(E,W) = \eta_c t\mu \frac{(1+\rho)}{2(\mu^2-\sigma^2)} \frac{2(\sigma+\mu\beta)-(\mu+\sigma)(1+\beta)e^{-(\mu-\sigma)W}+(\mu-\sigma)(1-\beta)e^{-(\mu+\sigma)W}}{(1+\beta)(\rho+\beta)e^{\sigma W}-(1-\beta)(\rho-\beta)e^{-\sigma W}} \quad (1),$$

where σ and β are the phosphor self-absorption and light scattering coefficients for the luminescent light, μ is the energy dependent mass attenuation coefficient[4], E is the incident photon energy, η_c is the light conversion efficiency, and t and r are the plate transmissivity and reflectivity. The term ρ is defined as $\rho = (1-r)/(1+r)$.

In the second model the phosphor screen is considered to be thin and microscopically non-uniform, as for the case when a screen is prepared by sedimentation. With this method phosphor grains are deposited randomly over the plate surface. For modelling purposes it is

Diagnostics for Experimental Thermonuclear Fusion Reactors 2
Edited by Stott *et al.*, Plenum Press, New York, 1998

assumed that the grains are identical and are laid down randomly over the surface so that grain stacks of different heights are built up on the plate surface (see Fig. 2). It is also assumed that the light emitted from individual grain stacks does not interact with other grain

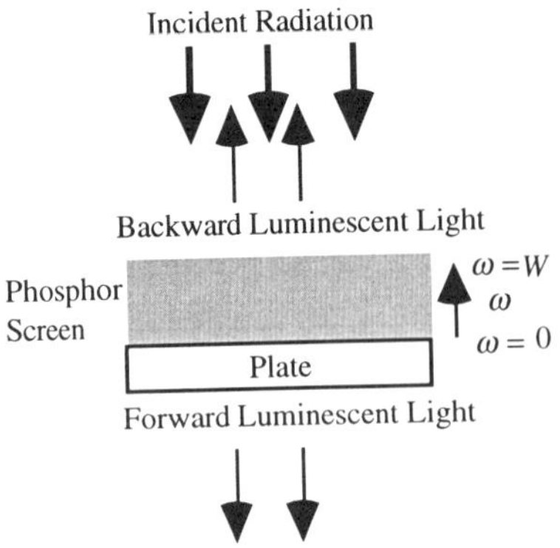

Fig. 1. Excitation of a phosphor screen with radiation

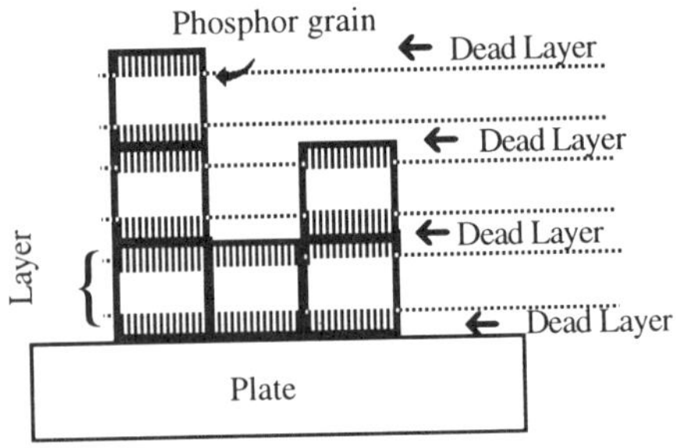

Fig 2. Dead layers in the phosphor grains. The grains are assumed to be rectangular.

stacks (i.e. scattering of light is not considered). The screen efficiency in transmission mode at energy E is estimated by summing the efficiencies over all possible grain stack heights weighted by each stack height fractional probability. This can be estimated by[5]

$$\eta_g(E,\omega) = \sum_{k=1}^{N} \eta_u(E,k\omega_{layer})(N_k)\alpha^k(1-\alpha)^{N-k} \qquad (2),$$

where $\eta_u(E,k\omega_{layer})$ are the transmission mode efficiencies of a layer $k\omega_{layer}$, (k being the layer thickness in grains and ω_{layer} the mass thickness of a single layer). Here, N is the total number of particles on the screen surface, α is the ratio of a single grain effective area to the total phosphor surface area under consideration, and (N_k) are the number of combinations of k grains taken from N grains.

In the granular model the presence of a dead layer on the phosphor grains is also considered, i.e., every grain has a superficial sublayer (see Fig. 2) that does not emit luminescent light. The thickness of this layer is an inherent phosphor parameter and it has been estimated[6] to be of the order ~ 0.1 μm.

When the incident radiation is not monochromatic, i.e., a continuous spectrum such as the output from an X-ray tube, then the screen efficiency is given by

$$\eta_{spectrum}(E_0,\omega) = \int_0^{E_0} \eta(E,\omega) f(E)dE \,/ \int_0^{E_0} f(E)dE \qquad (3),$$

where $\eta(E,w)$ is the screen efficiency at energy E. For the case of an X-ray tube with a thin aluminium filter between the source and screen, the flux incident on the screen $f(E)$ at energy E is given by[7]

$$f(E) = (1 - E/E_0)e^{-\mu_{Al}\omega_{Al}} \qquad (4),$$

where μ_{Al} is the mass attenuation coefficient for aluminium, ω_{Al} is the filter mass thickness, and E_0 is the X-ray lamp peak energy.

EXPERIMENTAL PROCEDURE AND RESULTS

Measurements to check and compare the models have been made using a tube source of X-rays and the phosphor $Y_3Al_5O_{12}$:Ce (P-46). This phosphor has been selected for use in the TJ-II stellarator[8] and has an emission peak at 530 nm (yellow-green), a density of 4.15 g/cm^3, and a mean grain size of 6.64 μm. Screens of mass thicknesses between 1 and 22 mg/cm^2 were deposited on quartz plates having $t = 0.94$ and $\rho = 0.89$. The samples were excited using lamp peak voltages of 30 kV and 52 kV. A 1 mm thick aluminium filter was used to shape the X-ray spectrum, see Fig. 3. The output from the phosphor screen was collected using an optical fibre and was analysed using a monochromator and a cooled photomultiplier tube to measure the luminescent light spectrum between 400 and 800 nm.

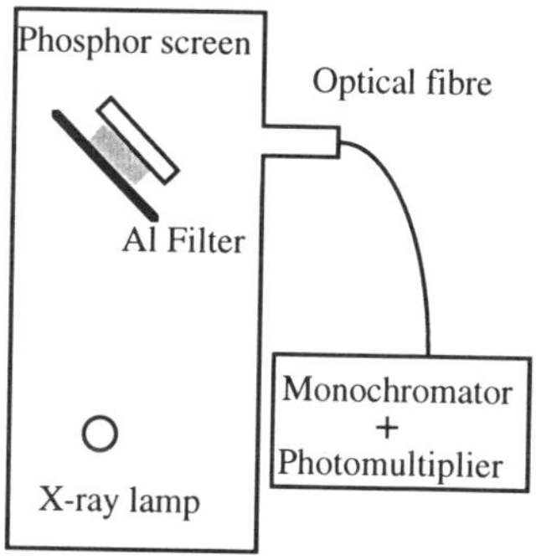

Fig. 3. Experimental set-up to measure the phosphor screen efficiency.

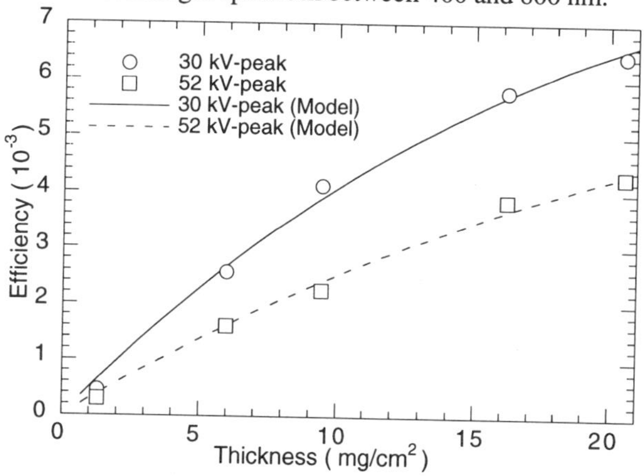

Fig. 4. Results from X-ray lamp for several screen thicknesses of the phosphor $Y_3Al_5O_{12}$:Ce. Experimental and theoretical results are indicated.

In Fig. 4, we show the experimental results for light measured at 530 nm together with curves obtained using the granular model. For these curves, a continuous filtered spectrum was used and curve was fitted to the data points setting the parameters β =0.063 and σ = 12.6 cm^2/g. The granular dead layer was taken to be 0.07 μm. The efficiencies were normalised to[9] η_c=0.033, to compare the results for different lamp energies. A factor $\sqrt{2}$ was included to account for the beam angle of incidence to the screen, i.e., 45°. Similar curves (not shown) were generated using the uniform model, albeit with different values for the parameters β and σ. It might be expected that the form of the curves from both models should be similar for the range of screen mass thicknesses used, as X-rays in this energy range should undergo little absorption when passing through the phosphor layers. In this narrow energy range the phosphor X-ray mass attenuation coefficients for the phosphor components are low, so the granular and dead layer effects are not as critical as in regions with higher coefficients where all the radiation is absorbed in the first few phosphor layers.

Although the range of experimental energies studied is quite narrow the measurements and curves provide the fitting parameters needed to extend the models to lower X-ray and VUV energies, where the diagnostic would be operated. For this we have chosen the granular model as it provides a more realistic description of the actual structure of the thin phosphor screens needed for this application. The results of these computations are shown in Fig. 5 for a wide range of screen mass thicknesses and a broad range of energies. At the lowest energies, < 10 keV, efficiencies of the order 10^{-2} are achieved with screen mass thicknesses of the order 20 mg/cm^2. Furthermore the efficiency to hard X-rays, i.e., > 25 keV, is < 2×10^{-3} and decreases rapidly with increasing energy. Hence, the background signal from high energy photons will be considerably diminished.

CONCLUSIONS

Good agreement has been found between different models to compare the response of a phosphor screen over the X-ray range 30 to 50 keV. These measurements have allowed curves to be generated for a wide range of energies and to allow screen mass thickness to be tailored to the energies of interest and to provide an indication of the background signal. The models can be applied to other phosphors once their fitting parameters are determined.

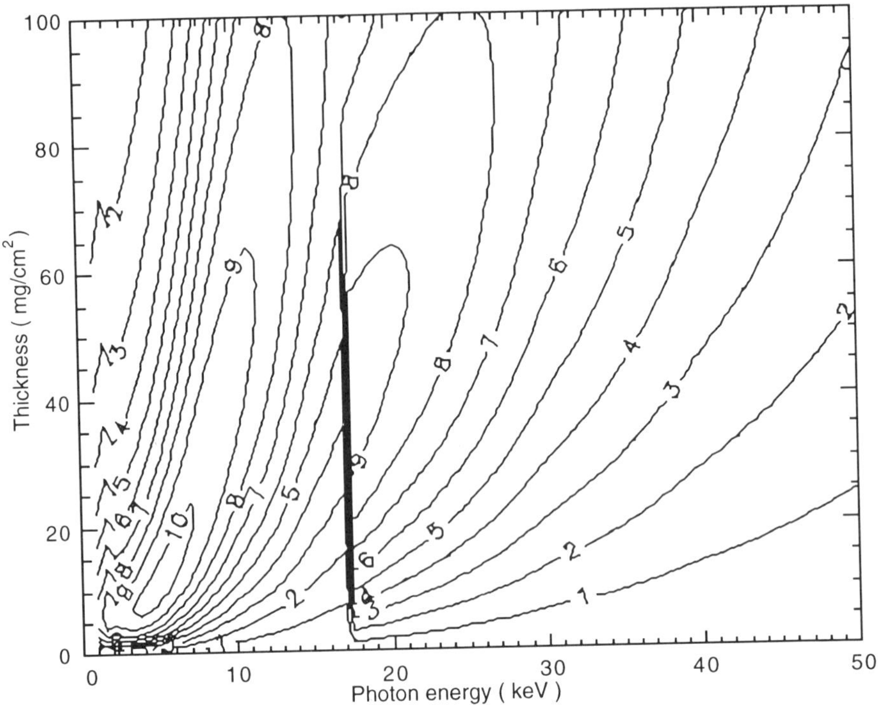

Fig. 5. Contour plot of $Y_3Al_5O_{12}$:Ce phosphor screen efficiencies (10^{-3}) modelled using the granular model for monoenergetic excitations. The large jump at 17 keV is due to yttrium.

ACKNOWLEDGEMENTS

This work was partially funded by the Spanish Ministry of Science and Education under contract DGCYT No. PB 94-1229.

A. B. is supported by a Scholarship from the Institute of Energy Studies.

REFERENCES

1. B. Zurro, C. Burgos, A. Ibarra and K.J. McCarthy, *Fusion Engineering and Design* , 34-35:353 (1997)
2. H. C. Hamaker, *Philips Res. Rep.*, 2:55 (1947)
3. G.W. Ludwig, *J. Electrochem. Soc.*, 118:1152 (1971)
4. J.H. Hubbell and S.M. Seltzer, *NISTIR 5632 -Web Version 1.02*, (1996)
5. G.E. Giakumakis, M.C. Katsarioti and I.E. Lagaris, *J. Appl. Phys.*, 69:6607 (1991)
6. S. Koboniwa, T. Hoshina and M. Kanamaru, *J. Electrochem. Soc.*, 120:1734 (1973)
7. E. Storm, *Phys. Rev. A*, 5:2328 (1972)
8. B. Zurro, C. Burgos, K.J. McCarthy and L.R. Barquero, *Rev. Sci. Instrum.*, 68:680 (1997)
9. G. Blasse, B.C. Grabmaier, Chapter 9, in: *Luminiscent Materials* , Springer-Verlag, Berlin Heidelberg (1994)

THERMONUCLEAR PLASMA X-RAY DIAGNOSTICS WITH COMPTON SCATTERING

D.V.Portnov, Yu.A.Kaschuck, A.V.Krasilnikov

Troitsk Institute of Innovation and Thermonuclear Researches,
Troitsk, Moscow area, Russia.

INTRODUCTION.

Our objective is to provide tools for X-ray spectral, time and spatial resolving measurements on ITER. Spectral range should be 2-5 T_e for good sensitivity to electron temperature changes. That means a spectral range up to 150 keV. The level of accompanying neutron radiation will be so high, that we need use some deflectors to separate neutron and X-ray fluxes. In the case, the crystalline reflectors cannot provide the wide range[1,2]. Kumakhov X-Ray optics[3] does not allow to deflect a radiation with an acceptable angle.

We suggest the other approach: to use a difference in scattering a X-Ray radiation and neutron on the light elements. Let's assume, that we have a deflector consisting of hydrogen. Neutrons cannot deviate on an angle more 90° by elastic scattering on protons. The X-Ray will be deflected according to Compton effect. In this ideal case the ratio of X-Ray and neutron powers will increase infinitely. However, hydrogen deflector is hard to use.

PRELIMINARY ESTIMATIONS.

Accepting an average differential Compton cross section of scattering equal 0.05 barn/steradian (varies ±30 %), we shall receive the following evaluations for above «amplification factor» (deflection angle 154°) for different light elements:

Table of 1 Interaction of neutrons with the light elements.

Element	Z	X-total	dXn-n1	Egamma	dX-ellastic	K
Li6	3	1.449	0.0001	1.5	0.002	75
Li7	3	1.442	0.0005	0.447	0.004	38
Be9	4	1.528	0.0070		0.039	5
C	6	1.27	0.0180	4.439	0.014	21

- Z - Number of electrons in atom,
- X-total - total cross section interaction with 14MeV neutrons, barn;
- dX-n-n1 - n-n'γ (the first level excitation cross section, is divided on 4π to compare with differential cross sections of X-ray and elastic neutrons scattering; we believe this channel of inelastic interaction is the main source of γ-noise from a reflector, for preliminary estimations the inelastic scattering of neutrons can be neglected, barn/ster;
- E_{gamma} - energy of a gammas produced at inelastic scattering, MeV;
- dX-ellastic - differential cross section of neutron elastic scattering, barn/ster;
- K - Arbitrary amplification factor

Diagnostics for Experimental Thermonuclear Fusion Reactors 2
Edited by Stott *et al.*, Plenum Press, New York, 1998

Li shows the greatest difference in Compton and neutron scattering. Besides, it is necessary to take into account a photo absorption in a reflector at measurements in low energy range (<10keV), e.g., it is necessary to select a material with small Z. And, at last, Li has the minimal inelastic cross section. Thus, it is necessary to apply for deflector substance containing a lithium, for example, LiH. Li6 isotope is preferable.

BACK SCATTERING METHOD APPLICATION FOR ITER.

In ITER the configuration with Compton deflector is much easier than configuration with application of X-Ray optics. Probably, it would be possible to use neutron collimator channels for these diagnostics. To do that, it is necessary to use Be entering window instead of steel and to locate a LiH deflector in the channel. The deflector is almost transparent for neutrons, so it'll not influence on neutron diagnostics. The detector of a deflected radiation can be arranged in additional collimator directed in the angle range 140-160° in respect to initial radiation direction.

Table 2 Parameters for estimations of potencies of streams on the detector

Parameter	Value	Units.
Volume of a reflector	400	cm^3
Square of a reflector	10	cm^2
Distance from plasma up to a reflector	15	m
Square of the detector	1	cm^2
Distance a reflector the detector	50	cm
Te, in a max.	25	keV
Ne, concentration	10^{14}	cm^{-3}

The application of a reflector gives improving a signal/noise ratio in dozens' time.

Table 3 Powers of streams

Position	First wall, W/cm^2*ster	On deflector, W/cm^2	On detecto W/cm^2
SXR	6.928	3.0E-05	1.1
neutron	35.7	1.0E-03	9.4
S/N Ratio	0.19	0.03	

The Detector Choice.

Tab.4-5 were calculated for Be entering window thickness 5mm, detector-2mm.

Table 4 Performances scintillating of detectors

Material of the detector	Z	Density, g/cm^3	Scintill. Effect. to NaI (Tl), %	Energy resolution for Eγ 662keV, %	Effectiveness of the detector Eγ 150 keV, %
NaI (Tl)	11,53	3.67	100	~ 6	30.8
CsI (Na)	53,55	4.51	85	~ 6	40.7
CsI (Tl)	53,55	4.51	45	~ 6	40.7
BaF_2	9,56	4.88	-	~ 8	39.2
$Bi_4Ge_3O_{12}$	32,8,83	7.13	10	~ 10	74.5
$CdWO_4$	8,48,74	7.90	20		67.7
$ZnWO_4$	8,30,74	7.33	28		65.6
$CaWO_4$	8,20,74	6.06	80		61.6

To improve a signal/noise ratio it is necessary to use scintillators with large Z and small sensitivity to neutrons. Such crystal can be $Bi_4Ge_3O1_2$ (BGO) - energy resolution <10% (for crystal 25 mm and thickness of 2.5 mm). A rather small highlighting time and absence of long-term components allow to receive a good time resolution.

Table 5 Performances of semiconductor detectors

Material of the detector	Z	Density, г/см3	Energy. A solution for Eγ 5.9 keV, keV	Energy. A solution for Eγ59.6 keV, keV	Energy. A solution for Eγ 122 keV, keV	Effectiveness of the detector for Eγ 150 keV, %
Ge	32	5.32				19.5
CdTe	48,52	6.06	1.1	3. 0	7.5	43.4
HgI_2	80,53	6.40	0.2	1.2	2.0	65.7

The detectors from HgI_2 are characterized by high radiation resistance. The energy resolution of detectors practically does not vary in an interval fluence 10^9 - $4 \bullet 10^{15}$ neutron/cm^2 (En = 8MeV).

Table 6 Interaction of neutrons with an energy 14 MeV with heavy elements

Material of the detector (presence, %)	S_{total} (barn)	$S_{inelastic}$ (barn)	$S_{ch.particles}$ (barn)	S_{SXR} 150keV (barn)	K
Na-23 (100)	1.72	0.75	0.199	5.16	26
Cs-133 (100)	4.86	0.16	0.013	163.727	12594
F-19 (100)	1.76	0.25	0.525	3.98	8
Ba-138 (71.7)	5.0	0.47	0.032	176.433	5514
Bi-209 (100)	5.35	0.39	0.005	722	144400
Hg	n.a.	n.a.	n.a.	590.293	.
Ge-74 (36.5)	3.70	1.22	n.a.	29.756	
O-16 (99.76)	1.61	0.43	0.220	3.5	16
W-nat	5.47	0.49	0.004	483.153	120788
I-127 (100)	4.79	0.55	0.019	147.159	7745
Cd-nat	4.62	0.37	0.021	103.688	4938
Te-130 (33.8)	4.67	1.50	0.002	134.567	67284

•S_{total} - total cross section of neutron interaction with substance;
•$S_{inelastic}$ - inelastic scattering cross section;
•$S_{ch.particles}$ - cross section of charged particles emission (n, p), (n, d), (n, a) etc.;
•S_{SXR} 150keV - cross section of interaction with a 150keV X-ray radiation
•K - Arbitrary sensitivity to SXR and to neutrons

Optimal choices are:

•From scintillators - BGO,
•From semiconductor detectors - HgI_2.

Spectrum Restoring Method Errors.

To evaluate errors the calculations for the targets of various shapes were carried out by a Monte-Carlo method. The events of a reflection (including multiple) and absorption in a reflector were played. The quants deflected under defined angles, were considered registered. The initial spectrum was selected from single line, Maxwell and uniform distribution.

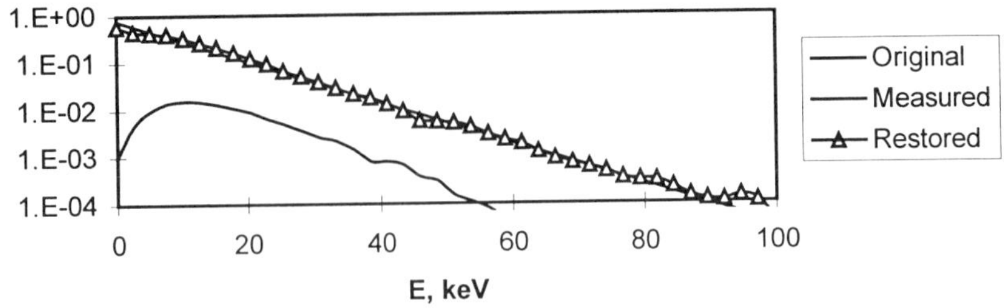

Figure 1. Restoring method evaluation.

The increasing of the slope of a «registered» spectrum with growth of an energy is due to dependency of Compton scattering from an energy and energy loss. The fall down at small energies is due to absorption in the deflector. At high energies the procedure of restoring does not meet problems. At small energies, the procedure of restoring requires to take into account an absorption in deflector. It is required to determine some function from an energy which is falling down from infinity for E = 0 up to 1 for large E. The registered spectrum should be multiplied on this function

Functions defined for one spectrum, appeared to be suitable for restoring all remaining. That is fitting parameters are completely determined only by geometry, and do not depend upon initial spectrum. The restoring does not increase the relative error.

Evaluation Of A Space And Time Resolution.

The spatial resolution cannot be better, than sizes of a reflector and collimator entering window, that is about several cm. From a point of view of the amplitude analysis the application Compton deflector gives a count rate exceeding possibilities of a modern electronics (more 10^6 s^{-1}). Probably, in a few years the progress of electronics will change the situation, but now, in case of the amplitude analysis, the time resolution is limited only by electronics and is of the 10ms order. For faster (<1ms) flux measurements the SXR-power is too small.

CONCLUSIONS.

Suggested X-ray diagnostics with Compton Scattering can be successfully used for electron temperature profile measurements in ITER burning plasma experiments with spatial and time resolutions of about 10 cm and 10 ms correspondingly. An application of Li deflector and HgI_2 semiconductor or BGO scintillator detectors in suggested diagnostic scheme will provide the increase (with respect to detector placement in strait collimator channel) of signal to noise ratio by ~ 40 times in ITER conditions.

REFERENCES

1 K.W.Hill, M.Bitter et al., *Proceeding of the International Workshop on Diagnostic for ITER*, edited by P.Stott, G.Gorini, E.Sindoni, p.341-351, (Varenna, Italy, 1995).

2 R.Barnsley, R.M.Giannella et al., *Ibid*, p.341- 351 (1995).

3 A.G.Alekseev, V.NAmosov et al., *Ibid*, p.365-368 (1995).

NPA at ITER: physical basis and conceptual design

A.I.Kislyakov, V.I.Afanassiev, A.V.Khudoleev, S.S.Kozlovskij*, M.P.Petrov

A.F.Ioffe physical-technical institute of Russian Academy of Sciences, St.-Petersburg, 194021 Russia
**State Technical University of St.-Petersburg, St.-Petersburg, 195251 Russia*

1. Introduction

To control the isotope density ratio of the hydrogen ions in the plasma is an important issue for optimization of burning plasma parameters in ITER. The measurement of the ratio of deuterium/tritium fluxes emitted by the plasma in the energy range of few maximal ion temperatures has been proposed to determine n_d/n_t ratio in the ITER plasma [1]. Along with the isotope ratio control the measurement of the energy distribution of He atoms in the energy range up to a few MeV can be used to diagnose the fast confined alpha particle behaviour [2]. In this report we will concentrate on the isotope ratio measurements. Possibilities of alpha particle diagnostics have been discussed earlier [3].

2. Physical bases

The hydrogen plasma ions can be neutralized due to charge exchange with the atoms penetrating into the plasma from the periphery (background neutrals), with the diagnostic beam atoms, with the not fully stripped impurity ions and due to radiation recombination with the electrons. For all the cases an intensity of the fluxes of the hydrogen, deuterium and tritium atoms produced by a unit of the plasma volume is proportional to the density of the corresponding ions:

$$\phi_{H,D,T}(E) = n_0\, n_{p,d,t}\, f_{p,d,t}(E) < \sigma v >_{cx} \qquad (1)$$

$$\phi_{H,D,T}(E) = n_e\, n_{p,d,t}\, f_{p,d,t}(E) <\sigma v>_r \qquad (2)$$

where n_0, $n_{p,d,t}$ and n_e - density of targets for charge exchange, density of hydrogen isotope ions and electron density, $<\sigma v>$ - rates for charge exchange and recombination, (1) is related to charge exchange and (2) - to recombination. Assuming that the distribution functions $f_{p,d,t}(E)$ of the different hydrogen ions are identical and measuring the ratio of the corresponding atomic fluxes we can determine the ratio of the hydrogen isotope ions in the plasma. Doing this

Diagnostics for Experimental Thermonuclear Fusion Reactors 2
Edited by Stott *et al.*, Plenum Press, New York, 1998

the differences of $<\sigma v>_{cx}$ and of absorption of atomic fluxes in the plasma for the different isotopes should be taken into account.

The atomic fluxes $\Gamma_{H,D,T}(E)$ emitted by the plasma and measured outside are different from $\phi_{H,D,T}(E)$ due to integration along line of sight and absorption by the outer layers of plasma

$$\Gamma_{H,D,T}(E) = \int_L \phi_{H,D,T}(E,l) \cdot \exp\{ -\int_0^l n\sigma_{abs}(E)dx \}dl \tag{3}$$

where l is coordinate along line of sight starting from the plasma edge and σ_{abs} is the total absorption cross section.

In the case of *active measurements* the detected atoms are produced by charge exchange with diagnostic beam atoms. The origin place of the detected fluxes is fixed by the geometry of the diagnostic set up and defined as a crossing area of the observation field and the beam. To obtain the isotope ratio of the plasma ions from measured ratio of the corresponding atomic fluxes we should take into account only the difference in the absorption and in charge exchange cross sections for the different isotopes.

In the case of *passive measurements* the plasma region in which emitted atoms are being born is not very well localized. Its position depends on a number of plasma parameters. It can be determined by modelling of the residual neutral density profile and calculating dependence of the emission function over the observation line. For the modelling and the calculating it is necessary to have detailed information about the plasma parameters, especially near the plasma periphery. As a rule it is better to operate with high energy particles which are emitted from the inner plasma area where they are neutralized mostly by recombination .

A radial dependence of the emission functions calculated for ITER plasma of one possible scenario with the flat ion temperature profile can be seen in fig.1a,b [4].

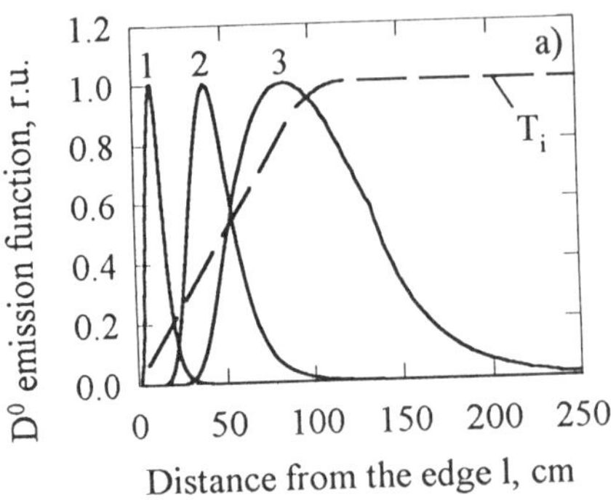

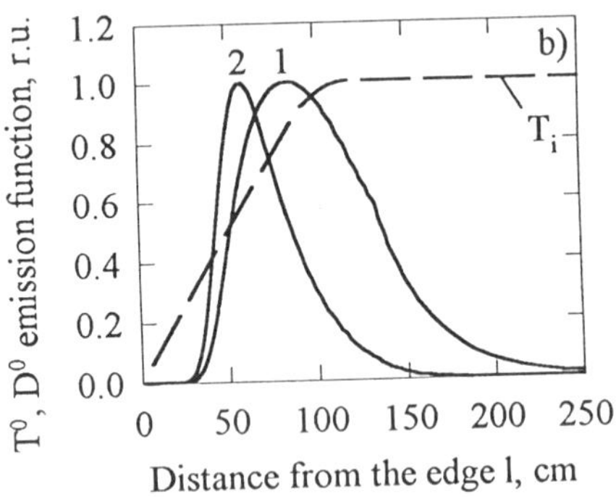

Fig.1. Emission functions of D and T atomic fluxes from ITER plasma for flat ion temperature and flat electron density profiles.
a - for D fluxes of 10 keV (1), 50 keV (2) and 150 keV(3).
b - for D (1) and T (2) fluxes of 150 keV.
Ti - ion temperature profile.

The passive fluxes of the atoms with higher energy are emitted from the deeper regions of the plasma (fig.1a). The atoms with the energy around 150 keV should be used to measure the isotope ratio of the more than 0.5 m deep plasma. Spatial resolution defined as width of the emission function on the half of the height is estimated as $\Delta x/x$ = 25 - 30 %, where x is a distance from the plasma edge till position of the maximum of the emission function. The emission functions of D and T atoms of the same energy have different radial distributions (fig.1b) and the isotope ratio measurements are possible only when plasma parameter profiles are very well known and when isotope plasma composition is assumed to be constant along line of sight of the NPA.

Above we considered the energy range up to 200-300 keV and assumed maxwellian distribution of the hydrogen isotope ions. In the burning plasma MeV energy tails of $f_{p,d,t}(E)$ are expected to be produced due to knock-on collisions of the hydrogen ions with alpha particles and with the other products of the fusion reactions [5,6]. Density of MeV energy tail d- and t-ions can reach 10^{-5} - 10^{-6} of the bulk ion density at the energy above 1 MeV. JET experiments [7] and later simulations [8] show a high probability for these ions to be neutralized due to charge exchange with H-like impurity ions such as C^{+5} and Be^{+3} and radiation recombination with electrons and we can expect a flux of MeV D/T atoms which will be high enough for n_d/n_t measurements [9]. Due to high permeability of ITER plasma for MeV D/T ions these measurement will provide an isotope ratio in the plasma core as well as an information about the distribution function of confined alpha particles because f_α should correlate with $f_{p,d,t}$ in this energy range. But more simulations taking into account different scenarios of the impurity behaviour in ITER are needed.

3. Description of the system.

Based on considerations presented above the following diagnostic system has been proposed (fig.2) [10].

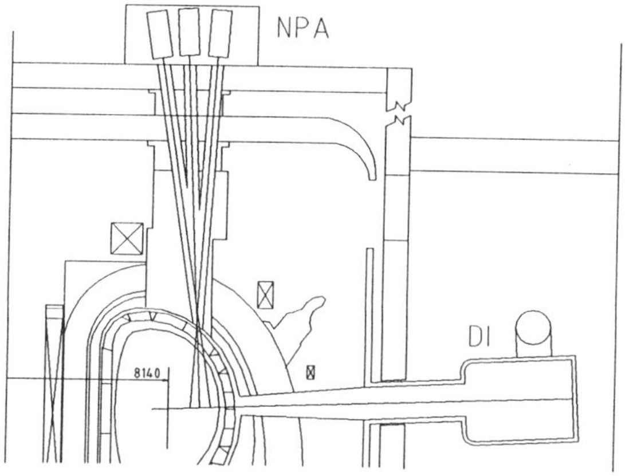

Fig.2. Set up of NPA diagnostics proposed in [10].
DNB - diagnostic beam injector.

It consists of the neutral beam diagnostic injector combined with 3 couples of neutral particle analysers (NPA). The diagnostic beam is injected into the plasma almost horizontally perpendicular to the plasma axis. Each NPA couple consists of one middle energy NPA and one MeV energy NPA. They are placed along the same axis one after another in such a way that the flux of neutrals passing through the first (middle energy) NPA enters the second (MeV energy) one. Three units of NPA couples view the plasma through a vertical port. Their lines of sight intersect the diagnostic beam at the distance of 0.56 m, 1.12 m and 1.62 m from the plasma edge to have the measurements at three different points along the radius. One more couple of NPA's with the line of sight along the major radius Crossing the plasma centre has been added.

Fig.3 presents the results of the computer simulation of the active and passive energy spectra of deuterium and tritium atoms for diagnostic beam of 10 MW beam power and 100 keV/a.m.u. particle energy and NPA which line of sight intersects the beam at 1.62 m from the plasma edge [4]. To provide the required 20% accuracy of the measurements with required time resolution 100 ms the counting rate should be not less than 10^3 kHz. For passive measurements it is possible in the energy range up to 130 keV and for active - in the energy range above 160 keV.

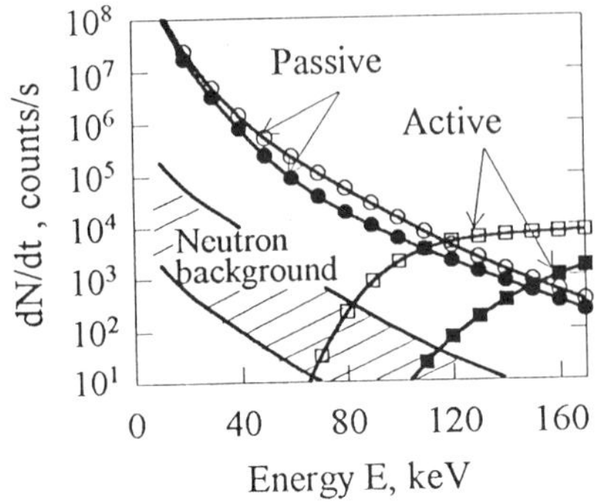

Fig.3. Active and passive fluxes of D (open symbols) and T (closed symbols) atoms from ITER plasma with $n_d=n_t$, flat ion temperature and flat electron density profiles.

The background level shown in fig.3 is calculated for NPA surrounded by additional concrete shielding of 1 m thickness with the input aperture 2×2 cm^2. The NPA is supposed to be equipped by advanced detectors with pulse height analysis system [11] which are assumed to be developed as a result of the proposed R&D (see paragraph 4). We should not expect problems with background counting rate but we should keep in mind that background produced by neutrons is strongly affected by surrounding diagnostic and technological equipment.

Computer simulation allows us to find a plasma parameter range where D/T flux ratio is less sensitive to the shape of the plasma density and temperature profiles for fixed isotope ratio of plasma ions. This is important for estimation of the reliability of the measurements. Some results are given in fig.4. We can see that situation improves with broadening of Ti profile and/or decreasing of the plasma periphery density.

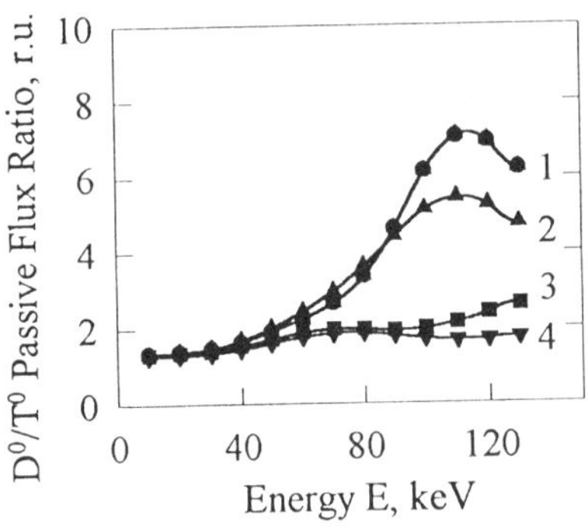

Fig.4. Energy dependence of Γ_D / Γ_T ratio from ITER plasma with $n_d=n_t$ for different energy and temperature profiles. 1,2,4 - $n_e \sim (1-(r/a)^{20})$, $R_{Ti(0)}$ =175 cm (1), $R_{Ti(0)}$ = 185 cm (2), $R_{Ti(0)}$ =200 cm (4) 3 - $n_e \sim (1-(r/a)^8)$, $R_{Ti(0)}$ =185 cm $R_{Ti(0)}$ - radius of plasma core where ion temperature Ti = Ti(0), for r> $R_{Ti(0)}$ Ti linearily decreases to 0 to the plasma edge r=a.

At present only the passive measurements are approved for ITER. In this case localisation and spatial resolution of the measurements are not determined by the diagnostic set up. They can be determined only by computer simulation. Besides data proceedings are not possible without information concerning density and ion temperature profiles, especially at the outer region of the plasma.

4. NPA prototype and R&D needed for its optimisation to ITER

GEMMA-type NPA [12] used for investigation of H and He neutral fluxes in the energy range 0.1 - 4 MeV at the largest tokamaks JET, JT-60U and TFTR is suggested as a prototype. Its operation is based on stripping atoms by means of a carbon foil, subsequent E || B analysis of secondary ions and detection of them by detectors that consist of thin (2-25 μm) scintillators optically coupled to photomultiplier tube (PMT). PMT's are chosen due to their high radiation resistivity. Thickness of scintillator is minimized according to the particle energy detected in the NPA channels to decrease the sensitivity of the detectors to the neutron and gamma background. The detectors can be used for detection of hydrogen and helium ions with the energy higher than 100 keV.

For implementation on ITER the prototype should be optimized to extend the low energy limit of the measurements to 10 - 20 keV, to decrease further the sensitivity to background radiation and to improve the signal to noise ratio. This is a purpose of our R&D program during which we are going to test

- new types of PMT;
- new types of scintillators;
- a more complicated version of the detectors using ion-electron converter with a high conversion ability;
- a system for acceleration of the secondary ions by additional voltage of order 100 kV which will transform the fluxes of the energy range 10 - 200 keV to the energy range 110 - 300 keV.

We hope results of the proposed R&D program enable us to make a design of the final NPA prototype for ITER probably covering the whole energy range from 10 - 20 keV till 4 MeV by one instrument.

5. n_p/n_d measurements at COMPASS tokamak and W7-AS stellarator.

Measurements of n_p/n_d were carried out with NPA ACORD-12 [13]which enables to detect simultaneously H and D fluxes and their energy spectra. Three pairs of its energy channels measure the fluxes of H and D atoms of the almost the same energy. It provides a possibility to follow for the correlation between $f_p(E)$ and $f_d(E)$ in the same energy range and it is convenient for n_p/n_d determination. The correction factors needed for obtaining n_p/n_d from the measured values of Γ_H / Γ_D are calculated by a computer program which simulates the production and emission of the active and passive fluxes of the fast charge exchange atoms.

At COMPASS tokamak the ohmically heated plasma with line density up to 2×10^{19} m^2 and ion temperature about 200 eV has been investigated [14]. A large portion of protons (up to 60%) has been observed in the plasma of deuterium discharges after switching the operation of tokamak from hydrogen to deuterium and wall boronization by $B(CH_3)_3$ being increased from shot to shot after cleaning the wall by glow discharge (fig.5). This result shows a sensitivity of the technique to changing of the isotope ratio of the hydrogen plasma ions.

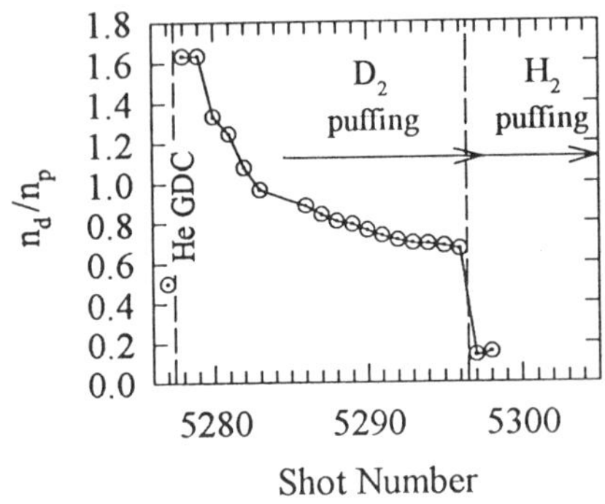

Fig.5. Isotope ratio n_d/n_p for a sequence of COMPASS shots in D_2 following 15 min He glow cleaning discharge.

At W7-AS stellarator a diagnostic neutral beam of 20 - 35 keV energy has been used to localize the measurements near the plasma centre [15]. By chopping the beam the active fluxes has been separated from the passive ones. It was found that active and passive measurements agree each other in the limit of the error bars (fig.6a,b,c). Data presented in fig.6 are averaged over 3 different energies and over duration of diagnostic beam pulse or time interval between pulses, normally 50 ms. In the case of electron cyclotron plasma heating a typical percentage of protons $n_p/(n_p+n_d)$ (ratio of the proton density to the total density of the hydrogen isotope ions) was found between 20% and 40% showing a weak tendency to the increase during a discharge (fig.6a). In the case of neutral injection heating a percentage of the protons increases during injection (fig.6b) which results from the injection of hydrogen by the heating beams.

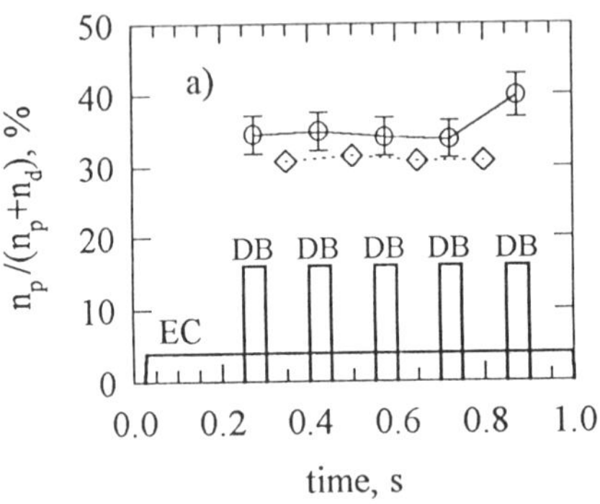

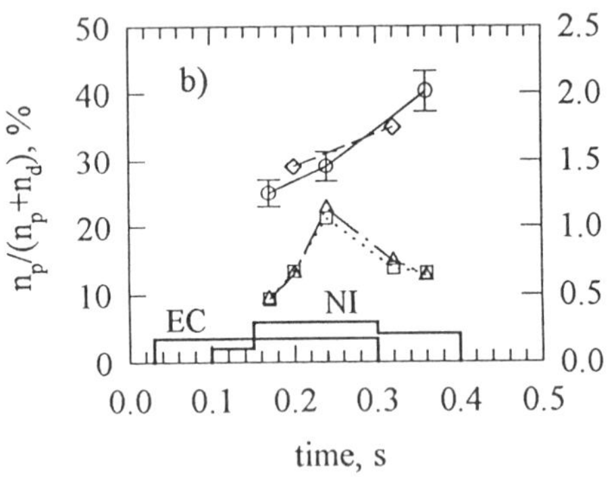

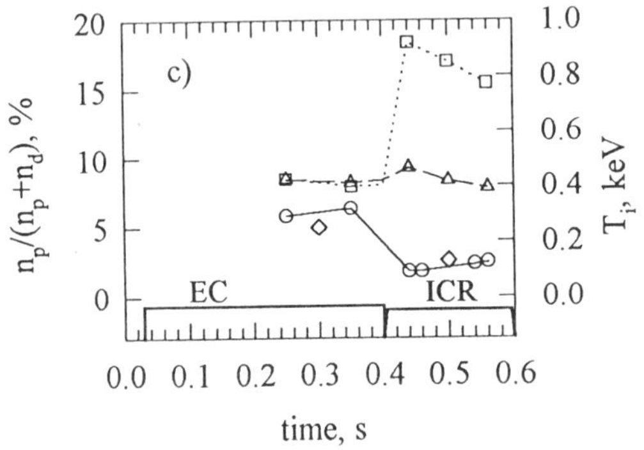

Fig.6. Relative proton content in deuterium discharges of W7-AS for different heating scenarios. Active (Φ) and passive (◊) measurements.
a - electron cyclotron heating.
b - neutral beam injection heating.
c - H minority ion cyclotron heating.
EC - ECR power heating, DB - diagnostic beam pulses, NI - neutral injection power as a function of time (.4 - 1.2 - .8 MW), ICR - ion cyclotron heating power. Ion temperature: protons (+), deuterons (×).

From diagnostic point of view measurements made in ion cyclotron H minority heating regime are of the most interesting. During ion cyclotron heating phase the effective temperature of the hydrogen minority increased more than twice compared to deuterium majority temperature (fig.6c). In spite of the large difference between T_H and T_D the measurements of the hydrogen percentage which has been made by three pairs of the NPA channels for three different energies agree very well if the difference between T_H and T_D is taken into account in calculations of the correction factors.

The measurements show a significant decrease of the hydrogen percentage in the ICR heating phase. There are two possible explanations of the decrease observed. Firstly, it is possible that the number of the protons in the plasma is decreased indeed due to enhanced losses of the energetic RF heated protons. Secondly, it is possible that the measured ratio n_p/n_d is decreased only due to the disturbance of the isotropy of the proton distribution function in the velocity space which results in a relative decrease of the particles with a velocity parallel to the line of sight of the NPA.

Thus, the measurements carried out for different heating scenarios which can influence the temperature ratio of protons and deuterons confirm a possibility to control the isotope ratio by charge exchange atomic fluxes during a plasma discharge. Unfortunately up to now the measurements has been made for the plasma with line density below 2×10^{19} m^2 which is much less than ITER parameters. The technique should be approved at the largest machines such as JET and JT-60U.

6. Conclusion

The isotope composition of the hydrogen plasma ions in ITER can be measured by measuring the ratio of D/T atomic fluxes. The ratio of the fluxes is proportional to the density ratio of the corresponding plasma ions. The coefficients of the proportionality are calculated by computer simulation using additional information about density and temperature radial distributions.

For measuring the density ratio of the hydrogen isotope ions in ITER plasma it was proposed to use the passive as well as the active diagnostics. If only the passive diagnostics is used the localization of the measurements will become worse and more intensive additional information will be needed to obtain required accuracy of the measurements.

Expected counting rates will provide accuracy of the n_d/n_t measurements equal 20% or better with time resolution 0.1s. Measurements in the energy range 1 - 3 MeV enable us to obtain n_d/n_t in the plasma core and the alpha particle distribution function.

Existing NPA should be optimized to extend its energy range to 10 - 20 keV and to decrease further its sensitivity to neutron and gamma background.

Efficiency of the techniques has been demonstrated at small scale plasma devices. The further testing at larger machines including JET and JT-60U is necessary.

References

1. A.B.Izvozchikov, A.I.Kislyakov, M.P.Petrov, A.V.Khudoleev, D/T/He composition diagnostics on ITER. Alpha Diagnostics, *ITER-IL-PH-7-9-S-07 (1989).*
2. M.P.Petrov, A.B.Izvozchikov, A.A Basalaev et al., Proposal for active charge-exchange diagnostics of charge fusion products in JET plasma, *Rep. A.F.Ioffe Physical-Technical Institute, Leningrad,* 1990; A.B.Izvozchikov, A.V.Khudoleev, M.P.Petrov et al., Charge

exchange diagnostics of fusion alpha particles and ICPF driven minority ions in MeV energy renge in JET plasma, *JET Rep. JET-R(91) 12, 1991.*

3. M.P.Petrov, R.K.Fisher, Charge exchange diagnostics of fast confined alphas: present situation and prospects for ITER, in *"Diagnostics for Experimental Thermonuclear Fusion Reactors"*, P.E Stott, G. Gorini and E.Sindoni, ed., Plenum Press, New York and London (1995) p.495.
4. A.V.Khudoleev, V.I.Afanassiev, F.V.Tchernyshev, Possibility to control an isotope composition of dense plasma by charge exchange atomic fluxes, *9th National Topical Conference on High Temperature Plasma Diagnostics, St.-Petersburg, Russia, June 2-4, 1997.*
5. D.Ryutov, Energetic Ion Population Formed in Close Collision with Fusion Alpha-Particles, *Physica Scripta 45 (1992) 153.*
6. R.K.Fisher, P.B.Parks, McChesney, M.N.Rosenbluth, Fast Alpha Particle Diagnostics using Knock-on ion Tails, *Nucl.Fus. 34 (1994) 1291.*
7. M.P.Petrov, V.I.Afanasyev, S.Corti *et al.*, Neutral particle analysis in the MeV range in JET, *Proc.19th EPS Conf.on Contr.Fus.and Plasma Phys., Innsbruck, 1992, vol. 16C, part II, p.1034.*
8. A.A.Korotkov, A.Gondhalekar, A.J.Stuart, Impurity induced neutralization of megaelectronvolt energy protons in JET plasmas, *Nucl.Fus., 37 (1997) 35.*
9. A.V.Khudoleev, V.I.Afanassiev, A.I.Kislyakov, M.P.Petrov, S.S.Kozlovskij, Neutral particle diagnostics in the MeV energy range at the large modern tokamaks, *9th National Topical Conference on High Temperature Plasma Diagnostics, St.Petersburg, Russia, June 2-4, 1997.*
10. ITER design description document (DDD), Neutral particle analyzers, WBS 5.5.E.08, RV4, 26 Nov.1996.
11. S.S.Kozlovskij, V.I.Mineev, V.V.Grebentschikov et al., Detector of particles emitted by plasma based on electron multiplier and converter with increased electron emission, *9th National Topical Conference on High Temperature Plasma Diagnostics, St.Petersburg, Russia, June 2-4, 1997.*
12. A.I.Kislyakov, A.V.Khudoleev, S.S.Kozlovskij, M.P.Petrov, High Energy Neutral Particle Analyzer, *Fusion Engineering and Design 34-35 (1997) 107.*
13. A.B.Izvozchikov, M.P.Petrov, S.Ya.Petrov, F.V.Chernyshev, I.V.Shustov. The Akord-12 multichannel analyzer for simultaneous recording of the energy spectra of hydrogen and deuterium atoms. Sov.Phys.Tech.Phys. 37 (1992) 201
14. H.G.Esser, S.J.Fielding, S.D.Hanks *et al.*, Boronization of COMPASS, *J.Nucl.Mat. 186 (1992) 217.*
15. A.I.Kislyakov, M.Kick and W7-AS team, Measurement of n_p / n_d ratio on W7-AS stellarator, *9th National Topical Conference on High Temperature Plasma Diagnostics, St.-Petersburg, Russia, June 2-4,1997.*

PROSPECTS FOR CORE HELIUM DENSITY AND RELATED MEASUREMENTS ON ITER USING ACTIVE CHARGE EXCHANGE

D.M. Thomas, K.H. Burrell, M.R. Wade,* R.T. Snider

General Atomics
P.O. BOX 85608
San Diego, CA 92186-5608

*Oak Ridge National Laboratory
P.O. Box 2009
Oak Ridge, TN 37831

ABSTRACT

The measurement of low-Z impurities, in particular He ash, in the core of International Thermonuclear Experimental Reactor (ITER) remains an outstanding diagnostic issue. The only credible candidate at present is active charge exchange recombination spectroscopy (CER) utilizing a diagnostic neutral beam (DNB) optimized for the dual requirements of beam penetration and charge exchange cross section, resulting in beam energies of ~100 keV/AMU. Using the existing ITER parameter profile and equilibrium data files and reasonable assumptions regarding viewing optics and DNB performance, we have employed a benchmarked multistep beam penetration code to yield signal-to-noise estimates for possible core helium concentration measurements. These studies confirm the importance of precise determination of beam intensities via accurate modeling and independent measurement, as well as the need for beam modulation, to satisfy the stated measurement precisions needed for ITER. Comparable calculations have been done for an intense pulsed neutral beam based on ion diode technology, as well as other candidate He-CER wavelengths, to assess the relative advantages of these techniques. Since any DNB-based diagnostic system actually deployed on ITER will likely be used for a variety of purposes, signal-to-noise calculations for the related active CER measurement of ion temperatures have also been performed and will be presented.

1. INTRODUCTION

Active plasma spectroscopy, or spectroscopy based upon local signal enhancement via a penetrating neutral atomic beam, is employed as a key diagnostic on almost all present-day tokamaks. This technique is utilized for a wide variety of measurements in the plasma edge and core, including ion temperature [via Doppler broadening of intrinsic impurity lines which are efficiently populated by charge exchange from beam atoms — charge exchange recombination spectroscopy (CER)], plasma rotation (via Doppler shift of the same impurity lines), and impurity profile information (via quantitative spectroscopy of the impurity

Diagnostics for Experimental Thermonuclear Fusion Reactors 2
Edited by Stott *et al.*, Plenum Press, New York, 1998

line intensities). Other important measurements that can be undertaken with active spectroscopy include plasma turbulence levels [via fluctuation analysis of the collisionally excited beam emission — beam emission spectroscopy (BES)] and local magnetic/electric field determination [via polarimetry of the beam emission polarization state induced either by the motional Stark or Zeeman effect (MSE)].

These measurements, all variations of a theme, have substantially increased our knowledge of the behavior and optimization of fusion plasma experiments. Our present picture of core and edge barrier transport formation based on shear suppression of turbulence owes its basic experimental underpinnings to well designed and deployed active spectroscopy systems.[1–3] Similarly, studies of core helium ash formation and transport, and the performance of pumped divertors for helium exhaust, have benefited from the precise, local, absolute measurements possible using this technique.[4,5]

Adoption of active spectroscopy for the International Thermonuclear Experimental Reactor (ITER) represents a challenging opportunity. As discussed in previous contributions to this workshop[6,7] there are numerous obstacles to deployment, the two most important of which are as follows:

1. The large size of ITER compared to present experiments implies much longer beam propagation path lengths resulting in severe attenuation of the neutral beam as it penetrates to the core region. One cannot simply go to higher energies to improve this situation. As pointed out,[6,7] the energy dependence of the beam attenuation and charge exchange cross sections results in an optimal beam energy of ~100–150 keV/AMU. The substantially longer path lengths through the plasma for the viewing chords imply any spectroscopic measurement, whether active or passive, will have to be made in the presence of a substantial bremsstrahlung background, typically several orders of magnitude larger than the desired charge exchange or emission line intensity. This implies that any usable system must have extremely good dynamic range and low noise.
2. The ignited phase of ITER will create an intense radiation zone around the device. Measurements in this environment will require diagnostic immunity/resistance to neutron and gamma radiation at a level substantially beyond present day experiments. Relevant issues include the radiation hardening of optical elements and detectors, including the survivability/distortion/erosion of first wall optical elements which will be required for any reasonable measurement. In addition, this same environment demands the utmost in reliability and remote maintenance capability from the diagnostic; aspects which are relatively less important (and hence less developed) on existing experiments.

Despite these obstacles, substantial design and development work continues on active spectroscopy systems for ITER because of the unique information they can provide on core plasma conditions. In the case of core helium density measurements — a crucial indicator of the success of ITER — no credible alternative to CER has emerged. In this paper we expand on previous work[6,7] to present a detailed performance estimate for a diagnostic system appropriate for measurements of core helium concentrations in the ITER environment, based upon active spectroscopy of the 4–3 He II transition. In Section 2 we review the desired measurement resolutions and relevant ITER geometry to arrive at a realistic layout for the diagnostic. Section 3 describes several ITER profile scenarios based upon the existing ITER parameter profile and equilibria data files and lists the assumed values for the diagnostic neutral beam (DNB) performance. Section 4 covers the calculations of expected beam intensities and signal-to-noise ratios, based on a benchmarked multistep beam attenuation code[8,9] and the various profile scenarios. Section 5 discusses the advantages associated with using a pulsed beam based on ion diode technology as opposed to the baseline DNB design. In Section 6 we investigate the possible benefit of using an alternative wavelength — the 5–4 He II transition in the infrared— for the helium concentration measurements. We discuss the prospects for ion temperature and rotation measurements using active CER in Section 7. In Section 8 a summary and some conclusions are presented.

2. ITER PARAMETERS AND REQUIREMENTS

The following diagnostic requirements are based upon the minutes of the ITER Diagnostic Expert Group Meeting (3/96):

These four parameters represent zeroth, first, and second order moments of the intensity distribution of the emitted He line. We note the fourth parameter listed — ion temperature — will be the most difficult measurement in terms of satisfying the quoted requirements, since the spectrum must be resolved sufficiently. The requirement for measuring plasma rotation demands orthogonal views of the beam, implying a vertical viewport must be considered a necessary part of this diagnostic system. The helium density profile measurement requires an absolute intensity calibration. This calibration must be maintained across the substantial number of viewing chords needed to satisfy the spatial resolution requirements. In Table 2 we lists what are believed to be reasonable parameters for a prototypical active CER system, based on the most recent engineering layouts and dimensions. The number of chords permits good coverage of the minor radius region 0.2–2.7 m (Fig. 1). The auxiliary chords for monitoring beam emission H_α serve as a crucial experimental cross-check of the beam attenuation calculations, which are sensitively dependent on local impurity densities and accurate cross sections. It may be desireable to have additional H_α channels having higher spatial resolution to accurately determine the lateral dimensions of the beam *in situ*. The auxiliary visible bremsstrahlung measurements at the He wavelength displaced laterally from the beam serve as a cross-check on the beam calculations.

Table 1. ITER requirements for active CER measurements.

Parameter	Range	Required (Spatial) (cm)	Resolution (Temporal) (ms)	Required Accuracy (%)
Core helium density (Category 2)	1%–20%	30	100	10
Toroidal plasma rotation (Category 1)	1–200 km/s	~5	10	30
Poloidal plasma rotation (Category 1)	1–50 km/s	~5	10	30
Ion temperature profile (Category 2)	0.5–50 keV	30	100	10

Table 2. Strawman active spectroscopy system parameters.

Number of optical penetrations	10 (5 vertical, 5 tangential)
Size of individual penetration	0.2 m × 0.2 m
Total estimated penetration area	0.4 m^2
Number of sightlines, total	128
Number of tangential sightlines	40 CER signal 14 H_α 10 Background
Number of vertical sightlines	40 CER signal 14 H_α 10 Background
Collection area of first mirror	0.2 m × 0.2 m
Characteristic chord length	~16 m
Presumed emission area-toroidal view	0.2 m vertical × 0.05 m radial
presumed emission area-vertical view	0.2 m toroidal × 0.05 m radial
Collection solid angle, first mirror	4.9×10^{-4} sr
Etendue through spectrometer (est.)	2×10^{-7} m^2-sr

3. PROFILES AND OPERATIONAL SCENARIOS

For the calculations presented here, we have used flat and peaked reference profiles from the ITER database[10] (Fig. 2). The relevant T_e, n_e, Z_{eff}, etc. profiles were first mapped to minor radius using the associated equlibria to provide suitable inputs for the beam attenuation and rate equation codes. We have assumed two cases for the study. The first is a steady state case represented by the flat profile, and assuming a constant (with density) 10% helium fraction. The second is an early ignition case, where we assume a 1% helium fraction at the core that falls off radially like the temperature profile.

DNB Projected Performance

A separate diagnostic neutral beam (DNB) design is crucial to the success of the above measurement. For our calculations we have relied on the provisional design parameters from Ref. 11, summarized in Table 3. Options to increase the design beam energy to 125–150 keV are being pursued. We also note the measurement time limitation, which serves to minimize the overall neutralizer gas flow into the plasma, will serve as a fundamental limit on any active spectroscopy measurements. Finally, the quoted footprint size may be somewhat optimistic based on present technology.

4. MULTISTEP BEAM PENETRATION MODELING

Modeling and estimation of the observed helium signal depends crucially on accurate cross sections and rates for beam attenuation and helium emission. We have adapted a previously benchmarked multistep collisional-radiative model[8,9] for the specific ITER geometry, profile conditions, and putative DNB parameters. The code uses the most recent collisional rates available from the Atomic Data and Analysis Structure (ADAS) database.[12] Initial runs verified the importance of the multistep corrections to beam attenuation calculations, with attenuation enhancement factors from 6–8 found compared to the earlier single-step estimates.[6] For the energy range of interest, attenuation factors of 10^{-3} to 10^{-4} are common for the core region $\rho_{min} \leq 0.5$ m ($\rho \leq 0.1$).

The helium emissivity based on the calculated source strength, along with the competing bremsstrahlung, is then calculated using the assumed geometrical factors from Tables 2 and 3. Figure 4 shows the results for the startup and fully ignited cases.

The signal-to-noise of the He II CER measurements may be estimated under the following assumption: for the no competing lines case we presume the fluctuations in the continuum background to be the dominant noise source. Then the SNR will be given by[6,7]

$$\mathrm{SNR} = \frac{N_{He}}{\sqrt{N_B + N_{He}}} = \frac{S_{He}\,\Delta t\,DF}{\sqrt{R\,S_{He}\,\Delta t\,DF}} = \frac{S_{He}\,\Delta t\,DF}{\sqrt{(S_{He}\,\Delta t)(R + DF)}} = \frac{\sqrt{S_{He}\,\Delta t}\,DF}{\sqrt{(R + DF)}} \quad (1)$$

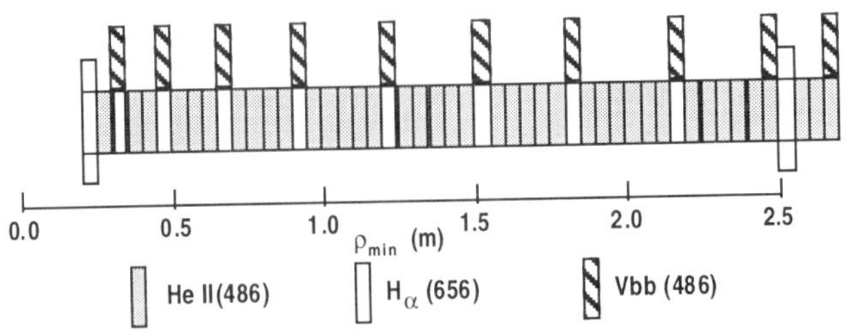

Figure 1. Possible layout of viewing geometry, showing the 64 He CER, Ha BES, and VBB (visible bremsstrahlung) locations. The basic element dimensions are 20 cm (lateral) by 5 cm (vertical), and are presumed superimposed on the beam trajectory. Two orthogonal arrays — one vertical, one toroidal — are presumed, for a total of 128 discrete spatial points.

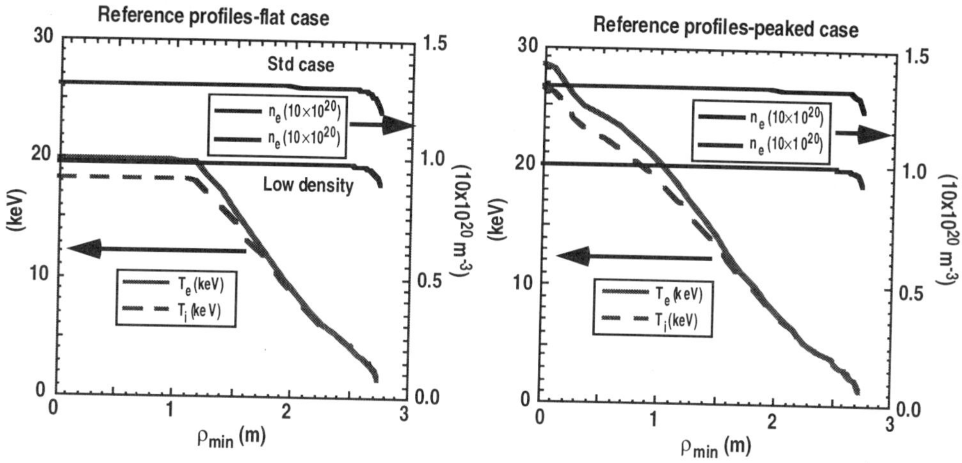

Figure 2. Reference profiles extracted from the ITER profile database, showing density, ion and electron temperatures for flat and peaked temperature profile scenarios. The "low density" (~1.0 × 20 m^{-3}) cases are most relevant to planned ITER operation. Z_{eff} profiles for all cases were reasonably flat with Z_{eff} ~1.5.

Table 3. Provisional parameters of diagnostic neutral beam[11]

Beam species	H^0
Beam energy (keV)	100
Neutral equivalent current (A)	50
Beam power (MW)	5
Beam divergence	0.2°
Beam footprint at beam focus	0.2 × 0.2 m^2
Beam focus	At r/a = 0.3–0.4
Equivalent H^0 density at focus (unattenuated)	1.78×10^{15} m^{-3} (~2 × 10^{-4} n_e)
Beam modulation frequency (Hz)	5 (0.1 s on, 0.1 s off)
Measurement time (s)	1–3 every 10–20

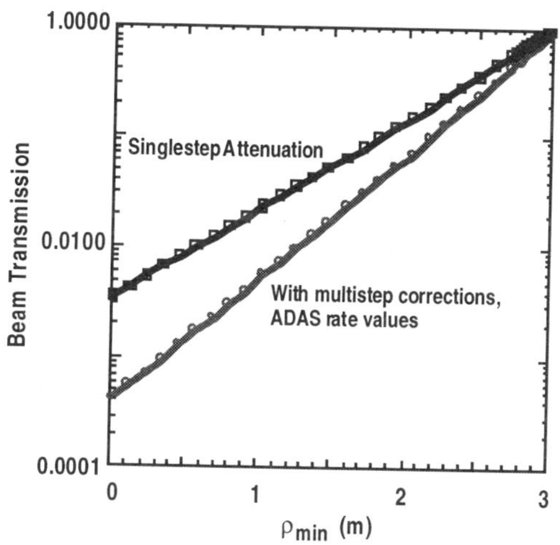

Figure 3. Calculation of 125 keV H^0 neutral beam transmission for the ITER flat profile case (scenario 1), using a simple single step and full multistep attenuation calculations.

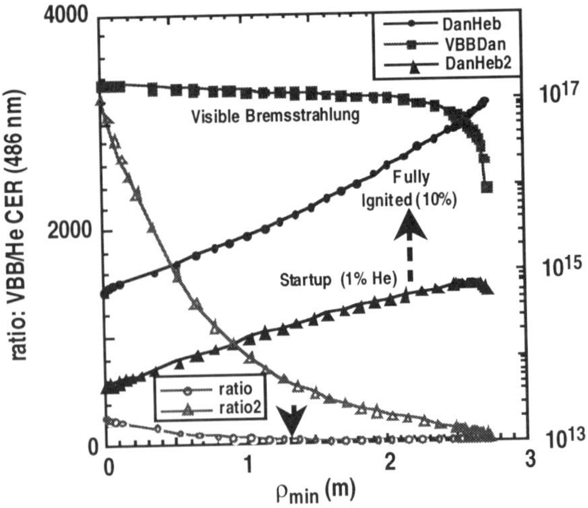

Figure 4. He CER and bremsstrahlung emissivities at 468 nm for the startup and fully ignited scenarios. Also plotted are the emissivity ratios for the two times. Beam energy = 125 keV, assumed neutral current density of 10^3 A/m^2 (40 A neutral equivalent current).

where S_{He} = helium intensity, R = VBB/He intensity ratio, Δt = integration time and DF = duty factor for beam. Calculations were performed for the two cases assuming a 50% duty cycle and 100 ms integration time — the cited time requirement. We note the 5 Hz DNB modulation frequency in Table 3 will permit four 50 ms timeslices (two signal and two background) per beam cycle. Longer integration times improve the SNR as $\sqrt{\Delta t}$. Figure 5 shows the results for the startup and fully ignited cases. For the calculation of absolute signal levels we have assumed the etendue and other optical parameters listed in Table 2.

The values of SNR found are marginal (for the desired time resolution) at the smaller radii, especially if we are interested in tracking the buildup of helium at the beginning of the ignited phase. The situation will be even more marginal for ion temperature and rotation measurements, or for plasma densities higher than those presumed in the modeling.

5. PULSED BEAM PROSPECTS

Because of the narrow limits set on the beam energy by the penetration and charge exchange requirements, the principal beam improvement that could be made would be an increase in current density. A substantial improvement in SNR would result from the deployment of an intense pulsed neutral beam based upon ion diode technology.[13] Figure 6 shows the calculated improvement from a hypothetical 100 keV, 50 kA IDNB, presuming the same beam footprint as for the conventional beam. Such beams are close (within a factor of about two) to satisfying the intensity and divergence requirements; remaining issues include beam repetition rate and effective neutralization. The use of these beams would improve the SNR at the stated time resolutions by factors of 10–20 in the core.

Especially for the temperature measurements, the increase in SNR will probably be necessary. A side benefit of the pulsed beam concerns the relative ease of fitting the lineshape: the technique of using background subtraction to reject edge He II emission will be far more effective on the 1 μs timescale, and the peaks may be better fit by a single Gaussian representing the hot core CER signal.

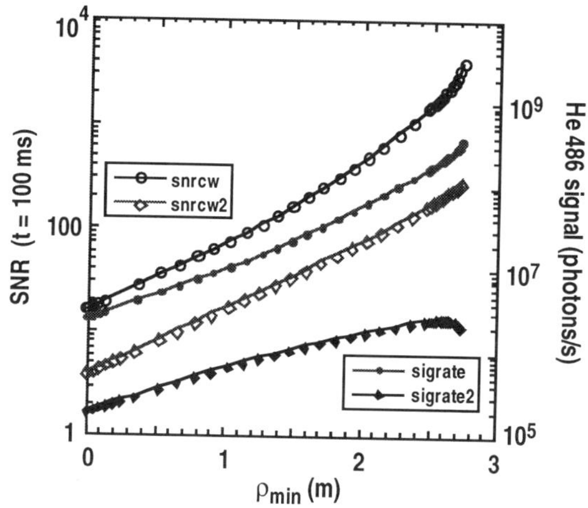

Figure 5. SNR estimates (and the associated signal levels in photons/s) based on the preceding beam calculations. Integration time 100 ms, 50% duty cycle. An overall system etendue of 2×10^{-7}, optical efficiency of 1×10^{-2}, and bandwidth of 2 nm is assumed in the calculations. Beam energy = 125 keV, assumed neutral current density of 10^3 A/m^2 (40A neutral equivalent current). Snrcw2 = SNR for startup, snrcw = SNR for fully ignited case. Sigrate2 = signal level for startup, sigrate = signal level for fully ignited case.

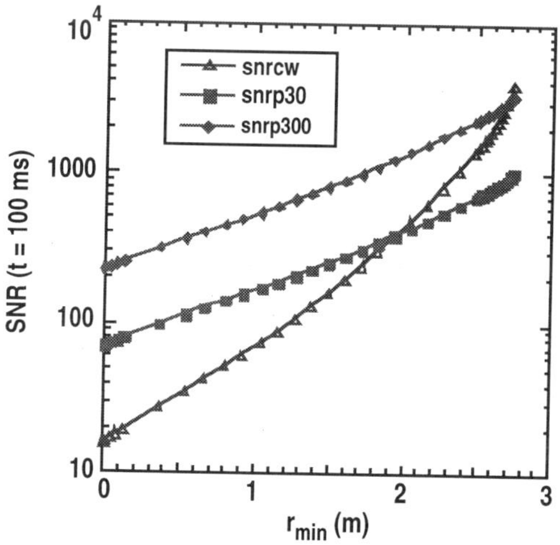

Figure 6. SNR estimates for a 1 μs 50 kA pulsed ion diode neutral beam (IDNB), compared to the conventional DNB for the fully ignited scenario. The two cases shown are for 30 pps and 300 pps rep rates. Integration time 100 ms, 50% duty cycle. An overall system etendue of 2×10^{-7}, optical efficiency of 1×10^{-2}, and bandwidth of 2 nm is assumed in the calculations. Conventional beam parameters: energy = 125 keV, assumed neutral current density of 10^3 A/m^2 (40A neutral equivalent current).

6. THE He-II 5–4 TRANSITION

One possible improvement for the helium intensity measurement involves the use of alternative wavelengths to the standard 486 nm $N = 4 \rightarrow 3$ transition of He II. In particular, the $N = 5 \rightarrow 4$ transition of He II offers several potential advantages. This near-infrared (λ = 1012.3 nm) line is also efficiently created by the charge exchange process and can be efficiently detected using present day technology. Figure 7 shows the two rate coefficients for Scenario 1 as a function of beam energy. For the energies of interest, the ratio of $\langle \sigma v \rangle_{4-3}$ to $\langle \sigma v \rangle_{5-4}$ is about 2.4 for the entire plasma profile.

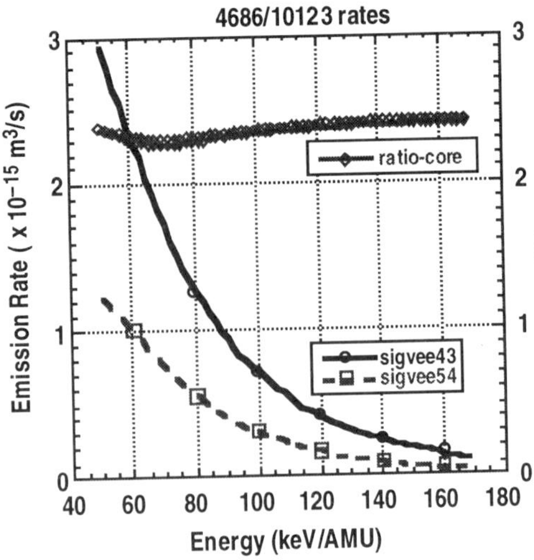

Figure 7. Emission rates for the 468.6 nn and 1012.3 nm transitions in He II as an injection energy function for the ITER flat profile case (Scenario 1). The ratio $\langle\sigma v\rangle_{4-3}/\langle\sigma v\rangle_{5-4}$ is also plotted.

Because of the λ^{-2} dependence of the bremsstrahlung intensity, the competing background light would be expected to drop by a factor of $(\lambda_{NIR}/\lambda_{VIS})^{-2} \sim 4.6$. Since the required Doppler measurements will also scale with wavelength, an improvement of $\lambda_{NIR}/\lambda_{VIS} \sim 2.2$ in the SNR should be obtained. This effect directly compensates for the sonewhat lower emission rate for the $5 \rightarrow 4$ transition.

Detection of the 1012.3 nm radiation can be accomplished using e.g., silicon APDs; at room temperature the quantum efficiency is ~70%.[14] These devices are presently used with great success for Nd:YAG Thomson scattering measurements on DIII–D at greater wavelengths ($\lambda < 1100$ nm).[15] Temperature stabilization of both detector and electronics is crucial for this application. We have observed both lines simultaneously on DIII–D discharges using cooled silicon CCD detectors.[16] In this case the system was not specifically designed for high NIR sensitivity and the quantum efficiency is about 10% at the 5–4 transition frequency. Germanium or other infrared detectors may be considered.

The use of the longer wavelength has two additional advantages related to the optical system. First, the transmission efficency through previously irradiated optical fibers is seriously diminished[17] in the visible. As it turns out the transmission is substantially better in the NIR[18] which would lead to relatively larger and less distorted measurements. Second, the longer wavelength will be more efficiently reflected from the metal mirrors used for the collection optics, and should be less affected by the surface coating, radiation damage and erosion which will inevitably occur to the plasma-facing mirrors.

For either transition wavelength, there are competing transitions due to the other light impurity ions, specifically C and Be (Table 4). Although we hope the absolute concentrations of these species will be low there may be enough to complicate the analysis of the He line. We are presently examining the relative scalings of these higher-n trasnsitions to see if this represents an additional advantage for the $5 \rightarrow 4$ transition.

Table 4. Relevant impurity transitions for helium measurements.

Ion Species	468.6 nm	1012.3 nm
He II	$4 \rightarrow 3$	$5 \rightarrow 4$
Be IV	$8 \rightarrow 6$	$10 \rightarrow 8$
C VI	$12 \rightarrow 9$	$15 \rightarrow 12$

7. ION TEMPERATURE/ROTATION MEASUREMENTS

As mentioned earlier, in contrast to the basic intensity measurement necessary for calculating the helium concentration, determining the ion temperature and rotation requires spectral analysis and fitting of the Doppler-broadened emission line — in effect taking higher order moments of the spectral distribution The degradation in SNR was earlier estimated[6] by presuming an increase in 10 in the required spectral resolution (to 0.2 nm) leading to a similar loss in the time resolution to recover the SNR. To follow the evolution of temperature and rotation during the plasma startup, or during a core barrier formation period where the temperature may be somewhat lower than the ignited case, it is clear we will require spectral resolutions substantially better than the factor of ten for adequate confidence in the line broadening and line shift values. For example, assuming a 20 km/s resolution requirement for the poloidal field measurement from Table 1 implies a needed spectral resolution $\Delta\lambda = \lambda(\Delta v/c) = 0.03$ nm in the case of the 486 nm line, a factor of seventy smaller that for the helium concentration determination. Similarly, a 1 keV resolution at 10 keV ion temperature will require differentiating between spectral halfwidths of roughly 0.90 and 0.86 nm, or 0.04 nm. Since the SNR scales with the square root of the signal intensity, it is clear that the requested time resolutions for temperature and rotation are incompatible with the capabilities of the conventional DNB. The only possible hope with the IDNB is to gate the detection in phase with the pulsed beam. As a reference point, in Fig. 8 we show the SNR estimated for the startup scenario with a 0.04 spectral window, assuming the imaging area has been increased by a factor of six to give 30 cm resolution, and 100 ms time resolution as before.

8. SUMMARY AND CONCLUSION

Using fairly detailed assumptions about ITER plasma conditions, geometry, and diagnostic neutral beam performance, we have tried to assess the prospects for measuring the core helium concentration profile, and to a lesser extent the ion temperature and plasma

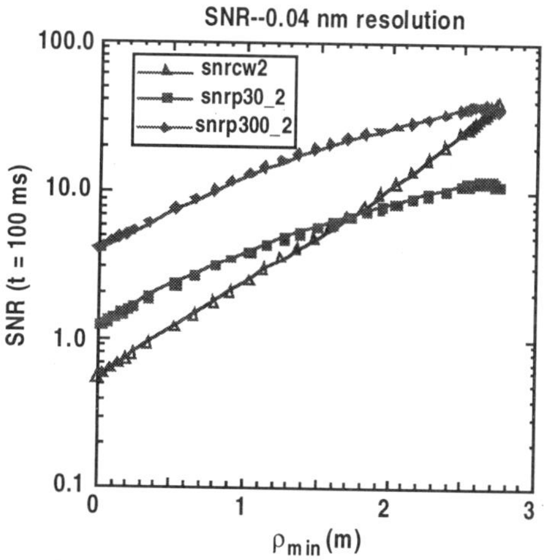

Figure 8. SNR estimates for a 1 μs 50 kA pulsed ion diode neutral beam (IDNB), compared to the conventional DNB for the startup scenario, using wavelength resolution suitable for temperature or rotation measurement. The radial resolution assumed is 30 cm. The two IDNB cases shown are for 30 pps and 300 pps rep rates. Measurement time 100 ms, 50% duty cycle. An overall system etendue of 2×10^{-7}, optical efficiency of 1×10^{-2}, and bandwidth of 0.04 nm is assumed in the calculations. Conventional beam parameters: energy = 125 keV, assumed neutral current density of 10^3 A/m^2 (40 A neutral equivalent current).

rotation profiles, using the visible He II 486.nm line. Using more accurate modelling of beam penetration and charge exchange we have performed signal to noise estimates for the desired measurement precision for the fully ignited case as well as a simple startup case. It is clear that measurements within r = 1 m will be difficult but doable for reasonable optical penetrations and collection efficiencies. However it is also clear that the stated measurement precisions for these parameters, as exemplified in Table 1, are not compatible with the expected DNB performance as listed in Table 3. This argues either for a relaxation of the desired resolution(s), or further work on a pulsed, high current alternative to the conventional DNB, if the more difficult (higher order moment) measurements are to be pursued. The short pulse length of the IDNB also lends itself to short detection gating strategies that should simplify the spectral analysis as well as increasing the SNR. We see that, even for the highest rep rate IDNB case, the Ti and V_{rot} measurements will be marginal for the very smallest radii at the highest expected densities. The use of other impurity species, e.g., C or Be is probably not of utility until their concentrations approach a few percent, which is unlikely for the presumed operation of ITER. Finally, the use of an alternate NIR He transition offers some advantages for making these measurements in the ITER environment, and we are continuing to investigate the relative merits of this option.

9. ACKNOWLEDGMENTS

Work supported by the U.S. Department of Energy under Contract Nos. DE-AC05-96OR22464 and DE-AC03-94SF20282, Subcontract ITER-GA-4002. The authors have benefited from numerous stimulating and informative discussions with C. Walker, G. Vayakis, V. Mukhovatov, K. Young, E. Marmar, R.C. Isler, and D.G. Whyte.

References

1. E.J. Strait, et al., Phys. Rev. Lett. **75**, 4421 (1995).
2. F.M. Levinton, et al., Phys. Rev. Lett. **75**, 4417 (1995).
3. K.H. Burrell, Phys. Plasmas **4**, 1499 (1997).
4. D.L. Hillis, et al., Phys. Rev. Lett. **65**, 2382 (1990).
5. M.R. Wade, et al., Phys. Plasmas **2**, 2357 (1995).
6. E.S. Marmar, *Diagnostics for experimental thermonuclear fusion reactors*, P.E. Stott, G. Gorini, and E. Sindoni, ed., Plenum Publishing Corp., New York, 281 (1996).
7. M.G. Von Hellerman, et al., *Diagnostics for experimental thermonuclear fusion reactors*, P.E. Stott, G. Gorini, and E. Sindoni, ed., Plenum Publishing Corp., New York, 321 (1996).
8. W. Mandl, Ph.D. Thesis, University of Munich, JET-IR (92) 05 (1992).
9. D.F. Finkenthal, Ph.D. Thesis, University of California, Berkeley (1994).
10. Profiles courtesy of D. Bouchet and G. Vayakis, ITER-JCT.
11. R. Hemsworth, A. Krylov, M. Hanada, *Proposal of a design of diagnostic neutral beam injector,* ITER Memorandum, December 1996.
12. ADAS documentation set, http://patiala.phys.strath.ac.uk/adas.
13. H.A. Davis, et al., Rev. Sci. Instrum. **68**, 332 (1997).
14. P.P. Webb, R.J. McIntyre, Properties of Avalanche Photodiodes, RCA Review, June 1974.
15. C.L. Hsieh, et al., Rev. Sci. Instrum. **61**, 2855 (1990).
16. D.M. Thomas, et al., Rev. Sci. Instrum. **68**, 1233 (1997).
17. D.V Orlinski, *Diagnostics for Experimental thermonuclear fusion reactors*, P.E. Stott, G. Gorini, and E. Sindoni, ed., Plenum Publishing Corp., New York, 51 (1996).
18. K. Young and A. Ramsey, *Application of fiber optics to present large tokamaks*, Proc. of ITER Technical Meeting of Irradiation Tests on Diagnostic Components, Garching, Germany (1996).

ACTIVE SPECTROSCOPY ON JET AND ITER

R.W. T. König, H. Anderson[2], P. Breger, W. G. F. Core, J. P. Feist, N. Hawkes[1], M. G. von Hellermann, A. C. Howman, A. C. Maas, P. D. Morgan, M. F. Stamp, H. P. Summers[2], J. Svensson, M. Tunklev[3], P. R. Thomas, K.-D. Zastrow

JET Joint Undertaking, Abingdon, Oxfordshire, OX14 3EA, UK
[1]UKAEA Government Division, Fusion, Culham, Abingdon, Oxon. OX14 3DB, UK (UKAEA/Euratom Fusion Association)
[2]University of Strathclyde, Scotland
[3]University of Lund, Sweden

INTRODUCTION

Ion temperature and velocity distribution functions, ion densities, plasma rotation velocities and magnetic fields have been identified by the ITER design team as key parameters[1], for which spatially-resolved profiles with medium time resolution are of vital importance in three major areas. These are: machine protection and plasma control, evaluation and optimisation of plasma performance, and investigation of the physical phenomena which might limit ITER's performance. Only a diagnostic based on active-charge-exchange (CXRS) and beam-emission spectroscopy (BES), using a dedicated neutral beam, can provide all these data with the accuracy and resolution required (table 1).

Table 1. Resolution and data accuracy requirements for those plasma parameters that can be provided by a CXRS diagnostic.

Parameter	Parameter range	Spatial resolution	Time resolution	Accuracy
v_{tor}	$1\ 10^3 - 5\ 10^4$ m/s	0.5 m	10 ms	30 %
Z_{eff} (steady state)	1 – 3	0.3 m	100 ms	10 %
Z_{eff} (transient)	1 – 3	0.3 m	0.1 ms	20 %
T_i	0.5 – 50 keV	0.3 m	100 ms	10 %
n_{He} / n_e	1 – 20 %	0.3 m	100 ms	10 %
$n_{im}p$ ($Z \leq 8$)	0.5 – 20 %	0.3 m	100 ms	20 %

As the specific requirements for a CXRS diagnostic have already been discussed at some length elsewhere, [2-4] we have concentrated here on a number of points in areas which still require further specification and where we have made considerable progress over the last two years.

Diagnostics for Experimental Thermonuclear Fusion Reactors 2
Edited by Stott *et al.*, Plenum Press, New York, 1998

BEAM REQUIREMENTS FOR AN ACTIVE CXRS DIAGNOSTIC ON ITER

A dedicated diagnostic beam with an energy of 120 keV/amu is required for ITER.[2] This specific energy results from the energy dependence of the CX excitation rates and the beam stopping cross-sections. As pointed out previously[2] it might be feasible to develop a neutral beam based on a negative ion source, which fulfils the requirements for a CXRS diagnostic. First experiments on JT-60U with this type of beam are very encouraging. The required low beam divergence of <3.5 mrad has already been demonstrated on this system.[5] Signal-to-noise considerations, however, indicate that the required time resolution of 100 ms cannot be achieved for central viewing lines. It might be possible to overcome this restriction with an intense pulsed ion beam from a diode, like the one presently under development at LANL.[6] Extrapolation from the existing source makes it appear feasible to develop one, which would improve the signal to noise ratio by about a factor of 30 compared to the CW beam, sufficient to fulfil the ITER requirements (tab. 1).

Table 2. Comparison of different neutral beam sources.

	[6]CX diagnostic beam required for ITER	ITER heating beam design (for comparison)	JT-60U	[7]Intense pulsed beam (under construction)	[7]Intense pulsed beam required for ITER
Beam source type	negative ion	negative ion	negative ion	Ion diode	Ion diode
Repetition rate	CW	CW	CW	1 – 30 Hz	300 Hz
effective beam power per source (MW)	10	16.7	11	3 kJ in 1.5 μs (2 GW)	10 kJ in 1.5 μs (10 GW)
energy (keV/amu)	120	1000	500	250	120
neutral beam current	80 A	40 A	22 A	12 kA	50 kA
beam divergence	0.2°	0.3 °	0.1° – 0.2°	0.8°	0.25 °
Dist. Source plasma centre / m	30				
beam dimensions in plasma centre (m)	$w_{\perp}$=0.1, $w_{l.o.s.}$=0.3				
central target density (m^{-3})	$1\cdot10^{20}$			$1.05\cdot10^{20}$	$1.05\cdot10^{20}$
beam path length (m)	2.8				
line integral to centre (m^{-2})	$2\cdot10^{20}$				
plasma composition	D(74%), He(10%)C(1%)				
Emission rates ($10^{15}m^3s^{-1}$)	He: 0.56, C: 5.09				
beam attenuation (and error)	0.013 (43%)				
integration time	500 ms				
S/N-ratio ($\tau_{exp.}$ = 100 ms) He-ash (line integrated T_i, v_{rot} (spectrally resolved)	~ 3 < 0.3			12 < 1	100 < 10

MODELLING AND MEASUREMENT OF THE BEAM ATTENUATION

It is the accuracy with which the local neutral beam density is determined that governs the accuracy with which the impurity density profiles can be measured. To control the helium ash density it is necessary to measure its central value to better than 10 %. This cannot be achieved by the current method of calculations of neutral beam attenuation. Extrapolating to ITER conditions – a central density of $1\cdot10^{20}$ m^{-3} and a 120 keV/amu

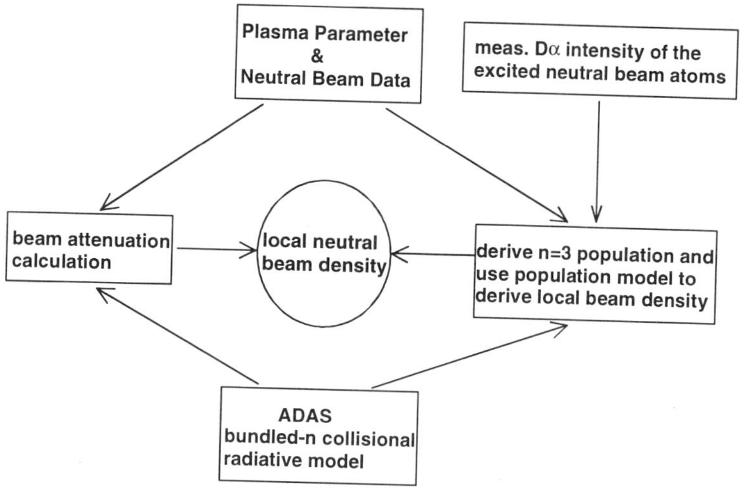

Figure 1. Methods for the determination of the local neutral beam density.

neutral beam with all measured plasma parameters and atomic data known to an accuracy of 10 %, an error of 40 – 50 % in the helium ash density is indicated.

Improved accuracy can only be achieved by measurement of the local neutral beam densities using BES. Such a diagnostic has been under development at JET for several years.[8] A major obstacle has been the complexity of the beam emission spectrum – up to 69 individual lines arising from 3 beam energy components, motional Stark splitting, normal and tangential beams, edge emission and impurity lines. However, reliable fitting routines have now been developed and Fig. 2 shows the result of such a fit.

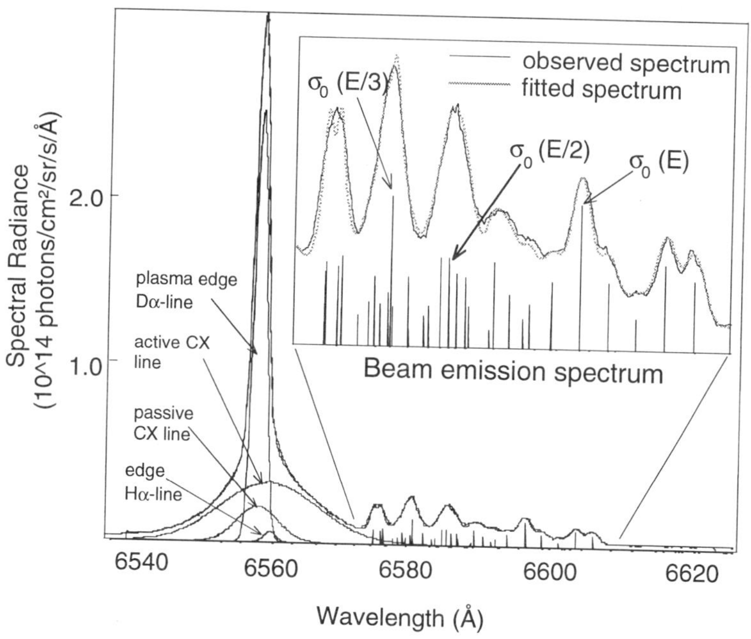

Figure 2. Neutral beam emission spectrum.

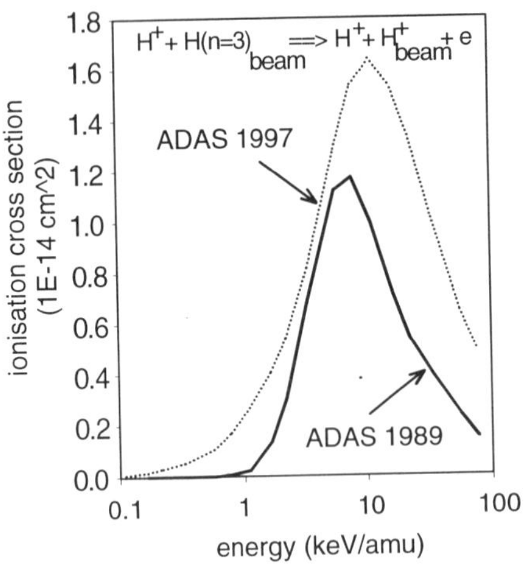

Figure 3. Cross sections for ionisation from n=3.

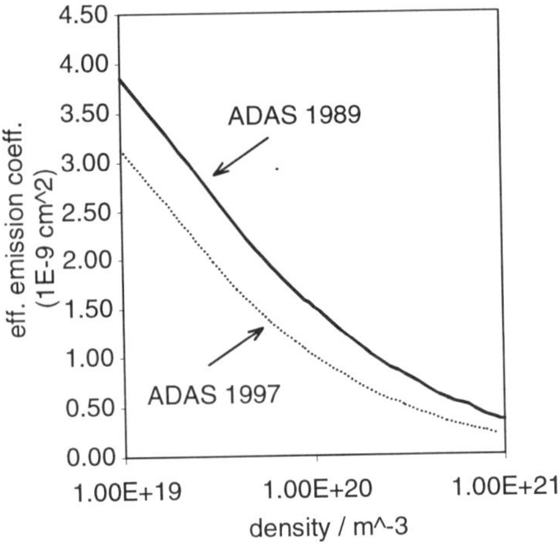

Figure 4. Effective Balmer-α emission coefficient as a function of the plasma density at an energy of 100 keV/amu.

In addition, early atomic data on ionisation cross sections were inaccurate, leading to differences of 100 % between beam neutral densities derived from BES and attenuation calculations. The adoption of improved atomic data, as implemented in the ADAS 'JET 1997 Ion Collision Database' has improved the agreement to better than 20 % (Figure 5). Figure 3 illustrates the old and current ionisation cross sections from excited H levels, while figure 4 shows the change in the effective Balmer-α emission coefficient for a 100 keV/amu neutral beam and for a plasma composition as stated in table 2.

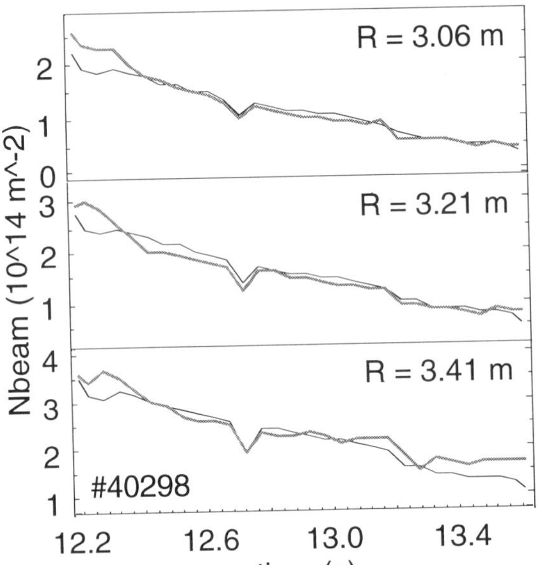

Figure 5. Measured (thick grey line) and calculated (solid line) local neutral beam density. The data are integrated along the l.o.s. through the neutral beam.

It is not only essential for ITER to have a real-time CX data analysis system, also at JET the need for such a system is ever increasing. In order to prevent disruptions, real time ion temperature gradients are for example required for the heating power control in optimised shear discharges. A neural network has been developed at JET which produces real-time ion temperature, toroidal rotation and CX line emissivity profiles.[9,10] The derivation of a 12 point radial profile takes about 5 ms on a conventional Pentium PC, which is well within the exposure times of 50 ms at JET and the anticipated 100 ms for ITER. The alternative scheme normally employed for CX data analysis is a comparatively slow interactive least-square minimisation procedure, which requires initial guesses by a skilled human operator and is therefore unsuitable for any real-time data analysis.

The neural network presently used at JET consists of a system of 22 (7x3+1) feed-forward networks with one hidden layer and sigmoidal activation functions (figure 6). Before a spectrum is passed on to one of seven groups consisting of three networks each, it has to pass a novelty detector test. The different groups are each trained for a particular range of radial positions in the plasma. Three networks were used in each group: one for ion temperatures, one for toroidal rotation velocities and one for the total integrated line intensity of the CX line. In the final step the results of the different networks are then put together to reconstruct real-time ion temperature, toroidal rotation velocity and CX line intensity profiles together with their error bars. A further separate network is used to predict the bremsstrahlung intensity underlying the CX spectra.

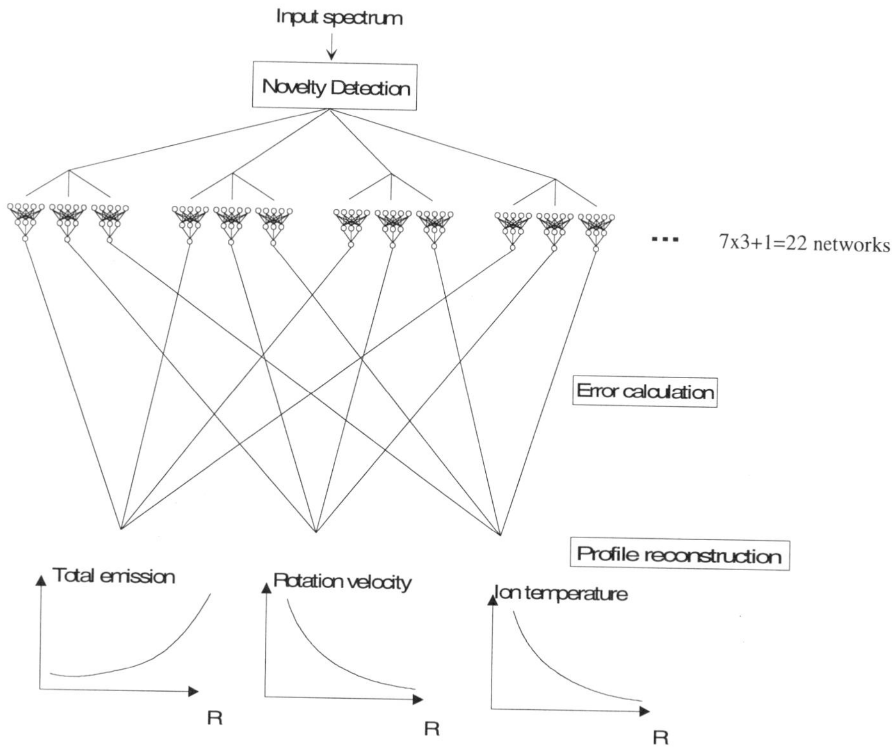

Figure 6. Structure of the neural network system presently used at JET.

For the training of the neural networks we did not follow the conventional way of directly using real data that had already been analysed with the least-squares method, since this would mean that after every change in the environment (change of the instrument function or dispersion of a spectrometer, new port location for the diagnostic causing a different Doppler-shift of the CX lines emitted from the rotating plasma, etc.) one would have to first analyse a large number of spectra with the least-squares method before one could re-train a neural network. (The training data set at JET consists of ~ 150000 analysed spectra analysed over several years.) We circumvented this problem by reconstructing the joint input space density in the space of the physical parameters that describe a full spectrum with all its spectral lines. Samples from this distribution are then used to produce synthetic spectra, which are likely to appear in a real JET experiment, with the correct noise characteristics and for any given new set-up of the diagnostic. The synthetic spectra can then be used to train individual neural networks for each parameter in the sample. This method has been successfully employed on JET to adapt the networks to a new viewing line geometry, without the need to first analyse a large number of spectra with the new system.

We have paid particular attention to the robustness of the neural network system since it is a major factor for its usefulness. A real-time novelty detector has therefore been developed, which prevents the network from analysing the data whenever novel spectra are presented to the input neurons, thereby allowing switching over to another control system. This novelty detector uses the Gaussian mixture estimate of the joint probability density in the space of the input neurons. [11] An input spectrum is regarded as novel when the probability density of the input vector is much smaller than for any point in the training data set. The advantage of using a novelty detector based on the Gaussian mixture method is that the calculation of the probability density is very fast; it takes only about the same amount of time as one feed-forward of the input data through the neural network. Figure 7 and 8 show an example of a case where a novel spectrum has been presented to the system at one time point (13.3 s).

The system also provides a real-time error estimate[10] for the output parameters T_i, Ω_{tor} and I_{CX} caused by the noise on the input signals (figure 9). A new method has been developed which allows to calculate the error to be calculated on the time scale of about one feed-forward of the net. This can be achieved by using a locally linear approximation to the feed-forward network function. The error of the output parameter can be calculated by using the values on the hidden layer of the previous feed-forward of the spectrum.

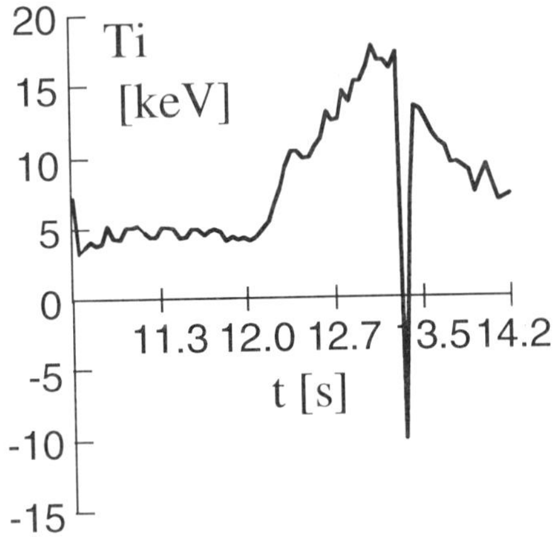

Figure 7. Network without novel spectrum detector.

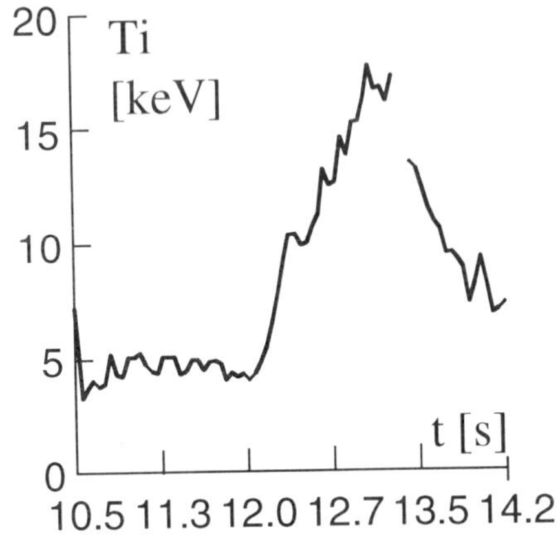

Figure 8. Novel spectrum detector prevents analysis at 13.3 s.

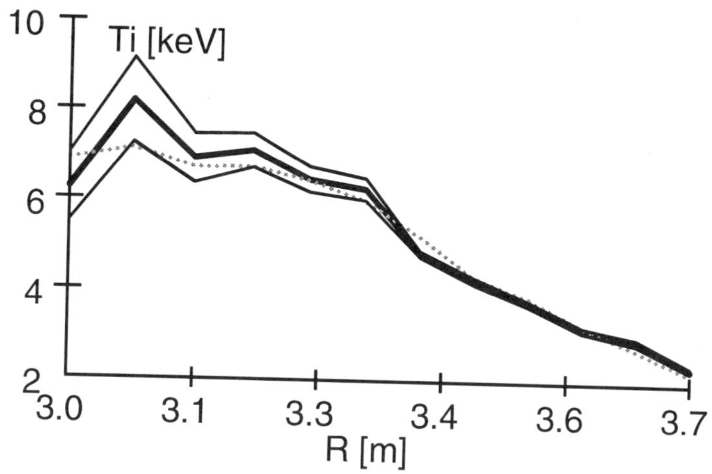

Figure 9. Neural network result (thick line) with error bars (thin lines) and least-squares fit result (dotted grey line)

IN-SITU CALIBRATION CONTROL WITH VISIBLE BREMSSTRAHLUNG

Maintaining the absolute calibration of the optical system will be a major challenge for an ITER CX diagnostic, since over long operating periods one has to expect coatings on optical components, as well as increasing transmission losses due to high neutron fluences. A possible solution for this problem could be the simultaneous derivation of Z_{eff} profiles from active CX measurements of all important low Z impurity ion density profiles and from Abel inversion of the bremsstrahlung signals underlying the CX spectra, as shown in figure 10.[12] On JET we can measure impurity density profiles of all relevant ions simultaneously, i.e. of He, Be, C, N and Ne. For the Abel inversion of the bremsstrahlungs signal we have so far used only the 12 l.o.s. that each of the three individual CX systems on JET provide.

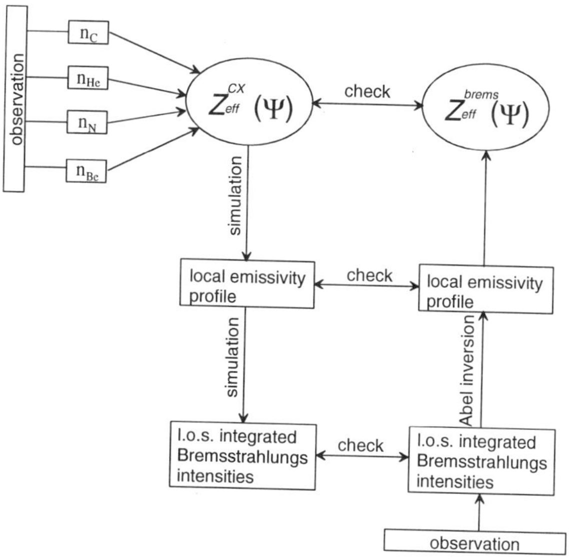

Figure 10. Principle of the self-consistency check for the channel to channel relative calibration control of a CX diagnostic system utilising the bremsstrahlung signals underlying each of the CX spectra.

With a least-squares fitting routine one could then vary the transmission factors for each l.o.s. until both methods give identical Z_{eff} profiles. Further cross checks can be undertaken at each of the intermediate steps shown in fig. 10. Implementation at JET has shown, that the main problem is that the Z_{eff} derived from bremsstrahlung is, in general, too large in the outer 1/3 of the plasma (figure 11). This phenomenon has been observed on many machines and is still not understood.

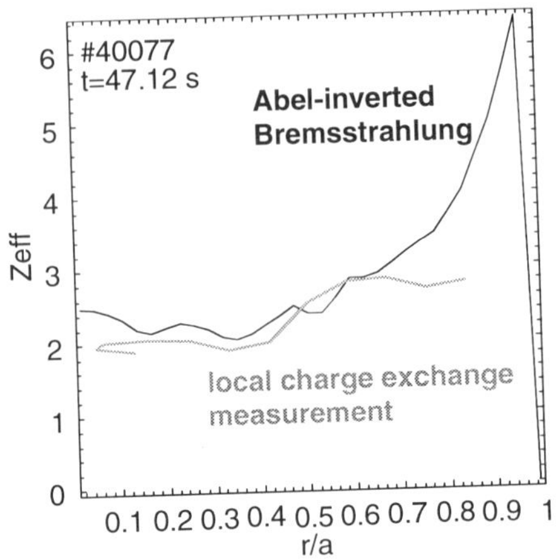

Figure 11. Z_{eff} profile comparison

RADIATION EFFECTS ON HEATED AND UNHEATED OPTICAL FIBRES

Fibre loops and blind fibres running right up to the optical heads of several diagnostics on JET are used to measure the transmission losses and the luminescence produced in these fibres by the neutron bombardment during tritium experiments.[13-15] So far the aluminium clad, wet fibres used at JET show no detectable long term transmission losses provided they are kept at ~250 °C. On JET the fibres inside the optical periscope are at about 250 – 300 °C determined by the vessel temperature of 300 °C. From the end of the periscope to outside the biological shield a special heated hose is used, which keeps the fibres at 250 °C. It will be difficult to extrapolate the transmission losses to ITER, since the neutron fluence inside the JET ports is 10^4 – 10^8 times smaller than on ITER behind a periscope or inside the diagnostic channel. We may, however, learn in the course of the DTE1 (Deuterium Tritium Experiment) experiments what the highest acceptable neutron fluences for heated optical fibres are, i.e. from which location onward one can use on ITER optical fibres instead of complicated mirror links. Moreover, JET is suitable to study the effect of fibre luminescence on the background light level, since the neutron fluxes at the JET periscopes are only 2 – 10 times smaller than inside the ITER diagnostic channels (fig. 12). The effect on Z_{eff} derived from bremsstrahlung certainly can not be neglected.[16] Since the intensity of the Cerenkov radiation, which is the main source of the fibre luminescence for heated fibres, is proportional to $1/\lambda^2$, it might be worthwhile to investigate whether the He CX-line at 1013.6 nm is more suitable for He-ash density profile measurements than the currently used line at 468.6 nm, because the additional 'background intensity' due to fibre luminescence would be ~5 times smaller there.

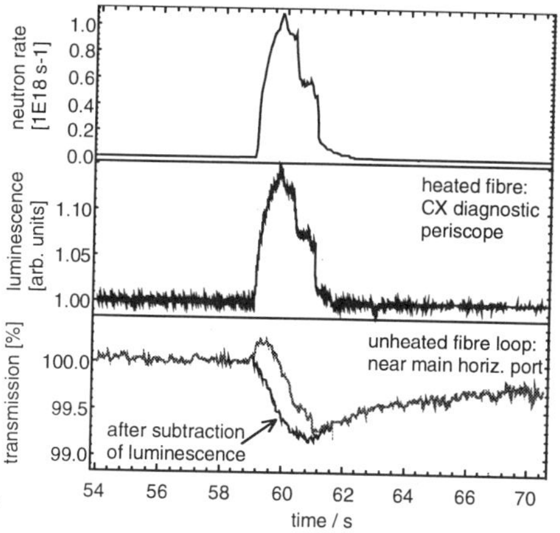

Figure 12. Observed neutron induced luminescence and transmission losses for a JET tritium pulse.

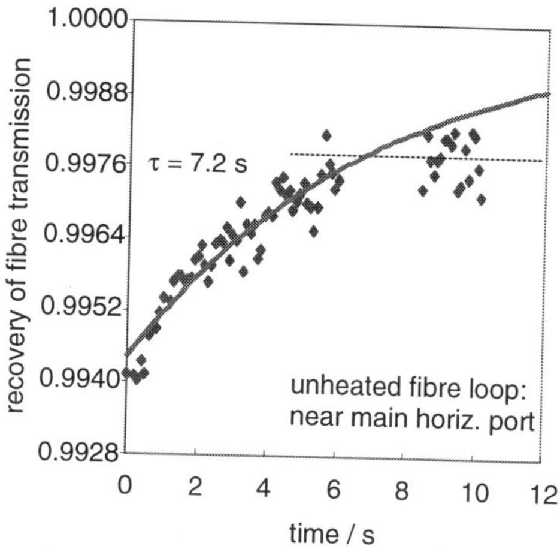

Figure 13. Recovery of an unheated fibre from induced absorption.

First experiments during the DTE1 campaign with unheated fibres running up to a position near a main horizontal port show only a small luminescence signal due to the relatively large distance to the plasma, i.e. to the neutron source (fig. 12). Nevertheless, the unheated fibre shows clearly increasing transmission losses in the course of the pulse reaching a maximum of ~0.7 % towards the end of the pulse (fig.13). At room temperature it takes about 7 s until the transmission losses have reduced to about 0.2 %.

SUMMARY

An active charge exchange and beam emission diagnostic will be absolutely vital for the control and analysis of ITER plasmas. There exists no alternative diagnostic that can monitor the helium–ash and the other low-Z impurity density profiles. Ion temperature and rotation velocity profiles can, in principle, also be measured by passive emission spectroscopy, but it will be difficult to achieve the required spatial resolution of $\Delta r/r < 0.05$. Any passive technique, like X-ray spectroscopy, neutral particle energy analysis or neutron spectroscopy, though very valuable as additional source of information, can only provide line-of-sight integrated parameters, i.e. non-localised quantities. Local values can only be derived indirectly from these measurements by tomographic techniques or by calculating the location and radial width of emission shells. For example, for an X-ray diagnostic system one requires a large fan of l.o.s. each dedicated to a specific ionisation stage of a specific ion, this often requiring the seeding of the plasma with some amount of high-Z impurities. The radial resolution will then depend on the width of the emission shells and therefore vary in the course of a pulse with changes in the plasma parameters, making the whole system not only technically difficult to install, but also the derivation of ion temperature and rotation velocity dependent on other parameters like electron temperature and density as well as atomic physics data. The charge exchange diagnostic has an overwhelming advantage over the other techniques, since it provides a direct local measurement of these parameters with good spatial resolution.

A CX diagnostic with a modulated negative ion source beam could just about meet the requirements on spatial and time resolution as well as accuracy over a considerable part of the plasma. The strong neutral beam attenuation towards the plasma centre would however mean that one had to accept integration times of up to 1 s for central viewing lines. It seems feasible to accomplish all requirements on time and spatial resolution as well as accuracy, if an intense pulsed ion diode based diagnostic beam were used. This would require some investment in R&D.

It is essential to combine the CX with a BES diagnostic since He-ash and other impurity densities can only be derived with the required accuracy if the local neutral beam densities are directly measured with a BES diagnostic. The recently reached accuracy of about 20 % at JET raises hopes that a further improvement to about 10 % within the next few years is achievable.

It has recently been demonstrated on JET that CX data can be analysed in real-time and can therefore be used for machine feedback control. A robust neural network system has been developed at JET, which includes real-time novelty detection and error analysis.

Experiments on JET show that in-situ calibration control with visible bremsstrahlung might be possible, but that the cause for the strong emission from the outer part of the plasma has to be understood first.

During the initial phase of the DTE1 experiments the heated optical fibres used at JET have so far shown no radiation induced transmission losses. If any long term transmission losses are observed later in the program they will give an indication of what the maximum permissible neutron fluences would be for ITER and therefore from which point onward one can use optical fibres instead of mirror links.

The effect of fibre luminescence can be studied on JET, since the neutron flux is nearly comparable to levels expected in the ITER diagnostic channels. The luminescence can lead to a considerable apparent increase of the bremsstrahlung background that needs to be corrected for.

REFERENCES

1. A.E. Costley et al., in “Diagnostics for Experimental Thermonuclear Fusion Reactors”, ed. by P. E. Stott, G. Gorini, E. Sindoni, Plenum Press, New York & London, 1996, p. 23
2. M.G. von Hellermann et al., in “Diagnostics for Experimental Thermonuclear Fusion Reactors”, ed. by P. E. Stott, G. Gorini, E. Sindoni, Plenum Press, New York & London, 1996, p. 321
3. E. S. Marmar, in “Diagnostics for Experimental Thermonuclear Fusion Reactors”, ed. by P. E. Stott, G. Gorini, E. Sindoni, Plenum Press, New York & London, 1996, p. 281
4. D. M. Thomas et al., Progress Meeting on the Design of ITER Diagnostic Systems, San Diego , 27 Feb. 1997
5. H. Horiike, private communications, 1997
6. M. G. von Hellermann et al., Plasma Phys. Contr. Fusion **35**, 799 (1993)
7. C. W. Barnes and H. A. Davis, Progress Meeting on the Design of ITER Diagnostic Systems, San Diego , 27 Feb. 1997
8. W. Mandel et al., Plasma Phys. Contr. Fusion **35**, 1373 (1993)
9. R. König,et al., 21st EPS Conference on Controlled Fusion and Plasma Physics, Montpellier, p. 1264 (1994)
10. J. Svensson, M. von Hellermann, R. W. T. König, in ‘*Neural Networks in Engineering*, ed. by A. B. Bulsari and S. Kallio, p. 321 (1997)
11. C. M. Bishop, ‘Novelty detection and neural network validation’ in IEE Proc.-Vis. Image Signal Process., Vol. 141, No. 4, August 1994
12. J. P. Feist, R.W. T. König, P. Breger, M. G. von Hellermann, A. C. Howman, ‘Selbstkonsistente Messung von Verunreinigungsionen in einem Fusionsplasma’, Verhandlungen der DPG, Mainz , 1997
13. A. T. Ramsey, H. G. Adler, and W. Tighe, ‘The use of optical fibres on the next generation of tokamaks’, 36th Annual Meeting, APS Division of Plasma Physics, Minneapolis, USA, 7 - 11 November 1994
14. A. T. Ramsey, Rev. Sci. Instrum. **66**, 871 (1995)
15. A. T. Ramsey, W. Tighe, and J. Bartolick, Rev. Sci. Instrum. **68**, 632 (1997)
16. P. D. Morgan, ‘Irradiation of optical fibres at JET through 14 MeV neutron production’, Proceedings of the 17th Symposium on Fusion Technology, Rome, 14- 18 September 1992, p. 722

MEASUREMENT OF THE H/D/T MIXTURE BASED ON DOPPLER-FREE TWO-PHOTON SPECTROSCOPY AND NEUTRAL BEAM INJECTION

Dietrich Voslamber and Wilhelm Mandl

Ass. EURATOM-CEA, DRFC-SPPF, CEN Cadarache
F-13108 St. Paul-lez-Durance

INTRODUCTION

Knowledge of the density ratios of the hydrogen isotopes is of prime importance in magnetic confinement fusion (MCF). The measurement of n_{D+}/n_{T+} in the plasma core is listed in "category (1)" of the diagnostics required for ITER, i.e. in the category that is considered to be indispensable for machine protection and plasma control. The determination of n_{H+}/n_{D+} in the plasma core (plasma dilution) and of n_D/n_T at the plasma edge is included in "category (2)", i.e. in the list of measurements for performance evaluation and optimisation.

For measurements at the plasma edge a local and time-resolved diagnostic, based on Doppler-free two-photon laser induced resonance fluorescence (in Ly-α), has been suggested previously[1,2]. Here we address the question whether this method can be extended farther into the plasma bulk by using a neutral diagnostic beam and exploiting the thermal neutrals that are created by charge exchange (CX) between the beam neutrals and the plasma ions.

DOPPLER-FREE TWO-PHOTON INDUCED FLUORESCENCE

In MCF plasmas the Doppler widths of the hydrogen lines are much larger than their isotope shifts. Isotope-selective resonance fluorescence can thus only be achieved by eliminating the linear Doppler broadening in the excitation process. This can be done using two-photon excitation in two anti-parallel laser beams, each having half the atomic resonance frequency and providing one photon to an atomic excitation. The Doppler shifts of the two photons have then opposite sign and cancel each other in the frequency sum. Hence all atoms, irrespective of their velocity, are in resonance with the two laser beams. While the two-photon absorption lines are thus Doppler-free, the fluorescence lines are not; these, however, can be separated by introducing a short time delay between the two laser shots for D and T.

Among the various possible transition schemes, we consider here the case where both excitation and fluorescence occur between principal quantum numbers $n = 1$ and $n = 2$ (Lyman-α). The $n = 2$ levels form a triplet due to the Stark-Zeeman effect that is caused by

Diagnostics for Experimental Thermonuclear Fusion Reactors 2
Edited by Stott *et al.*, Plenum Press, New York, 1998

the magnetic field $\vec{B}$ and the electric field $\vec{v} \times \vec{B}$ ($\vec{v}$ being an atom's velocity). Only the central component of this triplet can be excited efficiently, because its position is nearly unperturbed by the external fields and gives rise to a narrow peak in the two-photon absorption profile. Spectral profiles of this peak are shown in Fig.1 for D and T atoms in a plasma with equal portions of D and T. Their normalisation is such that the surface under the curves is proportional to the frequency-integrated two-photon transition strength and to the fluorescence intensity perpendicular to the magnetic field. The profiles are shaped by quadratic (relativistic) Doppler broadening, by motional Stark broadening (effect of the $\vec{v} \times \vec{B}$ field on the fine-structure components), by plasma Stark broadening (predominantly due to atom-ion collisions) as well as by distorting effects arising from fine and hyperfine splitting. Fig.1 does not include possible broadening effects arising from the laser (finite angle divergence and finite bandwidth). The total line width is seen to be at most a few GHz. This is much smaller than the 113 (335) GHz in laser frequency which separate the D from the T (H) lines.

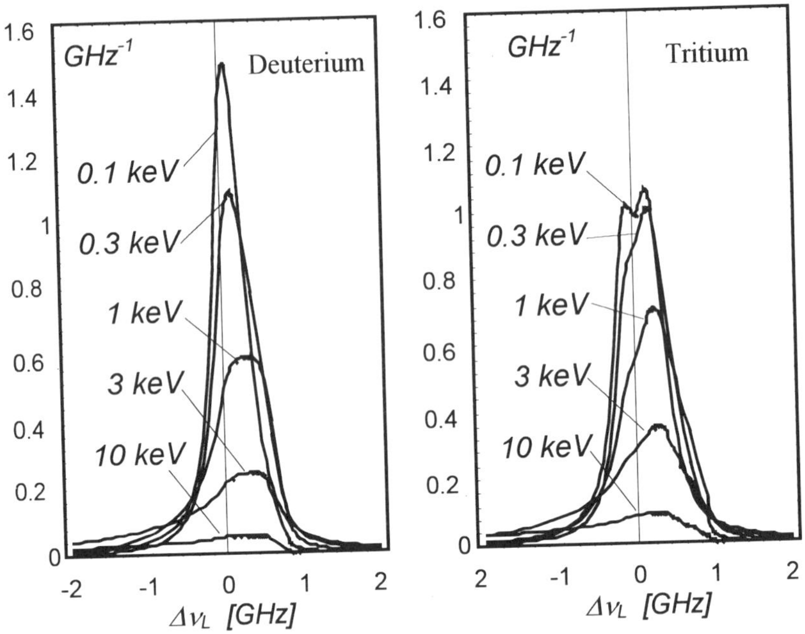

Figure 1. Spectral profiles (as functions of laser detuning) of the central Stark-Zeeman component of D and T atoms in a D/T plasma for various temperatures. ($B=4T$, $n_e=3\ 10^{19}m^{-3}$, $n_{D+}/n_{T+}=1$, $Z_{eff}=1$)

The peak intensities of the profiles are seen to decrease noticeably with increasing temperature. The reason is that for larger temperatures one has both a lowering of the two-photon transition strength and a rise of the quadratic Doppler width. Unlike the peak intensities themselves, their ratios $I_H/I_D/I_T$ taken from H, D and T profiles at identical plasma conditions depend only weakly on the plasma parameters. Therefore, rough knowledge of these parameters from independent diagnostics suffices for the determination of the mixture $n_H/n_D/n_T$.

DETECTION LIMIT AND EXPERIMENTAL REQUIREMENTS

Assuming the signal-to-noise ratio to be dominated by photon statistics, the detection limit n_c for the neutral densities n_H, n_D and n_T is found to be of the order of $10^{14}\ m^{-3}$. This number is based on customary optical parameters and on the assumption that the background radiation around the Lyman-α wave length is of the order of 10^{19} *photons* $m^{-2}s^{-1}sr^{-1}$, which is characteristic of the cool edge emission of tokamaks of any size (far from limiters, antennas and cold gas injection ports). Bremsstrahlung usually remains below this level, even in a tokamak like ITER. The same is true for the background radiation from excited beam-created neutrals (in case a neutral beam is used), except for such high neutral densities where the signal-to-noise ratio is so large that the background radiation can be ignored anyway.

While n_c is relatively independent of the plasma parameters, the laser characteristics that are required for producing sufficient fluorescence signals strongly depend on them. In particular, the pulse energy has to increase with temperature (roughly as $T^{3/2}$) so as to compensate for the decreasing peak intensities of the two-photon absorption profile (see Fig.1). Up to about *1-2 keV*, *30-50 mJ* pulse energy suffice to attain signal-to-noise ratios of the order of two at neutral densities of the order of n_c. However, nearly *1 J* laser energy is necessary to achieve the same at *15 keV*.

With the laser spectrometer developed at PTB Berlin[3] (tunable at 243 nm, bandwidth $\leq$ *0.5 GHz*, beam divergence $\approx$*0.3 mrad*, pulse duration *2.5 ns*, pulse energy up to *50 mJ*) the diagnostic should be applicable in edge and divertor regions of large tokamaks, such as ITER; it should further be applicable in the main plasma up to temperatures of *1-2 keV* if neutral densities in excess of n_c can be generated with the help of a neutral beam (see next section). Further progress in laser technology (*100 mJ* in each laser beam are foreseen for the near future) should allow measurements at *5-10 keV*.

OPERATING WITH A DIAGNOSTIC NEUTRAL BEAM

The diagnostic can be extended well beyond the edge and information can be won on the isotopic ion mixture $n_{H+}/n_{D+}/n_{T+}$ if a neutral particle beam is used whose characteristics are such that they assure efficient creation of thermal neutrals via CX reactions between the beam atoms and the plasma ions[4]. Choosing, for instance, a co-axial or nearly co-axial arrangement of the neutral particle beam and the laser beams, one would be able, with a single beam geometry and different viewing lines, to probe all those points along the beam line where the detection limit n_c is attained. However, for radial beam injection this necessitates a retro-reflector at the inner wall, which may be problematic for ITER. Different geometrical choices are, of course, possible. A co-axial arrangement of the viewing line with the laser beams[5] may be suitable when signal maximisation and background minimisation are important issues.

To study the potential of such diagnostics, we have done a model calculation for a D beam penetrating a D/T plasma that may contain some C^{6+} impurities ($Z_{eff}=1$ and 2)[4]. The beam attenuation was evaluated accounting for ionising collisions (including excited atomic states) with electrons and ions and for CX. From the birth rate of thermal neutrals due to CX and their loss rate due to ionisation and diffusion out of the neutral beam the neutral density profiles $n_D(r)$ and $n_T(r)$ were calculated and maximised to find the optimal beam particle energy $E_{opt}(r)$. The beam power $P_{crit}(r)$ needed for attaining n_c was then calculated as a function of r. The cross sections for the various reactions were taken from ADAS and Aladdin[6].

We have considered two ITER scenarios (A and B) with density profiles $n_e=n_{el}+(n_{e0}-n_{el})(1-(r/a)^2)^p$, where $n_{e0}=1.4\ 10^{20}\ m^{-3}$, $n_{el}=0$, p=.5 for case A, and $n_{e0}=1.3\ 10^{20}\ m^{-3}$, $n_{el}=5\ 10^{18} m^{-3}$, p=.1 for case B. For the temperature, $T=20\ keV\ (1-(r/a)^2)^2$, was chosen for case A, while for case B it was taken from a model calculation for ITER[7] (flat rise to *13 keV* at $r/a=1$, then steep rise to *27 keV* at $r/a=0$). A beam radius of *10 cm* was chosen in both cases.

Figure 2 shows the optimal beam particle energy E_{opt} and the required beam power P_{crit} as a function of the penetration depth $l = a - r$ of the diagnostic (distance from the edge)*. It is seen that with the *5 MW* power planned for the diagnostic beam of ITER the diagnostic

*Differences with respect to Ref.8 are due to the more accurate optimisation procedure and refined beam halo model used in the present calculations.

would cover about *1 m* of the outer plasma radius (assuming *60-80 keV* beam particle energy). In order to achieve this, the pulse energy per laser beam would have to be at least *100 J*. If *1 GW* beam power (at ≈100 keV) were available (short-pulse (≈ *1 μs*) multi-GW beams are under development but not yet accomplished[9]), the diagnostic might cover about *2 m* of the outer plasma radius of ITER, provided that the laser pulse energy attains *700 mJ*.

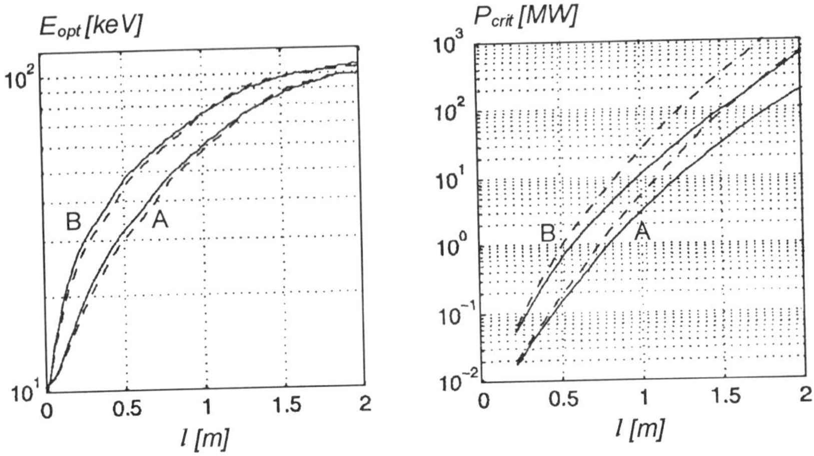

Figure 2. Optimal beam particle energy and required beam power as a function of the penetration depth *l* of the diagnostic. Curves A and B represent two different ITER scenarios. ——— $Z_{eff} = 1$, — — — $Z_{eff} = 2$.

FURTHER DIAGNOSTIC OUTPUT

Besides the isotope mix, further diagnostic output can in principle be obtained by tuning the laser over the atomic resonances (assuming narrow laser bandwidth) and also by exciting the beam atoms in addition to the atoms generated by CX. Information on the ion temperature *or* Z_{eff} (according to whether quadratic Doppler or plasma Stark broadening prevails) can be obtained from the line profiles. By crossing the two laser beams under a small angle, the ion velocity distribution perpendicular to the laser beams (implying temperature and plasma rotation) can be inferred from the linear Doppler broadening thus introduced. Excitation of the beam neutrals (in the π components of the Stark triplet) induces fluorescence lines largely Doppler-shifted from those of the CX-created neutrals. Their polarisation (parallel to the electric field $\vec{v}_{beam} \times \vec{B}$) would yield information on the q-profile analogously to the classical MSE experiments, while their intensity would be a measure of the beam attenuation.

REFERENCES

1. D. Voslamber, report EUR-CEA-FC-1387 (1990)
2. D. Voslamber, in Spectral Line Shapes, Vol.8 (AIP Conf. Proc. 328), Ed. A.D.May, J.R. Drummond, E. Oks, AIP Press, New York 1995
3. K. Grützmacher *et al*, in Diagnostics for Experimental Thermonuclear Fusion Reactors, Ed. P.E. Stott, G. Gorini, E. Sindoni, Plenum Press, New York 1996
4. D. Voslamber and W. Mandl, report EUR-CEA-FC-1607 (1997)
5. K. Grützmacher *et al*, these Proceedings
6. H.P. Summers, report JET-IR (94) 06 and Aladdin data base, ORNL-6086/V1 (1985)
7. I. Voitsekhovitch, private communication
8. D. Voslamber and W. Mandl, Proceedings of the 23rd EPS Conference on Controlled Fusion and Plasma Physics (Kiev 1996), Vol. 20C, part III, p.987
9. D.J. Rej *et al*, Rev. Sci. Instrum. 63, 4934 (1992)

QUANTITATIVE MEASUREMENTS OF TWO-PHOTON INDUCED Lα FLUORESCENCE: A PROOF FOR TOKAMAK EDGE DIAGNOSTICS

K. Grützmacher, A. Steiger, Ch. Seiser, M.I. de la Rosa and U. Johannsen

Physikalisch-Technische Bundesanstalt (PTB), Berlin, Germany

ABSTRACT

Future application of two-photon induced Lyman-α fluorescence (2γ-Lα) measurements as an isotope selective determination of small neutral densities (H, D, T) in magnetically confined fusion plasmas require quantitative studies of this novel technique. A first feasibility study has been terminated recently in a close cooperation between PTB and Max-Planck Institute of Plasma Physics (IPP), Ass. EURATOM, at the Plasma Generator[1] PSI-1 at IPP. Quantitative 2γ-Lα measurements were mainly carried out in order to proof the detection sensitivity achievable at PSI-1. Deuterium densities as small as 10^{10} cm^{-3} were determined in a measurement volume of only 0.1 cm^{-3}, while the background radiation was like commonly stated for tokamak plasmas. Relying on theoretical predictions[2,3,4], we are therefore convinced,that measurements of the n_T / n_D ratio from 0.1 to 10 will be possible in the edge plasma of ITER, with single shot accuracy better than 20% and with the rquired spatial resolution. The detection limit for each isotope is expected to be well below 10^9 cm^{-3}.

EXPERIMENTAL EQUIPMENT AND MEASUREMENT CONDITIONS

Although the principle of 2γ-Lα fluorescence is very simple,quantitative measurements of small neutral densities of hydrogen isotopes in the magnetically confined plasmas beam of PSI-1 required extensive experimental developments in order to achieve the demanded measurement performance which is depending on:

- the quality of the laser-spectrometer generating pulsed tunable narrowband radiation around 243 nm required for two-photon excitation,
- the properties of the detection system for measuring the 2γ-Lα fluorescence,
- the VUV background emission of the PSI-1 plasma in the spectral region around 121 nm.

Because of the small two-photon absorption probability, the method requires pulsed tunable laser radiation of high spectral quality at 243 nm with much more power than provided by common laser systems. A new type of solid-state laser-spectrometer has been developed

Diagnostics for Experimental Thermonuclear Fusion Reactors 2
Edited by Stott *et al.*, Plenum Press, New York, 1998

based on a commercial injection seeded Q-switched Nd:YAG laser, a single mode OPO/Ti:Sapphire system pumped by the second harmonic of the Nd:YAG,and a BBO crystal for the final generation of 243 nm radiation by frequency mixing. This table-top system delivered typically 30 mJ pulse energy in 2.5 ns with a bandwidth of about 300 MHz at a repetition rate of 10 Hz. Pointing stability (about 10^{-5}) and shot to shot reproducibility are excellent,and the spatial laser beam profile is sufficiently good. Frequency control of the laser output at 243 nm is provided by Doppler-free optogalvanic measurements in a reference cell containing atomic hydrogen or deuterium due to thermal dissociation.

The detection system was optimized to achieve high selective sensitivity for Lα-radiation at 121 nm, high temporal resolution, reduction of the dc background signal by gating in order to avoid degradation of the photomultiplier. Properties of the system are given below:

imaging:	1:1, f = 125 mm MgF_2 lens, ø = 50 mm
photomultiplier:	particle detector + MgF_2 window (special design, custom made)
	temporal resolution: 2.5 ns, gain of about 10^6
	electronic gate: 8 ms on, 60 times suppression of the dc signal
spatial filter:	aperture in the image plane : 3 mm x 16 mm
spectral filtering:	the photomultiplier itself, sensitivity range: 115 nm - 140 nm
	Lα-interference filters, optional for stepwise attenuation
	O_2 gas-filter, pressure dependent variable attenuation

The O_2 filter provided excellent suppression (i.e. two orders of magnitude) of the strong molecular background radiation emitted by the PSI-1 plasma, while the Lα-radiation was attenuated only by a factor of about 3. The absolute sensitivity of the entire detection system was determined several times, and was found to be sufficiently stable and reproducible. For this, a Lα-lamp was used, which has been calibrated as Lα-radiation transfer-standard by the PTB laboratory for VUV radiometry at the electron storage ring BESSY I.

The laser-spectrometer was set up in a nearby laboratory and the 243 nm radiation was directed to the PSI-1. After beam preparation, finally two counterpropagating circularly polarized beams of equal pulse energy - typically 10 mJ each - were directed into the PSI-1. The measurement volume, at a distance of 20 m from the laser, was as small as 0.1 cm^3 (Ø 3 mm x 15 mm). The detection system was situated on top of the target chamber viewing downwards.

The plasma generator PSI-1 is providing a magnetically confined horizontally propagating plasma beam of 100 mm in diameter. For the 2γ-Lα measurements, the discharge was operated in pure deuterium or various hydrogen-deuterium mixtures at currents from 70 A to 400 A. The corresponding parameters of the plasma beam were: Ne in the range of 10^{12} cm^{-3} to $5 \cdot 10^{12}$ cm^{-3}, Te in the range of 5 eV to 10 eV, magnetic field strength about 0.1 T, total pressure in the target chamber about 10^{-3} mbar.

DATA RECORDING, EVALUATION AND RESULTS

Typical measurements consisted of 600 recordings of the detector signal for subsequent laser pulses (10 Hz rep. rate). Such measurements were performed with the laser frequency either fixed at the center of the resonance or tuned across the resonance in order to obtain spectral profiles. For each laser pulse the Lα-detector signal at 50 Ω was recorded at 2 GHz sample rate with a fast digital storage oscilloscope for a time interval of 100 ns. Additionally the optogalvanic signal was recorded as frequency monitor. The recorded 2γ-Lα signal represents the 2γ-Lα fluorescence and the Lα radiance of the stationary plasma, see Fig.1.

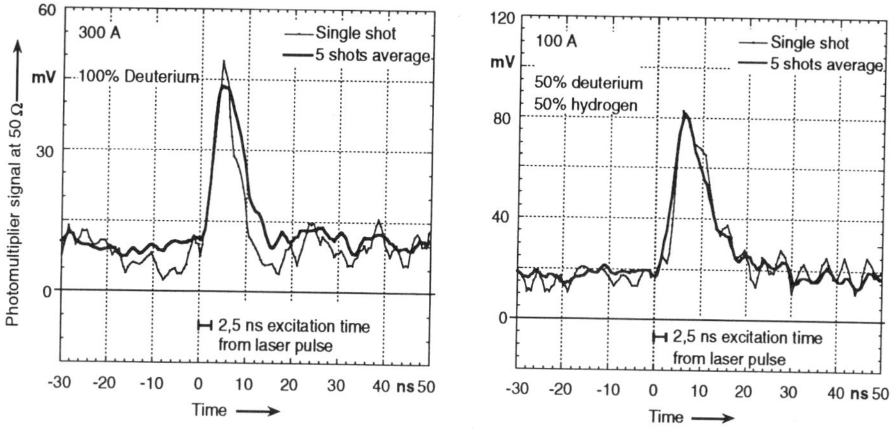

Fig. 1: Single shot signals compared to 5 shot averages, obtained for deuterium at different discharge conditions and different detector configurations. Immediately with the laser excitation appears well pronounced the 2γ-Lα signal. Left figure: deuterium discharge at 300 A. Right figure: discharge with 50% hydrogen and 50% deuterium at 100 A, corresponding to an atomic number density of about $5 \cdot 10^{10}\,cm^{-3}$ and Tokamak Lα background radiance.

From the Lα signals recorded, various results were evaluated as described below, e.g.:

- A statistical analysis exhibited, that the standard deviations of the signals were close to the limit of the Poisson statistic corresponding to the numbers of detected photons. Photon numbers corresponding to 2γ-Lα signals and the background were obtained from integrated photomultiplier signals and the photomultiplier gain (measured to be $8.5 \cdot 10^5$).
- The background radiance emitted from the stationary PSI-1 plasma was determined via the calibrated Lα sensitivity of the detection system. At currents of 100 A the radiance is equal to the Lα radiance commonly stated for tokamak plasmas, i.e. 10^{15} photons/(s·sr·cm^2). However, at the highest discharge current of 400 A investigated, the radiance of PSI-1 is twenty times larger compared to tokamak plasmas. This was a severe difficulty at the beginning of the measurements.

Beside of the calibrated detection sensitivity the determination of absolute atomic number densities requires the knowledge of the 2γ-Lα fluorescence yield Y_α, i.e. the probability per atom to emit a two-photon induced Lα photon.

- Y_α depends on the 2γ absorption profile and on the spectral, temporal and spatial distribution of the irradiances provided by the laser beams in the excitation volume. After the correct treatment of the spectral profiles (laser and 2γ absorption) rate equations accounting for two-photon excitation and de-excitation, as well as for the detrimental photo-ionization due to a third photon were time integrated for the temporal laser pulse shape. For the conditions at PSI-1, the maximum value of Y_α was found to be 10% for an time averaged (2.5 ns) irradiance of 100 MW/cm^2. This fits well with the chosen excitation conditions: about 10 mJ in 2.5 ns for each beam with 3 mm ø. Finally, the spatial integration was done for the beam profiles, which were recorded with a CCD camera.
- Compared to the ideal case of homogeneous irradiance, the distributions in the laser beam profiles led to a loss of about 50% in the integrated Y_α due to areas, where quenching by photo-ionization was dominant, e.g. for irradiances exceeding 150 MW cm^{-2}. However, this is partially compensated by the advantage, that - for higher irradiances and corresponding pulse energies - Y_α is fairly insensitive with respect to pulse energy fluctuations.

- Finally, atomic number densities of D and H were determined and found to range from 10^{10} cm^{-3} to 10^{12} cm^{-3}, depending on plasma parameters and isotope mixtures.

In addition to Doppler-free excitation by exactly counterpropagating laser beams, a small crossing angle of 52 mrad for the two laser beams was used. This causes a residual Doppler width of 2.6 % which dominates the 2γ absorption profile. From the residual Doppler width, a kinetic temperature of 1.6 eV was derived for atomic deuterium.

CONDITIONS AND SET UP FOR ITER EDGE AND BULK n_T/n_D RATIO

Considering tokamak edge diagnostic, the real advantage of 2γ-Lα measurements turns out for the determination of isotope density ratios, because ratios of the predicted two-photon transition strength for H, D and T are almost independent[2,3] on variations of the temperature (0.3 keV to 3 keV) and of Z_{eff} (1 to 3). 2γ-Lα fluorescence signals for two different isotopes represent therefore quite well directly the ratio of the isotope densities.

In order to obtain best 2γ-Lα fluorescence yield Y_α for ITER edge diagnostic[5], the laser setup should provide for each isotope two times 100 mJ pulse energy in 1.5 ns at 243 nm, 1 GHz bandwidth. The beams have to enter counterpropagating at the Mid Horizontal Port 8 (or Port 7) and Port 3. The detection system (observation angle 100 msr) has to be positioned in Port 8 or 7 and views side on the excitation volume (10 mm ø, e.g. 50 mm long). This setup will provide single shot n_T/n_D ratios, with a detection sensitivity well below 10^9 cm^{-3}.

Concerning the demanded n_T/n_D ratios in the bulk plasma of ITER, Voslamber[6] investigated the enhancement of ground state atom densities - which present the local ion densities - created by CX due to neutral beam injection. The study revealed promising results, i.e. the possibility of single shot n_T/n_D ratio measurements in the outer part of the bulk plasma. The laser beam handling described above for edge n_T/n_D ratios will allow also for n_T/n_D measurements in combination with the Diagnostic Neutral Beam (Port 4) at a penetration depth of 1.20 m. Under these excitation conditions, the laser beams are crossing the neutral beam nearly perpendicular. Together with a 2γ-Lα detection system viewing co-axially to the laser beams from Port 3, this arrangement is expected to combine the largest 2γ-Lα signal with the smallest Lα background signal and could provide therefore the best signal to background ratio achievable, i.e. the best detection sensitivity for small isotope densities.

ACKNOWLEDGMENT

This work was performed in a close cooperation with the Max-Planck Inst. of Plasma Physics (IPP), in order to develop 2γ-Lα measurement technique. We are very grateful to G. Fussmann as well as to W. Bohmeyer, H. Grote and H.D. Reiner operating the PSI-1 at IPP, Berlin Branch.

REFERENCES

1. W. Bohmeyer et al., in: *21st EPS Conference on Controlled Fusion and Plasma Physics*, E. Joffrin, P. Platz, P.E. Stott, eds., Europhysics Conference Abstracts Vol. 18 B, part III, p. 1328 (1994)
2. D. Voslamber, Determination of neutral particle density and magnetic field direction from laser-induced Lyman-α fluorescence, II - Two-photon excitation, Report EUR-CEA-FC-1387 (1990)
3. D. Voslamber, Some novel concepts for diagnostics in tokamaks, in: *Spectral Line Shapes, Vol. 8* (AIP Conf. Proc. 328), A.D. May et al. eds., American Inst. of Physics, New York (1995)
4. J. Seidel and D. Voslamber, to be published
5. K. Grützmacher , PTB Report, to appear in october 1997
6. D. Voslamber and W. Mandl, Determination of the D/T fuel mixture using two-photon laser induced fluorescence in combination with neutral beam injection, Report EUR-CEA-FC-1607 (1997), also appearing in this volume

PROGRESS OF THE REFERENCE DESIGN FOR ITER BOLOMETERS AND DEVELOPMENT OF A HIGH PERFORMANCE ALTERNATIVE

Roger Reichle[1], Marco Di Maio and L.Christian Ingesson

JET Joint Undertaking
Abingdon, Oxfordshire, OX14 3EA, UK
[1]new address: CEA Cadarache DSM/DRFC/SPPF/ITPP Bât. 508
13108 St-Paul-Lez-Durance CEDEX, France

INTRODUCTION

The reference design for the bolometer system of ITER, which had been outlined and justified against the ITER requirements [1] and to some degree detailed [2] before, has been critically reviewed against the first wall design and access possibilities [3]. The design iteration following this review is the main part of this contribution. As a separarate issue some initial results from a potential high performance alternative are also presented here. The reference design is based on ASDEX-Upgrade type detectors [4,5], modified for the JET high temperature in vessel bolometer [6,7] and draws on the experience in the European tokamaks [8,9,10,11]. The outline proposal was based on parallel lines of sights from 5 directions. Maintenance considerations, the progress of the first wall design, and the importance of the neutral particle contribution to bolometer measurements in the divertor drove the design towards its present manifestation, as will be detailed in the first chapter. The lines of sight are discussed and prioritised in the second chapter making use of tomographic methods. Some technological advances and radiation issues are discussed in chapter 3. Chapter 4 reviews a development project at JET which aims towards an alternative type of high performance bolometer. Finally the state of the different development lines is compared.

DESIGN CRITERIA

The main design criterium is to be able to produce a good tomographic reconstruction of the radiation distribution. The spatial resolution aimed for is 10 cm in the divertor and the edge and 50 cm in the bulk plasma. To achieve this spatial resolution by tomography without any a priory information requires a few thousand lines of sight. Since this is impractical, either the number of viewing directions in the case of parallel lines of sights, or the number of camera positions in the case of fan shaped arrangements has to be reduced from the optimum number, which than has to be recovered by mathematical means, like the anisotropic diffusion model tomography [12]. Nevertheless a few hundred lines of sight have to be regarded as the absolute minimum, with which this task can be fullfilled, and quite a few camera positions will have to be in the blanket[1], rather than in the port or the divertor cassette. This requires radiation hard high temperature detectors [1]. Tomographic artefacts should be kept to a minimum, and further developments along the lines already used [11] will be needed. The envisaged overall accuracy for spatially integrated measurements of bulk and divertor radiation should be better than 10%. The envisaged temporal resolution is 1 ms. With such a system, the lesser requirements for control and protection as well as the measurement of total radiation can be fullfilled by selection of a suitable subset of sight lines – especially when the full complement of lines of sight can be used extensively beforehand for gaining experience.

The best conditions for tomographic reconstructions are given when all data are sampled from the same poloidal plane. It is known [1], that the puffing of gas creates strong extra radiation locally at the gas inlet valve, mainly through charge exchange neutrals. Therefore it is advisable to select a poloidal plane which has no such disturbances of the main set of measurements. This should be complemented with a smaller subset of measurements in a sector with a gas inlet. In high density conditions, charge exchange neutrals give a large contribution to bolometer measurements in the divertor, which are not observed by measurements of the divertor radiation taken through the bulk of the plasma from the upper half of the blanket [10]. It has also become possible to determine the divertor neutral particle fluxes by making them an additional free parameter in tomographic reconstructions [11]. Closing the divertor increases neutral losses at JET and even larger effects are expected in ITER since it has an even more closed divertor. The consequence of this is, that a large number of lines of sight both in and out of the divertor are needed, to account for the loss power due to neutrals. Other reported neutral particle effects are neutral pressure related changes [1,8,13], which are being addressed by constructive means as outlined in the technology section later on. Measurements of the calibration constants as a function of the pressure at JET showed no significant change in the range of interest up to 0.1 mbar (see fig. 1), which should simplify potentially necessary corrections.

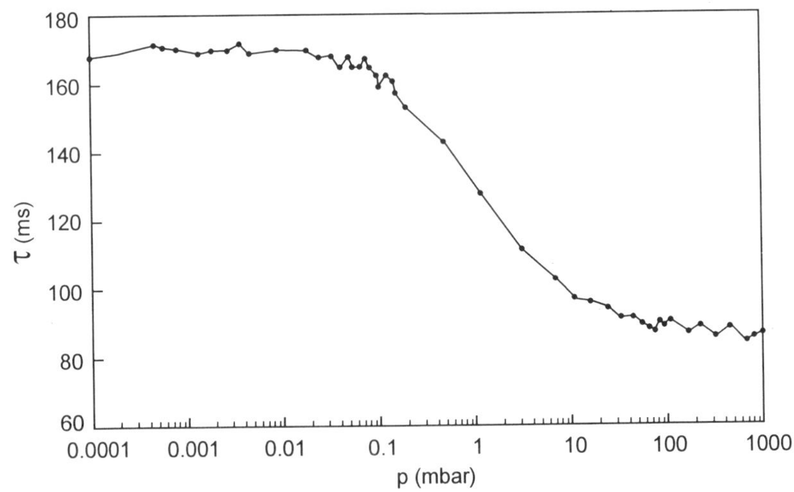

Figure 1. Calibration time constant as a function of the pressure for the present JET bolometers.

[1] The cameras look through poloidal gaps between blanket modules of 2 cm width, which gives sufficient signal strength [1].

The other main constraints in the design of the layout are technical concerns. For better maintenance it is advisable to concentrate as many of the important channels as possible in areas where they can be exchanged, i.e in the ports and the divertor cassette. The present first wall design features interlocking keys (see fig. 2) between toroidally adjacent blanket modules that are meant to withstand the shear forces induced by halo currents. This restricts the choice of viewing positions from the blanket to some degree and has to be accounted for. In the port area there is a different arrangement of the shear keys, such that they can be ignored there as far as the bolometers are concerned. The number of connectors and pins which are available at the back of the blanket restricts the number of sight lines.

The sectors to avoid with the main bolometer installation are 4, 8, 12, 16 and 20 since they contain gas inlets. Other potential disturbances by for example the pellets in sector 19 and the NBI in 4, 5, 6 - are considered negligible. Useable diagnostic ports are then the horizontal port in sector 9 and the vertical port in sector 10. Sector 10 has also 17 blanket connectors that can be used for bolometers. In sector 9 some more could be found by sharing with other diagnostics. The fact, that there is another set of blanket connectors available in sector 16 makes it attractive for the complementing detailed measurements in a gas inlet sector. There is a problem with the divertor cassette since the nearest sector with an instrumentation cassette is in sector 8, which causes a problem with the gas inlet. To solve this, it is here proposed to modify the cabling concept of the divertor cassettes, that the measurements can be taken in sector 10. This would then give the main set of measurements in sectors 9 and 10. For toroidal asymmetries a few wide angle channels in as many as possible sectors would be the best solution.

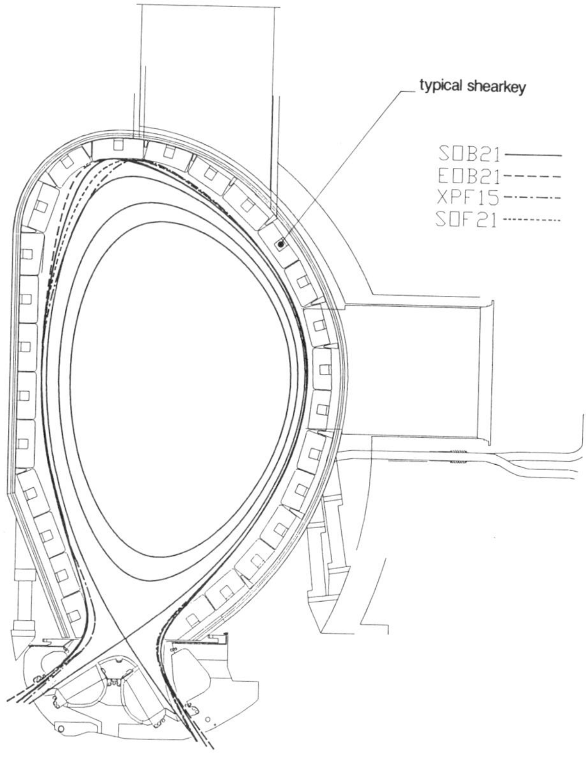

Figure 2. Various plasma shapes in the current first wall design, also showing the shear keys.

The expected radiation profile in ITER has a sharp localised peak at the edge and a relatively broad peak as central contribution [14], which justifies the high resolution in the edge and the low resolution in the bulk. There are various plasma shapes envisaged, the main difference of which - as far as the bolometers are concerned - is the shape in the upper inner corner [15] as fig. 2 shows.

LINES OF SIGHT

Fig. 3 shows the lines of sights which are proposed to be implemented in sector 10 and the horizontal port in sector 9 following the criteria outlined in the previous section. Fig. 4 gives an overview of all the lines of sights in a p-ξ diagram[2]. Both figures illustrate, that the the coverage will be dense in the edge and the divertor and relatively homogenous in the bulk area.

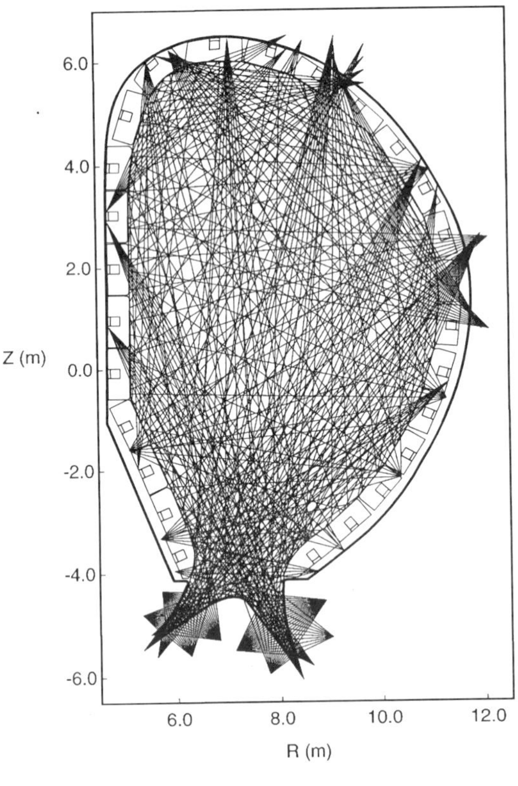

Figure 3. Overview of lines of sight.

The special emphasis given to the divertor coverage expresses itself in the implementation of a total of 8 divertor viewing fans, evenly distributed along the usable

---

[2] In such a diagram each line of sight is represented as a point defined by two coordinates: the impact parameter p relative to a reference point, usually the plasma centre, and the angle ξ of this line of sight relative to a reference direction.

circumference of the blanket, each looking with about 10 channels at the divertor. They are highlighted in fig. 4. The most important ones are the 2 fans located in the ports and the lowest inner and outer extremes. Edge coverage is provided by 10 fans, 5 of which originate from the ports with 5 to 6 channels each, which all are important. With this kind of edge coverage one can expect to be able to use the bolometers for position control of the plasma edge around the bulk plasma. The most important bulk coverage is provided from the port views and again from 2 relatively low fans at the inner and outer blanket half, about 2 m below the midplane. The other 5 bulk fans are required to obtain an even coverage of p-ξ space.

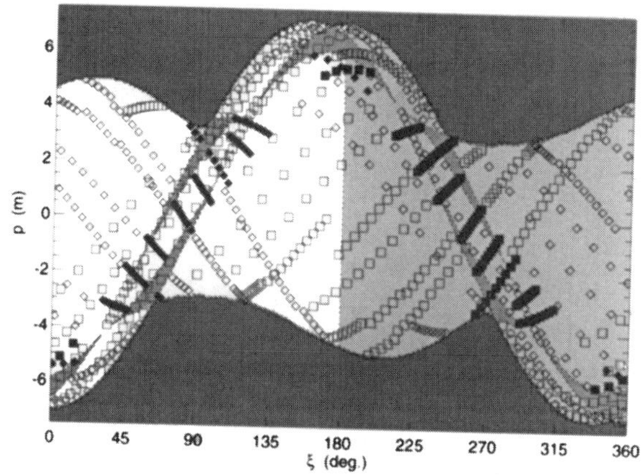

Figure 4. P-ξ Diagram of all sight lines with reference point in the centre of the plasma. The lines of sights looking from outside into the divertor are highlighted. The darkened regions within the plasma boundary are areas which are in the shadow of obstacles when looking from one of the two directions. The squares are the points corresponding to actual lines of sights, wheras the diamonds indicate virtual lines of sight that look in the opposite direction, where this view is not blocked by structures.

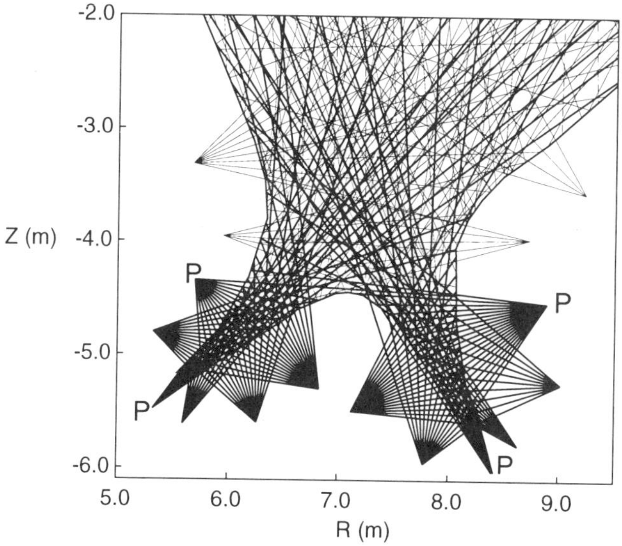

Figure 5. Divertor sight lines; priority fans are marked with 'P'.

The details of the divertor coverage are illustrated in fig. 5 and a special p-ξ diagram in fig. 6, for which the reference point has not been placed in the plasma centre as usual, but in centre of the outer divertor leg, to bring out the regularity of the coverage more clearly. There are 6 fans per leg proposed with on average 13 channels each, covering the same p-ξ area as the external cameras plus the areas that can not be accessed from outside. The highest priority have 2 orthogonal fans per divertor leg as indicated in fig. 5. To distinguish the neutral and the electromagnetic power loss, the complementing fans from outside the divertor are of course also necessary. Table 1 gives a summary of the lines of sights required and fig. 7 is a typical example of a tomographic reconstruction possible with the full complement of the lines of sight proposed and a present day tomographic algorithm [11].

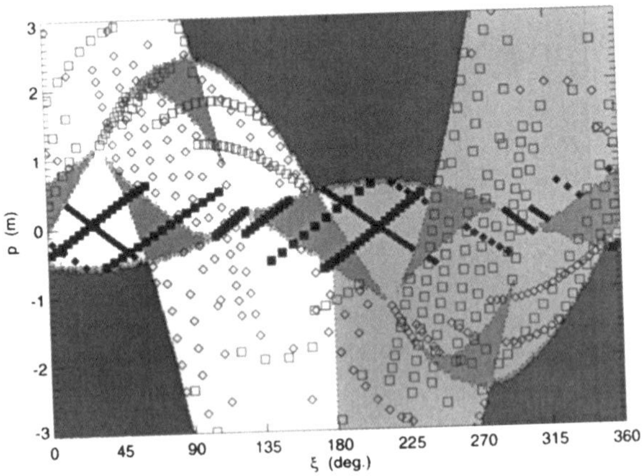

Figure 6. P-ξ Diagram of divertor sight lines in outer divertor leg with reference point in the divertor leg.

Table 1. Line of sight summary

Purpose	Detail	Number of channels
Priority fans for tomography		Total: 200
	All port fans for: divertor, edge and bulk	80
	2 orthogonal divertor fans per leg	55
	2 divertor viewing fans in the blanket	20
	5 edge viewing fans in the blanket	25
	2 bulk viewing fans in the blanket	20
For more detailed tomography		200
Gas inlet monitor		
	Subset of priority channels in gas inlet sector	50 ... 100
Toroidal asymmetries		
	Small subsets (10-15) of priority channels in as many as possible other sectors	100 ... 200

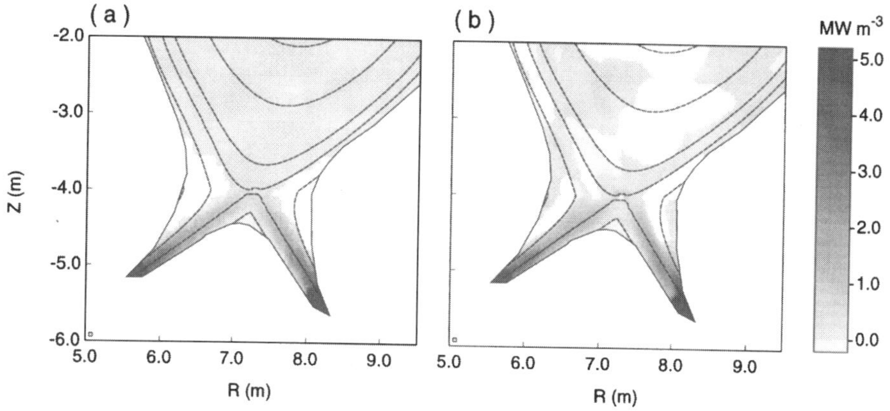

Figure 7. a) Phantom and b) reconstruction of a partially attached case study of the ITER radiation based on B2-Eirene results [16]. Note that the emission in the bulk plasma is more than one order of magnitude lower than in the divertor.

TECHNOLOGY

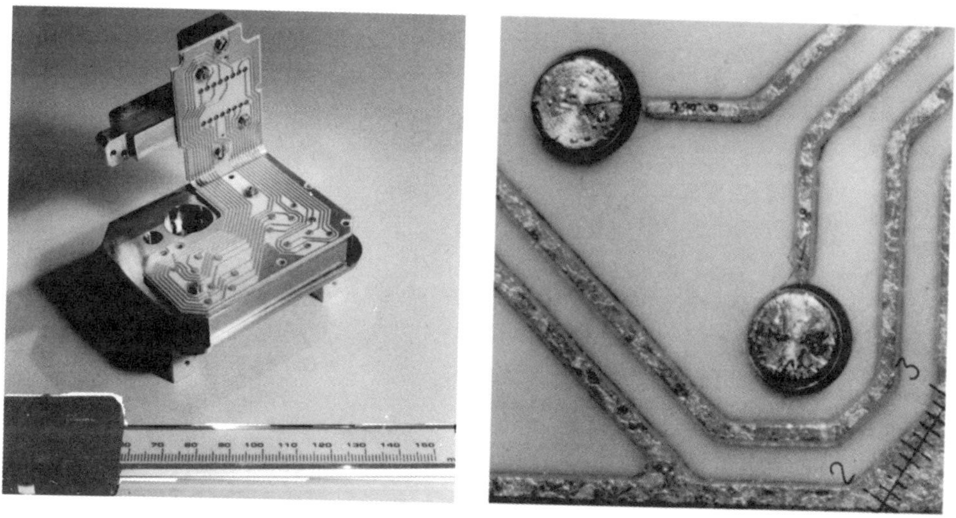

a) b)

Figure 8. a) Internal wiring of a JET gasbox divertor bolometer (small tick on scale: 1 mm); b) detail, showing ball bond connections (small tick on scale: 0.1 mm).

Neutral pressure effects have been minimized at ASDEX-Upgrade [17] and for the JET gasbox divertor bolometers by removing the copper shield in front of the reference window, to allow pressure equilibration. From the same design change it is also expected, that high frequency interference becomes less, since the conclusion from laboratory experiments at JET has been that this copper window actually allows an eddy current to flow which heats up the reference bolometer. On the side of electrical connections, it has become clear through investigations of used JET divertor bolometers, that the weakest electrical connection is the crimp connection at the back of the detectors. This crimp connection has now been replaced by a bond connection for the JET gas box bolometers - see fig. 8.

Radiation doses at bolometer locations at the back of the blanket during the first 2...3 years of operation of ITER are expected to be minimal. In the following 10 years 0.03...0.1 MWa/m^2 or dpa are possible and the final dose after 20 years may reach 0.13...0.3 MWa/m^2 or dpa. This means it is prudent to look for materials for the insulators in particular, that survive 1 dpa. Simultaneous technical requirements for the substrate are a reasonable flatness on the side where the meanders of the resistors are. Good values would be 20 nm...50nm. The operational temperature range should reach up to 300°C and no electrical breakdown should occur under conditions of $3.5V_{rms}$ at 20kHz over a 30 μm gap. Materials considered are Mica, Al_2O_3, AlN, Spinel, glasses and Silicon coated with Si_3N_4. Mica has been tested up to 10^{-4} dpa without problem [17], but due to its hydrogen content it is suspected, that it may fail at more relevant doses. Since the most positive experience exists with Al_2O_3, and it seems to fullfill most technical requirements, apart from the flatness, which may be achieved by polishing, it is the favourite candidate to start a development of substrate samples with thin enough membranes (≈20 μm). Spinel is electrically doubtful [18] but otherwise interesting and some glasses look as if they might be radiation resistant enough, which should make the surface flatness requirement achievable. Once suitable substrates have been produced, radiation tests with in situ measurements need to be performed.

HIGH PERFORMANCE ALTERNATIVE

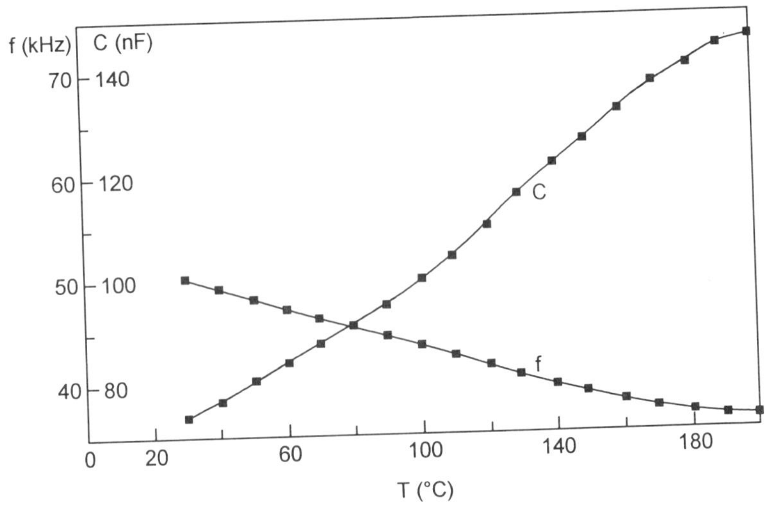

Figure 9. Capacity C and frequency f as a function of temperature.

Problems with the reference design are: noise pick-up on amplitude measurements, the relatively small change of the specific resistance with temperature and the large number of cables. At JET an alternative bolometer design has been started [19,20] which offers some interesting solutions to these problems. First of all the measurements taken are not amplitude measurements, but frequency measurements. This frequency is the resonance frequency of a tuned circuit, which has as a temperature sensitive element a ferroelectric capacitor. Near the Curie point, the dielectric constant rises very steeply with temperature and therefore the primary effect is larger as in resistor bolometers. In principle, many frequencies can be multiplexed on one cable, reducing the number of required feedthroughs. The potential problems of this alternative design are that the operational range of the temperature is limited to the region near the Curie point, which may be a useful range of 50...100 degrees. Fig. 9 shows a typical measurement of the change of the capacity and the frequency as a function of temperature. The electronics used to measure the change of the resonance frequency of the tuned circuit is a phase locked loop which presently is capable to track frequency changes well above 50 kHz, indicating that the electronics will not limit fast measurements. Tokamak tests are still outstanding. The design contains a coil in the vicinity of the temperature sensitive capacitor. This is inside the toroidal field coils. To avoid pick up problems a miniature toroidal design of the coils could be used, which is similar to low noise toroidal transformers used in audio equipment. This problem will be further alleviated by the fact, that the magnetic field varies only very slowly. A method for calibration is presently considered, which uses the intrinsic losses to heat up the detector foil. This still needs to be proven. The radiation hardness of ferroelectrica is reputedly good [18]. The cables have the additional specific requirement that they have to be able to transport 20 MHz. The present state of developments is that a proof of principle has been achieved. The next step is the improvement of the electrical Q value of the tuned circuit and the thinning of the membrane.

STATE OF DEVELOPMENTS

The steps required until installation on ITER differ in complication for the individual lines of developments. For the reference design the main issues apart from the technical detail of designing the individual assemblies are radiation related. Do mica, or the replacement substrate material and the mineral insulated cables survive?

In the case of the development of so called feedback bolometers [21, 22] new masks need to be made for the photolithographic processes that are used. Also new electronics are required. Both fields of work are presently addressed. After they have been solved, the final steps are the same as for the reference design bolometers. Ferroelectric bolometers first need to prove to work as bolometer system in the laboratory and then be installed on a real tokamak. Radiation hardness is the same concern as above.

CONCLUSION

The present line of sight proposal is compatible with the first wall. It allows good tomographic reconstructions from one poloidal position if the instrumented divertor cassette is moved. The neutral particles in the divertor and from the gas inlets are accounted for. Toroidal assymmetries will be recorded. Measurements of the total radiation can be done with a small subset of the channels, so no extra system is required. However the other requirements regarding spatial resolution in the divertor in particular make it indispensable to have the full system implemented. It looks as if the proposed system could be used to control the edge position. Initial steps regarding the radiation hardness of the reference design detectors are being taken. The high performance alternative to the reference detector looks promising, but still has a long way to go. The technical developments are progressing.

ACKNOWLED GEMENTS

The provision of relevant ITER data by C. Walker, L. De Kock, S. Gerasimow and D. Bouched as well as usefull discussions with K.F. Mast are gratefully acknowledged. We also thank P. Böcker and L. Horton for the help with the fig. 1 and 7 respectively.

REFERENCES

[1] 'Bolometer for ITER', R. Reichle, J.C. Fuchs, R.M. Giannella, N.A.C. Gottardi, H.J. Jaeckel, K.F. Mast, P.R. Thomas, in: Diagnostics for Experimental Thermonuclear Fusion Reactors, eds. P.E. Stott, G. Gorini and E. Sindoni (Plenum Press, New York, 1996) pp. 559

[2] 'Bolometry - core and divertor', R. Reichle et al., Progr. Meeting on Diagn. Systems, 15-17 Jan. 1997, Garching JWS, Germany

[3] priv. comm. C.I. Walker

[4] IPP 1/224 E.R. Mueller, G. Weber, K.F. Mast, G. Schramm, E. Buchelt, C. Andelfinger, Max-Planck Institut für Plasmaphysik, Garching, Oct 1985

[5] 'A low noise highly integrated bolometer array for absolute measurement of VUV and soft X-ray radiation' K.F. Mast et al., Rev. Sci. Instrum. 62(30) March 1991, p 744

[6] 'Divertor Radiation in JET', R. Reichle, N.A.C. Gottardi, R.M. Giannella, H.J. Jaeckel, A.C. Maas, Contrib. to the 36th ann. meet. of the DPP/APS 1994, 7-11 Nov. 1994, Minnesota, USA

[7] 'Radiation in JET's Mark I Divertor', R. Reichle et al., Contr. 22nd EPS on Contr. Fusion and Plasma Phys., Bournemouth, 3-7 July 1995, Poster R022

[8] 'Miniaturisiertes breitbandiges Bolometerarray', F. Mast et al., Technisches Messen: Sensoren, Geräte, Systeme, 4/97, Oldenburg Verlag, pp. 164

[9] 'Fast bolometric diagnostic in the RFX reversed field pinch experiment', K.F. Mast et al., Rev. Sci, Instrum. 63 (10), October 1992, pp. 4714

[10] 'Low energy neutral particle fluxes in the JET divertor', R. Reichle, et al., Contr. 12th Int. Conf. Plasm. Surf. Interact. in Contr. Fus. Devic., St. Raphael, France, 20-24 May, 1996

[11] 'Radiation distribution and neutral-particle loss in the JET MkI and MkIIa divertors', L.C. Ingesson, R. Reichle, G.C. Fehmers, H.Y. Guo, L. Lauro-Taroni, A. Loarte, R. Simonini, Contr. to 24th EPS Conference, Berchtesgaden, Germany, 9 -13 June 1997

[12] 'Two dimensional reconstruction of the radiation power density in ASDEX Upgrade', J.C. Fuchs et al., Contr. 21st EPS Con. Contr. Fus. Plasma Phys., Montpellier, France, June 27 - July 1, 1994

[13] 'Bolometric measurements in the ASDEX Upgrade divertor', J.C. Fuchs, K.F. Mast, G. Haas, ASDEX Upgrade team, et al., , Contr. to 24th EPS Conference, Berchtesgaden, Germany, 9 -13 June 1997

[14] S. Gerasimov, priv. comm. July 97

[15] D. Bouched, priv. comm. July 97

[16] 'Effect of light impurities on the divertor performance in ITER', A.S. Kukushkin, et al., Contr. to 24th EPS Conference, Berchtesgaden, Germany, 9 -13 June 1997

[17] K.F Mast, priv. comm.

[18] E. Hodgson, priv. comm.

[19] 'Ferroelectric Bolometer', M. Di Maio, R. Reichle, UK Patent application, provisional no: 9623139.4, Nov. 1996

[20] 'Design of a Ferroelectric Bolometer', M. Di Maio, R. Reichle and R. Giannella, Contr. to the 17th Symposium of Fusion Engineering IEEE/NPSS 6-10 Oct. 97 St. Diego, USA

[21] N.A.C. Gottardi, K.F. Mast, Euratom Patent

[22] K.F..Mast,et al., Progr. Meeting on Diagn. Systems, 15-17 Jan. 1997, Garching JWS, Germany

IMAGING BOLOMETRY DEVELOPMENT FOR LARGE FUSION DEVICES

G. A. Wurden[1], B. J. Peterson[2]

[1]Los Alamos National Laboratory
Los Alamos, New Mexico, 87545 USA

[2]National Institute for Fusion Science
Toki, 509-52 Japan

INTRODUCTION

As magnetic fusion experiments progress towards longer time-scales, more intense neutron and gamma backgrounds, larger physical dimensions, and use of diagnostics for real-time control, the measurement of total radiated power (including total neutral particle fluxes) becomes more challenging. The use of thin foils that absorb and convert the radiation and particle fluxes into a measurable temperature rise, is a standard tool for the plasma diagnostician. Both wide-angle, and tightly collimated foil detector arrays have been used on nearly every tokamak or alternative concept device. The temperature readout is usually accomplished via a DC or AC bridge circuits, to recover the millivolt-level signals, and then differentiated to get power from the time-integrated heat load on the foil and substructures. The challenge lies in making fast bolometers, responsive over the expected range in incident energies, which are also robust, reliable, compact, convenient to calibrate, and cheap enough to afford hundreds of channels for good spatial resolution using tomographic techniques. So-called "ASDEX" or JET-style bolometers manufactured by PTS GmbH, represent the present standard of performance. A new imaging bolometer[1], using infrared readout of hundreds of mini-foils in an actively cooled mask, has been built by LANL, and is presently being tested on the CHS and LHD plasmas in Japan. We describe work in progress towards a radically different approach to obtaining bolometer signals, which has many advantages (and a few disadvantages) over present conventional bolometer systems. We explore a system using technology available today, which simultaneously allows for a cheaper per/channel cost, a high degree of radiation-hardening, no wiring-harnesses in the vessel, no pre-amplifiers, good noise immunity, compatibility with long-pulse operation and real-time readout, and hundreds of channels to form a bolometric image. The drawbacks are the requirement for an optical path to the backside of the detector array, and some sacrifice of sensitivity. *

---

* This work is supported by the U. S. DOE contract #W-7405-ENG-36, and by Monbusho (Japan) through NIFS.

Diagnostics for Experimental Thermonuclear Fusion Reactors 2
Edited by Stott *et al.*, Plenum Press, New York, 1998

BACKGROUND

Numerous review articles have been written over the years on the topic of bolometry in fusion plasma devices[2,3,4]. The general theme is one of trading off various sensor types to achieve sufficient time and spatial resolution in measuring the radiation losses from the plasma. The most sensitive detectors (semiconductors, thermopiles, pyroelectrics, and thermistors) are all unacceptably sensitive to gamma and neutron damage as well. For this reason, metal foil bolometers, which are more radiation-hard and high temperature compatible, have been chosen by most large fusion devices in recent years (TFTR, JET, ASDEX, JT-60U, Tore Supra, etc.). A resistive element is bonded to the radiation absorber, and then wire leads (2 to 4 per unit, including a matched (blind) reference detector) are brought out of the vacuum vessel. Unfortunately, this resistive element is also subject to nuclear radiation-induced changes, in its temperature/resistance coefficient. The signal levels are very small, necessitating a pre-amplifier as close as is practical in the signal pathway. An AC bridge circuit is often used in the readout. With these techniques, sensitivity levels down to 1 μwatt/cm^2 using 10 millisecond averaging, have been achieved in the lab. Noise due to magnetic pickup or auxiliary heating typically raises this lower-bound number by several orders of magnitude. Arrays of 30 – 100 detectors have been employed to reconstruct the spatial profiles of the plasma radiation.

The next generation plasma fusion machines (LHD, W7X, KSTAR, and ITER), will offer a new (or at least more difficult) set of problems for implementing bolometry. The major difficulty will have to do with the issue of survivability in the face of neutron and gamma bombardment of all elements of the bolometer, including sensor, insulators, cabling, or any other close-in components. Furthermore, due to the fact that they will be capable of long-pulse (> 10 second) operation, bolometers which have up to now been inertially cooled, will need active cooling as part of their basic design. Although present-day analysis typically take into account the cooling of the sensor element, usually the cooling time is long compared to the desired time response, and the bolometer temperature rise is differentiated to yield the radiated power, with minor corrections for thermal losses.

Sources of Noise and Pickup

For the metal foil bolometer, the electrical bridge circuit which measures a change in resistance (of the resistor bonded to the metal absorber) is sensitive to a number of noise sources. At a very basic level, there is 1/f and shot-noise in the resistor. By using an AC bridge circuit measurement, at a carrier frequency of 100 kHz to several megahertz, the 1/f noise can be avoided. Although twisted pair or even coaxial leads are used, there still can be additional electrical pickup introduced in the relatively long leads coming from the diagnostic to the location of the discrete amplifiers. This pickup can include effects from magnetic field coils, RF Heating, and other auxiliary systems around the machine. In the case of ECH, it is even possible for actual thermal (resistive) heating of the foil to occur due to absorption of the microwave radiation itself. Consequently, the laboratory lab bench noise level limit of a working bolometer system is often much lower than the effective noise limit of the bolometer when used on a plasma device. The use of "blind" reference detectors, co-located with the plasma facing detectors is one way to "null out" a significant part of the magnetic pickup and nuclear radiation induced effects.

Sensitivity of Historical Metal Foil Bolometers

Metal foil bolometers have been the bolometer of choice for the present generation of large magnetic fusion plasmas, due to their moderately good radiation resistance, vacuum and bakeout compatibility, good time response (~10 msec) and reasonable sensitivity. The TFTR metal foil bolometers (derivatives which are also used on DIII-D) have an equivalent sensitivity of 60 mW/cm^2 with a 10 millisecond time resolution, and were packaged individually. The Physikalisch-Technische Studien GmbH (or PTS for short), Freiborg, Germany, has successfully developed a bolometer head with 2x4 sensors (four that are active, and four that are reference). The detectors are in a compact (33 mm x 20 mm x 15 mm) frame, which is bakeable to 300°C, and have a claimed sensitivity of 1 $microwatt/cm^2$ with a time resolution of 10 milliseconds, while responding to photon energies of up to 5 keV. These units have been sold to JET (England), RFX (Italy), ASDEX-UG (Germany), JT-60U and LHD (Japan), where they are employed in a pinhole/slit camera arrangement to form 1-D arrays.

At JET, the kapton substrate/gold resistor versions[5] (in the KB1 camera) have an equivalent noise level of 70 $microwatt/cm^2$ at 20 Hz, with a comparable level of magnetic pickup. Careful matching with blind detectors reduces both thermal drift and magnetic pickup noise problems. More recently, the new KB3D and KB4 cameras[6], employ a mica substrate/gold resistor, and a much smaller thermal capacity, leading to improved signals. A noise level of 20 $microwatts/cm^2$ is reported, but with substantially larger magnetic interference problems due to more demanding mounting locations, at 400 microwatts/ cm^2.

PTS units purchased for eventual use on LHD[7], are also measured at NIFS to have a 2 $microwatt/cm^2$ equivalent noise level on a lab test stand. On the CHS plasma, due to residual magnetic pickup and ECH interference, this number is also about 10x higher.

IMAGING BOLOMETER DEVELOPMENT

A new type of bolometer for magnetic fusion research is under development[1] at Los Alamos, using a state-of-the art infrared imaging video camera to image the temperature rise on a segmented foil, thereby making hundreds to thousands of "channels" of bolometry possible without wire leads to each foil element. In the late 1970's and early 1980's, single-channel, single detector infrared bolometers were first used on plasma experiments (TFR[8] and ZT-40M[9]). In the intervening 15-20 years, infrared detector technology has progressed to the point where 10's of thousands of detectors are available in one camera chip. Advantages of using an IR imaging bolometer for ITER were pointed out two years ago.[10] The sensitivity achievable by IR readout techniques (rather than resistive electrical measurements) will be a key tradeoff. Indeed, previous authors[2] have dismissed so-called "thermographic bolometers" because of sensitivity limitations (real or perceived) at that time. On the other hand, present specialists point out the need for 3000-4000 channels[6] for adequate reconstruction of ITER radiation profiles, and then dismiss this out of hand as too expensive. We hope to demonstrate that the IR bolometer can indeed answer these issues, and at the same time obtain unprecedented imagery of the plasma radiation profiles.

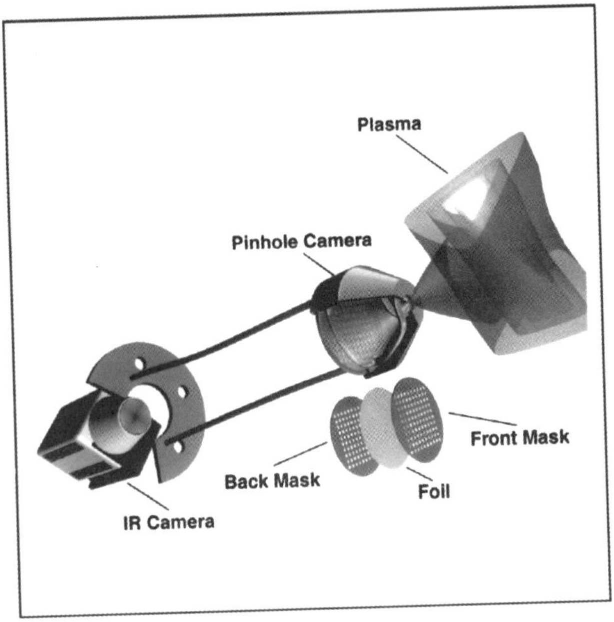

Figure 1: Artist's view of imaging bolometer concept for LHD. The pinhole-camera arrangement which holds the masks, is water-cooled.

The bolometer is very simple (as shown as an artist's concept in Figure 1, and in a CAD wire-frame in Figure 2, and actually realized in Figure 3). It uses an IR camera to view a segmented foil created by sandwiching a thin foil in-between heat sink material (copper) which has a suitable pixel pattern drilled into it. The pixel array views the plasma through a water-cooled pinhole camera. The plasma radiation heats the individual pixel elements on the foil, each of which is thermally isolated from its neighbor by the "thick" copper mask.

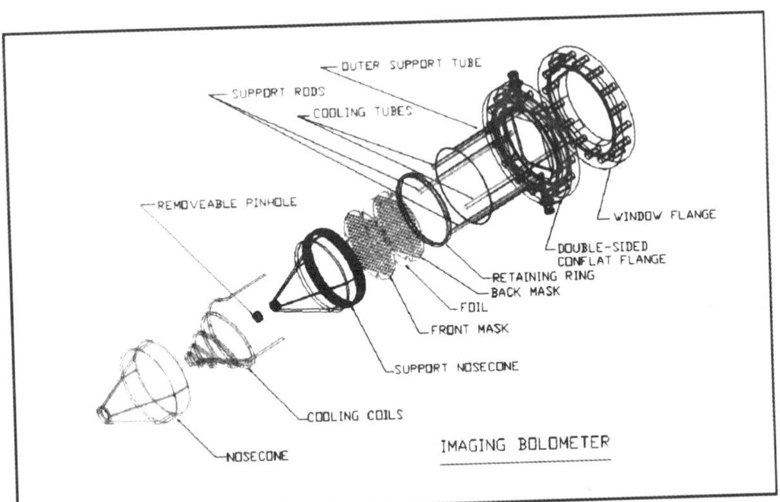

Figure 2: Wire frame schematic of imaging bolometer re-entrant head.

For a 1 μm-thick gold foil, coupled with a 25 m°K IR camera sensitivity, the imaging bolometer would have a 250 μwatt/cm^2 equivalent noise level at a 60 Hz image rate. Tests in the lab with actual windows, emissivity, and camera geometries suggest an actual sensitivity somewhat worse, at ~0.5 mW/cm^2 . If the foil is 10x thicker to stop higher energy photons, then the sensitivity will be correspondingly reduced.

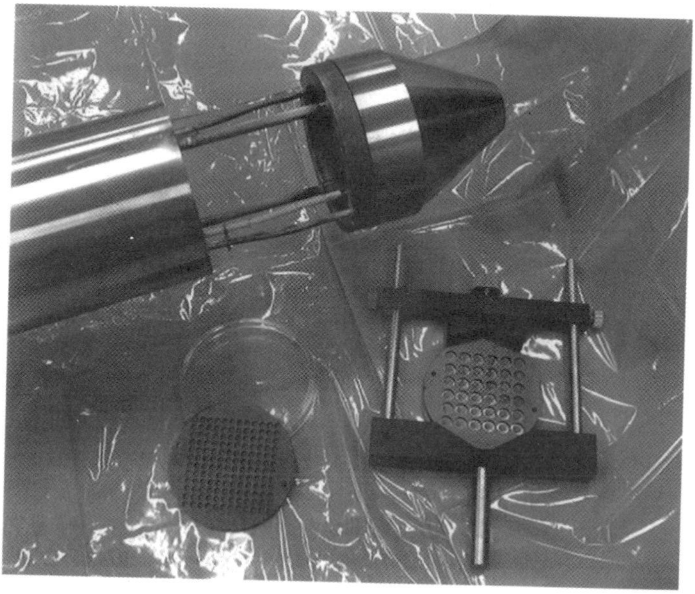

Figure 3: Front assembly of LHD imaging bolometer prototype, with two mask/foil assemblies shown. A mask/foil assembly is placed at the back of the water-cooled pinhole camera nosecone. Two (copper-constantan) thermocouples, and tubes carrying water, are also attached to the copper housing.

Figure 4: 12 x 12 copper mask, with gold foil which has been blackened with graphite spray.

For the small and relatively cold CHS plasma[11], we expect radiation ranging from 50 kW to 1 Megawatt levels. Thin foils (0.5 - 2.0 microns of gold) are sufficient to absorb the radiation of interest. In hotter (10-30 keV) plasmas, thicker foils are required (~10 microns gold). Even the most sensitive IR camera has only 10 milliKelvin sensitivity limit, at 30 frames/second near room temperature (the Amber Engineering Model 5128 research camera can do 1000 frames/sec but with only 128x128 pixel resolution). The infrared camera shown in Figure 5 costs approximately $100k as configured, including remote computer control and long lens. Consequently thicker foils will mean a loss of sensitivity for the bolometer measurement as well. Dynamic range of these camera systems is 12-bit in the infrared flux, which must then be converted to a temperature via Planck's Law for blackbody emission. Fortunately, ITER makes up for the need for thicker foils, by giving off more total radiated power (up to ~600 Megawatts during normal operation). However, it has been pointed out[12] that nuclear heating of the foils will compete with the desired plasma radiation signal. Blind pixels will be needed for compensation.

Figure 5: 3-5 micron band, 12-bit, infrared video camera, "Radiance 1" from Amber Engineering, with SiGe 75-250 mm f2.3 telephoto lens and internal filter wheel, all controlled by an RS-232 port. Standard 50 mm f2.3 lens is next to some of the filters.

Limitations on Sensitivity of IR Bolometer Systems

The sensitivity limitations come from several intrinsic issues: The noise equivalent power of the thermal imager in the IR wavelength band of interest sets a lower bound. Secondly, background object radiation (due to reflections or stray background thermal radiation) onto the foil of interest will contribute heat or IR emission in competition to that from the plasma radiated power deposited in the foil. Thirdly, IR photon statistics: You need to collect enough photons per desired observation time interval. That is to say, you can't determine a temperature change, if you don't have adequate IR photon statistics. Consequently, the optical throughput of the IR optics will effect the minimum ΔT that can be resolved. Next, in order to get the largest ΔT, for a given power flux, you want the lowest heat capacity for your absorber. This is in direct contrast to the requirement that the foil be thick enough to absorb a significant fraction of the soft x-ray flux that hits it. So, the hotter the plasma, the thicker the minimum thickness foil will need to be. This means that (strength issues aside) you can't make an arbitrarily thin foil, (in order to minimize the heat capacity of the target) and thereby boost up the ΔT for a given radiation flux onto the foil. Finally, the bolometer sensor must be shielded against microwaves, in a microwave-heated plasma, such as on CHS or LHD. A free-standing metal mesh must be placed over the entrance pinhole to preclude heating of the foil by microwaves. Cooling for this mesh must come from the boundary of the pinhole camera pinhole.

Calibration Tests (on the bench and with a plasma)

The imaging bolometer prototype, which employs a water-cooled copper pinhole-camera, shielded by a stainless-steel nose-cone, has been vacuum tested in the lab at the National Institute for Fusion Science, in Toki, Japan. A simplified (non-cooled) imaging bolometer has been installed on the CHS plasma in August 1997. To facilitate visible light He-Ne laser calibration, both gold and aluminum foils have been blackened with "Aerodag-G", commercially available micron-sized graphite spray from Acheson Corporation. Typical temperature rises are no more than 1°C for expected plasma sources. Response to a calibration laser is shown in Figure 6.

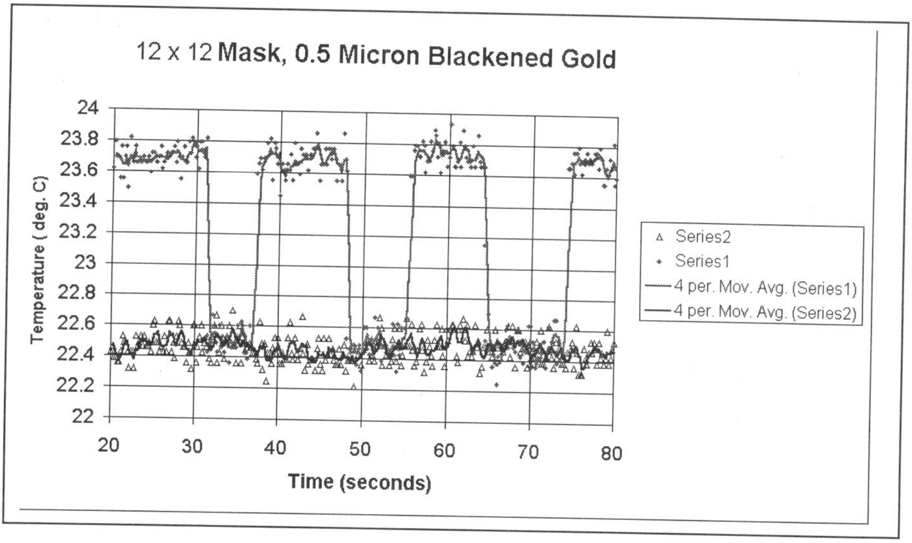

Figure 6: Point temperature measurement made by Agema 9600 IR camera of a foil pixel being illuminated by a 1.5 mW calibration laser (in air).

1). Difference between in-air, and in-vacuum response. The bolometer is observed to be approximately a factor of *2x* more sensitive to radiation in vacuum than it is in air, due to the additional cooling pathway represented by the air hitting the foil. This pathway is absent in vacuum below 100 milliTorr. The magnitude of observed signals in vacuum is at the same time a factor of ~2x larger than our simple model of a cylindrical absorber of conducting material surrounded by a perfectly clamped temperature boundary condition would predict. The peak to average temperature ratio across each pixel, is also observed to be different in air than in vacuum, due to the edge-cooling dominating in vacuum, but with surface-cooling also contributing in air.

2). Degree of thermal cross-talk observed between nearest-pixel neighbors. First generation masks and foil assemblies are observed to have approximately 10% crosstalk levels between nearest neighbor pixels. This is most likely due to imperfect bonding (or no bonding in some cases) between the thick cold mask, and the very thin foil laid across it. The mask must be very flat, and the foils smooth, for good thermal contact. In principle, the foil/mask assembly should be done in a clean room, to keep dust from getting between the surfaces.

3). Decay time of signal. The decay times have not yet been systematically compared to theoretical predictions. Some of the initial foils have an additional complication of a comparable thickness mylar backing film (for strength), which undoubtedly alters the thermal properties of the foil. For these foils, the decay times appear to be somewhat longer (~2-3 *x*) than initial simple design estimates, which don't include mylar.

4). Observation of air-currents across surface of the mask. When a mask/foil is heated by several degrees, and then allowed to cool, it is possible to see "waves of cooling" roll over the mask, which is the signature associated with air-currents. Similarly, spatial isolation of the signals from two ~1mW laser spots onto the array (in air) is easily achieved, as shown in Figure 7.

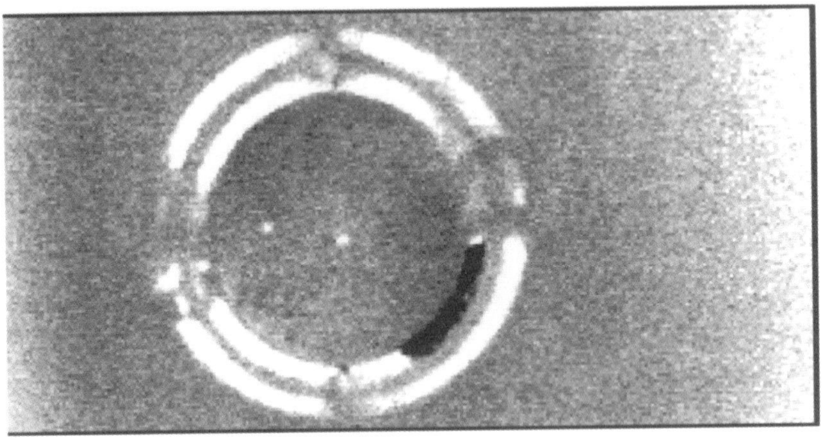

Figure7: Two laser spots illuminate the front surface of a 144-element array, and the IR image is taken from the back side of the foil/mask assembly. The two concentric bright rings are shiny copper and stainless sections of the 1.3 meter LHD prototype. An Agema 900 Thermovision 900 long wavelength series IR camera made the image.

6). Effect of a nearby hot filament ionization gauge: We installed a filament hot cathode ionization gauge about 20 cm to the side of the simple imaging bolometer for CHS. We could easily notice a several degree heating effect from this source. During plasma operations, the filament is off. We can report here that good signals have been seen during shots with from 50 to 400 kW of total radiation, although data is still being analyzed for possible ECH pickup. We use soft iron shielding to protect the IR camera from ~ 1 kG magnetic fields under the CHS machine.

7). Necessity to keep hot objects out of the optical IR field of view. The optical IR field of view between the backside of the mask, and the IR camera, should be isolated from objects hotter than the operating temperature of the mask, to prevent un-wanted IR reflections.

8). Problem of reflection of the cold camera dewar into the center of the field of view (FOV), for normal incidence alignment. This suggests the use of a background subtraction technique to allow for easier analysis.

9). The thermal emissivity of the backside of the foil (the side viewed by the IR camera) should be close to unity over the wavelength band being observed. Even though shiny gold is a good VUV and soft x-ray absorber, having a shiny gold on the backside of the same foil is bad, as this is an IR mirror!

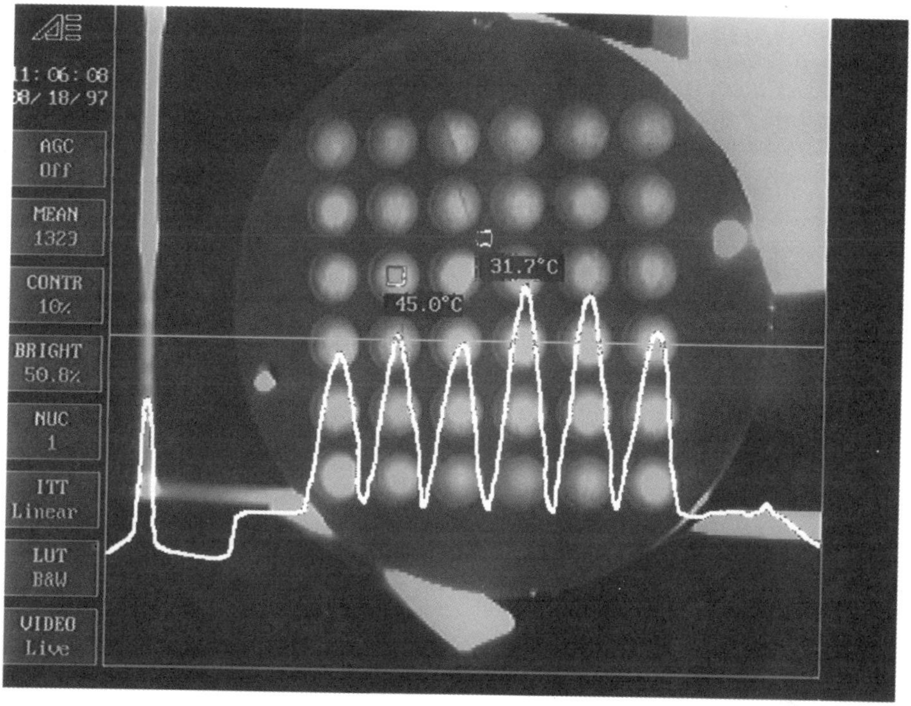

Figure 8: Heating of a masked 6x6 foil in air caused by a 100 watt heat lamp, 20 cm away from the mask, is easily seen by the Amber IR Video camera. The IR intensity profile on a line across the middle of the image, is graphically displayed near the bottom of the image. The temperature near the edge of each pixel is clamped by the thermal inertia of the thick copper mask.

SUMMARY

Testing of prototype infrared bolometers is in progress at NIFS in Japan, and preliminary indications are that there are no intrinsic problems to fielding such devices on the next generation of large fusion devices. At the cost of ~20x reduction in useful sensitivity (relative to standard metal foil bolometers), the IR imaging bolometer will eliminate all wiring harnesses inside the vacuum vessel, provide hundreds to thousands of channels (again, trading off sensitivity, spatial resolution, # of viewing positions) at a much lower cost per channel (<$1k). Furthermore, it will be immune to magnetic pickup, be compatible with steady-state data requirements, and be compatible with both high vacuum and high nuclear radiation environments. This assumes that the sensitive optical detector is shielded from magnetic and nuclear radiation fields. The transport optics will require metal mirrors, but they will not directly view the plasma. The penetrations through the ITER blanket will need to be of order ~1 cm, expanding quickly to ~ 30 cm diameter for the wide-angle "pinhole camera". A 20-meter stand-off distance to the IR video camera is assumed. Tangential views obtained with a bolometer head followed immediately by metal turning mirrors, may prove to be the most useful for real-time plasma position control during long-pulse operation. Unfortunately, we see no convenient way to position this type of system on the inner wall of ITER, although it may also be useful in a divertor port.

REFERENCES

[1] G. A. Wurden, B. J. Peterson, Shigeru Sudo, Design of an imaging bolometer system for the large helical device, *Rev. Sci. Instrum.* **68** (1) 766 (1997).

[2] P. E. Stott, Bolometry, pg. 187-195 in: *Diagnostics for Fusion Reactor Conditions*, Vol 1, Proceedings of the Intl. School of Plasma Physics, Varenna (1982).

[3] D. V. Orlinskij, G. Magyar, Plasma diagnostics on large tokamaks, *Nuclear Fusion*, **28** (4), 611 (1988).

[4] M. A. Ochando, Bolometry, in: *Diagnostics for Contemporary Fusion Experiments* ISPP-9 Varenna, Italy, P. E. Stott, et al. ed., Bologna (1991).

[5] K. F. Mast and H. Krause, Bolometric diagnostics in JET, *Rev. Sci. Instrum.*, **56** (5), 969 (1985)

[6] R. Reichle, J. C. Fuchs, et al., Bolometer for ITER, pg. 559-570, in: *Diagnostics for Experimental Thermonuclear Fusion Reactors,* P. E. Stott et al., ed., Plenum Press, New York, 1996.

[7] O. Motojima et al., *Fusion Engineering Design, **20** (1993).*

[8] TFR Group (presented by A. L. Pecquet), Bolometric techniques on TFR 600, *J. of Nuc. Materials*, **93&94**, 377 (1980).

[9] J. C. Ingraham and G. Miller, Infrared calorimeter for time-resolved plasma energy flux measurement, *Rev. Sci. Instrum.*, **54** (6), 673 (1983).

[10] G. A. Wurden, A rad-hard, steady-state, digital imaging bolometer system for ITER, pg. 603-606, in: *Diagnostics for Experimental Thermonuclear Fusion Reactors,* P. E. Stott et al., ed., Plenum Press, New York, 1996.

[11] K. Nishimura, et al., *Fusion Technology*, **17**, pg. 86 (1990).

[12] V. Mukhovatov, et al., ITER Diagnostics, pg. 65-68, *ITER Documentation Series No. 33*, IAEA, Vienna, (1991).

OVERVIEW OF FUSION PRODUCT DIAGNOSTICS FOR ITER

L. C. Johnson,[1] Cris W. Barnes,[2] K. Ebisawa,[1] A. V. Krasilnikov,[3] F. B. Marcus,[4] T. Nishitani,[5] G. J. Sadler,[4] C. I. Walker,[6] and the ITER Joint Central Team and Home Teams

[1]ITER Joint Central Team, San Diego, CA, USA
[2]Los Alamos National Laboratory, Los Alamos, NM, USA
[3]TRINITI, Troitsk, Moscow region, Russia
[4]JET Joint Undertaking, Abingdon, Oxfordshire, UK
[5]Japan Atomic Energy Research Institute, Naka, Ibaraki-ken, Japan
[6]ITER Joint Central Team, Garching, Germany

INTRODUCTION

In order for ITER to meet its operational and programmatic goals, it will be necessary to measure a wide range of plasma parameters.[1-3] Some of the required parameters – e.g., neutron yield, fusion power and power density, ion temperature profile in the core plasma, and characteristics of confined and escaping alpha particle populations – are best measured by fusion product diagnostic techniques. To make these measurements, ITER will have dedicated diagnostic systems,[4] including radial and vertical neutron cameras, neutron and gamma-ray spectrometers, internal and external fission chambers, a neutron activation system, and diagnostics for confined and escaping alpha particles. Engineering integration of many of these systems is in progress,[5-11] and other systems are under investigation.

This paper summarizes the present state of design of fusion product diagnostic systems for ITER and discusses expected measurement capability. Details are given in other papers in these proceedings.

MEASUREMENT REQUIREMENTS AND SYSTEMS

An extensive list of plasma parameters required for ITER, together with target measurement accuracies and resolutions, has been presented.[2,3] Measurement specifications for parameters associated with fusion reactions are shown in Table 1. Most of the listed parameters can be measured by fusion product diagnostic techniques, and nine systems based on these techniques are currently included in the ITER EDA Work Breakdown Structure (WBS) in the fusion product diagnostic group, as shown in Table 2. The table also lists the parameters for which each system can provide measurements and indicates the

Diagnostics for Experimental Thermonuclear Fusion Reactors 2
Edited by Stott *et al.*, Plenum Press, New York, 1998

Home Team primarily responsible for leading the system definition and design activity. For some systems, other Home Teams provide support for the lead party.

Note that there is not a one-to-one correspondence between systems and parameters. This is generally true for all ITER diagnostic systems. For some of the parameters, more than one system is needed to fulfill measurement requirements. In some cases, requirements can not yet be met.

Table 1. Target measurement specifications for parameters related to fusion products.

Parameter	Parameter range	Spatial resolution	Time resolution	Accuracy
1. Total neutron source strength	10^{14}–10^{21} n s^{-1}	integral	1 ms	10%
2. Neutron/α source profile	10^{14}–4×10^{18} s^{-1} m^{-3}	30 cm	1 ms	10%
3. Fusion power	≤2 GW	integral	1 ms	10%
4. Fusion power density	≤10 MW m^{-3}	30 cm	1 ms	10%
5. Ion temperature profile	0.2–50 keV	30 cm	100 ms	10%
6. n_T/n_D in plasma core	0.1–3	30 cm	100 ms	20%
7. Confined α energy spectrum	0.1–3.5 MeV	30 cm	100 ms	20%
8. Confined α density profile	10^{16}–2×10^{17} m^{-3}	30 cm	100 ms	20%
9. Escaping α flux (steady state)	≤2 MW m^{-2}	30 cm	100 ms	10%
10. Escaping α flux (transients)	≤20 MW m^{-2}	...	10 ms	30%
11. Neutron fluence on first wall	0–3 MW a m^{-2}	~10 locations	10 s	10%

Table 2. Fusion product diagnostic systems for ITER.

WBS #	Name of system	Parameters[1]	Lead party	Support party
5.5.B.01	Radial Neutron Camera	1,2,3,4,11	EU, JCT	...
5.5.B.02	Vertical Neutron Camera	1,2,3,4,11	US, JCT	...
5.5.B.03	Microfission Chambers[2]	1,3,11	JA	...
5.5.B.04	Neutron Flux Monitors	1,3,11	US	...
5.5.B.05	Radial Neutron Spectrometer	5	EU	RF
5.5.B.07	Gamma-ray Spectrometers	2,8	RF	...
5.5.B.08	Neutron Activation System	1,3,11	US	EU, RF
5.5.B.09	Lost Alpha Detectors[2]	9,10	US	...
5.5.B.10	Knock-on Tail Neutron Spectrometer[2]	7,8	US	...

[1]Parameter numbers refer to item numbers in Table 1.

[2]Requires new concept development.

Most of the fusion product diagnostic systems presently planned for ITER are based upon methods commonly used in contemporary large tokamaks. However, the systems must be tailored to a much more severe nuclear environment than that encountered in present-day experiments. Although central fusion power densities in ITER will be comparable to those observed in high power D-T operation on JET[12] and TFTR,[13] neutron flux on the first wall will be ten times higher, fusion power will be 100 times higher, and the neutron yield per pulse will be more than 10^5 times as large. Design of the fusion product diagnostic systems must not only account for the higher fluxes and fluences but also adjust to constraints imposed by the attendant massive radiation shielding. For example, radiation shielding around the flight tubes for systems which require line-of-sight diagnostic access to the plasma must be sufficient not only to prevent unacceptable heating of the superconducting magnetic field coils but also to limit nuclear activation of other components inside the bioshield. In addition, the aspect ratio of the vacuum vessel port nozzles and their extension ducts to the cryostat boundary will restrict the angular field of view of neutron cameras and spectrometers. This affects the accuracy of the fusion power calibration as well as the achievable resolution of spatial profile measurements. The thick blanket and vacuum vessel also complicate the design of material activation systems and limit the applicability of conventional calibration techniques for neutron source strength monitors. These issues will be addressed below in discussions of individual systems.

NEUTRON CAMERAS

Neutron cameras for ITER are intended to fulfill a number of measurement requirements. In an ideal implementation, at least two fan-shaped arrays of sight lines view the entire plasma, and absolutely calibrated detectors measure the uncollided flux and spectrum of neutrons emitted along those sight lines. Summation of the chordal signals from either array, together with knowledge of the plasma position, gives the global neutron source strength, hence the total fusion power. This provides a secondary standard for calibration of other neutron flux and fusion power monitors. Combining data from all the measurements allows tomographic reconstruction of the spatial distribution of neutron emissivity, $n_D n_T \langle\sigma v\rangle$, which determines the alpha particle source profile and fusion power density and constrains inferred values of fuel concentrations and effective ion temperature. Analysis of high-resolution neutron spectra provides chord-averaged ion temperature profiles in an ignited plasma.

As mentioned above, however, the ITER environment imposes limitations which prevent us from achieving an ideal configuration. In particular, only two arrays can be installed, and neither array can view the entire plasma. The neutron cameras are patterned after systems used on JET[8] and TFTR.[13] The principal challenges in implementing the cameras on ITER are (a) the design of satisfactory interfaces with the blanket, backplate, vacuum vessel, cryostat, and bioshield; (b) the selection of detector/collimator combinations which enable coverage of a very wide range of neutron fluxes with adequate accuracy and energy resolution; and (c) the development of suitable calibration methods.

Interfaces between the neutron cameras and major tokamak components must provide unencumbered diagnostic access to the neutron emitting region of the plasma while preserving the heat load handling capability of the first wall, the radiation shielding properties of the blanket, backplate, vacuum vessel, and bioshield, and the tritium containment characteristics of the vacuum vessel and cryostat. Figure 1 shows an exploded view of the present configuration of the Radial Neutron Camera[9] and illustrates design approaches for resolving interface issues. The arrangement provides three separate collimated flight tubes and detector housings for each poloidal angle, thus offering a variety of choices of collimator/detector combinations. A massive diagnostic shield plug in the vacuum vessel port, with an integrated first wall and shield blanket, surrounds the sight lines and reduces neutron streaming. A re-entrant cryostat door allows the collimator assembly to extend inward from the biological shield while remaining outside the secondary vacuum volume of the nozzle extension. Additional shielding blocks in the secondary containment volume surround the neutron beams emerging from the diagnostic shield plug and prevent unacceptable heating of the superconducting coils and unnecessary nuclear activation of neighboring components.

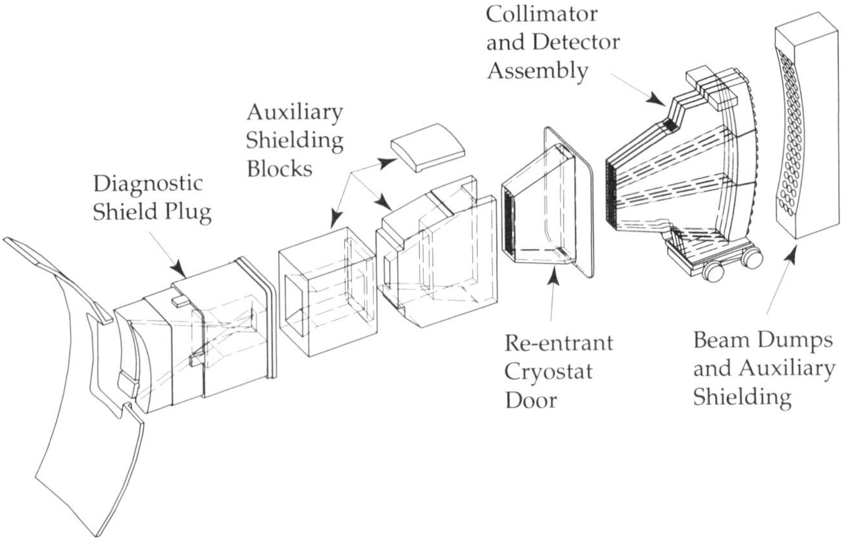

Figure 1. Exploded isometric view of the Radial Neutron Camera.

Design of the Vertical Neutron Camera is particularly challenging because of difficult interfaces with the blanket and backplate at the top port and with the cryostat lid. Unlike the situation at equatorial ports, ITER will not be equipped with large-diameter secondary vacuum extensions between the top port nozzles and the cryostat boundary. Shielding around flight tubes for the vertical camera must be either inside the primary vacuum region of the port or inside the cryostat vacuum. In addition, the width and length of the viewing slot through the blanket and backplate will be restricted, although precise limits have not yet been established, and the design must accommodate relative displacements between the upper port nozzle and the cryostat lid.

Figure 2 shows the present arrangement of the Vertical Neutron Camera. In this configuration, the front collimators are integrated into the design of the vacuum vessel port plug in order to make maximum use of the permissible aperture through the blanket and backplate and to maintain system alignment during thermal and operational excursions of the nozzle position. As a result, the diameters of the front collimators can not be changed after initial installation. In order to span the required range of flux with fixed front collimators, multiple detectors with different sensitivities will sample each sight line, as was the case in the TFTR neutron camera.[13] The detectors and rear collimators are in the space between the cryostat lid and the top bioshield. Additional shielding on the cryostat lid and around the detectors and rear collimators is not shown in the figure. A passive auxiliary shielding assembly encloses the sight lines inside the cryostat between the primary vacuum boundary and the cryostat lid. The height and angular position of this assembly can be adjusted from outside the cryostat vacuum space in order to account for displacement of the upper support structure under vacuum and static weight loads. Except for the support frame, the entire assembly can be installed and removed through a standard port in the cryostat lid.

Figure 3 shows expected line-integral power density profiles for the present designs of both neutron cameras. The profiles were computed from a simple model which approximates the magnetic flux surfaces with concentric ellipses. Two cases are shown, corresponding to a flat emissivity profile and one more peaked on axis. The dashed-line portions of the curves will not be observable because of limitations on the angular fields of view. Clearly, analysis of data from the neutron cameras will be subject to errors arising from the inability to observe the entire emission profiles. Measurements with the vertical camera, extending almost to the outer edge of the plasma, should be useful in constraining the profiles in the unobserved areas. The situation may be further ameliorated by invoking other information (e.g., knowledge of the position of the separatrix), but some errors will persist. The problem will be most severe for detectors which require corrections to the raw data because of their inability to completely distinguish uncollided neutrons from the background of gammas and scattered neutrons. A proper estimate of the expected errors under various operational conditions has not yet been made.

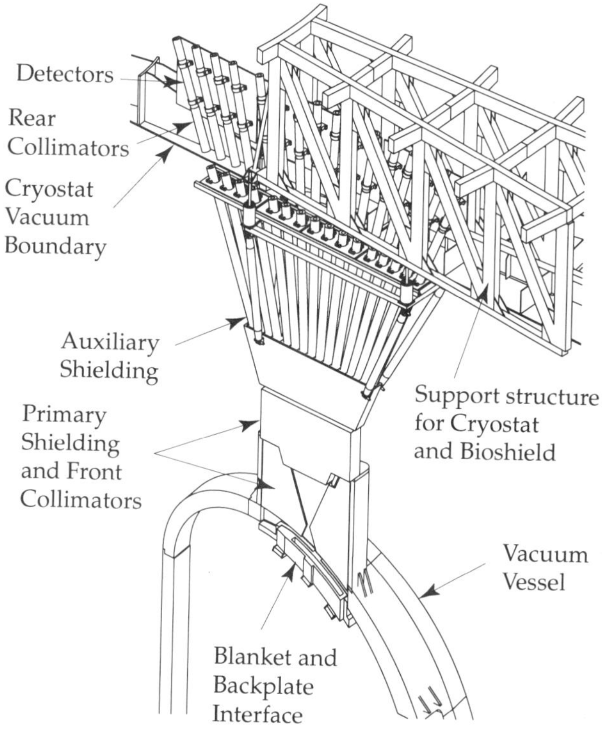

Figure 2. Isometric view of the Vertical Neutron Camera. For clarity, shielding on top of the cryostat lid and surrounding detectors and rear collimators is not shown.

NEUTRON SPECTROMETERS

The Radial Neutron Spectrometer is an adjunct to the Radial Neutron Camera, namely, a set of high resolution neutron spectrometers fitted to some of the camera flight tubes. In order to measure ion temperatures from the Doppler width of the 14.1 MeV emission over a reasonable fraction of the required parameter range, an energy resolution of less than 3% is needed. Among systems which may have sufficient resolution are natural diamond detectors[14] and various types of proton recoil and time-of-flight spectrometers.[15] Some of the spectrometers are quite bulky and not suitable for deployment around ITER except in small numbers. It is therefore not feasible to install enough large, high resolution spectrometers to measure the ion temperature profile with 30 cm spatial resolution. The natural diamond detectors are very compact and may serve as general purpose spectrometers in conjunction with flux detectors in the neutron cameras.

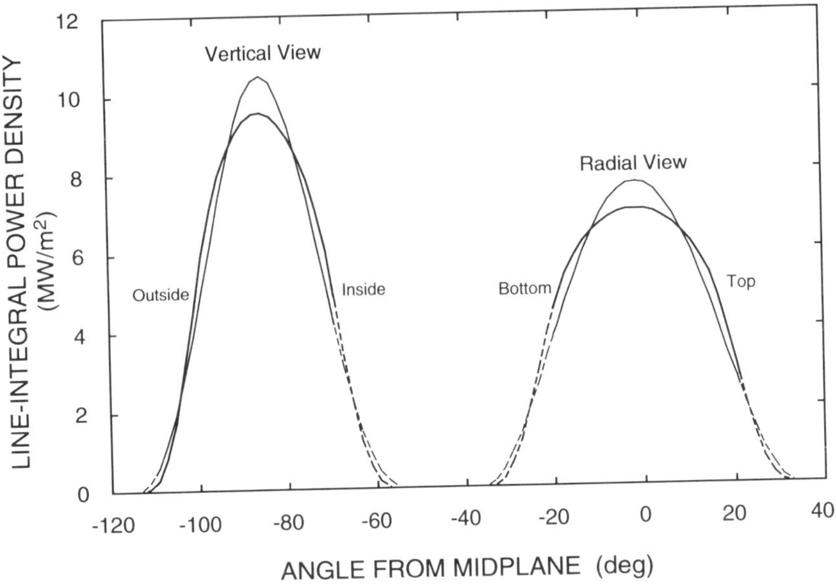

Figure 3. Computed line-integral power density profiles for the Radial Neutron Camera (right) and the Vertical Neutron Camera (left). The thick lines correspond to a flat emissivity profile, and the thin lines represent a profile more peaked on axis. The dashed portions of the profiles will not be observable because of limitations on angular fields of view.

Gamma-ray spectrometers[16] may be installed on some of the flight tubes of the Radial Neutron Camera. The purpose is to detect and analyze high energy gamma rays produced by fusion reactions and by interactions of fast ions with impurities such as Be. In order for the technique to be successful, the spectrometers must be able to distinguish the fusion gammas from a large background of prompt gammas produced in the first wall and blanket. The measurement capability of such spectrometers in the ITER environment is under study.

FISSION DETECTORS

The Neutron Flux Monitor System[5,7] will provide time-resolved measurements of neutron source strength. The proposed system for ITER will consist of a set of conventional ^{235}U fission chambers, similar to those used on JET, JT-60U, and TFTR. Each counter will be housed inside a moderator, to give flat energy response, and will be shielded against gamma rays.

As indicated in Table 1, the system is required to have a dynamic range of seven orders of magnitude with 1 ms time resolution and absolute accuracy of 10%. Since no single detector can satisfy these requirements, counters containing different amounts of fissionable material, and therefore having different sensitivities, will be chosen in order to span the necessary parameter range. The sensitivity may also be varied by locating some detectors near the plasma, e.g., in the pre-shield of the Radial Neutron Camera, while others will be situated in shielded locations outside the vacuum vessel. One complete set of detectors will consist of six fission chambers, with effective sensitivities decreasing in factors of no more than 25 to allow adequate overlap of linear operational ranges. This overlap is essential so that the six detectors of a set may be cross calibrated to one another.[18] In addition, "blank" proportional counters, containing no fissionable material but otherwise identical to the fission chambers, will be used to identify and correct for spurious signals generated by noise, etc. At least two complete sets of detectors will be needed to provide redundancy and protect against single-point failures.

The shielding effects of the blanket and vacuum vessel will make *in situ* calibration of flux monitors more difficult in ITER than in present large tokamaks. An important design goal is to situate the most sensitive detectors such that (a) they are close enough to the plasma so that they may be directly calibrated and (b) they view enough of the plasma volume so that their sensitivity is not strongly dependent on plasma position or shape. The calibration source will be a neutron generator mounted on an articulated arm attached to the standard remote handling transport system inside the vacuum vessel. Therefore, the most sensitive fission chambers and all machine components that affect their detection efficiency must be fully installed before the transporter is removed. Several possible detector locations are under study.[7]

The least sensitive detectors will be designed for operation in count rate mode during ignited D-T conditions. Instrumentation for all the fission chambers will operate in different electronic modes (count, Campbell, current) to extend the dynamic ranges of individual detectors and to allow temporal resolution of 1 ms. Provisions will be made for periodic insertion or permanent installation of standard calibration sources near the fission chambers to verify long term stability of the detectors and electronics.

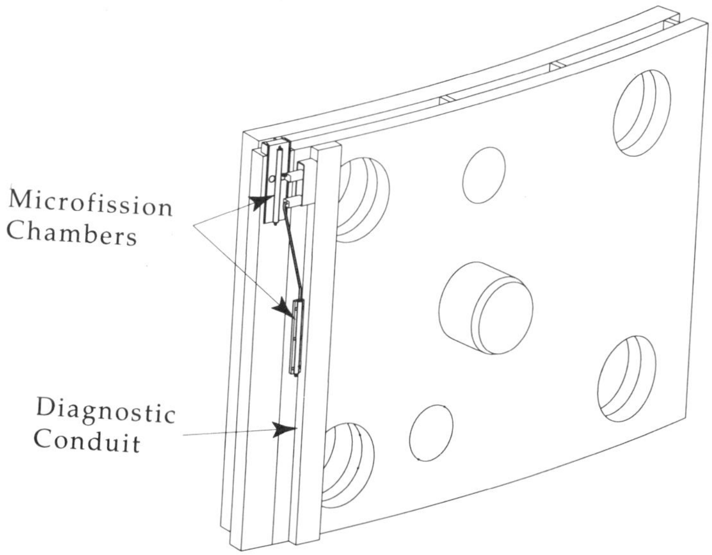

Figure 4. Possible locations of a pair of microfission chambers. The figure shows a portion of the backplate on the inboard side of the plasma. Signal cables are routed through a diagnostic conduit integrated into the backplate.

A set of Microfission Chambers[10] will augment the Neutron Flux Monitor System. These are small fission chambers, similar to those used for flux measurements inside fission power reactors. They will be installed in poloidal arrays at two toroidal locations inside the ITER vacuum vessel. Figure 4 shows possible locations for a pair of detectors on a portion of the backplate on the inboard side of the plasma. One detector is positioned behind a blanket module, and the other is in the gap between adjacent modules. Signal cables for the detectors are routed through diagnostic conduits integrated into the backplate.

Microfission chambers can use either ^{235}U or ^{238}U for the fissionable material, although they will differ in detector sensitivity, spectral response, and burn-up characteristics. The counters appear to be insensitive to gamma radiation and magnetic field effects. Extensive two-dimensional Monte Carlo calculations show that the detection efficiency is only modestly dependent on changes in plasma position and shape, and these profile effects may be further reduced by combining signals from opposite detectors in a poloidal array. The toroidal field of view will be limited for detectors located in the gap between modules, but *in situ* calibration should be possible.

NEUTRON ACTIVATION SYSTEM

A traditional and fundamental technique for determining neutron fluence is the measurement of radioactivity induced in some materials by neutron interactions. If the mean-free-paths of activating neutrons and decay radiation in the material are large compared with the sample size, the process is intrinsically linear and may be applied over a range of many orders of magnitude in neutron flux. Activation cross sections are accurately characterized for a number of materials whose nuclear and physical properties make them suitable for dosimetry. By choosing materials with reactions such as *(n,p)*, *(n,α)*, and *(n,2n)*, which usually require a minimum or threshold energy, thermal neutrons may be rejected and some degree of energy discrimination is possible. The method is insensitive to gamma flux at the sampling station, and decay radiation measuring equipment and analysis software are well developed. The principal problem in applying the technique to the measurement of neutron yield in fusion plasmas is in establishing the relationship between the plasma neutron source strength and the flux and spectrum at the sample irradiation station. This usually requires detailed Monte Carlo neutron transport calculations, and the resulting uncertainties depend critically upon the types and amounts of materials separating the exposed sample from the plasma and on the accuracy of the geometrical modeling of those materials in the vicinity of the sample.

The Neutron Activation System[6,7] for ITER will be similar to those used successfully on JET and TFTR. Encapsulated foils will be transferred pneumatically to an array of irradiation stations, deployed poloidally at two toroidal locations. After irradiation for about 100 s, the samples will be retrieved to remote counting stations, where gamma rays from the induced radioactivity will be measured.

Because of the high neutron and heat fluxes on the first wall and the long pulse lengths expected in ITER, special care will be required in the design of irradiation stations and re-entrant transfer tubes and in the handling of the extremely radioactive exposed samples. The samples should have wide-angle views of the neutron emitting region of the plasma with as little intervening material as possible, but the irradiation stations and foil capsules must be protected from excessive nuclear heating, and there can be no sharp bends in the transfer tubes. It does not appear to be feasible to embed the irradiation stations in the removable blanket modules. The present design approach is to locate irradiation stations in some of the permanent filler modules and view the plasma through the natural gaps between neighboring removable blanket modules. Routing of transfer tubes will make use of standard diagnostic conduits in the blanket backplate. Uncertainties in measurements of neutron yield arising from transport calculations can not be evaluated until the irradiation stations are designed in detail, but it is unlikely that the activation system for ITER can achieve the degree of accuracy demonstrated on JET and TFTR because of the less favorable local geometry.

CALIBRATION OF NEUTRON SYSTEMS

Calibration methods for neutron systems on ITER will be similar to those used on JET[17] and TFTR.[18] A variety of techniques will be utilized, including *in situ* calibration of flux monitors and neutron cameras with a neutron generator, activation measurements coupled with neutron transport calculations, and laboratory calibration of individual detectors. As discussed above, the thick blanket and vacuum vessel will impose limitations on some of the traditional methods. A comprehensive calibration program, which uses the advantages of some of the techniques to offset the limitations of others, is under development.[19]

ALPHA PARTICLES

A consideration of alpha particle physics issues in ITER indicates the need for measurements of alpha heat losses, alpha instabilities, confined alphas, and lost alphas and suggests measurement requirements and possible techniques.[20] The alpha birth profile should be well characterized by the neutron diagnostic systems. Under the assumption of classical confinement, this and other basic plasma parameters may be used to calculate the alpha heating profile, the confined alpha density and energy spectrum, and losses to the walls. Alpha heat losses and instabilities manifest themselves indirectly by giving rise to local hot spots on the first wall, observable with infrared cameras and thermocouples, and to plasma fluctuations, observable by magnetic loops, reflectometry, etc. A new detector using thin foil Faraday collectors to measure escaping alphas is being tested on JET[21] and may be suitable for installation on ITER. The direct measurement of confined alphas would be of great interest, but the present status of confined alpha diagnostics only allows a discussion of possible candidates and requires further R&D. Among the candidate techniques for monitoring the steady state alpha population are measurements of neutron and triton knock-on tails.[22] Measurements of the redistribution of alphas due to MHD events might be performed in the range $r/a>0.4 - 0.5$ through the use of neutral particle analyzers (NPA) with a helium diagnostic beam (200-400 keV, ~3 MW)[23] or by means of Pellet Charge Exchange (PCX),[24] but such systems are not presently planned for ITER.

CONCLUSIONS

The fusion product diagnostic systems required for the ITER device have been identified, and the principal characteristics of most of them have been established. Engineering of a number of the systems is well advanced, and solutions to many of the integration problems have been found. It appears that target measurement requirements for fusion power and emission profile can be met. The ion temperature profile can be measured, but it will be difficult to satisfy requirements on spatial resolution and parameter range. Methods for measuring properties of confined alpha particles are under study, but no technique is yet available that can satisfy requirements.

ACKNOWLEDGMENT

This report was prepared as an account of work performed under the Agreement among the European Atomic Energy community, the Government of Japan, the Government of the Russian Federation, and the Government of the United States of America on Co-operation in the Engineering Design Activities for the International

Thermonuclear Experimental Reactor ("ITER EDA Agreement") under the auspices of the International Atomic Energy Agency (IAEA).

REFERENCES

1. K. M. Young, A. E. Costley, T. Matoba, D. Orlinski, and P. E. Stott, The ITER diagnostic programme, in: *Diagnostics for Experimental Thermonuclear Fusion Reactors*, P. E. Stott, G. Gorini, and E. Sindoni, eds., Plenum Press, New York (1996).
2. A. E. Costley, R. Bartiromo, L. de Kock *et al.*, Requirements for ITER diagnostics, in: *Diagnostics for Experimental Thermonuclear Fusion Reactors*, P. E. Stott, G. Gorini, and E. Sindoni, eds., Plenum Press, New York (1996).
3. V. S. Mukhovatov, R. Bartiromo, D. Boucher *et al.*, Role and requirements for plasma measurements on ITER, *These Proceedings* (1997).
4. L. C. Johnson, C. W. Barnes, A. Krasilnikov *et al.*, Neutron diagnostics for ITER, *Rev. Sci. Instrum.* 68:569 (1997).
5. C. W. Barnes and A. L. Roquemore, Neutron source strength monitors for ITER, *Rev. Sci. Instrum.* 68:573 (1997).
6. C. W. Barnes, M. J. Loughlin, and T. Nishitani, Neutron activation for ITER, *Rev. Sci. Instrum.* 68:577 (1997).
7. C. W. Barnes, D. L. Jassby, G. LeMunyan, A. L. Roquemore, and C. Walker, Design considerations for neutron activation and neutron source strength monitors for ITER, *These Proceedings* (1997).
8. F. B. Marcus, O. N. Jarvis, M. Loughlin *et al.*, A neutron camera for ITER, *Rev. Sci. Instrum.* 68:514 (1997).
9. F. B. Marcus, J. M. Adams, P. Batistoni *et al.*, A radial neutron camera design for ITER with integral shielding, *These Proceedings* (1997).
10. T. Nishitani, K. Ebisawa, L. C. Johnson *et al.*, In-vessel neutron monitor using micro fission chambers for ITER, *These Proceedings* (1997).
11. C. I. Walker, T. Ando, A. E. Costley *et al.*, Diagnostic access for ITER, *These Proceedings* (1997).
12. The JET Team, Fusion energy production from deuterium-tritium plasma in the JET tokamak, *Nuclear Fusion* 32:187 (1992).
13. L. C. Johnson, C. W. Barnes, R. E. Bell *et al.*, Fusion product measurements in D-T plasmas in TFTR, in: *Diagnostics for Experimental Thermonuclear Fusion Reactors*, P. E. Stott, G. Gorini, and E. Sindoni, eds., Plenum Press, New York (1996).
14. A. V. Krasilnikov, E. A. Azizov, A. L. Roquemore *et al.*, TFTR natural diamond detectors based D-T neutron spectrometry system *Rev. Sci. Instrum.* 68:553 (1997).
15. T. Elevant *et al.*, The role of neutron spectrometers for ITER, *These Proceedings* (1997).
16. V. G. Kiptily, I. A. Polunovskii, V. O. Naidenov *et al.*, ITER gamma diagnostics: 2-D neutron and gamma camera, in: *Diagnostics for Experimental Thermonuclear Fusion Reactors*, P. E. Stott, G. Gorini, and E. Sindoni, eds., Plenum Press, New York (1996).
17. O. N. Jarvis *et al.*, Use of activation techniques at JET for the measurement of neutron yields from deuterium plasmas, *Fusion Technology* 20:265 (1991) and references therein.
18. L. C. Johnson, C. W. Barnes, H. H. Duong *et al.*, Cross calibration of neutron detectors for deuterium-tritium operation in TFTR, *Rev. Sci. Instrum.* 66:894 (1995) and references therein.
19. G. J. Sadler, J. M. Adams, C. Barnes *et al.*, Calibration issues for neutron diagnostics, *These Proceedings* (1997).
20. S. J. Zweben, S. Putvinski, M. P. Petrov *et al.*, Alpha-physics and measurement requirements for ITER, in: *Diagnostics for Experimental Thermonuclear Fusion Reactors*, P. E. Stott, G. Gorini, and E. Sindoni, eds., Plenum Press, New York (1996).
21. M. J. Loughlin, F. E. Cecil, M. Hone *et al.*, evaluation of ITER compatible thin foil Faraday collector as a lost alpha particle diagnostic for high yield d-t plasmas, *Rev. Sci. Instrum.* 68:361 (1997).
22. R. K. Fisher, C. W. Barnes, A. Gondhalekar *et al.*, ITER Alpha particle diagnostics using knock-on ion tails, in: *Diagnostics for Experimental Thermonuclear Fusion Reactors*, P. E. Stott, G. Gorini, and E. Sindoni, eds., Plenum Press, New York (1996).
23. M. P. Petrov and R. K. Fisher, Charge exchange (C-X) diagnostics of fast confined alphas: present situation and prospects for ITER, in: *Diagnostics for Experimental Thermonuclear Fusion Reactors*, P. E. Stott, G. Gorini, and E. Sindoni, eds., Plenum Press, New York (1996).
24. S. S. Medley *et al.*, *Rev. Sci. Instrum.* 67:3122 (1996).

A RADIAL NEUTRON CAMERA DESIGN FOR ITER WITH INTEGRAL SHIELDING

F B Marcus, J M Adams[1], P Batistoni[2], D S Bond[1], T Elevant[3], N Hawkes[1], G Janeschitz[4], O N Jarvis, L Johnson[5], L de Kock[4], M Loughlin, M Rapisarda[2], G Sadler, P Stott, N Watkins[1]

JET Joint Undertaking, Abingdon, Oxon. OX14 3EA, United Kingdom
[1]AEA Technology, Harwell, Abingdon, UK
[2]ENEA Frascati, Italy
[3]Royal Inst of Tech, Stockholm, Sweden
[4]ITER, Garching, Germany
[5]ITER, San Diego, USA

INTRODUCTION

This paper presents the reference design of the radially viewing horizontal neutron profile monitor for ITER. It incorporates integral neutron shielding, and with full machine integration, should provide control and plasma diagnostic information based on the deduced neutron emission profile. Information can be obtained on total neutron emission (fuel burn-up rate), alpha-particle birth profile, plasma position and the time dependent effects of plasma instabilities such as sawteeth.

Previous papers on initial designs have been presented (Marcus et al., 1995,1997), which had been constrained by the cryostat and bio-shield boundaries to have separate pre-shield and collimation shielding. This design permitted some neutron scattering inside the bio-shield region, although at levels which would have been acceptable from the point of view of induced coil heating (Marcus et al., 1997). In this paper, the design has been evolved to include a re-entrant cryostat and integral diagnostic shielding, which greatly reduces neutron scattering outside the diagnostic. This has been accomplished by lengthening the collimator channels which are imbedded in a long re-entrant shield plug, and replacing rotating collimator adjustment with parallel collimators with different sized inserts to adjust the neutron flux. The camera detectors are placed outside the cryostat and bio-shield, allowing relatively easy access. The neutron fluxes at the detectors are expected to be about the same as in the JET neutron profile monitor, as discussed below.

The modular concept of camera design is retained, allowing changes in detector systems and collimation geometries up to the time of final construction for ITER. The reference design currently proposed has fixed collimation on each channel, with inserts to

Diagnostics for Experimental Thermonuclear Fusion Reactors 2
Edited by Stott *et al.*, Plenum Press, New York, 1998

adjust the neutron flux. Options involving adjustable collimation are also under consideration.

RECENT OPERATIONAL EXPERIENCE FROM JET D-T EXPERIMENTS

The JET Neutron Emission Profile Monitor (Upgrade) (Marcus et al., 1997) has been used during the initial "DTE1" experiments with deuterium-tritium (DT) fuelling in JET, and is providing a demonstration of the collimation geometry, detector response, and analysis techniques suitable for the ITER camera. Two other camera systems have also been used in DT experiments: the first version of the JET Profile Monitor (Adams et al., 1993) during the PTE experiments and the neutron camera on TFTR (Roquemore et al., 1990).

Neutron emission from recent experiments with DT in JET has been measured with cylindrically shaped solid organic scintillators (Bicron type BC418) with a small diameter (15 mm) and thickness (10 mm), and compared with NE213 detectors. The advantage of the Bicron detectors, when used in conjunction with discriminators that detect recoil protons with energies of about 9 MeV or greater, is that gamma rays are not recorded as they cannot deposit enough energy in the small detector volume. Count rates in the MHz range are acceptable. The geometry used during high power DT experiments has 10 mm diameter collimation for all channels in both the vertical and horizontal camera. In the horizontal camera, the front and rear collimator channels are both 250 mm long. The maximum separation of the channels occurs in the edge channels of the viewing fan, with the entrance to the front collimator and the exit to the rear collimator separated by 1.466 m. On the camera midplane, the separation is 1.390 m.

Fig. 1 shows the measurements and deduced results obtained from the neutron profile monitor during DT operation. The diverted, ELMy H-mode plasma is heated by up to 10.7 MW of tritium beams, producing near steady conditions with ion and electron temperatures of a few keV. Three of the channels from the horizontal camera are shown, centrally viewing channel #4, and intermediate position viewing channels #7 and #8, out of the 10 channels total. The count rates on the central channel reach about 2×10^5/s at a fusion power level of about 2.5 MW, and the detectors and electronics should comfortably function at an order of magnitude higher rates. The time variation in the central channel is different than in the intermediate channels, and is not linearly proportional to the global neutron emission. The data from both cameras are analysed with a program which determines a least squares fit to the emissivity function, and contours of constant neutron emissivity are plotted between a major radius through the maximum emissivity of 2m to 4m versus time. The oscillations in central (predominantly beam-thermal) neutron emission are due to variations in the beam power and to core sawteeth. The integral of the emissivity is then plotted on the same scale as the global neutron emissivity deduced from fission chamber measurements. The agreement of the time-dependent variations is excellent.

THE REFERENCE DESIGN

The machine integration aspects of the reference design for the horizontal neutron camera are shown in Fig. 2. A monolithic structure contains a fan of 18 sets of 3 flight tubes with different collimator inserts, each backed by a detector module and a neutron

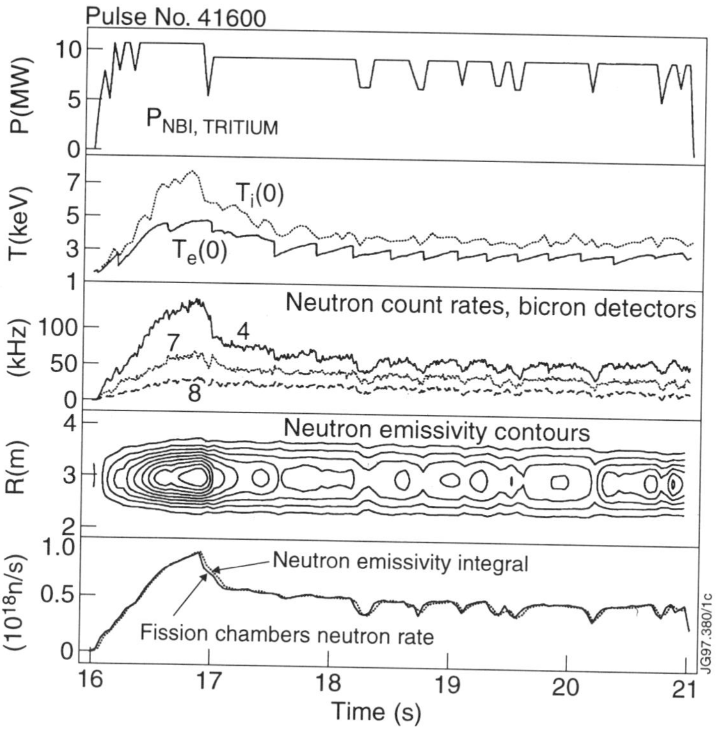

Fig. 1. Tritium beam heating power, central ion and electron temperature, horizontal neutron camera line integral measurements for channels 4,7,8, contours of constant neutron emissivity versus major radius, and fission chamber global neutron rates compared to the integral of the emissivity profiles.

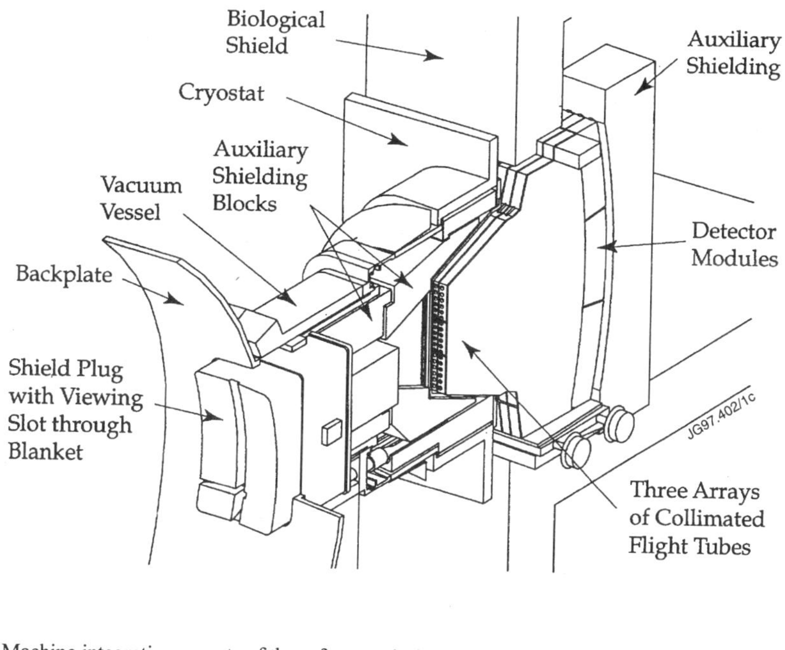

Fig. 2. Machine integration aspects of the reference design for the radial neutron camera.

beam dump imbedded in the auxiliary shielding. The neutron flight tube channels are sized and placed so as to enable a minimisation of the width of the slot in the blanket. The structure mounts on the bio-shield, with the exit ends of the collimators flush with the outside of the bio-shield. The plasma facing side of the structure nests into a re-entrant cryostat flange surrounded by auxiliary shielding blocks, with shielding inside the cryostat nozzle. That shielding structure, in turn, mates to a shield plug with a viewing slot through the main neutron blanket. Appropriate gaps are designed in to allow for relative movement of the various components. The arrangement is designed with the re-entrant part of the cryostat flange as narrow as possible, consistent with shielding requirements, in order to minimise the weight of the cryostat flange and the weight of the cantilevered portion of the collimator structure. Other diagnostic systems are integrated into the same port. This arrangement greatly reduces neutron levels outside the diagnostic.

Collimator and Viewing Geometry

Fig. 3 shows the fan geometry of the 18x3 viewing channels that have fan angles of 21.3° upwards (on the plasma side) and 20.0° downwards, with the focal point at major radius R and vertical height Z (R,Z) = (14.160 m, 1.590 m). The plasma axis is at (8.140 m, 1.440 m), so the sight-lines above the plasma axis now cover a vertical height (at the major radius of the plasma axis) of 2.347 m, i.e. about half of the distance of 4.381 m from the plasma axis to the separatrix surface, which should enclose much of the neutron emission (depending on plasma profiles). This is less favourable than with the JET situation, where the (vertically) lowest channel (in the plasma region) of the horizontal camera effectively views outside the neutron emitting region. Channels which view regions further out would therefore be highly desirable in either the horizontal or vertical neutron camera. The current viewing extent of the collimator fan is determined by the ITER standard size of the blanket opening on the plasma side, and by the ITER standard cryostat flange opening at the rear. Non-standard flanges openings and flanges are not allowed.

Fig. 4 shows a top view of the collimators with the three channels which are present at any given fan angle. The focal point of these collimators in this plane is R=11.100 m, with an angular separation of 0°, +1.4°, -1.4°, to fit through the slot in the blanket. Each flight tube has an inner diameter of 50 mm, allowing inserted collimators with inner diameters up to 40 mm. The detector assemblies themselves mate into the rear of the collimator assembly, and are backed by polythene beam dumps imbedded in the rear auxiliary shielding. There are two thin steel vacuum boundaries, the first an extension of the vacuum vessel, the second an extension of the re-entrant part of the cryostat, with 5 mm thickness.

The 18 channel spacing corresponds to slightly better than a 7% resolution in radius. The improved resolution compared to the 10 channels per camera of TFTR or JET, made possible due to the larger system size, is highly desirable, especially if there are strong emissivity gradients.

On the midplane, the viewing cone from a collimator of 10 mm (and 20 mm, 40 mm) diameter for both front and rear has its apex half way down the 3.700 m length at major radius 18.391 m, and a full opening angle of 0.31° (0.62°, 1.24°). The total opening of this cone at the focal point R=14.160 m is 23 mm (46 mm, 92 mm), at the shield plug slot R=11.100 is 39 mm (78 mm, 156 mm) compared to a slot width of 180 mm, and at the plasma axis R=8.140 m is 55 mm (110 mm, 220 mm), or about 1/83 (1/41, 1/20) of the viewing height in the plasma covered by the 18 channels. The 40 mm maximum diameter

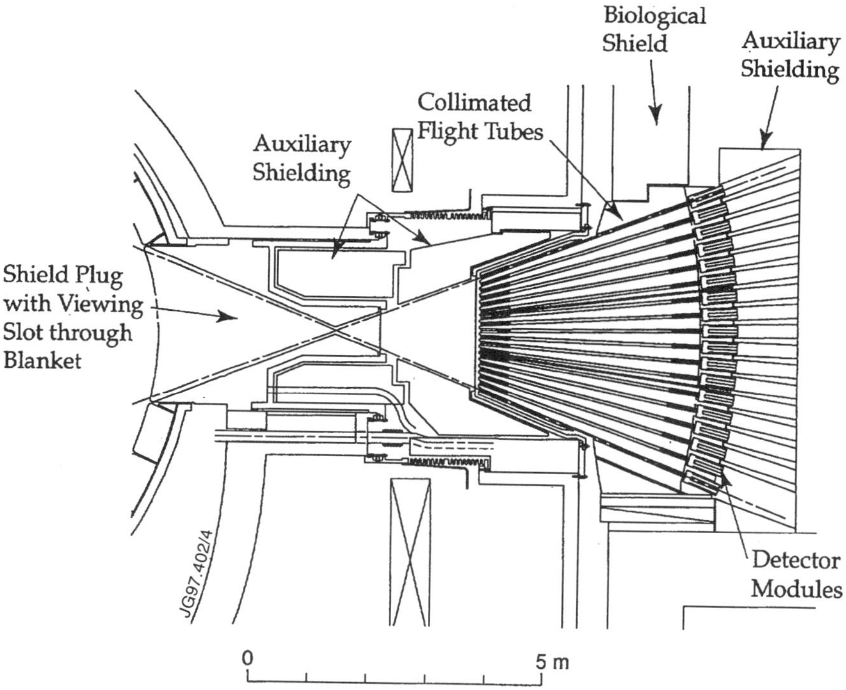

Fig. 3. A side view of the collimator viewing geometry showing 18 collimator groups in a fan configuration.

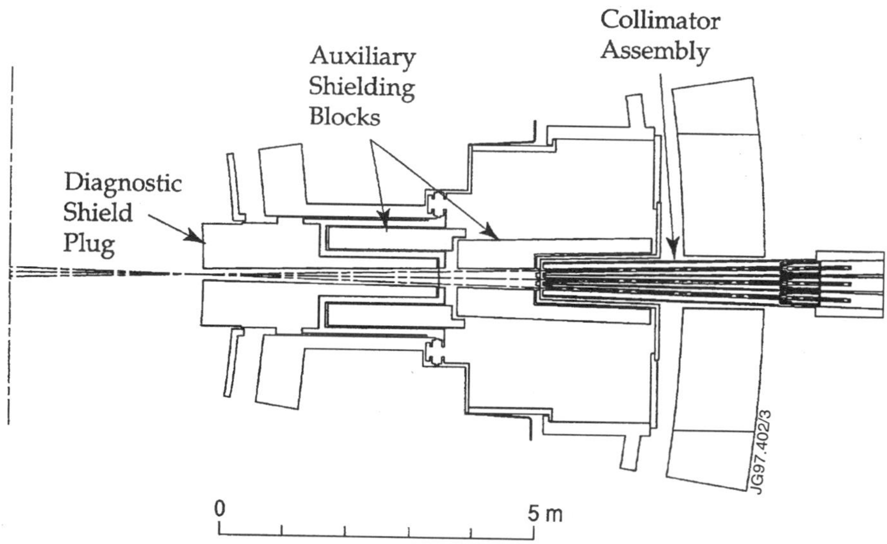

Fig. 4. A top view of the collimator viewing geometry showing 3 flight tubes at a constant vertical fan angle, with the possibility of different collimation inserts.

collimator therefore fits within the blanket slot opening, and makes the width of the cone nearly equal to the separation between channels at the plasma centre.

Neutron Flux and Rate Detectors

An initial collimator and detector design is based on the JET camera, with apertures of 10 mm and 21 mm. For ITER, we choose the collimator diameter sizes of 10, 20 and a third size which could be as large as 40 mm, to be able to allow for the fluxes expected in either d-d or d-t operation, for low intensity in peripheral channels, or for large aperture detectors for maximum count rates. The 10 and 20 mm apertures are suitable for the types of detectors used on JET as shown in Fig. 5, with 25 mm diameter NE213 detectors for 2.5 MeV and low rate 14 MeV neutron operation, and 15 mm diameter Bicron detectors, suitable for high rate 14 MeV operation. The collimator length is chosen to be 3.700 m to reduce the flux to the detectors. It is assumed that ITER will have a d-d phase. The variation in neutron rate (4th power of aperture diameter) is 1:16:256 (compared to a typical ratio of 88 between 50-50 d-t and equivalent 100% d-d plasmas.

The use of the largest allowable collimator with a high count rate detector such as those used on TFTR (Roquemore et al., 1997) would allow the observation of details of MHD fluctuations. However, the neutron flux is so large, 256:1, compared to that of the 10 mm collimator, that there is a real danger that gamma production and neutron scattering between adjacent detector boxes would swamp the signals on the detectors with a lower flux. When using massive scintillators in current mode to observe fluctuations, those camera channels would no longer function usefully as a flux monitor for neutrons, since we could no longer distinguish between gammas and neutrons. Detailed neutronics scattering calculations of the lack of interference between the three sets of detectors at high yields will be required to demonstrate that the full size apertures can justifiably be used in conjunction with smaller apertures. Similarly, calculations are required to determine the best relative placement of the three apertures, especially whether the smallest aperture should be in the middle or in an edge channel. An adjustable collimator option (see below) may need to be developed.

The line integrated neutron emissivity from a 1.5 GW ITER plasma at a detector viewing the plasma axis along a horizontal sight-line is 2.4×10^{18} neutrons/m^{2}s of 14.1 MeV neutrons, based on a simple model (Adams et al., 1991). (The line-integral emissivity deduced (Johnson et al., 1995) for the horizontal camera was about the same for a standard plasma profile, and could double with different assumptions.) ITER also allows for up to 20% power excursions. For a 10 mm diameter, 3.7 m long collimator, the maximum flux on a centrally viewing detector is 1.1×10^{12} neutrons/m^{2}s (1.8×10^{13} neutrons/m^{2}s for a 40 mm collimator). An efficiency of 1.5 % for typical discriminator bias settings of a 10 mm diameter, 10 mm thick active area Bicron detector would lead to a count rate of about 1.3 MHz, compatible with the operating range of the fast organic scintillators being used on JET. For edge channels (especially), background scattering needs to be calculated and subtracted, and it sets a lower limit to the measurements near the plasma edge of about 1% of the peak emission in DT, or higher in DD.

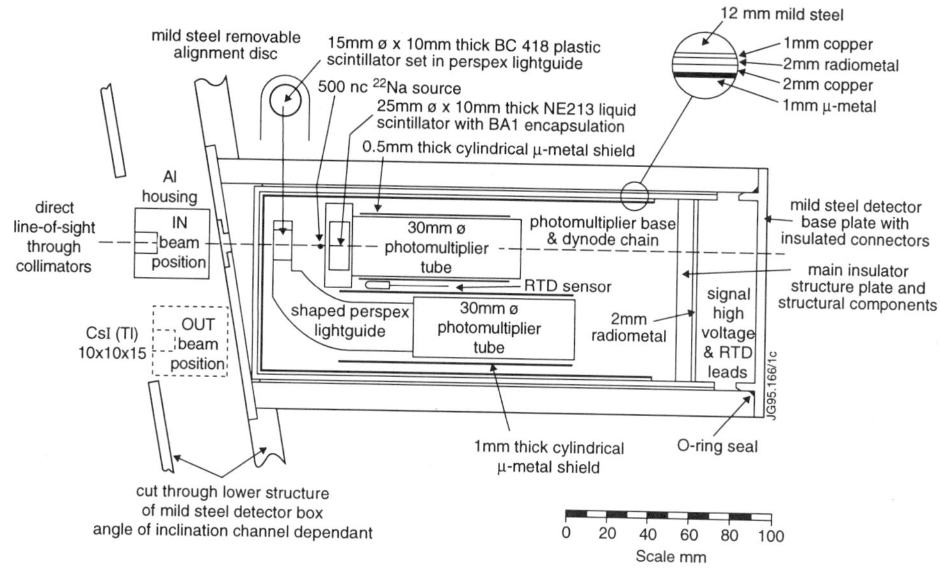

Fig. 5. Detector system for the JET neutron profile monitor, comprising a 25 mm diameter NE213 detector and a 15 mm diameter Bicron detector.

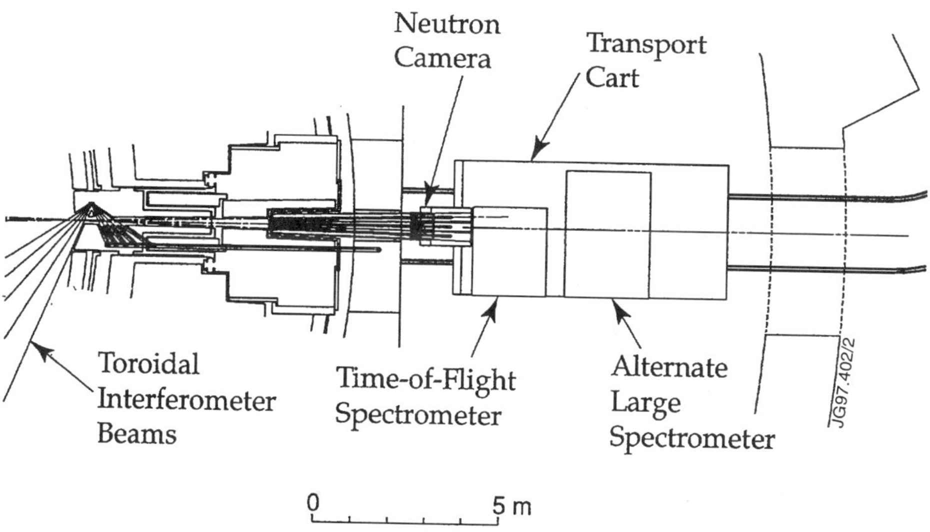

Fig. 6. Large spectrometer design integration - time-of-flight neutron spectrometer.

NEUTRONICS

A neutronics assessment of the radial neutron camera has been made (Shatalov et al., 1997). ITER Port #16 (now #9) containing the radial neutron camera and toroidal interferometer systems was analysed using the MCNP-4A code. The key findings are that the heat deposition and other radiation effects in the TF coil are well below the permissible level and that the interference between the two systems in the port is very low. The shielding in general is very effective. For the neutron emitting model of the plasma used in the study, the full energy neutron flux reached very low levels in the edge channels, a few percent of the central channels.

NEUTRON SPECTROMETERS

The various options for spectrometers were discussed previously (Marcus et al., 1997), with details in the referred papers in that article, and these options are still viable for the present design. For example, the third flight tube of each group, instead of a large diameter opening, could be adapted to each hold a compact spectrometer of various types such as scintillators, diamond, scintillating fibre, and silicon diodes.

It is also possible to incorporate larger spectrometers. There is space behind the rear auxiliary shielding, which can be used for a few of these larger spectrometers as shown in Fig. 6, with the space envelope for a time-of-flight (TOF) neutron spectrometer. The spectrometer can be moved in to just behind the auxiliary shielding, and removed to make way for maintenance or for other spectrometers. The 3 m TOF neutron spectrometer shown in Fig. 6 allows the measurement in thermal plasmas of ion temperatures greater than 7 keV in DD plasmas and with minor modifications, greater than 9 keV in DT plasmas. A 4 m TOF would allow DT temperatures down to 5 keV to be determined. Fig. 6 also shows the space requirement for such larger spectrometers. At present, large spectrometers have more operating experience and significant advantages over more compact spectrometers, and probably a few large spectrometers will be required.

"ADJUSTABLE" DESIGN OPTION FOR COLLIMATORS

From a historical perspective, the original design of the camera with a front and rear rotatable collimator solved the problem of adjustable collimation. Its problems were that there were a large number of mechanical systems to rotate the collimators, and that neutron streaming occurred through the large pre-shield opening towards the bio-shield. The reference design solves this problem by having a re-entrant cryostat wall filled with shielding that contains flight tubes with inserts to define the collimation, which leaves no flexibility for easy adjustment. It was necessary that the modules be lengthened to reduce the size of the opening angle of the sight line, so that the sight line fits through openings in the blanket. There is no way of adjusting the neutron flux throughput on individual channels, or indeed of shutting off the flux altogether on some channels to reduce neutron streaming into the detector area. The idea of using inserts only allows changes over maintenance periods, whereas it would be desirable to change the collimation in front of a detector between shots or even during a shot.

The "adjustable" option uses the philosophy from the original design to allow for altering the flux reaching the detector system. The actual design depends on how much space is available around each channel at different positions, and whether or not the number of channels needs to be reduced. The same access requirements and integral shielding size are maintained as in the reference design. One detector per module instead of two as in the previous design would allow smaller diameter modules, and the modules could be lengthened to reduce the size of the opening angle so that the sight line fits through openings in the blanket.

There is a limit on the minimum size of the rotatable collimator. If the diameter of the rotatable collimator is compact to fit each channel, for example 8 cm in diameter, containing holes of, say, 2.5, 1.0, and 0. (blank) cm diameter, then the holes might provide streaming paths in the unused channels for neutrons, even if the entrance is blocked off. Careful evaluation is required.

The idea that each channel might have only a single compact rotatable collimator just in front of the detector and a fixed front collimator based on inserts in the flight tube was rejected, because the dynamic range is too limited. In this case, the flux on the rear aperture remains unchanged, and the total neutron throughput scales as the square of the diameter of the rear aperture. If the detector (for example a diamond detector) is smaller than the aperture, the count rate of the detector is almost independent of the size of the final collimator.

"COMPACT" OPTION FOR ADJUSTABLE COLLIMATORS

Another option for the design for a radial neutron camera was considered, which was very similar to the existing horizontal neutron profile monitor at JET. After the pre-shield, it has two massive, rotatable collimator cylinders with 10 mm and 25 mm holes at right angles, followed by a detector housing block inside the cryostat and bio-shield. The detectors have been moved inward to allow a wider plasma coverage. This option has the apparent virtue that it is compact and fits inside the cryostat, where it does not perturb the cryostat or the bio-shield. This option was not retained for further study, after a number of problems were identified. These problems were as follows:

It is very important to have flexibility for using a whole range of neutron detectors, and that we need to have access to the collimation and detector system. Massive monolithic collimators would lead to a total lack of flexibility for instrumentation and throughput adjustment. Having detectors inside the bio-shield is unacceptable, since they will need periodic replacement due to radiation degradation, and the massive magnetic shielding could cause problems with magnetic forces and tokamak field perturbation. Signal transmission from the detectors is also a major problem, especially for NE213 with PSD detection. Hundreds of very high quality and large diameter cables may be required for the various neutron diagnostics. The design complicates the installation of large spectrometers (Time-of-flight, MPR, etc) located behind the detector block, since the required space may not be available. The production of gamma rays in the block and the magnetic shield may be a serious problem that has to be considered once the magnetic shield has been specified.

It is also unacceptable to consider a system where the front and rear collimators are mechanically separate. Each channel needs to be in a mechanically rigid cylinder containing all systems.

CONCLUSIONS

A reference design has been determined for ITER of a radially viewing neutron profile monitor, which is constructed with integral neutron shielding which greatly reduces stray neutron fluxes. This design is fully consistent with the overall ITER requirements including standard flange sizes and blanket interfaces, but as a result the zone of the plasma viewed is reduced compared to the previous design. Viewing of the plasma outer half by additional channels or by a vertical camera would be highly desirable. The radial camera is highly flexible, allowing modification of collimator geometries with inserts and different detector systems. Instrumental flexibility would be improved further if the collimation on each channel were adjustable by means other than inserts. Two collimator design options have been considered: the "adjustable" collimator option was kept for future consideration, since it would only involve a modification to the shield block holding the collimators in the reference design; the "compact" option was rejected, which which be a major modification of the reference design, and which has major problems of accessibility. With the insert system and multiple collimators in the reference design, a large range of neutron fluxes can be measured by a variety of flux detectors and spectrometers, compatible with those being tested in existing experiments.

ACKNOWLEDGEMENTS

We gratefully acknowledge the contributions to configuration control, design and CATIA drawing by Carl Lippert at the ITER San Diego Site, and to profile monitor data interpretation codes by S Conroy.

REFERENCES

J.M. Adams et al., 1991, Neutron diagnostics for NET, *AEA Technology 411/90.2/FU-UK,* Harwell, UK.

L.C. Johnson, 1995, "Study of generic access for neutron cameras in ITER", *Report ITER/US/95/PH-07-13.*

F.B. Marcus et al, 1995, "A Neutron Camera for ITER: Conceptual Design," *Diagnostics for Experimental Thermonuclear Fusion Reactors,* Edited by P. Stott, G. Gorini and E. Sindoni, Plenum Press, New York, 1996, Proceedings of the International Workshop on Diagnostics for ITER, Varenna, Italy

F.B. Marcus et al., 1997, "A Neutron Camera for ITER", *Rev. Sci. Instrum.* 68, No. 1, pp. 514-519

A.L. Roquemore et al., 1997, "Recent Expansion of the TFTR multichannel neutron collimator", *Rev. Sci. Instrum.* 68, No. 1, pp 544-547.

G.E. Shatalov et al., 1997, "Neutronic Calculations," in *Diagnostic Design Work Performed under Contract with Ministry of the Russian Federation on Atomic Energy (#09-97), Report for 2nd Quarter 1997*, Moscow.

ROLE OF NEUTRON SPECTROMETERS FOR ITER

T. Elevant[+,*] and J. Scheffel[*]

[+]JET Joint Undertaking, Abingdon, Oxfordshire, OX14 3EA, U.K.,
[*]Alfvèn Laboratory, Royal Institute of Technology, Stockholm, Sweden,
EURATOM/NFR Fusion Association.

ABSTRACT

Conditions for retrieval of ion temperature profiles through measurement and evaluation of neutron energy spectra emitted in ITER-like fusion plasmas are discussed. Spectra from deuterium- and deuterium-tritium plasmas heated by 100 MW, 1 MeV neutral deuterium beams, and ignited H-mode plasmas are simulated by means of a Monte-Carlo Fokker-Planck code. Regions are defined where information on ion temperature clearly is preserved. The correction from line-of-sight integrated measurement to peak temperature is found to be 4-10% for deuterium plasmas of all temperatures and for deuterium-tritium plasmas of temperatures below approximately 5 keV. For higher temperatures the correction equals 10-30% depending on the density and temperature profiles. Ion temperatures can be determined with an accuracy better than 10% in this region only if profile data is made available.

Several spectrometers have been proposed. Operating ranges in ion temperature and beam power, spatial- and temporal resolution, space requirements and life expectancy are compared with ITER measurement requirements. Necessary R&D work is also discussed.

1. INTRODUCTION

For evaluation of energy and particle transport in magnetic fusion experiments, information on ion temperature profiles, $T_i(r)$ are used as input data. Measurements of ion temperature profile will also be used for optimisation of burn in ITER[1] plasmas. For Maxwellian plasmas the ion temperature will cause a Doppler broadening of the Gaussian shaped neutron energy peak[2]. A direct method for evaluation of ion temperatures is based on accurate measurements of the width of this peak using neutron spectrometers[3]. However, using deuterium beams for heating deuterium (DD)- or deuterium-tritium (DT)-plasmas will generate reactions between beam- and thermal ions. This will cause a deviation from Gaussian shaped spectrum; thus more complex analysis and better statistics will be required to evaluate ion temperatures.

Radiation constraints inside the ITER cryostat restrict the neutron flux in various positions around ITER. Spectrometers located in or behind the horizontal neutron camera[4] planned for ITER are thus exposed to a well defined maximum neutron flux. Taking the sensitivity of the specific spectrometer and data collection time into account give a measure of the spectrometer statistics obtainable for a given fusion reaction rate. Apart from the statistics, the accuracy in the evaluated ion temperature depends on the spectrometer energy resolution and its accuracy, the width of the measured spectrum, fraction of events from thermal reactions, and accuracy in the number of background events including non-thermal reactions. Thus, selecting spectrometer(s) matching ITER measurement requirements is a multi-parameter choice that depends on priorities given.

In an attempt to enlighten the complex situation, T_i-operating ranges deduced for various spectrometers, located in or behind the ITER horizontal neutron camera, are here compared with ITER measurement requirements[5]. The required accuracy, spatial and time resolution, are briefly described in section 2. A 3-D Monte-Carlo neutron kinematics code[6] used for calculation of neutron energy spectra during Maxwellian and non-Maxwellian conditions is described in section 3. Data from standard operating scenarios in ITER[7] are used. The geometry is given by radiation constraints and standard ITER port components[8]. Calculated energy spectra from neutral beam injection (NBI) heated- and Maxwellian plasmas are presented in section 4. In section 5 statistics and background requirements are discussed and the proposed spectrometer candidates described. Spectrometer operating ranges during ITER NBI-heating scenarios with increased T_i and fuel density are also shown. The R&D status and suggested work is briefly discussed. Spectrometer selection will be influenced by the individual merits to match the physical, technical and economical priorities given as discussed in the summary and conclusions.

2. ITER MEASUREMENT REQUIREMENTS

Requirements for ITER measurements have previously been identified; ion temperatures from 0.5 to 50 keV have to be measured[5]. Restriction in port dimensions means that low ion temperatures outside r = 0.75a have to be measured by non-nuclear methods. Reference density and temperature profiles for design of diagnostics are given by[9] $n_e(r)=n_e(0)[1-(r/a)^2]^{\alpha}$, ($\alpha \approx 0.5$) and $T_i(r)=T_i(0)[1-(r/a)^2]^{\beta}$, ($\beta \approx 1.0$). The horizontal camera with the opening in the first wall, pre-shield and collimated cones-of-sight is shown in fig. 1. Radiation constraints restrict the pre-shield opening to $13{\cdot}23$ cm^2. Avoiding any intersection between the collimated cones-of-sight and the pre-shield and allowing for thermal movements yields a collimator cross-section area of $3{\cdot}3$ cm^2. This corresponds to a viewed plasma cross-section area of $16{\cdot}16$ cm^2 at the magnetic axis.

The ion temperature gradients[7] may be as large as 10 keV/m. Having a line-of-sight width equal to the required spatial resolution of 30 cm with T_i ranging from 10 to 30 keV is thus inconsistent with an accuracy of 10% required in T_i measurement[5]. A line width of 5 cm is therefore also considered. The requirements considered on T_i measurements using neutron spectroscopy are: *T_i ranging from 3 to 50 keV, accuracy $\Delta T_i/T_i < 10\%$, time resolution Δt = 100 ms, spatial coverage $r \leq 0.75a$, spatial resolution from shielding constraints Δr = 16 cm and spatial resolution from T_i gradient Δr = 5 cm.*

Although the assumed profiles and adopted measurement requirements may have to be revised they serve as a basis for providing an overview of the operating range of the spectrometers in ITER and the relative merits of the instruments will remain valid.

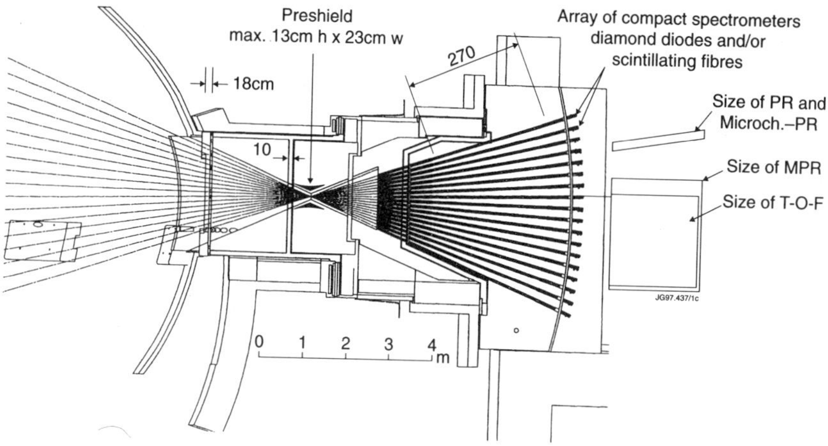

Fig.1 Side view of neutron horizontal camera and spectrometer location is shown. Maximum aperture in the precollimator is 13 cm high and 23 cm wide corresponding to a collimator dimensions of 3·3 cm^2 and a viewed plasma cross-section area (at $R=R_0$) of 16·16 cm^2. Measures are given in cm.

3. CODE FOR CALCULATION OF NEUTRON ENERGY SPECTRA

There are four basic heating scenarios during which neutron diagnostics may be used: purely ohmic, ignited, NBI- and radio frequency heating. In the first two scenarios reacting ions have approximately Maxwellian distribution functions. Beam injection physics is classical in the sense that Fokker-Planck models are consistent with measurements of fast beam ion slowing down in the background plasma. RF-heating deposition and effect on velocity distribution is more complex. In this study we calculate neutron energy spectra emitted from Maxwellian and NBI-heated plasmas.

The tool used is a 3-D Monte-Carlo neutron kinematics code, NSPEC[6]. The plasma is specified by its dimensions, physical parameters, reacting species and type of heating. The magnetic flux geometry is modelled using the geometric centre R_0, inner boundary R_{min}, outer boundary R_{max}, plasma height above magnetic axis Z_{axis}, the Shafranov shift Δ, the elongation κ, as well as the toroidal current I_p and toroidal magnetic field B_ϕ. The model computes approximate flux surfaces, which are labelled monotonically (going outwards) in the range $0 \rightarrow 1$. Thus plasma parameters are calculated at each position where a L.O.S intersects a magnetic flux surface. Parameter used unless otherwise stated are: $R_0 = 8.06$ m, $a = 2.94$ m, $I_p = 24$ MA, $B_\phi = 5.7$ T, $\kappa = 1.7$ and $T_e \approx T_i$.

Along the specified L.O.S. through the plasma a discrete number (typically 9-19) of volumes are defined with their circular cross-sections defined by the length and diameter of the neutron collimator. Within each of these sub-volumes, the fusion reactivity of typically 10^5 events are computed. Only those neutrons that fall into the collimator are accounted for, rendering the calculations highly efficient. Fusion cross-sections used are those of Bosch[10]. The Fokker-Planck distribution for beam-heated plasmas is computed in ref. 11.

Neutron energy spectra for NBI-heated ITER plasmas are calculated using L.O.S. passing through the magnetic axis. The details of the semi-tangential beam geometry are given in ref. 12. The neutral beam deposition is assumed to have an exponential profile with H(0) = 2.5 (see fig.2). This is an overestimate of the deposition in the plasma centre. The ratio of neutron fluxes from thermonuclear and beam-plasma reactions obey approximately the scaling $\Phi_{pp}/\Phi_{bp} \propto n^2\ T_i^1\ P^{-1}\ H^{-1}$, where n is the plasma density, T the temperature, P the beam power and H the (central value of) normalised beam deposition profile. Simulations[6] suggest a temperature dependence T^{η} where $0.5 \leq \eta \leq 1.0$. At the high beam energies considered (low DT fusion reactivity) the dependence may be even stronger. Thus, the neutron flux scaling is favourable for diagnosing hot and dense fusion plasmas.

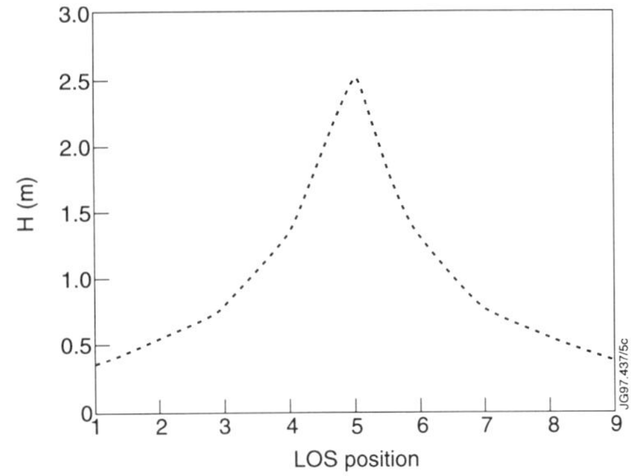

Fig 2. The neutral beam deposition profile denoted H(2.5) is shown.

Neutron energy spectra for NBI-heated ITER plasmas are calculated using L.O.S. passing through the magnetic axis. The details of the semi-tangential beam geometry are given in ref. 12. The neutral beam deposition is assumed to have an exponential profile with H(0) = 2.5 (see fig.2). This is an overestimate of the deposition in the plasma centre. The ratio of neutron fluxes from thermonuclear and beam-plasma reactions obey approximately the scaling $\Phi_{pp}/\Phi_{bp} \propto n^2\ T_i^1\ P^{-1}\ H^{-1}$, where n is the plasma density, T the temperature, P the beam power and H the (central value of) normalised beam deposition profile. Simulations[6] suggest a temperature dependence T^{η} where $0.5 \leq \eta \leq 1.0$. At the high beam energies considered (low DT fusion reactivity) the dependence may be even stronger. Thus, the neutron flux scaling is favourable for diagnosing hot and dense fusion plasmas.

4. INFORMATION PRESERVED IN ENERGY SPECTRA

Calculations of neutron energy spectra are performed with deuterium NBI-power up to 100 MW and DD- and DT- plasmas. In the following results from heating H-mode scenarios with rapidly increasing temperatures and slowly increasing density rise are presented. It is concluded in ref. 13 that analysis with better than 30% accuracy in reaction fractional contributions can be obtained provided the thermal reaction fraction is at least 70% of total. In the following we require minimum 80% thermal reactions.

4.1 Deuterium NBI-heated Plasmas

Neutron energy spectra from 100 MW of 1 MeV deuterium NBI-heated DD-plasmas are calculated. Fig.3a shows a spectrum from a plasma with the central $T_i = 10$ keV and central $n_D = 5\cdot10^{19}$ m^{-3} and fig. 3b shows a spectrum with the corresponding parameters equal to 15 keV and $7\cdot10^{19}$ m^{-3}. The thermal fractional contribution is approximately 68% and 90% of total contribution in the thermal peak area and the peak width correction for the integrated measurements along the L.O.S. is equal to 4% and 7% for the two cases. Applying the scaling laws and the required thermal fractional contribution to be larger than 80%, interpolation suggests that a range for evaluation of T_i in deuterium NBI-heated plasmas is to be found above approximately 12 keV. The relatively small correction also suggests that central T_i can be evaluated from a single L.O.S. measurement.

Neutron energy spectra from 100 MW of 1 MeV deuterium NBI-heated DT-plasmas are shown in fig.4. The ion temperature is 5.0 and 7.5 keV and electron density equal to $0.45\cdot10^{20}$ and $6.0\cdot10^{20}$ m^{-3} respectively. In the low temperature case the fractional

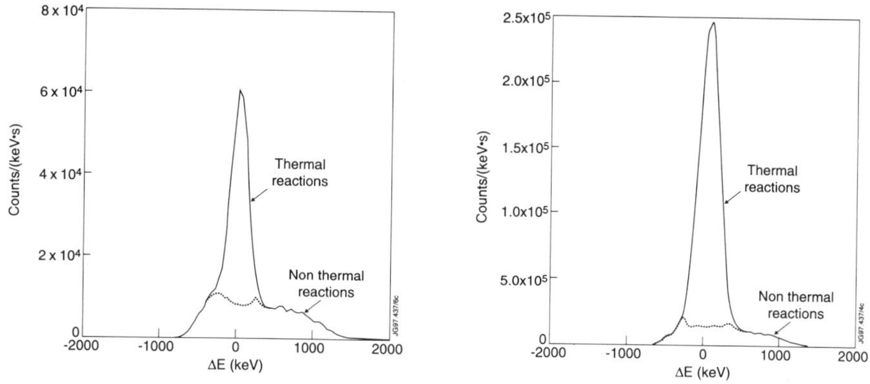

Fig.3 Calculated neutron energy spectra from a reference H-mode scenario with 100 MW, 1 MeV deuterium NBI-heated DD-plasmas are shown. In fig.3a) is the central T_i = 10 keV, central $n_d = 5\cdot10^{19}$ m^{-3} and the resulting thermal reaction fraction in peak area equal to 70%. The corresponding numbers in fig. 3b) are T_i = 15 keV, $n_d = 7\cdot10^{19}$ m^{-3} and 90%.

contribution from thermal reactions is 80% of the number of events in the peak area and 73% of total number of counts. In the high temperature case the corresponding numbers are 93% and 96% respectively. This demonstrates the feasibility to evaluate ion temperatures during high power NBI-heating provided T_i exceeds approximately 6 keV.

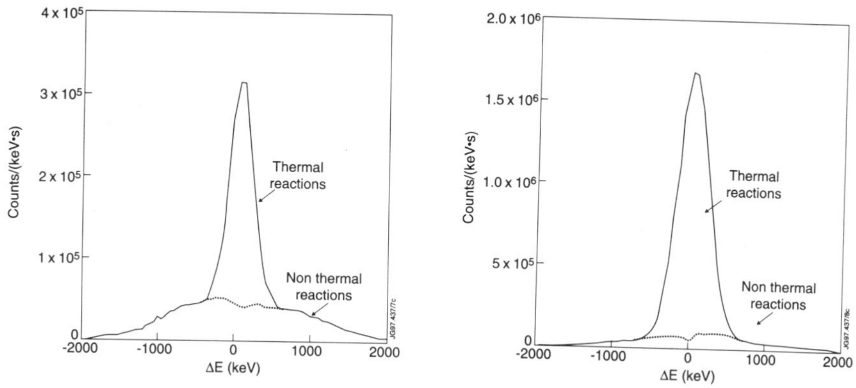

Fig.4 Calculated neutron energy spectra from a reference NBI-heating scenario with 100 MW, 1 MeV deuterium NBI-heated DT-plasmas are shown. In fig.4a) is the central T_i = 5 keV, central $n_e = 4.5\cdot10^{19}$ m^{-3} and the resulting thermal reaction fraction in peak area equal to 76%. The corresponding numbers in fig. 4b) are T_i = 7.5 keV, $n_e = 6\cdot10^{19}$ m^{-3} and 90%.

4.2 Maxwellian deuterium-tritium plasmas

The feasibility to evaluate ion temperature from neutron energy spectra in Maxwellian DD-[13] and DT-plasmas[3] has previously been demonstrated. For a single L.O.S. the applied correction is 5-10% for DD-plasmas and low temperature DT-plasmas. For DT-plasmas of high temperature, i.e. $T_i > 6$ keV, the correction is 10-30% and depends on temperature and density profiles. Unless profile information is obtained the accuracy will be of the same order. However, using an array of spectrometers will provide the ion temperature profile shape and thereby improve the accuracy in T_i measurement to $\approx 10\%$. The correction was shown to be relatively insensitive to broad variations in magnetic flux surface profile shapes.

5. NEUTRON SPECTROMETERS

Although information on T_i may be preserved in the energy spectrum it is not always readily retrievable. The flux at the spectrometer position, spectrometer sensitivity and energy resolution must be consistent with the statistics required to measure the spectrum shape with necessary accuracy, spatial and temporal resolution. Fulfilling the ITER measurement requirements this forms an operating range in T_i for each spectrometer.

5.1 Statistics and Resolution

For an ideal and well defined case where the ion distribution is Maxwellian and the spectrometer resolution function is Gaussian shaped the accuracy in ion temperature measurements can be expressed in terms of spectrum- and spectrometer parameters[14]:

$$\Delta T_i/T_i = 2\{ [1+R^2/W^2]^2 [1/(2 N_S)^{0.5} + |\Delta N_B|/N_S]^2 + (R^2/W^2)^2 (\Delta R/R)^2 \}^{1/2} \tag{1}$$

N_S is here the number of useful counts, ΔN_B the uncertainty in background counts below peak, W ($\propto T_i^{1/2}$) the actual width of the neutron spectrum, R the instrumental resolution and ΔR its accuracy. Calibrating the instrument using a neutron generator providing a Gaussian energy distribution with $\Delta E_n \approx$ 50 keV [FWHM] known to 1% accuracy and a spectrum containing $6 \cdot 10^4$ events will bring down $\Delta R/R$ to 1%. Furthermore, by reducing the uncertainty in background counts through good statistics eq. (1) gives the minimum required number of counts in the thermal peak as a function of ion temperature.

5.2 Spectrometer Proposed

Several neutron spectrometers have been proposed for measurements of T_i in ITER. Principal reaction and measurement, physical weight and size of the spectrometers are shown in Table 1. Some details are presented elsewhere[3] and only a brief description is given here. For DD-plasmas, only a time-of-flight (ToF) spectrometer similar to the one used at JET for several years[15] has been proposed. Seven spectrometers for analysis of DT-plasmas have been proposed. For all but the diamond diode is the well characterised (n,p) elastic scattering process utilised. For the proton-recoil (PR) variants the proton energy is measured. In the case of the scintillating fibre spectrometer the proton range in the fibres is measured and the corresponding proton energy evaluated. For the proton-recoil-time-of-flight (PR-ToF) also the flight-time for the scattered neutron is measured. In the case of the pure ToF spectrometer only the neutron flight-time is measured. For the diamond diode is the total energy of the charge particle products from the intrinsic reaction measured.

Some advantages like the small size of Diamond Diodes and Scintillating Fibers are obvious and enable these to be used in the collimators of the horizontal neutron camera. As many as 40 spectrometers could be used yielding good profile coverage. Furthermore, a figure of merit related to the radiation damage is given. Of importance is the number of spectra that can be produced between replacement of any vital component. The MPR-, Scintillating Fibre-, the PR- and the two ToF- spectrometers all have expected life times of more than 5000 full power (1.5 GW and 1000s duration) discharges.

Table 1. Spectrometer candidates

Spectrometer type	Principal reaction and measurement	Weight	Width x Height x Length	Radiation damage[A)] No. of[B)] discharges	Replacement of
DD-operation:					
Time-of-Flight, (ToF)[15]	$(n,p)_{el.}$, n-ToF	< 300 kg	1 x 1 x 2 m^3	$1\cdot10^6$	Scintillators
DT-operation:					
Compact spectrometers with detectors in neutron beam					
Diamond Diode[16,17]	$^{12}C(n,\alpha_0)^9Be$, α_0+ ^9Be energy	< 1 kg	≈ 1 cm^3	500	Diodes
Scintillating Fibre[18]	$(n,p)_{el.}$, p-range	< 1 kg	≈ 2 cm^3	10^5	Fibers
Large spectrometers with scintillators outside neutron beam					
Time-of-Flight, (ToF)[14]	$(n,p)_{el.}$, n-ToF	< 300 kg	1x 1 x 3 m^3	$3\cdot10^5$	Foil
Magnetic[19,20] proton recoil,(MPR)	(n,p), dispersive p-detection	21 tons	3 x 1 x 2 m^3	$7\cdot10^4$	Foil
Medium sized spectrometers with Si-diodes off neutron beam					
Proton Recoil, (PR)[21]	$(n,p)_{el.}$, p-energy	< 200 kg	0.5 x 0.5 x 2 m^3	10^4	Diodes
Proton Recoil, Time-of-Flight, (PR-ToF)[22,23,24]	$(n,p)_{el.}$, p-energy and n-ToF	< 400 kg	2 x 2 x 1 m^3	10^3	Diodes
Microch.[25,26] Proton Recoil (PR)	$(n,p)_{el.}$, p-energy	< 600 kg	0.5 x 0.5 x 2 m^3	10^3	Diodes

[A)] Fluence 15 cm from neutron beam is $3\cdot10^{-4}$ fluence in beam. [B)] Assuming 1.5 GW and 1000 s.

5.3 T_i Operating Ranges for Spectrometers

Efficiencies, energy resolutions and T_i operating ranges for DD- and DT-spectrometers are shown in fig.5. Considered are two different plasma cross-section areas, 5·5 and 16·16 cm^2 corresponding to collimator cross-section areas of 1·1 and 3·3 cm^2 respectively, see section 2. Furthermore, following the plasma parameter evolution for a NBI-heating scenario, densities and corresponding neutron fluxes for a 3·3 cm^2 collimator are indicated. Count-rates are also shown for each spectrometer and collimator used.

The DD-ToF spectrometer proposed for ITER is a modified version of the one used at JET. It can operate above approximately 12 keV, as seen in fig. 5a, limited by the large amount of neutrons from beam-thermal

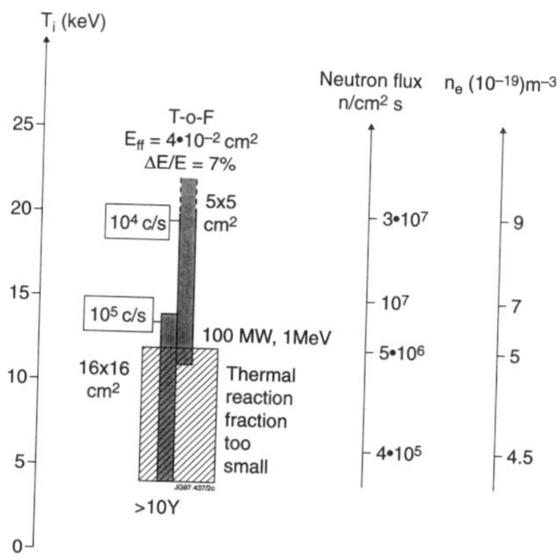

Fig.5 DD-Time-of-Flight spectrometer maximum T_i operating range is shown for two collimator dimensions corresponding to 5·5 and 16·16 cm^2 plasma cross-section area respectively. Also shown are the maximum countrates.

reactions below this temperature. Only 7% energy resolution seems necessary. The spectrometer will deliver central ion temperatures with a time resolution of 50 ms.

The operating ranges for the seven proposed spectrometers to be used for DT-operation are shown in fig. 5b. The Microch.-PR and PR-ToF spectrometers have essentially the same basic specifications and are shown in the same column. Efficiency and energy resolution (in FWHM) for each spectrometer are given. The maximum count-rate over the operating range is shown within squares. Due to its high efficiency combined with the good energy resolution the ToF spectrometer can operate at temperatures as low as 6 keV. The Scintillator Fibre-, MPR- and PR-spectrometers are useful above approximately 9-10 keV and the Diamond Diode, the Microch.-PR and PR-ToF above approximately 17 keV. Below 6 keV does the large fraction of non-thermal reactions prevent any spectra to be evaluated during NBI-heating. No spectrometer has the capability to measure spectra during ohmic heating (T_i ranging from 4 to 5 keV) fulfilling the measurement requirements. However, relaxing on the time resolution to 1 s makes the ToF spectrometer useful also for ohmic heated plasmas.

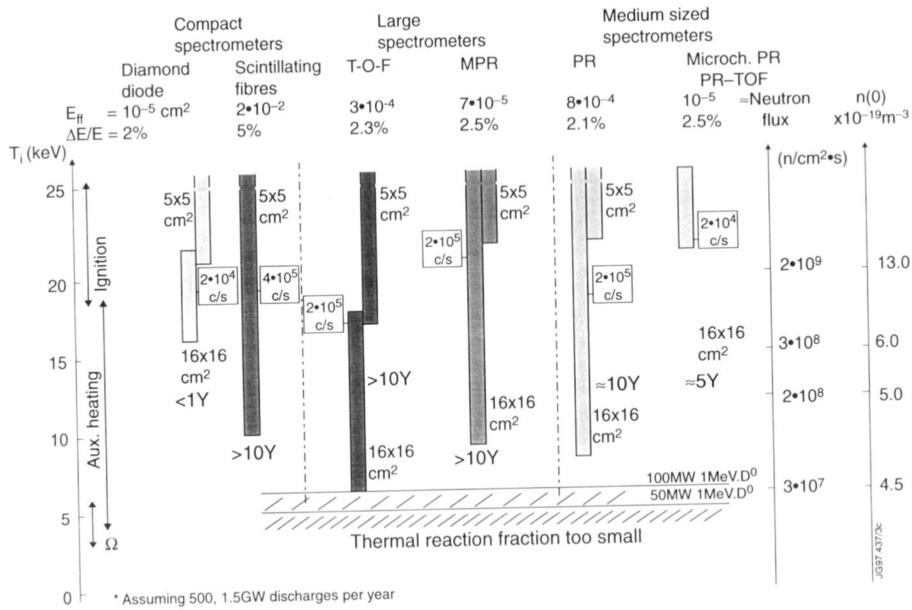

Fig.6 Maximum T_i operating ranges are shown for proposed DT-spectrometers. Collimator dimensions corresponding to 5·5 and 16·16 cm^2 plasma cross-section area respectively are considered. Also shown are the maximum countrates and the expected life length in years (y) between replacement of vital components assuming 500 discharges per year generating 1.5 GW each.

5.4. R&D Status and Needs

The spectrometers are in different development stages with the Diamond Diode-, MPR-, PR-, PR-ToF and Microch.-PR spectrometers developed into prototypes that have been tested, either at a neutron generator or at a Tokamak. This is also the case for the DD-ToF spectrometer that has been in operation at JET for several years. The Scintillator Fibre spectrometer is on the other hand in an early stage of development where a conceptual design has been carried out but no prototype has been constructed. To explore the potential of this small and efficient spectrometer it is proposed that a prototype also of this will be constructed and tested with respect to energy resolution, efficiency and separation of neutron from gamma events also at high count-rate.

In the case of Diamond Diode, development of high count-rate (>1 MHz) electronics for pulse high analysis is needed. Also radiation resistance, annealing methods and an increase of the relatively low efficiency (10^{-5} cm^2) need to be investigated.

Neutron transport calculations have been performed for the DT-ToF spectrometer[14] to an extent where further information only can be obtained through test of a prototype. Due to its large size this is particularly justified if measurement in the low temperature region is a prime and other smaller and simpler methods can not be extended to this region. For the PR spectrometers, only adjustments to conditions expected in ITER will be needed and no particular R&D is foreseen [27].

6. SUMMARY AND CONCLUSIONS

A 3D Monte-Carlo neutron kinematics code is used for calculation of neutron energy spectra during high beam power NBI-heating. Matching the ITER measurement requirements the results show that information on T_i is preserved in the neutron energy spectra even with 100 MW, 1 MeV of deuterium beams injected into DD-plasmas with $T_i > 12$ keV and DT-plasmas with $T_i > 7$ keV. Central T_i can be evaluated from one L.O.S. measurement for DD-plasmas and $T_i < 5$ keV in DT-plasmas while multi-L.O.S. measurements are needed to determine $T_i > 5$ keV in DT-plasmas.

To perform measurements of ion temperature profiles, several spectrometers have been proposed. For DD-plasmas there is only one spectrometer candidate, ToF. No further R&D is needed. For DT-plasmas there are seven different spectrometers proposed with different features. Depending on the physical-, technical- and economical priorities given, different spectrometers would be preferred as shown in Table 2.

Table 2. Most suitable spectrometer

Priority issues	Most suitable spectrometer
Physical	
Low T_i range ($T_i < 9$ keV)	ToF
Good profile cover	Diamond Diode
Good spatial resolution (5·5 cm^2)	Scintillating Fibres
High count-rate	MPR
Technical	
Small size and weight	Diamond Diode
Long life time	Scintillating Fibres
Economical	
Low cost	PR

The need for R&D work is obvious for the small spectrometers to be fully investigated and tested. A prototype of the Scintillating Fibre spectrometer needs to be designed, constructed and tested. For Diamond Diode there is a need to develop MHz electronics for analysis of pulse amplitudes. Furthermore, radiation damage and annealing properties need to be studied and the efficiency needs improvement.

REFERENCES

1. Tombaechi, J.R. Gilleland, Yu.A. Sokolov, R. Tosci and the ITER team, ITER conceptual design, *Nuclear Fusion*, Vol. 31.No.6 (1991). ITER Joint Central Team, Parameters of the ITER EDA design, *Plasma Phys. Control. Fusion* 35 (1993) B23.
2. Brysk, Fusion neutron energies and spectra, *Plasma Physics,* Vol. 15, (1973),611-617.
3. Elevant, et.al., Ion Temperature profile Measurements by means of Neutron Spectroscopy, *Diagnostics for Experimental Thermonuclear Fusion Reactors*. P.E. Stott et.al., Plenum Press, New York, 1996.
4. Marcus, et.al., A neutron camera for ITER: conceptual design, *Diagnostics for Experimental Thermonuclear Fusion Reactors*. P.E. Stott et.al., Plenum Press, New York, 1996.
5. Minutes of 1st Meeting, Physics Expert Group on Diagnostics. July 18-22, 1994; ITER Joint Work Site, San Diego, USA. SCX M1 94-08-03 F1.
6. Scheffel, Neutron spectra from beam-heated fusion plasmas, *Nucl. Instr. and Meth.* 224 (1984) 519-531.
7. ITER-EDA-JCT, ITER information, July 1995, S 55 RE 1 95-07-11 F1.
8. Private communication Dr.s. L. Johnsson and C. Walker, ITER-JCT, San Diego and Garching, June 1997.
9. Mukhovatov, H. Hopman, S. Yamamoto, K. M. Young, P. Stott et.al., ITER Diagnostics, *ITER Documentation series, No. 33*, International Atomic Energy Agency, Vienna, 1990.
10. Bosch, Review of data and formulas for fusion cross-sections, *Report IPP* I/252, Max-Planck Institut für Plasmaphysik, Garching bei Munchen 1990.
11. Scheffel, Solution to the Fokker-Planck Equation for High-Energy Ions, *Nucl. Fusion* 27(1987)1173.
12. Hemsworth et.al., ITER neutral beam injector design, *Sixteenth IAEA Fusion Energy Conference*, IAEA-CN-64/FP-18, Montréal, Canada (1996).
13. Olsson et. al., Analysis of neutron energy spectra from neutral beam heated plasmas in the JET tokamak, *Plasma Phys. and Controlled Fusion* 35, (1993) 179.
14. Elevant et.al., On ion temperature profile measurements in ITER by means of neutron spectroscopy, to appear in Fusion Technology (1997).
15. Elevant et.al., The new JET 2.5-MeV neutron time-of-flight spectrometer, *Review of Scientific Instruments*, Vol. 63, No.10 (Part II), October (1992).
16. F. Borchelt et. al., First measurements with a diamond microstrip detector, *Nucl. Instr. and Meth.*, A 354 (1995) 318-327 and A. V. Krasilnikov, Diamond detector based DT neutron spectrometer for ITER, *Diagnostics for Experimental Thermonuclear Fusion Reactors*. P.E. Stott et.al., Plenum Press, New York, 1996.
17. A. V. Krasilnikov, Natural diamond neutron spectrometer, Workshop on *Diagnostics for experimental fusion reactors,* Varenna, Italy, September 4-11, (1997).
18. Elevant et. al., Scintillating Fibre Neutron Spectrometer, *Diagnostics for Experimental Thermonuclear Fusion Reactors*. P.E. Stott et.al., Plenum Press, New York, 1996.
19. J. Kallne and H. Enge, Magnetic Proton recoil spectrometer for fusion plasma neutrons, *Nucl. Instr. and Meth.* A311 (1992) 595.
20. J. Kallne et.al., Neutron spectrometry for ITER, *Diagnostics for Experimental Thermonuclear Fusion Reactors*. P.E. Stott et.al., Plenum Press, New York, 1996.
21. N. Hawkes, Harwell Laboratories, U.K. Private communication, June 1995.
22. G. Grosshog et.al., The use of the neutron-proton scattering reaction for D-T fusion spectrometry, *Nucl. Instr. Meth.*, A249 (1986), 468.
23. K. Drozdowicz, M. Hoek and D. Aronsson, Energy calibration of neutron detectors for neutron spectrometer TANSY, *Nucl. Instr. and Meth.*, A306 (1991), 315.
24. M. Hoek, N. S. Garis, G. Grosshog, Simulation of the neutron and proton transport in the 14 MeV neutron time-of-flight spectrometer TANSY, *Nucl. Instr. and Meth.* A322, 248 (1992).
25. T. Iguchi et.al., Conceptual design of neutron diagnostic systems for fusion experimental reactor, *Fusion Engineering and Design* 28 (1995) 689.
26. T. Nishitani et.al., Design of radial neutron spectrometer array for the International Thermonuclear Experimental Reactor, *Rev. Sci. Instrum.* 68 (1), January (1997).
27. N. Hawkes, Harwell Laboratories, U.K. Private communication, June 1997.

NATURAL DIAMOND NEUTRON SPECTROMETER

Anatoli V. Krasilnikov

Troitsk Institute for Innovating and Fusion Research
142092 TRINITI, Troitsk, Moscow region, Russia

INTRODUCTION

The fundamental importance of neutron diagnostic for determination and control major parameters of plasma ion component on future fusion reactor are well known. Two-dimensional multi-chord neutron flux measurements on International Thermonuclear Experimental Reactor ITER will give an information for deuterium - tritium (DT) plasma ignition, burn and fueling control, disruption avoidance and instabilities study. Multi-cone of views neutron spectrometry will give an information about deuterium, tritium and α-particles birth energy distribution functions on different magnetic surfaces of reactor's plasma during ignition and burning. This information will be very important for plasma burning optimization and studies of high energy ions transport mechanisms. Tangential neutron spectrometry could provide information about core plasma rotation velocity and averaged tangential deuteron and triton velocities during asymmetric neutral beam injection (NBI) or ion cyclotron range of frequency (ICRF) heating. The multi-cone of view systems of compact detectors are preferable for spectrometry and flux measurements on plasma fusion reactor.

Here, I focus on the results of application of developed in Russia natural diamond detector (NDD)[1,2] based neutron spectrometers (DNS)[3,4] in experiments on modern large tokamaks and most intensive deuterium - tritium (DT) neutron sources. These results demonstrate the potential of NDD application for DT neutron spectrometry and flux monitoring in tokamak - reactor, and in particular as was discussed previously[4] in ITER perpendicular neutron cameras[5,6] and tangential neutron collimator.

NATURAL DIAMOND DETECTOR PROPERTIES AND PRINCIPLE OF OPERATION.

Natural Diamond Detector is one of the best choice for DT neutron spectrometry and flux measurements due to number of it's outstanding properties[4].

NDD is the metal-insulator-metal system, where metal is deposited contact (in our case Au or Ti/Pt/Au) and insulator is pure selected natural diamond of group IIa. The mechanism of detection is based on collection of charge carriers which are created by the

Diagnostics for Experimental Thermonuclear Fusion Reactors 2
Edited by Stott *et al.*, Plenum Press, New York, 1998

interaction of the radiation with diamond. Charge produced within the sensitive volume of a diamond detector is collected by means of an external applied electric field. Any radiation generating charge carries in NDD can be detected. This includes UV, X-rays, γ-rays, high energy particles such as α-particles, electrons, atoms, neutrons and others. In the case of full charge collection the amplitude of detector electrical signal is proportional to kinetic energy of ionizing charged particle and detector operate as spectrometer.

The possibility of using NDD for DT neutron spectrometry and flux monitoring bases on the shape of NDD pulse height spectrum in DT neutron flux[3,4,7,8] shown on Fig.1.

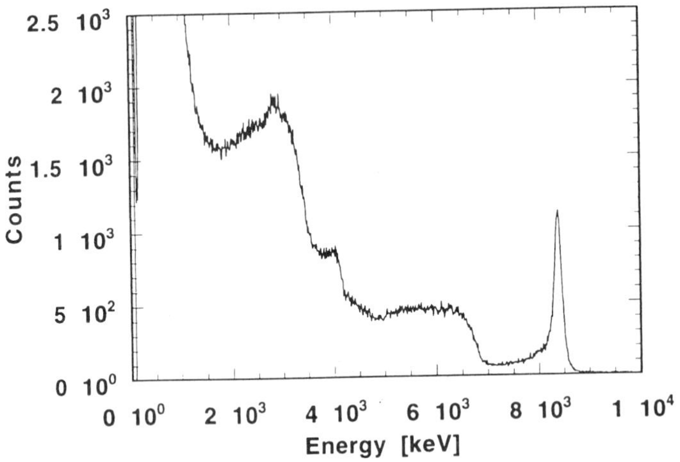

Figure 1. Total NDD pulse height spectrum in 14.1 MeV neutron flux of FNS.

This pulse height spectrum is defining by high energy charged particles (α, ^{9}Be and ^{12}C) created inside diamond by nuclear reaction and neutron elastic and inelastic scattering. Measured pulse height spectra are in an excellent agreement with results of their calculations made using double differential cross sections for reactions taking place under DT neutrons interaction with diamond[3,4]. The $^{12}C(n,\alpha_0)^9Be$ reaction is responsible for the useful for spectrometry E_n-5.7 MeV peak in the NDD pulse height spectrum. The absence of contribution from the other nuclear reaction in the pulse height spectrum around the $^{12}C(n,\alpha_0)^9Be$ peak, namely in the range $E_{\alpha 0+9Be} \pm 2$ MeV, makes NDD very useful for DT neutron spectrum measurements of hot (up to ion temperatures ~ 50 keV) burning plasma[3,4].

A high band gap, resistivity and break down resistance, large saturation velocity, and mean free drift time (10-15 ns)[9] are very important properties of pure group IIa natural diamond[10,11] as materials for radiation detectors. NDDs have very high radiation resistance (maximum acceptable fluence about $5x10^{14}$ n/cm^2)[12] and can operate as spectrometer in temperature range 100 - 600 K. High radiation resistance of diamond is a very important property for it's application for neutron measurements. Study of NDD radiation resistance in the fluxes of 14 MeV neutrons is in progress now. We have not seen any degradation of NDD energy resolution when last received fluence of 14 MeV neutrons equal to $6x10^{12}$ n/cm^2. The radiation resistance of NDD to gammas should be essentially higher then to fast neutrons. This is important for it's possible application as the gamma flux monitor.

A charge collection distance is an extremely important characteristic of spectrometric detector material. It is equal to a product of charge carriers saturation velocity and mean free drift time and represents the distance inside detector sensitive element material passing which the number of charge carriers decreases e times. The large enough charge collection distance of the best (from group IIa) natural diamond crystals makes it possible to create on their base detectors for high resolution particle spectrometry. Best detectors produced from artificial diamonds (chemical vapour deposition (CVD) or DC arc jet)[13] have essentially lower charge collection distance and so can't be used for spectrometry.

RESULTS OF NDD APPLICATIONS.

Three NDDs have been successfully used for DT neutron spectrum measurements on SNEG-13[3] (Russia), FNG[7] (Italy), FNS[8] (Japan) and PPPL's[14] (USA) neutron generators and tokamaks TFTR[14-17] and JT-60U[18]. One NDD was successfully used for energetic charge exchange atoms spectrometry on TFTR. Four detectors are using now on JET for DT neutron spectrometry and two on TORE-SUPRA for unconfined charged particles spectrometry. Some results of these applications will be present in this review.

Six NDDs shown 2-3% energy resolution (FWHM) in these experiments.

NDD with dimensions: area - 16 mm^2, thickness 0.2 mm and area of thin (50 nm) gold contacts - 9 mm^2 has sensitivities:

- 5×10^{-4} counts/(n/cm^2) total for DT neutrons;
- 10^{-5} counts/(n/cm^2) due to $^{12}C(n,\alpha_0)^9Be$ reaction for DT neutrons;
- 10^{-3} counts/(n/cm^2) for DD neutrons;
- 10^{-3} counts/(γ/cm^2) for ^{137}Cs and ^{60}Co gammas.

Pulse height spectrum of NDD irradiated by 2.5 MeV neutrons of FNG, ^{137}Cs (0.66 MeV) and ^{60}Co (1.25 MeV) gamma sources[7] shown on Fig.2.

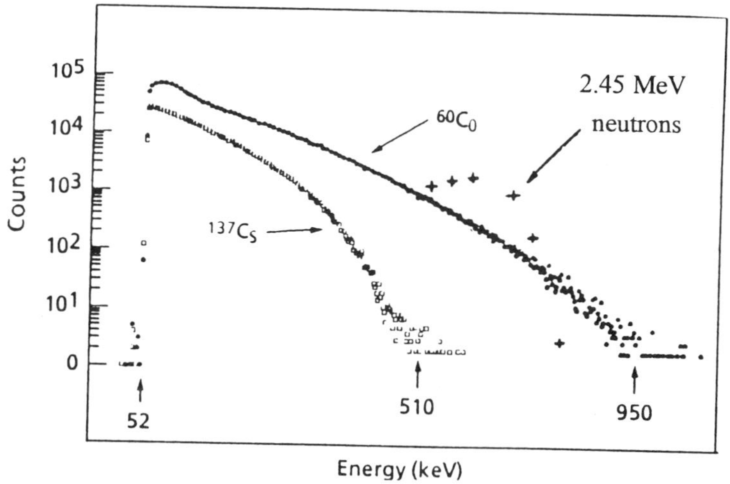

Figure 2. NDD pulse height spectrum in fluxes of 2.5 MeV neutrons, ^{137}Cs and ^{60}Co gammas.

Measurements on DT Neutron Sources

Application of NDDs with energy resolution equal 2-3 % for DT neutron spectrometry on neutron sources SNEG-13[3], FNG[7] and FNS[8] gave possibility to measure detail neutron energy distribution in positions placed around tritium target at angles 0, 95, 100 and 160 degrees with respect to deuterium beam direction. The comparison of measured on FNS neutron spectra (parts of NDD pulse height spectra under $^{12}C(n,\alpha_0)^9Be$ peak) with results of calculations shown in Fig.3. The peak position for 95^0 is show in Fig.3.b by arrow. The energy scale there corresponds to the sum of the kinetic energy of the charged products created inside NDD by DT neutrons. Due to high (1.95%) energy resolution of used NDD it was possible to measure not only the broadening (like in experiments on SNEG-13 and FNG) of DT neutron spectra for 0^0 and 160^0, but also the shape of distribution.

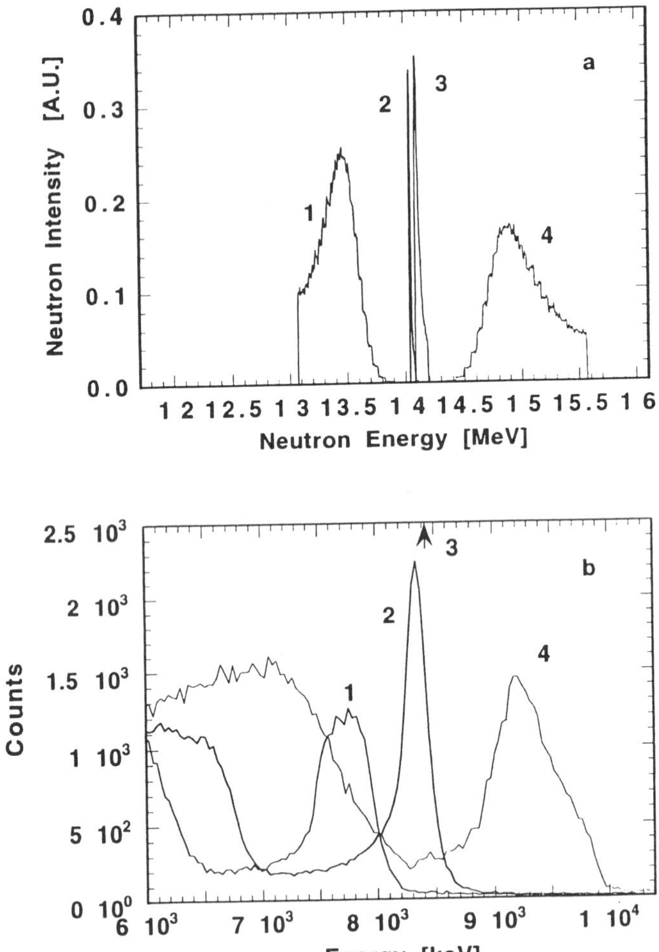

Figure 3. The calculated FNS DT neutron energy spectra (in arbitrary units - A.U.) - a, and measured NDD pulse height spectra - b for points located 20 cm from the target at 1 - 160^0, 2 - 100^0, 3 - 95^0, and 4 - 0^0 with respect to 350 keV deuteron beam.

High energy resolution of used in these experiments NDD gave possibility to detect the difference between perpendicular (95^0) neutron spectra for the cases of deuteron energies equal to 200 and 380 keV. This difference is determined by deuteron scattering inside titanium target before it slowing down to energy 110 keV - peak energy of DT reaction cross section. The restored from these measurements mean projected angle of angle distribution of participating in nuclear reactions deuterons equal to $\pm$ 7^0 and is in a good agreement with calculated using TRIM code value - $\pm$ 5^0.

DT Neutron Spectrometry and Flux Monitoring on TFTR and JT-60U

Three DNS have been used for DT neutron spectrometry and flux monitoring during NBI and ICRF heating deuterium-tritium experiments on TFTR. During these studies, up to six NBI sources were aimed in the counter-direction and up to six in co-direction, with total power of 36 MW of 105 keV DT injection. Up to 3 MW of ICRF power at 30 MHz was injected with wave phasing 60^0 (counter), 180^0 (π), and 300^0 (co with respect to plasma current) during ICRF heating at fundamental triton cyclotron frequency. The high (up to $3*10^{18}$ n/s) yield of 14 MeV neutrons during TFTR experiments has made it possible to use neutron spectrometry to study the ion energy distributions under reactor-like conditions. The small size of NDD housing (~ 1 cm^3) allowed the detectors to be inserted into the TFTR multichannel neutron collimator and fusion gamma shield without interfering with other detectors operating there. The first detector (NDD1) was placed inside a center channel of the TFTR vertical neutron collimator (see Fig.4) to provide a view perpendicular to the plasma current axis. A second detector (NDD2) was placed inside TFTR fusion gamma shield as shown on Fig.5 and measured neutrons with emission angles in the range of 110 - 180^0 with respect to plasma current. The third detector (NDD3) was placed inside a small specially designed shield in the TFTR test cell (see Fig.5) and provide measurements of neutrons having angles of emission of 60 - 120^0 with respect to plasma current. The sensitivities of the detectors and the speed of the signal processing electronics drive the location and shielding required for each detector[14].

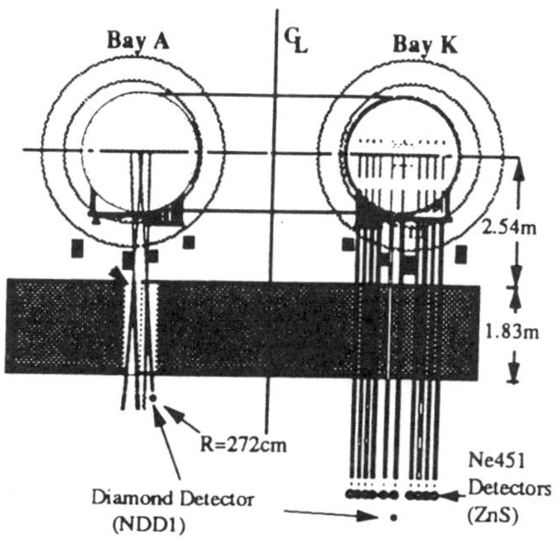

Figure 4. Vertical viewing location in TFTR Multichannel Neutron Collimator

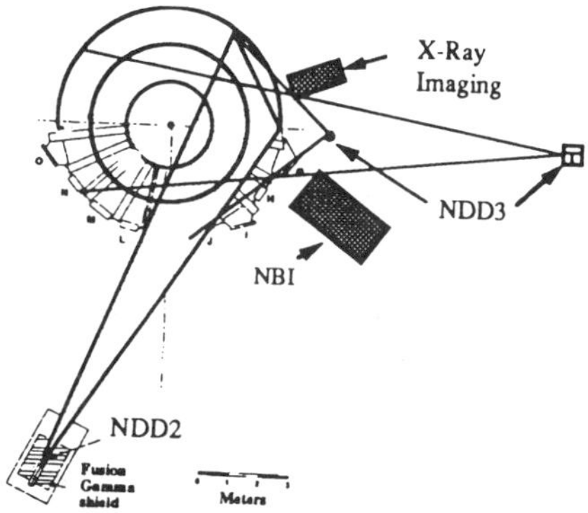

Figure 5. NDD positions inside fusion gamma shielding (NDD2), in compact shielding enclosure (NDD3), and in open space (NDD3) in test cell.

For a Maxwellian DT plasma with ion temperature T in the range 30-50 keV the FWHM of DT neutron energy distribution is given by

$$\Delta E_n = 180\ \sqrt{T_{eff}} \qquad \text{(where } \Delta E_n \text{ and } T_{eff} \text{ in keV } ^{15}) \tag{1}$$

The ion component in TFTR contains both thermal and hot beam ions. Energetic ions also appear during ICRF heating. The emitted neutrons result from three channels: beam-plasma (55-75%), beam-beam (0-20%) and thermal neutrons (10-30%). The anisotropy of TFTR hot ions produces the anisotropy of DT neutron energy distribution. Examples pulse height spectra accumulated by perpendicular and tangential DNS during 1.5 seconds of 3

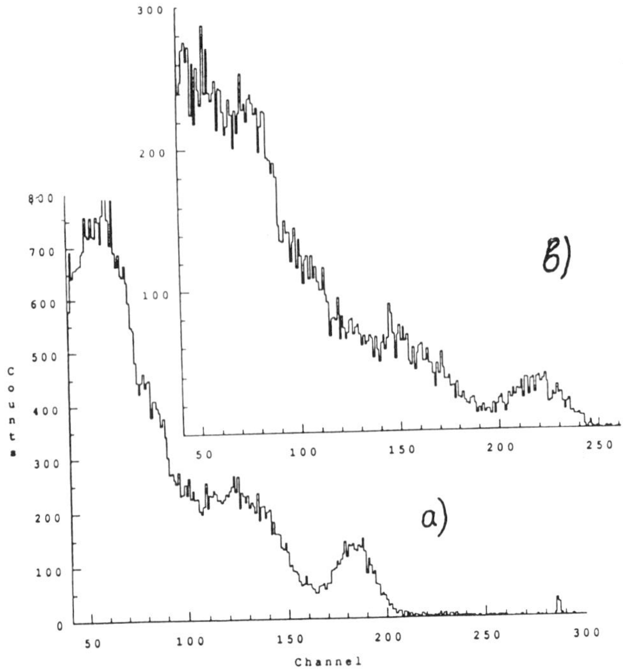

Figure 6. Pulse height spectra accumulated during 1.5 seconds of balanced 18 MW deuterium NBI by perpendicular a) and tangential b) DNSs

TFTR shots with balanced deuterium NBI (18 MW) into deuterium-tritium plasma are shown in Fig. 6. Stored in plasma energy, electron density, ion and electron temperatures during these experiments reach 2-3 MJ, 4.5×10^{13} cm^{-3}, 20-30 and 5-7 keV correspondingly. Dynamics of tangential and perpendicular DT neutron spectra broadening (FWHM) measured in these experiments with pure deuterium and pure tritium injection are shown in Fig.7.

NST code and semi-analytical model were created[15] for calculation DT neutron spectra in all detector positions for anisotropic plasma of TFTR. For the case of TFTR experiments with pure deuterium and pure tritium injection semi-analytical model gives[17] for

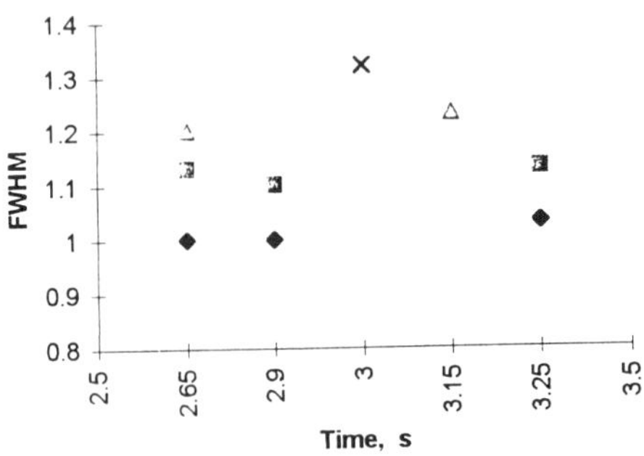

Figure 7. Perpendicular (△,◆) and tangential (×,■) DT neutron spectra broadening during pure deuterium (■ , ◆) and pure tritium (× , △) balanced injection on TFTR.

broadening of tangential and perpendicular neutron spectra from one side and corresponding averaged energies of beam ions and plasma temperatures from another side next relations.

$$FWHM_{||,d} = 180 * (0.2*(4*\langle E_{||,d}\rangle + 3*T_i))^{0/5} \qquad (2)$$

$$FWHM_{||,t} = 180 * (0.2*(6*<E_{||,t}> + 2*T_i))^{0/5} \quad (3)$$
$$FWHM_{\perp,d} = 180 * (0.2*(2*< E_{\perp,d} > + 3*T_i))^{0/5} \quad (4)$$
$$FWHM_{\perp,t} = 180 * (0.2*(3*< E_{\perp,t} > + 2*T_i))^{0/5} \quad (5)$$

In experiments shown in Fig.7 during pure tritium injection FWHM of measured perpendicular and tangential neutron spectra are equal to 1.22 and 1.33 MeV, that in accordance with (5) and (3) for T_i = 30 keV gives $<E_{\perp,t}>$ = 57 keV and $<E_{||,t}>$ = 35 keV. During similar shots with pure deuterium injection FWHM of tangential and perpendicular neutron spectra are equal to 1.13 and 1.03 MeV, that in accordance with (2) and (4) for T_i = 20 keV gives $<E_{||,d}>$ = 34 keV and $<E_{\perp,d}>$ = 52 keV.

Measurements of tangential neutron spectra during pure co (5 sources of total power ~ 10 MW) and pure counter (4 sources, ~ 7 MW) deuterium beam injection into deuterium-tritium plasma have been done in experiments with flow shear (shots 103260-103280). Peak of tangential neutron energy distribution moves on 0.44 MeV down to low energies during co injection and on 0.43 MeV up to high energies during counter injection (see Fig.8). Accuracy of these measurements is ± 60 keV.

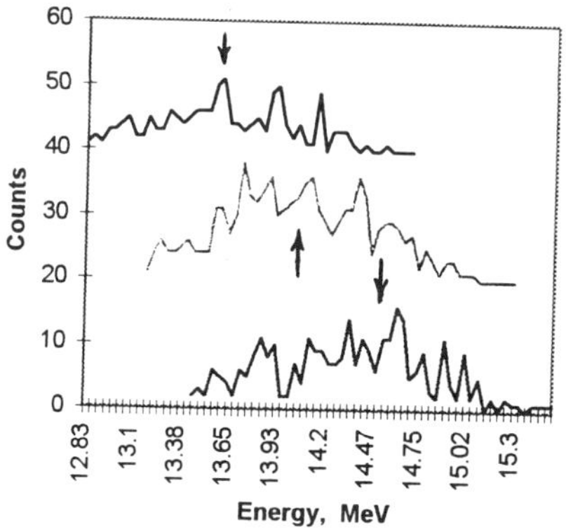

Figure 8. Tangential DT neutron energy distribution during co (1), balanced (2) and counter (3) deuterium injection into TFTR DT plasma. Spectra (1) and (2) shift on 40 and 20 counts for better view. Arrows show energy distributions peaks positions.

Taking into account that energies of beam deuterons are essentially higher than energies of plasma tritons we can write next equation for relation between averaged deuteron tangential velocity and energy of neutron spectrum peak[17]:

$$E_n = 14.1 + 0.56* E_d + 2.13*\cos\theta \sqrt{E_d} \quad \text{(where all E in MeV)} \quad (6)$$

Averaged deuteron tangential energies estimated using equation (6) from presented in Fig.8 neutron spectra are 43 keV and 41 keV for co and counter injection. These values are in a good agreement with averaged deuteron tangential energy received from measurements of neutron spectra broadening during balanced deuteron injection.

The system of three DNSs was used to study the dynamics of DT neutron distribution functions and fluxes during TFTR fundamental tritium ICRF heating experiments. During these experiments plasmas with major radius 277 cm and minor radius 96 cm were heated during 1 second (from 2.1 s to 3.1 s) by 17 MW NBI and during 0.8 second (from 2.3 s to 3.1 s) with up to 3 MW ICRF heating at a frequency of 30 MHz with wave phasing 60^0 (counter), 180^0 (π), and 300^0 (co with respect to plasma current). The resonance region for the first harmonic of tritons was in the range of major radius 221-249 cm depending upon $k_{||}$ - tangential wavenumber and $v_{||}$ - tangential component of triton velocity. The dynamics of tangential and perpendicular effective ion temperatures were measured during these experiments[16]. The effect of beam penetration changing due to electron density increase during NBI was noticed by DNSs[16]. The triton slowing down time was estimated[16] from the decrease of tangential effective temperature after deuterium and tritium beams were turned off. The derived value of about 100 ms is in reasonable agreement with calculations. Measurements made by perpendicular DNS demonstrate that the highest heating of tritons occurred in experiments with low triton concentration[16].

During experiments with ICRF heating resonant tritons increase not only perpendicular but also tangential velocity. Additional triton tangential velocities were directed along or opposite plasma current in accordance with sign and value of $k_{\|}$. Tangential DNS measured corresponding shifts of tangential neutron energy distribution peak positions presented in Fig.9. During the first half of the heating time the measured shift corresponded to averaged triton tangential velocities of $\pm 10^7$ cm/s and during second half to $\pm$ 7×10^6 cm/s. The decrease in the averaged triton tangential velocities can be explained by the increase of plasma density during NBI.

Some resonant counter-moving tritons with energy higher than 150 keV can be lost during TFTR off-axis triton ICRF heating[16]. The influence of the triton first orbit losses on the ICRF heating efficiency is shown in Fig.10. NBI in these experiments was pure deuterium. Tritium was injected into plasma by gas puffing. Both tangential and perpendicular effective ion temperature measured by corresponding DNSs are higher for the case of ICRF heating with π phasing and power 1.2 MW than ICRF heating with counter phasing and power 1.5 MW.

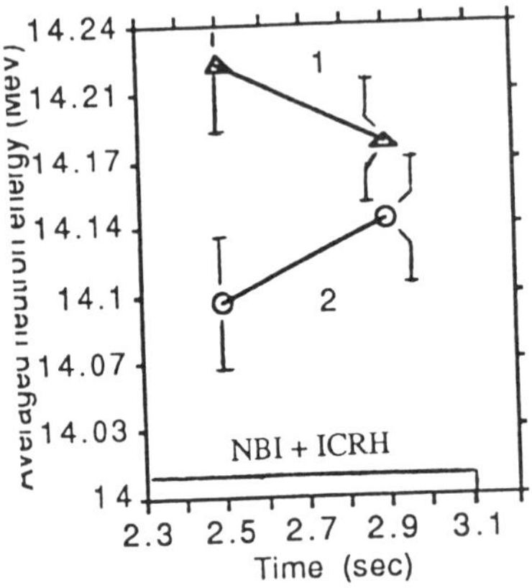

Figure 9. Peak energy of tangential neutrons energy distribution during tritium ICRF heating experiments with counter (curve 1) and co (curve 2) wave phasing.

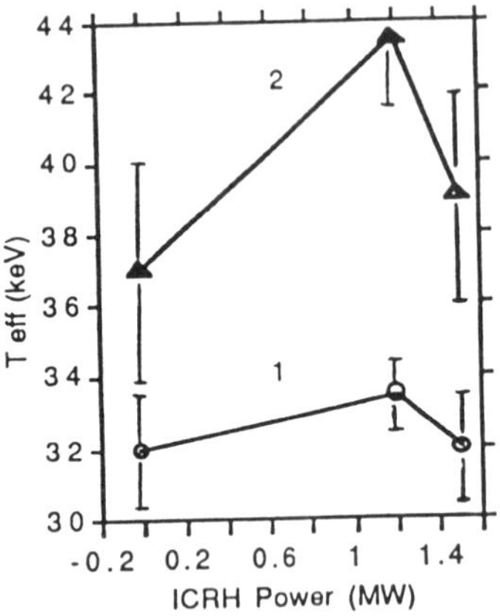

Figure 10. Perpendicular (curve 1) and tangential (curve 2) effective ion temperature of plasma with few % of tritium due to gas puffing during 16.8 MW deuterium NBI and ICRF heating with π (1.2 MW) and counter (1.5 MW) wave phasing

In parallel with spectrometric channels NDD's electrical pulses which had an amplitude corresponding to energy 1.5 MeV and higher were sent to counters. This provides parallel NDD application for DT neutron flux monitoring with sensitivity of the order 2×10^{-4} counts/n/cm^2. Dynamics of neutron fluxes measured by NDD in experiments with almost balanced 17 MW NBI of six deuterium sources and one tritium source injected in co (shots 93336 and 93337) directions and 2 MW ICRF heating with co (shots 93335 and 93336) and counter (shots 93332 and 93337) wave phasing shown in Fig.11.

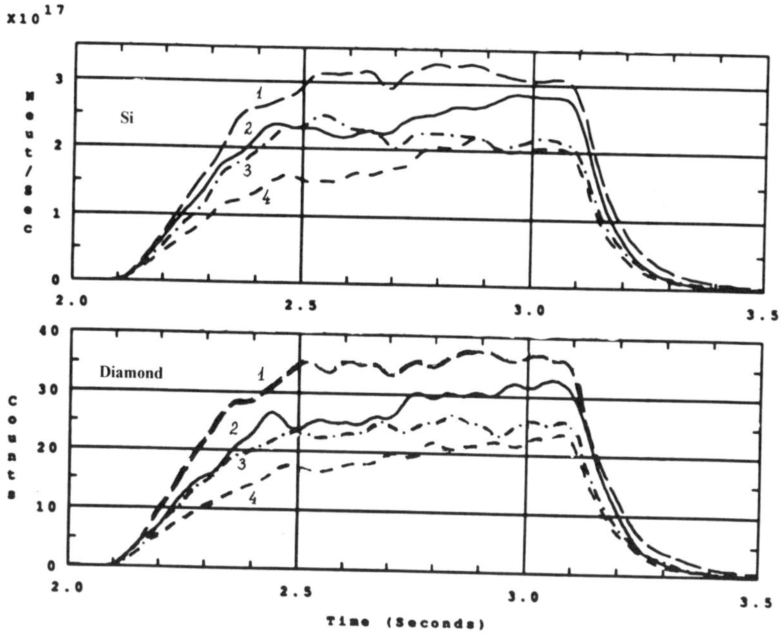

Figure 11. Dynamics of DT neutron fluxes measured by NDD during shots with balanced 17 MW NBI with six deuterium and one tritium sources and 2 MW ICRF heating. The tritium was injected in co (curve 1 and 2) and counter (curve 3 and 4) directions. ICRF wave has co (curve 1 and 3) and counter (curve 2 and 4) phasing.

NDD application for DT neutron spectrometry during experiments on JT-60U gave possibility to measure[18] the averaged energy of tritons appearing inside plasma with energy 1 MeV due to nuclear reaction during and after high power deuterium NBI. Averaged triton energy was measured as 410 keV during beam injection and 250 keV 200 ms after beam was turned off. The triton slowing down time was estimated from this measurements as 400 ms. Dynamic of parts of NDD pulse height spectra corresponding to DT and DD neutrons gave information about dynamics of their fluxes after beam was turned off.

R&D DEMANDED FOR DNS APPLICATION ON ITER

To meet ITER requirements to neutron spectrometry time resolution DNS should operate at count rate 10^6 counts/s. Development of NDDs with useful for DT neutron spectrometry sensitivity $\sim 3*10^{-5}$ counts/n/cm^2 and fast NDD optimized electronics are necessary.

In discussed here experiments we used fastest standard spectroscopy amplifiers (ORTEC-673 and CANBERRA-2024) which could provide spectrometry at input count rate up to 3×10^5 counts/s. Output count rate in this case are not higher then 1.2×10^5 counts/s. (NDD itself can successfully operate at count rates up to 10^8 counts/s). Our studies[8] shown that energy resolution of combined system based on NDD and a 0.25μs shaping time amplifier changes from 1.95% to 3.58% due to pileup effect when input amplifier count rate changes from 10^4 to 3.8×10^5 counts/s. Special development of a fast pulse signal processor[19] will provide DNS successful operation at count rate 10^6 counts/s in ITER conditions.

NDD radiation resistance will provide it operation during 500 full power ITER DT shots. The development and studies of the different annealing procedures for NDD could essentially increase it's operation time.

CONCLUSION

DNSs have been successfully used for DT neutron spectrometry and flux monitoring on neutron generators, in TFTR DT and JT-60U triton burn-up experiments. On TFTR DNSs provide measurements of the effective ion temperatures, averaged tangential and perpendicular beam ion energies, triton slowing down time and averaged tangential velocity of resonant tritons during co and counter wave phasing ICRF heating. Spectrometry and flux monitoring made on TFTR by DNSs shown the decrease of effective temperatures and DT neutron fluxes during ICRF heating with counter triton NBI and counter wave phasing.

The excellent energy resolution, high radiation resistance, and very small size make NDDs one of the best choices for creation ITER multichord neutron spectrometer on the base of vertical and horizontal neutron cameras and tangential neutron collimator in the case of it's approval and design. DNS could provide measurement of ITER core plasma ion temperature with a relative arrow of 10%, a time resolution 0.1-1 s, and spatial resolution ~ 30 cm. These measurements will be very important for burn optimization, determination of α-particles birth energy and spatial distributions, transport study. Tangential DNS channel can be used for measurements of plasma rotation velocity and averaged tangential triton and deuteron velocities during additional heating.

In parallel with spectrometry NDDs will provide multichord DT neutron flux monitoring and core ($r<2/3*a$) emission profile measurements with time resolution 1 ms, spatial resolution ~ 30 cm, and accuracy 10%. This measurements will be very useful for burn and fueling control, disruption avoidance, and instabilities study.

ACKNOWLEDGMENTS

I am grateful to Drs. E.A.Azizov, V.S.Khrunov, K.M.Young, A.L.Roquemore, L.C.Johnson, R.V.Budny, N.N.Gorelenkov, M.Pillon, M.Angelone, T.Nishitani, M.Isobe, J.Kaneko, F.Maekawa for participating in original works cited in this revue.

REFERENCES

1. E.A.Konorova, S.F.Kozlov, *Sov. Phys. Semiconductors* 4, 1600 (1971).
2. A.E.Luchanskii, S.S.Martynov, V.S.Khrunov, V.A.Cheklaev, *Sov. Atomic Energy* 63, 639 (1987).
3. À.V.Krasilnikov, *Sov. Voprosi Atomnoy Nauki i Tehniki*, 1, 36 (1995).
4. A.V.Krasilnikov in *"Diagnostics for ITER"* P.E.Stott, et.al. Editors, p.435, Plenum Press, New York and London (1995).
5. L.C.Johnson et. al. Overview of fusion product diagnostic, *These Proceedings* (1997).
6. F.B.Marcus et. al. A radial neutron camera design for ITER, *These Proceedings* (1997).
7. M.Pillon, M.Angelone, A.V.Krasilnikov, *Nucl. Instrum. Methods B*, 101, 473 (1995).
8. A.V.Krasilnikov, M.Isobe, J.Kaneko, et.al. *Rev.Sci.Instrum.* n.8 (1997).
9. E.A.Konorova, S.F.Kozlov, V.S.Vavilov, *Sov. Phys. Solid State* 8, 1 (1966).
10. V.S.Vavilov, *Phys. St. Solidi (a)* 31, 11 (1975).
11. *The Properties of Natural and Synthetic Diamond*, J.E.Field Editor, Academic Press, London (1993).
12. S.F.Kozlov, R.Stuck, M.Hage-Ali, P.Siffert, IEEE *Trans.Nucl.Sci. NS-22*, 160 (1975).
13. L.S.Pan, S.Han, D.R.Kania, M.A.Plano, M.I.Landstrass, S.Zhao, M.Kagan, *3rd Int. Symp. on Diamond Materials* (Honolulu Hawaii, May 16-21) p.735 (1993).
14. A.V.Krasilnikov, E.A.Azizov, A.L.Roquemore, et. al. *Rev.Sci. Instrum.* 68 (1) 553 (1997).
15. A.V.Krasilnikov, N.N.Gorelenkov, R.V.Budny, L.C.Johnson, and A.L.Roquemore in *Controlled Fusion and Plasma Physics* (Proc. of 23th EPS Conference, Kiev, 1996).
16. A.V.Krasilnikov, A.L.Roquemore, N.N.Gorelenkov, R.Budny, in *Proceedings of "Sixteenth IAEA Fusion Energy Conference"* Montreal, 7-11 October 1996.
17. A.V.Krasilnikov, A.L.Roquemore, to be published in *Plasma Physics Reports* 1 (1998)
18. M.Isobe, T.Nishitani, A.V.Krasilnikov, et.al. *Fusion Engineering & Design* 34-35, p.573 (1997).
19. S.V.Dudin, O.V.Ignatyev, and A.D.Pulin, *Nucl. Instrum. Methods* A, 352, 610 (1995).

MPR NEUTRON SPECTROMETRY AT JET AND ITS NEXT STEP IMPLICATIONS

J Källne, E Traneus and the NPF team*

Department of Neutron Research
Uppsala University
S-751 21 Uppsala, Sweden

INTRODUCTION

The first MPR (magnetic proton recoil) spectrometer for fusion neutron measurements has been constructed and recently installed in a diagnostic system at the Joint European Torus (JET) of EURATOM. The MPR on JET experiment[1] has been in production operation since December 1996 and is used to measure the neutron emission from plasmas of both pure deuterium (D) and mixed deuterium-tritium (DT). The objectives are to perform plasma diagnostic studies and to exploit the unique characteristics of the new technique to push back the observational limits of fusion neutron spectrometry. At the same time, it is important to acquire operating experience and to check the diagnostic performance as projected from instrumental specifications against the results obtained. The latter is motivated by the need for input to the ongoing work on the new generation neutron diagnostics for the next step fusion experiments entailing burning plasmas with high neutron fluxes[2]. In this perspective, the MPR on JET experiment, which will be discussed here, represents neutron spectrometry under flux conditions approaching those of pre-ignition plasmas. As the MPR method can be used with enhanced performance also at the highest fluxes foreseen for burning tokamak plasmas[3,4], the results are relevant for the diagnostic planning for the International Thermonuclear Experimental Reactor (ITER).

The MPR technique is based on the conversion of neutrons to same energy (recoil) protons[5,6]. This is done in a thin CH_2-foil placed at the end of a neutron collimator viewing the plasma. The target acts as a recoil proton source that is imaged onto the focal plane of a magnetic spectrograph. The image is spatially dispersed according to proton momentum and the momentum distribution of such images is in turn a reflection of the incoming neutron energy distribution. The relationship between the neutron spectrum and the focal plane proton distribution (measured) is defined by a well known instrumental response function which includes the spectrometer ion optics.

The ion optics was designed to give a small first order image in momentum as this is an inescapable contribution to the energy resolution, $\Delta E/E_n$. Moreover, ther are the higher order aberration terms connected with the use of a large target area (A_T) and solid angle recoil aperture (Ω_A). These add to $\Delta E/E_n$ but they are not determinable with same certainty as that of the first order term so they are kept relatively small by design. The neutron (flux) detection efficiency (ε) increases proportionally with A_T and Ω_A so they are optimized within the limit set

* L Ballabio, S Conroy, J Frenje, G Ericsson, P Prandoni, M Tardocchi, PU Renberg, and G Gorini.

Diagnostics for Experimental Thermonuclear Fusion Reactors 2
Edited by Stott *et al.*, Plenum Press, New York, 1998

by the finite $\Delta E/E$-value. The performance of the MPR spectrometer is represented by the combination of ε and $\Delta E/E_n$ (say $\varepsilon=5\cdot10^{-5}$ cm^2 for $\Delta E/E_n\approx2.5$ % or $\varepsilon=2\cdot10^{-4}$ cm^2 for $\Delta E/E_n\approx10$ %) with respect to time (ε value) and energy resolution; the requirement on energy resolution is set relative to the thermal Doppler broadening in the neutron spectrum where, for instance, T_i=4 keV gives a Doppler broadening of $\Delta E_D/E\approx2.5$ % (FWHM). This is referred to as the basic diagnostic resolving power of the instrument. It applies to observations of the principal component of the neutron emission while the weak components require high sensitivity which comes from the combination of resolving power and good control of extraneous radiation. The MPR instrument possesses both these properties.

During the first half year of operation, sufficient neutron flux has been received from JET (up to $F_n\approx5\cdot10^8$ n/s) to make the MPR pass the 100 kHz count rate level, a new record for neutron spectrometry. This means that plasma phenomena, manifested in the neutron emission spectrum, can be studied at the time resolution level of a few tens of milliseconds. Extrapolated to ignited plasmas, the MPR method would afford count rates up to the 10-MHz level with corresponding enhancement in resolving power. Moreover, it has been possible to observe minority components of the neutron emission thus testing the handling of background at very low signal/noise (S/N) values. In addition, the *ab initio* calibration of the MPR has been used for the first absolute measurement of the fusion neutron spectrum.

This paper discusses the development of fusion neutron spectroscopy, the principles of the MPR on JET experiment, the outcome of the plan to install the MPR diagnostic as a turn-key-ready instrument and some examples of results that illustrate attainments as compared to aims and envisions. The empirical knowledge acquired will be pointed out for some projected rôles that neutron diagnostics can or will play in the next step fusion experiments. The next step involve new high power fusion test reactors (e.g., ITER) to conduct experiments with no less aim than the first controlled thermonuclear ignition of matter.

NEUTRON SPECTROMETRY DEVELOPMENTS

The first important fusion neutron spectrometry measurement dates back some 40 years when a cloud chamber was used on the ZETA device[7]. The results disappointingly disproved the hope that the neutron production observed was of thermonuclear origin. The neutron spectrum of thermonuclear $d+d\rightarrow{}^3He+n$ reactions in a tokamak were measured in 1982 with a ^3He ion chamber; the count rate was a modest few Hz as limited by low neutron flux[8]. The same type of detector was later used on JET which produced sufficient flux to surpass the count rate capability limit ($C_n^{CAP}\approx10^2$ Hz) of this instrument. Yet low, this count rate was sufficient to show, for instance, a higher than expected fuel depletion[9]. Later, a time-of-flight spectrometer was used to raise the count rate capability into the kHz range making it possible to study, for instance, the relative thermonuclear to fast ion reactions, the latter being connected with strong neutral beam injection (NBI) or RF-heating[10]. It is interesting to note that the presence of non-thermonuclear reactions is no longer seen as the research failure it was seen as in 1958, but is even considered a legitimate quantity to measure.

As tokamaks are entering DT plasma routine operation, the neutron flux is increased a factor of 10^4 relative to D-plasmas. This new situation is now being exploited with the MPR spectrometer as illustrated in the passing of the 100-kHz count rate limit which is still below its capability of $C_n^{CAP}\geq10$ MHz. Although the capability is fully tested first in post-JET ignition experiments, making fluxes in the 10^{11} n/s·cm^2 range available for neutron spectroscopy, the MPR will bring neutron spectrometry into the category of the routine diagnostics already now.

Coupled with the raised count rate is the capability of the new instrument to obtain data of high calibration accuracy. The thus raised quality of the experimental data means that the requirements on the theoretical description of the fusion neutron production have risen, too.

Formulas relating the Doppler width (FWHM) of the spectrum (i.e., the second moment), as well as the amplitude, to the ion temperature were derived in the 60's [11]. The underlying dd and dt reaction cross sections, which are needed to describe the neutron spectra, have been improved successively so that they are now at the few percent level in the aspects relevant for fusion applications[12]. With the advent of the new measurements, it is necessary to know the shape of the neutron energy distribution, due to thermal ion reaction, over several orders of magnitude, both with regard to amplitude and absolute energy. This means that a relativistic formalism has to be used to calculate fusion reactivities and neutron emissivities. Such results have recently been presented relating plasma conditions to measured neutron parameters such as the spectral moments[13].

Besides the principal neutron emission, there are several low amplitude contributions that can be used for plasma diagnostic purposes. One such example is the so called α-particle knock-on signature ($S_\alpha(E_n)$). This arises due to head on collisions between α:s of the slowing down population and fuel ions of the bulk plasma (d or t). The resulting fast d and t ion populations have recently been calculated taking full account of the nuclear and the Coulomb scattering amplitudes as well as their interference terms resulting in an accurate determination of $S_\alpha(E_n)$ and its plasma parameter dependencies; contrary to initial estimates, the nuclear contribution is not insignificant[14].

The instrumental development in fusion neutron spectrometry has up till now been done through adaptation of techniques already existing in other fields of nuclear physics. It seems that this will continue while the dedicated main development area will concern the interfacing of the neutron spectrometers in view of two principal aspects. One is that the performance of the spectrometers will depend on how well one can handle the effects of the tokamak environment so that they do not curtail the inherent instrumental sensitivity and precision. The other is that the highest performance is obtained for maximum neutron flux which generally complicates the interfacing having an impact on diagnostic design as well on that of the machine. For instance, with regard to the magnetic interference between the machine and the MPR, i.e., a 'magnetic stealth' technique would need to be developed in the sense that the diagnostic does not 'see' the machine and also that it is 'invisible' to the machine. The diagnostic performance of neutron spectrometry can only be maintained at their theoretical limits provided the interfacing can be handled instrumentally and technologically. The theoretical limits will be verified empirically by the MPR experiment at JET that has now begun.

The proton detector is one part of the MPR system that presents scope for a new development. This is likely to be driven from within the fusion research field as the application is unique with regard to the radiation to be detected and rejected, The problem has already been formulated and a first study suggested that a factor of 20 improvement can obtained by choice of scintillator material[15]. The objective of the detector R&D work should be to eliminate detector limitations on MPR measurements altogether. The performance of the 14-MeV MPR prototype spectrometer now at JET is within a factor of about 3 of the conceivable upper limit; a dedicated design for 2.5-MeV neutrons has yet to be done. Other known spectrometer types have theoretical efficiency limits of $\varepsilon >> 10^{-4}$ but with compromised performance in other essential respects, such as count rate capability, so that the diagnostic resolving power of the MPR is unmatched for plasmas producing neutron fluxes in the high range of 10^9 n/s·cm^2.

PRINCIPLES OF THE MPR TECHNIQUE

The MPR technique (Fig. 1)can be seen as involving three main functional steps . The first step is the neutron-to-proton conversion which is facilitated by the neutron collimator (C), the target (T) where the neutron produce proton recoils and the recoil acceptance aperture (A). These are defined by their; solid angles (Ω_C, Ω_T, and Ω_A); the maximum angle deviation from the central axis (θ_C, θ_T and θ_A,) and the related lateral deviations are noted as x and y; the target is also defined by its thickness, t_T, and area A_t. The second step is provided by the magnet with its beam optics which produces an image of the target on to the magnet focal plane (x'_F and y'_F) with a magnification of 0.6 and 3.2, respectively; one image is produced for each value of proton momentum. The third function is provided by the focal plane detector which records the x_T-distribution of recoils.

The MPR energy resolution is defined by three main factors: Kinematic broadening because of the angular dependence $E_r{=}E_n \cdot \cos(\theta_n{+}\theta_r)$ for angles within the range limited by θ_C and θ_A; the proton energy loss difference in the target being proportional to t_T; the ion optics contribution being proportional to $g_M \cdot x_T \cdot [1+\Gamma(\theta_R,\theta_T)]$ where g_M is the magnet magnification and Γ represents the aberration terms.

The flux detection efficiency (ε) increases proportionally with Ω_T, Ω_A and t_T, and depends also on the magnet transmission factor T_M. The count rate is obtained as $C_n=\varepsilon \cdot F_n$ where $F_n \propto \Omega_C \cdot n_d \cdot n_t \cdot <\sigma \cdot v>$ is dependent on an instrumental parameter (Ω_C), and the plasma parameters of ion densities and reactivity. The measured recoil spatial distribution on the focal plane is a convolution of the instrumental response function, including both the broadening and imaging aspects mentioned above, and the incoming neutron spectrum.

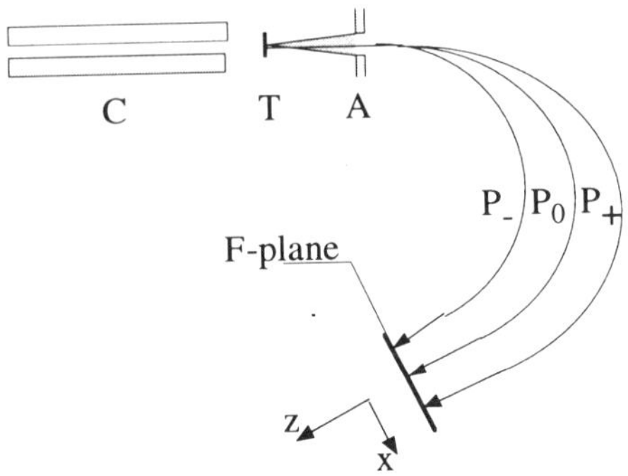

Figure 1. Sketch of the MPR spectrometer showing the neutron collimator (C), the target (T) and the proton aperture (A); dispersed recoil proton θ=0° orbits of low, central and high momentum (p_-, p_o and p_+); the focal plane and detector location.

A unique feature of the MPR technique compared to other recoil techniques is that the recoil energy is nearly the same as the neutron energy (forward scattering). This benefits the discrimination against extraneous neutrons and minimizes the kinematics broadening effects. Another unique feature is the energy measurement which is performed passively while the only active detector element used refers to the counting of protons. In this way, the control of the background interference in the measurement and the instrumental calibration are both rate independent. The rate independence is crucial for the reliability of the determination of the instrumental specifications .

A consequence of the MPR properties is that the instrumental specifications can all be determined with certainty within assigned errors. They are based on calculations but with empirical input in the crucial aspects. This also means that it is easy to establish the theoretical specification limits of the MPR and thus to define the performance limits. For instance, given a required resolution of $\Delta E/E$=2.5 %, the solid angles can not exceed about 35 msr (all three being the same) and the target thickness must not exceed 10 mg/cm^2. As mentioned, the present 14-MeV MPR spectrometer is close to the specification limit for this type of instrument. A 2.5-MeV version for studies of the d+d$\rightarrow$^{3}He+n means other specifications and design criteria. The detector used on the present spectrometer is a plastic scintillator hodoscope. This serves the purpose for the JET experiments but is not an optimized focal plane detector for the MPR as it is desirable to improve both the background immunity and spatial definition for use in future measurements where the statistical sensitivity is expected to be 1 to 2 orders of magnitude higher. Both the detector and the 2.5-MeV spectrometer design are R&D tasks for the future.

PERFORMANCE OF MPR AT JET

The MPR instrument built has the (maximum) solid angles Ω_p=53 msr and Ω_t=33 msr which together with A_T defines the maximum ε–value. The collimator solid angle is Ω_C=6 msr as limited by interface restrictions. This means that the count rate for given neutron brightness for a homogenous plasma source is a factor of 3 lower than the value for an ideal MPR. For a discharge of 2.4 MW fusion power, the MPR recorded a count rate of 120 kHz at a resolution of about 4 % (FWHM); at 2.5 %, which is another resolution setting used, the count rate would have been about 50 kHz.

The calibration accuracy of the MPR can be described by noting the following. With regard to efficiency ε, the error is at the few percent level. With regard to channel variations along the focal plane, the recorded data suggest differences from expectations at the same level. No corrections to the *ab initio* specifications are made. With regard to the energy calibration, the accuracy expressed in terms of ability to predict the spatial distribution on the focal plane for a given energy distribution of the neutron source is determined to be $\delta E_n/E_n = 2 \cdot 10^{-3}$. The calibration stability is estimated to be at the 10^{-4} level or better. All calibrations, as the determination of operating settings of the MPR, are done in the laboratory before installation. Only charged particles are used in these preparations and neutrons are only used to examine the sensitivity of the focal plane detector to background. The MPR spectrometer is designed to be a turn-key-ready device and this has essentially proven to be true by the JET installation.

The sensitivity of the instrument with regard to its ability to detect and measure weak spectral components in the neutron emissions can be illustrated by indicating two quantities. With a count rate of 10^5 n/s, the statistical sensitivity level is at about $\Sigma = I_{max}/I_{min} > 10^4$ (as represented by spectral intensity ratios), and a signal/background(noise) ratio at the level of $S/N > 10^5$.

JET MEASUREMENTS

The MPR experiment at JET has so far (August 1997) included measurement of 14-MeV neutrons from D plasmas and from some DT plasmas. The D plasmas could be observed in the case the primary neutron production rate was in the range $Y_n = 1$ to 5 times 10^{16} n/s where the latter number represents a JET record. The observed count rate of the triton burnup neutron (TBN) emission reached only 100 Hz or less so that statistics was acquired by accumulating data for successive discharges of similar characteristics all with NBI heating where the beam ions are involved in most of the neutron producing reactions. The DT plasmas could be observed in the neutron production range 10^{16} to $8 \cdot 10^{17}$ n/s where the latter value refers to the JET record corresponding to a fusion power value of 2.4 MW. These plasmas range from those being only ohmically heated to those with NBI and RF heating, or, both. Spectra were acquired with an intrinsic time resolution of 10 ms corresponding to the scaler read-out frequency. Besides the logic data containing the spectral information of the measurement, analog pulse height information was also acquired for the purposes of monitoring the detector performance and setting the electronics working points (e.g., discriminator levels). The analog information was also used for making background assessments.

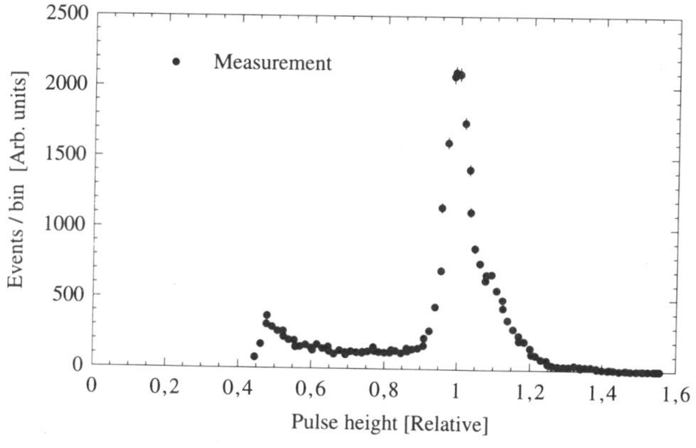

Figure 2. Pulse height spectrum recorded with one scintillator channel of the MPR hodoscope; the data from three JET discharges were added to improve statistics.

An example of a pulse height spectrum recorded for a DT discharge is shown in Fig. 2. The recoil protons give a pulse height distribution of the shape expected from photon statistics (≈8 %, FWHM) and the scintillators' longitudinal response variation (the slope); all scintillators are 100 mm long and the center ones are 2.8 mm thick and 8 mm wide. Some recoils give fractional energy deposition in the scintillators resulting in a flat low energy distribution extending to 0 pulse height. The background radiation is responsible for the rest of the spectrum which can be explained as due to neutron leakage through the shielding, γ:s from local neutron capture leading to γ→e conversion in the scintillators or the surrounding material; both electron sources cause detector background. The discriminator level is typically set at 70 % of the signal peak which cuts out ≈10 % of the recoil count rate and leaves a background admixture of ≈0.9 %; these values are known with sufficient accuracy so that the correction applied introduces an error of less than 1 % including drifts; this correction is count rate independent.

EXAMPLES OF PLASMA RESULTS

The MPR spectrometer provides data on the neutron emission spectrum based on absolute calibration which is illustrated with measurement on ohmic DT discharges at JET (Fig. 3). These discharges can be assumed to produce neutrons under thermal ion reaction conditions so the emission spectrum is accurately known from reaction kinematics and the Doppler broadening (given by T_i) but no Doppler shift (e.g., no rotation). This neutron source function distribution has been calculated[13] and was convoluted with the known MPR response function at given magnetic field setting as a parameter (known to within 10^{-5}). The result of the predicted neutron energy distribution is shown in Fig. 3 compared with the observation; T_i was used as a free fitting parameter with T_i=1.4 keV found to give the best fit together with a position shift adjustment amounting +0.4 channels (equal to 3 mm on the focal plane or +30 keV in energy). This is consistent with MPR *ab initio* calibration having an error of ±3 mm (corresponding to $\Delta E/E \approx 2\cdot 10^{-3}$) as limited by survey uncertainties. The adjusted calibration error is $\Delta E/E = 2\cdot 10^{-4}$ (statistics limited) so the limit for absolute neutron energy determinations is, e.g., ±4 km/s in terms of toroidal rotation assuming the ohmic JET plasmas are not rotating; no additional experimental uncertainty is expected at this level.

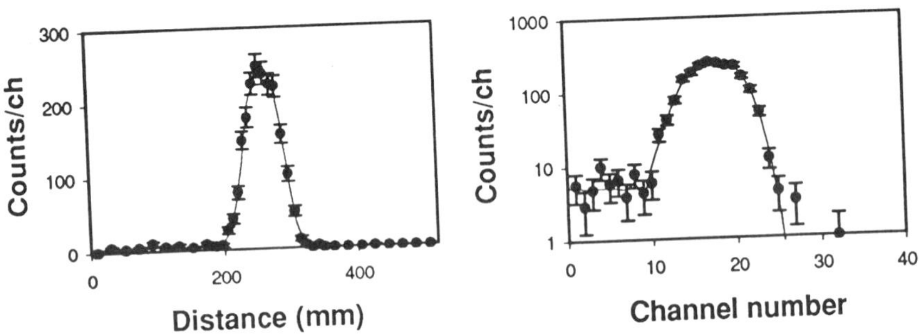

Figure 3. The spectrum for neutron emission for ohmic JET discharges measured with the MPR spectrometer as function focal plane position (left) and channel number. The curve shown is calculated based on folding the adjusted (see the text) with the Gaussian neutron distribution for T_i=1.4 keV (best fit value).

The close agreement obtained between the calculated spectrum and data for the thermal plasma condition case affirms the response function determination for the MPR. Therefore, the instrumental resolution can reliably be subtracted from the measured spectral broadening (e.g., the FWHM value) should one wish to deduce T_i for the plasma this way being the conventional one. This can be done even if $\Delta E_I/E$ is not negligibly small compared to $\Delta E_D/E$ but instead $\Delta E_I/E$ can be chosen so that the statistical uncertainty in the T_i determination is minimized, i.e., $\Delta E_I/E \approx \Delta E_D/E$. One should also note that the shape of the response function is verified to be correct over several orders of magnitude in intensity which is an important basis for a number of new diagnostic tasks to be performed with the MPR.

The maximum count rate recorded with the MPR so far at JET is 120 kHz. It was reached for a DT discharge (#41760) with a 2 s sharp NBI pulse at 10 MW and E_d=80 keV combined with an RF pulse with a slow up/down ramp with a flat top at 4 MW. An example of the spectra recorded each 10 ms is shown in Fig. 4. Examples of the time dependent information collected are also shown in the form of the measured number of counts per time bin, and the extracted mean value of the peak position, $\underline{X}(t)$, and RMS-width $\Delta X(t)$. The statistics is sufficient to follow the time evolution of the plasma expressed in these parameters over a dynamic range of about 5 and with the shown intrinsic time binning of 20 ms .

In these raw data are embedded, for instance, the plasma rotation in $\underline{X}$. For this particular discharge, the measurement suggests a maximum Doppler shift in the neutron emission of ≈100 keV. During the NBI a shift can arise both due to passing ion reactions and rotation but

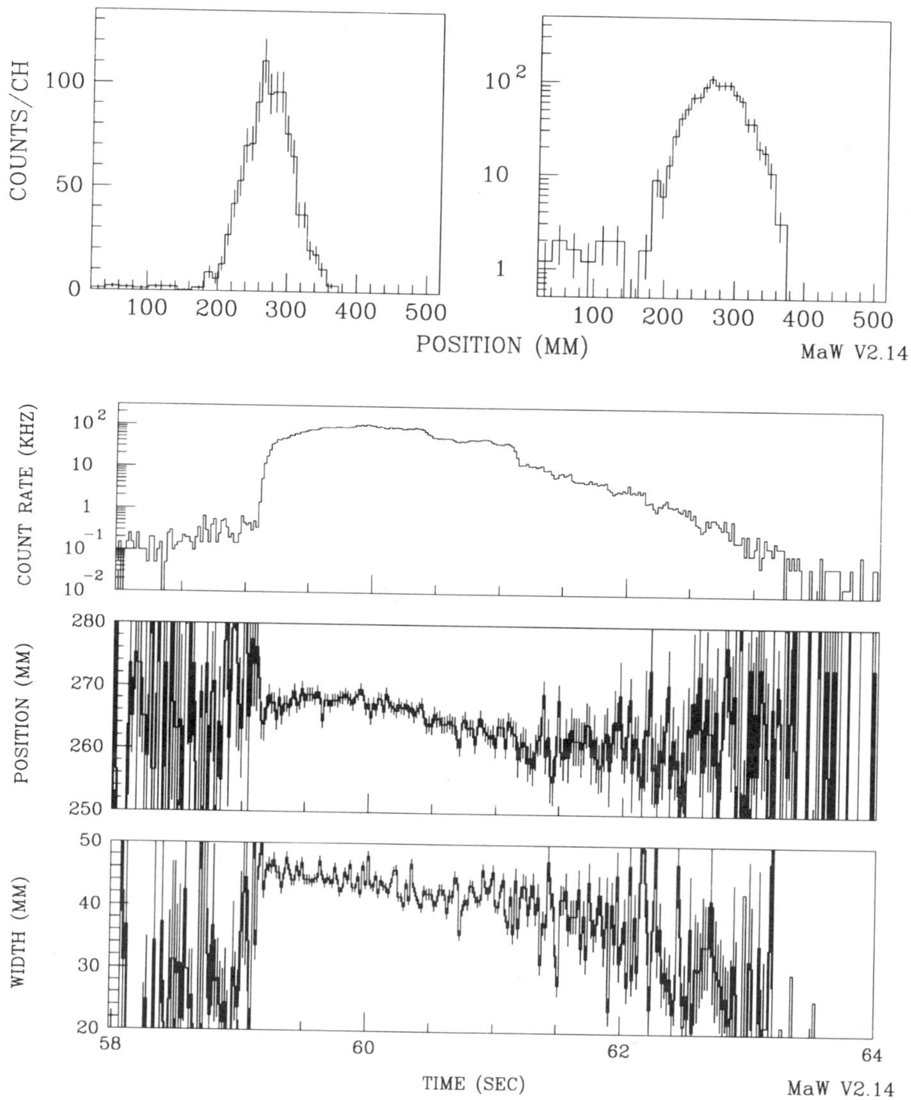

Figure 4. Examples of results from the MPR spectrometer in high count rate operation for JET DT discharge (#41760) from the top are: Neutron spectrum in terms of focal plane distribution recorded during 10 ms; time traces with 20 ms time binning for the count rate, the position and the width of the d+t→α+n peak.

only the latter when beam ion reactions are absent. It can be noted that the data suggest some rather fast changes in toroidal rotation during the RF heating at the 0,1 s time scale; the maximum Doppler shift observed during this discharge corresponds to a core rotation of 200 km/s.

The results on the peak width show that this parameter decreases monotonically from a high level at the beginning of the NBI pulse and continues so until it approaches the value of the instrumental resolution $\Delta \underline{X}_I$ =26 mm. It is interesting to note that the results on the width suggest oscillations during certain time periods, e.g., the RF ramp-up phase, and that there may be concurrent changes in $\underline{X}(t)$. If and how they are correlated remains to be studied.

D-plasmas produce a minority component of 14-MeV triton burn-up neutrons (TBN) whose spectrum has been cleanly measured with the MPR for the first time. The intensity is reduced a factor of 100 relative to the dd neutron rate and about a factor 10^4 for same condition neutron production in DT plasmas. The observation of the TBN is thus a test of the spectrometer sensitivity. In order to get sufficient statistics for time variation observations some 15 discharges were added. Examples of measured spectra with 500 ms time bins are shown in Fig. 5 (for transient plasma conditions of the leading and trailing regions of the NBI pulse) while some of the time dependent information obtained is also illustrated in terms of the measured number of TBN counts per time bin and the energy distribution width, C(t) and $\Delta X(t)$. It can be noted that the variation in $\Delta X(t)$ can be connected with the fact that the TBN production is due to reactions between bulk deuterons and fast tritons of the slowing down population where the population is lacking low and high energy components, respectively, in the leading and trailing regions[15].

Another objective will be to use the high sensitivity of the MPR to measure weak spectral components, especially, in the tail regions. Such observations are within reach as the count rate approaches the 100 kHz range. An example of such observations is shown in Fig. 6 which concerns an RF heated discharge. The main spectral component is simply fitted with a Gaussian which gives a reasonable account of the data but for the high and low energy tail regions. On the high energy side, an apparently exponential tail that breaks off from the Gaussian where the amplitude has fallen by almost a factor of 10^3 which is also near the statistical sensitivity limit in this measurement. On the low energy side, there is a flat level at 10^{-2} of the peak amplitude which is not a reflection of the neutron spectrum but extraneous effects which must be treated before this low energy region can provide information; there is a suggestion of such information at the foot of the peak. The data obtained so far suggest that the production of fast ion tails when RF power is injected into the plasma can be studied with the MPR. One issue to address in these studies is response of the plasma to the RF power injection in terms of heating and fast ion production as detectable in the Doppler broadening (T_i) and 'anomalous' tails of the neutron emission, the latter measuring tail to main component amplitudes.

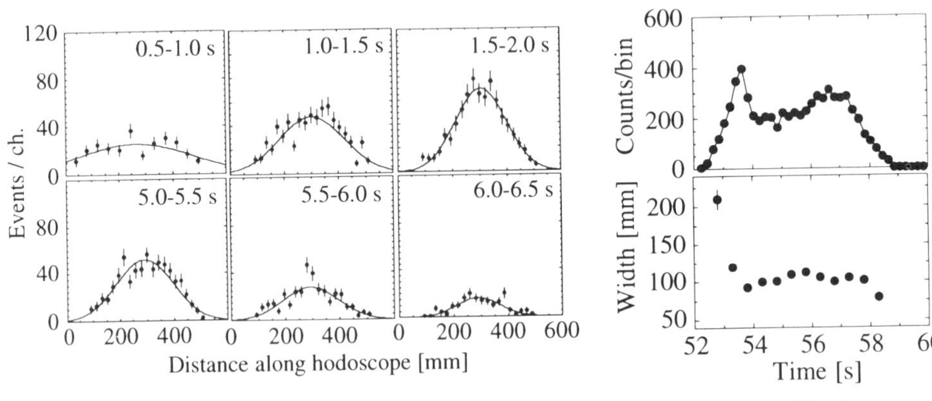

Figure 5. The triton burnup neutron (TBN) spectrum measured with the MPR during JET D-plasma operation with pulsed NBI heating in 0.5 s time bins from the beginning of the pulse and at the end (left panel). Time traces of the number of counts (per 0.2 s) and the width (per 0.5 s) of the TBN peak (right panel).

OPERATING EXPERIENCE AND IMPLICATIONS FOR ITER

The MPR diagnostic was fully calibrated and system tested before being transported in parts to JET. At JET, the system was assembled, interfaced (physically and in terms of cable connections) and commissioned over a 3 months' time period; it uses its own local data acquisition and control system so the interface to the JET CODAS system is minimal. An important feature of the MPR system is a pulse testing and monitoring system so that working points for the PM tube gains and discriminator levels can be set while possible drifts, cable noise pickup and other perturbations can be monitored. While all hardware is as originally installed (apart from electronics which has been gradually complemented), the procedures have developed with operating experience over time. Especially, the use of the momentum analyzed proton recoils from discharges was hard to predict but has turned out to be more useful than anticipated. The installation and initial operating experience of the MPR on JET suggest that diagnostics based on this technique can be delivered for turn-key-ready operation.

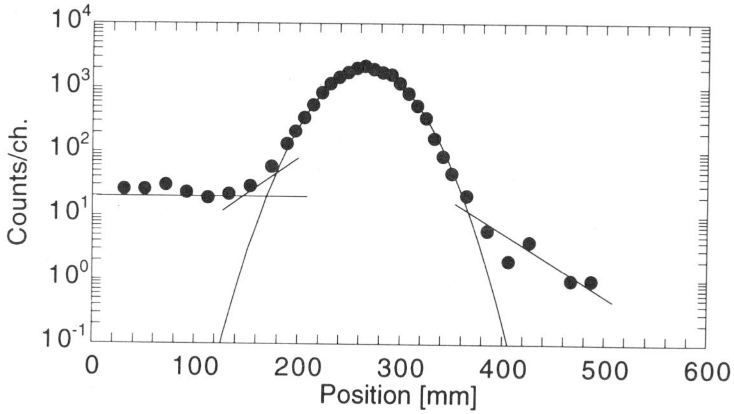

Figure 6. Measured spectrum of the neutron emission for JET DT discharge with RF injection (#41735); the results are presented as counts per channel as function of focal plane position. The curve is a Gaussian fit to the main component and the lines are guides for the eye.

What concerns the ability to clarify uncertainties about features in the data whether they are due, for instance, to malfunctioning hard ware, statistical fluctuations in the data intermittent noise interference, it has been found to be absolutely essential that the energy determination is separate from the means used to discriminate background and noise from the genuine signal pulses. From this experience we tend to draw the conclusion that neutron diagnostics based on techniques that do not allow clear separation of energy determination and background separation (including noise rejection) will have an exceedingly hard time to establish working points, let alone verifying that they are not subject to off-drifts. The latter presents itself as an even greater challenge if some of the neutron measurement functions have an intrinsic rate dependence. Most alternative techniques that have been proposed for use in fusion neutron spectrometry (as motivated, for instance, by increasing the efficiency) suffer from the mentioned problems; in fact, these problems are known to have afflicted existing neutron diagnostics, too.

CONCLUSION

This paper has reported on progress on neutron spectrometer diagnostics on D and DT fusion plasmas in tokamaks. The magnetic proton recoil (MPR) spectrometer was described and its use in the MPR on JET experiment from which results were reported on the first high performance neutron diagnostic studies of fusion plasmas. It has been proven that the experiment could quickly be put into productive operation. This should be considered in view of the importance of demonstrating 'turn-key-ready' start-up operation for diagnostic equipment of potential use on ignition experiments as this has been a declared requirement, for instance, for ITER. The MPR on JET experiment was also installed with an *ab initio* absolute calibration which has been verified to be correct within specified uncertainty limits. Moreover, the instrumental stability allowed the establishment of an adjusted calibration to an accuracy of $\Delta E/E=2\cdot10^{-4}$ permitting plasma toroidal rotation measurements down to the ±2 km/s level (assuming that the ohmic plasmas at JET are rotation free to the same level); this demonstrates neutron spectrometry as a rotation diagnostic to be comparable to x-ray spectroscopy as conventionally used. A record count rate of 120 kHz has been attained demonstrating the first use of neutrons spectrometry as a time resolving plasma diagnostic. The high sensitivity of the MPR spectrometer has been demonstrated in the first clear recording the triton burnup neutron spectrum whose amplitude is down by four orders of magnitude relative to the emission in similar DT discharges. In the continued JET campaign on DT plasma operation it is expected even higher neutron fluxes will be produced so that the performance limits can be further tested. As statistics is the limitation for many of the envisaged diagnostics tasks of the MPR and its use for plasma physics studies, the coming years can be expected to be fruitful in three respects: To prove diagnostic principles; contribute new diagnostic information to the JET experimental program; potentially gaining new insight into the physics of extreme sates of matter. From the point of view of the planning of the next step neutron diagnostics, the continued MPR on JET experiment will provide information on the theoretical limits of neutron spectrometry.

ACKNOWLEDGMENT

This work was performed under the auspices of Association NFR (Sweden) of EURATOM in collaboration with Association ENEA-CNR (Milano, Italy). It has been financially supported by the Swedish Research Councils NFR, FRN and NUTEK; EURATOM; JET. We are grateful for stipends from The G Gustafsson foundation (SC and LB) and for travel grant from the Swedish institute (JF). We gratefully acknowledge significant personal contributions: Prof. H Enge on spectrometer design work; S Hultqvist and B Trostell of our home team on the instrument; G Leman of NFR for administration and liaison with EURATOM and JET; S Scott, N Hammond and J Bird of the JET team for engineering the experimental interface to the tokamak; John Reid and his crew for assembly and the installation at JET.

REFERENCES

1 J Källne and G Leman, 'Neutron spectrometer for high power DT plasmas; proposal for installation and measurements at JET'. Uppsala University Neutron Physics Report, UU-NF 93/#1, March 1993.

2 L Johnson, 'Overview of fusion product diagnostics', This proceedings.

2. J.Källne et al, "Neutron Spectrometry for ITER", in Proc. of Workshop on Diagnostics for Exp. Thermonuclear Fusion Reactors (Varenna, 1995), Plenum Publ. Comp, New York (1996), p.397.

4 J Källne, Comments on Plasma Phys. Contr. Fusion, **12**(1989)235

5 J Källne and H Enge, Nucl. Instr. Meth. A311(1992)595

6 G Ericsson et al, "The MPR neutron spectrometer project", in Proc. on Workshop on Diagnostics for Experimental Thermonuclear Fusion Reactors, Plenum Publishing Company, New York (1996), p.413.

7 B Rose , AE Taylor and E Wood, Nature **181**(1958)1630.

8 W Fisher, SH Chen, D Gwin and R Parker, Phys Rev A 28(1983)3121

9 ON Jarvis, G Gorini, J Källne, J Merlo, G Sadler and P van Belle, Rev. Sci. Instr. **57**(1986)1717 and ON Jarvis, Plasma Physic. Contrl Fusion **36**(1994)209, and ON Jarvis, In proc. of Int. School of Plasma Physics, Varenna 1991 (ISBN 88-7794-043-3), p. 555

10 T Elevant et al, Rev Sci. Instr. **63**(1992)4586

11 G.Lehner, F.Pohl, Zeit. Phys. **207**(1967) 83.

12. H.S.Bosch, G.M.Hale, Nucl. Fusion **32**(1992) 611.

13 L Ballabio, G Gorini, and J Källne, 'Relativistic calculation of fuison product spectra for thermonuclear plasmas', Uppsala Univ. Neutron Physics Rep., UU-NF 97#4 (July 97) ISSN 1401-6269, subm. to Phys. Rev. E.

14 L Ballabio, G Gorini and J Källne, 'The α−particle knock-on signature in the neutron emission of DT plasmas. Phys. Rev E 55(1997)3358.

15 J Frenje, S Conroy, G Ericsson, J Källne, P-U Renberg, and E Traneus, Nucl. Instr. Meth. **A376**(1996)462.

HIGH PRECISION CALIBRATION OF NEUTRON SPECTROMETER DETECTORS WITH α-PARTICLES

P. Prandoni[1], L. Ballabio, S. Conroy, G. Ericsson,
J. Frenje, J. Källne, M. Tardocchi, E. Traneus

Department of Neutron Research, Uppsala University, Uppsala, Sweden* and
[1] INFM and Department of Physics, University of Milan, Milan, Italy*

INTRODUCTION

A neutron spectrometer for the diagnosis of tokamak plasmas has been built and is now operating at the Joint European Torus (JET). The instrument is of a new type being based on the magnetic proton recoil (MPR) technique. As has been described earlier[1] it makes use of a CH_2 target in the collimated neutron flux from the plasma where some neutrons collide with protons transferring approximately all kinetic energy so that there is a nearly one-to-one correspondence between neutron and proton recoil energy. The protons that pass within a physical aperture are deflected in the magnet, imaged onto the focal plane according to their momentum and recorded by a 1-D position sensitive detector (hodoscope).

The position determination is an essential aspect of the MPR spectrometer because it is closely related to the protons' energy. The MPR instrument is also equipped with a target wheel where 4 targets of different thickness are mounted together with a ^{241}Am source. The former provides different combinations of energy resolution and n→p conversion efficiencies, while the latter is used in general for the MPR calibrations and testing of the hodoscope. The geometrical properties of the MPR hodoscope were studied in a dedicated experiment based on a scan with a well collimated α-source. This paper describes the α-scan study aimed specifically at determining the geometrical tolerances of the detector as compared to design values, namely, the position of the individual scintillators. The experimental set up used together with the results obtained in this test are presented and discussed. Special consideration is given to the applicability of the method for calibration of future focal plane detectors where the detector tolerance should be pushed by at least a factor compared to the present MPR prototype spectrometer requiring corresponding testing facilities.

THE DETECTOR

The MPR hodoscope is composed of 37 plastic scintillators which each is connected through a light guide to a PM tube. The scintillators are of three different sizes for detecting ions of low, medium or high energy; normally, they span the energy range of the focal plane band width (±20%) normally centred around E_o = 14 MeV. The dimensions are (width x thickness) 20x2, 8x2.8 and 20x3.8 mm^2 respectively, all of length 100 mm.[2] To avoid light cross talk between adjacent scintillators, they were wrapped with 10 μm thick aluminum foils which leaves 14-MeV proton recoils practically unaffected. However, this is not the case for the 5.5-MeV α-particles used for the testing as we shall discuss later. The scintillators are positioned with gaps of ζ_G=0.5 mm. This means that the overall detector efficiency is reduced by a few percent (escape factor) which is accounted for in the detector response function.[3] Another

* Part of the collaboration between the Association NFR-EURATOM, Sweden, and the Association ENEA/CNR(Milano)-EURATOM, Italy .

consequence of the gap is that the channel position can vary slightly across the hodoscope, the upper limit being of ±0.25 mm. In order to evaluate the actual scintillators' positions, and the gaps between them, the present α-particle test was performed as described in the next section.

EXPERIMENTAL SET UP

The experimental set-up of the α-test is schematically shown in Fig. 1. It consists of a collimated α-source which was used to irradiate the detector hodoscope at well determined positions. The source was placed in a holder very near the hodoscope surface, the typical clearance being 2 to 4 mm. Moreover it could be moved across the hodoscope (the x-direction in Fig. 1) with high accuracy. The translation was provided by a stepping motor which turned a screw through a threaded hole in the α-source holder; the motion was 1.988 μm/step over the length of 250 mm. The holder moved on guide rails to keep it securely positioned in the y-z plane. The stepping motor and the guide rails were placed in a rigid frame which was secured to the detector and aligned to desired accuracy. The source assembly could be moved to different x-position to cover the entire hodoscope. The whole set-up was evacuated in order to avoid having the α's pass through air and the beam definition degraded.

The source was ^{241}Am whose main α-emission has an energy of E_α=5.5 MeV. These α's have a range of 4 cm in air which made it necessary to use evacuated test conditions. The range in Al is 25 μm which is only 2.5 times the foil thickness so that range effects cannot be ignored.

The scanning process, as well as the data collection, was under full automatic computer control. Data consist of histograms for each channel as a function of source position (x).

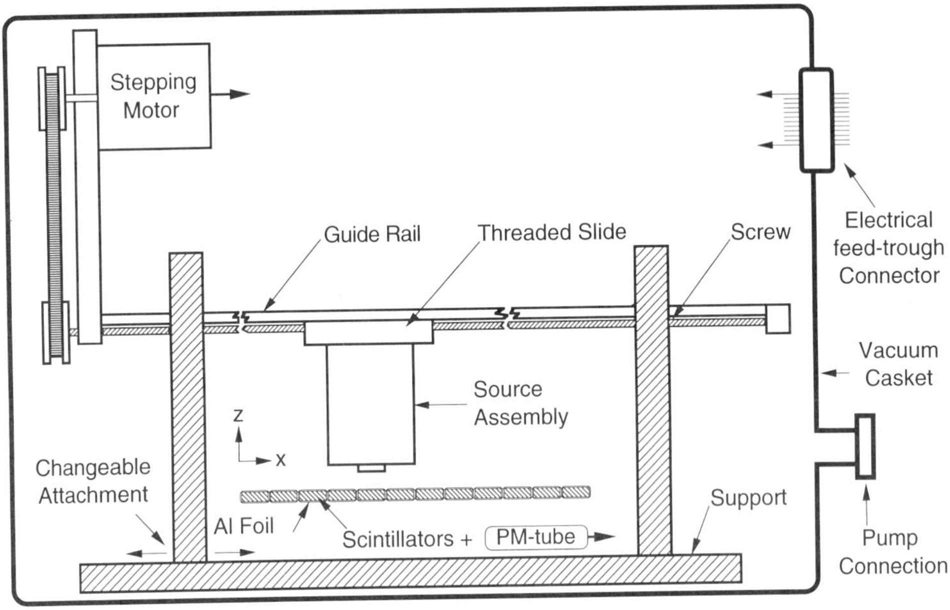

Figure 1. Schematics of the experimental set-up showing the stepping motor, the guide rail, the source assembly and the scintillators all enclosed in a vacuum casket with pump connection and electrical feed-through connectors. The guide rail system could be set to cover different parts of the hodoscope.

DATA AND RESULTS

Two series of measurements were performed. First, the hodoscope was scanned in steps of 0.5 mm in the regions with wide scintillators and in step of 0.1 mm in the region with narrow scintillators. The larger step length was chosen in order to decrease the measurement time. Second, the gap regions between the wide scintillators were studied in detail using 25 steps of 0.1 mm as a complement to the full scan. The acquisition time was fixed and chosen so as to obtain N=10^4 counts giving a statistical counting error of ±1%.

The data were recorded in the form of histograms and reduced to make them suitable for analysis. The first step was the subtraction of background which generally meant a correction of 2-6 % of the total counts depending on the dimension of the scintillator.

Several approaches were used to analyse the data; the most direct one was to determine the scintillator positions and the gap sizes from the half maximum height (HMH) points in the recorded x-scan histograms which gave information on the edges of each channel in the recorded histograms. Two adjacent scintillators with no gap in between would, when the source is positioned over the gap, record the same number of events in both scintillators equalling one half of the events produced by the source. Generally, we define x_A and x_B as the HMH positions of two adjacent channels in a histogram. The difference, $\zeta_G = (x_A - x_B)$, is a direct measure of the width of the gap. The center position of the gap is $X_G = (x_A + x_B)/2$ and this together with ζ_G was determined for all scintillator pairs of the hodoscope. Each pair was identified by the highest channel number in the pair with X_G-value of gap #1 the zero reference point.

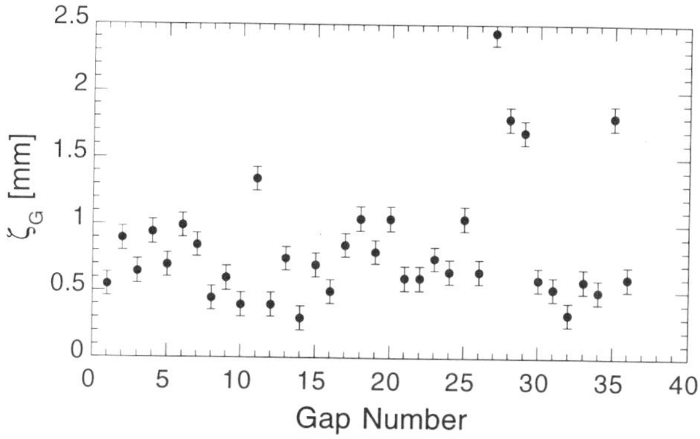

Figure 2. Summary of results on gap sizes for all channels. The error bars shown represent the estimation of typical errors.

Fig. 2 shows the ζ_G values as a function of the gap number. Several of the data points exceed the designed value of 0.5 mm and a few are even larger than 1 mm thus apparently exceeding the physical bound. The observed X_G-values, however, are not only showing the physical gap size, but are also sensitive to α-absorption in the foil which can show local differences. The error bars represent an estimation of the errors induced by this kind of analysis. Fig. 3a shows the gap positions X_G as a function of the design position X_D; a linear fit of data was also done giving a slope of 0.99961. This means that the real position of the gaps, and therefore the position of the scintillators, is very close to those of the design. Differences between the fitted line and the data points are not appreciable on this plot.

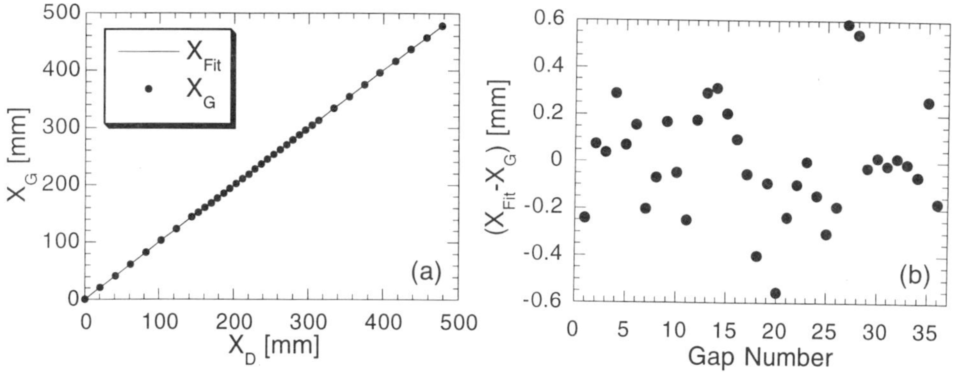

Figure 3. The measured gap positions (X_G) for the hodoscope channels plotted versus the theoretical positions according to design (X_D); the line shown is a linear fit (a). Results on the gap position showing the difference between measured (X_G) and projected (X_{Fit}) as a function of the gap number (b).

Fig. 3b shows the difference between X_G and the fitted line as a function of the gap number. One observes that all points are within a deviation interval of ±0.6 mm. This, in turn, provides evidence that the construction of the MPR hodoscope is faithful to the design to within set construction and assembling tolerances. The cases with apparent deviations is ascribed to α-particle foil absorption as foil wrinkles could be seen to concur with such measurement deviations.

IMPLICATION FOR NEXT STEP NEUTRON SPECTROMETER

The nominal precision in the α-scan of the detector of the MPR focal plane hodoscope is about 2 μm which corresponds to an accuracy of $\Delta E/E < 7 \cdot 10^{-6}$ which is compatible with the aim of attaining an overall accuracy of 10^{-5}. This aim could be reached in neutron spectrometry measurements of ignited plasmas where the statistics of the measurements would be sufficiently good so as not to be a limitation at the 10^{-5} accuracy level. The observation in this experiment that the foil effects interfere with the detector characterisation would be eliminated by considering the α-particle testing already in the design of future high precision measurements. Indeed, this would removed the foil effects as an error source, which is the dominant one in the present exercise mainly representing a test of the calibration principle.

CONCLUSION

The detector system of the MPR neutron spectrometer was studied with the use of a collimated α-beam from an ^{241}Am source. The results provide information on the separations between the scintillators (gaps) and their location. These show that the mechanical properties of the MPR hodoscope can be characterised with the help of α-particle measurements. Specifically the obtained results confirmed that the hodoscope construction held the tolerances of the design values. Some of the determined gaps appeared to exceed the expected upper limit value of 0.5 mm width. This is ascribed to the presence of local effects due to irregularities in the Al-foil wrapping. This problem will not affect the use of the MPR spectrometer for 14-MeV recoil protons but would have some effect on 2.5-MeV protons. Finally the effectiveness of the used method of analysis was assessed giving an estimate characterisation accuracy of the MPR hodoscope corresponding to about 10 microns and compatible with the aim of reaching better than 10^{-5} in the neutron energy determination; this is the requirement set even for the most demanding diagnostic applications of the MPR spectrometer as projected for the use on ITER.

ACKNOWLEDGEMENT

This work was performed under the auspices of the NFR association of EURATOM; it was financially supported by the Swedish Natural Science Research Council (NFR), and the ERASMUS student exchange of EU (for PP), and the staff mobility program of EU (for JK).

REFERENCES

1. J. Källne and H. Enge, Magnetic proton recoil spectrometer for fusion plasma neutrons, *Nucl. Inst. and Methods* A311(1992), and G. Ericsson, L. Ballabio, S. Conroy, J. Frenje, G. Gorini, J. Källne, P-U. Renberg, M. Tardocchi, E. Traneus, The MPR neutron spectrometer project in: *International Workshop on Diagnostics of ITER,* Plenum press, Varenna (1995).
2. J. Frenje, S. Conroy, G. Ericsson, J. Källne, P-U. Renberg, *Design of the MPR focal plane detector,* Uppsala University Neutron Physics Report UU-NF 94/#11.
3. S. Conroy, L. Ballabio, G. Ericsson, G. Gorini, J. Frenje, J. Källne, E. Ramström, P-U. Renberg, M. Tardocchi, E. Traneus. *Simulation of the MPR spectrometer performance,* Uppsala University Neutron Physics Report UU-NF 95/#15.

FUEL DENSITY MEASUREMENT IN BURNING PLASMAS USING NEUTRON SPECTROMETRY

G Gorini[1] , L Ballabio, J Källne, S Conroy, J Frenje, G Ericsson,
P Prandoni, M Tardocchi and E Traneus

Department of Neutron Research, Uppsala University, Uppsala, Sweden
[1]INFM, Department of Physics, University of Milan, Milan, Italy and
Istituto di Fisica del Plasma, CNR, Milan, Italy

INTRODUCTION

The use of neutron spectrometry as a density diagnostic was first proposed in 1990 [1] and has been the objective for a preliminary assessment [2]. A major concern for this diagnostic is the presence of wall emission (WE) neutrons in the measured spectrum causing a background to the plasma emission (PE) signal. An initial study of the WE component was performed with the aim to identify which neutron scatterings and reactions dominate the WE spectrum and which changes can occur in magnitude and shape due which causes (for instance, and especially, in the plasma conditions) [3]. We have now performed neutron transport calculations with the MCNP code [4] in a model representing the latest ITER geometry and a plasma facing wall of composite multiple-nuclide material and multi-layered structure. The results are used here to predict the signal and background intensities, as well as the signal-to-noise ratio (S/N) that can be expected in a proposed fuel ion density diagnostic based on neutron spectrometry for ITER. The predictions are based on the use of neutron spectrometers of the magnetic proton recoil type (MPR) a prototype for which is in use in the JET tritium phase operation.

As the signal intensity as well as the S/N ratio is important for obtaining optimum diagnostic quality, a tangential sight line is significantly better than a radial one. It is true that a tangential sight line is more difficult to accommodate because of anticipated port space limitations on ITER. It is also a concern that the MPR is a rather large installation as far as diagnostics go. On the other hand, the MPR fuel density diagnostic is rather small compared to the NBI-systems that are planned for ITER. The NBI systems do use tangential beam lines and share other interface problems with MPR such as those concerning effects from the magnetic stray fields from the machine, and, conversely, effects on the plasma field. These problems seem to have been solved for the NBI-systems and they should be solvable for an MPR diagnostic. This will be the subject of future studies.

Diagnostics for Experimental Thermonuclear Fusion Reactors 2
Edited by Stott *et al.*, Plenum Press, New York, 1998

This paper is focused on the more principle question of the effect of the WE background on the diagnostic quality of the fuel density determination.

DIAGNOSTIC PRINCIPLE AND SIGNAL STRENGTH

The diagnostic basis of the measurement is given in Ref [2] and can be summarized as follows. The neutron spectrum from a DT plasma consists of the two peaks, one at 2.5 MeV due to $d+d \rightarrow {}^3He+n$ (1) and another at 14 MeV due to $d+t \rightarrow \alpha+n$ (2) besides a distribution at $E_n<9.4$ MeV due to $t+t \rightarrow \alpha+n+n$ (see Fig. 1). The diagnostic basis is the peak intensities I_1 and I_2, and the peak widths Γ_1 and Γ_2 . For the reference case of a thermal plasma, the peaks are of Gaussian shape and their widths are directly related to the (same) ion temperature T_i. The intensities are products of a density factor ($n_d^2/2$ or $n_d \cdot n_t$) and the core reactivities ($\rho_{dd}(T_i)$ or $\rho_{dt}(T_i)$) besides the effective plasma volume seen by the spectrometer neutron collimator (being proportional to the chord length Λ) and a proportionality constant:

$$I_1 = 0.5 \cdot n_d^2 \cdot \rho_{dd} \cdot \Lambda_1 \cdot C_1 \qquad (1)$$

$$I_2 = n_d \cdot n_t \cdot \rho_{tt} \cdot \Lambda_2 \cdot C_2 \qquad (2)$$

The fuel densities derive from

$$n_d = (2I_1/\rho_{dd}\, \Lambda_1 C_1)^{1/2} \qquad (3a)$$

$$n_t/n_d \equiv r = 0.5 \cdot (I_2/I_1) \cdot (\rho_{tt}/\rho_{dd}) \cdot (\Lambda_2/\Lambda_1) \cdot (C_2/C_1) \qquad (3b)$$

$$n_t = r \cdot n_d \qquad (3c)$$

The intensity I_2 is measured directly while the measurement of I_1 includes a background contribution that must be subtracted from the spectral data $I(E_n)$. The background is given by an amplitude factor B_1 (representing the region around E_n=2.5 MeV) and a shape factor $f_B(E_n)$. The $I_n(E_n)$ spectrum consists of signal neutrons (I_1) and background neutrons (B_1) besides contributions from I_3 (belonging to the t+t reaction). The ratios I_1:I_2:I_3 are a function of temperature as given by the reactivities. A summary of the quantities of Eq. 3 is given in Table 1 which also specifies their role in the determination of the densities.

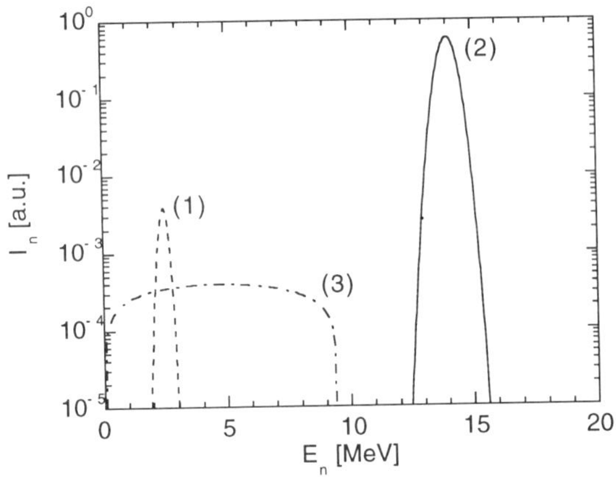

Fig. 1 The neutron emission from a DT plasma at T_i=20 keV showing three separate contribution. (1) The 2.5-MeV peak due to dd-reactions; (2) The 14-MeV peak due to dt reactions; (3) The distribution (kinematic phase space) due to tt-reactions.

Table 1. Parameter summary

PARAMETER	INFORMATION SOURCE AND USE.
I_2	measured
Γ_2	measured and used to determine T_i
I_1	derived from the measured spectrum $I(E_n)$
Γ_1	(i) determined based on $\Gamma_2(T_i)$ and used for data reduction (ii) measured and used to determine T_i (applicable if $r<<1$)
B_1	measured
$f_B(E_n)$	obtained from computation and cross checks
Λ_1	from external information source (neutron camera)
Λ_2	as Λ_1; $\Lambda_1 \approx \Lambda_2$
C_1	given by geom. dimensions and spectrometer efficiency (ε_1)
C_2	given by geom. dimensions and spectrometer efficiency (ε_2)

The feasibility of this diagnostic is connected with the possibility of obtaining tolerably small background and sufficiently high signal intensity I_1. The latter determines the statistical uncertainty and is proportional to the fusion power (P_f). At peak power level (P_f^{MAX}=1.5 GW, corresponding to a total dt neutron yield of $Y_n^{MAX}=6\cdot10^{20}$ n/s), the maximum 2.5- and 14-MeV neutron fluxes would be $F_1^{MAX}\approx4.4\cdot10^8$ n/cm^{2}s and $F_2^{MAX}\approx1.4\cdot10^{10}$ n/cm^{2}s, respectively. These fluxes are estimated at the spectrometer targets, assumed to be located at 10 cm and 150 cm distance from the collimator exit with the 2.5-MeV spectrometer being closest to the collimator. The collimator is assumed to be of cylindrical shape (aperture A_c=10 cm^2, length l_c=70 cm) viewing the plasma tangentially from a location at the biological shield wall (R=20 m). Interface restrictions may require a smaller collimator solid angle with proportionally lower neutron fluxes (see Discussion). The reference plasma conditions of T_i=20 keV and n_t/n_d=1 are used in this estimate. As a reference model for the plasma emissivity geometry we take a toroidal volume (major radius $R_m=R_o$=813.7 cm and horizontal minor radius a_m=178.9 cm) with elliptical cross section (elongation κ=1.5) and uniform emissivity. The model radius values represent w=0.6 where $w=a_m/a$ and a is plasma radius (a=298.1 cm); this width corresponds to the case of a predicted broad profile for an ignited ITER plasma. At the maximum neutron fluxes, the signal count rates in the two spectrometers (assumed to have flux efficiencies of $\varepsilon_1=3\cdot10^{-5}$ and $\varepsilon_2=5\cdot10^{-5}$ cm^2) would be $C_1^{MAX}\geq10$ kHz and $C_2^{MAX}\approx0.7$ MHz, respectively. These values pertain to instruments of the MPR type similar to the prototype in use at JET [5]. The statistics in I_1 is sufficient to obtain results at the 10 % accuracy level (not counting background subtraction effects) with a time resolution of Δt=100 ms assuming the neutron flux from plasma burning at the power levels of $P_f>0.1\ P_f^{MAX}$. If the plasma changes profile, this can have a noticeable effect on the signal as it varies roughly as $I\propto w^{-3/2}$ and can be significantly higher for peaked profiles.

WALL EMISSION BACKGROUND

The background B_1 is dominated by scattered (energy degraded) neutrons from the dt-reaction. It has three major components, namely, WE, in-scattering in the neutron collimator and window scattering by the windows in the neutron flight path between the plasmas and the spectrometer converter foil. Our estimates suggest that WE dominates by an order of magnitude assuming the window to be 5 mm steel and having a suitably formed viewing/collimator channel. Therefore, the WE-level is the factor that checks the feasibility of this diagnostic with respect to the issue of background. Hence, only results with respect to the WE part are presented here. Although the tt-reaction can also interfere with the measurement of I_1, its relationship to the dt and dd reactions is known appearing as a smooth background under the dd peak. Finally, there can be effects on the diagnostic from the ambient radiation level around the detector. It is estimated, based on the prototype design calculations for the JET experiment, that this can be kept at a level that gives a count rate that is insignificant relative to WE (see below).

Table 2. ITER wall dimensions and composition used for the MCNP model.

Outboard radius of layer [cm]	Material	Rel. atomic composition [%]
1111.8	Be	Be(100)
1112.8	Cu+SS[1]+H_2O	Cu(83), Fe(3), Cr(0.8), Ni(0.6), H(8), O(4)
1114.8	SS+H_2O	Fe(53), Cr(14), Ni(11), H(15), O(8)
1137.3	SS+H_2O	Fe(37), Cr(10), Ni(8), H(30), O(15)
1155.3	Void	-

[1] SS316L

The WE background has been calculated with a model of the ITER plasma facing wall consisting of layers of composite materials as shown in Table 2; the relative nuclide abundance of each layer is also given. The viewing geometry and the WE source spot on the wall are indicated in Fig. 2. This together with the reference plasma geometry described above provided input to the neutron transport calculations with the MCNP code and the prediction of the WE spectrum that the spectrometer records. The signal spectrum (i.e., the dt and dd peaks) was calculated assuming a thermal ion velocity distribution and isotropic emissivity from the plasma volume intersected by the viewing cone defined by the neutron collimator. The 3-body kinematical phase space distribution of tt neutrons was calculated in the same way. The latter contribution is in effect part of the background for r-values around 1 or smaller. The results are presented in the form of spectra as they would be recorded by the spectrometer, namely, with the instrumental response function (resolution) incorporated by means of smoothing of the WE and plasma emission (signal) spectra; the latter incorporate the thermal Doppler broadening given by T_i.

The results of the calculations for the reference case are shown in Fig. 3a. This is the spectrum the spectrometer would record if set to place the dd peak in the center. The signal (I_2), the background (B_2) and the tt component are shown individually as well as all superimposed (total). We find that the s/b ratio is about 1/2.5 (using peak value). It is estimated that this model calculation should be within about 10 % of a full 3D model including more tokamak details (e.g., diagnostic apertures and the divertor region). The present results are in line with those of previous calculations taking into account differences in input conditions. It is believed that the result is a fair basis for judging the feasibility of this diagnostic for the ITER conditions as of present.

The s/b will vary depending on the r-value. If the r-value is decreased from r=1 (the case shown in Fig. 3a), the s/b-ratio will closely vary as 1/r as the tt-contribution is vanish-

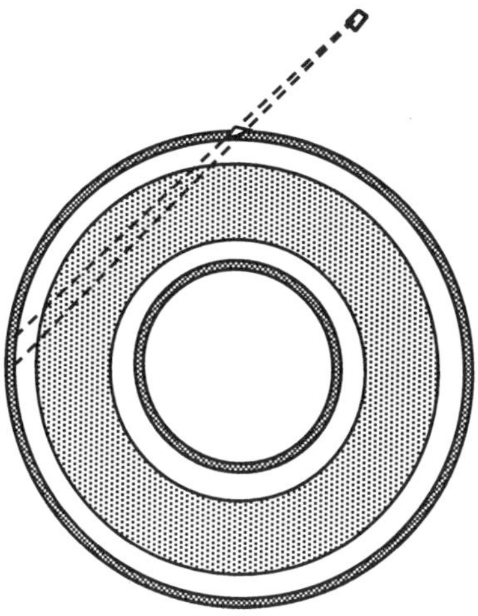

Fig. 2 The model used for the MCNP neutron transport calculations to mimic the conditions of ITER. Shown is the top view cross section in the equatorial plane; the cross section of the torus is elliptical and has the same elongation as the plasma neutron source. The neutron collimator of the spectrometer is located at the apex of the tangential viewing cone (dashed lines) at a distance of about R = 20 m.

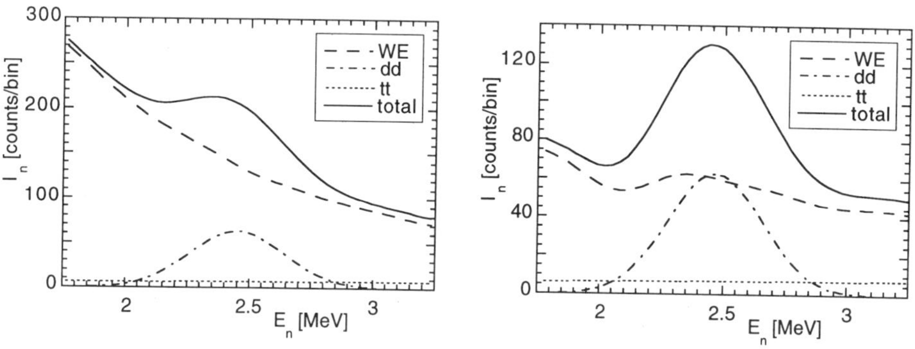

Fig. 3 **(a)** Simulated ITER neutron spectrum recorded by a tangentially viewing neutron spectrometer for reference plasma conditions (see text). A time resolution of Δt=100 ms and $P_f \approx P_f^{MAX}$ are assumed giving about 1,000 signal counts. The spectrum has been convoluted with the instrumental function of the spectrometer assumed to be of Gaussian shape (FWHM=10 %). The binwidth is 30 keV. The density ratio is n_t/n_d=1.**(b)** Same as Fig. 3 but assuming that the viewing region consists only of water.

ing. For instance, the case r=0.1 gives s/b≈5/1. If r is increased to above r=1, the s/b will have the same 1/r dependence with regard to the WE background but the tt-contribution will come into play further lowering the s/b value. For the extreme case r=10 we predict s/b<1/30. These results suggest that the dd peak can be used as a diagnostic basis for r-values below an upper limit (r≲2 for the reference case). For r>2 the statistical requirements become more severe and would certainly imply long integration times. For r>>1 it is possible to use the tt-component as the diagnostic base. This is a potential that would be best exploited with the spectrometer set to measure in the energy range 4-8 MeV where the s/b for this reaction would be highest. This possibility will not be considered here.

The WE spectrum was calculated for a few different cases of wall compositions namely, the reference case (Table 2) besides the cases with either Be+Cu or H_2O removed.

The results show that the Be+Cu layers act in the same way as the bulk of the wall material, i.e., the WE spectrum does not change when it is removed. This can be understood as the neutronics properties of Cu are quite similar to those of the bulk and the amount of Be is relatively small to have any quantitative impact although its neutronics features are different. Removing H_2O has a detectable impact on the WE spectrum, especially for low energies, showing the effectiveness of hydrogen as a neutron moderator. A high water content in the wall reduces the WE background as illustrated by the calculations assuming that the view spot of the wall is all water (Fig. 3b). This increases the s/b-ratio by a factor 2.5 (to s/b=1/1) and gives a much gentler background slope around the 2.5-MeV peak. This result is interesting as it puts the issue of a 'viewing dump' in a new perspective. In contrast, the WE background is not sensitive to the width of the plasma profile but the s/b ratio will vary because of the w-dependence in s. It is also very insensitive to other plasma changes [3].

ERROR EFFECTS DUE TO THE BACKGROUND

The effect of background on the statistical uncertainties in the determination of I_l has been studied by statistical analysis of simulated neutron energy spectra. The results obtained, for the reference case (Fig. 3a) and using I_l=1000 counts, show a statistical error of $\Delta I_l/I_l$=6 %. The background free case is of course smaller $\Delta I_l/I_l$=3.2 %, so the increase is a factor of 1.9 when the s/b-ratio is 1/2.5. If the s/b ratio goes down by a factor of two, the error is increased by a factor of 1.37 (to 8.2 %). If the s/b ratio is increased, using the case of water enhanced viewing dump (Fig. 3b), the s/b-ratio is 1 and the error is reduced by a factor of 1.27 (to 6.0 %). The error in I_l enters with full weight into the determination of n_t and n_t/n_d, and half weight into n_d (Eq. 3).

Another error effect due to background occurs if the shape $f_b(E_n)$ of B_l deviates from the one used in the fitting. We have reason to assume that this shape is a quantity characterized by great inertia towards change. Therefore, any deviation is most likely to be of a systematic nature. Previous studies [3] have shown that the calculated $f_b(E_n)$ responds very little or not at all to changes in input conditions, such as plasma conditions, which can occur fast and uncontrolled. With regard to the correctness of the initial $f_b(E_n)$ function used, this would be based on predictions which can be verified empirically with neutron scattering measurements simulating the diagnostic situation or crucial elements thereof. Moreover, the f_b function would be verified through its use in the data reduction analysis and corrected based on the goodness of fit attained. Similarly, as the data base of the diagnostic measurements builds up to very high statistics, one can analyze the acquired fitting experience and ascribe a lack of fit to a deviation between assumed and actual $f_b(E_n)$ shape. The acquired fitting experience is thus a base for verification and corrections of f_b. The main effect of an error in the shape function is therefore a systematic one and it should be possible to reduce this error well below the 10 % level.

DISCUSSION

It must be pointed out that neutron spectrometry can measure the individual n_d and n_t densities as well as their ratio. In fact, the statistical accuracy is better for absolute density measurements than for their ratio. The availability of two independent fuel density signals makes this measurement more robust for control applications. The use of the measurement for control application should be tested on the basis of simulations of envisaged plasma scenarios in ITER requiring the intervention of the control system. In this way the ability of the measurement in detecting variations in the control signals at the 5% accuracy level can be assessed.

The use of this diagnostic for control applications depends on the online availability of other plasma diagnostic information for the elaboration of the measured data. The neutron emissivity profile (in terms of its moments: horizontal position, horizontal width, and elongation) is the most important information needed. It is therefore important to design a neutron camera system equipped with detectors offering the best performance in terms of accuracy and stability. Ion temperature information is also needed for the elaboration of the data but this will be provided by the spectrometry measurement itself. A high dt countrate is essential for this purpose.

The instrument to be used in this diagnostic is a double neutron spectrometer of the MPR type. A prototype is in use at JET [5]. The performance specifications needed for the density diagnostic are essentially the same as those of the prototype spectrometer. The main modifications refer to the spectrometer for the dd neutrons which is a scaled down version of the prototype with increased energy bite (±30 % instead of ±20 %); it would normally be run at an energy resolution setting of $\Delta E/E$=10 %. The JET experiment will test the diagnostic in most respects for the ITER application but for the fact that the magnetic stray fields will be an order of magnitude higher at ITER. The operation and the acceptability of magnetic instruments like MPR spectrometers is an important issue for ITER and an interesting technological topic for further study of principle solutions and their practicality.

Interface issues are the ultimate factor determining the feasibility of the measurements. Here we only point out that the tangential sight line required for the measurement would be similar to one for a NBI installation. Other interface issues would also be similar to those of NBI including magnetic interference effects between the machine and the diagnostic. It is therefore possible that interface solutions common to the NBI system and the tangential neutron spectrometer will be found.

There is no alternative to a tangential view for the measurement. A radial view would give a s/b ratio that is worse by a factor of 3 compared to the reference case with a tangential view which is unacceptable. A vertical view would be intermediate between a tangential and a radial view in terms of s/b ratio. However the penetration through the blanket would probably have to be narrow and would represent a limiting factor on the signal intensity.

Assuming a tangential view can be accomodated, there would still be engineering restrictions on the size of the aperture through the blanket. For instance, if the allowed aperture diameter is reduced to 20 cm, the dt and dd countrates are reduced by a factor of 2.5 and 10, respectively, relative to the values given before. This means the dd countrate would be $C_1^{MAX} \approx 1.5$ kHz which falls somewhat short of the target measurement requirements in terms of time resolution. A diameter of 30 cm would improve the performance of the spectrometer by a factor >2 as measured in terms of the dd and dt signal countrates.

The space resolution of the measurement is limited by the fact that the measurement is line integrated; its resolution depends on the effective width of the emissivity profile intercepted along the viewing line more than on the aperture of the viewing line itself. In the case of a tangential view the dependence of the resolution on the profile width is weaker that for a radial view. Another comment is that the measurement is intrinsically coupled to the plasma volume where the fusion reactions occur, which is where the fuel densities need to be controlled. Therefore, the space resolution is not a serious reason preventing this measurement from being used for tokamak control functions.

CONCLUSION

This study supports the findings of the earlier study [2] that fuel densities can be measured with a tangential neutron spectrometer for burning plasma conditions with typical values for statistical and systematic errors of ±10 %. The systematic error is lower (±5 %) for the density ratio mainly because it does not depend on knowledge on the plasma chord length Λ (Eq. 3 and Table 1). One may note though, that input information is essential for extracting n_d and n_t, and it is important that this can be provided by the neutron camera for the wide profiles now predicted for ITER operation. As a reference for the performance assessment the case of balanced fuel densities ($n_t/n_d=r$) has been used here. In general, there is nothing preventing this density diagnostic being used for plasma conditions of low densities $r<1$ down to the $r<<1$ values of D plasmas with only traces of tritium. In principle, it can also be used for conditions of $r>>1$ should it be of interest to operate in tritium plasmas (using the tt-component as indicated above). For $r>2$ the measurement is still possible but rather long sampling times are required to achieve a 10% statistical accuracy. The ultimate feasibility check is believed to be the interface in two respects: (1) the accommodation of a tangential sight line, possibly connected with a water enhanced viewing dump; (2) the accommodation of the magnetic material of the MPR neutron spectrometers on ITER requiring a stealth technique to make the instrument invisible to the plasma magnetic field and the protection against the influence of the stray field on the diagnostic.

ACKNOWLEDGEMENT

This paper has been prepared as an account of work assigned to the European Home Team under Task Agreement number G 55 TD 06 within the Agreement among the European Atomic Energy Community, the Government of Japan, the Government of the Russian Federation, and the Government of the United States of America on Cooperation in the Engineering Design Activities for the International Thermonuclear Experimental Reactor ("ITER EDA Agreement") under the auspices of the International Atomic Energy Agency (IAEA).

REFERENCES

1. J. Källne, Neutron diagnostics for burning DT plasmas of ITER/JET, Uppsala University Report UU-NF 91/1, Uppsala University, Uppsala, Sweden (1991).
2. J. Källne, P. Batistoni and G. Gorini, *Rev. Sci. Instrum.* 62:2871 (1991).
3. P Antozzi et al, *Nucl. Instrum. Meth.* A368:457 (1996).
4. Documentation for CCC-200A/MCNP Code Package, NEA Data Bank, Gif-sur-Yvette, France (1991).
5. J. Källne, E.Traneus and the NPF Team, Neutron Spectrometry in JET and Next Step Implications, these Proceedings.

IMPROVEMENT OF THE NEUTRON ENERGY SPECTROMETER, COTETRA (COUNTER TELESCOPE WITH THICK RADIATOR)

M. Osakabe[1], M. Sasao[1], M. Isobe[1], S. Itoh[2], T. Iida[3],
A.L. Roquemore[4], L.C. Johnson[5]

[1]National Institute for Fusion Science, Toki, Gifu-509-52, Japan
[2]Dept. of Nuclear Engineering, Nagoya University, Nagoya 464-01, Japan
[3]Dept. of Nuclear Engineering, Osaka University, Suita 560, Japan
[4]Princeton Plasma Physics Lab., Princeton, N.J., USA
[5]ITER Joint Central Team, San Diego, CA, USA

INTRODUCTION

A new type of neutron spectrometer, named COTETRA (counter telescope with thick radiator), had been developed, and used to measure the 14MeV neutron energy spectra on the DT experiment of TFTR. It showed the energy resolution of about 9.5% on the laboratory test[1], but the peak width of the energy spectrum from a NBI heated plasma of TFTR was 23%, wider than the expectation (11%)[2].

Following problems have been pointed out as causes of the resolution degradation,
a) fluctuation of the DC levels of signal lines,
b) degradation of the photomultiplier tube (PMT) gain due to high counting rates,
c) change of the PMT gain due to the stray magnetic field.
d) the effect of the back foil of the ΔE-scintillator
e) proton scattering on the side wall of the holder

The fluctuation of DC levels was monitored simultaneously during the shot, and it was found to be negligibly small. In the present paper, the results of the PMT improvement and the evaluation of the effects d) and e) are reported.

MAGNETIC-FIELD RESISTIVE PMT AND ACTIVE DIVIDERS

The COTETRA is a ΔE-E type proton recoil telescope and uses a relatively thick plastic scintillator as a radiator. The light output from the ΔE detector is detected by two PMT's at the side, and that from E detector is also received by a PMT. The energy of the incoming neutron is obtained by the sum of the three PMT outputs. Original version of COTETRA uses HAMAMATSU R372-02, which has high counting rate capability with excellent linearity .

During TFTR discharges, the degradation of PMT gain due to the effect of magnetic field and the saturation of PMT was observed. To overcome this defect, a magnetic-field resistive PMT, HAMAMATSU R2490-07 was tested. An active divider[3,4] was developed for this PMT to avoid the saturation.

Figure 1 shows the electronic circuit of the devider. The divider current was driven by seven pairs of inverted darlington transistors at the last seven stage, and each dynode voltage was kept by reference resistors (R3). When the high voltage of -2kV was applied to the divider without any light source, the current of 3.63mA was supplied.

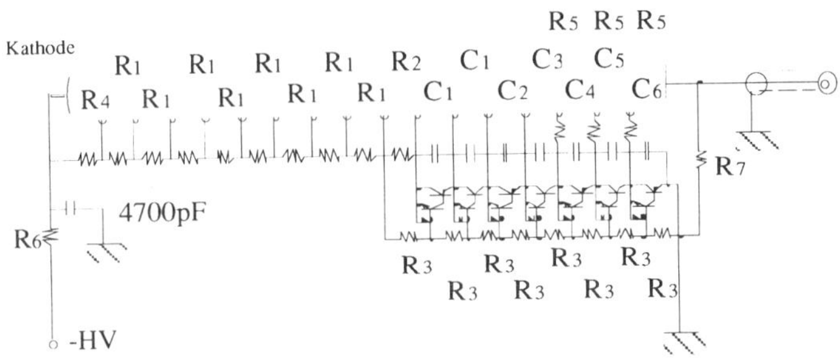

Fig.1 Schematic drawing of the electronic circuit for the active divider. The character R and C express resistors and capacitors, respectively. The value of each resistors and capacitors are; R1 (33kW), R2 (34kW), R3 (1MW), R4 (84kW), R5 (50W), R6 (3.9kW), R7 (10kW), C1 (0.01mF), C2 (0.022mF), C3 (0.047mF), C4 (0.1mF), C5 (0.22mF), and C6 (0.47mF).

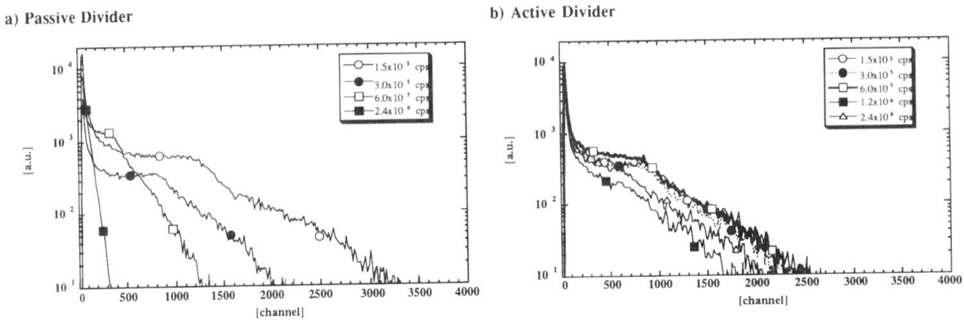

Fig.2 Pulse height spectra of NE102A for D-T neutrons.
a) Spectra obtained by a passive divider. (b) Spectra obtained by the active divider of Fig. 1.

Using OKTAVIAN D-T neutron generator at Osaka University, the performance of this active divider was tested and compared with the normal (passive) resistive chain divider, which is available from HAMAMATSU Photonics. Fig. 2(a) shows the pulse height spectrum (PHS) of NE102A plastic scintillator bombarded with D-T neutrons, using a passive divider, and Fig.2(b) shows that obtained by an active divider. The counting rate of the detector was measured when it was low, and those for higher cps were estimated by comparing the D-beam current at the T-target of OKTAVIAN. As the counting rate of the detector increases, the edge position of the PHS is decreasing from $\approx$1500 channel to $\approx$300 channel in the case of the passive divider as shown in Fig.2(a). On the other hand, those of the active divider stay at $\approx$1000 channel up to $6x10^5$ cps and the deterioration of edge position is small even for the case of $2.4x10^6$ cps. This shows the effectiveness of the active divider for the use of high counting rate environment. The neutron spectrum from the DT generator (OKTAVIAN) measured with these new PMT's and deviders are shown in Fig3. The peak shows an asymmetric shape with a tail on the low energy side. The FWHM estimated from the high energy side is 15% , larger than the that of the original COTETRA with R237-01. The larger value of FWHM is due to the poorer linearity of the new magnetic resistive PMT.

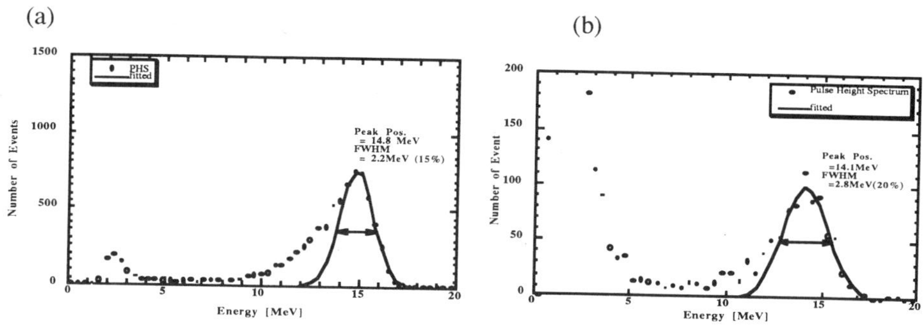

Fig.3 The 14 MeV neutron spectrum of COTETRA with R2490-07, and active deviders of Fig.1, from the D-T generator (a), and from the NBI-heated DT plasma (b).

The new set of PMT's were used on a series of TFTR-DT experiments of 1995. In Fig. 3(b) is shown the neutron spectrum from a NBI-heated DT plasma. The FWHM of 2.8 MeV can be explained by the expected Doppler broadening of beam-plasma interaction and the peak shape in Fig. 3(a).

BACK FOIL AND SIDE WALL EFFECTS ON THE LOW ENERGY TAIL

In an ideal case, 14MeV neutron events form a straight line on a 2-dimensional map of ΔE-E pulse height. However, in the case of Fig. 3, events of proton energy deposit in the E-detector higher than 8.8 MeV show a different slope from those of $E \leq 8.8$ MeV, and the latter form the tail on the low energy side. The peak shape which was composed with events of $E \geq 8.8$ MeV is symmetric, as shown in Fig. 4.

Possible cause of the tail is effect of the back foil thickness of ΔE scintillator, or the proton scattering on the side wall of the holder cylinder of ΔE- detector. The effect of the foil thickness on the peak shape was tested with the DT generator, as shown in Fig. 5. The difference between the spectrum without the backfoil (open circles) and that with it (open squares) is small.

A new holder has been designed so as to diminish the proton scattering effect and tested on TFTR. The DT neutron spectrum measured with this new holder is shown in Fig.6. The spectrum is symmetric and the FWHM is 2.4 MeV, which is consistent with the expected Doppler broadening.

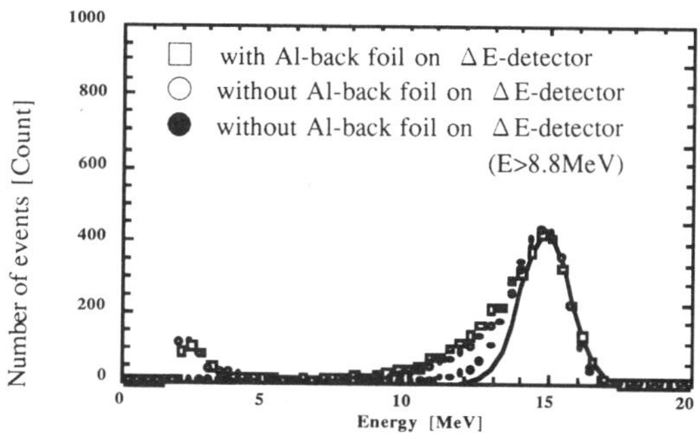

Fig.4 COTETRA neutron spectra with differnct back foil thickness, and without it measured with a DT generator.

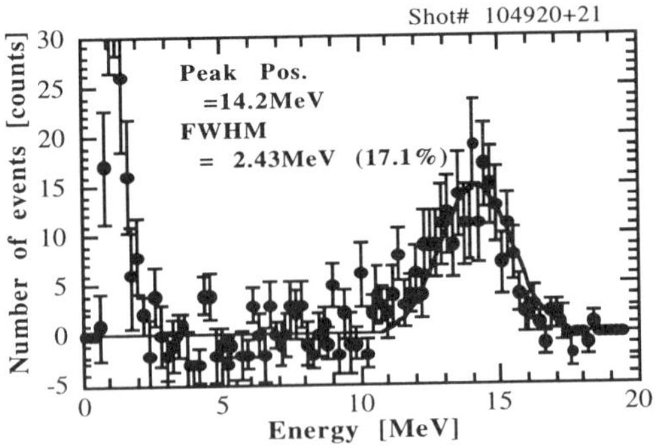

Fig.4 D-T neutron spectrum from the NBI-heated DT, with new ΔE detector holder.

CONCLUDING REMARKS

Several problems, such as the counting capability of PMT, Magnetic field effects, proton scattering on the sideholder, of COTETRA have been solved. The peak shape of 14 MeV neutrons from TFTR is symmetric, and consistent to the theoretical prediction and the laboratory test. Present version of COTETRA has an intrinsic resolution of 17 %, which is mostly determined by the PMT linearity. By improving the circumstance against magnetic stray field, the resolution less than 8% can be expected. Although this detector has worse energy resolution resolution than a diamond detector, it can be considered to be an economical, radiation resistive detector for a profile measurement on ITER.

Authors wish to thank all the stuff of TFTR for encouragement, and for their help and adequate suggestion on the TFTR experiment, specially to Dr. K. M. Young and Dr. D Johnson. We also wish to Dr. N.Sasao at Kyoto Univ. and Dr. G.A.Wurden at LANL for their useful suggestions on the active divider, and to Prof. T.Iguchi and Dr. J. Kawarabayashi and the stuff of the OKTAVIAN for their help at the test of the dividers.

REFERENCES

[1] M.Osakabe, et. al., Rev. Sci. Instrum. **66**, 920 (1995)
[1] M.Osakabe, et. al., Rev. Sci. Instrum. **66**, 920 (1995)
[3] G.A. Wurden et al., Rev. Sci. Instrum. 66 (1995), 901
[4] Y.Takeuchi, Mas. Thesis, Kyoto Univ. (1996)

EXPERIMENTAL ESTIMATION OF THRESHOLD PROPERTIES OF THE BUBBLE CHAMBER FOR KNOCK-ON NEUTRON MEASUREMENTS

V.A. Agureev,[1] V.P. Romanikhin,[1] V.G. Rumyantsev,[1]
G.N. Cheryomukhin,[1] S.V. Trusillo,[1] V.S. Zaveryaev[2]

[1]Russian Federal Nuclear Centre - All-Russia Research Institute of Experimental Physics, 607190 Sarov
[2]RNC "Kurchatov Institute" - Nuclear Fusion Institute, 123182 Moscow

INTRODUCTION

The problem of the high energy DT neutron spectrum tail measurement has been already discussed[1,8]. The importance of this problem for ITER plasma diagnostics caused us to search for an effective threshold detector capable of coping with this problem. The fundamental difficulty of this task lies in the necessity to detect a weak knock-on neutron flux close to the extremely high peak of thermonuclear DT neutrons.

Earlier we proposed a special bubble chamber[2,6] having the adjustable energy threshold level E_T. The threshold has been predicted theoretically[2], and its value can be easily controlled with the main parameters of a bubble chamber such as a temperature and a lower pressure.

The detector energy threshold steepness defines the extent to which DT neutrons detection may be suppressed. It is the reason why the detector threshold properties should be well known. In this paper we present the main results of experimental investigation of bubble chamber threshold properties.

EXPERIMENTAL

To get a trustworthy answer on whether a bubble chamber-based detector can be applied for the knock-on neutron measurement it was decided to estimate the DT neutron detection efficiency if the energy threshold is 14 MeV or over. DT neutrons were generated with a pulse neutron source based on the plasma focus device[7]. To achieve reliable results there were produced about 100 neutron pulses with the duration of 40 ns and the output from 10^7 to 10^{11} n/pulse.

Since both DT and DD neutron sources were available, the next ratio could be measured rather accurately

$$R(E_T) = \varepsilon(E_T)/\varepsilon(E_{2.45}),$$

Diagnostics for Experimental Thermonuclear Fusion Reactors 2
Edited by Stott *et al.*, Plenum Press, New York, 1998

where R and ε are relative and absolute detection efficiencies, E_T is the threshold energy.

In fact the rate of increase in the working liquid optical density was measured instead of the absolute detection efficiency because they are proportional. All the measurements were carried out using a bubble chamber previously designed for neutron imaging of plasma focus devices[7]. The chamber had a temperature stabilizing system and a fast valve to release a pressure inside the chamber; its working liquid was Freon-12 (CF_2Cl_2).

Unfortunately, the bubble chamber construction did not allow for the precise setting of the lower pressure defining the energy threshold level, therefore to control it the following was used:

1. Adjustment of the initial pressure levels P_1 and P_2 which corresponded to two different periods of the release valve opening;
2. Adjustment of the gas volume dM released during the valve opening.

Several series of measurements with different variants of energy threshold control allowed the results to be compared, and hence provided reliability of measurements. In order to use these parameters practically one has to study first how the energy level depends on them.

First of all a theoretical relationship between the relative detection efficiency and the threshold levels $R(E_T)$ was calculated within 2.5 - 14 MeV. It was done based on recoil nuclei spectra[3], energy losses of the recoil nuclei in the bubble chamber working liquid[4], and the bubble growth rate as a function of the temperature and the pressure[5]. This theoretical relationship $R(E_T)$ is shown by the solid line in Figs.1 and 2.

Further the bubble chamber was irradiated with a pulse DD neutron flux. Adjusting a control parameter, we achieved the zero rate of the optical density increase. It meant that the energy threshold was 2.45 MeV.

Leaving the energy threshold unchanged, the bubble chamber was irradiated with a pulse DT neutron flux, and the optical density increase rate was measured. Then the control parameter was changed and the optical density increase rate was measured again. In such a way the relative detection efficiency R of the bubble chamber was obtained.

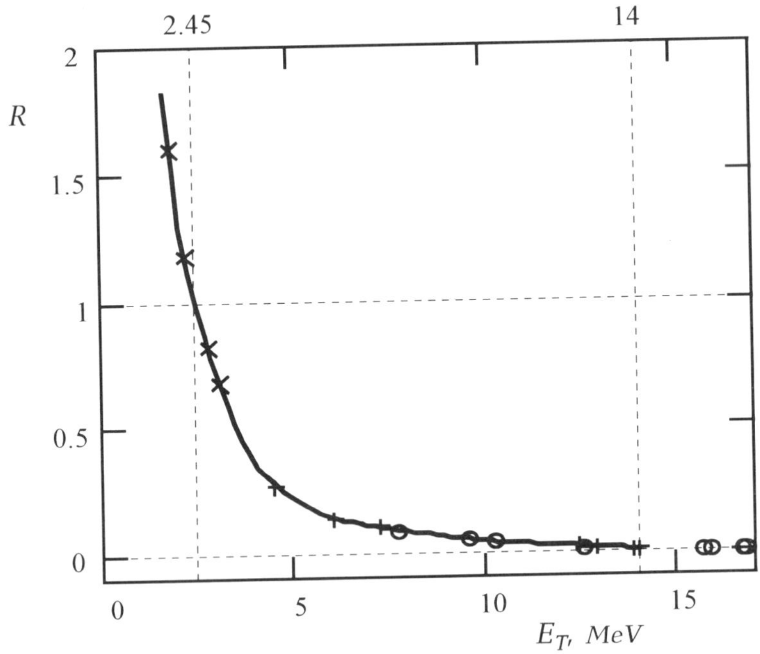

Figure 1. Theoretically calculated relative efficiency $R(E_T)$ with experimental points.

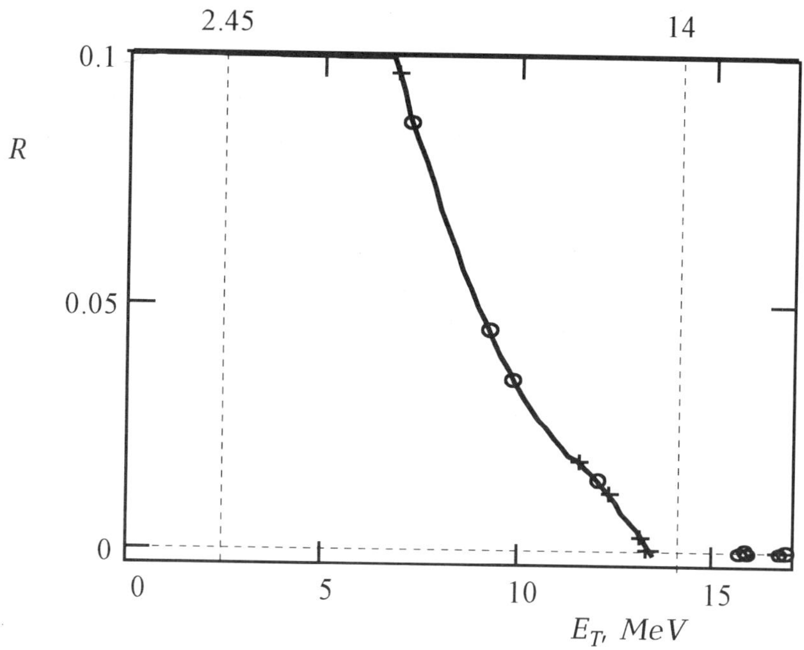

Figure 1. Theoretically calculated relative efficiency $R(E_T)$ with experimental points taken close to 14 MeV.

Making the same procedure for other control parameters, we found all the functions $P_1(R)$, $P_2(R)$ and $dM(R)$ versus the detection efficiency R down to $R < 10^{-2}$. In order to derive the information about the energy thresholds out of these functions, the experimental points were marked on the mentioned theoretical function $R(E_T)$. Applying the function as a scale we could get the series of energy threshold values. The series was treated with a Least Squares Technique and smooth functions $P_1(E_T)$, $P_2(E_T)$ and $dM(E_T)$ were obtained. Such a method provided a way of setting the energy threshold of the bubble chamber with a standard deviation less than 0.6 MeV.

To evaluate the suppression degree of DT neutrons, the detection threshold levels were set in the range from 15 to 18 MeV. After that the relative detection efficiency R was measured with the maximal neutron flux and the utmost value of sensitivity. The latter was set in such a way that it was possible to measure the rate of the optical density increase caused by only one bubble.

DISCUSSION

No DT neutron signal has been detected starting with the energy threshold greater than 16.5 MeV, which corresponds to the relative detection efficiency R less than 10^{-7}. This result has been obtained with the neutron flux density up to $10^{15}/cm^2s$, which is six orders higher than the DT neutron flux density inside ITER neutron cameras [6].

It is known that the knock-on neutron spectrum extends to 20.5 MeV and its intensity is five orders less than that of thermonuclear neutrons[6]. In the range from 16.5 to 20.5 MeV the average calculated efficiency is about 10^{-3} if the threshold level equals 16.5 MeV, which provides the bubble density in the chamber to be close to $100/cm^2s$. This is quiet sufficient for successful detection of knock-on neutrons on ITER using the bubble chamber designed according to the proposal [6].

REFERENCES

1. R.K. Fisher, et al., *Nucl. Fusion* 34, 1291 (1994).
2. V.A. Agureev and S.V. Trusillo, *Proc. Int. Workshop on Diag. for ITER*, Varenna, Italy, 1995.
3. A.M. Baldin, et al., *Nuclear Reactions Kinematics*, Gosizdat, Moscow, 1959 (in Russian).
4. M.S. Nemets, *The Heavy Ions Ranges*, Naukova Dumka, Kiev, 1984 (in Russian).
5. Y.A. Alexandrov, et al., *Bubble Chambers*, Indiana University Press, Bloomington, Indiana (1967).
6. R.K. Fisher, S.V. Trusillo, V.S. Zaveriaev, *Rev. Sci. Instr.* 68 (1997) 1103.
7. S.V. Trusillo, et al., *JET Lett.* 33 (1981) 148 (in Russian).
8. R.K. Fisher, et al, *Diagnostics for Experimental Thermonuclear Reactors*, ed. by P.E. Scott et al., Plenum Press, New York (1996).

DESIGN CONSIDERATIONS FOR NEUTRON ACTIVATION AND NEUTRON SOURCE STRENGTH MONITORS FOR ITER

Cris W. Barnes

Los Alamos National Laboratory
Los Alamos, NM USA 87545

D. L. Jassby, G. LeMunyan, A. L. Roquemore

Princeton Plasma Physics Laboratory
Princeton, NJ USA 08543

Chris Walker

ITER Joint Central Team, Garching Joint Work Site
Garching, Germany

INTRODUCTION

The International Thermonuclear Experimental Reactor[1] will require highly accurate measurements of fusion power production in time, space, and energy. Spectrometers in the neutron camera could do it all, but experience has taught us that multiple methods with redundancy and complementary uncertainties are needed. Previously, conceptual designs have been presented for time-integrated neutron activation[2] and time-dependent neutron source strength monitors[3], both of which will be important parts of the integrated suite of neutron diagnostics for this purpose.[4] The primary goals of the neutron activation system are: to maintain a robust relative measure of fusion energy production with stability and wide dynamic range; to enable an accurate absolute calibration of fusion power using neutronic techniques as successfully demonstrated on JET[5] and TFTR[6,7]; and to provide a flexible system for materials testing. The greatest difficulty is that the irradiation locations need to be close to plasma with a wide field of view. The routing of the pneumatic system is difficult because of minimum radius of curvature requirements and because of the careful need for containment of the tritium and activated air. The neutron source strength system needs to provide real-time source strength vs time with ~1 ms resolution and wide dynamic range in a robust and reliable manner with the capability to be absolutely calibrated by *in-situ* neutron sources as done on TFTR[8], JT-60U[9], and JET[10]. In this paper a more detailed look at the expected neutron flux field around ITER is folded into a more complete design of the fission chamber system. Overall issues of neutron calibration for ITER will be presented elsewhere[11].

Neutron Activation

Irradiation End Location

The locations of the neutron activation irradiation ends are a critical choice in creating a working, maintainable system that can be accurately calibrated using neutronics modeling of the local geometry. In the previous Phase II design[2] a location in the front surface of the blanket/shield module (BSM) was desired. However, pneumatic access to any such location through the BSM would be very difficult, and the resulting "unique" BSMs were rejected because of the heavy requirements on remote handling and installation procedures that made special modules extremely undesirable. As the design of the BSMs continues to evolve, the need for "toroidal filler modules" to plug gaps between BSMs from the curvature of the torus has been recognized. Placing neutron activation irradiation ends in such special toroidal filler modules at the intersection of four BSMs should allow remotely maintainable units with a good view of the neutrons from the plasma.

The primary design concern for the neutron activation system is achieving the necessary wide angle of view of the neutron emitting region. If the field of view to a significant number of neutrons is cut by the edge of a BSM, then the geometry of the activation location relative to that edge becomes the dominant effect in the neutronics uncertainty. This was the case on TFTR where the location of the RF limiter relative to the re-entrant irradiation end determined the toroidal cut-off of the field of view and created a 5% uncertainty.[6] The field of view should clearly include the center of the plasma and out to approximately the half-radius, and also a considerable distance toroidally. If one wants to see ±1.4 m at 2.8 m distance one needs an angular field of view of 26°. If the activation location is set back a distance d from a gap of radius 10 cm with corner of radius e the viewing angle is min(atan(10/(d-e)),atan((10+e)/d)). For e~7.5 mm, d must be less than 29 mm to achieve this 26° angular field of view; for e~10 mm then d must be less than 32 mm. This simple calculation says the activation location needs to be within 2-3 cm of the front face of the BSM.

Figure 1 is a scale drawing of a typical proposed irradiation end in a toroidal filler module. An actual Catia CAD working model of the BSMs was used as the source. One nice aspect of this drawing is that the rounded corners of blanket are significant and give a better field of view than the above calculation would imply. As in the previous Phase II design, the pneumatic transport for ITER will be 1.1 cm inner diameter tubes carrying 1 cm diameter capsules. A coaxial design at the irradiation end as used on TFTR with a 2 cm outer diameter allows the return air to be transmitted to the inner tube. Figure 2 shows a close-up of the irradiation end. While the inside of the irradiation end does not need to be in vacuum, a secondary vacuum seal and tritium containment is needed further down the pneumatic system (see the next section).

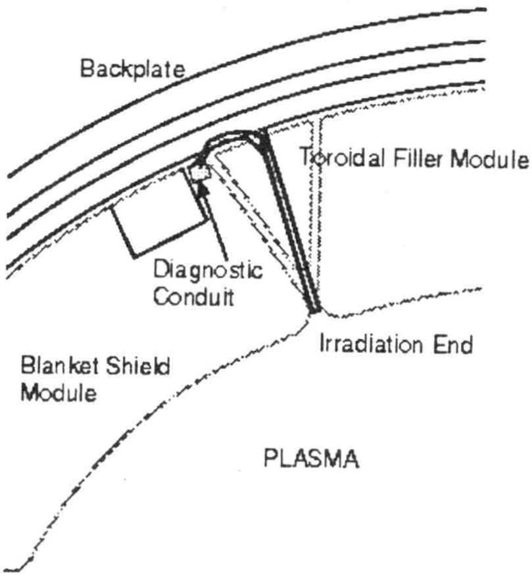

Figure 1: Schematic of a Neutron Activation Irradiation Location in a Toroidal Filler Module. A tube of radius-of-curvature greater than 10 cm is drawn going into the backplate diagnostic conduit.

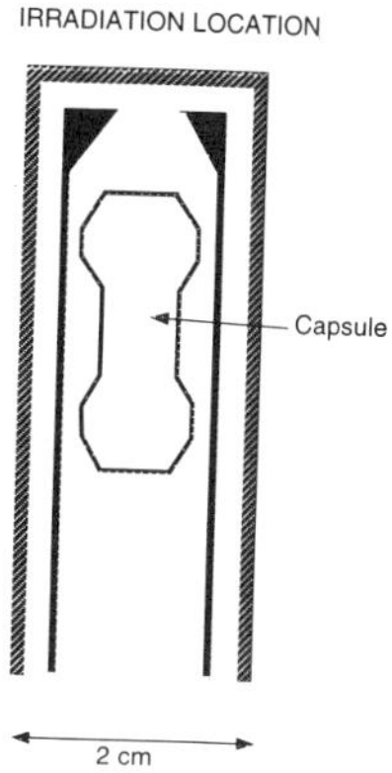

Figure 2: Close-up view of neutron activation capsule inside of irradiation end. The capsule self-plugs the return air hole at the end, providing a pressure change signal used to determine arrival of the capsule.

A critical issue will be heating of irradiation end components and cooling. The nuclear heating into the toroidal filler module region has been analyzed.[12] Nuclear heating will be 10 W/cm^3 in this region and the plasma heating probably only 10%-20% of that. The inner wall of the irradiation end is made quite thin for low total nuclear heating. Thick connecting "fins" from the inside tube to the outside transfer heat; the outer tube is then mounted to the BSM for thermal contact. There exist "lucky" and "unlucky" BSMs depending on whether the neighboring BSM also needs to be removed during remote handling because of the system of keys. The irradiation ends would be attached to "unlucky" BSMs to provide thermal contact. Then during remote handling procedures the neighboring BSM will have to be removed anyway, allowing access to the toroidal filler modules with irradiation ends. There is still some worry about the nuclear heating of the capsule itself which will be in poor thermal contact with anything. Presently we do not plan to have any significant flow of pneumatic gas past the capsule; on the contrary, we hope to use plugging of the channel by the capsule and resulting pressure changes as a robust radiation-insensitive monitor of capsule arrival at the irradiation end. We still may be able to use polyethylene since it may not get too hot. Any capsule components will need to be made of low activation material. The tube itself could be 316 SS or Inconel or Tungsten or even possibly copper; eventually the relative thermal coefficients of expansion will need to be considered in a detailed design.

Which exact filler modules are to be used needs to await the final design of the BSMs. The poloidal angles of the four irradiation locations at each of two toroidal angles would be driven by the sensitivity issues raised in Ref. 2. The toroidal angles will be chosen within constraints of use of the backplate diagnostic conduits by other systems and routing concerns (see next section).

Routing of Pneumatic System

We have considered whether the system should be hydraulic (using water to push the capsules) rather than pneumatic. Water does gives one better thermal conduction. But air moves the capsules much faster, can operate at lower pressure, and is easier to "airlock" to provide secondary vacuum and tritium containment. Our actual preference is to use helium which has good heat transfer, reduced problems on the vacuum if the system fails, low activation, and good availability in the ITER plant.

The critical issue in routing the pneumatic system is the minimum radius of curvature allowed and available. Consider a capsule of diameter d and length l in a hose of inner diameter d+ε. Choose a radius of curvature R such that the hose just touches the ends of the capsule. For ε/d~10% one can support about a 10 cm radius of curvature for a 1 cm capsule. (This translates to minimum 10 inch radius of curvature for 1 inch capsules in 1-1/8" I.D. tubes on TFTR, which agrees with experience.) Thus our pneumatic routing scheme will be plausible as long as we can negotiate any necessary turns with greater than 10 cm radius of curvature.

The initial problem is getting out of the irradiation end in the toroidal filler modules. On Figure 1 is illustrated a path with the minimum radius of curvature defined above. The irradiation end will be centered over a feature in the backplate known as the backplate diagnostic socket (see paper by C. Walker in these proceedings for figure and details[13]). The tube can then curve into the backplate wiring conduit[13] (actually a conduit in the BSM) and following its routing. This backplate wiring conduit was initially designed for electrical diagnostics (such as magnetics) and is also used by the bolometer system. Thus our principle is to use features already existing in the backplate, and then don't be affected by small changes in its design.

The route can get out of the wiring conduit at water manifolds either at the top or midplane of the machine. Figure 3 shows the present backplate design, both in cross-section and 3-D view, illustrating that there is plenty of room to pass the 1 cm tubes through. An extra pipe can be placed through the double wall construction for the pneumatic tubes. If we have 20 mm diameter pipes (12 mm ID) then we can have two of them (transfer and return air) in a 50 mm diameter sheath; it could probably be even smaller.

At least 2 toroidal locations are desired to have poloidal arrays of irradiation ends for redundant backup. We propose to put this system in toroidally at locations of the so-called "in-situ welds of multi-sectors"; that is, sector joints that are welded together in the pit rather than welded before being lowered into place. Electrical diagnostics should not be on

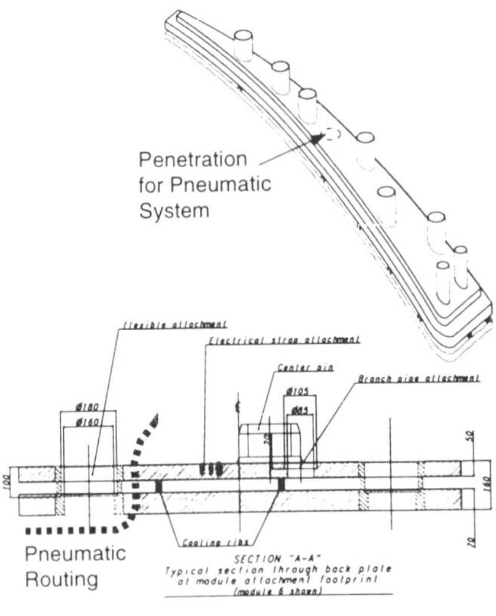

Figure 3: Views of backplate near vertical pipe assemblies showing where activation system pneumatic routing exits from primary vacuum vessel.

this critical path of the construction, but our pneumatic system can be. Eight out of the 20 sector welds are like this. We could conceptually use any pair of in-situ weld sectors and collect the tubes toroidally in the Heat Transfer System (HTS) Vault and thence to Diagnostic Hall (on south side). Sectors 13 and 17 on the south side are our initial preferred choice.

Routing of the pneumatic system into either the HTS vault or the neutral beam injection (NBI) vault provides a location for airlocks and secondary tritium containment. The need for two different toroidal locations implies they (probably) both cannot use NBI vault and also be at in-situ welds, besides which the NBI vault only applies to routing out near the midplane which is problematic for inboard locations. Hence the primary design route is along vertical water manifolds to the HTS vaults. Once into the vertical pipe extensions our pneumatic tubes can follow the same routing as the water pipes[14], using their primary and secondary containment.

Once the tubes penetrate into the HTS vault the problems of providing secondary containment and removing activated air from inside the pneumatic tubes must be solved. Figure 4 is a schematic of the pneumatic system. Capsules are routed to and from loaders, drop outs, and counting systems in the Neutron Activation Laboratory in the Diagnostic Hall to a special room with a "carousel" multiple-in / multiple-out queuing device. The carousel room in the tokamak building would be "off-limits" usually, but the analysis room in the diagnostic hall probably only needs partitions to provide controlled access and distance from the counting stations. All the lines from the carousel to the tokamak go through a single penetration into the HTS vault, and are then routed toroidally[15], all using insulated (plastic) piping to prevent shorting sectors. The radiation dose inside the HTS vault should not be any greater than the TFTR Test Cell itself, and this non-metal piping will not fail from radiation damage. Inside the HTS vault are located "shoebox-size" airlock devices which provide primary containment and ability to purge the air lines. We

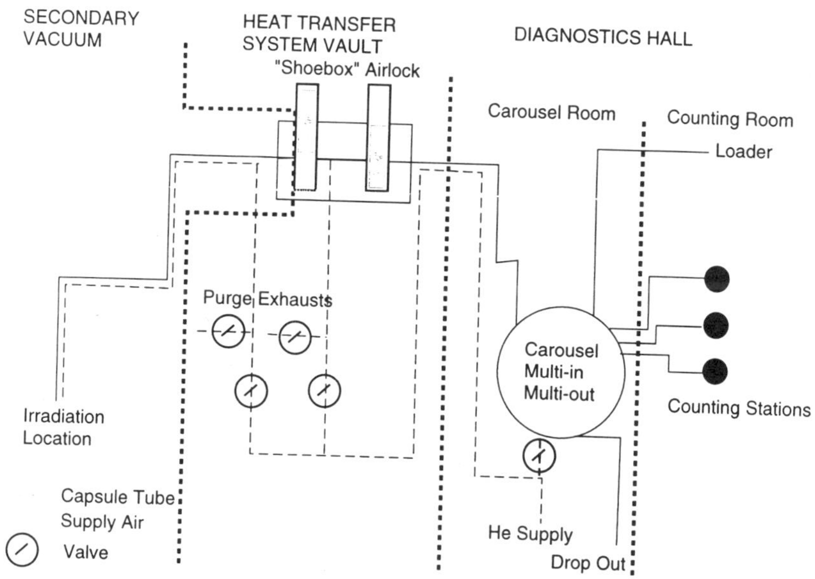

Figure 4: Generalized layout of pneumatic system from counting room to carousel to one irradiation location illustrating tritium containment and activated air control. The "shoebox" airlocks feature two valves each with three positions: open, closed with air holes, and fully closed and sealed. Activated air is purged via exhausts in the Heat Transfer System vault.

thus plan to vent the activated air in the pneumatic system into the HTS vault.

There are other options available for variations of the pneumatic routing if necessary. There are "HTS upper and lower communication shafts" or "pipe chases" which run between the upper and lower HTS vaults (to allow twice the volume in loss-of-coolant accident). These provide communication between top and bottom cooling systems. Cooling water from the "outer blanket limiter" (outboard below midplane port) and from the divertor go *down* to the lower HTS vault, but the piping at the midplane goes to the pipe chases and then down (or for the neutron activation system up to connect to pipes at top). A possible alternative location for the airlock shoeboxes would be on the walls of the pipe chase at the cryo(pump) handling level or diagnostic level and thence out. Exact details can be worked out during construction as long as the concepts are sound and space for the airlock shoeboxes is available.

An issue recognized during the conceptual design was how to measure the presence of capsules in the pneumatic system in a way that could withstand the radiation and heat of the environment. Operation of the pneumatic system on TFTR has led us to recognize that pressure changes occur when the capsules reach the irradiation ends and "stopper" the flow of gas. This method can provide a very robust measurement technique for arrival and departure timing and determination of the presence of the capsule at the irradiation end. More usual electrical and optical techniques can be used at the airlocks and in the HTS vault and beyond where the environment is not as intense.

A similar system needing pneumatic access to the inboard side is the pellet injection system. However, that system uses a smaller pipe, probably needs a much larger radius of curvature, and only needs one or two specific routings and cannot be generalized to multiple poloidal locations. Its primary vacuum seal must be pulled out to the pellet injection cask, and not just its secondary seal. Thus the pellet system is more difficult and not as general, and cannot be used as a template for the neutron activation system.

Key Outstanding Issues

Before final design and construction of the neutron activation system, there are still several outstanding projects that must be addressed.

1. The basic design must be modeled in the MCNP neutronics code and its sensitivity to location determined. The field of view and its uncertainties must be carefully considered.
2. A thermal model should be developed to confirm that sufficient heat transfer to the BSMs and the toroidal filler module itself and their cooling is achieved so that no active cooling of the irradiation end is needed.
3. Each irradiation location will need to have its pneumatic system routed through the final backplate design. No "jogs" or bends of small radius of curvature can be allowed, and there may need to be a special circumstance or two.
4. In particular, it is not clear from present drawings whether irradiation ends on the outboard lower side can be routed past the midplane ports to be collected in the upper vertical pipe assembly. A special routing out at the midplane ports and up through the pipe chases may be necessary.
5. Neutron Activation Laboratory space for both the analysis / counting room and for the carousel needs to be allocated by Naka Joint Work Site team in the Diagnostics building.

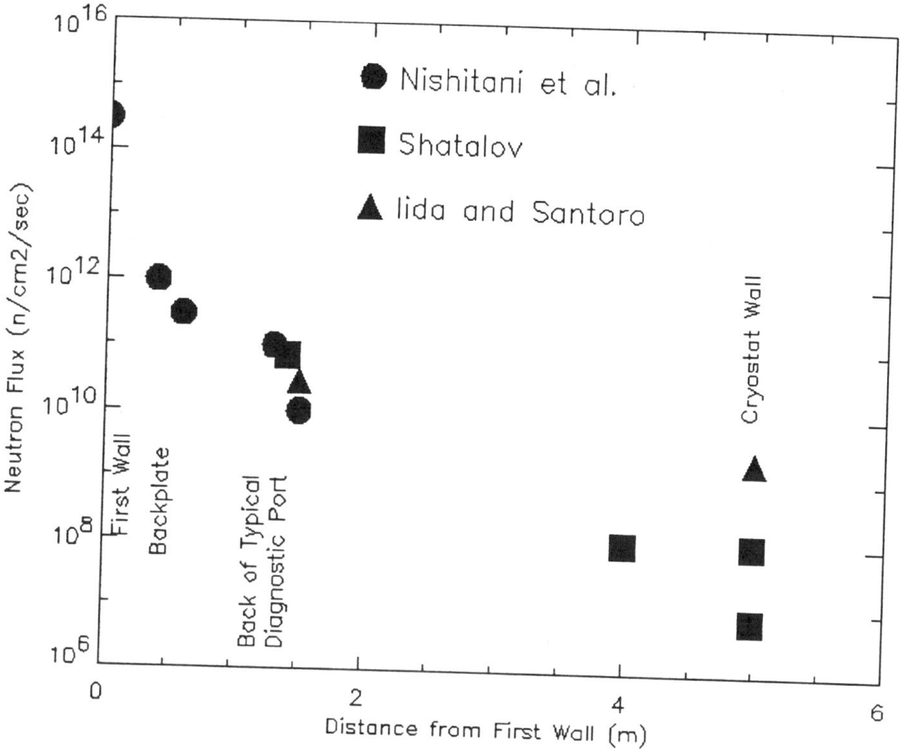

Figure 5: Summary of expected neutron fluxes at full power (typically 1 MW/m^2 at wall) at different distances from the ITER plasma.

Neutron Source Strength Monitors

Location of Detectors in Neutron Field

Further detailed design of the neutron source strength monitor system has required some estimate of the local neutron flux within the bioshield of ITER. Figure 5 summarizes a variety[16,17] of MCNP neutronics calculations done for ITER. There is an outstanding question of whether there is more than 1 order of magnitude drop in (total) neutron flux from outside of the vacuum vessel to the bioshield and whether the source strength monitors will need to be placed behind toroidal field coils to get more shielding. However, we believe that we will *not* need the shielding of the TF coils and can use a limiter (remote handling) port shield plug instead.

As argued previously[3], 7 orders of magnitude detector efficiency dynamic range is needed in the system with a factor of no more than 25 between sensitivity of detectors, and hence 6 detectors in a complete set. We need three different locations because we cannot get the needed dynamic range of detectors in one physical location. One detector (the most sensitive) must go "close" to the plasma to allow *in-situ* calibration. As long as a detector responds to neutrons from at least one toroidal field "bay" or more (and is not effectively collimated to significantly less than one bay) the calibrations of that detector should be meaningfully related to the total plasma volume emission. We plan to place two 200mm x 50mm fission chamber detectors in the neutron collimator pre-shield[18] following lines of a

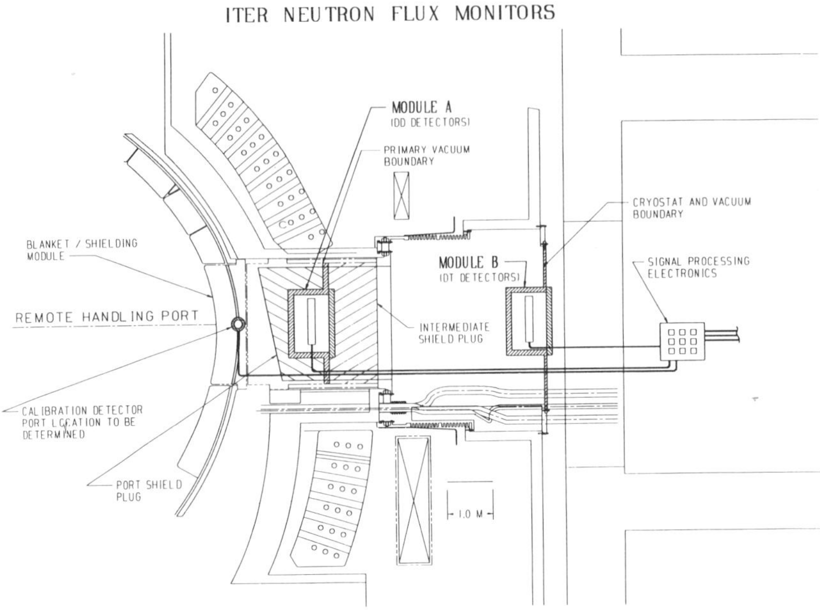

Figure 6: Elevation view of remote handling port depicting locations of one set of neutron source strength monitors. Module A contains 2 fission chambers for DD operation and is re-entrant into the primary vacuum shield block. Module B contains 3 fission chambers for DT operation and is re-entrant into the cryostat vacuum boundary. A calibration detector located near the plasma is the sixth fission chamber in the set.

bolometer channel also to be placed there. We may also try to use one of the micro-fission detector[17] locations at the back of the blanket shield module (BSM). Our "calibration detectors" are thus not in "limiter/Remote handling" (RH) ports and they can be calibrated by a source held *in-situ* by standard RH equipment. No special track inside the vessel would be needed. Three detectors (the least sensitive for DT operation) can go into a module "topologically outside the interspace vacuum" (see Module B in Figure 6) on a coverplate of the secondary vacuum wall in a re-entrant configuration. They are thus outside of any vacuum (but inside bioshield). A variation of sensitivity of a factor of 25x25 can be achieved between these three detectors by changes in fissionable mass. Commercially available fission detectors with up to 10 grams now exist.

This leaves two detectors (the middle range of sensitivity, for ohmic and/or DD operation) which must go into a different module (Module A of Figure 6) with less shielding in front of it than the least sensitive ones in the module B outside the interspace vacuum. The module of two detectors of intermediate sensitivity would need to be in the interspace vacuum, just outside the primary vacuum shield block. The need for three locations (most sensitive, least sensitive, and middle sensitivity) is dictated by the desire to use count mode over the entire dynamic operating range of ITER and by the requirement to use machine shielding (and distance) to provide some of the dynamic range. The least sensitive and middle sensitive modules can hopefully be part of the plug assembly for the limiter/RH ports.

Two similar sets of six detectors are then required at two different toroidal locations. In addition to basic redundancy, two arrays of monitors are needed to: control for local changes in shielding and hence detector efficiency over time; control for local noise sources like RF interference; and control for localized neutron production from runaway electrons or disruptions.

Details of Fission Chambers in Modules

Each detector is inserted into a double-walled stainless steel module box with hydrogenous material for moderation (polyethylene, about 8 cm thick between the walls), attached to the lid of the module for easy removal by remote handling. The stainless steel provides fireproofing as well as radio frequency shielding. The modules need not be vacuum tight. From each module, through a vacuum penetration in the cryostat, runs a single 10-cm diameter electrical conduit containing the electrical connections. No electronics are needed in the detector modules, because modern fission chambers operate with low-capacitance, low-leakage cables many hundreds of feet long.[19] The electronics can be outside the reactor hall in the diagnostics building. In each module there will be one "blank" or fissionable-material-free detector to help identify noise issues. In general, the renormalization (radioactive) sources can just be built-in to the modules, and not have to have any technique to insert or remove them. They provide a constant background of a low counting rate that can be subtracted from the signals. However, each detector may require its own source because of the wide variation in sensitivity of detectors in a module.

Module Name	Detector	Mass (g)	Maximum Source Strength (n/sec)	Efficiency (counts/ neutron)
Cal	NE-1	10	4.0e14	2.5e-10
A	NE-2	10	1.3e16	8e-12
A	NE-3	5	2.6e16	4e-12
B	NE-4	6	6.6e17	1.5e-13
B	NE-5	0.25	1.6e19	6e-15
B	NE-6	0.01	4.0e20	2.5e-16

Table 1: Representative set of 6 detectors for the ITER Neutron Source Strength Monitors.

Table 1 shows a representative set of six detectors for the neutron source strength monitor array. The maximum source strength is that which provides 100-kHz count rate at the detector. The 100-kHz count rate is chosen as a rate where pileup and dead time effects are small and easily accounted for. This set will measure a range of source strength all in counting mode from 4×10^{20} n/sec to below 10^{14} n/sec with lower counting rates in NE-1. Following Figure 5 the flux is assumed to decrease by a factor of 30 between the calibration detector location and Module A, and by another factor of 30 between Module A and Module B. The actual detector array for ITER will be different, depending on final design and exact location and neutronics analysis. This table does reflect how the design conditions can be met.

In practice the efficiencies of some detectors may have to be adjusted in the field to achieve the required ranges of sensitivity. Such adjustments are readily made with thin sleeves of cadmium, which has an exceptionally high absorption cross section for very slow (<0.4 eV) neutrons. The detector efficiency can be reduced as required by wrapping the detector body in a cadmium sleeve of appropriate thickness. With the expected degree of neutron thermalization in the modules, a factor of 2 reduction in efficiency is achieved with 0.5 mm of cadmium.

It will be important also to have time-dependent monitors of the DT neutron emission rather than just the total emission, for initial triton burnup work but more importantly for trace-tritium shots and tracking tritium clean-up in DD shots. Possible threshold detectors for this purpose are radium-based fission chambers, silicon surface barrier diode detectors, scintillating fiber detectors, and natural diamond detectors. Additional types of threshold detectors are presently under development.[20] The fission threshold of ^{226}Ra at 3.0 MeV makes a radium-based fission detector all but impervious to D-D neutrons that have nominally 2.5 MeV. (One would still detect the extremity of the D-D tail in the case of beam-target reactions.) Such a detector could in principle measure only the D-T neutrons in a mixed DD-DT flux. However, the fission cross section $\sigma_f = 0.02$ barn at 14 MeV[21] is very small (more than 5000 times smaller than the thermal cross section for ^{235}U), so that a large mass or huge neutron flux is needed for adequate sensitivity. Because ^{226}Ra is an intense alpha emitter (half-life = 1620 yr), the noise discrimination problems would be much more severe than with uranium-based detectors. Alpha pileup is also a serious issue. A study at ORNL concluded that it is not practical to put more than 10 mg of radium in a single detector, both because of alpha pileup and because of safety problems in fabrication. However, several radium-based detectors could be used in parallel for increased sensitivity. A 0.1-g radium detector array at the location of Module A would require a neutron source strength of $2x10^{19}$ n/s (56 MW fusion power) to generate a minimal count rate of 500 c/s. That means that the detector would be useful only when the D-T neutron flux is so large that there is no need to discriminate against D-D neutrons. Silicon surface barrier diodes[22] are sensitive to neutrons with E_n>4 MeV and generally use a system threshold set at 5.5 MeV. The efficiency to neutron detection is typically 0.001 counts per unit flux. At the front of module A this would enable useful operation with a D-T source strength of $5x10^{16}$ to $5x10^{18}$ n/s which is ideal for trace-tritium operation in ITER. One would like a factor of 10-100 higher sensitivity for D-D triton burnup which would imply a location near the close-in calibration detectors. SBDs have a lifetime of about 10^{12} n/cm^2; at the cryostat wall with a neutron flux of 10^8 n/cm^2/s they would only last 10 full-powered 1000-sec long discharges. Such detectors would still provide valuable information on ITER during the period prior to full-power operation; they could be installed in the source strength monitor module boxes and sacrificed when full-power tritium operation begins. Scintillating fiber detectors[23] use a directional array of fibers and pulse-height analysis to reject Compton recoils from gamma interactions and low-energy proton recoils from DD neutron interactions. They have been successfully used for triton burnup and 14-MeV neutron dectection with better efficiency and time response than silicon diodes. Diamond detectors[24] are also radiation resistant; in addition to use in the neutron cameras they might be useful as DT source strength monitors.

Key Outstanding Issues

1. Different neutronics calculations of the ITER environment[16] have reached conclusions about the drop off in neutron flux to the bioshield that differ by a couple orders of magnitude. This difference needs to be resolved to better determine the location of the source strength monitors. Also, the effect on the neutron flux of the shield plug in the limiter port needs to be ascertained.
2. The exact location of the calibration detectors needs to be negotiated.
3. The selection of complementary energy-sensitive (DT only) flux monitors needs to also be made.

Acknowledgments

We thank Ken Young (PPPL) for support and encouragement for this design work. The ITER Neutron Working Group has been a source of ideas and comments for our efforts. This report has been prepared as an account of work assigned to the United States Home Team under Task Agreement number G55 TD 09 with the Agreement among the European Atomic Energy Community, the Government of Japan, the Government of the Russian Federation, and the Government of the United States of America on Cooperation in the Engineering Design Activities for the International Thermonuclear Experimental Reactor (ITER EDA Agreement) under the auspices of the International Atomic Energy Agency (IAEA). This work was supported by DOE Contracts No. W-7405-ENG-36 and DE-AC02-76-CH03073.

REFERENCES

[1] R. R. Parker *et al.*, in: *Diagnostics for Experimental Thermonuclear Fusion Reactors*, page 1, edited by P. E. Stott *et al.*, Plenum Press, New York (1996).

[2] Cris W. Barnes, M. J. Loughlin, and T. Nishitani, "Neutron activation for ITER", *Rev. Sci. Instrum.* **68** (1997) 577.

[3] Cris W. Barnes and A. L. Roquemore, "Neutron source strength monitors for ITER," *Rev. Sci. Instrum.* **68** (1997) 573.

[4] L. C. Johnson *et al.*, "Overview of fusion product diagnostics for ITER," these proceedings.

[5] M. Pillon, K. A. Verschuur, O. N. Jarvis, J. Källne, and M. Martone, "Calibration of neutron yield activation measurements at JET using MCNP and FURNACE neutron transport codes," *Fus. Eng. Des.* **9** (1989) 347; M. Pillon, M. Martone, O. N. Jarvis, and J. Källne, "Calibration of neutron yield activation measurements at Joint European Torus," *Fusion Technology* **15** (1989) 1420.

[6] Cris W. Barnes, A. R. Larson, and A. L. Roquemore, "Calculations of neutron activation response for the Tokamak Fusion Test Reactor, and absolute calibrations of neutron yield," *Fusion Technology* **30** (1996) 63.

[7] L. C. Johnson *et al.*, "Cross-calibration of neutron detectors for deuterium-tritium operation in TFTR," *Rev. Sci. Instrum.* **66** (1995) 894.

[8] H. W. Hendel *et al.*, "In situ calibration of TFTR neutron detectors," *Rev. Sci. Instrum.* **61** (1990) 1900.

[9] T. Nishitani *et al.*, "Absolute calibration of the JT-60U neutron monitors using a ^{252}Cf neutron source," *Rev. Sci. Instrum.* **63** (1992) 5270.

[10] B. J. Laundy and O. N. Jarvis, "Numerical study of the calibration factors for the neutron counters in use at the Joint European Torus," *Fusion Technology* **24** (1993) 150.

[11] G. Sadler *et al.*, "Calibration issues for neutron diagnostics on ITER," these proceedings.

[12] V. Khripunov, technical memo IDoMS NO:NA NAG-13-11-30-96.

[13] Chris Walker, "Diagnostics access for ITER," these proceedings.

[14] Yuri Utin, private communication (June, 1997). Refer to ITER Catia CAD model 10.0001.0002 and .0003 for view of vertical route to HTS vault.

[15] See Catia CAD Model 26.0009.0001, recent 10 loops option of the upper HTS vault.

[16] T. Nishitani *et al.*, "Design of radial neutron spectrometer array for the International Thermonuclear Experimental Reactor," *Rev. Sci. Instrum.* **68** (1997) 565. V. S. Zaveriaev, G. E. Shatalov, and N. N. Vasiliev, "Final Report for 1996," S 55 RI 1 96-07-03 F 1. H. Iida and R. Santoro, private communication of 1-D calculation, June, 1997.

[17] T. Nishitani et al., "In-vessel neutron monitor using micro fission chambers for ITER," these proceedings.

[18] F. B. Marcus *et al.*, "A radial neutron camera design for ITER with integral shielding," these proceedings.

[19] For example, from Gamma-Metrics, 5788 Pacific Center Blvd., San Diego, CA 92121.

[20] B. Esposito, private communication of paper "Application of micro strip gas chambers (MSGC) as neutron detectors in controlled thermonuclear fusion experiments," presented at 7th Pisa Meeting on Advanced Detectors on May 25-31, 1997.

[21] Babenko, *et al.*, *Sov. J. Nucl. Phys* **7** (1968) 186.

[22] E. Ruskov *et al.*, "Measurement of 14 MeV neutrons at TFTR with Si-Diode Detectors," *Rev. Sci. Instrum.* **66** (1995) 910.

[23] W. C. Sailor *et al.*, "Conceptual design for a scintillating fiber neutron detector for fusion reactor plasma diagnostics," *Rev. Sci. Instrum.* **66** (1995) 898; G. A. Wurden *et al.*, "Scintillating-fiber 14 MeV neutron detector on TFTR during DT operation," *Rev. Sci. Instrum.* **66** (1995) 901.
[24] A. V. Krasilnikov *et al.*, "TFTR natural diamond detectors based D-T neutron spectrometry system," *Rev. Sci. Instrum.* **68** (1997) 553; A. V. Krasilnikov *et al.*, "Fusion Neutronic Source deuterium-tritium neutron-spectrum measurements using natural diamond detectors," *Rev. Sci. Instrum.* **68** (1997) 1720.

IN-VESSEL NEUTRON MONITOR USING MICRO FISSION CHAMBERS FOR ITER

Takeo Nishitani, Katsuyuki Ebisawa†, Larry C. Johnson †, Chris Walker‡, Yoshio Kita# and Satoshi Kasai

Japan Atomic Energy Research Institute, Naka-machi, Naka-gun, Ibaraki 311-01, JAPAN
†ITER Joint Central Team, San Diego, CA 92037, USA
‡ITER Joint Central Team, Garching, GERMANY
#Toshiba Corporation, Fuchu, Tokyo 183, JAPAN

INTRODUCTION

The measurement of the absolute neutron source strength is very important for controlling the fusion power in a fusion experimental reactor such as ITER. In present large tokamaks such as JET[1], TFTR[2] or JT-60U[3], the neutron source strength measurement has been carried out using ^{235}U or ^{238}U fission chambers installed outside the vacuum vessel. Detection efficiencies of those detectors are easily affected by surrounding equipment such as other diagnostics or heating systems. ITER has a thick blanket and vacuum vessel, so that detectors outside the vacuum vessel can not measure the neutron source strength with sufficient accuracy without cross-calibration to more sensitive detectors. We are designing micro fission chambers, which are pencil size gas counters with fissile material inside, to be installed in the vacuum vessel as neutron flux monitors for ITER[4-6]. By installing the detectors at several poloidal angles, this neutron monitor may remove or reduce the error of the neutron source strength caused by changes of the plasma position and/or neutron source profile. This paper provides the conceptual feasibility for this neutron monitor system.

ITER REQUIREMENTS FOR NEUTRON MONITOR

The neutron monitors have important functions in providing measurements for machine protection and plasma control, and they are also needed for performance evaluation and optimization, and physics understanding. The primary function is measurement of the total neutron source strength, hence the fusion power. Table 1 specifies parameter ranges, temporal resolutions, and measurement accuracy. The neutron monitors have to have wide dynamic range and fast response. From the technical point of view, it should be robust in the ITER environment such as radiation, electro-magnetic noises, and

Diagnostics for Experimental Thermonuclear Fusion Reactors 2
Edited by Stott *et al.*, Plenum Press, New York, 1998

mechanical vibrations, and it should not be sensitive to gammas. The detection efficiency should be stable over the ITER operation life, and should be calibrated easily. In ITER, conventional neutron monitors installed outside the vacuum vessel and in-vessel neutron monitor using micro fission chamber are proposed for the neutron source strength measurement. Neutron cameras[7] with their neutron source profile measurement are also available. However, they need Abel inversion or tomography techniques to get the neutron source profile.

Table 1. Requirement for neutron monitor.

Parameter	Parameter range	Spatial resolution	Time resolution	Accuracy
Neutron source strength	10^{14}-10^{21} n s^{-1}	integral	1 ms	10 %
Fusion power	≤2 GW	integral	1 ms	10 %

DETECTOR DESIGN

Micro Fission Chamber

A micro fission chamber is a pencil-size gas counter with fissile material inside, which was developed as an in-core monitor for fission reactors. Figure 1 shows the schematics of the typical micro fission chamber with wide dynamic range which is commercially available. In this detector, about 12 mg of UO_2 is coated on the cylindrical electrode and 14.6 atm of Ar + 5% N_2 gas is filled between the electrodes. The micro fission chamber can be operated with pulse counting mode at low neutron flux, Campbelling (mean square voltage) mode[8] at medium flux and current mode at high flux. Combination of those operational modes may provide a wide dynamic range of 10^{10} with the temporal resolution of 1 ms from a single detector, which satisfies the ITER requirement. The most popular candidates of the fissile material in the micro fission chamber are ^{235}U and ^{238}U, the former has a large fission cross-section for thermal neutrons and the later has fission events only for fast neutrons higher than ~0.8 MeV. The signal is transferred by a MI (mineral insulated) cable, so that we can install micro fission chambers inside the vacuum vessel in the same manner as magnetic probes. The operational temperature can not be higher than 300°C, so that cooling of the detector is necessary due to the nuclear heating inside the vacuum vessel.

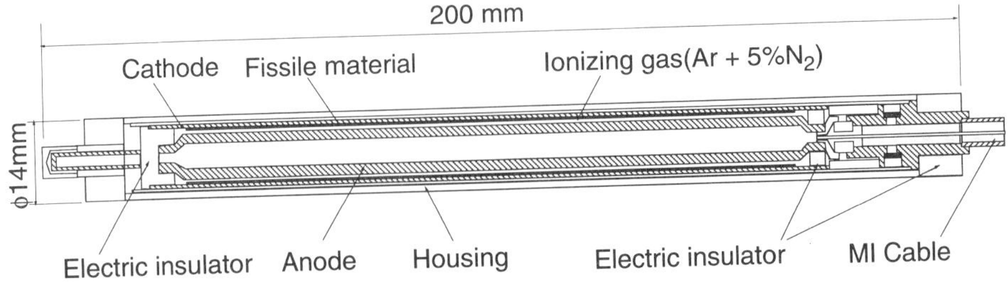

Figure 1. Schematics of typical micro fission chamber. Fissile material such as ^{235}U is coated on the cylindrical electrode. Ionizing gas of Ar + 5% N_2 (14.6 atm) is filled between the electrodes.

Installation Position

In ITER CDA(Conceptual Design Activity), we has a plan to install micro fission chambers just under the first wall. However, this would be rather difficult due to the intensive nuclear heating larger than 10 W/cc and the burn-up of the fissile material resulting in a short life time of the detector. A neutronics calculations using the two dimensional neutron transport code DOT 3.5 was used to fined suitable positions around the shielding blanket and to determine the kind of fissile material. The model includes the first wall, shielding blanket with gap between adjacent modules, back plate of blankets and vacuum vessel. Figure 2 shows the distributions of neutron and gamma flux, and the nuclear heating rate for stainless steel along the radial direction in the shielding blanket. From an engineering point of view, installation of a detector inside the shielding blanket is rather difficult due to the extremely tight space among the cooling pipes. A possible candidate for the installation position is in the gap between adjacent blanket modules and behind the blanket on the back plate. Figure 3 shows the neutron and gamma spectra for those two positions. The neutron flux in the gap is about 10 times larger than that behind the blanket. On the other hand, the gamma flux in the gap is only three times larger.

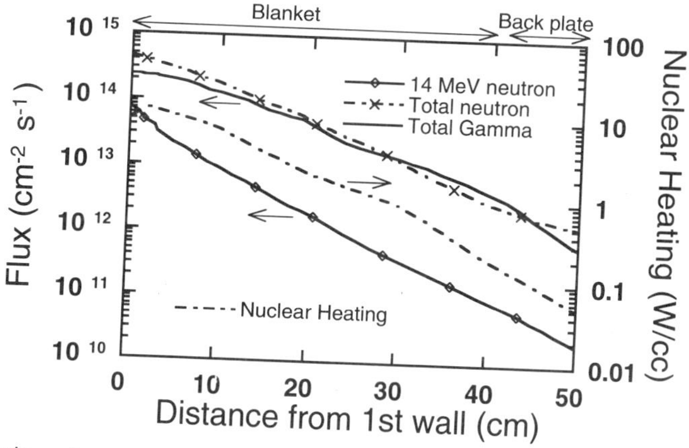

Figure 2. Distributions of neutron and gamma flux, and the nuclear heating rate for stainless steel along the radial direction in the shielding blanket.

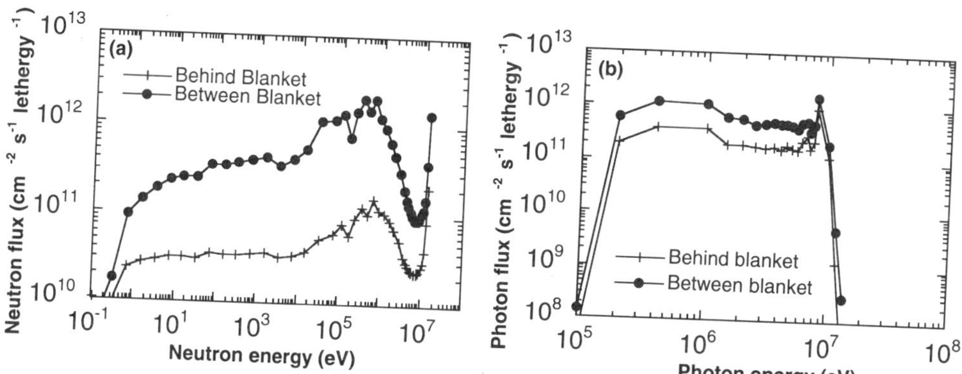

Figure 3. (a)Neutron and (b)gamma spectra in the gap between the adjacent blanket modules and behind the blanket on the back plate.

Total neutron and gamma flux, and expected fission reaction rates of ^{235}U and ^{238}U micro fission chambers with 10 mg uranium in the ITER 1.5 GW operation are listed in Table 2. In order to get wide dynamic range, the fission reaction rate should be as high as the operation limit in the maximum power operation of ITER. Maximum fission reaction rate of this chamber is 10^{10} s^{-1} for the Campbelling mode. The ^{235}U chamber can be positioned both the behind blanket and in the gap. The ^{238}U chamber can only be used in the gap.

Table 2. Fission reaction rates in ^{235}U and ^{238}U micro fission chambers.

Position	Total neutron flux (cm^{-2} s^{-1})	Total gamma flux (cm^{-2} s^{-1})	Fission reaction rate ^{235}U (s^{-1})	Fission reaction rate ^{238}U (s^{-1})	Nucl. heating (w/cc)
Behind blanket	9.0×10^{11}	1.2×10^{12}	9.9×10^{8}	2.3×10^{6}	~0.05
In gaps	9.7×10^{12}	3.5×10^{12}	2.3×10^{9}	1.3×10^{7}	~0.5

Life Time

The change of the detection efficiency due to the burn-up of the fissile material is a worry. ^{235}U is burned up mainly due to the fission reactions, because the thermal neutron flux is relatively small, and the fission cross-section is larger than the capture cross-section even in the low energy region. The change of the detection efficiency is only 0.1 % and 0.2 % behind blankets and in the gap, respectively, for the ITER life time of 1.5 GW•year. Thus we can use ^{235}U chambers without replacement in the ITER life time.

Burn-up of ^{238}U is more complicated. ^{238}U will be burned up mainly by the neutron capture reaction in the thermal energy range, but the capture reaction will produce ^{239}Pu via following chain;

$$\begin{aligned} &^{238}U + n \rightarrow {}^{239}U + \gamma \\ &\qquad {}^{239}U \rightarrow {}^{239}Np + \beta \\ &\qquad\qquad {}^{239}Np \rightarrow {}^{239}Pu + \beta . \end{aligned} \tag{1}$$

The detection efficiency of the ^{238}U chamber may increase due to ^{239}Pu which has large fission cross-section for thermal neutrons. We calculated the fission reaction rate of ^{238}U chamber for the neutron spectrum in the gaps of adjacent blankets. After the 1.5 GW•year operation the burn-up of ^{238}U is only 0.2 %, however, the detection efficiency will increase 50 % due to ^{239}Pu. The change is too large to use ^{238}U chamber as a stable neutron monitor in ITER.

Gamma-ray Effect

This micro fission chamber can be operated in the pulse counting, Campbelling and current modes. We evaluated the noises from gammas for those three operational modes. In the pulse counting mode, the current pulse generated by the fission fragments or gamma reaction in the ionizing gas is measured. The fission reaction gives ~100 MeV to the ionizing gas, while on the other hand deposited gamma energy is less than 10 MeV (see

Fig.3b). So we can eliminate the gamma pulses by conventional pulse discrimination technique.

The Campbelling mode sensitivity is given from the following formula,

$$I_{rms}^2 = N q^2 \int f^2(\omega)\, B(\omega)\, d\omega/\pi \qquad (2)$$

where N is the fission reaction rate, q is electric charge generated by the fission, f(ω) is the frequency spectrum of the detector output, and B(ω) is the filter function of the amplifier. The gamma sensitivity in Campbelling mode is 7.7×10^{-29} $A^2/Hz/(R/h)$ for those chambers. The expected Campbell currents due to neutrons and gammas are listed in Table 3, where gamma dose rates are converted from the gamma spectra shown in Fig.3b. Thus the we can neglect gamma effects in Campbelling mode. Also we calculated the outputs in the current mode as shown in Table 3. The gamma sensitivity in current mode is 2.2×10^{-12} A/(R/h). Current mode can not used due to gamma noises.

Table 3. Outputs from neutrons and gammas in Campbelling and current modes.

Detector/location	Gamma dose rate (R/h)	Campbelling mode (A^2/Hz)		Current mode (A)	
		Neutrons	Gammas	Neutrons	Gammas
^{235}U behind blanket	3.5×10^6	2.3×10^{-17}	2.7×10^{-22}	2.6×10^{-4}	8.0×10^{-6}
^{235}U in gaps	8.6×10^6	5.3×10^{-17}	6.6×10^{-22}	6.0×10^{-4}	1.9×10^{-5}

Dynamic Range

Wide dynamic range of 10^7 in single fission chamber has been demonstrated in the JT-60U neutron monitor[3] using both pulse counting and Campbelling modes, which meets the ITER requirement for the neutron monitor. The linearity was calibrated using a fission reactor with output power range of 10^7 before installation on JT-60U. We confirmed two decades of overlap in pulse counting and Campbelling modes. From this experience, the expected dynamic range of a micro fission chamber with 10 mg ^{235}U is shown in Figure 4. Thus this system can cover the 10^7 dynamic range of ITER operation.

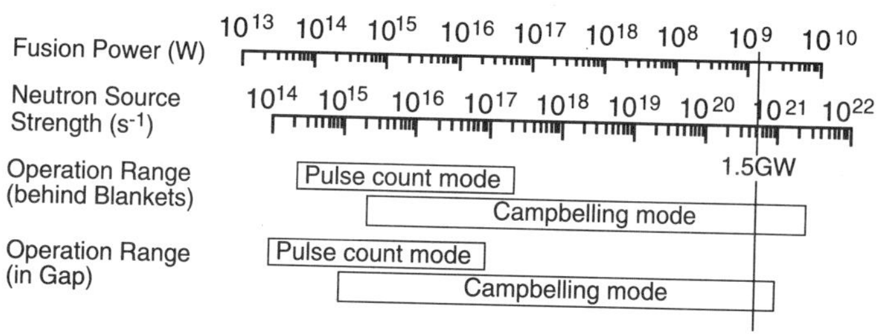

Figure 4 . Operation range of ^{235}U micro fission chambers behind blanket and in the gap of adjacent blankets.

Magnetic Field Effect

The effect of the strong magnetic field on the fission chamber is another problem. In the fission chamber, we measure the electron induce current from ionization by fission fragments. So we calculated the electron drift orbit in the magnetic field. The electron drift velocity **u** is represented by

$$\mathbf{u} = \frac{\mu_e}{1 + \omega_c^2/\nu^2}\left[E + \frac{\mathbf{E} \times \mathbf{B}}{B}\frac{\omega_c}{\nu} + \frac{(\mathbf{E} \bullet \mathbf{B})}{B^2}\frac{\omega_c^2}{\nu^2} \right] \tag{3}$$

where μ_e is the electron mobility, ν is the collision frequency of the electron to neutral atoms, and ω_c is the electron cyclotron frequency in the magnetic field **B**. If we assume $\mathbf{E} = (E_x, 0, 0)$ and $\mathbf{B} = (0, 0, B_z)$, then the components of the velocity are

$$\mathbf{u} = \left(\frac{\mu_e E_x}{1 + \omega_c^2/\nu^2}, \frac{\mu_e E_x}{1 + \omega_c^2/\nu^2}\frac{\omega_c}{\nu}, 0 \right) \tag{4}$$

The angle between **u** and **E**, the Lorentz angle α, is represented by $\tan \alpha = \omega_c/\nu$. In the case of the micro fission chamber with 1 atm Argon gas, are applied voltage of 100 V across 0.25 mm electrode gap, in a magnetic field of 5.7 T, the Lorentz angle is evaluated to be $\tan \alpha \approx 0.1$. Thus the magnetic effect on the electron drift velocity is negligible.

EFFECTS OF PLASMA POSITION AND NEUTRON SOURCE PROFILE

The detection efficiency of a neutron detector inside the vacuum vessel is affected by changes of plasma position and the neutron source profile. However, there is a possibility of removing or reducing these effects by installing the detectors at several poloidal angles. We investigated the responses of the micro fission chambers by a neutron Monte Carlo calculation using the MCNP version 4A code[9]. An 18 degree sector of the first wall, shielding blanket, blanket back plate and vacuum vessel is modeled by a simple torus with elliptic cross-section as shown in Figure 5. The divertor cassettes, ports and coils are not included. Only one poloidal gap between adjacent blankets is modeled. We use the neutron cross-section set based on JENDL 3.2[10]. The neutron source is a toroidally symmetric source with 14 MeV monoenergetic energy. The source has a poloidal distribution as follows:

$$S = \left[1 - \left\{ \frac{(R-R_P)^2}{a_P^2} + \frac{(Z-Z_P)^2}{a_P^2 \kappa^2} \right\} \right]^m \tag{5}$$

where R_P is the major radius, a_P is the minor radius, Z_P is the vertical shift of the plasma center, κ is the ellipticity, and m is the power of the parabolic profile. The angular emission is isotropic. The reference parameters are $R_P = 8.14$ m, $a_P = 2.3$ m, $Z_P = 1.4$ m, $k = 1.6$ and $m = 0.8$. In this case, the neutron source peaking factor is represented by

$$\frac{S_n(0)}{\langle S_n \rangle} = m + 1 \tag{6}$$

where $S_n(o)$ and $<S_n>$ are the central and volume averaged neutron emissivity, respectively.

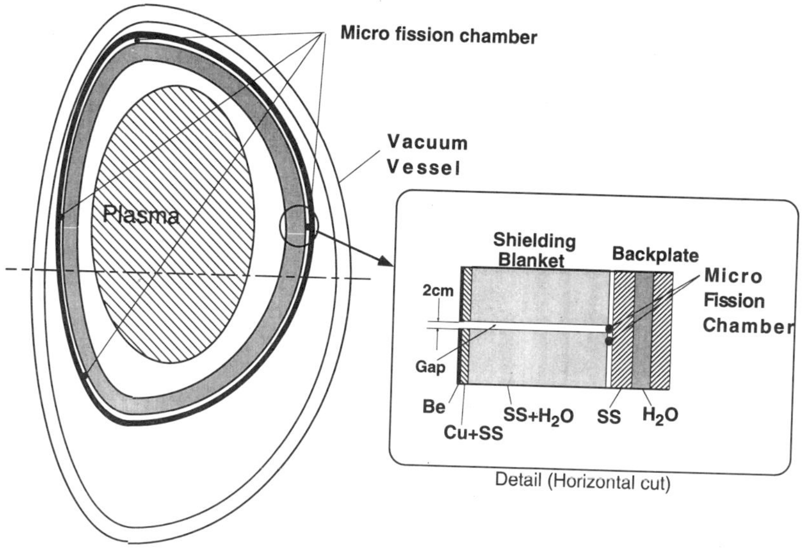

Figure 5. Model for a neutron Monte Carlo calculation. An 18 degree sector of the first wall, shielding blanket, blanket back plate and vacuum vessel is modeled by a simple torus with elliptic cross-section.

We calculated the detection efficiencies of the ^{235}U micro fission chambers at four typical poloidal positions in the gap as shown in Figure 5. We have not obtained sufficient statistics for those positions behind the blanket. Figure 6 shows the plasma position dependence of the detection efficiencies. For the horizontal shift of plasma position in the range ± 50 cm, the detection efficiency of the bottom detector is almost constant. The efficiency of the outside detector increases only 1.5 % with 50 cm increase in horizontal plasma position, while that of the inside one decreases 12 % and the top detector's efficiency decreases 3%. For a vertical shift of the plasma position in the range ± 50 cm, detection efficiencies of inside and outside detectors are almost constant. The efficiency of the top detector increases 8 % with a 50 cm increase in vertical plasma position, while that of the bottom one decreases 18 %.

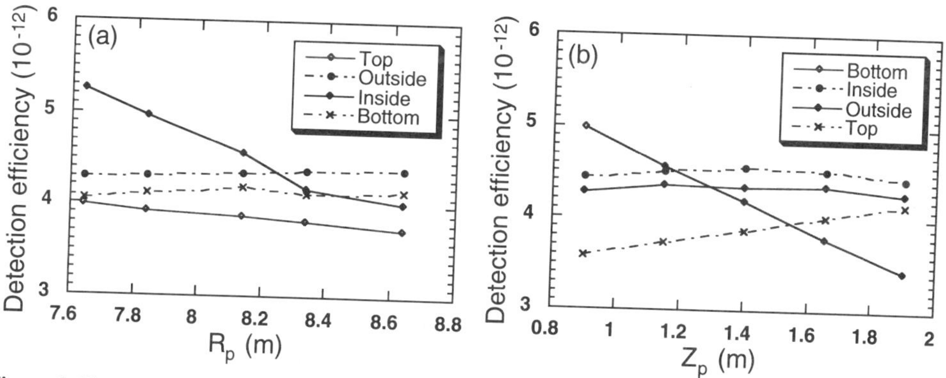

Figure 6. Plasma position dependence of the detection efficiencies; (a) is for a horizontal shift and (b) is for a vertical shift.

Figure 7 shows the neutron source profile dependence of the detection efficiencies. Detection efficiencies of both inside and outside detectors increase 10 % with the increase in the peaking factor of the neutron source profile from 1.8 to 3. However, the efficiencies of the top and bottom detectors are less sensitive to the peaking factor. The inside and outside detectors are so close to the plasma that they are easily affected by the plasma position and/or neutron source profile. There is a possibility of reducing those effects by combining the outputs of detectors. For example, the average of the top and bottom detector signals derived only ± 2 % from the neutron source strength for both changes is plasma position and profiles.

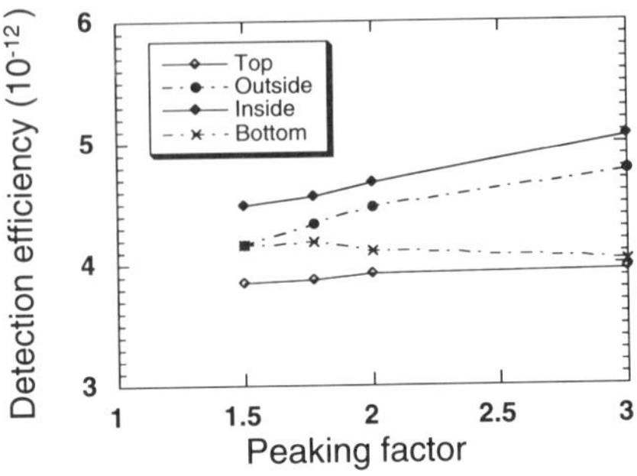

Figure 7. Neutron source peaking factor dependence of the detection efficiencies.

CALIBRATION

Absolutely calibrating the source strength monitors is their most critical issue. In present tokamaks, neutron monitors are calibrated by moving a neutron source such as ^{252}Cf radioactive source or a DT neutron generator. In ITER, an *in-situ* calibration by moving a DT neutron generator remotely inside the vacuum vessel should be performed. Figure 8 shows the detection efficiency (of the outside detector) for the point source on the plasma axis at R = 8.14 m plotted against the toroidal angle of the source. A DT neutron generator with emission rate of ~10^{10} neutrons/s to get sufficient counting rate during the *in-situ* calibration. Such an intense neutron generator which is movable by the remote handling apparatus should be developed.

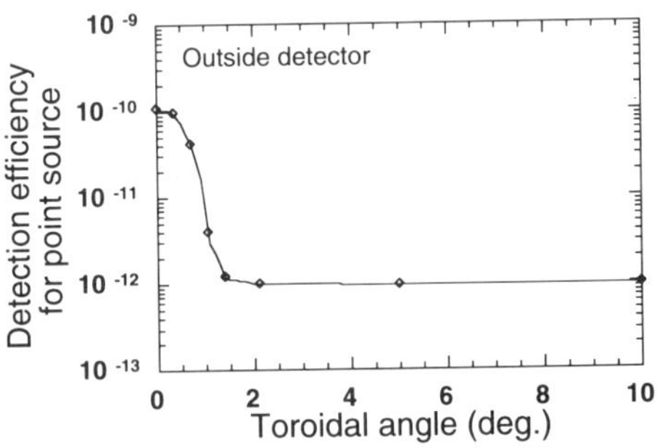

Figure 8. Detection efficiency of outside detector for a point source on the plasma axis at R = 8.14 m plotted against the toroidal angle of the source.

ARRANGEMENT ON ITER

Figure 9 shows the proposed arrangement of micro fission chambers on ITER. Ten pairs of ^{235}U chambers are installed at 10 poloidal locations. One chamber is behind the blanket and another is in the gap of adjacent blankets in each poloidal location. Another chamber is installed in the divertor cassette just under the dome. MI cables of chambers are wired via generic diagnostics conduit.

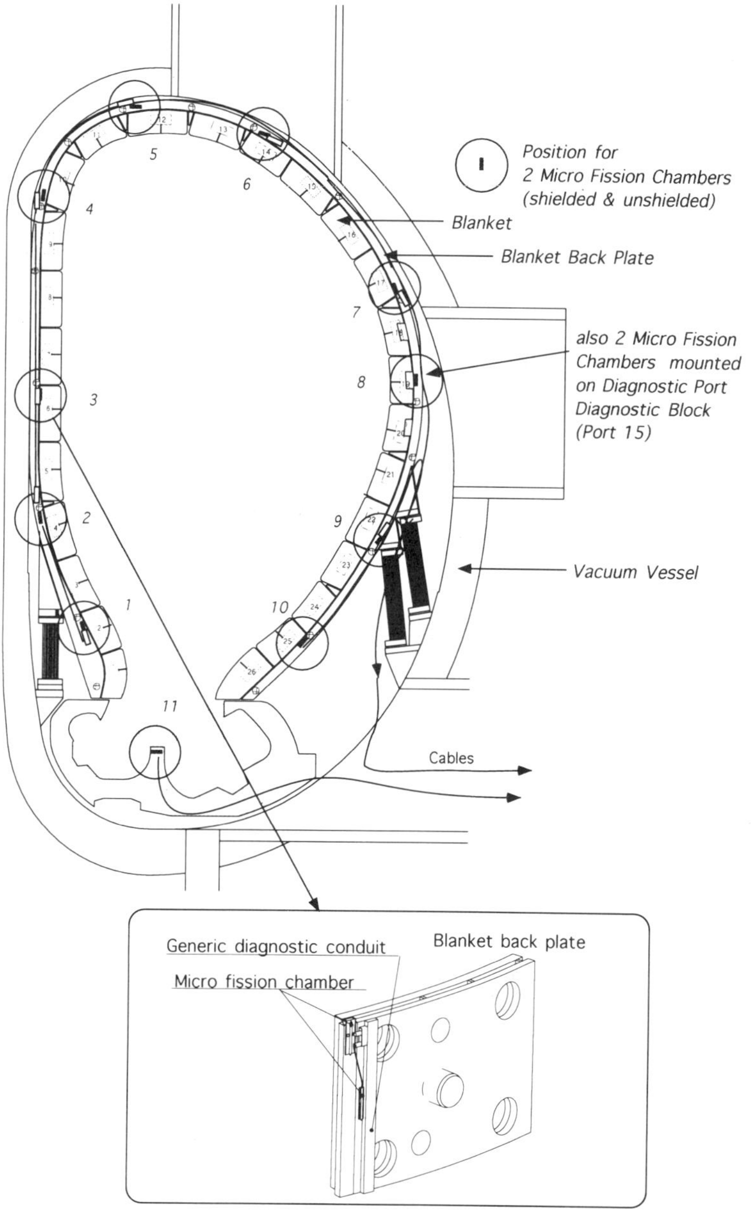

Figure 9. Arrangement of micro fission chambers on ITER.

ACKNOWLEDGMENTS

The authors appreciate Dr. S. Sato and Dr. H. Iida for their support on the neutronics calculations. We appreciate Dr. S. Yamamoto for his pioneer work on the micro fission chambers in ITER CDA. We thanks Drs. Cris W. Barnes and G.A. Wurden for comments and suggestions. This report has been prepared as an account of work assigned to the Japanese Home Team under Task Agreement number S 55 TD 02 FJ within the Agreement among the European Atomic Energy Community, the Government of Japan, the Government of the Russian Federation, and the Government of the United States of America on Cooperation in the Engineering Design Activities for the International Thermonuclear Experimental Reactor ("ITER EDA Agreement") under the auspices of the International Atomic Energy Agency (IAEA).

REFERENCES

1. O.N. Jarvis, G. Sadler, P. van Bell and T. Elevant, In-vessel calibration of the JET neutron monitors using a ^{252}Cf neutron source: Difficulties experienced, *Rev. Sci. Instrum.* 61: 3172 (1990).
2. H.W. Hendel, R.W. Palladino, Cris W. Barnes, et al., In situ calibration of TFTR neutron detectors, *Rev. Sci. Instrum.* 61: 1900 (1990).
3. T. Nishitani, H. Takeuchi, T. Kondoh, et al., Absolute calibration of the JT-60U neutron monitors using a ^{252}Cf neutron source, *Rev. Sci. Instrum.* 63: 5270 (1992).
4. V. Mukhovatov, H. Hopman, S. Yamamoto, et al., in: *ITER Diagnostics, ITER Documantation Series, No.33*, IAEA, Vienna (1991).
5. T. Iguchi, J. Kaneko, M. Nakazawa, T. Matoba, T. Nishitani and S. Yamamoto, Conceptional design of neutron diagnostics system for fusion experimental reactor, *Fusion Eng. Design* 28: 689 (1995).
6. T. Nishitani, K. Ebisawa, T. Iguchi and T. Matoba, Design of ITER neutron yield monitor using microfission chambers, *Fusion Eng. Design* 34-35: 567 (1997).
7. F.B. Marcus, J.M. Adams, P. Batistoni, et al., A neutron camera for ITER: Conceptual design, in: *Diagnostics for Experimental Thermonuclear Fusion Reactor,* P.E. Stott, G. Gorini and E. Sindoni ed., Plenum Press, New York (1996).
8. Y. Endo, T. Ito and E. Seki, A counting-Campbelling neutron measurement system and its experimental results by test reactor, *IEEE Trans. Nucl. Sci.* NS-29: 714 (1982).
9. LANL Group X-6, *MCNP-a general Monte Carlo code for neutron and photon transport version 3A, Report LA-7396-M, Rev.2*, Los Alamos National Laboratory, Los Alamos (1986).
10. T. Nakagawa, K. Shibata, S. Chiba, et al., Japanese evaluated nuclear data library version 3 revision-2: JENDL-3.2, *J. Nucl. Sci. Technol.* 32: 1259(1995).

CALIBRATION ISSUES FOR NEUTRON DIAGNOSTICS

G. J. Sadler, J. M. Adams[1], C. W. Barnes[2], P. Batistoni[3], D. Bond[1], N. Hawkes[1], O .N. Jarvis, L. C. Johnson[5], F. B. Marcus, M. Loughlin, A. L. Roquemore[4], P. van Belle, N. Watkins[1]

JET Joint Undertaking, Abingdon, Oxon. OX14 3EA, UK
1 AEA Technology, Harwell, Abingdon, UK
2 Los Alamos National Laboratory, Los Alamos, NM 87545, USA
3 ENEA, Frascati (Roma), Italy
4 PPPL, P.O. Box 451, Princeton, N.J. 08543, USA
5 ITER, San Diego, USA

1. INTRODUCTION

The performance of diagnostic systems are limited by their weakest constituents, including their calibration issues. Neutron diagnostics are notorious for problems encountered while determining their absolute calibrations, due mainly to the nature of the neutron transport problem. Although neutrons penetrate large amounts of material and are difficult to shield against, they lose essential information (energy and direction) whilst interacting with shielding materials or other machine components. Detectors, fully characterized in the laboratory, respond only to the local neutron field, which is determined by radiation transport in the tokamak and the diagnostic itself. In general, detectors will not distinguish between virgin (unscattered) neutrons coming directly from the plasma and neutrons that have undergone scattering events, although some discrimination is possible if energy-sensitive detectors are used. In order to facilitate the determination of an accurate and precise calibration, the diagnostic design should be such as to minimize the scattered neutron flux. In addition, the largest possible fraction of the neutron emitting volume has to be viewed so as to avoid errors caused by profile and collimation effects, which is best achieved by choosing a location as close to the plasma as possible. The general issue of obtaining accurate calibrations for the neutron diagnostics on present-day tokamaks were discussed at a specialist meeting in 1989[1].

ITER will use a comprehensive set of neutron diagnostics[2] - comprising radial and vertical neutron cameras, neutron spectrometers, a neutron activation system and internal and external fission chambers - to provide accurate measurements of fusion power and power densities as a function of time. The calibration of such an important diagnostic system merits careful consideration. Some thoughts have already been given to this subject during the

Diagnostics for Experimental Thermonuclear Fusion Reactors 2
Edited by Stott *et al.*, Plenum Press, New York, 1998

conceptual design phase in relation to the time-integrated neutron activation[3] and time-dependent neutron yield monitors[4]. However, no overall calibration strategy has been worked out so far.

This paper represents a first attempt to address this vital issue. Experience gained from present large tokamaks (JET, TFTR and JT60U) and proposals for ITER are reviewed. The need to use a 14-MeV neutron generator as opposed to radioactive sources for in-situ calibration of D-T diagnostics will be stressed. It is clear that the overall absolute determination of fusion power will have to rely on a combination of nuclear measuring techniques, for which the provision of accurate and independent calibrations will constitute an ongoing process as ITER moves from one phase of operation to the next.

2. ITER REQUIREMENTS

The complexity of the calibration issue can be appreciated from a brief consideration of the range of ITER machine operations during which neutron diagnostics are required: starting with hydrogen (H) plasmas, pure deuterium (D) and trace tritium (D(T)) will next be employed before moving on to 50:50 deuterium-tritium (D-T) plasmas. Hydrogen plasmas will generate a small neutron yield due to the natural abundance of D in H and the usual photo-neutron contribution is expected. It will be important to have absolutely calibrated 14-MeV neutron monitors even during the D-D phase to study triton burn-up. The same monitors cannot not be used during the D-T phase as the 14-MeV neutron yield will then be more than 4 orders of magnitude higher. Calibration for the D-T phase probably presents the least difficulty.

The use of different fuels causes large variation in neutron yield and important changes in neutron energy spectra, as does the employment of different plasma heating techniques. Ideally, the instantaneous neutron yield diagnostic would be designed to have:

- a linear response to neutron yield over some seven orders of magnitude,
- no sensitivity to the neutron energy spectrum,
- no sensitivity to neutron emission position, shape, profile, and anisotropy,
- no sensitivity to gamma-rays,
- long-term intrinsic stability,
- no sensitivity to changes to the machine structure.

It may seem over-ambitious to specify all these factors but it should be borne in mind that if the main diagnostic is lacking in a particular attribute then it implies the need for a subsidiary diagnostic which does not. Our present understanding is to meet the mission of ITER the calibration issue *alone* implies the need for an activation system, a profile monitor and neutron spectrometers to complement the neutron yield monitors.

3. METHODS USED ON PRESENT DAY MACHINES

Greatest emphasis will be placed in this article on experience obtained from JET since the material from this machine is most readily available to the authors, although it should be remarked that the exchange of information between the main fusion laboratories has always been excellent so that the experimental techniques have been developed in common; the main procedural differences can often be related to differences in construction of the different tokamaks and in their perceived missions.

3.1 JET

At JET, the instantaneous neutron yield is monitored with three sets of fission chambers arranged around the machine, each set comprising a ^{235}U chamber and a ^{238}U chamber to provide the required dynamic range. These chambers are operated in pulse-counting and current modes. The fission chambers are checked against each other on a discharge by discharge basis and an alarm is raised if the response of one of the detectors deviates by more than a preset value (~6%) from their average. Apart from being triggered occasionally by photo-neutrons, it allows early detection of drifts (not observed at JET, so far) and changes to the disposition of equipment around the machine which influence the detector calibration (this inevitably happens during long shut-downs).

The absolute calibration of JET neutron yield monitors was first obtained in 1984, following the example of TFTR, from an extensive series of measurements[5] where the response of the fission chambers to a ^{252}Cf radioisotope source was mapped as a function of source location inside the vacuum vessel. At the same time, the insensitivity of the measurement to the neutron energy spectrum was demonstrated using additional sources (Am-Be and a pulsed 14 MeV neutron generator). The overall accuracy of the calibration was determined to be ± 10%. The calibration was checked at various occasions using the same ^{252}Cf source but never repeated in its entirety due to access limitations.

Later, foil activation measurements (see below) were used to derive the absolute calibration with an improved accuracy of around ± 7% [6]. Unfortunately, as time passed more and more bulky and heavy equipment was installed inside and outside the main access ports near to the fission chambers, inhibiting neutrons scattering from the access port windows from reaching the fission chambers but not affecting those arriving at the fission chambers after penetrating the concrete-filled mechanical shell. Consequently the fission chamber response changed considerably over the first 7 years of operation. It also transpired that their response had become dependent on the source energy spectrum[7] so the use of ^{252}Cf was renounced and, in the absence of convenient D-D and D-T generators, in-vessel calibrations were abandoned altogether. The causes of the neutron energy dependence and the associated changes in calibrations were investigated in a series of neutron transport simulations[8].

3.1.1 The activation method

At JET, the preferred method of calibrating the neutron yield monitors involves determining the induced activation of small samples of carefully selected materials exposed to the neutron field close to the vacuum vessel. Although a total of 8 irradiation stations were originally provided, the most accurate results were obtained with the two irradiation stations mounted inside the vacuum vessel (the others are outside but close to the vacuum vessel wall). The neutron transport calculations, required to relate the measured activation to the source, are most reliably performed for the inside locations. The resulting calibration is insensitive to hardware outside the vacuum vessel but it does depend on the position of the plasma within the vessel and it is only weakly dependent on internal machine components. Transport calculations are performed with two codes based on contrasting methods: FURNACE[9], a ray tracing code using double differential albedos, and the standard 3D Monte Carlo Code, MCNP. After extended investigations and iterations, agreement between the two codes was achieved and accuracies of ±7% obtained[6]. The inside of the machine has subsequently been changed by the installation of a divertor, the up-down symmetry of plasma position in the vacuum vessel has been lost and only one of the previous two inside irradiation ends is now available. The computer models have been updated and the profile monitor is needed to measure the up-shift (~30 cm) in the plasma position. The present (DTE1) calibration from the neutron activation system provides an accuracy ~ ± 10%.

The choice of material for irradiation and the operational aspects of the activation technique are standard practice and will not be elaborated here, other than to note that a very considerable dynamic range is readily achievable. The induced activity can usually be determined to an accuracy of about 3%, comparable with the uncertainty in activation cross-sections for dosimetry standard materials. The greatest uncertainty derives from the neutron transport calculations. Of special relevance for JET (but possibly not for ITER) is the use of delayed neutron counting using thorium or uranium samples for D-D discharges and silicon for D-T discharges (or for triton burn-up measurements in D-D discharges) because their associated decay half-lives are suitable for repeated use of the same samples in successive discharges. The delayed neutron measurement technique was calibrated very accurately at a research (fission) reactor installation. The silicon activation cross-sections are not well known so this material had to be cross-calibrated against dosimetry standard materials.

3.1.2 The neutron profile monitor

The new, upgraded, neutron profile monitor presently in use at JET is very similar to the one under design for ITER. It consists of two fan-shaped arrays ("cameras"), one viewing the plasma radially and the other viewing the plasma vertically. Absolute calibrations are provided for the two cameras, as described below.

First, a short description of the neutron detection system used in the profile monitor is presented. The JET neutron detectors operate in the pulse height mode and are energy sensitive. The photo-multiplier pulses are amplified in a fast linear amplifier after having traveled over 60-m-long cables between the torus hall and the signal processing cubicles. For the BC418 scintillators used for operation in D-T, the amplified pulses are fed to a fan-out unit, from which five of the outputs are fed to pulse height discriminator units - 4 for measuring neutrons, the fifth for checking long term stability from in-built radioactive sources. A sixth output is fed to a 'charge' integrating preamplifier followed by a pulse shaping amplifier, so that fully shaped pulses can be presented to the input of an multi-channel analyzer (after having passed through a linear gate). Pulse height spectra can thus be accumulated in order to determine end-points and thresholds by gating the linear gate with the outputs from the various pulse height discriminators. Pulses from the NE213 detectors (used for D-D and trace tritium plasmas) can be treated in a similar way with the difference that we have to consider 2 branches: a D-D branch and a D-T branch. Fortunately, each branch is handled by one single, double-width, NIM pulse shape discriminator unit providing properly shaped output pulses for neutrons and gamma-rays already suitable for multi-channel analysis.

Setting up the low-energy branch of the NE213 detectors using the 1275-keV γ-rays from the in-built ^{22}Na sources posed few difficulties. Initially, the same procedure was used in setting up of the NE213 high energy (D-T) branch but, due to the large extrapolations involved, significant energy discrimination variations resulted and the measured profiles were not smooth. A subsequent attempt, using an Am/Be neutron source, proved very helpful while setting up the n/γ discrimination for the low energy branch but was of no assistance for the high energy branch due to the lack of neutrons with energies above 9 MeV.

Real progress was made with both types of 14-MeV neutron detector when a D-T generator was borrowed from PPPL. As the output of the generator (~1-2 10^7n/s) was too low to be of use when placed at the focal point of the camera, the rear camera shields were removed and all detectors units irradiated simultaneously from a short distance. Spectra for all detectors of a given type from one camera at a time were obtained in parallel by multiplexing the shaped pulses to a fast ADC of a Canberra PC-based multi-channel analyzer. Proceeding this way allowed statistically meaningful data to be obtained on an hourly basis for all detectors simultaneously. The high voltages and discriminator thresholds were adjusted until 'identical' spectra were observed. Only approximate settings could be obtained by proceeding as described above due to neutron scattering in the rather bulky components of the detector box.

The final adjustments were made using dedicated plasma discharges containing some trace tritium at the beginning of the DTE1 campaign. ELMy H-mode plasma discharges yielding relatively flat neutron profiles were used so as to obtain acceptable statistics in the wing channels within a single discharge. The relative efficiencies for all channels lie within less than 5% of the mean value.

Applying absolute, solid-angle-corrected detector efficiencies should lead to an absolutely calibrated system. However, a thorough accelerator characterization of the BC418 scintillator-detector units, as required here, has yet to be undertaken and the light output functions used for the NE213 detectors had to be utilized. A concern here is the inherent non-linearity of light output functions coupled with the fact that the experimentally determined NE213 light response functions do not agree well with published data.

Finally, corrections for back-scattering from the vacuum vessel wall regions viewed by the camera, attenuation, in-scatter and out-scatter all have to be made. The absolute magnitude of the backscatter correction is only weakly channel dependent. For D-T neutrons, the fractional backscatter contribution to the line-of-sight data is ~10% for a wing channel whereas, for D-D neutrons, this correction is ~10% for a central channel, despite the order of magnitude higher unscattered fluxes in central channels. Corrections for attenuation and scattering of neutrons directed towards the detectors, as determined by Monte Carlo calculations, are typically about 30% for D-T neutrons and 50% for D-D neutrons.

Our first D-T results are encouraging as the instantaneous 14-MeV neutron yields from the profile monitor agree to within 8 % with those obtained using silicon diodes calibrated using the activation system (when the discharges are analyzed using both camera simultaneously). However, the imbalance of ~12% between vertical and horizontal camera remains a concern. The calibration provided by the horizontal camera on its own is preferred because of the fortunate circumstance of the radial position of the focus of the fan-shaped arrangement of the lines-of-sight being almost exactly twice the radius of the major plasma axis. This means that the line-of-sight density of the viewing arrangement on the inboard side of the machine is significantly lower than on the outboard side. The good match between the radial dependence of the line-of-sight density and radial dependence of toroidal volume means that the radial-viewing neutron camera is, in principle, able to measure accurately the total volume integrated neutron emission strength without recourse to information regarding the spatial distribution of the neutron source profile along the camera's lines-of-sight (i.e. information from the vertical camera is not required). A weighted summation of the lines-of-sight data can therefore be used for deducing the total neutron yields. The viewing extent of the horizontal camera is such that at maximum an error of 5% is introduced for extreme, almost unrealistic, profiles. Unfortunately the location of the focus of the radial camera proposed for ITER does not have the property described here and information on the spatial distribution along the lines-of-sight (from a vertical camera and/or model calculations) will be required in order to deduce the total neutron yield.

3.2 TFTR

The early D-D neutron production measurements with fission chambers at TFTR relied on direct calibrations performed in-situ with radioisotope sources, as at JET. The sensitivity of the TFTR fission chamber responses to source energy spectra is less than at JET since the TFTR support structure is considerably less massive. Nevertheless, in later work particular emphasis was laid on the use of neutron generators as these possess the great advantage of providing the correct neutron source energy spectrum. Despite careful characterization of the directional emission from the D-D generator, only moderately accurate results (> 15%) were obtained from the first attempt[10]. Calibration for D-T operation involved a D-T neutron generator; however, such key issues as the number of spatial points that were necessary and the effects of anisotropic sources were addressed with ^{252}Cf sources[11]. Activation techniques were used to determine the output of the D-T neutron generator used for in-situ calibrations; the $^{27}Al(n,\alpha)^{24}Na$ reaction was chosen and a NIST-traceable gamma-ray source was used for calibrating the efficiency of the HPGe detectors.

In order to cover the required range of neutron intensities from TFTR, a variety of neutron detectors with differing sensitivities was employed. The absolute calibrations using portable neutron sources were performed for the highest sensitivity detectors; transfer of the calibrations to the low-sensitivity detectors used at high fusion powers required considerable care to deal with issues of non-linearity and detector drift over time. With proper attention to detail and the demonstrated stability of key detectors, the uncertainties from this process were reduced to a few percent at most[12, 13].

The use of the neutron activation technique for calibrating the fission chambers was given lower priority at TFTR than at JET and, indeed, as at JET, some initial inconsistencies were experienced while employing an irradiation position outside the vacuum vessel of the tokamak. Subsequent work with an irradiation position well inside the vessel led to reliable results being obtained. The internal consistency of the activation technique was demonstrated using materials with a range of threshold energies, those with low reaction thresholds being most sensitive to details of the neutronics modeling.

The TFTR neutron profile monitor provides only a vertical view through the plasma, considered sufficient as the flux surfaces are nearly circular (there being no divertor). The radiation detectors in the collimated sight-lines were calibrated[14] using neutron sources positioned inside the tokamak. A special feature of TFTR is the possibility of translating the plasma column radially through substantial distances without substantially altering the discharge characteristics; this was used[15] to test the relative efficiencies of the different collimation channels and to determine experimentally the magnitude of the backscatter corrections.

The fission chamber calibrations obtained from (i) activation measurements performed on high yield discharges coupled with neutron transport calculations[16], (ii) extrapolations using the absolute efficiencies of the most sensitive detectors determined with in-situ neutron generators[17], (iii) the profile monitor[18] and also (iv) from a "wide-angle collimated view" of a NE213 liquid scintillator spectrometer[19], were all very similar. The final "official" TFTR D-T fusion powers were derived from an independent-uncertainty-weighted average[20] of all these methods, giving a 1 s.d. uncertainty of ± 7%.

3.3 JT60U

The calibration of the fission chambers at JT60U was first derived from an extensive and carefully planned calibration[21] in which an in-situ ^{252}Cf neutron source was moved to 92 different toroidal locations using a purpose-built automated source carrier. Spectrum effects were examined using MCNP neutron transport calculations, showing no difference within statistical errors in the detection efficiency for ^{252}Cf or D-D neutrons. After integrating over the plasma volume, a final error bar of about ± 10% was deduced.

JT60U, as at TFTR, suffered problems with a "noisy" detector which created a problem during their calibration studies which was overcome by careful analysis. The need for redundant and complementary systems is emphasized by such experiences.

Calibrations of the fission chambers using foil activation measurements coupled with neutron transport calculations have recently been performed[22], providing an overall accuracy of ± 15-20% for D-D discharges, in good agreement with the in-situ source calibrations. The estimated error for 14 MeV neutron measurements (for triton burn-up studies since it is not proposed to employ tritium fueling) is ~ ± 20%.

4. ITER

4.1 Yield monitors

The US proposal for instantaneous neutron yield monitoring is to provide two arrays of 6 neutron detectors operating in the count rate mode with varying sensitivities (defined by the mass of ^{235}U in the fission chambers, location and extent of shielding) to measure over 7 orders of magnitude in neutron intensity. These detectors are to be located at two positions within the remote handling ports: two detectors will be embedded in the inboard shielding block in a chamber re-entrant into the primary vacuum (but actually in the bioshield) to measure primarily D-D and trace tritium yields while three less sensitive detectors devoted to D-T plasmas will be located in a re-entrant chamber at the outside wall of the cryostat vacuum (but actually at atmospheric pressure). This system will be duplicated in a second RH port for redundancy as well as to provide discrimination against local events such as photo-neutrons from runaway electrons and RF pickup. For calibration purposes, a sixth detector per array, will be located up against, and possibly embedded in the inner shield wall/blanket module of the vessel. These detectors will be considered "sacrificial" as no provisions will be made for maintenance unless there is a major vessel reconfiguration. Their primary function is to facilitate calibration using an "in-vessel" neutron source. They will be placed close to the plasma to provide the widest possible solid angle of view so that the calibration source best represents a plasma. The remaining detectors can be cross-calibrated to this detector using the plasma as a neutron source. One criticism of this proposal is the inconvenience of merging the count-rate signals provided by 6 detectors of differing sensitivities and limited ranges when, by using them in current mode, far fewer detectors are required. However, current and Campbelling modes have been observed to exhibit non-linearities on TFTR. Thus, a conservative approach would be to use count-mode over the entire dynamic range, backed by the other electronic modes for operational simplicity and to achieve better time resolution.

4.2 Proposed neutron yield monitor calibration techniques for ITER

The calibration issue for ITER has already been considered briefly[23] and is elaborated below.

4.2.1 In-situ calibration using in vessel neutron generators

Performing this task is evidently a non-trivial problem, since a very strong source of neutrons is required. A serious limitation is the fact that any extensive in-situ calibration can only be performed before start-up and/or during major interventions before the machine becomes radio-active; thereafter, the use of a neutron generator requires remote handling access and will occur infrequently. Only the most sensitive of the neutron yield monitors and the most sensitive detectors in the profile monitor can be calibrated in this manner. Because conventional neutron generators are bulky, a non-negligible fraction of the emitted neutrons interact with components of the generator itself, leading to spectrum degradation and anisotropic emission. The main source of error thus lies with the characterization of the neutron output of the generator. Careful monitoring with activation foils as demonstrated at TFTR and JET provides a measure of the neutron emission strength from the generator. The directionality of the neutrons emission and the energy spectrum, which also may be a directional property, need to be established.

The U.S. and Japanese proposals are that ITER should be provided with a robust, high-intensity D-T neutron generator capable of operation at the end of a remote handling arm inside the torus. The optimum generator would have an isotropic neutron output of over 10^{10} neutrons/s with a "fusion" (~14-MeV) energy and the capability of accurate "associated particle" detection (that is ability to measure 100% of the charged fusion products in some known solid angle) for direct absolute calibration of the emission strength. A comparable scheme (without the remote handling involvement) using a small Van de Graaff generator was briefly considered for JET but was rejected as being too complicated. The Inertial Electrostatic Confinement devices that are currently under development would be far more practicable.

4.2.2 Calibration using a pneumatic transport activation system

This is the preferred method in use at JET, where all other techniques are relegated to a supporting role. Its advantages are clear: it is intrinsically linear and can therefore be used for detecting possible non-linearities in neutron yield monitor responses. It is also stable in time, provided the necessary care is taken to maintain the calibration of the decay radiation detectors with standard radio-isotope sources. Reactor dosimetry materials, i.e. well chosen materials with evaluated cross sections yields, should be employed. Suitably high threshold reactions add energy discrimination, thereby reducing the importance of low energy, scattered neutrons. Its main weakness lies in the essential neutron transport calculations that involve a detailed modeling of the machine, especially the region close to the irradiation ends. The closer the irradiation station is to the plasma, the less difficult and time consuming is the modeling. At JET and TFTR, the irradiation ends protruding into the main vacuum vessel proved the most useful and are used to inter-calibrate supplementary ex-vessel stations.

At ITER, the irradiation stations would have to be protected from excessive heat flux and cannot be located too close to the first wall. Establishing a precise location is fraught with difficulty. Confidence in the transport calculations can be gained by using activation materials offering a range of threshold energies.

As a learning exercise at TFTR, a foil placed in an irradiation end was activated using an in-situ neutron source. Agreement between predicted and measured activities was obtained to about ± 20% accuracy, the limitation being the low strength of the source. On ITER, even with the planned source strength of 10^{10} neutrons/s, it is not clear that this direct in-situ approach would be possible due to the low of sensitivity of foils, even those with high cross-sections such as copper. Alternatively, a small well-characterized scintillator/photomultiplier tube could be placed in the irradiation end during the calibration period. This might even have the sensitivity to provide a full line-of-sight calibration of the irradiation end as a well-characterized generator is moved through a full transit around the vessel, thus providing a meaningful benchmark for the neutronics codes.

A variant of the activation technique would be to replace the pneumatic transport of small solid samples by the use of a fluid flow loop, where the active material is either the fluid itself or dispersed within it. This type of activation system has been used in fission reactor applications but has not yet been employed at a fusion device. It may have advantages for ITER and should be elaborated further. The calibration of such a system would be performed in a manner comparable to that for the pneumatic transfer system.

Instead of measuring the induced activation in a flowing fluid, it would be feasible simply to measure the temperature rise of the fluid due to nuclear heating between entry and exit from the machine as this should be related linearly to the fusion power production. However, it would be necessary for the pipework to be thermally insulated from the machine components, otherwise the calculation of temperature rise would have to include a consideration of the effects of all cooling systems, cryogenic as well as main coolant flows. Also, the full calculation would involve a careful assessment of deep penetration of neutrons and gamma-rays to *all* regions crossed by the pipework. To provide an accurate independent calibration this measuring system would be very difficult and the temporal response would be very poor. It has recently been suggested[24] that the temperature rise in the main coolant could also be used as a measure of the fusion power production. The task of providing an independent absolute calibration would be daunting indeed, due to the need to compute the nuclear heating in all machine components traversed by the pipework (no longer thermally insulated). The accuracy of such a calibration could never be competitive with that provided by activation techniques using high threshold reactions.

4.2.3 Profile monitors

A diagnostic comprising a pair of properly designed neutron profile monitors offering orthogonal full coverage views across the ITER plasma would provide a wealth of important information concerning plasma performance. The degree of compromise needed to achieve a practicable solution will be discussed elsewhere. We address here only the issue of providing an absolute calibration for the detectors. There are two alternative approaches that can be followed, although perhaps both should be pursued:

(i) In-situ calibration of cameras with generators:

The disadvantages of using neutron generators inside a tokamak were considered above, although for calibrating the profile monitor some of the constraints may be relaxed. Specifically, the neutron generator does not have to be exceptionally strong since the detectors view mostly uncollided neutrons and the angular emission is not an issue provided the foil for the activation calibration is placed at the same orientation as the collimated sight-line. This approach was used successfully at TFTR[25].

(ii) Calibration of the profile monitor from first principles:

As explained earlier, the practice at JET is to determine the detector using a first principles approach. The detectors are characterized at an accelerator laboratory, which actually means determining their light responses as a function of energy of incident monoenergetic neutrons. (Light response functions are available in the literature but are specific to the type of detector and, we suspect, production batch). A series of neutron transport calculations are then performed to determine the effective solid angles of the collimators, any attenuation and in-scattering effects and, separately, the neutron back-scattering from the vacuum vessel wall that intercepts the lines-of-sight.

The main problem with the JET D-D neutron detectors has always been the need to discriminate gamma-rays from neutrons, an essential requirement since the lines-of-sight unavoidably intercept the vacuum vessel walls and supporting structures. This is achieved through pulse-shape analysis, a conventional but inconvenient procedure. The pulse-shape discrimination electronics need careful setting up and are not very stable. Accordingly, gamma-ray spectra from in-built sources are acquired between discharges for monitoring purposes. This is not sufficient, however, and a new system is currently being assembled that will permit Am-Be neutron sources to be moved into position behind each detector in turn, also between discharges. For D-T measurements, on the other hand, it is possible to reduce the detection of gamma-rays to a negligible level by good detector design. Nevertheless, determining the neutron energy discrimination level remains a problem. Accordingly, it is strongly recommended for ITER that a set of small (switchable) 2.5 and 14-MeV neutron generators be installed behind each detector for between-discharge monitoring.

5. CONCLUSIONS

In order to fulfill the parameter ranges, target measurements resolutions and accuracies as laid down by ITER[26], particular attention has to be given to calibrating the neutron diagnostics. This could be the largest source of error in the total fusion power and power density measurements. Previous experience from present-day large tokamaks dictates that several different methods should be used. The use of a neutron generator for direct calibration of the total yield monitors and of profile monitor detectors is practical only before start-up and on the rare occasion of a major intervention. An activation system for routine calibration checks of the yield monitors is strongly recommended; apart from yielding an independent calibration, it is the only way of proving the linearity and stability of the various systems. In addition, a neutron profile monitor is required for neutron emissivity measurements and this can be calibrated (from first principles, if necessary) and should constitute a second independent calibration method available at all times. Experience with existing large machines indicates that the activation method is more accurate than the first-principles calibration of the profile monitor but the problems anticipated for ITER could result in this situation becoming reversed .

---

[1] J.D.Strachan *et al*, *Neutron calibration techniques for tokamak results, Rev. Sci. Instrum*. 61: 3501 (1990).

[2] L.C. Johnson, Cris W. Barnes, A. Krasilnikov *et al., Neutron diagnostics for ITER, Rev. Sci. Instrum*. 68: 569 (1997) and L.C. Johnson *et al, in these proceedings.*

[3] Cris W. Barnes, M.J. Loughlin, and Takeo Nishitani, *Neutron activation for ITER, Rev. Sci. Instrum.* 68: 577(1997).

[4] Cris W. Barnes and A. L. Roquemore, *Neutron source strength monitors for ITER, Rev. Sci. Instrum.* 68: 573 (1997).

[5] O.N. Jarvis *et al., Calibration of the neutron yield detectors on JET*, JET Report JET-IR(84)02 and O.N.Jarvis *et al., Further calibrations of the time resolved neutron yield monitor*, JET Report JET-IR(85)06.

[6] O.N. Jarvis *et al., Use of activation techniques for the measurement of neutron yields from deuterium plasmas at the Joint European Tokamak, Fusion Technology* 20: 265 (1991).

[7] O.N.Jarvis, G.Sadler, P.van Belle and T.Elevant, *In-vessel calibration of the JET neutron monitors using a* ^{252}*Cf neutron source: difficulties experienced, Rev. Sci. Instrum.*, 61: 3172 (1990).

[8] B.J. Laundy and O.N. Jarvis, *Numerical study of the calibration factors for the neutron counters at the Joint European Torus, Fusion Technology*, 24: 150 (1993).

[9] K.A. Verschuur, *Neutronic calculations for JET,* Netherlands Energy Research Foundation Report ECN-C-96-053 (1996).

[10] E.B. Nieschmidt *et al., Effects of neutron energy spectrum on the efficiency calibration of epithermal neutron detectors, Rev. Sci. Instrum*. 56: 1084 (1985).

[11] H.W.Hendel, R.W. Palladino, Cris W. Barnes *et al., In situ calibration of TFTR neutron detectors, Rev. Sci. Instr.*, 61: 1900 (1990).

[12] Cris W. Barnes, M.G. Bell, H.W. Hendel *et al., Absolute calibration of neutron detection systems on TFTR and accurate comparison of source strength measurements to transport simulations, Rev. Sci. Instr.*, 61: 3151 (1990).

[13] Cris W. Barnes, H H. Duong, D. L. Jassby *et al., DT neutron measurements and experience on TFTR* in: *Diagnostics for Experimental Thermonuclear Fusion Reactors*, P.E. Stott, G. Gorini, and E. Sindoni, eds., Plenum Press, New York, (1996), 379-383.

[14] J.D.Strachan, Cris W. Barnes, M. Diesso *et al., Absolute calibration of TFTR helium proportional counters, Rev. Sci. Instr., 66: 1247 (1995).*

[15] J.D.Strachan, A.L.Roquemore, M.Diesso, S.L.Liew and J.Roberts, *Influence of neutron scattering on measured TFTR neutron profiles, 7th EPS Conference on Controlled Fusion and Plasma Heating, Amsterdam, June 1990. Europhys. Conf. Abstracts* 14B, part IV, 1548.

[16] Cris W. Barnes, A. R. Larson *et al., Calculations of neutron activation response for the Tokamak Fusion Test Reactor and absolute calibrations of neutron yield, Fusion Technology*, 30: 63 (1996).

[17] D. L. Jassby, Cris W. Barnes and L. C. Johnson *et al., Absolute calibration of Tokamak Fusion Test Reactor neutron detectors for D-T plasma operation, Rev. Sci. Instr.*, 66: 891 (1995).

[18] L.C. Johnson, *Validation of spatial profile measurements of neutron emission in TFTR plasmas, Rev. Sci. Instr.*, 63: 4517 (1992).

[19] Cris W.Barnes and J.D.Strachan, *Calibration of the TFTR neutron spectrometers, Rev. Sci. Instr.*, 61: 2383 (1990).

[20] L. C. Johnson, Cris W. Barnes, H. H. Duong *et al., Cross-calibration of neutron detectors for deuterium-tritium operation in TFTR, Rev. Sci. Instr.*, 66: 894 (1995).

[21] T. Nishitani, H. Takeuchi, T. Kondoh *et al., Absolute calibration of the JT-60U neutron monitor using a* ^{252}*Cf neutron source, Rev. Sci. Instr.*, 63:5270 (1992).

[22] M. Hoek, T. Nishitani and Y. Ikeda, *Neutron yield measurements by use of foil activation at JT-60U, Rev. Sci. Instrum*. 66: 885 (1995).

[23] L.C. Johnson *et al.*, in *ITER Diagnostic Review; Neutron Systems*, March 3-7 1997.

[24] J.Wesley, *Fusion power measurement by cooling water calorimetry, ITER Document*: S19 RI 8 97-07-10 F1

[25] A.L. Roquemore, T.C. Chouinard, M. Diesso *et al., TFTR multichannel neutron collimator, Rev. Sci. Instrum*. 61: 3163 (1990).

[26] A.E. Costley *et al., Requirements for ITER,* in *Diagnostics for Experimental Thermonuclear Fusion Reactors*, P.E. Stott, G. Gorini, and E. Sindoni, eds., Plenum Press, New York, (1996), 23-37 and V.S. Mukhovatov *et al., Role and Requirement for Plasma Measurements on ITER, in these proceedings.*

GAMMA RAY SPECTROMETRY IN ITER: CONCEPTUAL DESIGN

V.G.Kiptily[1], T.Kondoh[2], V.O.Naidenov[1], I.A.Polunovski[1], Yu.V.Tuboltsev[1], I.N.Chugunov[1], A.E.Shevelev[1]

[1]A.F.Ioffe Physical Technical Institute
194021 St.Petersburg, Russia
[2]Japan Atomic Energy research Institute
Naka-machi, Naka-gun, Ibaraki-ken, 311-01, Japan

INTRODUCTION

This paper presents the result of conceptual design of Gamma Ray Spectrometry system, integrated in the ITER Neutron Camera, to provide plasma diagnostic information, based on measurements of line-integral gamma-ray emissivity and deduced yield profile. Information can be obtained on alpha-particle birth profile and the profile of confined particles, impurity density in plasma core, as well as different effects of plasma instabilities such as sawteeth and TAE modes. Design of the system presented here is based on the Radial Neutron Camera[1]. It is assumed that for Vertical Camera the system will have the similar construction and capabilities.

Besides of the conceptual design, this paper presents the outline of diagnostic R&D works that required to fulfill this design phase. A prototype of GAMMACELL spectrometer is manufactured and tested. Neutron sensitivity of BaF_2 scintillator which used in the spectrometer is examined and neutron shield testing is performed at neutron generator. Cross sections of some diagnostic nuclear reactions are measured, and special nuclear database is developed. A new diagnostic approach to measure escaping alphas is developed.

BASIS OF ITER GAMMA DIAGNOSTICS

Gamma-ray spectrometry in ITER is based on registration and analysis of radiation from nuclei excited in reactions that can take place during plasma discharge. There are three type of nuclear reactions that may be noted for diagnostics: reactions between fuel nuclei, interaction between fast charged particles and plasma impurities, and reactions with specially doped impurities. Some of the nuclear reactions are given in Table 1. It is supposed that gamma-ray measurements will be performed during both Zero Activation and High Activation phases of the ITER operation.

Gamma-ray emissivity of fusion plasma is defined by a value of the reaction rate

$$R(r) = \frac{n_1(r)\, n_2(r)}{1+\delta_{12}} \langle \sigma v \rangle ,$$

where $\langle \sigma v \rangle$ is rate parameter averaged over the Maxwellian distribution of fusion particles; n_1 , n_2 are the densities of reacted ions and δ_{12} is Kronecker's symbol. In the case of fast ion reactions the rate is expressed as

$$R_i(r) = \left(\frac{2}{m_i}\right)^{1/2} \int n_t(r)\, \sigma(E)\, E^{1/2}\, F_i(E,r)\, dE ,$$

where $F_i(E,r)$ is the fast ion distribution function; $n_t(r)$ is a density of the impurity target.

Table 1. Main diagnostic reactions and associated gamma rays (energy in MeV)

FUSION REACTIONS		BACKGROUND REACTIONS		"PELLET" REACTIONS	
$D(p,\gamma)^3He$	5.5	$^9Be(p,p'\gamma)^9Be$	1.7, 2.4, 3.05, 4.7	$^7Li(p,p'\gamma)^7Li$	0.478
$T(p,\gamma)^4He$	20	$^9Be(p,\gamma)^{10}B$	5.1, 6.7, 7.5, 7.8	$^7Li(p,\gamma)^8Be$	17.6
$D(d,\gamma)^4He$	24	$^9Be(p,\alpha\gamma)^6Li$	3.56	$^{11}B(p,p'\gamma)^{11}B$	2.4, 4.44, 5.02
$D(t,\gamma)^5He$	17	$^9Be(d,n\gamma)^{10}B$	2.8, 2.15	$^{11}B(p,\gamma)^{12}C$	4.44, 11.7
$D(^3He,\gamma)^5Li$	17	$^9Be(d,p\gamma)^{10}Be$	3.4	$^{19}F(p,\alpha\gamma)^{16}O$	6.1, 6.9, 7.1
		$^9Be(^3He,p\gamma)^{11}B$	7.3, 4.44	$^{10}B(\alpha,p\gamma)^{13}C$	3.1, 3.7, 3.9
		$^9Be(^3He,n\gamma)^{11}C$	6.3-6.9, 4.3		
		$^9Be(\alpha,n\gamma)^{12}C$	4.44		

A spectrometer, placed in a collimator views a plasma volume along the line of sight. The gamma-ray count rate can be written as

$$P = k(E_\gamma)\, \varepsilon(E_\gamma) \int R\, \Omega\, dV_{LOS} ,$$

where $k(E_\gamma)$ is a gamma-ray attenuation factor in the LOS, $\varepsilon(E_\gamma)$ the detector efficiency; Ω the detector solid angle, V_{LOS} the plasma volume viewed by the spectrometer. A temporal resolution of the spectrometer, that is, an expected time of measurement at the required accuracy δ and the background count-rate B , is calculated by means the expression

$$t_m = \left(\frac{\sqrt{P+B}+\sqrt{B}}{\delta P}\right)^2 .$$

The Gamma Ray Spectrometry system is installed in the Radial (Vertical) Neutron Camera that consists of fan-shaped array of flight tubes, viewing the ITER plasma through slots in the blanket/shield, intersecting at a common aperture defined by a specialized shielding plug in a midplane/vertical port, and penetrating the vacuum vessel, cryostat, and biological shield through stainless steel windows. A set of gamma detectors are chosen to provide the required range of sensitivity, and temporal and spectral resolution.

FUNCTIONS

The Gamma Ray Spectrometry system in coupling with the neutron cameras has important functions in providing measurements included in ITER diagnostic categories (i), (ii) and (iii) given in Design Description Document[2]. Plasma parameters for which the gamma spectrometry may provide measurements are given in Table 2.

The primary functions are measurement of the *dt* alpha particles. Sum of the chordal 17-MeV signals, together with knowledge of the plasma configuration, gives the fusion power in plasma core. Combined with data from the vertical and radial cameras, these measurements allow tomographic reconstruction of the spatial distribution of fusion reaction rate, $n_D n_T \langle \sigma v \rangle$, which determines the alpha-particle source profile and fusion power density. Signals from 4.44-MeV gamma rays, created when fast alphas react with the main plasma impurity ions beryllium allows to determine the 2-MeV alpha-particle distribution[3], using additional data from the impurity monitor and x-ray crystal spectrometers. In the case of knowledge of the alpha-particle distribution the system yields the beryllium concentration in plasma core.

Signals from the $D(t,\gamma)^5He$, $T(p,\gamma)^4He$, $D(d,\gamma)^4He$ capture reactions can provide information on fuel density; especially high gamma-ray yield is expected in the case of beam-fuel reactions. Because of the known functional dependence of the reaction cross sections on temperature, knowledge of T_i gives ratios n_H/n_D and n_D/n_T in plasma core.

Table 2. System Functions

Category & Measurement	Plasma Parameters
(i)#9	m=2 MHD Modes, Sawteeth
(i)#11	n_D/n_T in Plasma Core
(ii)#27	Fishbones, TAE Modes
(ii)#30	Confined Alpha-Particles (in Plasma Core)
(ii)#31	Escaping Alphas
(ii)#32	Impurity Density (Be in Plasma Core)
(ii)#44	n_H/n_D in Plasma Core
(iii)#49	MHD Activity in Plasma Core

In Zero Activation Phase (*He, H* plasma) and in *d-d* plasma of High Activation Phase measurements of time-dependent gamma-ray emissivity profile of plasma core allow to determine the fast ion distribution function or ion termalization, localization of fast ions and a response to plasma instabilities (sawteeth, *TAE* modes)[4,5].

DESIGN DESCRIPTION

Geometry of System.

The design for Gamma-ray Spectrometry system based upon the horizontally viewing Neutron Camera[1] designed by EC team and spectrometers developed in A.F.Ioffe Physical Technical Institute. Gamma detector modules are connected with neutron ones outside the ITER biological shield and have the same fan-shaped viewing geometry. Each module contains special pair of sight-lines with adjustable collimation, allowing to change the detector for gamma-ray spectrometry over a wide operating range and different tasks. The system measure line-integral gamma-ray emissivity along chords, viewing inwardly through both midplane and vertical ports, and spanning central part of the plasma cross section. The gamma detectors are housed in an auxiliary shielded enclosure outside the biological shield, and additional shielding along and between flight tubes provides necessary collimation.

The modular system, where neutron and gamma detector sub-system has its own precision machined collimator, allows individual units to be maintained or altered as necessary. A module consists of a cylindrical plug 0.4 m in diameter and 3.5 m length containing collimators and detectors. Each module has neutron and gamma detector housings separated by a beam dump module with 3 cm diameter aperture for the gamma-ray measurements. The suggested 10 sightlines cover ~ 50% of total vertical plasma height and this satisfies the 30 cm spatial resolution requirement.

For the gamma spectrometry the collimator module has two additional apertures of 4.5 cm and 1 cm diameters with ^{6}LiH plugs for neutron flux attenuation. The optimal total size of the plugs is 80 cm (can be extended up to 140 cm) that sufficiently to reduce flux for a 1.5 GW plasma. The internal diameter of each collimator can be remotely set to 1.0 cm or 4.5 cm, to be able to adjust for the fluxes expected in either *d-d* or *d-t* operation, or for low intensity in peripheral channels.

The gamma viewing cone from the aperture of 4.5 cm (1 cm) diameter has a full opening angle of 1.43^{o} (0.57^{o}), the same as in the Radial Neutron Camera. The opening of viewing cone is 29 cm (12 cm) at the plasma axis.

The line integrated gamma-ray emissivity from a reference 1.5 GW ITER plasma at the detector viewing along a horizontal sight-line would lead to a count rate of about 2 MHz. In the collimator with 80 cm ^{6}LiH plug, the neutron intensity at the detector is ~ 2 10^{4} n/s, while the 14.1-MeV neutron flux to first wall is ~ 10^{14} n/cm^{2}/s. This neutron flux generates ~ 5 10^{3} cps in the detector.

Individual Collimator and Detector Module.

In an individual module, the front and rear collimators consist of two cylinders, 36 cm outer diameter (allowing for wall and bearing thickness) and 40 cm in length. Each cylinder has two additional holes drilled on a circle of diameter 18 cm: hole of 4.5 cm diameter with ^{6}LiH plugs and hole (rotated 180° from the first one) of 1 cm diameter. The collimators are individually rotatable to bring the appropriate collimators in line with the detectors, allowing adjustments for all phases of operation. The intermediate flight tube assembly stops most of the neutrons that do not enter the rear collimator hole and reduces the radiation outside the bioshield. An additional bioshield extension is required to reduce the effect of neutron streaming through the various apertures of the system.
Cross sectional views of a collimator/detector module are shown in Fig.1.

Gamma-ray Spectrometers.

Two types of detectors have been suggested for measurement of gamma-ray energy spectra. The *GAMMACELL* spectrometer[6] is developed by the RF Home Team. It intends for all phases of ITER operation. Other one is a conventional high purity germanium detector with best resolution which is suggested to use during Zero Activation Phases only. The *GAMMACELL* spectrometer consists of nine optically independent scintillation detectors (42x42x142 mm) assembled in a single housing; the general assembling draft of it is shown in Fig. 2. The spectrometer will be located inside of the collimator in such a manner that investigated gamma radiation impinges only on central section of array. All the detectors are included in a summarizing circuit which yields full energy of gamma ray, falling on a central

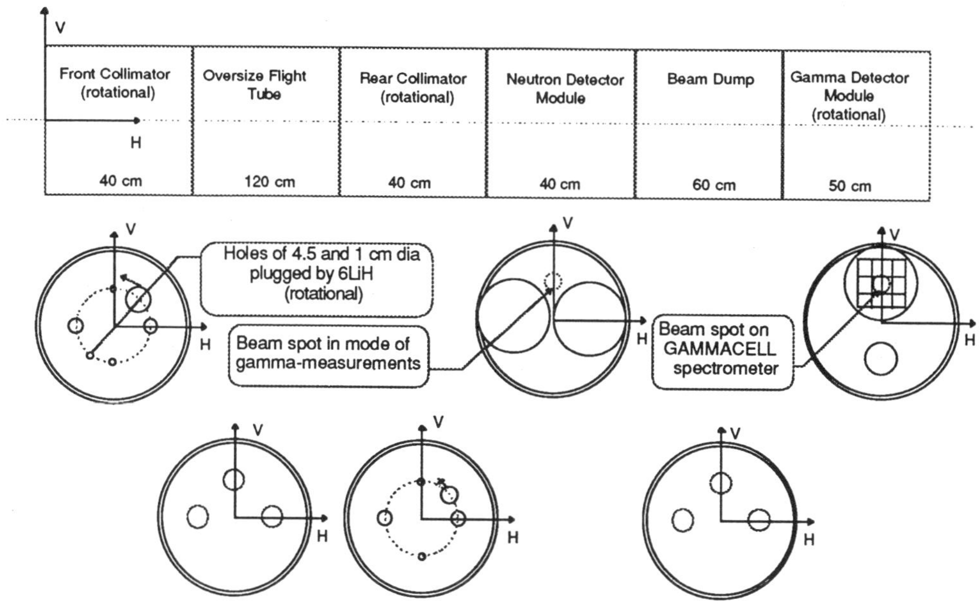

Figure 1. A Collimator and gamma detector module, integrated in the Radial Neutron Camera.

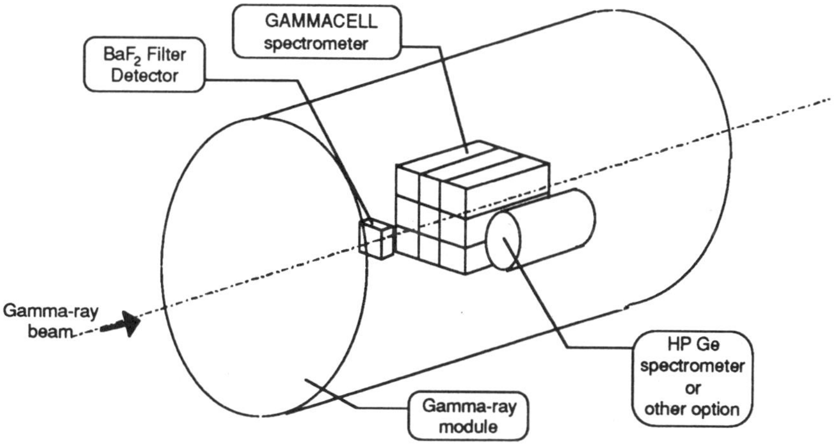

Figure 2. A gamma detector array in the collimator module.

crystal. Before the central crystal, it is supposed to establish one more detector (passive-active gamma-ray filter) with the sizes 60x60x40 mm, which must functionate as an absorber of low energetic gamma radiation and, simultaneously, as a rejecter of scattered radiation. This ensemble of detectors allows to reduce considerably the background load to the basic spectrometer.

Pilot version of the data acquisition and processing system allows to accumulate gamma spectra at count-rate more than 1 MHz. Required time for processing of a single event is equal to 150 ns. The system is manufactured as two unit in *CAMAC* standard.

DIAGNOSTIC REACTION RATES

Most part of nuclear reactions presented in Tab. 1 are examined for feasibility using in the ITER plasma diagnostics. Geometry of gamma-ray measurements outlined here is used for the calculations.

Results of estimations for fast proton nuclear reactions at resonance energies are given below in Table 3. Measurements of these diagnostic reactions can provide information on distribution function and a response to plasma instabilities in Low Activation Phase of the ITER operation. The $^{9}Be(p,\alpha\gamma)^{6}Li$ reaction is very promising with the HPGe spectrometer to study the distribution anisotropy during ICRF heating.

Table 3. Calculation results for Low Activation Phase (He, H)

Resonance Energy, MeV	Gamma-Ray Energy, MeV	Reaction Rate[1], $cm^{-3}s^{-1}$	Count Rate, cps
	$^{9}Be(p,\gamma)^{10}B$		
0.139	5.13	$3\ 10^{4}$	10^{3}
0.938	6.71	$3\ 10^{4}$	10^{3}
0.992	7.48	$2.8\ 10^{5}$	$8\ 10^{3}$
1.083	6.84	$3.0*10^{4}$	10^{3}
1.290	7.75	$1.4\ 10^{5}$	$4\ 10^{3}$
	$^{9}Be(p,\alpha\gamma)^{6}Li$		
2.57	3.57	$1.2\ 10^{7}$	$1.3\ 10^{4}$

[1]All calculations were performed at fast ion density equal to 10^{9} cm^{-3} keV^{-1}

At steady state plasma in the High Activation Phase the radiative capture reactions $D(t,\gamma)^{5}He$, $T(p,\gamma)^{4}He$ and $D(d,\gamma)^{4}He$ would be very useful to measure the value of fuel density ratio, but only first one can be measured in ITER with sufficient temporal resolution ~ 0.5 ms and accuracy 10% . Other two reactions can be measured with the temporal resolution ~ 200 ms and ~ 25 s, correspondingly, with accuracy 20%. Nevertheless these reactions can provide information during NBI heating, if the LOS of detectors cross the beam.

Study of the confined alpha-particles can be carried out by means of the $^{9}Be(\alpha,n\gamma)^{12}C$ reaction[3]. Information on profile of 2-MeV alphas in steady state plasma is deduced from 4.44-MeV gamma-rays with temporal resolution ~ 2 ms at the 2% Be density and the 10% accuracy.

A contribution of background reaction $^{12}C(n,n'\gamma)^{12}C$ that occur in plasma at the carbon density ~ 0.2 % is less then 10% of the true signal. The same contribution, arisen from 4.44-MeV wall emissivity will be observed in the case if the 5 g of total carbon mass is in the detector LOS.

New diagnostic reaction $^{10}B(\alpha,p\gamma)^{12}C$ that induces the 3.09, 3.68, 3.85-MeV gamma-rays would be very prospective for alpha-particle diagnostics[6]. However, a special pellet injector is required to bring in the additional impurity in plasma. Other problem is a high background radiation in the energy range 3 - 4 MeV that predicted by neutronic calculations. This reaction may be very promised for alpha-particle diagnostics in future *d-*^{3}*He* fusion reactors.

CRITICAL DESIGN ISSUES

The main drawback for all diagnostics that designed for ITER, including the system presented here, is a very high neutron and gamma-ray background during ignition. As to gamma-ray spectrum measurements, optimization of the shield and detector system improvement are needed to reach of it feasibility and the best results for peak to background ratio and counting rate.

A wide and useful experience of gamma spectrometry is accumulated at the JET[4,7] and JT-60U[5,8] for fast ion study at ICRF and NBI heating of *d-d* and *He-H* plasma. These results allow to hope in feasibility and reliability of the gamma-diagnostics of fusion plasma.

RESEARCH AND DEVELOPMENT

GAMMACELL Spectrometer

A prototype of diagnostic gamma spectrometer for ITER is developed. The pilot detector is produced and tested using accelerator. Results of the examination and parameters projected for advanced data acquisition system of the spectrometer are given in Table 4. The goal of the activity is to reach a high efficient registration of gamma rays with energies up to 30 MeV at high counting rate, and to provide the best peak-to-background ratio.

Table 4. Test results and projected parameters of the spectrometer

GAMMACELL Parameters	Achieved	Projected
Energy range	0.4 - 10 MeV	1 - 30 MeV
Energy resolution @ 0.662 MeV	27%	15%
@ 4.44 MeV	15%	5-7%
Efficiency @ 4.44 MeV	53%	60%
Counting rate	not examined	> 2 MHz

Neutron Sensitivity of the BaF_2 Scintillator

The BaF_2 crystal, used in the spectrometer has been examined on sensitivity to 2.8-MeV and 14.8-MeV neutrons. The measurements have been performed using a calibrated neutron generator. The neutron yield was controlled with an accuracy ~ 3%. The spectra induced by the neutrons are obtained and sensitivity of the scintillator is evaluated.

Neutron Shield

To suppress neutron flux in the collimator a shielding plug is developed on the base of ^{6}LiH powder. Tree samples of the plug with dimensions D30xL300 mm have been manufactured. The reducing factors of the plugs are measured using the neutron generator with 2.8-MeV and 14.8-MeV beams. These results have been used for conceptual design of the system.

Diagnostic Nuclear Reactions & Database

Gamma-ray spectra and cross sections of diagnostic reactions ^{9}Be(α,nγ)^{12}C, ^{10}B(α,pγ)^{13}C, ^{9}Be(p,$\alpha\gamma$)^{6}Li, ^{19}F(p,$\alpha\gamma$)^{16}O have been measured at accelerators. These data were utilized for the reaction rate estimations. In result of the analysis of experimental and evaluated nuclear reaction data the diagnostic database *FUSION* has been developed.

Escaping Alpha-particle Technique

To provide information on escaping alpha-particles an approach is developed based on the using investigated reaction ^{10}B(α,pγ)^{13}C. Intensity of lost alphas is deduced from gamma-ray emission from ^{10}B targets mounted on first wall in detector LOS. A result of preliminary estimation for diagnostic escaping alphas is given in Tables 5. It is suggested that ^{10}B targets have the dimentions ~ 20 cm in diameter and 20 - 30 μm thickness.

Table 5.

	Steady state: < 2MW/m^2		Transients: < 20 MW/ m^2	
Alpha-particle Energy Range, MeV	Temporal Res. @ 20% accuracy, ms	Count Rate, cps, and B/P ratio	Temporal Res. @ 10% accuracy, ms	Count Rate, cps, and B/P ratio
> 1.2	> 200	$2*10^4$ (> 30)	> 7	$2*10^5$ (> 3)

CONCLUSIONS

This paper presents a conceptual design for gamma-ray measurements in ITER. The design is based on the construction of the Radial Neutron Camera. Gamma-ray detectors are incorporated behind the neutron detectors. The system can provide plasma diagnostic information on alpha particles and fast ions, the response to MHD instabilities and time-averaged fuel density ratios. Moreover, it could be used at the beginning of Low Activation phase of the ITER operation. A set of R&D works are carried out for successful fulfillment of the conceptual design. Implementation of the results in large operated tokamaks JT-60U, JET, D-IIID is needed for further development of the ITER Gamma Spectrometry system.

REFERENCES

1. F.B.Marcus, J.M.Adams, P.Batistoni, T.Elevant, O.N.Jarvis, L.C.Johnson, L.de Kock, G.J.Sadler and P.E.Stott. A neutron camera for ITER: conceptual design, in: *Diagnostics for Experimental*

Thermonuclear Fusion Reactors, P.E.Stott, G.Gorini, E.Sindoni, eds., Plenum Press, New York and London (1996).

2. K.M.Young, A.E.Costley, T.Matoba, D.Orlinski, P.E.Stott. The ITER diagnostic programme, *ibidem*
3. V.G.Kiptily. Capabilities of gamma spectroscopy for fast alpha-particle diagnostics, *Fusion Tech. 18:454 (1990).*
4. G.J.Sadler, S.W.Conroy, O.N.Jarvis, P.van Belle, J.M.Adams, M.A.Hone, *Fusion Tech.18:5 (1990).*
5. H.Kimura, M.Saigusa, S.Moriyama, T.Kondoh, Y.Neyatani, T.Ozeki, T.Nishitani, Y.Kusama, T.Fujii, M.Sato, M.Nemoto, K.Tobita, C.Z.Cheng, Excitation of high *n* toroidicity-induced Alfven eigenmodes and associated plasma dinamical behaviour in the JT-60U ICRF experiments, *Phys. Lett. A199:86 (1995).*
6. V.G.Kiptily, I.A.Polunovskii, I.N.Chugunov, V.S.Zaveriaev, S.V.Popovichev, A.V.Khramenkov, S.N.Abramovich, A.G.Zvenigorodskii, M.V.Savin, ITER gamma diagnostics: 2-D neutron and gamma camera, in: *Diagnostics for Experimental Thermonuclear Fusion Reactors,* P.E.Stott, G.Gorini, E.Sindoni, eds., Plenum Press, New York and London (1996).
7. O.N.Jarvis, J.M.Adams, P.J.A.Howarth, F.B.Marcus, E.Rigni, G.J.Sadler, D.F.H.Start, P.van Belle, C.Warrick, N.Watkins, Gamma-ray emission profile measurements from JET ICRF-heated discharges, *Preprint JET-P(95)53.*
8. T.Kondoh, H.Kimura, Y.Kusama, A.Morioka, S.Moriyama, M.Sigusa, K.Nagashima, R.Yoshino, H.Harano, Gamma-ray measurements in JT-60U ICRF heated plasma, *J. Plasma Phys. Fusion Res. 72:1397 (1996).*

DATA BASE FOR ENSURING THE WORKS ON GAMMA-DIAGNOSTICS OF TOKAMAK PLASMA

V.G.Kiptily,[1] S.M.Taova,[2] A.G.Zvenigorodskii [2]

[1]Physics and Technology Institute by A.F.Ioffe - FTI
[2]All-Russia Scientific Research Institute of Experimental Physics - VNIIEF

INTRODUCTION

The determination of ion temperature and the investigation of fast charged fusion products behaviour represent the significant task of hot thermonuclear plasma diagnostics. Among various methods proposed by now and suitable for different temperature intervals, considerable role is played by nuclear-physics methods. The last are based on various types of interaction between plasma ions and fast nuclear particles, produced as the result of thermonuclear fusion, with specially admixture additives with further registration of secondary products of this interactions.

There were proposed various approaches to the problem of hot plasma behaviour study by nuclear-physics methods, from which there should be set off the following ones:

Activation methods

These methods are based on the reactions between charged particles and specially chosen admixtures. Owing to their high sensitivity and efficiency the given methods can be used in the most interesting temperature interval - from 1 to 10 keV /1/. In this case it is possible to place a thick target just on the first wall of reactor chamber and to register the events of charged particles yield from the region of plasma confinement.

Spectrometry of high-energy γ-quanta

The possibility of hot plasma diagnostics with the aid of resonance (z,γ) reactions between the confined fast ions (H, D, T, He-3, α) and light admixtures which are present in plasma or artificially introduced to it, was considered in detail in papers /2, 3/. Thus, in case of α-particles, most suitable for diagnostic purposes are D, ^{6}Li, ^{7}Li (α,γ)-reactions, in which narrow resonances are observed.

The method which take advantage of the Doppler gamma-spectra shape for the obtain of information on the function on particles distribution in plasma, opens fresh opportunities for the investigation of fast ions and, specifically, thermonuclear α-particles. In case of DT plasma ^{9}Be$(\alpha, n\gamma)^{12}$C and ^{10}B$(\alpha, p\gamma)^{13}$C reactions are most effective for this purpose. The application of these reactions for TOKAMAK plasma diagnostics is described in more detail in papers /3,4/.

Neutron diagnostics of TOKAMAK plasma

Neutron spectroscopy was used at the dawn of thermonuclear researches as a method for measuring ion temperature in deuterium- deuterium or deuterium-tritium plasma. There were analyzed the neutrons from the following reactions:

$$^{2}H(d,n)^{3} + 3.269MeV,$$
$$^{3}H(d,n)^{4} + 17.589MeV,$$

as well as the neutrons from the "on flight" reactions.

Diagnostics for Experimental Thermonuclear Fusion Reactors 2
Edited by Stott *et al.*, Plenum Press, New York, 1998

In terms of the three approaches specified, there was formulated the list of reactions (see table 1), the data on which present particular interest for the activities on hot plasma diagnostics, using nuclear-physics methods.

Table 1. List of reactions included into the FUSION data base.

Reaction type	Reaction energy, MeV	Energy range, MeV	Data type	Possible usage
1-H-2(p,γ)2-He-3	5.494	0.024-22.6	sig	γ-diagnostics
1-H-2(d,γ)2-He-4	23.847	0.026-18.92	sig	γ- diagnostics
1-H-2(d,n)2-He-3	3.269	0.005-20.0	sig,dif	n- diagnostics
1-H-2(d,p)1-H-3	4.033	0.005-14.0	sig,dif	ch. part.-diagn
1-H-3(p,γ)2-He-4	19,814	0.1-17.8	sig	γ- diagnostics
1-H-3(d,γ)2-He-5	16.696	0.042-5.0	sig,res	γ- diagnostics
1-H-3(d,n)2-He-4	17.589	0.005-19.0	sig,dif	n- diagnostics
1-H-3(t,2n)2-He-4	11.332	0.01-4.0	sig	n- diagnostics
2-He-3(d,γ)3-Li-5	16.387	0.048-16.8	sig	γ- diagnostics
2-He-3(d,p)2-He-4	18.352	0.01-20.0	sig,dif	ch. part.-diagn.
2-He-4(d,γ)3-Li-6	1.475	0.1-12.0	sig,res	γ- diagnostics
2-He-4(t,γ)3-Li-7	2.468	0.01-1.51	sig	γ- diagnostics
2-He-3(τ,γ)4-Be-7	1.588	0.86-23.83	sig	γ- diagnostics
3-Li-6(p,γ)4-Be-7	5.606	0.157-1.174	sig	γ- diagnostics
3-Li-6(α,γ)5-B-10	4.460	0.4-1.7	sig, res	γ- diagnostics
4-Be-9(p,γ)5-B-10	6.586	0.102-1.194	sig, res	γ- diagnostics
4-Be-9(p,αγ)3-Li-6	2.126-E_{lev}	0.028-11.58	sig,dif, dop	γ- diagnostics
4-Be-9(d,γ)5-B-11	11.815	0.5-11.9	sig	γ- diagnostics
4-Be-9(α,nγ)6-C-12	4.439-E_{lev}	1.9-6.5	sig,dop	γ- diagnostics
5-B-10(α,pγ)6-C-13	4.063-E_{lev}	1.25-4.25	sig,dif, dop	γ- diagnostics
5-B-10 (p, γ)6-C-11	8.691	0.1-12.0	sig	γ- diagnostics
5-B-11 (p, γ)6-C-12	15.957	0.16-30	sig	γ- diagnostics

sig - integral cross-section; dif - differential cross-section;
res - resonance parameters; dop - γ-quantum spectra with account of Doppler broadening.

DESCRIPTION OF FUSION DATA BASE

To create the Fusion data base there was chosen the relation data base management system (DBMS) Ms Access, which stores and extracts the data in accord with the established between them connections defined by the user. Ms Access functions within the Windows frames and, having a graphical interface, uses the advantages of the graphical environment in the full scale and presents visual access to data.

The Fusion data base ensures the execution of the following procedures:

- arrangement of the required information search for the given reaction definition;
- calculations of the values of integral, differential and resonance cross-sections in the given energy range;
- output of the table of γ-lines characteristics for residual nuclei;
- output of the values of reaction rates;
- output of graphical presentation of the desired data to a screen;
- viewing the tables for the given reaction on the display screen;
- output of selected in the course of visual scan data to one or another medium (disk, print);
- output of the level schemes for the given nuclide;
- possibility of adding data to the base;
- possibility of getting reference for the data considered.

In the Fusion program the dialogue with the user is implemented through the use of window mode of information display. The input of initial values is performed immediately in the dialogue window. Bibliographical and other reference data are presented in the form of a table. The calculation data can be presented in the form of a separate value, table and chart.

Provision is made for the modes of editing table data and chart, for their output to print. The table data can be recorded on the external file in ASCII format and used for further calculations.
There are three main processing modes in data base Fusion:

Search data	- search and extraction of data on the given reaction name.
Edit data	- editing of the values introduced immediately in he tables themselves.
Add data	- introduction of new values into the basic tables with the use of manual input and from file data input modes.

Mode of data searching

To operate in the mode "Search data" it is necessary to specify the required reaction. For this purpose the user is to select from a set of data types the required ones. The modes of data processing, accessible for the reaction chosen, will be specified in the dialogue window.

Mode of data editing

This mode makes it possible to correct the data just in the tables themselves. One can remove the data in the records, add some new data, destroy the entire records.

Mode of data adding

The process of adding new data to the base is an automatically operated process. There are several modes of adding in the data base: data replacing, data adding and input of new data. To enter data templates are used.

CONCLUSION

The fusion data base gives the user wide possibilities for obtaining exhaustive data on the reactions, described in the papers related to hot plasma diagnostics. The availability of amicable interface makes the process of communication with the data base convenient and pleasant.

The growing necessity in information systems of this type provides for further development of the Fusion data base. Not only periodic replenishment of data base with new information, but the creation of additional modes connected with the processing of the existing ones, are expected. The possibility of coupling with the program modules of Dynamic Linking Libraries (DLL) allows to perform sufficiently complicated mathematical processing of data within the frames of the Fusion system itself. The usage of such procedures will ensure a powerful instrument, assisting in the solution of many science - research and practical tasks in the sphere of thermonuclear synthesis.

REFERENCES

1. Young F.C., Golden J. and Kapetanakos G.A., *Diagnostics for intense pulsed ion beams.* Rev.Sci. Instr., v.48, n4, (1977), pp.432-443.
2. Cecil F.E., Zweben S.J., Medley S.S., *A method for determining fast alpha particle confinement in TOKAMAK plasmas using resonant nuclear reactions.,* Nucl. Inst. and Meth., A245(1986) pp.547-552.
3. Kiptily V.G., *Fusion Technology*, v.18, N 4 (1990), p.583, v.22, N4 (1992) p.454.
4. Kiptily V.G. et.al, *ITER gamma-diagnostics: 2D neutron and gamma camera "Diagnostics for experimental thermonuclear fusion reactors,* Edited by P.E.Stott, Plenum Press, New York and London, pp.463-466.

DEVELOPMENT OF A THIN-FOIL FARADAY COLLECTOR ARRAY AS A LOST ALPHA PARTICLE DIAGNOSTIC FOR HIGH YIELD D-T FUSION PLASMAS

F.E. Cecil[1], J. C. Barbour[2], P. v.Belle[3], D. Darrow[4], M. Hone[3], O.N. Jarvis[3], S . Kern[1], M. Loughlin[3], S.S. Medley[4], A. Nowak[1], A.L. Roquemore[4], G. Sadler[3], Y. Takimoto[1]

[1]Colorado School of Mines, Golden CO 80401, USA; [2]Sandia National Laboratory, Albuquerque NM, 87131, USA; [3]JET Joint Undertaking, Abingdon OX14 3EA, UK; [4]Princeton Plasma Physics Laboratory, Princeton NJ 08540, USA.

INTRODUCTION

Conventional charged particle detectors, such as surface barrier Silicon detectors or scintillation detectors suffer from the inability to operate effectively at high temperatures and high background radiation fields. Alternatively, a stack of parallel thin metallic foils, electrically insulated from one another and connected to separate ammeters may serve as a charged particle spectrometer capable of operating under fairly extreme conditions of temperature (very low temperatures as well as high temperatures) and background radiation. Such devices may therefore be utilized on tokamaks such as JET or ITER as first wall lost alpha particle diagnostics. We have been examining this new concept in charged particle spectrometry by i) accelerator based testing of various designs to evaluate energy resolution and ii) installation of prototype devices at JET and TFTR to evaluate the radiation and temperature insensitivities and to insure that the electro-mechanical noise levels in the measured currents are well below the expected signal levels of lost alpha particles.

ACCELERATOR BASED TESTING

Our first accelerator based tests of this concept were carried out at the Univ. of Birmingham, U.K.[1]. More recently we have tested several device designs using He^{+} and He^{++} beams with energies from 1 to 7 MeV from the tandem Van de Graff accelerator at Sandia National Laboratory. U.S.A. The tests consisted of bombarding the detectors with monoenergetic beams while measuring the current in each foil. One detector consisted of four foils of 2.5 µm thick Ni separated by 2 mm thick ceramic rings. This detector was assembled from components fabricated at JET and is identical to the the detectors installed in JET for the 1995 and 1997 run periods described below. The results of this test are shown in Figure 1 where the *fraction* of the total beam current in each of the four foils is plotted veritcally for He^{++} beams from 1 to 6 MeV

Diagnostics for Experimental Thermonuclear Fusion Reactors 2
Edited by Stott *et al.*, Plenum Press, New York, 1998

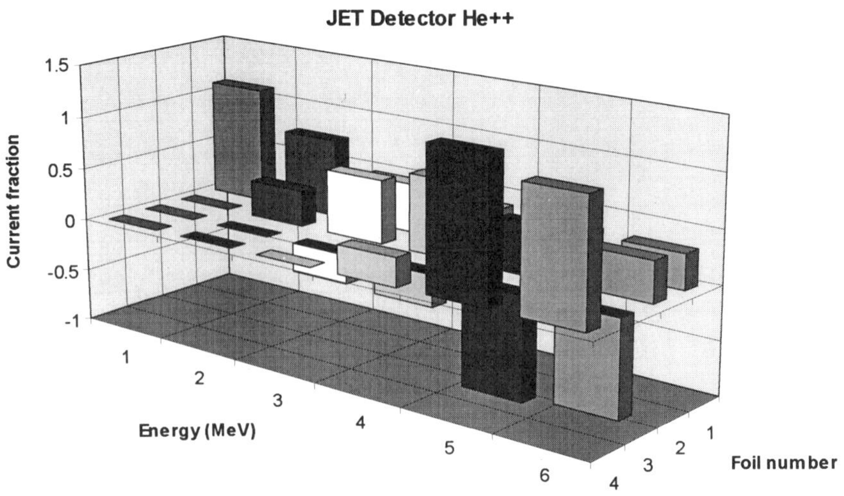

Figure 1. Fraction of total α-particle bombarding current in each foil for α-particle energies from 1 to 6 MeV. The *negative* current fractions in certain foils are the result of secondary electrons emitted from *adjacent* foils. The *positive* current fractions in the front foils at the higher energy α-particles are due to secondary electrons emitted from the front foils as the α-particles pass through these foils. The influence of the secondary electrons was partially reduced when the detector was placed in an external magnetic field

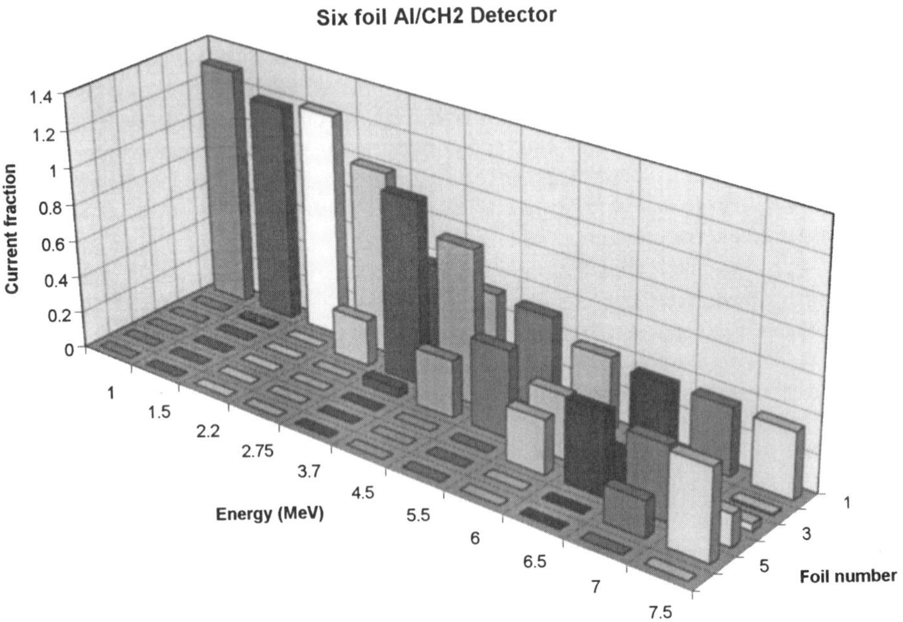

Figure 2. Measured current fractions in each foil during the bombardment of the Aluminum/polyethylene detector with α-particles with energies from 1.0 to 7.5 MeV. We are able to distinguish between beams of, for example, 7.0 and 7.5 MeV indicating an energy resolution *better* than 7%. The current fractions in the front foil which are in excess of unity at the lowest energies are due to the emission of secondary electrons from the front face of this foil in addition to *all* of the α-particles stopping in this foil. The negative current fractions which were apparent the the four foil Ni detector in Figure 1 are effectively suppressed by the insulating polythelene layers between the conducting Al foils. The positive current fractions in the first foil at the higher energies are due to the emission of the secondary electrons from the front foil as the alpha particles pass through this foil. An insulating layer *in front of* the first foil would suppress this positive current.

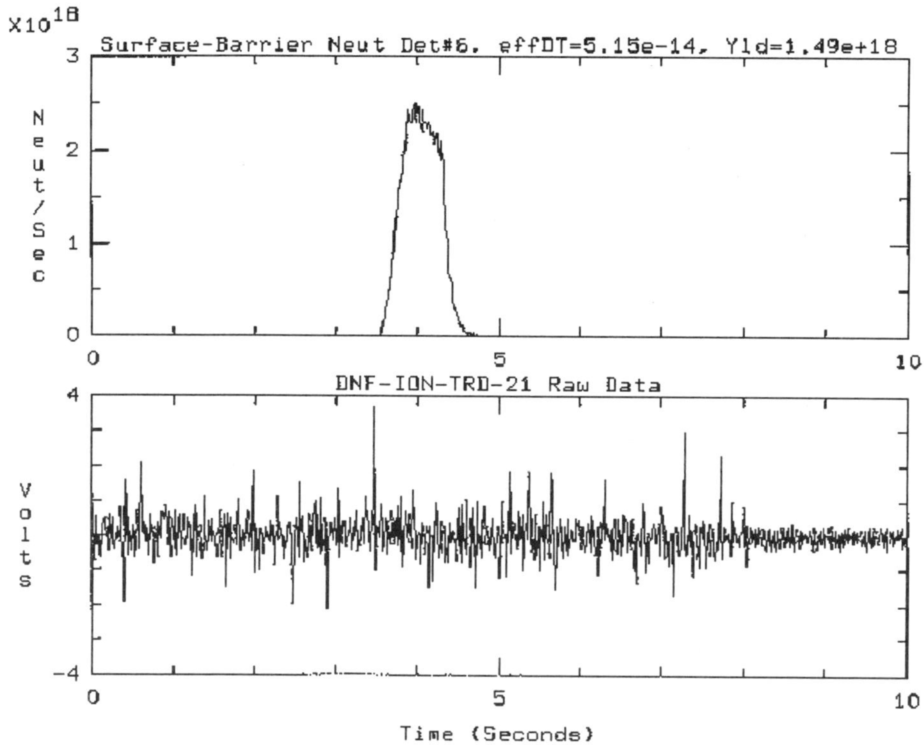

Figure 3. Comparison of neutron yield during TFTR d-t shot No. 76748 (top) and digitized current from foil immediately outside TFTR vacuum vessel (bottom).

A second detector which was tested consisted of a sandwich of 6 foils of 6 μm thick Al each separated from one another by 1 μm of polyethelene. Although this particular design is *not* capable of operating at high temperatures due to the low melting point of the insulating polyethelene layers, the design was included in the present study by virute of the ready availablilty of the materials. The results are shown in Figure 2.

PROTOTYPE TESTING AT TFTR AND JET

Prototype devices were installed at JET and TFTR to evaluate the radiation and temperature insensitivities and the electro-mechanical noise levels in the measured currents. These noise levels are to be compared to an expected current of alpha particles due to classical first orbit losses on TFTR or JET during high yield d-t plasmas of about 1 $\mu A/cm^2$[1]. At TFTR a device was mounted immediately *outside* the tokamak vacuum vessel during the 1994 run period to measure the response during moderately high yield d-t plasmas. The response during the highest yield shot of this period is shown in Figure 3. A signal of 2 V would have corresponded to an alpha particle current of about 100 nA/cm^2 and thus the noise level *during* the plasma production was much less than that due to the expected alpha particle flux.

On JET, detectors similar to the one discussed above were installed for the 1995 and 1997 run periods. The results from the 1995 run period are described elsewhere[2]. For the present run period, two detectors are installed slightly below the torus midplane and a few cm beyond the last closed flux surface. These detectros were monitored during a few low level d-d plasma pulses near the start of this run period. The noise levels in all observed plasma pulses corresponded to a current of about 10 nA/cm^2. We expect to monitor these detectors during the continuing d-t experiments in 1997 and through the next series of d-t experiments presently expected in 1999.

This work is supported in part by the U.S. Dept. of Energy Office of Fusion Energy Sciences

REFERENCES

1. F.E. Cecil, P. van Belle, O.N. Jarvis and G.J. Sadler, "Development of a thin foil Faraday collector as a lost alpha particle diagnostic for high yield d-t tokamak fusion plasmas. Proc, 21st European Physical Society Conf. on Controlled Fusion and Plasma Physics (Montpelier) European Conf. Abstracts, **18B** (1994) 1340.

2. M.J. Loughlin, F.E. Cecil, M.Hone, O.N. Jarvis,S. Medley and L. Roquemore,G.J. Sadler, P. Van Belle, G. Whitfield,"Evaluation of ITER compatible thin foil Faraday collector as a lost alpha particle diagnostic for high yield d-t tokamak plasmas. Reviews of Scientific Instruments , **68**,(1997) 361

MAGNETIC DIAGNOSTICS FOR ITER

L. de Kock[1], S. Bender[2], A. Costley[3], J. Leuer[4], P.J. McCarthy[5], F.C. Morabito[6], A. Portone[7], I. Senda[8], T. Shoji[8], R. Snider[4], C. Walker[1]

[1] ITER Joint Central Team, Garching
[2] Efremov Institute, St. Petersburg, RF
[3] ITER Joint Central Team, San Diego
[4] GA, San Diego, USA
[5] University College, Cork, Ireland
[6] CREATE, Napoli, Italy
[7] ITER Joint Central Team, Naka
[8] JAERI, Naka, Japan

INTRODUCTION

The measurement and the control of the plasma boundary is one of the fundamental aspects of a magnetically confined plasma. Traditionally, inductive magnetic measurements have been used for discharges of limited duration. For ITER with its projected pulse duration in excess of 1000s with a possible extension to steady state, classical integrators exhibit unacceptable drifts caused by small thermo-electric voltages and the operational amplifier offset current resulting in an erroneous boundary identification by tens of cm. Long term drift compensation schemes have recently been developed based on the intermittent observation of a dummy channel with the same integrator. These schemes have the potential to reduce the drift to < 1 cm for a 1000 s pulse. Therefore the magnetic diagnostic in ITER is based on inductive measurements with voltage loops, saddle loops and pick up coils.

The magnetic diagnostic must produce the input measurements for the PF control system and for operation and interpretation. The required accuracy of the reconstruction for both control and physics interpretation is 2 cm and is related to the power scrape-off thickness of 1-2 cm. For the PF control system, average quantities such as plasma current and average position of the current channel are required during the start-up, and the size of 6 gaps between the plasma and the separatrix during the flat top. For diagnostic purposes the entire boundary and internal profile information will be available for interpretation of other diagnostic measurements and analysis of the plasma-wall interaction.

The location of the plasma boundary can be derived only from numerical calculations. Several methods are available and all of them are well developed. In ideal conditions the reconstruction can be as accurate as 2 cm, which is the target for ITER.

Diagnostics for Experimental Thermonuclear Fusion Reactors 2
Edited by Stott *et al.*, Plenum Press, New York, 1998

MEASUREMENT REQUIREMENTS

The physics and technical requirements originate from the role of the magnetic measurements in establishing and controlling the plasma configuration. This requires both accuracy and reliability. Once the plasma is established the magnetic configuration is required for evaluation of diagnostics and physics. MHD mode analysis from locked modes to very fast fluctuations give essential information for judging stability.

For PF Control

The most important application of the diagnostic is its input to the current and shape control. The target accuracy of 2 cm for diagnostic use exceeds the required accuracy for control. Also, the time resolution is an order of magnitude better. There are several candidate reconstruction codes to provide the input parameters, which are basically a subset of the overall equilibrium. For following the development of small bore plasmas and movements during disruptive events it is useful to have the opportunity to construct the average properties of the plasma from the moments of the current distribution. MHD mode analysis will provide the means for disruption evasion. A very important aspect of the control application is that it must be reliable and robust. For the purpose of recovery from unwanted situations or to start protective or evasive action this diagnostic must remain operational far outside the nominal scenarios.

For Evaluation and Optimisation

The full MHD equilibrium reconstruction will form the basis of the diagnostic and physics interpretation. The plasma boundary shows the degree of plasma wall contact and the matching of the divertor configuration to the dome and target plates. From the equilibrium reconstruction the internal parameters β_p and l_i can also be obtained. MHD fluctuations in the range from DC, corresponding to locked modes, to 200 kHz, corresponding to fast TAE modes, can be detected. In addition the magnetic diagnostic should have the capability to measure the static error field generated by asymmetries in the TF and PF coil sets, if large enough.

The Shafranov integrals are obtained as part of the measurement of the diamagnetism from which the energy content is obtained. The energy content can also be obtained from the magnetic reconstruction as β_p.

For equilibrium measurements, the plasma should be followed on a <100 ms time scale, smaller than the L/R times of the conducting shells, requiring compensation of eddy currents.

From the technical side, the sensitivity of the coils and loops must be determined in relation to the input circuits and the bit resolution of the data acquisition system. Further requirements originate from the ITER environment: vacuum, temperature, radiation, remote replacement and reliability.

PROPOSED SYSTEM CONFIGURATION

The sensors, voltage and saddle loops, pick up coils, diamagnetic and Rogowski loops are supported from the structural elements of the ITER machine. The accuracy of the mounting directly reflects on the absolute accuracy of the plasma boundary. Effects of thermal expansion can be taken into account, but mechanical deformations during the pulse are difficult to compensate for.

Equilibrium

The main set of sensors for the calculation of the boundary is located on the plasma side of the back plate and on the sides of the instrumented divertor cassettes. The locations of the

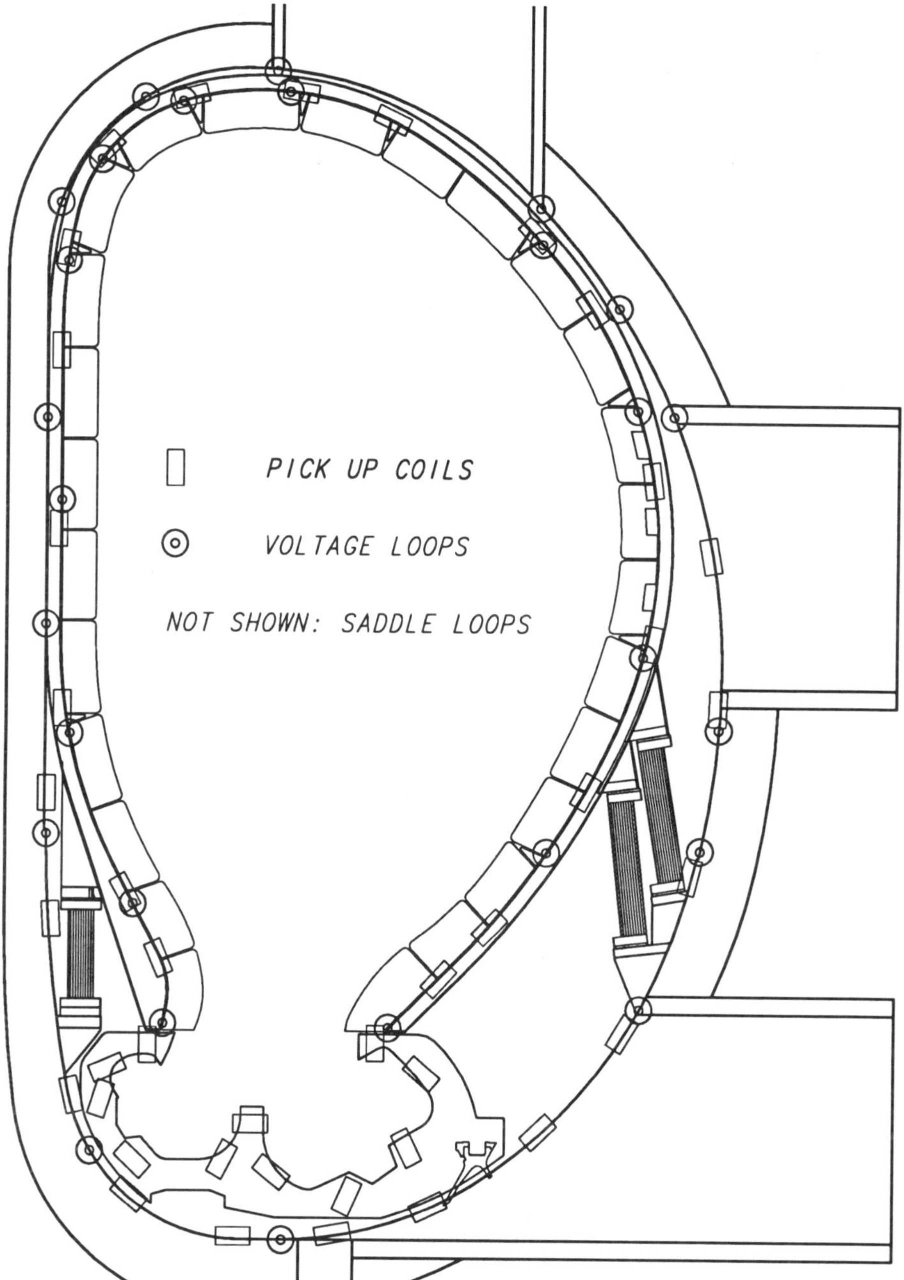

Figure 1. The locations of the flux and field measurements on the inner wall of the vessel and on the plasma side of the back plate are indicated by open circles and small boxes respectively (size not to scale). The flux measurements are a mixture of voltage loops (full flux measurements) and saddle loops (differential flux measurements). The poloidal field measurements on the vessel wall are a mixture of tangential and normal field measurements; on the back plate there are tangential equilibrium and h.f. coils at all locations. Distribution to be optimised.

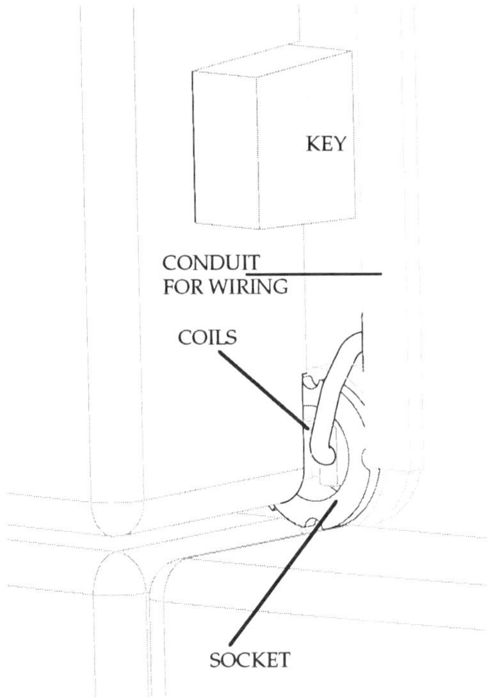

Figure 2: Location of the coils on the plasma side of the back plate. The coils are in the shadow of the shield module to reduce the radiation flux. The keys, electrically insulated, compensate the twisting moments on the shield modules.

equilibrium pick up coils, the voltage loops are given in Figure 1. The set of saddle loops giving intermediate flux values e.g. in the port areas, are not shown.
On the back plate there are 19 flux measurements of which 13 are voltage loops and 12 saddle loops. There are 18 tangential coils for poloidal field measurements and in the divertor cassette there are 11 coils. The voltage loops together with the resistivity of the back plate will give the poloidal distribution of axisymmetric currents in the back plate. The details of the mounting of the coils behind the shield modules in a removable socket is shown in Figure 2. This set is not magnetically shielded and has a response time limited by the construction of the pick up coils. For the equilibrium coils the limit is 10 kHz caused by the short circuit of the sheath of the MI cable. For the high frquency coils the limit is caused by the presence of the conducting structural materials, and the limit is 200 kHz. There is no major eddy current correction necessary if only the back plate and divertor sets are used. The sets have a 5-fold toroidal symmetry with 4 sets at 90°.

The sensors on the vessel wall have a supporting role. The 14 voltage loops give the poloidal distribution of the vessel current. 12 sets of tangential and normal pick coils are located in the lower part of the vessel where space is available. The flux and field measurements in this area can be used for finding the configuration around the X-point from a local field expansion method. These sets are well shielded from radiation and therefore a two-fold toroidal symmetry will be sufficient for redundancy.

Rogowski Loops

Four continuous Rogowski loops are planned outside the vessel. They will measure the total current and need correction with the back plate and vessel current as will be available from the voltage loops. This gives a back up for the basic plasma current measurement by the summation of the 18 poloidal field measurements on the back plate and one under the dome of the divertor cassette.

Diamagnetism

The diamagnetic loops will be mounted on the inner wall of the vessel. The response will be slowed down by the presence of the back plate, which is electrically connected to the divertor cassette forming a poloidal shell. This requires a major correction with the L/R time constant of this structure if a 10 ms response time is required. The reference toroidal flux will be provided by a large pick up coil near the outer equator. The compensation of erroneous pick up has to be prepared in the initial operation stages by measuring the mutual inductance between the diamagnetic loop and the various coils and the plasma current loop.

Halo Currents

Each shield module has two earth straps. In principle there is space to mount a local Rogowski coil on the back plate connection avoiding an extra connector. This allows the measurement of the toroidal and poloidal extent of the halo currents limited only by the practicality of the number of coils required.

A supporting measurement is proposed on the earth straps between the lower edge of the back plate and the divertor cassette. This method will miss the fraction of the current flowing directly into the vessel wall, but will show the toroidal distribution.

MHD modes

The system of coils and loops provides also the means to measure the MHD mode activity from DC, i.e. locked modes, to 200 kHz, fluctuations and TAE-modes. Locked modes are measured by poloidally extended saddle loops with a fixed spatial Fourier analysis with m=1,2,3 and n=1, low frequency oscillations by the equilibrium coils (<10 kHz) and the highest frequencies (>10 kHz) by special high frequency coils. The high frequency coil is shown in figure 2. It is placed in the gap between the back plate and the module just in the shadow of the shield module. The poloidal and toroidal extent of the coils is such that sufficient mode analysis capability ($m \leq 9$ and $n \leq 5$) is offered. It is the intention to use the fluctuation measurement system as a monitor of dangerous MHD activity.

DETAILED DESIGN CONSIDERATIONS

Inductive magnetic measurement system have been successfully used since magnetic fusion started in the 50's. The underlying physics and the mathematical techniques are fully developed and validated. From this vast experience we treat here a limited number of aspects which are important or essential in the context of ITER.

Reconstruction: Accuracy and Optimisation

The factors that affect the lowest detectable signal are: the effective area of the coils and loops, the gain of the front end integrator or amplifier related to the bit resolution of the analogue to digital conversion, the expected noise and uncompensated drift (off-set) of the input circuit, and finally the interference. Examination of these factors suggests that the plasma current can be measured at 1% accuracy from 1 MA and used for control at 0.1 MA albeit at reduced accuracy. The equilibrium reconstruction reaches its full potential at 2.5 MA with its accuracy impaired by the small size of the plasma (large extrapolation).

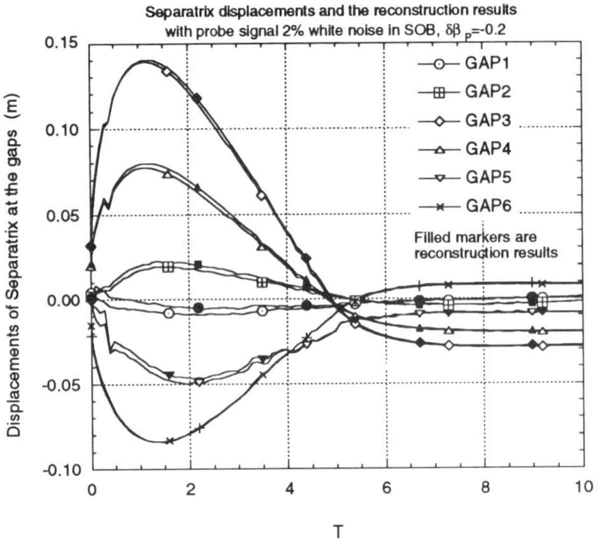

Figure 3: The evolution of the separatrix displacements and the reconstruction result with 2% white noise in probe signals (from [2]).

The reconstruction methods used for investigating the proposed ITER magnetic diagnostic are: EFIT (US), a filament method (RF and JA), a Local Field Expansion method (RF), and statistical methods such as Artificial Neural Networks (ANN) and Function Parameterization (FP) (EU). The numerical codes are used to test the accuracy of the reconstruction under imposed statistical errors on the simulated measurements, to study the weight of the contributions of individual sensors to arrive at an optimised set, and to assess the influence of eddy currents.

Tests on the previous configuration [1] have shown that under 2% statistical errors the gap measurements are accurate to ±2 cm rms. There is no reason to believe that the present proposed set will be different. Two preliminary results of reconstructions using the proposed arrangements have become available.

The first is a steady state calculation using the ANN and FP methods [2]. Only the sets on the back plate and in the divertor have been used, thus no major eddy current effect needs to be included (but can, if required). The imposed errors cover a range of 0.2 to 1.0 % of the average poloidal field. As an example we quote here some results of the calculation with 1% error (10 mT). For the flux differences the error level of the field is used which leads to 500 mVs flux error using the average area of the differential flux loop area of 50 m^2. With these values reconstruction errors of the outer equatorial gap of 12 mm and of the top gap of 23 mm have been obtained. Further details can be found in [2].

The second study uses a current filament method which follows dynamically the plasma shape and position. Coupled with this model is a finite element description of the eddy currents in the axisymmetric conducting shells formed by the back plate and the vessel. The main use of this code is for the study of the PF control action [3]. For this second study results are only available which use the complete set of sensors i.e. inclusive of those on the vessel wall. This introduces some large eddy current effects into the reconstruction. The study of the response to a decrease of β_p=0.2 for all 6 gaps shows that eddy currents can efficiently be corrected using the model (see next section). In this study the imposed errors are different. On the input 2% white noise is used on the time derivative and the resulting gap deviations are <1 cm. For this case the plasma current and the PF currents area assumed to have no errors (Figure 3). If we allow a 1% error on the plasma current and 0.1% on the PF currents with 2% white noise on the measurements, as used before, the gap deviation increases to ~2 cm. This result is shown in Figure 4. The noise has a large effect on the

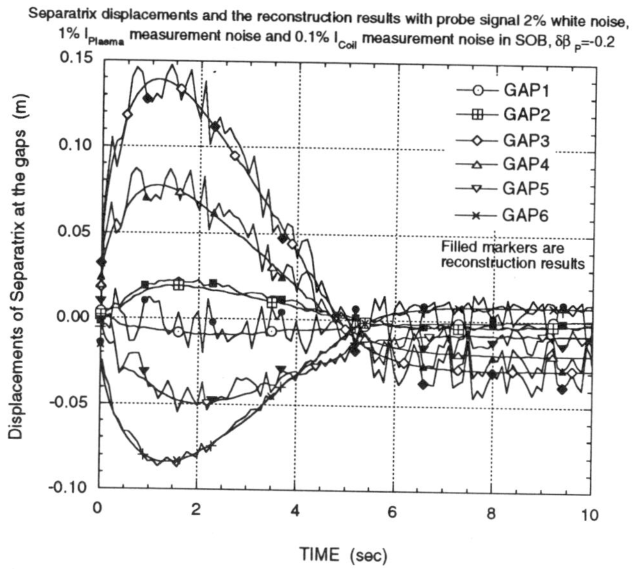

Figure 4: The evolution of the separatrix displacements and the reconstruction result with 2% white noise in probe signals, 1% white noise on the plasma current and 0.1% on the PF currents (from [2]).

reconstruction and indicates that plasma current and the currents in the PF coils must essentially be noise free. For these cases the eddy current correction is incorporated and the full bore plasma is under feedback control.

Optimisation studies have not been completed on the new set. The procedure is to switch off the contribution of each individual probe with imposed noise present and record the changes in accuracy. If the deletion of a sensor causes large increases one must consider increasing the local sensor density. When this test is done one needs to keep in mind that different plasma shapes have to be treated e.g. small initial limiter plasma, large limiter plasma, limiter plasma in the upper part of the vessel, divertor plasma, etc. One also needs to consider different reconstruction methods: EFIT, filament codes, Local Field Expansions and statistical methods (ANN and FP).

Eddy Currents

No major eddy current corrections, due to axi-symmetric currents in the back plate and vessel are necessary to the measurements on the back plate and the divertor when they are used for equilibrium reconstruction. However, during start up and plasma movements, concentrated current channels will occur near the lower part of the back plate (divertor gap). These can affect the plasma current integral and the accuracy of the reconstruction in this area. The time scale of these eddy currents is 100's of msec. The contributions can only be quantified by numerical calculations and corrected. Examples of a major eddy current correction are shown in Figure 5 where a small initial plasma has been examined during the current rise with substantial currents in the vessel and back plate structure. These contributions can be removed to a high accuracy by invoking the finite element eddy current model as in [3] bringing the reconstruction accuracy back to the required specification. In a similar way rapid movements caused by a drop in β_p of 0.2 of the full bore divertor plasma have been investigated with the PF feedback switched on. As an example the effect on the outer divertor leg gap is shown in Figure 6. Compensation for eddy currents leads to an improvement from 2 cm to 1 cm error, the original accuracy. The response of the other gaps with various noise contributions are shown in Figures 3 and 4. Once the relation between the eddy current channels and the correction has been established one can also use the *measured* back plate currents for a direct on-line correction to the field measured by the pick up coils.

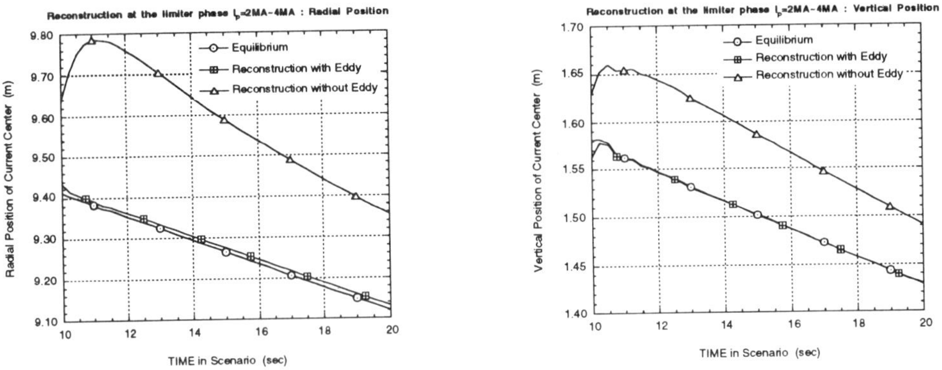

Figure 5: The radial and vertical position of the plasma current centre in the reconstruction of the limiter plasma.

The voltage loops on the back plate are not sensitive to the local eddy currents in the shielding modules, but the poloidal coils will pick up the stray fields of the modules and the penetrations. The estimated time scale is several 10's of ms and therefore not visible in the reference [3]. Further work is underway to assess this contribution.

High Frequency Fluctuation Measurements

Two factors determine the ability to measure the high frequency fluctuation field: the limitations due to the coil construction and the penetration of the fields to the location where the coil is. The coil consists of a ceramic body with 2 layers of ceramic coated wire. Provisional calculations on the frequency response suggest a roll-of frequency at 200 kHz as the required specification. The penetration is through the poloidal gap of 2 cm between the shield modules as shown in Figure 7. However this gap is blocked by the keys between the shield modules to compensate the moments occurring during disruptions. These form a wave guide in cut-off. The

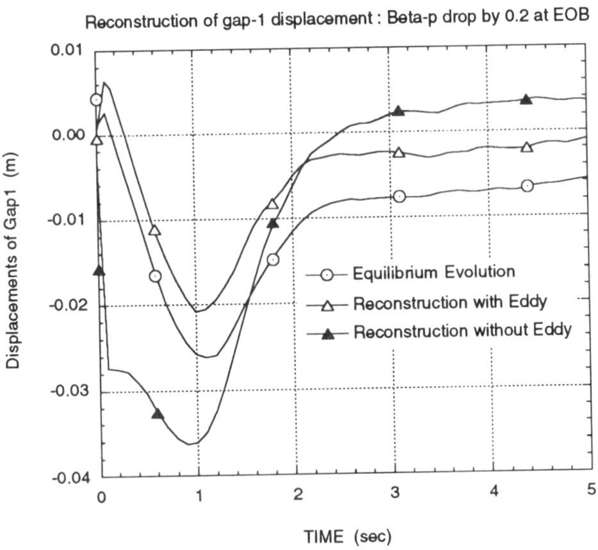

Figure 6: The displacement of the separatrix at the gap 1 (at the inner divertor leg) in the drop of the poloidal beta by 0.2 at EOB plasma.

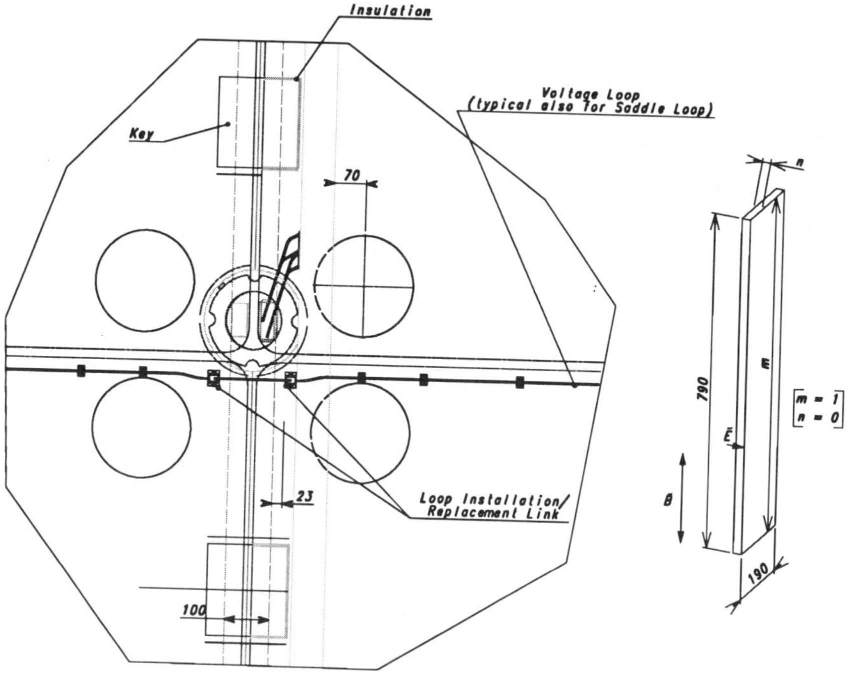

Figure 7: Front view of the blanket modules with the diagnostic socket and the coils. The poloidal gap between the modules and the vertical limitations by the keys is clearly visible. The dimensions of the wave guide in cut-off are given on the right hand side.

wave guide dimensions are 2x790 mm^2 and 190 mm deep. The cut-off frequency is approximately 38 MHz. At the frequencies of interest, <500 kHz, the attenuation becomes frequency independent given by $\exp(-\pi x/a) \approx 0.47$, where x is the depth and a the height.

At very low frequencies the capacitive short (formed by the electrical insulator of the key in the wall of the wave guide in cut-off will open up and allow the equilibrium field to reach the poloidal field pick up coil without attenuation.

Measurement Noise

The most dominant direct cause of the noise is the movement of the plasma boundary. For the local loop voltage V_l we can write:

$$V_l = -2\pi \, d\Psi/_{dt} = -2\pi\left(\partial\Psi/_{\partial t} + (\bar{u}.\nabla)\Psi\right)$$

showing that there is a contribution of the change in flux and a change due to the movement **u** of the boundary. Any change in the integrated loop voltage leads to a gap variation.

The measurements in [5] show clearly that the flux and field derivatives have a white noise character. This means that the gap noise has a 1/f frequency dependence. As a working hypothesis we take the reference of 2 mm gap noise at 1 Hz.

From earlier numerical experiments it is known that the average rms accuracy of the reconstruction shows a behaviour like in Figure 8. The general opinion is that we have a 2% error on the input measurements which leads to a 2 cm rms error on the gap values. Obviously this is an extra noise that is not correlated with the autonomous noise on boundary fluctuations and much larger, but it is the gap that is fed into the controller. Every new time point is therefore subjected to this noise and the controller is continuously working to adjust the boundary to this stochastically wrong value.

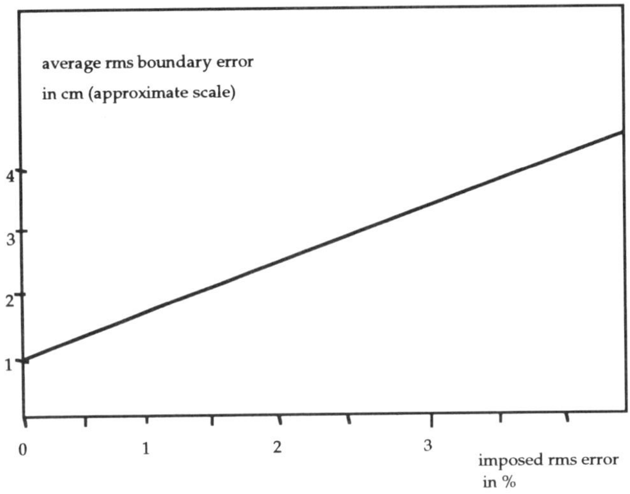

Figure 8: Boundary error versus imposed rms error on simulated measurements

If we do a completely new reconstruction for each time point and there is large statistical error which causes the gap values to fluctuate, we have three remedies:

a) We take an average of each sample at intervals of 2 ms and do a running average over 10-20 ms or even 100 ms.

b) We take an approach as with EFIT as used for control [6]. This means that we keep the current density profile constant for a longer time, say 100 ms, thus restricting the amount of fluctuations.

c) Use ANN or FP which do not have this problem.

Radiation Effects

The allowable radiation induced conductivity (RIC) in the standard equilibrium coil is 10^{-6} S/m. On the vessel wall all sensors are well shielded and RIC is many orders of magnitude below the limit. The coils on the back plate are close to the gap between shield modules and scattered radiation can reach the coils. With the common materials the coils will operate at the given limit. With the choice of high purity alumina as insulator in the Mineral Insulated (MI) cable the limit can be improved by a factor 100. Permanent damage effects to insulators have not been observed up to 3 dpa in the common candidate insulators. Since the shielding factor for the coils on the back plate is around 10, the most exposed coils will survive the lifetime of ITER.

The nuclear heating of the coils and loops on the back plate is considerable. The design aims at cooling by conduction to the back plate. No detailed calculations have been carried out, but explorative calculations have shown that conductive cooling should be possible for coils in the proposed location (see Figure 2).

Steady State Considerations

The steady state conditions during the flat top suggest a direct measurement of the magnetic field or a measurement of the location of the boundary by non-magnetic means. An example of the first category is the Hall probe, well shielded from radiation, and of the second category the reflectometer. In the latter case only a full bore plasma can be measured, smaller bore plasmas as occur during start up can only be measured by magnetic means. An attractive method proposed independently by Hara and Woolley [7] combines a classical pick up coil and a jxB method to measure the steady state component. This method is investigated further in the Japanese HT [7]. In ITER some of these techniques will be used to gain experience and to have the possibility to detect drifts of the inductive system.

CONCLUDING REMARKS

The new design of the blanket/shield structure and the back plate give the opportunity to obtain the magnetic equilibrium without being magnetically shielded by eddy currents. Additionally the fast fluctuations can be measured through the poloidal gap between the modules only attenuated by transmission through a wave guide in cut-off, which is frequency independent. Only the sensors on the back plate are subjected to considerable radiation levels, but this is just below the damage levels for the full life time of ITER. Further design must resolve the conduction cooling of these coils and loops. Confirmation of the number and locations of the sensors through a formal optimisation is still outstanding.

The study of the noise on the measurements has just started and the first results are available [2,3]. The largest source of noise is the movement of the plasma boundary and the statistical errors in the reconstruction.

Acknowledgements:

This report is an account of work undertaken within the framework of the ITER EDA Agreement. The views and opinions expressed herein do not necessarily reflect those of the Parties to the ITER Agreement, the IAEA or any agency thereof. Dissemination of the information in this paper is governed by the applicable terms of the ITER EDA Agreement.

References:

1. L. de Kock et al., Overview of Magnetic Diagnostics Planned for ITER, Diagnostics for Experimental Thermonuclear Fusion Reactors, Varenna 1995, p. 71
2. P.J. McCarthy and F.C. Morabito, these proceedings
3. I. Senda and T.Shoji, Report on ITER Design Task D324, 1997 (in progress)
4. M. Matsukawa, et al, Fusion Eng. Des. 23 (93) 341.
5. P. Vyas and J.B. Lister, Report ITER Design Task D324-1 Phase II, Aug. 1997
6. J.R. Ferron et al., Real Time equilibrium reconstruction for control of the discharge in the DIII-D tokamak. APS, Denver, CO, USA, Nov. 1996
7. S. Hara, M. Abe and K. Moriyama, A Functional Test of a Magnetic Sensor Designed for Long Period Measurements in a Magnetic Fusion Device, J. of Plasma Fusion Res., vol. 73, p 501 (1997) and
   R. Woolley, Tokamak poloidal magnetic field measurements accurate for unlimited time durations., 16th IEEE/NPSS Symp. on Fusion Endineering (SOFE),1995, p. 1530-1533.

REALTIME IDENTIFICATION OF PLASMA SHAPE AND POSITION ON ITER USING AN EQUILIBRIUM DATABASE

Patrick J. Mc Carthy[1] and Francesco C. Morabito[2]

[1]Physics Department, University College Cork, Association Euratom DCU, Cork, Ireland
[2]Department of Electrical Engineering and Applied Mathematics, University of Reggio Calabria, Italy

1. INTRODUCTION

Realtime control of the plasma shape and position on ITER requires a fast and accurate determination of the equilibrium configuration. A suitable identification method must be robust in the presence of measurement noise and must also remain reliable when the plasma geometry makes possibly large deviations from the pre-programmed configuration. The realtime use of an iterative equilibrium code will, under benign conditions, yield the solution most closely satisfying a given set of input diagnostic measurements. However, apart from the relatively large computational requirement, the danger of uppredictable results due to faulty measurements and/or convergence to an incorrect solution by the minimization algorithm in such codes is always present. In contrast, a non-iterative identification of the equilibrium which naturally incorporates robustness against measurement noise is possible by methods based on offline analysis of a database of randomly generated equilibria chosen to span the entire parameter space accessible by ITER. We describe here the generation of one such database by the Garching Equilibrium Code[1] using the ITER FDR PF coil set and spanning a cigar-like volume in a high-dimensional (16-D) parameter space whose axis is a line segment connecting Start of Burn (SoB) and End of Burn (EoB) fiducial ITER equilibria. We investigate the identification of a variety of plasma position, shape and current profile parameters from magnetic flux and field measurements prescribed by the most recent conceptual design of the magnetic sensor layout.[2] We analyse a linear model, which is expected to be inadequate, and two leading identification techniques, namely Function Parameterization and Artificial Neural Networks.

2. DESCRIPTION OF THE EQUILIBRIUM DATABASE

A database of 1522 lower x-point equilibria with fixed $I_p = 21\,\mathrm{MA}$ and $B_o = 5.7\,\mathrm{T}$ was generated by partial Monte Carlo variation of 16 input parameters consisting of 9 external coil currents and a 7-parameter family of toroidal current density profiles within the plasma. The database was designed to contain equilibria clustered around a 1-D sequence spanning the fiducial SoB and EoB ITER equilibria "sob21-l09-b09-c1.t1" and "eob21-l09-b09-c1.t1". This was achieved by a part-deterministic part-random variation of the PF currents using the scheme

$$I_j = I_{\mathrm{SoB}} + (I_{\mathrm{EoB}} - I_{\mathrm{SoB}}) \times \left(\frac{j}{N_C} + \tilde{U}\{0,f\}\right)$$

where I is any PF current, I_{SoB} and I_{EoB} are its SoB and EoB values, j is the equilibrium index in the database, N_C is the number of cases and $\tilde{U}\{0,f\} \in [-f,f]$ is a random number chosen

Diagnostics for Experimental Thermonuclear Fusion Reactors 2
Edited by Stott *et al.*, Plenum Press, New York, 1998

from the uniform distribution with limits $\pm f$. The current density profile

$$j_t(R,\psi) = Rp'(\psi) + FF'(\psi)/(\mu_o R)$$

was randomly varied using a 7-parameter family given by ($\hat{\psi}$ is the normalized poloidal flux: $\hat{\psi} = 1$ at the magnetic axis, $\hat{\psi} = 0$ at the plasma boundary)

$$p'(\hat{\psi}) \;\propto\; \hat{\psi}^{a_P} \exp\left\{b_P(1-\hat{\psi}) + c_P(1-\hat{\psi})^2\right\}$$

where a_P, b_P, c_P are randomly chosen for each $p'(\hat{\psi})$ profile, a_{FF}, b_{FF}, c_{FF} specify the $FF'(\hat{\psi})$ profile, and the 7th parameter determines β_{pol}. Reversed shear and edge pedestal profiles are easily generated with this family. A total of 47 raw measurement inputs were calculated for each equilibrium. These consist of 18 flux differences and 29 pick-up coil field measurements, all located on the plasma side of the blanket backplate. Their geometry is described elsewhere.[2]

3. OFFLINE TRAINING METHODS

3.1 Linear Models

As a benchmark for the more sophisticated models considered below, the recoverability of each output parameter was investigated by carrying out a well-conditioned linear regression using the input signals as predictors and noting the rms fitting error averaged over the database.

3.2 Function Parameterization

In the method of Function Parameterisation[3] (FP) the dependence of a set of outputs (plasma parameters) on a corresponding set of inputs (magnetic measurements) is modelled by (typically) a second degree polynomial whose arguments are the most strongly varying statistically independent linear combinations of the inputs. These linear combinations, known as Principal Components (PCs), are determined by diagonalizing the correlation matrix **R** of the set of inputs. The eigenvalues of **R** are the variances of the PCs and those PCs with eigenvalues that are close to zero or below the expected noise level of the inputs can be discarded. Thus the PC transformation has the dual purpose of identifying independent linear combinations of the inputs and reducing the dimensionality of the problem. FP is a widely used technique within plasma physics research and is the routine method of realtime and between-shot equilibrium identification on the ASDEX Upgrade tokamak.[4] If we denote the ith PC variable by $\mathbf{\Phi}_i$ then $\mathbf{\Phi}_i = \boldsymbol{\gamma}_i \cdot \mathbf{V}_s$ where $\boldsymbol{\gamma}_i$ is the ith eigenvector (in order of eigenvalue magnitude) of **R** and $\mathbf{V}_s(1 \times m)$ is the vector of m standardized input variables (note $\mathbf{\Phi}_i$ and **V** take a different value for each case in the database) then the FP representation for each output parameter is given by

$$G_n \;=\; \sum_{\substack{i=0\\ j=i}}^{p} c_{i,j,n} \mathbf{\Phi}_i \mathbf{\Phi}_j$$

where G_n is the nth output parameter variable, p is the number of retained PCs and the $c_{i,j,n}$ ($c_{0,0,n}$ is the intercept and the $c_{0,1,n}$ are linear PC coefficients) are $(p+1)(p+2)/2$ unknown coefficients to be determined (for each G_n) by a *linear* least squares regression.

3.3 Artificial Neural Networks

Artificial Neural Networks (ANNs) are also widely used as a method of rapid data interpretation in plasma physics[5] The ANNs we consider here are of the multilayer type, in which the first layer is directly connected to the inputs and the last layer outputs the plasma parameter values. Here we use one hidden-layer of nodes (neurons) which nonlinearly transform a selected set of linear combinations of the inputs using the hyperbolic tangent function. Using similar notation to that of the FP model above, we have the following ANN model:

$$G_n \;=\; \sum_{i=0}^{p} c_{i,n} \tanh[\mathbf{w}_i \cdot \mathbf{V} - b_i]$$

where $\mathbf{w}_i \cdot \mathbf{V}$ is again a linear transformation of the inputs vector. Here, the $Np + N$ output weights $c_{i,n}$, the p biases b_i and the matrix of input weights $\mathbf{W}(p \times m)$ are simultaneously deter-

Table 1. Linear, FP and ANN parameter recovery statistics for 1522 ITER equilibria.

Parameter	Min	Avg	Max	Spread	Model	$\bar{\epsilon}$ 0 mT	$\frac{\bar{\epsilon}_{\bar{\delta}=0\,\mathrm{mT}}}{\mathrm{spread}}$	$\bar{\epsilon}$ 2 mT	$\frac{\bar{\epsilon}_{\bar{\delta}=2\,\mathrm{mT}}}{\mathrm{spread}}$
R geometric (mm)	7779	8084	8347	109	Linear	4	3.6%	5	4.1%
					FP	0	0.4%	2	1.8%
					ANN	2	2.1%	2	2.3%
Z geometric (mm)	328	689	1056	154	Linear	15	10.0%	16	10.3%
					FP	4	2.8%	6	3.8%
					ANN	5	3.0%	6	3.6%
Z current centre (mm)	893	1164	1499	119	Linear	1	0.6%	2	2.0%
					FP	0	0.3%	2	1.8%
					ANN	2	1.9%	2	1.9%
R x-point (mm)	6845	7136	7480	110	Linear	4	3.8%	5	4.1%
					FP	0	0.4%	2	1.8%
					ANN	3	2.8%	3	2.7%
Z x-point (mm)	-4399	-4165	-3837	122	Linear	8	6.7%	9	7.2%
					FP	2	1.3%	5	4.1%
					ANN	3	2.8%	5	3.8%
horizontal minor radius (mm)	2188	2674	2932	112	Linear	4	3.6%	5	4.4%
					FP	1	0.5%	2	1.8%
					ANN	2	1.7%	2	2.2%
elongation ×1000	1580	1820	2358	107	Linear	7	6.8%	8	7.1%
					FP	2	1.5%	2	2.3%
					ANN	3	2.6%	4	3.4%
R inner (mm)	5163	5410	5840	142	Linear	5	3.8%	7	4.8%
					FP	1	0.4%	3	2.2%
					ANN	3	1.8%	3	1.8%
R outer (mm)	10135	10757	11054	170	Linear	6	3.4%	7	3.8%
					FP	1	0.4%	3	1.5%
					ANN	3	1.9%	4	2.2%
Z top (mm)	5019	5543	5999	253	Linear	28	11.0%	29	11.3%
					FP	8	3.2%	10	3.9%
					ANN	6	2.3%	10	4.1%
R(Ztop) (mm)	6254	7133	7819	244	Linear	44	17.9%	44	18.1%
					FP	21	8.8%	22	9.0%
					ANN	12	5.0%	19	7.7%
R magnetic axis (mm)	8164	8442	8662	123	Linear	18	14.7%	20	15.9%
					FP	4	3.1%	14	11.4%
					ANN	8	6.7%	11	9.0%
Plasma Volume (m^3)	1587	1834	2035	75	Linear	4	5.1%	5	6.1%
					FP	1	1.0%	2	2.5%
					ANN	2	2.8%	4	5.3%
Bottom Triang. ×1000	-455	-355	-284	30	Linear	3	8.5%	3	8.9%
					FP	0	1.2%	1	3.8%
					ANN	1	3.8%	2	6.9%
β_{pol} ×1000	613	875	1330	172	Linear	11	6.4%	16	9.5%
					FP	2	1.2%	6	3.7%
					ANN	4	2.6%	9	5.3%
l_i ×1000	524	951	1574	170	Linear	18	10.7%	29	17.0%
					FP	3	1.7%	10	5.7%
					ANN	8	4.6%	15	8.9%
q_0 ×1000	390	1007	2664	296	Linear	112	37.9%	113	38.1%
					FP	44	14.9%	85	28.7%
					ANN	87	29.5%	112	37.8%
$q_{95\%}$ ×1000	2536	2966	3566	176	Linear	27	15.6%	29	16.6%
					FP	8	4.3%	14	8.2%
					ANN	10	5.9%	21	12.0%
Gap Out (mm)	178	470	1087	170	Linear	6	3.3%	6	3.7%
					FP	1	0.6%	3	1.6%
					ANN	3	1.6%	5	2.8%
Gap Top (mm)	87	564	1200	263	Linear	21	7.9%	22	8.4%
					FP	3	1.2%	6	2.3%
					ANN	5	1.8%	7	2.6%

mined for all N output parameters in the ANN by the *backpropagation algorithm*, an iterative gradient descent technique designed to minimize the error between the true and ANN output parameter values. Thus unlike FP, where the PC transformation is done without regard to the parameters to be recovered, the ANN can optimize the choice of input signal mappings (i.e. **W**) for the vector of output parameters **G** that are to be recovered. Fifteen hidden nodes have been chosen here for the hidden layer and the parameters were modelled by four separate ANNs, each of which was trained on two-thirds of the cases and validated on the remaining third.

3.4 Simulation of Signal Noise

The magnitudes of the flux difference and magnetic field signals in the database are typcally 50 Vs and 1 T, respectively and Gaussian noise was optionally added to these input signals prior to training. For a given choice of noise magnitude $\bar{\delta}$ mT for the pick-up coils, 50 $\bar{\delta}$ mVs was used for the flux differences. Since the area subtended by each flux loop pair is typically 50-60 m^2, this gives (approximately) *equal uncertainties in the magnetic field for all measurements.*

4. RESULTS AND CONCLUSIONS

We report results for models of 15 PCs or hidden nodes, reflecting the number of degrees of freedom in generating the database. (Note this dimensionality is much higher than a previous database analyzed in a similar manner to here[6].) Table 1 gives database summary statistics and rms recovery errors, both absolute and expressed as a percentage of the parameter Spread (the standard deviation of the parameter about its mean value in the database) for Linear, FP and ANN models both for noise-free and noisy measurements (noise level: $\bar{\delta} = 2$ mT). For the 20 parameters tabulated, the noisy model recovery error for the three methods was, on average, in the ratio FP : ANN : Linear :: 1 : 1.3 : 2.2. Thus FP performed moderately better than ANN and clearly better than the Linear model. For geometric parameters, the typical noisy model absolute rms error is 4 mm for FP, 5 mm for ANN and 12 mm for the Linear model.

Of relevance for realtime control is the computational requirement of the identification algorithm. If it can be assumed that the plasma is of the single null type, then it is straightforward to show that for m inputs, p PCs or hidden nodes and N outputs, the number of multiplications per evaluation cycle for FP and ANN is given by

$$\begin{aligned} N_{FP} &= mp + p(p+1)/2 + N(p+1)(p+2)/2 \\ N_{ANN} &= \nu mp + \nu pt + N(p+1) \end{aligned}$$

where ν separate networks are required for the N output parameters and t is the number of equivalent multiplications to evaluate the tanh function. Using a test fortran program running on a 170 MHz sparc, the following timings for $m = 47$, $p = 15$, $N = 20$ and (for ANN) $\nu = 4$, were obtained: $\tau_{\text{Linear}} = 0.28$ ms, $\tau_{\text{FP}} = 0.84$ ms, $\tau_{\text{ANN}} = 1.27$ ms.

We conclude, on the basis of the results presented here, that there are only moderate differences in the accuracy and speed of realtime parameter identification between the FP and ANN models and that both methods would be well suited to this task on ITER.

REFERENCES

1. K. Lackner, *Computation of ideal MHD equilibria* Comput. Phys. Commun. **12** (1976), 33–44.
2. L. De Kock, S. Bender, A. Costley, J. Leuer, P. Mc Carthy, A. Portone, I. Senda, R. Snider, C. Walker, *Magnetic diagnostics for ITER* this workshop.
3. B.J. Braams, W. Jilge and K. Lackner, *Fast determination of plasma parameters through Function Parametrization* Nucl. Fusion **26** (1986), 699–708.
4. P.J. Mc Carthy, W. Schneider, K. Lackner, H.P. Zehrfeld, K. Büchl, J. Gernhardt, O. Gruber, A. Kallenbach, G. Lieder, R. Wunderlich, *MHD Equilibrium Identification on ASDEX Upgrade* in "19th European Conference on Controlled Fusion and Plasma Physics, Innsbruck, 1992", Europhysics Conference Abstracts **16 C-I** (1992) 459–462.
5. J.B. Lister, H.Schnurrenberger, Nucl. Fusion **31** (1991), 1291.
6. R. Albanese, E. Coccorese, O. Gruber, R. Martone, P.J. Mc Carthy, F.C. Morabito, *Identification of Plasma Equilibria in ITER from Magnetic Measurements via Functional Parameterization and Neural Networks* Fusion Technology **30** (1996) 219-236.

STEADY STATE MAGNETIC SENSOR IN A FUSION DEVICE

Shigemitsu Hara, Yuzuru Neyatani, Mitsushi Abe[1], Akira Nagashima and Satoshi Kasai

Naka Fusion Research Establishment, Japan Atomic Energy Research Institute, Naka-machi, Naka-gun, Ibaraki-ken 311-01, JAPAN
[1] Hitachi Research Laboratory, Hitachi Ltd., Hitachi, 319-12, JAPAN

INTRODUCTION

Magnetic measurement is the one of the most important issue for the magnetic confinement device like a tokamak. The magnetic measurement provides an accurate knowledge of the plasma position, shape, current and many parameters. In future machine like International Thermonuclear Experimental Reactor (ITER), the magnetic sensor is required to measure accurately in severe environment for a long time. Then, it will be necessary to measure magnetic field continuously in a long discharge during which radiation damage will occur[1]. Conventional sensors, which consist of a pick-up coil and integrator, undergo zero level drift and they are unfeasible for a long pulse operation. Hall elements have been used for a long pulse discharge[2]. However, their use is problematic in a high neutron and high temperature environment as in a fusion device.

A new hybrid magnetic sensor was proposed which consists of a mechanical sensor and a conventional magnetic probe[3]. A mechanical sensor is a new type of magnetic sensor for frequencies smaller than 0.1Hz and employs the measurement of electromagnetic force. This paper provides the practical feasibility of this hybrid sensor. The prototype of a new hybrid sensor was manufactured and tested on the JT-60U tokamak discharges. Our results show that the hybrid sensor is a promising candidate for the measurement of magnetic fields on next generation long-pulse fusion devices for the magnetic confinement of thermonuclear plasmas.

SYSTEM OUTLINE DESCRIPTION

Figure 1 shows a system diagram of the hybrid sensor, which consists of three parts; an electromagnetic force measurement part (mechanical sensor), a magnetic probe part (conventional sensor), and a signal synthesis part. In the first part, there is a transmitter

Diagnostics for Experimental Thermonuclear Fusion Reactors 2
Edited by Stott *et al.*, Plenum Press, New York, 1998

module which supplies *E* the bridge voltage of load cell, and amplifies *V* the signal voltage, which is proportional to the magnetic field. However the mechanical sensor can not measure high frequency components (>1 Hz) of the field, due to its own inertia. Therefore high frequency components are measured by a conventional magnetic probe.

The signal synthesis part combines the signals from the mechanical sensor and the conventional sensor. The low pass filter cuts the high frequency components off for the mechanical sensor. However, for the conventional magnetic probe, the low pass filter acts as an integrator in the high frequency region, producing a signal proportional to the magnetic field without the low frequency components. These two filtered signals are added to produce a signal proportional to the magnetic field.

This new hybrid sensor have two good features. One is that the hybrid sensor does not produce offset drifts of the analog integrator since only the high frequency component of the signal from the conventional magnetic probe is integrated. Another is that the hybrid sensor may be tolerant of high neutron fluxes since the magnetic field detection parts (i.e. the sensing coil, magnetic probe and load cell) are not made of semiconductors.

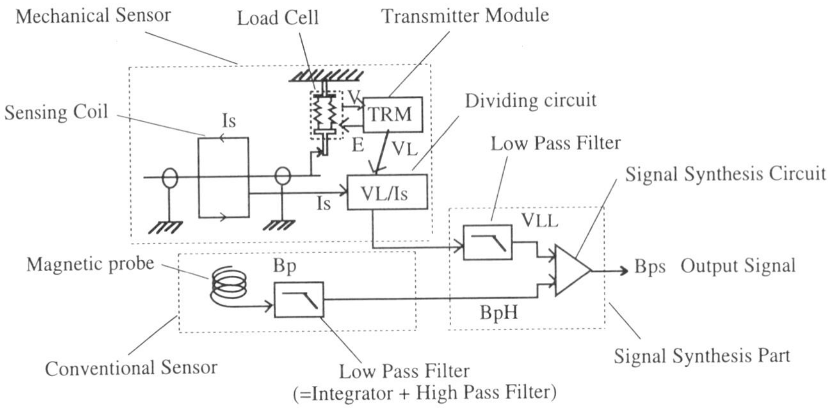

Figure 1. System schematics of the hybrid sensor, showing its three parts, the mechanical sensor, conventional sensor, and signal synthesis part.

TEST RESULTS ON JT-60U

Performance tests of the hybrid sensor were carried out by using JT-60U discharges. The hybrid sensor was set inside the toroidal field coil. The measurement test was enforced under the condition of strong toroidal magnetic field of higher than 1 Tesla. At the setting position, Br was larger than Bz in the typical discharge. We measured Br component of magnetic field and up to a 1×10^{-1}T magnetic field was measured.

Figure 2 shows the waveform of the magnetic measurement during the plasma discharge. The waveform of 120 second is shown in Fig. 2(a) and time-expanded one is shown in Fig. 2(b). The waveform of integrated pick up coil signal(Bp), the pick up coil signal through low pass filter(BpH), load cell signal(VL), load cell signal through low pass filter(VLL) and synthesized signal(Bps) obtained by adding both signals were shown. The toroidal field ripple was observed in Fig2(a). The hybrid sensor seems to be more sensitive for the toroidal field ripple than conventional magnetic probe. It is found that the waveforms of Bps and Bp are similar within several percent error in Fig. 2(b). The results shows that it is possible to measure real magnetic field by the hybrid sensor essentialy.

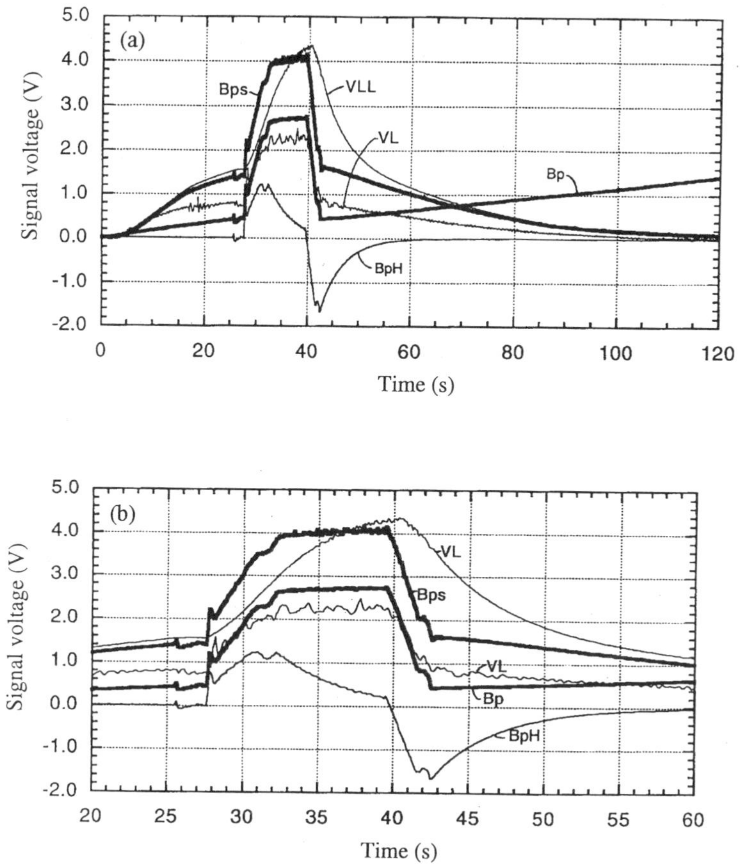

Figure 2. Example waveform of 120 second magnetic field measurement(a) and waveform(b) on an expanded time scale.

The long time measurement test was enforced by hybrid sensor. The hybrid sensor measured magnetic field for about 2 hours. There was no drift of in the measured signal.

The feasibility of dynamic range measurement was tested by changing sensing coil current during discharge. When the sensing coil current was changed violently from 0.1 A to 0.5 A, the hybrid sensor could measure magnetic field normally with no influence of sensing coil current changing. The results shows that it is possible to measure the magnetic field on wide magnitude range by adjusting sensing coil current.

FUTURE PLAN

The irradiation effect of the hybrid sensor is one of the most important issues. Especially, RIC (radiation induced conductivity) and RIED (radiation induced electrical degradation) of insulating materials are important issues under DC/AC electrical fields[4]. We are planing the irradiation test of the hybrid sensor. Figure 3 shows the schematics of the hybrid sensor for irradiation test. The load cell will be made without organic materials and the sensing coil will be wound with the ceramic coated wire. We will investigate the irradiation performance of the load cell and sensing coil using Co^{60} Gamma-ray irradiation facility in JAERI Takasaki around the end of 1997. In the next step, the hybrid sensor will be irradiated in the neutron irradiation condition using JMTR (Japan Materials Testing Reactor).

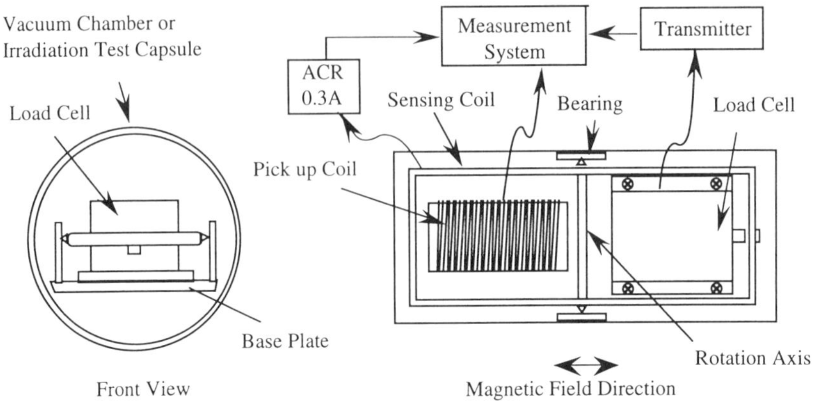

Figure 3. Schematics of hybrid sensor for irradiation test. Low frequency component of magnetic field can be found by measuring the electromagnetic force which is raised on a sensing coil . The hybrid sensor is set in a vacuum chamber or irradiation test capsule in order to prevent from oxidization.

REFERENCE

1. V.Mukhovatov, H. Hopman, S. Yamamoto, et al., in: *ITER Diagnostics, ITER Documantation Series, No.33*, IAEA, Vienna (1991).
2. S. Itoh, N. Hiraki, Y. Nakamura, K. Nakamura, A. Nagao, S. Moriyama, et al., in: *Proc. 13th Int. Conf. on Plasma Physics and Controlled Nuclear Fusion Research*, Washington D. C., 1990 1: 733, IAEA, Vienna (1991).
3. S. Hara, M. Abe and K. Moriyama, A Functional Test of a Magnetic Sensor Designed for Long Period Measurements in a Magnetic Fusion Device, *J. of Plasma Fusion Res.*, **73**: 501 (1997), Robert D. Woolley, Tokamak poloidal field measurements accurate for unlimited time duration, in *Proceedings of the16th IEEE/NPSS Symposium on Fusion Engineering (SOFE) 1995, 1530-1533.*
4. T. Nishitani, T. Iida, Y. Ikeda, et al, Irradiation Tests on Diagnostics Components For ITER in 1955, *JAERI Tech* 96-040, Japan Atomic Energy Research Institute, Tokai (1996).

PLASMA VIEWING IN JET USING ENDOSCOPES AND A DETAILED DESIGN FOR ITER

J Paul Coad, Johann Lingertat, *Jean-Blaise Migozzi, Peter E Stott, **Christopher I Walker and Colin H Wilson

JET Joint Undertaking, Abingdon, Oxon OX14 3EA, UK
*JBM Optique, F-91140 Villebon-sur-Yvette, France
**ITER Joint Work Site, D-85748 Garching, Germany

1. INTRODUCTION

For the efficient operation of a tokamak like JET, it is very important to provide immediate visual information on the position of the plasma within the vessel (and possible regions of interaction with the wall) from a wide-angle view showing the complete plasma cross-section, and of the strike-point positions in the divertor. It will be more important to have an comprehensive on-line viewing system in ITER from the very start of operations, since the much larger heating power involved has a greater potential for causing problems. The development of viewing systems at JET has highlighted the advantages and difficulties inherent in such systems, and has also indicated the possibilities of the technology.

Video cameras were fitted to the JET tokamak in 1994 to provide detailed views of the newly-installed divertor and a wide-angle view of the inside of the torus in real time. The detailed views of the divertor were provided by cameras on the upper main vertical ports (UMVP) at Octants V and VIII, looking down almost vertically (the central ridge of the divertor is at 7° to vertical) in the poloidal plane, as shown in Figure 1 (which also shows a typical plasma configuration). As can be seen in the figure, almost all the divertor floor is visible: the view of the inner and (especially) the outer wall of the divertor is very oblique and partially shadowed, but this was not considered important at the time. The viewing lines are shared by spectroscopic (1) and infra-red cameras (1,2). The wide-angle view installed in 1994 was from a camera in a re-entrant water-cooled tube mounted in the pumping box at Octant I. The camera position is limited to approximately 1.5 metres from the point of entry of the Oct I main horizontal port (MHP) through the first wall by the requirement to leave the port clear for viewing by other diagnostics and for personnel entry into the vessel: this restricts the view of the interior of the torus to that shown in the computer-generated (CAD) image in Figure 2.

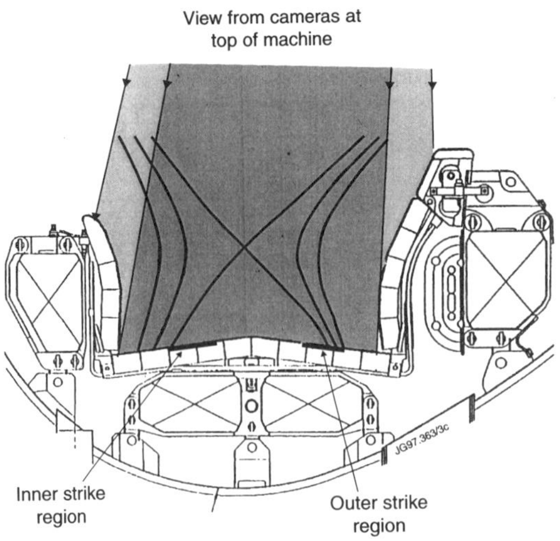

Figure 1 Poloidal section through the JET Mk I divertor used during 1994/5 and showing the lines of sight from the cameras mounted on the top of the torus and a typical plasma configuration.

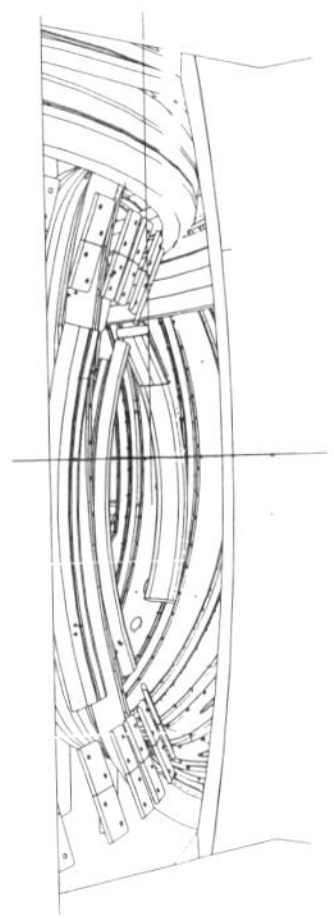

Figure 2 CAD drawing of the wide-angle view of JET from the camera in the Oct Pumping Box.

A number of problems were envisaged for subsequent JET campaigns. Firstly, the 1996/7 (Mk II) and planned future divertors have a more closed aspect and there is increasing interest in shaping the plasma so that the strike points are on the side walls, so a complete view is not possible from the top ports. Figure 3 (a) is a poloidal cross-section of the 1996/7 Mk IIA divertor, showing a typical plasma configuration. It is clear that the areas of interaction on the vertical plates cannot be viewed by the cameras at the UMVP, and part of the outer strike zone is also shadowed by the outer divertor wall. For the Mk II Gas Box (due to be installed during 1998) which has a shape similar to the proposed ITER design the situation is even worse (Figure 3(b)). Secondly, the wide-angle view is very restricted so that only about 5% of the vessel surface is seen, and the complete torus cross-section is not visible. Finally, none of the existing cameras can be used during plasma operations using D-T fuelling, because the neutrons (and associated γ-rays) will destroy the detectors.

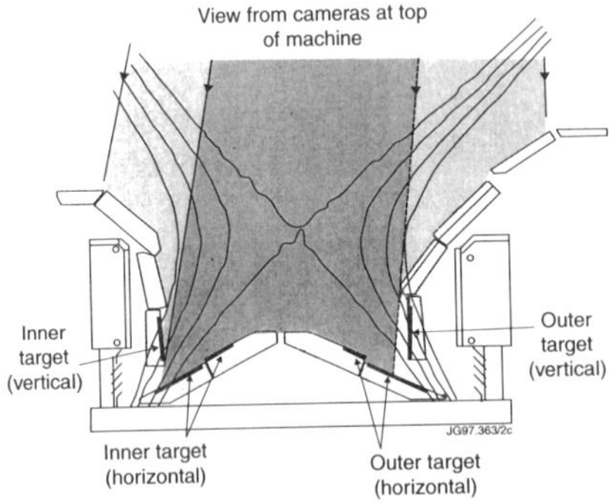

(a)

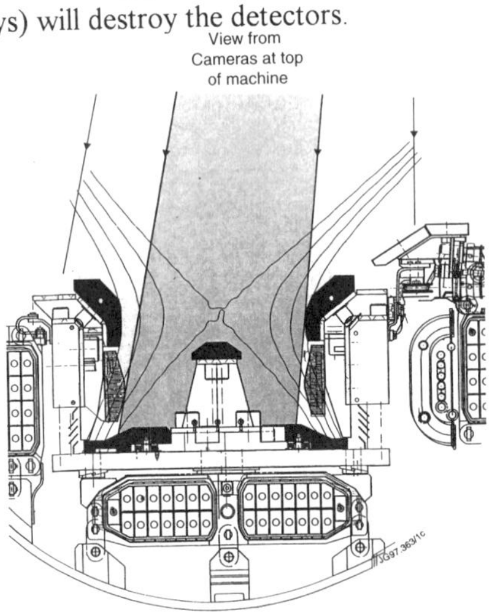

(b)

Figure 3 Cross-sections of the JET divertor showing a typical plasma configuration and the viewing lines from cameras mounted at the top of the machine, for (a) Mk IIA in use during 1996/7 and (b) Mk II Gas Box due to be installed in JET in 1998.

2 THE NEW ENDOSCOPE VIEWING SYSTEM ON JET

The ideal position from which to obtain both wide-angle and divertor views is just inside the vessel near the outer midplane. Since the JET vessel is operated with wall temperatures at 250 to 350°C (and of course under high vacuum conditions) cameras cannot simply be placed at such points. Instead re-entrant tubes have been fitted within two of the old Limiter Guide Tubes (LGT). The LGTs are horizontal vacuum ports welded into the outer wall which extend through the vessel support structure to a flange which is easily accessible: the length of each port is 1.8 metres from outer wall to flange. There are pairs of LGTs in each octant (in sectors B and D), one 280 mm above and one 280mm below the outer mid-plane, and although all these ports have been re-allocated to various diagnostics since the discrete limiters were removed in 1987, the upper ports in Octants IVB and VIIIB became available when one of the diagnostics was removed in 1995. The re-entrant tubes are mounted on the flange and end in a double sapphire window at the ideal viewing point just within the outer vessel wall. The window is in the shadow of an outer poloidal limiter which protects it from the plasma, and is covered by a shutter during vessel conditioning treatments such as glow discharge cleaning and Be evaporation. The window is angled at 45° to the vessel radius to give a clear view from inside the window in the anti-clockwise direction inside the torus: the poloidal limiters to the left of each port block the view in the clockwise direction. From each of these two vantage points approximately 30% of the vessel is visible.

An optical relay (endoscope) system has been constructed to allow cameras mounted outside this flange to view through the window. The endoscope allows three different views into the torus; a wide-angle view (WAV), and detailed views of the divertor in the visible and in the infra-red. Each of these views is accomplished via a separate set of relay lenses within the endoscope (contained within tubes of 44mm internal diameter), and the assembly is water-cooled to maintain the mechanical stability required to give high resolution images. The optical design of the WAV gives an included angle for the viewing cone of 80° at an aperture of f/4, and the objective lens of the video camera is chosen to give an image with the diagonal just equal to this value. Figure 4, which is a frame taken from a video of a JET pulse, shows the area of the torus that is visible, and the improvement over the previous view (Figure 2) is obvious.

The optics which provide a detailed view of the divertor in the visible spectral range image a circle of 14° centred in the divertor with an aperture of f/4. The camera is fitted with a filter so that light of one specific wavelength is recorded (currently the CII line at 658 nm). The optics for viewing in the infra-red are optimised for operation at 5 μm, and the view is split into two halves, one giving a close-up of the inner half of the divertor floor, and one of the outer half. However, at the present time the infrared camera is still viewing from the top of the torus whilst an improved image processing package is developed.

The glasses used in the construction of the WAV and divertor viewing systems are not radiation-resistant.Thus the endoscope was temporarily removed in March 1997 for the duration of the experiment with deuterium-tritium fuelling (code-named DTE1) to avoid possible neutron-induced damage. The endoscope is due to be replaced later in 1997, when once again a colour WAV and a close-up of the divertor will be recorded for every pulse.

3. EXAMPLES OF THE USES OF THE VIEWING CAMERAS

3.1 to verify plasma equilibrium and identify major hot-spots

$H\alpha$ radiation (mostly $D\alpha$ for deuterium-fuelled discharges) dominates the visible emissions from within the torus during plasma pulses. As the deuterium is ionised close to the surfaces interacting with the plasma, the edge of the plasma is the primary light source.

Figure 4 A frame taken from the video of the wide-angle view of pulse 38907 when the plasma was also in contact with the outer poloidal limiters.

Figure 5 A frame taken from the video of the wide-angle view of pulse 40466 which includes the light from a giant ELM.

Thus the wide angle view clearly shows all points of interaction with the plasma, and is a primary check on the position of the plasma within the vessel.

During the divertor phase of the discharge the only surfaces with significant plasma interaction should be the target tiles in the divertor. However, in order to develop the optimum plasma conditions plasma shape (especially elongation) and size are important variables: the WAV gives an immediate indication of the available working space for these experiments. The plasma shape is also varied to try to provide the optimal coupling between the ICRH antennas and the plasma for each heating scenario. Figure 4 is an example of what happens when there is too much power loading on the outer poloidal limiters during ICRH heating. Strong interaction is observed with the limiters to the right of the antenna banks at just above the outer midplane. The limiter surfaces have reached sufficiently high temperature for carbon atoms to sublimate, and as these atoms are ionised they move around the torus along the magnetic field lines. Emission from the carbon can be seen as streaks angling from lower left to upper right across the interaction areas, and in the colour original continues as blue light around the top of the inner wall (top right of image). For sublimation to occur a temperature in excess of 2500°C is required, so some of the light from the limiters must be thermal in origin, but the Dα predominates. However, if the plasma is suddenly moved away from the outer wall the Dα light source there is "switched off", and the thermal emission from the limiter tiles can be seen decaying with time.

A further useful diagnostic tool visible in Figure 4 concerns the clearance to the inner wall. The scrape-off layer (SOL) is clearly illuminated by emission of Dα from particles streaming up from the inner leg of the divertor, so that the distance of the last closed flux surface (LCFS) from the wall can be seen to well above the midplane. (In this example the separation is rather large and the illumination of the inner SOL correspondingly weak.)

3.2. Studies of ELM interactions outside the divertor

During an ELM up to a few per cent of the energy content of the plasma is expelled in about 100 microseconds. It is known that some of the energy goes to the divertor (but usually deeper into the SOL regions (2)), but little is known of interactions elsewhere. A frame is recorded from the WAV camera every 20 milliseconds, and the information within that frame represents an integration over a period of between 1 and 20 milliseconds (depending on the camera settings). Thus if an ELM occurs within the active period for the frame, all the interactions associated with the ELM will be integrated in the picture: if the ELM occurs during the gap between the active periods it will be missed. Many examples of ELMs have thus been recorded on the WAV videos. Figure 5 shows a video frame that captures the interaction effects of a giant ELM. The strong interaction within the divertor increases in area (and intensity) and normally extends to the tiles at the entrance to the divertor. Light is also invariably seen from some other points in the vessel - in this instance there was power deposited near the tops of the outer poloidal limiters, on the protection tiles for the upper saddle coils (inner) and at various points on the roof tiles. However, each ELM has a different set of interaction points, and no part of the first-wall is immune from these effects. Future fusion machines are likely to operate in ELMy regimes, so it is important to learn more about the dynamics of the power deposition to different parts of the first wall during ELMs before defining the design criteria for these surfaces.

3.3. Monitoring UFOs and looking for first wall component failures

Flashes of light (UFOs) are often seen in the plasma during a pulse. The flashes are usually caused by a particle burning up after entering the plasma. Occasionally showers of particles are seen after material is dislodged from the wall (eg following a partial disruption), and sometimes the source of particles can be seen due to interactions within the field of view of the camera. For example, particles have been emitted from tiles during strong interaction with the outer poloidal limiters such as shown in Fig. 4.

An example of a previous component failure can be seen in Figure 4. There is a dark rectangle at the midpoint of the outer poloidal limiter between the two which are interacting at this time, which is due to a gap in the limiter where one tile pair is missing. These tiles were damaged during a previous experiment when the limiters were overheated during several pulses. On another occasion edges of the protection tiles for the upper saddle coils (inner) were overheated whilst a plasma quench procedure was being developed, until one tile was seen on video to break and two pieces of the tile dropped down to the bottom of the vessel: one fell between the divertor and the inner wall, and the other into the divertor. Neither piece nor the damaged wall tile could initially be seen by In-Vessel Inspection System (IVIS), but the damaged tile was eventually located after directions were taken by comparison with the video footage. It was decided to improve the mountings of these tiles at the next vessel entry, and the broken pieces were retrieved at the same time.

3.4. Detection of high levels of visible bremstrahlung emission from the core associated with high performance discharges (e.g. optimised shear regime)

During normal operation there is no emission seen in the visible range from within the plasma, all light emanating from the periphery of the plasma. However, during experiments on the optimised shear regime (OSR), which is a possible operating scenario for ITER, low levels of yellowish light can be seen in the plasma core due to bremstrahlung emission. The initial experiments on OSR have been encouraging, and further work is planned wherein some attention may be given to the camera information.

3.5. Diagnosis of density limits due to MARFE formation and disruptions

Plasma instabilities register as fluctuations on many diagnostics, but it is often difficult to visualise what is actually happening. Some instabilities such as MARFEs may last long enough to follow by video, so that the 3-D complexity can be seen readily.

Such instabilities may end in a disruption. Almost all plasma diagnostic signals are lost/overloaded at a disruption, so characterisation is poor. Video recordings of pulses ending in a disruption provide some useful information, such as which surfaces take the energy released at the disruption. In JET the majority of disruptions go to the top of the machine, however disruptions to the inner or outer walls are not uncommon. If a disruption occurs at high plasma current (as opposed to minor disruptions that frequently occur well into the ramp-down phase of the discharge) all the edges/asperities in the contact area are seen to glow red-hot in several video frames after the disruption, indicating that surfaces can be heated to the order of 2000°C.

3.6. Applications of the divertor viewing camera

A typical frame from the divertor-viewing camera is shown in Figure 6. The strike points are clearly visible, and radiation can also be seen from the inner divertor leg extending to the X-point. The narrowness of the bands of light at the strike points indicates the zone width is only of the order of 1 cm. The divertor view is principally used to study the positions of the strike points. There has been much interest in recent months in the comparison of the effects of horizontal and vertical divertor targets on divertor recycling and detachment (3). The divertor view has been valuable in providing a direct visual confirmation of whether the strike points are indeed on the planned targets and to provide a check for the Langmuir

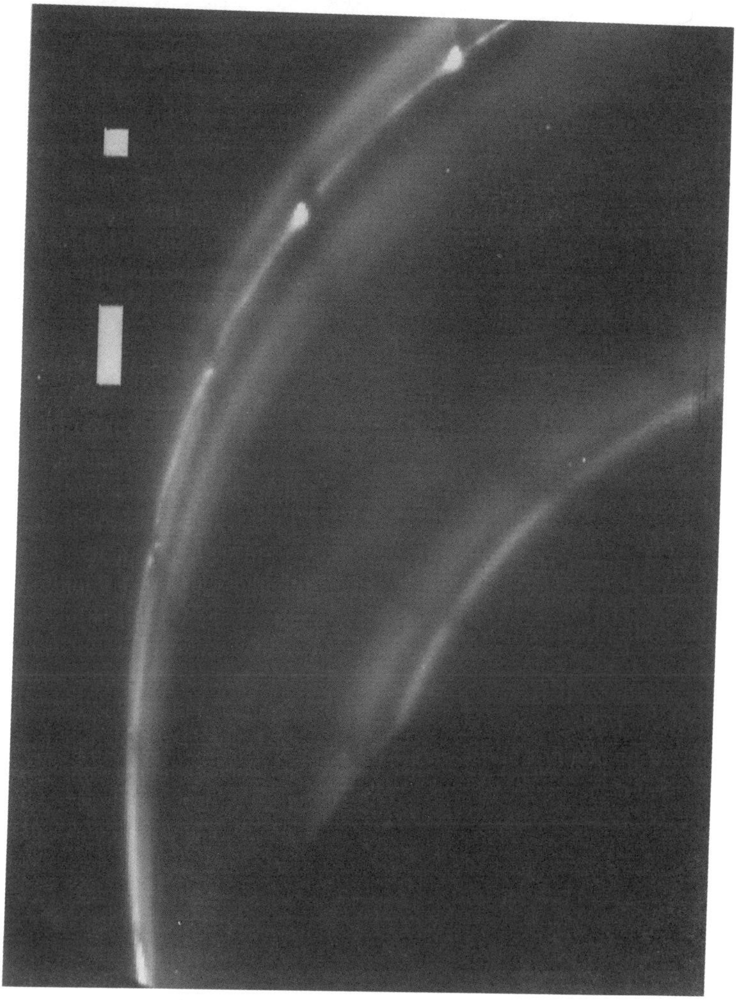

Figure 6 View of the divertor recorded in CII light during JET pulse 38267.

probe data, since the magnetic field measurements cannot give the strike-point position to the accuracy required. The plasma is also swept during portions of many discharges to distribute the power, and this movement (which is usually done at a frequency of a few Herz) of the strike points can be precisely corellated with the divertor tile structure using the divertor viewing camera.

The divertor view has proved useful for studies of detachment, when not only is the reduced interaction at the target surface seen (as also observed by Langmuir probes (3)) but also radiation can be seen moving up the divertor leg to the X-point region: the 3-D nature of the divertor view gives a unique visualisation of the X-point in addition to the targets. The integrated picture of ELM activity in the divertor can also be seen to complement the WAV image of elsewhere in the vessel (the WAV image sensitivity is such that the image is saturated in the divertor region).

The divertor viewing system is being modified before the endoscope is replaced on the JET machine later in 1997 so that three cameras can view through the optical system at the same time by means of beamsplitters. Each camera will have a filter changer, so that a number of experiments will become possible, such as measuring divertor temperatures by comparing helium spectral lines during He-puffing experiments and using filters for impurity lines, as well as studying various D and C spectral lines.

4. A DESIGN OF AN ENDOSCOPE SYSTEM FOR ITER

All the attributes of the JET viewing system are required for ITER. In addition a complete visual coverage of the vessel is required in case there is any toroidal non-uniformity in the plasma power load ings which could result in erosion or melting at the first wall of the main chamber (which is expected to be contructed of beryllium). Positions in two mid-plane blanket modules (on opposite sides of the machine) have been allocated for wide-angle viewing, with viewing in both clockwise and anti-clockwise directions from each position to allow complete visual coverage of the vessel. Furthermore, divertor viewing from these positions as well as from two other mid-plane modules would also be possible.

A major difference between viewing systems for ITER and the existing JET endoscope is that for ITER systems must be able to withstand the large neutron (and associated γ-ray) fluxes. Naturally the design must also be able to withstand the thermal and mechanical loads anticipated, but these are similar to those experienced at JET.

An outline design for a system to view the interior of ITER from the outer midplane has been produced which uses reflective optics for the first 2 metres (approximately) from the first-wall, changing to refractive optics to relay the images ~10 metres through the primary and secondary vacuum containment, the intermediate structure and the biological shield to the cameras. The scheme is shown in Figure 7. Two independent optical systems are portrayed with a cross-over just behind the first-wall: one sytem views in the clockwise direction, the other anti-clockwise.

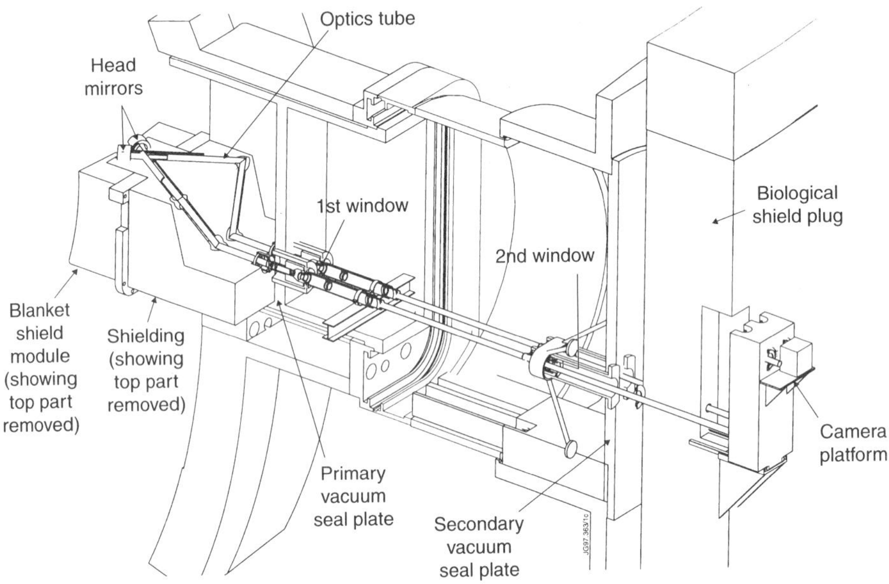

Figure 7 Overall scheme for wide-angle viewing from the outer mid-plane of ITER.

The design uses a concave aspheric mirror as the first element (Figure 8), and a catadioptric telescope (Figure 9) to produce a planar image to relay via conventional diffractive optics to the cameras which are located behind the biological shield. A high resolution wide-angle view inside the vessel can be obtained for the assessment of the plasma, and the same optics

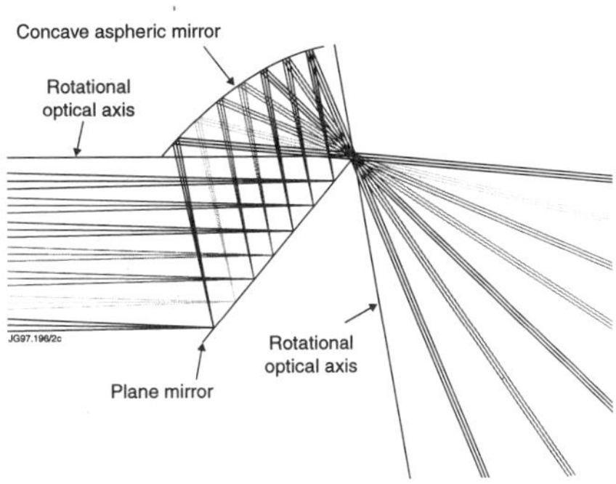

Figure 8 Concave aspheric mirror proposed for use at the first-wall of ITER to provide a wide-angle view of the interior of the vessel.

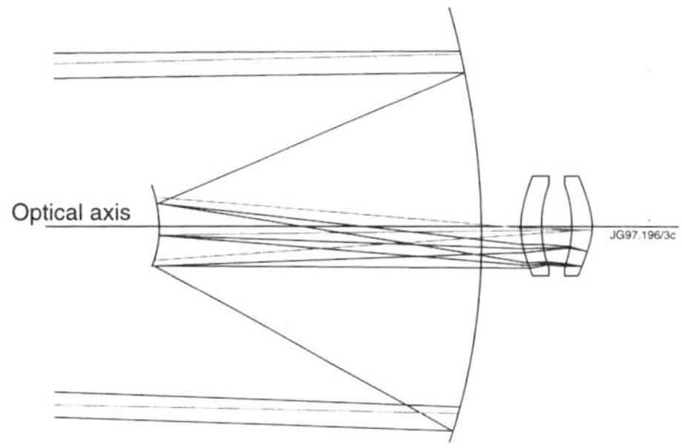

Figure 9 Catadioptric telescope stage to convert the spherical image of the interior of ITER to a planar image for relaying through the shielding surrounding the machine.

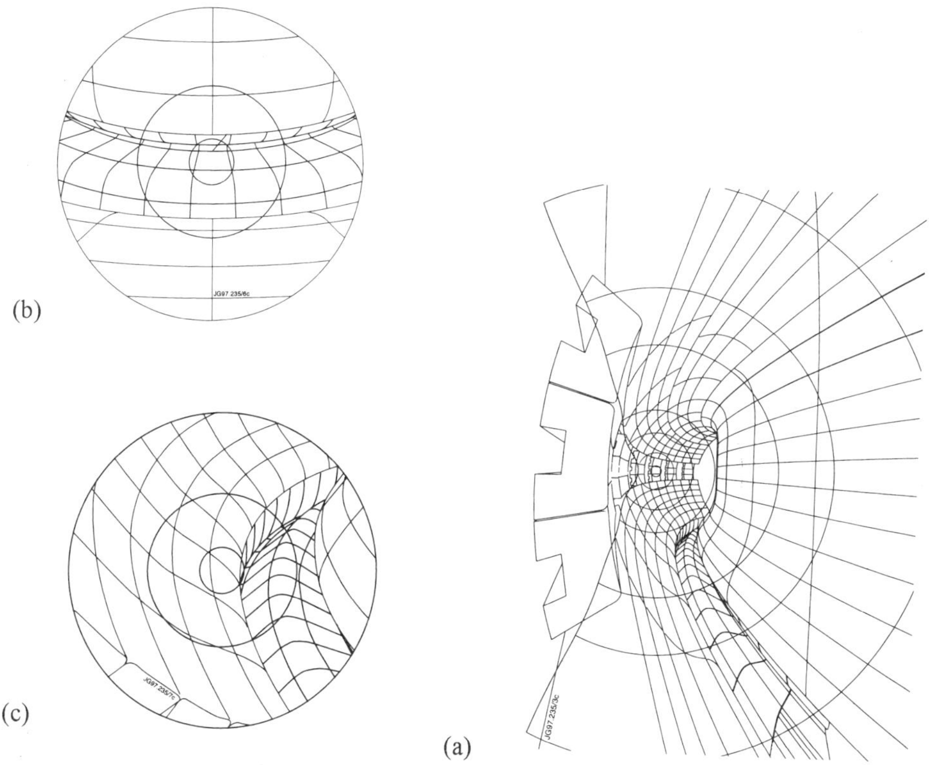

Figure 10 (a) Simulated wide-angle view inside the ITER vessel. (b) and (c) optimum angles for viewing into the inner and outer legs of the divertor (respectively).

can be used to examine closely any parts of the first-wall that appear to interact strongly with the plasma by means of a zoom lens, and to view into the outer leg of the divertor. Figure 10(a) is a simulated image of the interior of ITER, and the approximately semi-circular lines show the view obtained at different included angles for the aspheric mirror: figure 10(b) shows an expansion of the optimum angle for viewing the outer divertor leg. The optimum view into the inner leg shown in Figure 10(c) is along a line of declination of approximately 50° in the radial plane. Dedicated optics will be necessary for viewing into the inner divertor leg, and greater resolution may be obtained in the visible range if dedicated optics are also available for the outer leg: in any case viewing in the infra-red for temperature measurement will require separate optics to viewing in the visible for each divertor leg.

At the diffraction limit, the spatial resolution for viewing items in the ITER vessel of the proposed wide-angle optical system is 2-3 mm at areas 10 metres from the mirror at the first wall. This resolution should allow any significant damage at any point on the first wall to be visible, and since the precursor to any damage event (i.e. strong plasma interaction) will show up clearly on the live video, the operators will already know precisely where to search for any signs of damage. If the included angle in the viewing cone is reduced by a factor of perhaps five to give detailed viewing into the divertor, there is a commensurate improvement in the resolution limit. Erosion of the target tiles is a major concern for high power operation of ITER: with sub-millimetre spatial resolution there is the possibility of designing an on-line indicator of the extent of erosion in these areas (e.g. the shadow cast by grooves or steps in the target surfaces). However, this assumes certain values for the manufacturing and assembly tolerances, and for the mechanical and thermal stability of the system. It is very important to check that components can be made and assembled in materials suitable for ITER to the standard required, and to test the design in a suitable environment. Accordingly, a prototype is being designed to be tested on JET which incorporates as many of the features of the proposed ITER system as possible, including a dog-leg for radiation screening and a three-point focussing camera with zoom capability. The prototype will be of similar size to the ITER design, and will be subject to at least as severe thermal conditions as expected on ITER. It is planned for the prototype to be built and installed by the end of 1998, so that if there is an extended period of DT operation on JET in 1999 the radiation-resistance of the device can be thoroughly tested.

CONCLUSIONS

A new endoscope system has been installed on JET which gives greatly improved wide-angle and divertor plasma viewing at high resolution. Similar real-time viewing of the ITER plasma may be possible using an aspheric mirror as the first optical component. The critical issues which control the resolution obtainable in ITER are the mechanical tolerances and stability that can be acheived on assembly and maintained long-term in the high radiation levels, and the thermal stability of the endoscope system. The next step is therefore to build a prototype system which is as close as feasible to the ITER configuration that can be tested in the laboratory and then installed on JET.

REFERENCES

1. J Lingertat, A Tabasso et al *J Nucl. Materials* 241-243 (1997) 402
2. E Gauthier, A Chankin et al, in *24th EPS Conf. on Controlled Fusion and Plasma Physics*, Berchtesgaden, Germany, June 1997.
3. R Monk, A Loarte et al, in *24th EPS Conf. on Controlled Fusion and Plasma Physics*, Berchtesgaden, Germany, June 1997.

PRESSURE GAUGES AND NEUTRAL PRESSURE MEASUREMENT IN ITER

G. Haas[1], H.-S. Bosch[1], D. Coster[1], L. de Kock[2], R. Maingi[3], J. Neuhauser[1], R. Schneider[1]

1) Max-Planck-Institut für Plasmaphysik
IPP-EURATOM Association
D–85748 Garching, Germany
2) ITER JCT, JWS Garching, Germany
3) Oak Ridge National Laboratory
Lockheed Martin Energy Research,
presently assigned to General Atomics, San Diego, USA

ABSTRACT

The neutral gas environment in a poloidal divertor tokamak like ITER plays an important role for the global machine performance.
In order to investigate the local neutral gas dynamics with sufficient time resolution, specific neutral pressure gauges have been developed originally for ASDEX, and are now applied in modified form in most existing experiments. Application of these gauges for ITER seems to be feasible (Diagnostic Design Document DDD 5.5.G.03). In this paper, recent progress in gauge design and understanding is described and experimental results from several divertor tokamaks are presented. The results are used to check numerical edge-divertor codes (B2-EIRENE), which, in turn, can be applied for ITER predictions and neutral diagnostics optimization.

INTRODUCTION

The parameters of the neutral gas environment of the divertor and the main plasma have a crucial influence on the overall operation of ITER as of any tokamak. Therefore two years ago at the ITER Diagnostic workshop in Varenna a neutral gas diagnostic for ITER was proposed [1], based on so called ASDEX gauges [2]. The proposal was adopted by the ITER design team and was incorporated into the Diagnostic Design Document DDD 5.5.G.03. It is foreseen to install 50 to 100 ASDEX gauges at the walls of the instrumented standard divertor cassettes (See Fig. 1).
These gauge heads are small and versatile and with appropriate shield structures robust enough to be installed even in positions of high recycling and power load like divertor

Diagnostics for Experimental Thermonuclear Fusion Reactors 2
Edited by Stott *et al.*, Plenum Press, New York, 1998

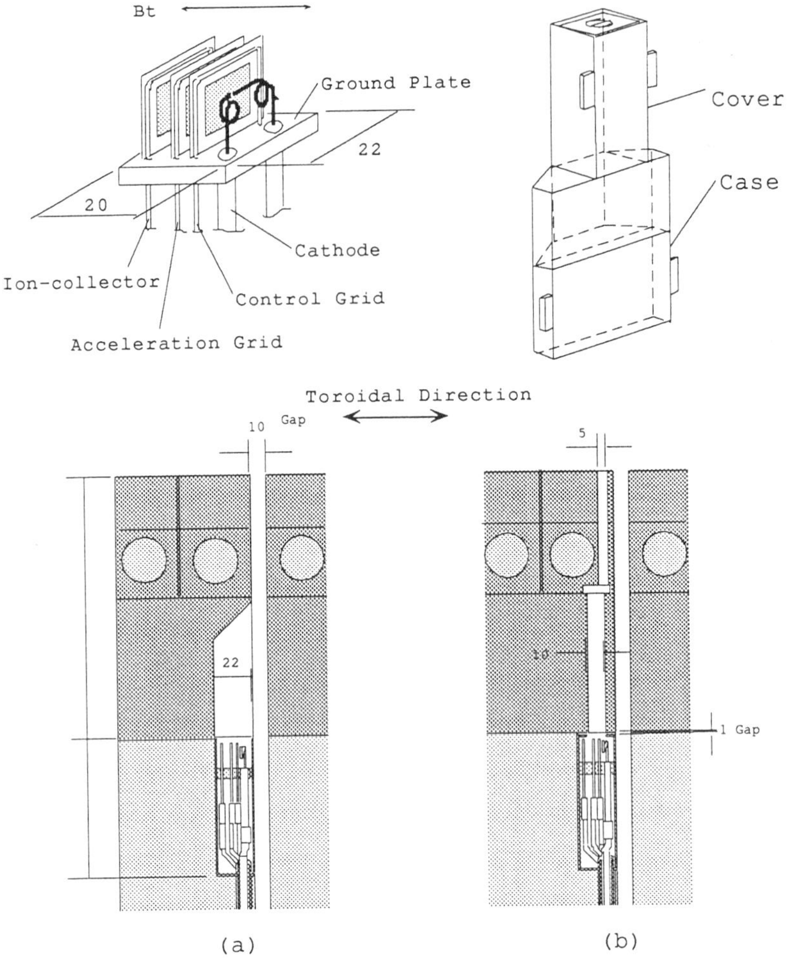

Figure 1: Two versions of attachment of a pressure gauge head to a target tile using either the gap between the tiles or a small hole in the tile as duct to access the plasma. The inserts show the gauge head and its housing.

targets or scoop limiters in existing tokamaks [2, 3]. Only those positions have to be excluded where the gauge head or its shield structure would cause additional recycling and would influence the neutral gas or the boundary plasma. With this exception no problems are expected with installation and operation of the gauges at any desired position in ITER. The final positions have to be chosen in accordance with predicting model calculations. Also the in-vessel cabling and electrical connections and the feedthroughs do not pose specific problems except those which have to be solved in general for other diagnostics too. Therefore it is still considered as premature to develop a complete and detailed design of the whole neutral gas diagnostic system, but one has to ensure that the components are compatible with the constraints set by the overall design of ITER and its divertor systems and by the operation of ITER. In the next chapter we deal with some new developments of the gauges and the implications on their use in ITER. In chapter 3 and 4 we describe some examples how the gauges are used in present tokamaks phenomenologically and in connection with model calculations, respectively.

TECHNICAL PROGRESS AND DEVELOPMENTS ON THE ASDEX GAUGES

The ASDEX gauge [2, 4, 5] is used as a diagnostic tool on many fusion experiments and it is subject to ongoing development to meet new requirements. Some examples relevant for ITER are given below.

The Al_2O_3 base plate has been replaced by a grounded metal plate with inserted Al_2O_3 insulators. This is similar to what was used until about 1990. It completely avoids leakage currents between the electrodes even for long term operation. This was possible with the all ceramic base plate only by conducting, grounded strips on the ceramic surface, but it does not work as well as a complete metal barrier. Like the gauge head with all ceramic base plate, the new version is fully brazed in contrast to the oldest design. This one could be dismantled for repair, but that is not a useful option for tokamaks like JET or ITER, anyway. The change became possible by progress in metal to ceramic brazing.

A diameter reduction at the central part of the filament allows to operate the gauge with reduced heating current without impairing the stability. Beside the advantages of the reduced power (lower gauge temperature, smaller feedthrough and cable cross section) it avoids the danger of deforming the filament by the Lorentz force especially if the feedback loop fails and the heating current rises to the limit necessary for conditioning.

It was shown during tests performed in a superconducting magnet at Forschungszentrum Karlsruhe, that startup and normal operation of the gauge for hours in strong magnetic fields of up to 9 Tesla do not present any problems. A limitation in the achievable emission current and a non-linear behaviour of the gauge were probably the consequence of the poor vacuum condition as recent experiments in IPP Garching with a better vacuum system indicated. The same experiments in Garching cast some light on the dominating processes in the gauge immersed in a strong magnetic field [5]:

The main result is that the strong enhancement of the sensitivity due to even small magnetic fields is caused by electrons which return to the filament after one passage through the gauge. These electrons contribute to the ionising electron density in the ionisation volume, but not to the measured emission current.

The Radiation Induced Electrical Degradation (RIED), considered two years ago still as a severe problem for all components containing ceramic insulators in an electric field, seems to be avoidable if appropriate ceramics are used. Therefore tests of a gauge in a neutron radiation field will be postponed for the moment.

SELECTED RESULTS FROM DIVERTOR TOKAMAKS

In this chapter we will demonstrate on a few examples the importance of the high time resolution of the in vessel pressure gauges.

Generally, only 10 or 20 % of the gas injected into a (divertor) tokamak are found in the plasma and the neutral gas [2, 6]. The rest is stored on the walls, most of it, however, as transient inventory participating in the particle recycling. If one removes this wall inventory or tries to build up the plasma without it, achievement of stationary plasma density proves impossible [7]. The gas is probably rather unevenly distributed on the walls and hardly accessible to a direct measurement, but only to a global particle balance. Such balances are determined on a slow time-scale, i.e. by taking into account the particle input by fueling, the changes of the particle contents of plasma and neutral gas and the particle flux pumped by the external pumps [6, 8, 9]. This gives inventory changes averaged over the whole vessel, and integrated over the full duration of a discharge it yields changes in the permanent wall inventory.

Neutral gas balances, however, are sometimes useful also on a much faster time scale, and one such example is a disruption, i.e. the sudden breakdown of the plasma which frequently leaves behind a rather uneven distribution of the neutral gas in the vacuum vessel. During a disruption elongated tokamak plasmas loose the vertical stability and move towards the vessel walls. In ASDEX Upgrade the plasma is mostly dumped in the lower part of the vessel (See Fig. 2 for a poloidal cross section of the vessel).

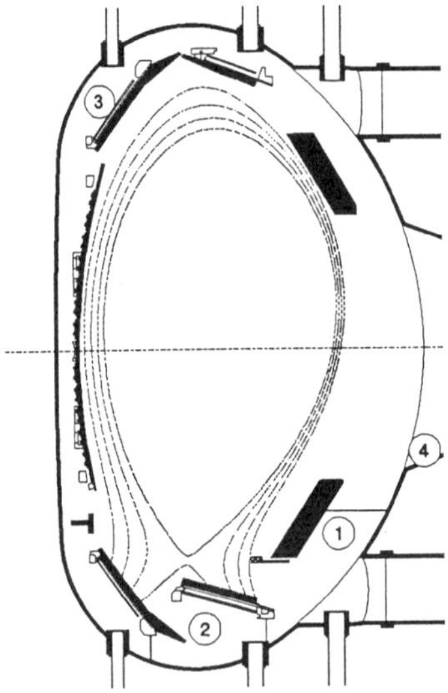

Figure 2: Poloidal cross section of the ASDEX Upgrade vessel. The numbers denote typical positions of gauges. For the particle balances gauge 2 (private flux region) and 3 (vessel) are used.

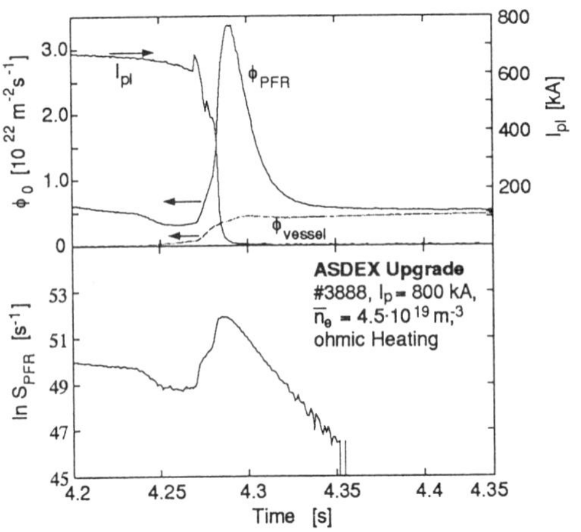

Figure 3: Neutral particle flux densities and plasma current during a disruption in ASDEX Upgrade (upper figure). The bottom curve displays the exponential decay of the fast portion of the particle source in the private flux region. The slowly decaying portion is deducted.

This raises the neutral flux density ϕ in the divertor much more and much faster than in the upper part of the vessel. The neutral flux density is connected with the neutral density n_0 by

$$\phi = \frac{n_0 \cdot \bar{v}}{4}$$

with $\bar{v}$ being the mean velocity of the neutrals.
In the neutral gas flow range typical for tokamaks the neutral flux density is especially for non homogeneous temperature distribution or non-Maxwellian velocity distributions the better defined quantitiy than the particle density or the pressure of the neutral gas which may change locally much more. Since moreover the quantity primarily measured by the encased gauges is the neutral flux density, we prefer for quantitative discussions the neutral flux density rather than the neutral pressure or the neutral density.
The particle balance of the volume below the outer target plate (private flux region) as an example is:

$$\frac{d\phi_{PFR}}{dt} + \frac{\phi_{PFR} - \phi_{vessel}}{V_{PFR}/L_{PFR}} = \frac{S_{PFR} \cdot \bar{v}/4}{V_{PFR}}$$

with ϕ_{PFR} and ϕ_{vessel} being the neutral flux density in the private flux region and in the upper part of the vessel, respectively, and $\bar{v}$ the mean velocity of the neutrals in the private flux region. V_{PFR} and L_{PFR} are the volume of the private flux region and the conductance to the main plasma chamber. De- or adsorption of particles at the wall appear as additional particle source or sink S_{PFR}. It is composed of a small, slowly decaying part ($\tau\ \approx 0.5$ to 1 sec) and a large one which decays exponentially with a time constant of 10 to 15 msec over more than 2 decades. The neutral flux densities and the fast portion of the source S_{PFR} are plotted in Fig. 3. The bottom part of this figure clearly shows the exponential decay with a time constant of $\approx$ 10 ms. The total amount of gas stored in the walls of the private flux region can be derived by integrating the particle source. It exceeds the gas content of the private flux region at the maximum of ϕ_{PFR} (at 4.29 s) by more than a factor 3. In the rare event of a plasma moving upwards during a disruption this term appears as particle sink with the same time behaviour.
The neutral flux density is modulated by ELM's, too. The absolute and relative amplitude, the rise time and some other details of the wave form change from tokamak to tokamak and also depend on the discharge parameters. In Fig. 4 examples from DIII-D and ASDEX Upgrade are shown. In both cases the neutral flux density in the private flux region during a type I ELM is plotted together with the D_α intensity as time marker and indicator for the ELM. The neutral flux starts to rise just after the ELM indicated by the D_α signal is nearly finished.

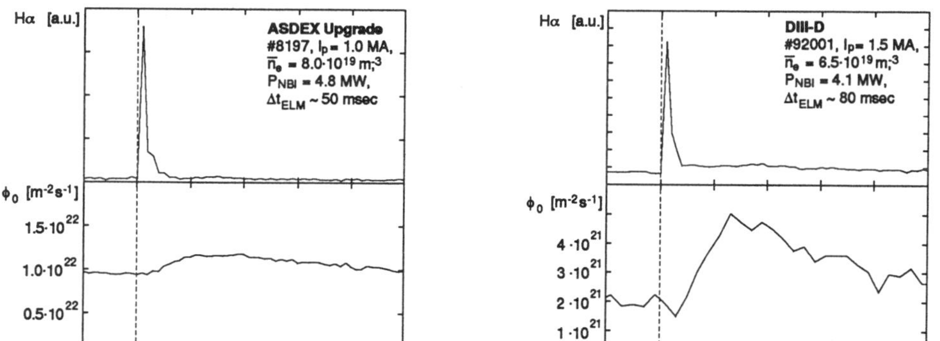

Figure 4: ELM history in ASDEX Upgrade (left) and DIII-D (right). Time is the time relative to the onset of the ELM, determined from the H_α signal (upper traces).

Sometimes a slight decrease of the neutral flux density is observed during the ELM. Maybe it is caused by the intensified pumping ability of the divertor plasma which has been heated up by the increased energy flow during the ELM. On the other hand, however, indications have been occasionaly found that gas is desorbed due to the ELM [8]. The absolute increases of the neutral fluxes are comparable in the examples of figure 4, but due to the lower base neutral flux the relative modulation is larger in DIII-D (In JET still larger modulations have been observed up to a factor of 5 to 10). The most surprising fact is the almost identical rise time of about 10 msec in both tokamaks which is much longer than the ELM and coincides with the decay time of the source term derived from the fast particle balance in a disruption (See Fig. 3). The large differences in the volumes of the private flux regions and in the connections to the divertor plasma in both tokamaks and the fact that even in ASDEX Upgrade the vacuum time constant of the PFR is by a factor of about 3 shorter than the rise time excludes pumping times as explanation. A more likely explanation is a large reservoir of adsorbed gas which needs this time constant to come into an equilibrium with the gas in the volume.

THE ROLE OF MODELLING

The neutral gas has a rather complicated distribution which shows no symmetry according to the magnetic field and may change especially in the divertor on a short length scale. To get experimentally the full information on the neutral flux density would require the installation of more gauges than is technically possible and partly on positions where it is impossible at all. Therefore one has to restrict the number of gauges and to use their output for cross checks of Monte Carlo model calculations rather than as input for them. Being constrained by experimental plasma data, i.e. with the transport parameters properly adjusted, model calculations can deliver all desired parameters of the neutral gas.

To assess the relative merits of different divertor schemes the numerical models and codes used have to be able to treat all the atomic physics processes as well as the complicated geometries involved. Due to the strong non-linear plasma-neutral interaction a sophisticated coupling between fluid plasma code and kinetic neutral transport code is necessary to get converged solutions. The basic setup is the exchange of sources for the plasma fluid code due to neutrals and plasma background for the neutral transport code.

On ASDEX Upgrade the B2-EIRENE code package [10, 11, 12] is used as a tool for modelling the plasma boundary, the scrape-off layer plasma and the neutral gas. In EIRENE [13] a rather complete set of atomic and molecular data is available, including elastic collisions of neutral atoms and molecules and ions [14]. Neutral recycling sources are defined by the target plasma fluxes, accounting also for the sheath acceleration of the ions. Plasma fluxes to the side walls are fed back locally as thermal molecules or fast neutrals, sampled from a Maxwellian energy flux distributrion of 2kTi. Pumping is simulated with an appropriate albedo at the pump duct entrance (for external pumps) and/or at the walls (for wall pumping). Various experimental scenarios for ASDEX, ASDEX-Upgrade and other tokamaks including ITER have been successfully modelled using the B2-EIRENE package [10, 15, 16, 17, 18]. Fitting the coefficients for the anomalous radial plasma transport to experimental profiles (usually at low density operation) comparison with neutral gas measurements show very good agreement even for neutral fluxes different by more than four orders of magnitude (outer divertor volume, private flux region, main chamber).

A neutral flux density ϕ_0 can only be converted directly into a neutral density n_0 if the mean velocity of the molecules $\bar{v}$ is known, e.g. if the gas is fully thermalized, according to $n_0 = 4 \cdot \phi_0/\bar{v}$. Near to a plasma this condition is usually not fulfilled,

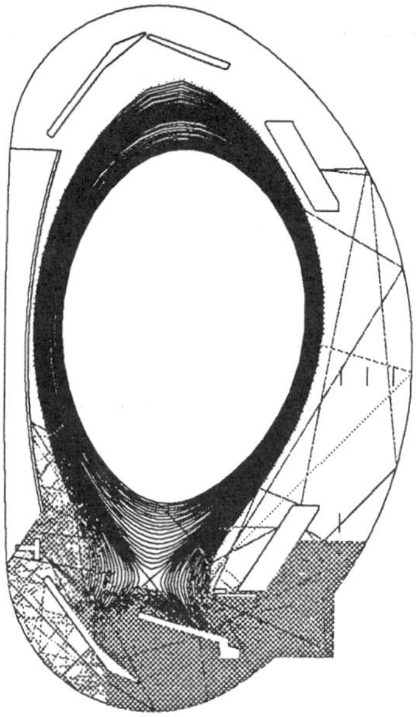

Figure 5: Trajectories of particles started at the inner and outer target plates. For each plate 100 particles are launched according to the particle and energy flux distribution of a self-consistent B2-Eirene result.

but an essential portion of the neutral flux is carried by energetic atoms [19]. Even molecules may be heated up to 1 eV by elastic collisions with deuterium ions and atoms. Fortunately, for the majority of physical questions the neutral flux density is a more important quantity than the neutral particle density. Therefore, the conversion is frequently unnecessary for the analysis. Only detailed particle balances for which also the number of particles in the gas is needed require such a conversion.

The overall power load and the power density on the target plates have to be reduced in ITER to a safe value of below 5 MW/m^2. This can be achieved only by operation under partially or totally detached conditions. The physics of detachment, which has to be understood therefore in detail, is closely related to the physics of the neutrals. Candidates for power loss and wider energy distribution are neutral recycling and impurity radiation. It would be desirable to avoid the addition of impurities. Two energy loss channels - CX- and elastic collisions with and atomic radiation from neutral hydrogen - are in principle available [20]. Both processes, however, constrain strongly the parameters, and can, even in connection with other loss channels, transmit only a minor fraction of the total power for realistic ITER parameters. This has been demonstrated by modelling calculations [21, 22, 23] but can be understood also from simple analytical estimates [24].

Due to the localized reionization of neutrals the ion flux to the target plate increases faster than the midplane density, whereas the temperature decreases at the target plates. Below a temperature of about 3 eV CX- and elastic collisions with neutrals prevent a further increase of the ion flux to the target and produce a pressure drop along the field-lines. At temperatures of about 1 eV volume recombination (especially three-body recombination) stops the ion flow completely before reaching the target and the target plasma flow drops at least by one order of magnitude.

A proof for this mechanism is the observation, that detachment experimentally manifests itself as a drop of the plasma flux (or ion saturation current) onto the target plate and of the plasma density and pressure at the target plate combined with an increase of the D_α signal and of the neutral flux densities in the divertor. The explanation of this contrary behavior of D_α and neutral flux on the one hand and the plasma density and flux to the target on the other forces the introduction of strong volume-recombination [25]. Modelling with the volume-recombination suppressed shows that without this effect no significant drop of the particle flux onto the plate is possible. Elastic collisions between molecules and ions represent another important process in these low temperature plasmas resulting in an additional momentum loss and creating a hot (about 1 eV) population of molecules as measurements in the private flux regions of JET and DIII-D have shown. Neutral particle flux densities measured in the private flux regions could be brought into accordance with D_α intensities measured simultaneously at the adjacent gas/plasma interaction zone only if elastic collisions were assumed. Otherwise the particle fluxes were by far too large [26]. In accordance with this analysis of the detachment process divertor geometries which are closed by the SOL plasma are tested in several tokamaks to maximize the neutral flux density in the divertor, to enhance energy and momentum losses and to allow easier access to detachment. It may, however, result in a reduced SOL density limit which is determined by complete divertor detachment [27].

CONCLUSIONS

Neutral gas measurements are of vital importance for ITER, especially in the divertor since neutrals play a key role for safe divertor operation (compatible with good confinement). The application of ASDEX-type neutral flux gauges for ITER seems to be technically feasible. For the analysis of the experimental results modelling with sophisticated codes like B2-Eirene is a key element, because it helps to understand them in detail and (in an iterative process) it allows improvement and validation of the model for extrapolation to ITER.

References

[1] G. Haas, H.-S. Bosch, and L. de Kock, Neutral Gas Diagnostics for ITER, in *Diagnostics for Experimental Thermonuclear Fusion Reactors*, edited by P. E. Stott, G. Gorini, and E. Sindoni, pages 571–579, New York, 1996, Plenum Press.

[2] G. Haas, J. Gernhardt, M. Keilhacker, E. B. Meservey, and the ASDEX Team, Measurements on the Particle Balance in Diverted ASDEX Discharges, J. Nucl. Mater. **121**, 151–156 (1984).

[3] T. E. Evans, J. S. DeGrassie, G. L. Jackson, N. Ohyabu, A. J. Wootton, K. W. Gentle, W. L. Hodge, S. C. McCool, P. E. Phillips, T. L. Rhodes, et al., Experiments to Test an Intra-Island Scoop Limiter on TEXT, J. Nucl. Mater. **145–147**, 812–818 (1987).

[4] C. C. Klepper, T. E. Evans, G. Haas, G. L. Jackson, and R. Maingi, Neutral Pressure Studies with a Fast Ionization Gauge in the Divertor Region of the DIII-D Tokamak, J. Vac. Sc. Techn. **A11**(2), 446–450 (1993).

[5] G. Haas and H.-S. Bosch, In Vessel Pressure Measurements in Nuclear Fusion Experiments with ASDEX Gauges, Vacuum, 1997, in press.

[6] G. Haas, H.-S. Bosch, A. Kallenbach, C. Klepper, R. Maingi, D. Meisel, W. Poschenrieder, ASDEX Upgrade Team, and NBI Team, Particle Exhaust and

Wall Pumping Studies in ASDEX Upgrade, in *Europhysics Conference Abstracts (Proc. of the 22th EPS Conference on Controlled Fusion and Plasma Physics, Bournemouth, 1995)*, edited by B. Keen, P. Stott, and J. Winter, volume 19C, part I, pages 321–324, Geneva, 1995, EPS.

[7] G. Haas, M. Kaufmann, R. S. Lang, ASDEX Team, and Pellet Team, J. Nucl. Mater. **145–147**, 519–552 (1987).

[8] J. Ehrenberg, P. Andrew, L. Horton, G. Janeschitz, L. de Kock, and V. Philipps, Material Related Aspects of Density Control in JET, J. Nucl. Mater. **196–198**, 992–996 (1992).

[9] R. Maingi, B. Terreault, G. Haas, G. L. Jackson, W. Zuzak, P. K. Mioduszewski, M. A. Mahdavi, M. R. Wade, S. Chiu, M. J. Schaffer, J. T. Hogan, and C. C. Klepper, Comparison of Wall/Divertor Deuterium Retention and Plasma Fueling Requirements on the DIII-D, TdeV, and ASDEX Upgrade Tokamaks, J. Nucl. Mater. **241–243**, 672–677 (1996).

[10] R. Schneider, D. Reiter, H. P. Zehrfeld, B. Braams, M. Baelmans, J. Geiger, H. Kastelewicz, J. Neuhauser, and R. Wunderlich, B2-EIRENE Simulation of ASDEX and ASDEX-Upgrade Scrape-off Layer Plasmas, J. Nucl. Mater. **196–198**, 810–815 (1992).

[11] D. Reiter, Progress in two-dimensional Plasma Edge Modelling, J. Nucl. Mater. **196-198**, 80–89 (1992).

[12] D. Reiter, C. May, D. Coster, and R. Schneider, Time dependent Neutral Gas Transport in Tokamak Edge Plasmas, J. Nucl. Mater. **220–222**, 987–992 (1995).

[13] D. Reiter, The EIRENE Code, Version: Jan. 92 Users Manual, Technical Report Jül-2599, KFA Jülich, Germany, 1992.

[14] D. Reiter, Neutral Gas Transport in Fusion Devices: Atomic and Surface Data Aspects, in *Atomic and Plasma-Material Interaction Processes in Controlled Thermonuclear Fusion*, pages 243–266, Elsevier Science Publishers, 1993.

[15] K. McCormick, S. Fiedler, G. Kyriakakis, J. Neuhauser, D. Reiter, R. Schneider, J. Schweinzer, and N. Tsois, Particle and Energy Transport in the ASDEX Scrape-Off Layer, in *Proc. of the 20th EPS Conference on Controlled Fusion and Plasma Physics, Lissabon, 1993*, volume II, pages 587–590, Petit-Lancy, 1993, EPS.

[16] H.-S. Bosch, J. Neuhauser, R. Schneider, A. Field, A. Herrmann, G. Lieder, W. Junker, C. S. Pitcher, D. Reiter, B. Braams, and ASDEX-Upgrade Team, 2D-Modelling of the ASDEX-Upgrade Scrape-off Layer and Divertor Plasma, J. Nucl. Mater. **220–222**, 558–562 (1995).

[17] H. Kastelewicz, R. Schneider, J. Neuhauser, D. Reiter, B. Braams, U. Wenzel, K. Büchl, M. Laux, and V. Mertens, Numerical Marfe Simulations at ASDEX and ASDEX Upgrade, in *Proc. of the 20th EPS Conference on Controlled Fusion and Plasma Physics, Lissabon, 1993*, volume II, pages 807–810, Petit-Lancy, 1993, EPS.

[18] R. Schneider, D. Reiter, D. Coster, J. Neuhauser, K. Lackner, and B. Braams, Analysis of Cold Divertor Concepts for ITER, J. Nucl. Mater. **220–222**, 1076–1080 (1995).

[19] H.-S. Bosch, R. Dux, G. Haas, A. Kallenbach, M. Kaufmann, K. Lackner, J. Neuhauser, S. de Pẽna Hempel, W. Poschenrieder, D. Coster, R. Schneider, ASDEX Upgrade Team, and NI-, ICRH- and ECRH Teams, Particle Exhaust Studies in ASDEX Upgrade, Plasma Phys. Contr. Fusion, in press.

[20] J. G. Watkins, R. A. Moyer, D. N. Hill, D. Buchenauer, T. N. Carlstrom, R. Conn, J. Cuthbertson, R. Doerner, R. Lehmer, M. A. Mahdavi, R. T. McGrath, L. Schmitz, and R. Stambaugh, Scrape-off Layer Measurements in DIII-D, J. Nucl. Mater. **196–198**, 829–832 (1992).

[21] R. Schneider, D. Reiter, J. Neuhauser, J. Igithkanov, H. Kastelewicz, M. Baelmans, and B. Braams, Analysis of Cold Divertor Concepts by 2-D Simulations, in *Proc. of the 20th European Conference on Controlled Fusion and Plasma Physics, Lisbon, 1993*, volume II, pages 775–778, Petit-Lancy, 1993, EPS.

[22] A. Kukushkin, Feasibility of Gas Target Mode of Divertor Operation in ITER, Contrib. Plasma Physics **34**(2/3), 282–292 (1994), 4th Workshop on Plasma Edge Theory, October 1993, Varenna, Italy.

[23] S. Weber, R. Simonini, and A. Taroni, Numerical Simulation of Hydrogenic Atomic Losses in a Divertor Plasma, Contrib. Plasma Physics **34**(2/3), 374–379 (1994), 4th Workshop on Plasma Edge Theory, October 1993, Varenna, Italy.

[24] K. Lackner, Figures of Merit for Divertor Similarity, Comments Plasma Phys. Controlled Fusion **15**(6), 359–365 (1994).

[25] K. Borrass, R. Schneider, and R. Farengo, A Scrape-Off Layer Based Model for Hugill-Greenwald Type Density Limits, Nucl. Fusion **37**, 523–537 (1997).

[26] G. Haas, P. Bachmann, D. Düchs, R. Reichle, D. Reiter, and R. Schneider, Elastic Backscattering of Hydrogen Molecules from a Plasma, J. Nucl. Mater. **196-198**, 481–484 (1992).

[27] K. Borrass, D. Coster, and R. Schneider, Numerical Study of the Impact of Divertor Closure on Detachment, EPS 1997.

HIGH RESOLUTION IR TEMPERATURE MONITORING SYSTEM FOR THE ITER TARGET PLATES

A. Herrmann and H. Salzmann

MPI für Plasmaphysik, EURATOM Association, 85748 Garching, Germany

INTRODUCTION

Monitoring the surface temperature of the divertor target plates of ITER is needed for machine protection. In this study no complete 2D coverage of the divertor is considered but a measurement in a single poloidal plane with the front optics being placed in the dome of the diagnostic cassette.

In addition to machine protection, the diagnostic serves for measuring the actual position of the strikepoints on the target plates to be compared with the prediction of the magnetic reconstruction codes, and for determining the heat flux profile in the SOL.

The IR measuring technique is well established on present day tokamaks and the difficulties in applying it to ITER are concerned with: the capability of the optical front end to endure, the complexity of the optical system inside the divertor cassette, a possible luminescence of the vacuum window and the movement of the vessel relative to the building. In this paper an unconventional optics is proposed which should reduce the number of optical elements inside the divertor cassette by wavelength-multiplexing.

Additionally, the more basic problem of the error in the temperature measurement due to the change of the emissivity of the divertor material as a consequence of particle and neutron impact is discussed.

UNCONVENTIONAL COLLECTION OPTICS

In competition with a complex reflective optics system transmitting the light collected from a wide angle of view, an unconventional optics approach is proposed. This method is based on the fact that the information about the temperature of the radiating surface is contained in a wide range of wavelengths. A collection optics for the divertor diagnostic cassette was designed which wavelength-multiplexes the collected thermal radiation from different points of the target plates into a single beam. Thus the number of optical elements inside the vacuum vessel is drastically reduced and only a small optical window is needed.

Diagnostics for Experimental Thermonuclear Fusion Reactors 2
Edited by Stott *et al.*, Plenum Press, New York, 1998

The mentioned wavelength-multiplexing of the beams from different target locations is achieved by using a low-frequency grating close to the front mirror, but shielded from the divertor plasma (Fig. 1). We propose a kind of inverse Rowland circle spectrometer: When an approx. 3 cm wide strip of a large poloidal section of the target plates is imaged, the image surface will be curved, and, to first order can be approximated by a circle. Thus, when imaging an input stop A onto A' by an off-axis ellipsoidal mirror, the images of points of the target plate can be taken as the 'exit slits' on a Rowland circle spectrometer defined by these images and A', the position of the curved grating. Light at different wavelengths, determined by the grating equation, entering from these 'exit slits' then emerges at the 'entrance slit' B as a single beam of broadband radiation. This beam has a smaller étendue than the beam of collected light in a conventional collection optics imaging the same area of the target plates and can therefore be transmitted more easily. The information on the spatial origin of the collected radiation is encoded in the wavelength and can be recovered passing the light once more through a spectrometer, similar to ECE measurements of the plasma electron temperature. Note that the chosen wavelength interval is such that the first and second diffracted order do not overlap. The shown Rowland circle spectrometer is folded to fit into the dome structure of the diagnostic cassette using additional mirrors[1].

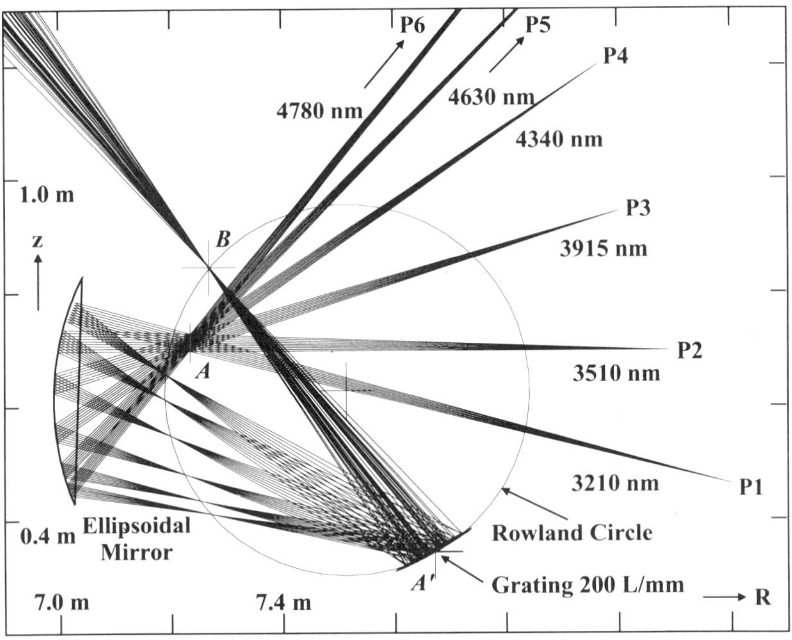

Figure 1. Principle of the proposed optics. P1 - P6 are points of the contour of the outer target plate. R: radius from torus centre; z: height above the centre line of the horizontal viewing duct of the diagnostic divertor cassette.

The drawback of the method obviously is the linkage between spatial and spectral resolution: E.g., viewing the outer target plate with a single system (field of view 75°) requires a spatial resolution of about 210 points to achieve 5 mm resolution on the target surface. Assuming a reasonable spectral range of 3200 nm - 4800 nm for the measurements, the information on a spatial point has to be transmitted in a bandwidth of about 7 nm. Thus the SNR achievable with the proposed optics is certainly less than the one for the conventional optics where the bandwidth can be chosen independently from the spatial resolution. Further, the spectral resolution of the Rowland circle spectrometer in combination with the ellipsoidal mirror used as front optics has to match this requirement. First optical ray tracing calculations assuming a spherical grating with constant groove separation (200 L/mm) along the surface (R = 654 mm) indicate a spectral resolution of about 12 nm combined with a spatial resolution of about 8 mm.

The question of how to determine the position being observed in a certain wavelength interval is solved by the planned measure to cover one part of the target plate with carbon and the other one with tungsten. The distinct change in spectral emissivity between the two areas can serve as a 'landmark'.

ACCURACY AND CALIBRATION

Thermography systems used for monitoring the surface temperature of in-vessel components in addition to an IR detector with imaging optics comprise of a vacuum window and a relay optics. Although modern commercially available IR cameras are calibrated, the optical transmission of the additional components has to be taken into account for the overall calibration. If the transmission or reflection of the components is known, the new calibration factor can be calculated. Alternatively the whole IR system used on the experiment has to be calibrated against a black body source either in the laboratory before installation or in situ at the experiment. Once the linearity of the detector has been checked the calibration is a two point calibration with A and B acting as calibration factors. The detector signal U_{out} is given by

$$U_{out} = \varepsilon \tau A M_p^b(\lambda, T) + B \qquad \text{(equ. 1)},$$

where $M^b_p(\lambda, T)$ is the emittance of the target, $A M^b_p(\lambda, T)$ is the detector signal caused by a black body at temperature T via loss-free optics, and ε and τ account for the emissivity of the target and the transmission of the optical system, respectively. B represents the background radiation. Calibrating an IR camera or detector against a black body source (ε=1, τ=1), the measuring error at high temperatures is dominated by the error of the black body calibration and is about 1%. At low temperatures the error due to the background temperature (temperature of the optics) is dominating. If we consider such an ideal thermography system it matches by far the ITER specifications of ±10 % at 2500 C. However, in practice additional uncertainties of temperature mesurements are caused by other reasons, such as:

i) uncertainties of the emissivity of the target as function of the temperature and the surface structure which may be changed by particle and neutron impact as well as by deposited layers.

ii) degradation of the optical transmission of the system by coatings or neutron damage.

iii) hot spot radiation from other locations of the inner wall reflected at the target.

iv) Bremsstrahlung.

The contributions i) and ii) reduce the received signal from the target and have to be taken into account when evaluating equ. 1, whereas iii) and iv) influence the background term. Assuming Wien's law for the emittance of the target, the error of the temperature calculation is

$$\frac{\Delta T}{T} = \frac{\lambda T}{C_2}\left(\frac{\Delta\varepsilon}{\varepsilon} + \frac{\Delta\tau}{\tau} + \frac{\Delta A}{A}\right) \qquad \text{(equ. 2)},$$

where C_2 is the second constant of Planck's law. To meet the ITER requirements of 10 % temperature error (±250 K at 2800 K) the sum of the uncertainties of emissivity, transmission, and the calibration accuracy should be about 50 % for a system measuring at 1 µm and 10 % at 4.5 µm, respectively. A typical change of the emissivity of graphite with temperature at a fixed wavelength is about 8%, and the variation of the emissivity for different graphites, which may be an indication for the influence of graphite layers building up, is of the order of 10% [3]. Thus, the ITER requirements are met easily for a system measuring at 1 μm. However, the spectral range of 3.2 μm to 4.8 μm is needed to meet the requirement of measuring down to 200 C. Then, considering only the uncertainty in emissivity it is already difficult to meet the ITER requirements without performing a periodical in situ calibration of the system. This in situ calibration can be done by monitoring in between discharges the temperature of the target plates heated by the cooling water. To avoid a deterioration of the calibration by background radiation it is necessary in this case to keep the temperature of the target well above the temperature of the optics. If this is not possible, an additional measurement must be performed using a heated target of known emissivity placed at the position of the entrance stop.

The consequences for the three different tasks, machine protection, strike point recognition and power density measurement are different. Systems for measuring the power density are operating at wavelengths above 3μm and thus can be calibrated in situ. The location of the strike point may be derived from relative data by searching the temperature maximum, as long as the emissivity and the transmission vary little across the surface. The consequences for the machine protection depend on the operational scenario. If the divertor is carbon dominated the change in emissivity can be assumed to be in the range of 10 % resulting in a temperature error of 170 K at 3.2 μm and 260 K at 4.8 μm which can be taken into account for machine protection by decreasing the upper temperature limit. The temperature measured for metallic targets such as tungsten is strongly affected by carbon deposition layers which increase the emissivity from about 20% to 70%. As a result the measured temperature is higher than the real one. From the point of machine protection it is an inherently safe scenario.

The contribution of bremsstrahlung to the temperature evaluation is dicussed in [4]. Its effect to the measured temperature as well as the effect of reflected radiation of hot spots results in a measured temperature higher than the true temperature which again is favourable for machine protection.

SUMMARY

The merit of the proposed unconventional optics is the smaller étendue compared to conventional optics which reduces the number of optical elements inside the vacuum vessel and the necessary diameter of the optical window. The spectral region of 3.2-4.8 μm anvisaged for the measurements allows a in situ calibration reducing the measuring errors of temperature due to a change of emissivity and system transmission.

REFERENCES

1. ITER DDD 5.5.G.06, IR Thermography Divertor, (1997)
2. E. Ishitsuka, Sample No. JA-W-T246-(1)-01, *ITER R&D Task T246 'Irradiation Effects on Diagnostic Components'* (1997)
3. G. Neuer, R. Kochendörfer, F. Gern, High temperature behaviour of spectral and total emissivity of CMC materials, *High Temperatures - High Pressures* 27/28:183-189, (1995/96)
4. A. Herrmann, Optical surface temperature measurement, in: *Diagnostics for Experimental Thermonuclear Fusion Reactors*, Edited by P.E.Stott et al., Plenum Press, New York, (1996)

THE USE OF X-RAY DIAGNOSTICS IN PLASMA CONTROL

A. W. Edwards, B. Alper, K. Blackler, R. D. Gill, M. Lennholm

JET Joint Undertaking, Abingdon, OXON, OX14 3EA, United Kingdom

INTRODUCTION

Plasma control is normally performed using analogue signals from magnetics diagnostics. However, with the increase in performance of digital signal processors (DSPs) more use is being made of digital systems for plasma control. For the pumped divertor phase of JET a new VME based Plasma Position and Current Control system (PPCC) was developed[1]. PPCC consists of two separate subsystems; the first for vertical stabilisation of the plasma position (VS), the second for plasma shape and current control (SC)[2].

When pumped divertor operation began in JET, it was found that the bottom pick-up coils were corrupted by noise at 600 Hz (together with higher harmonics) from the ripple in the divertor power supply. The input to the VS control algorithm was therefore limited to signals from a reduced number of magnetic coils positioned in the upper part of the vessel.

A JET working party was set up, therefore, to make recommendations on improving the performance of the VS system. The principal recommendation of the working party was to introduce high power filters for the 4 divertor coil power supplies. These it was estimated would reduce the 600 Hz noise by a factor of 10, allowing the use of the pick-up coils in the bottom of the vessel near and around the divertor.

In addition, it was proposed that the use of a control signal measuring the plasma vertical velocity derived from a soft X-ray diagnostic system should be used in the VS system as an alternative to the signal from the magnetic pick-up coils.

One of the requirements was that this new soft X-ray system should provide such a signal through the planned JET D-T experiments, as well as during D-D plasma operation. A proposal was made, therefore, to design two D-T compatible soft X-ray cameras which could be used together with a modified version of the JET Central Acquisition and Trigger System (CATS) to provide a real-time measurement of the vertical velocity of the plasma. This signal could then be used within PPCC to stabilise the vertical position of the plasma.

It is important to emphasise how the design of the hardware and data acquisition in this new soft X-ray system complement each other. The CATS design not only made provision for a significant degree of real-time processing to produce multiple event triggers, but in addition, the design allowed for data to be routed in real-time to processors where real-time control signals could be derived. Although this feature was not used in the initial phase of CATS operation, its inclusion in the original design was an important factor in determining the feasibility of providing additional real-time plasma control signals.

In this paper some further background on PPCC will be given first. This will be followed by a brief description of the design of the new soft X-ray system and a review of CATS and the additional changes implemented to calculate the control signal in real-time. Some examples of the use of the system for real-time control of the JET plasma position will then be given. Finally, future considerations and improvements will be outlined, together with how this work can impact on the ITER design: for plasma control; for diagnostic data acquisition; and for soft X-ray imaging.

THE JET PLASMA POSITION AND CURRENT CONTROL SYSTEM

The intended control algorithm for use in the VS system was based on constructing a full current moment signal from pick-up coils measuring the field around the entire plasma. However, the corruption of the bottom pick-up coils by the 600 Hz noise and the elimination of these signals from the measurement meant that the control algorithm had to be based on the derivative of the radial flux. The VS feedback system was not, therefore, driven by a true measurement of the vertical plasma velocity.

The consequences of this power supply noise and the change in control algorithm were firstly that it was difficult to understand the behaviour of the VS system and secondly, in most of the plasma disruptions, the VS circuit applied the wrong FRFA polarity leading to vertical displacements and increased forces on the JET vessel.

In addition, the shots with the highest neutron fluxes during the JET MKI pumped divertor phase were seen to have a high probability of ending in a disruption following the giant elm at the end of the high performance phase. In these cases the plasma current was often still above 3.5 MA. When preparing for the MKII pumped divertor phase, the 3 high performance shots in the MKII simulation experiments all disrupted at 3.8 MA following a giant elm. A sudden upward jump in the position of the plasma X-point was inferred from magnetic signals and divertor probe measurements. If this jump exceeded a critical amplitude, the VS system was found to be unable to stop the vertical displacement.

Further exploration of the JET high performance regime required that these problems be overcome. Hence the recommendation to install filters to the divertor power supplies and to propose a new soft x-ray system to provide an alternative vertical stabilisation control signal. In principle this signal would: be de-coupled from any edge effects such as the divertor power supply ripple; provide a control signal derived predominantly from a measurement made over the plasma core; provide valuable insight into the operation of plasma control systems using multiple control signal inputs derived from different sources.

THE JET D-T COMPATIBLE SOFT X-RAY SYSTEM - CAMERA DESIGN

It is difficult to determine the soft X-ray emission from a plasma in the presence of high fluxes of energetic neutrons. Firstly the neutrons induce in the X-ray detector a background signal which needs to be corrected for. Secondly, the neutrons give rise to a disproportionately large noise level because the incident radiation has much higher energy per particle compared with soft X-ray radiation. Finally, if Si pin diodes are used as detectors, these will start to suffer from radiation damage at approximately 200 rads (Si) and the detectors will cease to work at radiation levels 100 times larger.

The camera design is based on cautious estimates of the predicted performance of the proposed system during the second planned JET D-T Experiment (DTE2). It is assumed that DTE2 will consist of 1000 D-T shots each producing 10 MW of power of which 80% will be in the form of 14 MeV neutrons. The total soft X-ray power observed for each pulse is taken to be 100 kW. The detectors to be used are Si diodes with a thickness of 250 μm and a detection efficiency of 100% for soft X-rays in the region of interest.

The results of this design study[3] showed that with a suitable shield it was possible to satisfy the criteria of a low neutron background signal, a signal to noise ratio of better than 1% and the ability to withstand the total DTE2 dose of 6×10^4 rads.

Two cameras have been constructed and a schematic cross-section of one is shown in Figure 1. Each views a vertical plasma cross-section through a JET main horizontal port. One camera is displaced by 180° toroidally around the machine from the other in order to overcome m=n=1 plasma instabilities such as the sawtooth crash, which would otherwise lead to a spurious VS control signal being generated. In addition, the cameras also have the capability of measuring soft X-ray signals from mhd events in D-T plasmas.

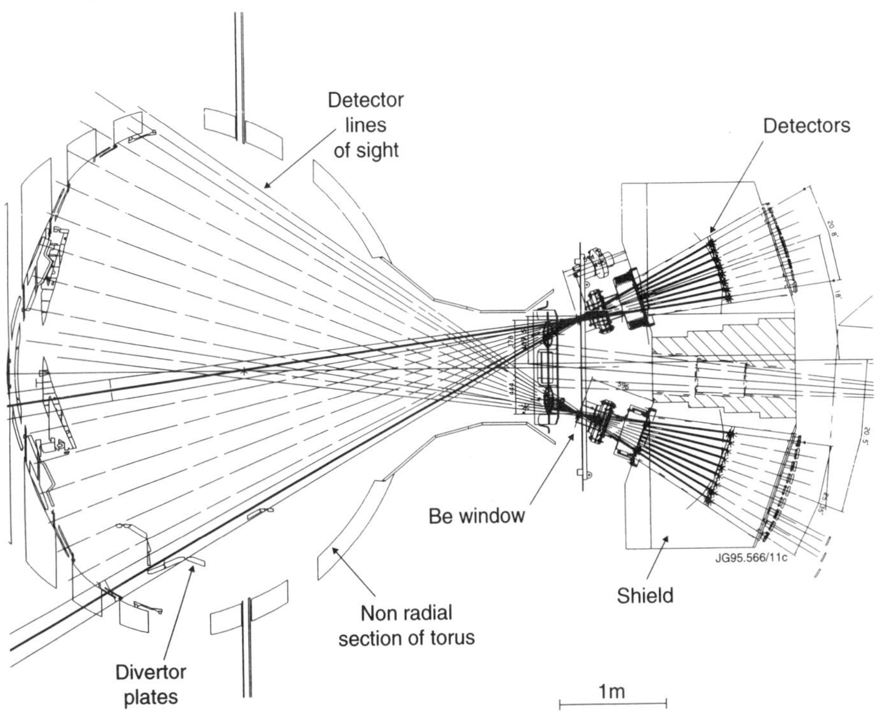

Figure 1. The outline design of one of the JET D-T compatible soft X-ray cameras.

The two cameras each contain 17 detector assemblies embedded into a large shield made of barytes concrete. This type of concrete has a high Ba content. The shield is 75 cm deep in front of the detectors in the plasma direction and roughly half that thickness in all other directions. The detectors view the plasma through precision machined stainless steel collimators and 250 μm Be windows which separate the machine vacuum from that of the diagnostic. The detector vacuum is maintained at 10^{-2} torr.

Each detector assembly has a stack of different elements as follows:

1. a Centronics Si diode detector (type OSD196-OG(CER)) which is 14x14 mm square and has an efficiency of 100% for X-rays with an energy less than 10 keV, but a much lower efficiency for 14 MeV neutrons which are mainly transmitted through the detector with very little attenuation and provide a noisy background signal;

2. a Si diode detector, identical to the first, which is used to measure the neutron induced background signal;

3. a plastic scintillator (Bicron BC 400) 20cm long and 20mm diameter which has a high efficiency for the detection of neutrons. The light subsequently emitted by the charged particles produced by the neutrons is measured by a third identical Si diode.

THE JET D-T COMPATIBLE SOFT X-RAY SYSTEM - ELECTRONICS DESIGN

The Central Acquisition and Trigger System[4] (CATS) has been under continual development at JET since 1991. This system now provides advanced data acquisition with real time data selection and analysis for very large volumes of data acquired at high sampling frequencies in the range of 100's of kHz. CATS is designed to be modular and can easily be expanded to include new channels. The system design also allows for hardware failures without affecting the acquisition of other parts of the system. Such a system has great relevance to all long pulse machines, in particular ITER.

An overview of the CATS system is shown in Figure 2. The CATS system can be considered in four parts. The first is the signal conditioning electronics, responsible for taking in analogue signals and outputting digital data. Next the data collection system receives this data, holds it and then stores relevant data in response to events from the third part, the trigger system. Finally, in the fourth part, data is stored on high volume local disks and then centrally archived. The data from the local disk or the central archive can be examined using Windows NT workstations. In addition, the system can be controlled from Windows NT workstations in the JET Control Room.

Signal Conditioning

The CAT system was designed to provide a standard data acquisition solution for a large number of channels from several diagnostics systems with different signal characteristics. The specification of the CATS ADC card includes the following: the analogue signal input can be either uni- or bipolar current or voltage; the gain of the signal is switchable in software from 1 - 64; the signal offset can be adjusted from software; the analogue signal is sampled at either 1, 2, 4 or 8 MHz with a 12-bit simultaneous sample and hold ADC; the 12-bit ADC data is then digitally filtered to 250, 500, or 1000 kHz; the output is fully anti-aliased and has an inherent noise of less than 1-bit.

In addition, there is high clock coherency between all the widely distributed systems, only one single clock being used to generate all timing and clock signals throughout the system. There is also full electrical and optical isolation between channels and each channel is shielded against high electrical and magnetic interference.

Control signals are distributed to each rack of ADC cards via an Interface card in each cubicle. A Control card in each ADC rack allows monitoring and control of the ADC cards outside of a data acquisition sequence. During data acquisition, the Control cards packages the data from all ADC cards in a rack and transmits this data through an optical TAXI link to a remote data collection system.

Data Collection

The data collection system receives data, in parallel, from many serial links. Each element of the collection system holds the data in a 200 ms ring buffer while extracts are passed to the trigger system for analysis. Upon receipt of trigger events (see later), all data collection modules save relevant extracts of data into longer term memory. Following the pulse, this is written to disk and may be collected later for archival by the central systems. Currently there are approximately 35 data collection modules, receiving data from over 500 channels, resulting in a total data flow into the system in excess of 250Mbytes/second.

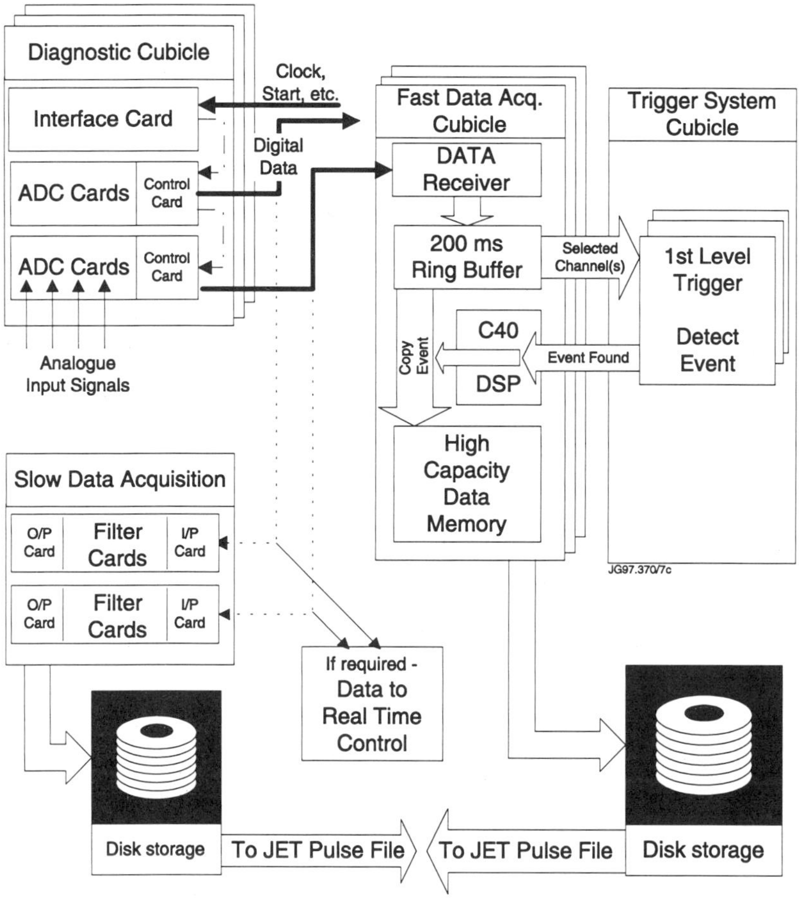

Figure 2. An overview of the JET Central Acquisition and Trigger System (CATS).

Triggering

A specific trigger module receives data in real-time from selected channels. After applying a certain algorithm, an event may be detected. If an event is found, the data window to be recorded for that event is broadcast to all data collectors. Up to 100 ms of pre-event data may be stored and any amount of post-event data, subject to the limits of the memory. Essentially, any number of trigger events may be detected during a pulse. Trigger algorithms have so far been developed to detect: threshold levels in neutron production rates (R_{DD} and R_{DT}); sawteeth; peaks in signals - in particular ELMs.

System Control

The CAT system is controlled by a PC running the Windows NT server operating system. Data is written to a large bank of disks, capable of holding data from many pulses. These are generally archived overnight to the JET IBM mainframe computer, while selected shots are written to a local Windows NT server to enable faster access for analysis.

THE VERTICAL STABILISATION CONTROL SIGNAL

In order to provide a real-time control signal to the VS system, the CAT system described above was extended to allow real-time processing of all the soft X-ray data from the new D-T compatible cameras. To ensure the system is as reliable as possible, the front-end signal conditioning electronics are not positioned close to the detectors as with the existing soft X-ray system[5]. Rather, a pre-amplifier was designed based upon that used in the CATS ADC card. These are placed as close as possible to the detectors and the current input signals are then amplified and converted to differential voltage signals. These signals are brought to racks of CATS ADC cards positioned in the JET Diagnostic Hall where they are readily accessible should any faults develop in this part of the system.

At present the ADC cards used in the D-T compatible soft x-ray system are configured to sample at 500 kHz. This allows the physics studies of D-T plasmas undertaken with the new cameras to be combined with the real-time calculation of the control signal. The data from each rack of ADC cards is transmitted to both a CATS data collection module, as described earlier for data recording and analysis, and to an equivalent system in which the real-time calculation of the plasma vertical velocity is made.

In order to derive a velocity from the measured soft X-ray data, the centroid of the soft X-ray profile for each camera is first calculated. At the first level, the data sent to the real-time system is dealt with in four processors which each evaluate the centroid for one half of each camera. These results are then passed to a final processor which combines the results to extract a centroid for each camera and then averages the results in order to remove (as far as possible) effects from the presence of m=n=1 modes. This value of the vertical position of the soft X-ray plasma centroid is then converted to a velocity by comparing the change in displacement from the previous value calculated.

An accuracy of better than a few millimetres is normally achieved when determining the centroid, this value being essentially limited by counting statistics. However, given that the signal bandwidth is 200 kHz, even an accuracy of a few millimetres can lead to velocities in the range of 100's of metres/s. However, it is not necessary for the VS system to determine the plasma vertical velocity on such a timescale. Indeed, whilst the VS system is designed to stabilise plasma growth rates of up to $1000s^{-1}$, typical growth rates are at least an order of magnitude below this. Further, the VS system itself uses 20 kHz ADCs to digitise the control signal transmitted to it from the soft X-ray system.

The plasma vertical velocity derived from the D-T compatible soft X-ray cameras is therefore filtered to a much lower frequency. First attempts at filtering this signal resulted in a significant propagation delay being introduced in to the signal. To avoid introducing such a delay, a modified approach was adopted where instead of applying the more traditional FIR filter, resulting in a significant delay, an effective rolling box car average was applied to the raw velocity signal.. This resulted in a both a reduced delay and a signal which was now suitably filtered and appropriate for use in the VS system.

FIRST RESULTS USING THE SOFT X-RAY CONTROL SIGNAL

The soft X-ray control signal (SXR CS) has so far been tested within the VS system during the restart of JET operations following a JET shutdown. Initially, the SXR CS was recorded and monitored within the VS system to assess the suitability of the SXR CS for use as a plasma control signal. This showed, for example, that the first filtered signal derived within the soft X-ray real-time system had an excessive delay. Hence, as described earlier, the filtering method was changed to reduce the propagation delay in the SXR CS.

Following this period of integration of the SXR CS in to the VS system, successful vertical stabilisation of the JET plasma using the SXR CS was achieved, firstly in ohmic plasmas and then in ICRH heated plasmas. For ohmic plasmas, the soft X-ray emission from the plasma above the Be filter cut-off is rather low, due to the low electron temperature of the ohmic plasma. The SXR CS derived under these conditions is noisier than that from the magnetics. For the ICRH heated plasmas a strong soft X-ray signal is measured which allows a very clean SXR CS to be determined and the quality of the SXR CS under these conditions is normally comparable with that from the magnetics.

Using the SXR CS plasma control has now been maintained for up to one second in ICRH heated plasmas, this being the maximum period allowed for testing. There are no reasons to believe that control could not have been maintained for much longer periods using the SXR CS, especially in ICRH heated plasmas.

Figure 3 shows signals measured during VS of the JET plasma position for a period of 1 second using the SXR CS for an ICRH heated plasma. At 20 seconds, a switch was made within the VS system from control using the magnetics signals to the SXR CS. A small jump in the plasma position is seen at that time. For the first ½ second a high gain was used in the VS controller, followed by a lower gain for the remaining ½ second through to a time of 21 seconds. At this time control switched back to the magnetics.

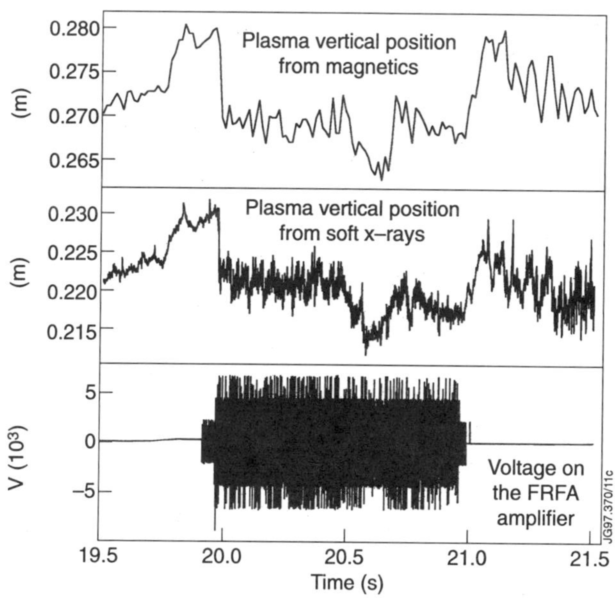

Figure 3. Vertical stabilisation of an ICRH heated JET plasma using the soft x-ray control signal.

Figure 3 shows clearly that the plasma position can be controlled using the SXR CS to the same level as that of the magnetics. Indeed, during the period when the high controller gain was used, control of the plasma position is kept within ±2mm. When the controller switches to a lower gain at 20.5 seconds, the variation in the plasma position is comparable with that during the period that the magnetics is used to control the plasma position. The main difference between the two control systems is seen in the voltage applied to the radial field amplifier (FRFA). During the period that the SXR CS is used, the voltage on the FRA is switched in larger steps than that used during the magnetics. However, although larger FRFA voltages are presently switched using the SXR CS, during normal operation this voltage does not saturate.

The tests have also conclusively demonstrated that the use of signals from the 2 cameras 180^0 apart toroidally has effectively suppressed any m=n=1 contribution to the signal, in particular from effects due to the sawtooth crash.

The system has not so far been tested using NBI heated plasmas to study the response of the VS system to the SXR CS measured during giant elms. However, studies of the system have been made during ICRH heated plasmas when small elms are present. Figure 4 shows the signals measured during an ICRH heated discharge when the magnetics were being used for plasma control. Away from the elm, excellent agreement is seen between the magnetics' control signal and that derived from the soft X-ray measurements. At the time of the elm the signals are observed to diverge. The sudden inferred change in the plasma position as derived from the magnetics is also clearly seen. When the JET DTE1 run is over, it is hoped that more extensive testing of the system will be possible, especially in the presence of large elm activity.

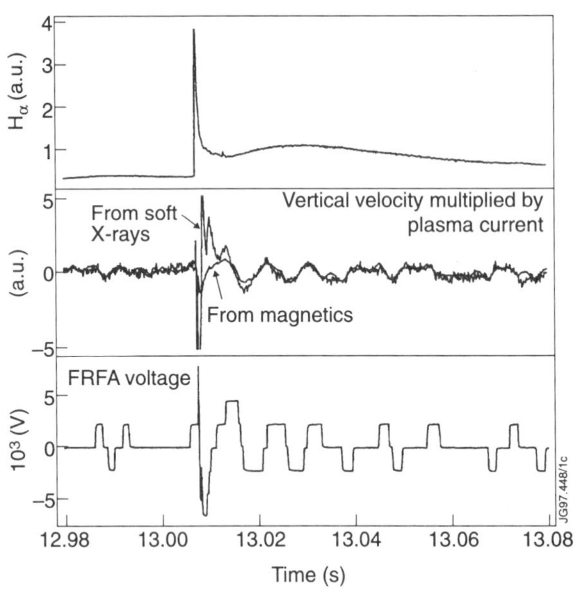

Figure 4. Vertical stabilisation of an ICRH heated JET plasma during elm activity using the magnetics control signal.

FIRST SOFT X-RAY MEASUREMENTS FROM D-T PLASMAS

First operation in tritium has already taken place at JET and this has provided an opportunity to compare the actual behaviour of the new D-T soft X-ray cameras with the anticipated performance. By comparing the neutron induced signals in the edge and central channels it is possible to conclude that over 85% of the neutron induced signal seen in the diode comes from neutrons viewed directly along the line of sight of a channel.

If it is assumed that D-T neutrons are produced within the inner one third of the plasma radius, then excellent agreement is obtained between the signal measured in the background detector and that predicted assuming this current is produced principally by 8 MeV recoiling alpha particles. This same level of noise will also effect the front detector, although it will be larger here as this detector has a higher bandwidth. Again, if the increased bandwidth is taken into account then good agreement is found between the measured and predicted value of the signal noise. The relative noise found is 1.8% at a bandwidth of 220 kHz. (This would be reduced to 0.7% if the bandwidth were limited to 30 kHz as had been assumed in the design calculations.)

When a comparison is made of the estimated neutron flux at the detectors in the new shielded D-T compatible soft x-ray cameras with that in the existing unshielded cameras, then it is seen that the neutron flux is reduced by a factor of 1700. This confirms both the design and that the detectors will continue to operate beyond the end of DTE2.

FUTURE CONSIDERATIONS

The present CAT system is limited in the amount of data that can be passed from one processor to another through the processor network. This means that at present any large amounts of data within CATS that are to be used for control purposes need to fanned out after the front-end ADC racks and before the data collection system. If certain data were required for more than one control algorithm this would require it to be fanned out a number of times. The introduction of a "ring" for passing large quantities of data at very high rates between large numbers of processors is being actively worked on at present[6]. This should permit CATS to further evolve such that in the next stage front-end data will be first brought into the back-end data collection system, from where it can be easily passed to all parts of both the trigger and control system that require it. Hence, it should be possible to develop ever more complex algorithms using ever larger quantities of data from many different diagnostics to either find and classify events, or to produce control signals.

Regarding plasma control using X-ray diagnostics, the work described in this paper has so far concentrated on bringing the JET D-T soft X-ray system into operation and producing a suitable signal for the control of the plasma vertical position. The flexible design of CATS allows further DSPs to be added with ease. The use of such DSPs provides the ability to perform error checking and correction of the input signals to a level that was not possible in previous hardware based systems. Confidence levels of the output control signals can also be derived such that an overall controller can easily verify that the control signal is acceptable to be used. Part of the future work will, therefore, concentrate on this question of error checking, correction and validation of the input and output signals in real-time. In particular, the use of neural networks within CATS is currently being investigated. If this preliminary work proves successful then one aim would be to extend this work into the area mentioned above.

In principle, the work described in this paper can be adapted to other diagnostics of a similar type. Nonetheless, soft X-ray detectors provide a simple, reliable and effective way of diagnosing the plasma with a high temporal and spatial resolution. Reliable control signals can be derived from the soft X-ray measurements, without the long pulse integration problems present in the magnetics signals on machines such as ITER. If soft X-ray systems are to be incorporated into the ITER design, then work is required on developing suitable radiation hardened soft X-ray detectors. It is hoped to be able to test various possible candidate detectors on JET, in particular Gallium Arsenide detectors, which could provide a simple alternative to existing silicon diodes, and vacuum photo-diode X-ray detectors.

CONCLUSION

This paper has described the work carried out at JET to provide an alternative control signal to the JET vertical stabilisation system to that normally used from the magnetic diagnostics. This work has involved the development of D-T compatible soft x-ray cameras integrated closely into the existing fast central data acquisition and trigger system developed at JET. It is important to stress this, that is that the hardware, electronics and data acquisition system are designed as a whole, not in isolation of each other.

The CAT data acquisition system is considered highly relevant to long pulse machines such as ITER. The modular nature of the design firstly allows new diagnostic systems to be easily "added in". In addition, whilst the functional elements of the design are relevant to all machines, the modular nature of the design means that the actual components can be chosen to suit the particular machine concerned, and further can be easily upgraded as new digital electronics components come onto the market.

The soft X-ray system described has been shown to be capable of vertically stabilising the JET plasma position.

Finally, it is intended to continue this work in the future to provide further information on the use of such signals in control systems, particularly on ITER.

ACKNOWLEDGEMENTS

The work described here includes valuable contributions from many people made over a long period of time. John Holm and Erik Van der Goot helped to start the work on real-time event recognition and data processing. Dennis Thompson, William Mulligan and Hunt Engineering are responsible for the design of the CATS electronics Richard Robins and Perry Beaumont have provided technical support and Robert Felton software support in bringing this work to its present status. Simon Dillon has been responsible for the hardware design of the D-T compatible soft X-ray cameras. Stefano Puppin and Filippo Sartori have been involved with much of the work within PPCC.

REFERENCES

1 M Garribba et al., The new control scheme for JET plasma position and control systems, in: "Proceedings of the 15th Symposium on Fusion Engineering (SOFE)", Hyannis, Massachuetts, USA, 1993

2 M Garribba et al., First operational experience with the new plasma position and current control system of JET, in: "Proceedings of the 18th Symposium on Fusion Technology", Karlsrhue, Germany 1994

3 R. D. Gill, B Alper, A. W. Edwards, S Dillon, A radiation hardened soft X-ray camera, JET-IR(95)03

4 K Blackler and A. W. Edwards, The JET fast central acquisition and trigger system, *IEEE Trans. On Nucl. Sc.* 41:111 (1994)

5 B. Alper et al., The JET soft X-ray diagnostic systems, *Rev. Sci. Instr.* 68:778 (1997)

6 Hunt Engineering, Brent Knoll, Somerset, UK at http://www.hunteng.demon.co.uk

STATUS REPORT OF PREPARATION FOR NEUTRON MEASUREMENT ON LHD

M. Sasao, M. Isobe, H. Yamanishi, M. Osakabe, and T. Iguchi[1]

National Institute for Fusion Science (NIFS), Toki, Gifu-509-52, Japan
[1]Dept. of Nuclear Engineering, Nagoya University, Nagoya 464-01, Japan

INTRODUCTION

Helical confinement systems have been recently revived with a strong interest. One of them which are operating or are expected to be operated in near future, is the Large Helical Device (LHD), of a 3.9 m major radius and a 0.6 m average plasma radius[1]. The LHD is also one of the largest, superconducting torus devices, and various R&D issues of ITER can be studied on LHD.

In LHD, DT experiments will not be conducted, and most of the operation will be made with hydrogen discharges in the first phase, while DD and D-^{3}He experiments are now under consideration for the second phase. Plasmas will be initiated by 400 kW ECH, and be heated by 20 MW NBI and/or 5 MW ICRF. Because the study of the confinement property of high energy ions is one of the main physics research issues on LHD, neutron diagnostics are expected to play an important role. Then it is important to prepare an efficient detection system, and the system with a moderate spatial resolution. Presently, the preparation for a neutron emission monitor located at the torus center, a neutron profile monitor, and an escaping high energy particle probe have started[2].

Among various R&D issues on neutron diagnostics presently being examined, the followings are shortly described in this paper,

a) the magnetic field effect on a gas chamber
b) the effect of the liquid helium on a calibration factor
c) a profile monitor viewing trough a port with a long throat.

MAGNETIC FIELD EFFECT ON A FISSION CHAMBER

The highest detection efficiency for toroidally extended neutron source can be obtained at the center position of the torus, which is available on a helical system. In Fig. 1 is shown the schematic view of the neutron emission monitor at the torus center of

LHD. An efficiency as high as 10^{-5} can be expected by using a commercially available 1"Ø x 10" long-^{3}He or a BF_3 counter surrounded by a 10 cm-hick moderator at this position. With this high detection efficiency, even in a case of 180 kV-hydrogen beam injection into a hydrogen plasma, the DD neutron signals of 10^4-10^5 counts/sec can be obtained[2].

However, the biggest concern is the magnetic field effect on a proportion chamber or an ionization chamber, because a residual vertical magnetic field at this position is predicted to be about 50 % of the toroidal field on the axis, which is 3T during the first phase. A test experiment was performed on CHS to prove that a BF_3-detector works at the torus center under the magnetic field as high as 1T. No effect was observed on the pulse height spectrum, and on the counting efficiency[3]. The results has been theoretically explained with the collisional diffusion of electrons toward the electrodes of the detector[3].

In this model, collisional cross sections of low energy electrons (about 50 eV) produced through the ionization process in a gas chamber by a high energy charged particles, are considered[4]. In a typical fission chamber, an Ar gas of 1-4 atom. is filled. The mean free path of 50 eV electrons in an 1 atom. Ar gas is about $3x10^{-4}$, while the electron gyro-radius under 1T is 2 x 10^{-3}. This ratio is about 10 times larger comparing with that in the CHS test case. Therefore, a higher filling gas pressure is desirable for LHD.

A micro fission chamber presently proposed is filled at 14 atom. The ratio of the collisional mean free path to the electron gyro-radius is about the same as in the CHS test case, and the detector will be tolerable under a 1-T magnetic field.

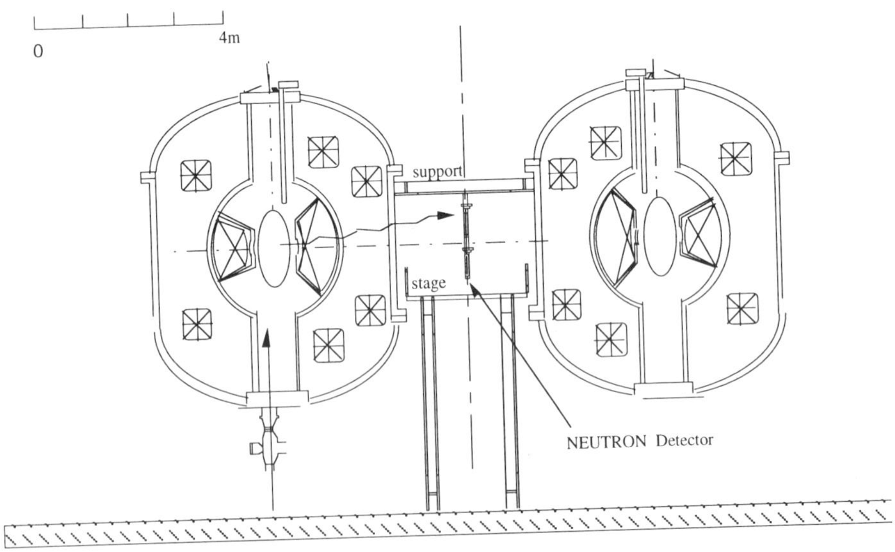

Fig. 1 The shematic view of the neutron emission monitor at the torus center of LHD.

EFFECT of LIQUID HELIUM ON A CALIBRATION FACTOR

The magnetic system of LHD consists of three pairs of poloidal field coils of 3.6m, 5.6 m, and 11m diameters, and one pair of helical coils. All of them are superconductive. It takes about a month to cool down the coils to the liquid helium temperature, 4.5 K, and takes about a month to bring them back to normal. Although the heat shield between coils and the vacuum vessel are enough to break the vacuum at 4.5 K, the number of calibration experiment with liquid helium in coil cases is very limited and an in-vessel man work is not allowed. Therefore, it is important to evaluate the effect of the liquid helium on the calibration factor.

Neutron spectra at the position of the neutron monitor at the center of the torus have been calculated, using a 2-dimensional neutron transport code, for cases of DD operation . In LHD, an elongated plasma cross section is rotating along the torus axis. At the angle where the central detector looks the vertically elongated palm, a massive helical coil with liquid helium in between exits, but at the angle where the central detector looks the horizontally elongated plasma, there is no coil to attenuate the neutron source. The neutron spectrum is approximated by taking mean values of these two spectra. The component of Helical coils(a) and Poloidal coils(b) are obtained from Hitachi Co.. For both of them, Cu and Al are main elements but the number density of He contributes about several % of the total.

The calculation was performed for two cases, one with liquid helium and one without. The calculated spectrum with He is shown in Fig. 2a. The difference of two cases are shown in Fig. 2b. In general, the effect of helium is within 10%, but the change depends on the neutron energy. So, it is important to measure the effect of helium on the calibration factor experimentally, and evaluate the validity of the transport calculation. After that, the factor calibration obtained with in-vessel source without helium can be used. On LHD, one of the possible method is a use of the in-vessel port which is prepared for foil activation.

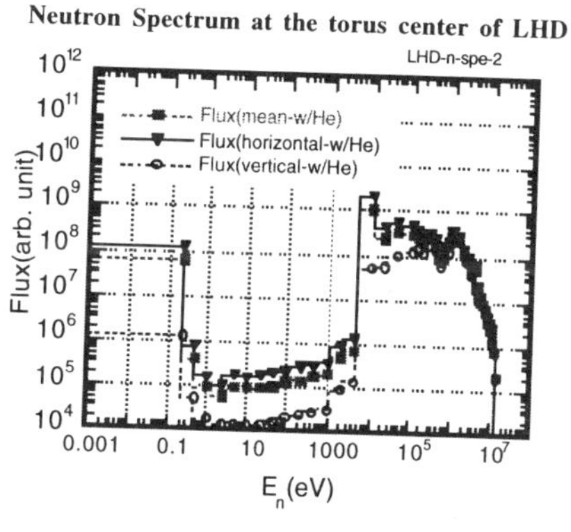

Fig. 2(a) Neutreon Spectrum at the torus center of LHD.(Data are provided from Safty / Enviroment Research Center of NIFS)

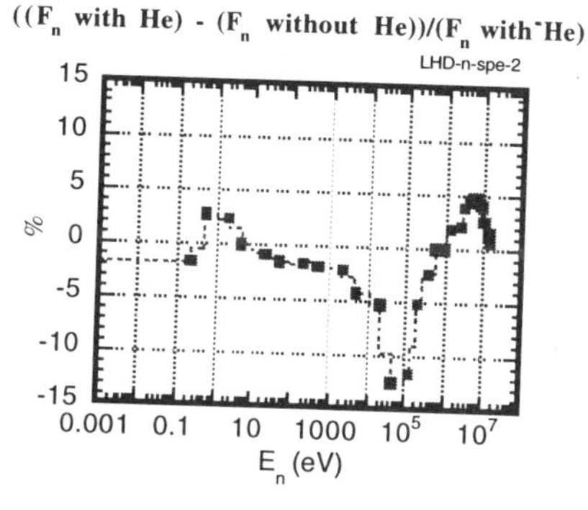

Fig. 2(b) Effect of Liquid Helium in %

PROFILE MONITOR THROUGH A PORT WITH A LONG THROAT

Because the whole plasma devise is set in the cryostat for superconductive coils, the diagnostic port has a long throat. We use a 2 m thick- concrete floor for the neutron shield of the profile monitor and collimators and detector boxes will be set in the floor. Fig. 3 shows the vertical cut of the neutron profile monitor. For helical plasma, the 3 dimensional information in one helical pitch is important. The neutron detector array will be tilted towards two opposite directions so that some information might be obtained. The plasma cross section is in an egg shape. However, the information will be very limited with this long port throat.

Authors wish to thank Mr. Saito of Hitachi Engineering Co. For the neutron transport calculation.

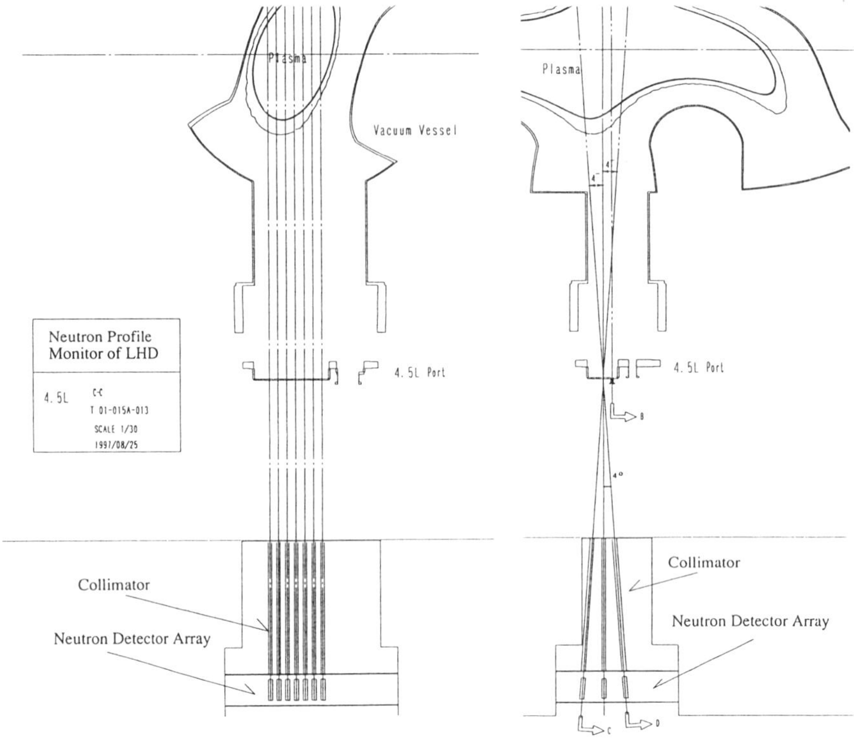

Fig. 3 Schematic view of the neutron profile monitor of LHD

REFERENCES

[1] A. Iiyoshi, et. al., Phys. Plasmas **2** , 2349 - 2356 (1995)
[2] M.Sasao et. al., Fusion Engineering and Design **34/35**, 595 - 598 (1997)
[3] M. Isobe et al., Rev. of Sci. Instrum. **68**, 532 - 535 (1997)
[4] S. Fukui, *Introduction to High energy Particle Measurement* (Kyoritu Press, 1992)

STATUS AND PROSPECTS OF PULSE RADAR REFLECTOMETRY ON THE START TOKAMAK

Vladimir F. Shevchenko[1]

UKAEA Fusion, Culham Science Centre, Abingdon
Oxfordshire, OX14 3DB, UK
(UKAEA/EURATOM Fusion Association)
[1]TRINITI, 142092, Troitsk, Moscow region, Russia

I. INTRODUCTION

The START experiment has exhibited both peaked and flat density profiles, with steep gradients either internally or at the edge. Time resolved profile information is therefore important to explore profile changes. Any diagnostics should be resilient to the transient MHD instabilities including edge fluctuations. The START tokamak has been in operation since 1991 and is the first spherical tokamak to have produced high temperature plasmas[1]. The diagnostics set on START includes multipoint Thomson scattering (TS), a neutral particle analyser, fast CCD video camera, multichord SXR camera, HCN interferometer, magnetic pick-up coils, flux loops and spectroscopy.

A multifrequency pulse radar reflectometer[2] (MPRR) is now one of the routine diagnostics on the START tokamak. The main feature of the technique used in this system is the direct time delay measurement of microwave signals reflected back from the plasma. The MPRR has been in operation on START since September 1994, and now the total operation time of the system is about 18 000 hours. The reflectometer has been fully commissioned and cross checked with the results of other diagnostics. The MPRR allows the study of the plasma density profile dynamics at different regimes, to follow the behaviour of the plasma boundary position and edge density gradient during the shot. In combination with other diagnostics it also permits the identification of toroidal and poloidal mode numbers of MHD perturbations and analysis of their structure, localisation and dynamics.

In this paper a description of recent MPRR modifications is presented in Section II. Sources of measurement errors and different types of MPRR experimental data filtering are discussed in Section III. A problem of plasma density profile reconstruction from reflectometry data is analysed with two different approaches in Section IV. A brief overview of recent experimental results obtained with MPRR is presented in Section V. A schematic of a novel frequency scanning pulse reflectometer proposed for GLOBUS-M[3] (Russia) is considered and possible system developments for various machines are discussed in Section VI, while a summary is presented in Section VII.

II. MPRR MODIFICATIONS ON START

The MPRR system has been designed to measure electron density profiles on the START tokamak using four probing frequencies: 19 GHz, 33 GHz, 47 GHz and 63 GHz which cover a plasma density range from $0.45 \cdot 10^{19}$ m^{-3} to $4.92 \cdot 10^{19}$ m^{-3} with o-mode plasma probing[2]. Initially, the antenna system consisted of a pair of separate horn-like transmitting and receiving antennas which produced a wide solid angle of plasma probing. In order to reduce the angle of plasma view the MPRR antenna setting has been modified by addition of two elliptical mirrors arranged in the periscopic system. This is illustrated in Fig. 1; probing beams are focused to the centre of the machine for all probing frequencies.

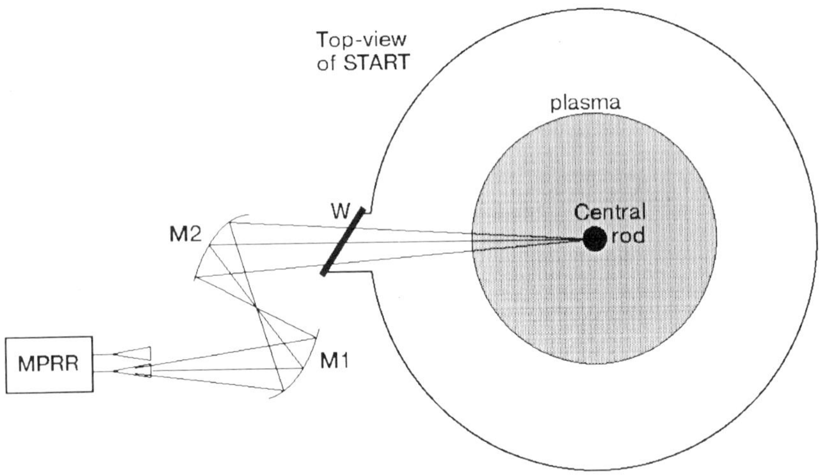

Figure 1. Periscopic antenna lay-out on the START tokamak.

This kind of focusing should provide very small losses of the reflected power both in plasma absence and in presence of the spherical plasma. Gaussian beam optics also predicts an improvement of the antenna performance with beams focused. In practice due to beam focusing, the signal-to-noise ratio in the receiving system of MPRR has been increased by approximately 2 times in comparison with unfocused beam reflection from the central rod of the START tokamak.

Another improvement of system performance due to probing beam focusing is dealing with the reduction of two-dimensional effect contribution in reflecting plasma layers. This is because of reduced sizes of the reflecting area and better spatial shape matching between the reflecting plasma layer and the spherical probing beam. As a result of partial suppression of two-dimensional effects the experimental MPRR data jitter has been decreased from 290 ps without focusing to 245 ps with beam focusing while the data jitter in the absence of plasma remains the same - about 80 ps (see Fig. 2). This data jitter reduction improves the spatial resolution of the system. But it should be mentioned that the

MPRR sensitivity to vertical plasma displacement has been increased with decreasing of the area of reflection in the plasma.

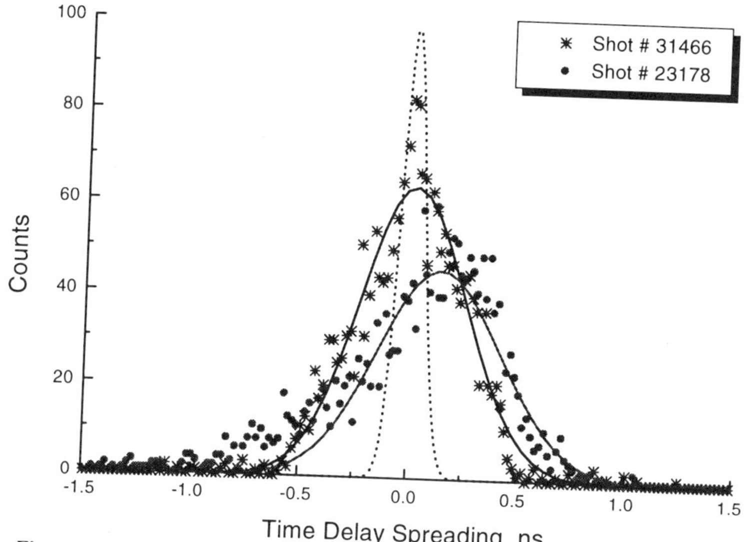

Figure 2. MPRR (33 GHz channel) data spreading due to density fluctuations with probing beam focusing - (*) and without - (•). Data jitter in the absence of plasma presented by dotted line.

The last MPRR modification concerns to the temporal locking circuit. The original circuit performed the temporal locking at a constant fraction of the reflected signal. The value of the fraction was selected close to 50% of the detected signal amplitude[4]. The new locking system is based upon a zero-crossing detector which locks to first zero crossing of a second derivative of the detected signal. The temporal position of this point is very close to the half amplitude fraction point for a wide range of reflected signal amplitudes and settling times. Having slightly improved the dynamic range of analysed signal amplitudes the new circuit performs locking two times faster in comparison with the old one, and as a result it has better temporal stability and protection against pickup and amplitude variations on top of the reflected signal.

III. EXPERIMENTAL DATA ERRORS

The main sources of experimental data perturbations in reflectometry are well known, they are: plasma density fluctuations, spurious reflections, o, x - mixed mode probing, reflected beam deflection, plasma emission and absorption. Nevertheless the data response to these perturbing factors is very dependent on the measurement technique and on peculiarities of the plasma machine. For this reason the sensitivity of reflectometry to perturbations of different origin has to be analysed for specific method of measurements on particular machine.

Plasma Density Fluctuations

Density fluctuations reveal themselves in MPRR data as strong broadening of the time delay jitter. Usually time delay jitter is increased in the presence of plasma by a factor of 3-

4 with respect to its vacuum value of 80 ps. As predicted by one-dimensional wave equation simulations[5] density fluctuations should introduce additional slowing-down for probing microwave pulses. This effect results in an asymmetrical form of experimental data jitter with dominant part corresponding to signals delayed in the plasma. Data jitter asymmetry observed in experiments with unfocused beam probing was stronger indicating that two-dimensional effects in the reflecting layer enhance the slowing-down of signals.

Fig. 2 illustrates time delay jitter measured in two similar shots: one with probing beam focused and another without focusing. Low frequency components of experimental data were removed by a high-pass digital filter (cut-off frequency of 1 kHz) after that time delay histograms were plotted for 1000 experimental points. Zero time delay shift corresponds to averaged experimental data while histogram maxima indicate most probable values of time delays. Histogram maxima are usually shifted from mean values to the positive side that is an agreement with slowing-down effect.

A special digital filter is applied to MPRR experimental data to calculate probability maxima for selected time intervals. Another digital filtering procedure applied has transfer performance of an arctangent function type which allows a very effective suppression of single short spikes of both signs (see Fig. 5 below). Such kind of filtering allows us to reduce the plasma fluctuation influence in density profile reconstruction.

Spurious Reflections and Mixed Mode Probing

In pulse reflectometry spurious reflections of probing signal can reveal themself in two different manner: large distance retroreflections inside the vacuum vessel with time separation greater than rise time of the probing signal and small distance reflections with time separation less than signal rise time. The temporal locking system of MPRR selects only the first class for further analysis. The time separation of retroreflections inside the START tokamak is defined by the distance between plasma boundary and vacuum vessel. Its value is about 3 ns that is greater than the rise time of probing signals which is 1 ns. Small distance spurious reflections can appear in the antenna-waveguide system having in a result an amplitude variations of the reflected signal due to interference between phase shifted reflections. Such kind of reflections has been reduced to a possible minimum by means of antenna system alignment.

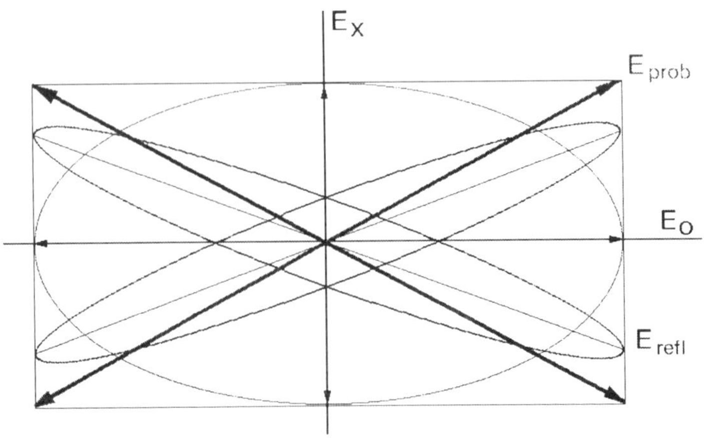

Figure 3. Polarisation effect with mixed mode probing for low q plasmas.

It is difficult to arrange pure o-mode probing on the START tokamak because the magnetic field direction on the plasma boundary can vary significantly during the plasma shot. So actually MPRR operates with mixed o, x - mode probing. The probing beam is split at the plasma boundary into two orthogonal components which have different points of reflection inside the plasma and different phase shift on the output of plasma. Reflected components can not interfere directly because they are orthogonal, so the resultant wave coming into the receiving antenna has elliptical polarisation. Because of the dependence on phase difference, the orientation of the ellipse will oscillate between two linear polarisations with respective changing of ellipticity, as shown in Fig. 3. The detector selects the linear polarisation with the same orientation as the probing beam. As a result, the detected signal will have amplitude variations with phase difference changing. These variations can reach up to 100% if the angle between the input polarisation and magnetic field is close to 45°. This effect was experimentally observed on START as a short time loss of reflected signal in one frequency channel while other channels show no evidence of plasma inclination or vertical displacement.

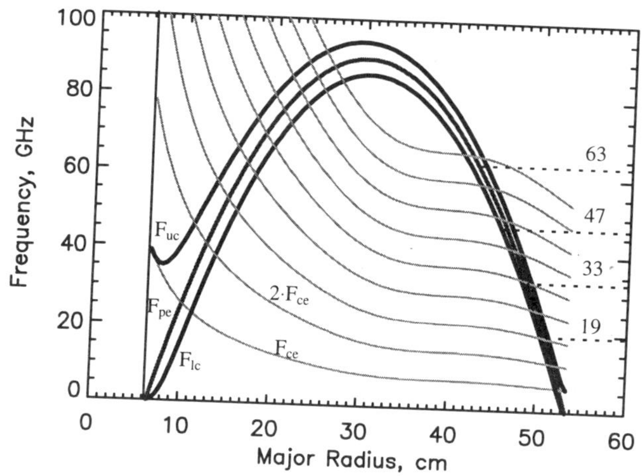

Figure 4. Cutoff and cyclotron frequencies on START at reference conditions (0.3 T central field, $1 \cdot 10^{20}\,m^{-3}$ central density, 0.1 MA plasma current).

Such kind of reflected signal behaviour is appropriate to the START tokamak because cut-off layers of both modes are localised quite close in the plasma (see Fig. 4), consequently the time separation between components is relatively small and does not disturb the signal shape significantly. In the case of large time separation between components reflected from plasma strong signal shape perturbations are expected because of mode scrambling[6].

Plasma Absorption and Emission

The START tokamak (spherical tokamak in general) has a very unusual topology of cutoffs and cyclotron frequencies because of the low toroidal field. Due to this fact a regular set of harmonics of electron cyclotron frequency is localised in transparent plasma layers in front of cutoffs as shown in Fig. 4. These harmonics should reveal a resonant absorption and emission of microwave power, but the optical depth of cyclotron harmonics decreases very fast with harmonic number increase. So in START only the first two harmonics can play a feasible role in absorption/emission processes. Typically they are located below the first probing frequency of MPRR and do not disturb the beam propagation in the plasma.

IV. DENSITY PROFILE RECONSTRUCTION

The principle of reflectometry is based on continuous or discrete frequency plasma probing from zero to a highest probed density with measurement of reflected signal time delay or phase shift. The density profile can be reconstructed from reflectometry data by consequent solving of integral equations involving measured values with plasma parameters from zero frequency to the highest. But in practice the lowest probing frequency is limited by plasma dimensions or by accessibility restrictions. Even with x-mode probing it is very difficult to diagnose close to zero densities because of interpretation problems. So an unknown portion of density profile below first probed density always exists for any reflectometry technique.

The standard solution of this problem is the estimation of the first reflecting layer position from other diagnostics and application of some density profile model below that layer[7]. Similar initialization methods were applied for START plasmas in ohmic shots where edge density profile was extrapolated by linear function with gradient estimated from data obtained from the two lowest frequency channels[2]. But in NBI heated plasmas edge density gradient can vary in a very wide range during the plasma shot and the previous edge density model becomes incorrect.

A new method of density profile reconstruction has been developed. The idea of the method is to solve the initialization problem starting not only from the plasma edge but from any point where additional information is available or some profile model is applicable. This method is much more flexible and allows us to apply any additional information or model in order to avoid the uncertainty due to unknown edge profile.

Let's suppose a piecewise linear approximation for probed part of density profile where the pieces meet at cutoffs. In this case for every cycle of measurements with k frequency channels (for simplicity let k=4) we can write the system of k linear equation for k+1 unknown positions,

$$\begin{bmatrix} - & - & - & - & - \\ a_{10} & a_{11} & 0 & 0 & 0 \\ a_{20} & a_{21} & a_{22} & 0 & 0 \\ a_{30} & a_{31} & a_{32} & a_{33} & 0 \\ a_{40} & a_{41} & a_{42} & a_{43} & a_{44} \end{bmatrix} \times \begin{bmatrix} x_0 \\ x_1 \\ x_2 \\ x_3 \\ x_4 \end{bmatrix} = \begin{bmatrix} - \\ \tau_1 \\ \tau_2 \\ \tau_3 \\ \tau_4 \end{bmatrix}, \qquad (1)$$

where x_0 is the plasma boundary position, x_i is the cutoff position and τ_i is the measured time delay of i-th frequency channel. In the case of o-mode probing matrix components can be calculated as follows:

$$a_{k,i} = \frac{4}{c} \cdot \left(\frac{1}{\sqrt{1-\frac{N_{i-1}}{N_k}} + \sqrt{1-\frac{N_i}{N_k}}} - \frac{1}{\sqrt{1-\frac{N_i}{N_k}} + \sqrt{1-\frac{N_{i+1}}{N_k}}} \right) \text{(for } 0<i<k\text{)}, \qquad (2)$$

and

$$a_{k,k} = \frac{4}{c} \cdot \frac{1}{\sqrt{1 - \frac{N_{k-1}}{N_k}}}, \tag{3}$$

$$a_{k,0} = \frac{4}{c} \cdot \left(\frac{1}{2} - \frac{1}{1 + \sqrt{1 - \frac{N_1}{N_k}}} \right), \tag{4}$$

where c is speed of light, N_i is the cutoff density of i-th probing frequency. In the case of x-mode probing it is necessary to apply some approximation to internal magnetic field as well in order to calculate $a_{k,i}$ matrix coefficients. Note that matrix coefficients (3) are greatest and responsible for particular cutoff coordinates, while coefficients (4) define the contribution of zero density coordinate to position of appropriate frequency cutoff. From here it is easy to find that the relative contribution of initial plasma layer decreases with increasing frequency channel number as fast as N_1/N_i.

In order to have single solution of system (1) we need to introduce one additional linear equation which can be some model of local or edge density profile or it can be the cutoff position of particular frequency channel obtained for example from Thomson scattering measurement.

V. EXPERIMENTAL RESULTS

As is well known, MHD perturbations in a tokamak are described with toroidal - n and poloidal - m numbers. Thus n corresponds to the number of periods of perturbation originating on a full turn of a torus in the toroidal direction, and m - the number of periods in the poloidal direction. If the probing frequency is fixed the reflecting "point" will draw a line of equal density while the plasma rotates either in poloidal or toroidal direction. So when MHD is localised near cut-off, the time delay of the reflected signals should have m periodical perturbations per full turn in the case of poloidal MHD rotation and n for toroidal rotation. The situation is different for interferometric measurements through the plasma centre: the number of phase shift perturbations per full toroidal turn equals n for even numbers of m, and - $2 \cdot n$ for odd.

The described effect has been rather clearly observed on the START tokamak. The signals of third MPRR channel and HCN interferometer are presented on Fig. 5. The reflectometer signal is shown in raw (points) and filtered (solid line) variants. Filtering was carried out with a digital filter which has transfer performance of an arctangent function type plus conventional low pass filtration with 20 kHz cut-off frequency. As well seen from Fig. 5 the main mode exhibits two times higher frequency on interferometer signal in comparison with reflectometer. This perturbation is accompanied by quite a regular structure of small phase oscillations which correlate with spikes on reflectometer signal with appropriate phase shift due to different toroidal location of those diagnostics. The dominated perturbation was identified as a toroidally rotated mode m/n = 1/1, while small scale oscillations are caused by higher order modes located in transparent plasma layers. The amplitude of mode oscillations in major radius direction can be estimated from MPRR data as big as 12 cm that is in a good agreement with SXR camera data.

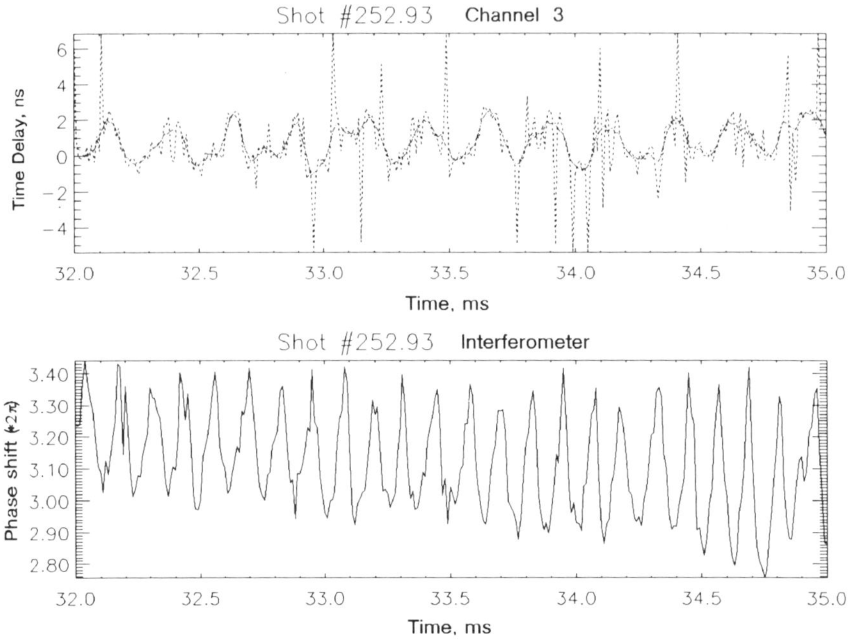

Figure 5. MHD activity registered by MPRR (47 GHz channel) and HCN interferometer.

In high β regimes[8] START plasma exhibits a very steep density gradient at the edge which is accompanied by a low density plateau starting at a density level approximately ten times lower then central. That complicates the interpretation of MPRR data with standard initialization method so the system approach described above was applied using the normalisation of highest frequency channel to the Thomson scattering profile, which gives the position of 63 GHz cutoff and allows the estimation of the plasma edge location. Fig. 6 illustrates the high β plasma density profile measured by multipoint Thomson scattering and MPRR. As is seen from Fig. 6 the profile is quite sharp down to the density level of $0.5 \cdot 10^{19}$ m^{-3} at major radius of 59 cm while low densities reach major radii about 70 cm.

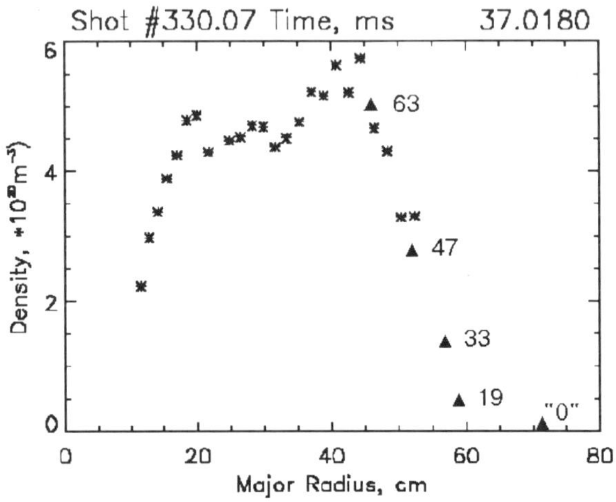

Figure 6. Plasma density profile measured by Thomson scattering (∗) and MPRR (Δ).

VI. FREQUENCY SCANNING PULSE REFLECTOMETER

MPRR on the START tokamak has a high temporal resolution and moderate spatial sampling limited by the number of channels employed in the system. Spatial sampling in principle defines the possible spatial resolution of the density profile measured with reflectometry. It is possible to increase spatial sampling with channel number build-up that leads to undesired hardware complexity. Another solution is the employment of frequency scanning pulse reflectometer for plasma density measurements. The main problem met in the designing of such kind of system is to provide fast frequency scanning over operational frequency band keeping desired level of the signal to noise ratio. With the appearance of fast swept solid-state microwave sources[9] this problem is significantly simplified. Such sources in combination with active frequency multipliers can cover now the frequency range up to 100 GHz with output power of tens mW that is quite close to desired values for most existing tokamaks.

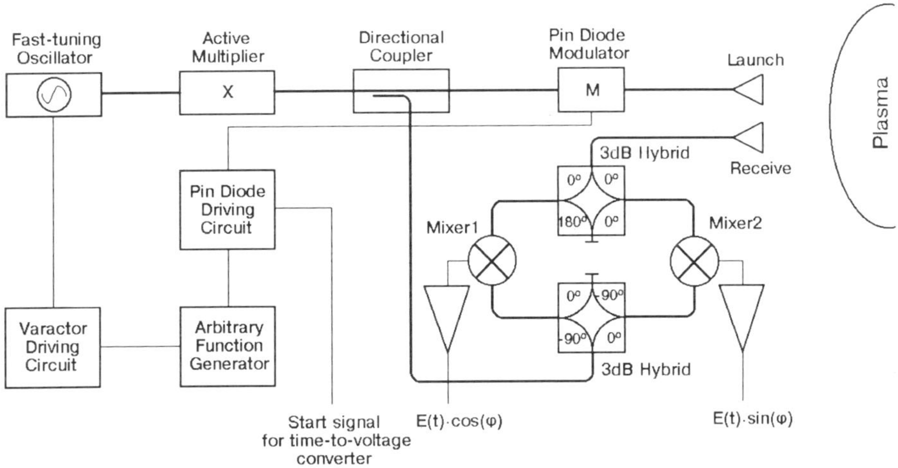

Figure 7. Schematic of the frequency scanning pulse reflectometer with simultaneous instant phase measurement.

The frequency scanning pulse reflectometer (FSPR) based on the fast swept solid-state technique has been proposed for GLOBUS-M[3] (Russia) and MAST (UK) tokamaks. The basic schematic of the single-band section of FSPR is shown in Fig. 7. It consists of a hyperabrupt varactor tuned oscillator followed by an active frequency multiplier, so as to cover a frequency range preliminary selected for GLOBUS-M of 26-40 GHz. During every step-wise frequency scan the fast p-i-n diode modulator cuts short microwave signals which are directed toward the plasma by antenna/waveguide system. The receiver is based on the quadrature balanced scheme allowing the separation of quadrature components of reflected signals. The amplitude of those components are registered by ADC to perform instant phase calculations for each reflected pulse. For time delay measurements quadrature components are summed in analogue quadrature summator to have an envelope of reflected power for further temporal analysis which is mainly the same as employed in the MPRR. Performing both the time delay and instant phase registration FSPR system allows us to improve the accuracy of time of flight measurements and to combine advantages of two different reflectometric techniques in one device.

An appropriate set of FSPR single-band sections can be applied separately or combined in a common antenna system with band-pass filter joints in order to cover the desired frequency range.

VII. SUMMARY

The multifrequency pulse radar reflectometer on START has shown itself as a highly reliable and powerful diagnostic tool for plasma density profile studies. Using pulse reflectometry data in cooperation with other diagnostics it is possible to investigate the dynamics and structure of MHD oscillations. MPRR has a high temporal resolution and moderate spatial sampling limited by the number of channels employed in the system. The frequency scanning pulse reflectometer is a logical extension of this technique.

ACKNOWLEDGEMENTS

The author would like to thank the START and COMPASS teams for their help and continuing support, especially T. Edlington, M. Gryaznevich, A.W. Morris and M.J. Walsh. This work was funded by the UK Department of Trade and Industry and EURATOM.

REFERENCES

1. A. Sykes, et al., *Nucl. Fusion* **32**, 694 (1992).
2. V.F. Shevchenko, M.J. Walsh, *Rev. Sci. Instrum.* **68**, 2040 (1997).
3. V.E. Golant, et al., *SOFE 1995*, Vol. II, 1430.
4. V.F. Shevchenko, et al., *Int. J. Infrared Millim. Waves* **14**, 1755 (1993).
5. V.F. Shevchenko, *Plasma Physics Reports* **23** (2), 169 (1997).
6. A.J.H. Donne, et al., *Rev. Sci. Instrum.* **68**, 473 (1997).
7. P. Varela, et al., *Rev. Sci. Instrum.* **66**, 4937 (1995).
8. A. Sykes, et al., *24 EPS Conf.*, Berchtesgaden, Germany, Topical Lecture 10 (1997), to be published in *Plasma Phys. Contrtol. Fusion.*
9. K.W. Kim, et al., *Rev. Sci. Instrum.* **68**, 466 (1997).

BOLOMETRIC AND SXR TOMOGRAPHY IN RFX

Piero Martin[1], Andrea Murari, Gianni Gadani, Gabriele Manduchi, Lionello Marrelli, Angelo Parini and Cesare Taliercio

Consorzio RFX
Corso Stati Uniti 4, 35127 Padova, Italy

INTRODUCTION

Among the various diagnostic systems which equip the RFX Reversed Field Pinch experiment (Rostagni, 1995; Bagatin et al., 1995) the new integrated bolometric and soft x-ray tomography is one of the most complex. It has been designed and to operate in a harsh environment following standards compatible with the requirements of large fusion experiments. The bolometric detectors which have been used are the same adopted in the major tokamaks (Mast et al., 1991)

This paper is dedicated to describe those features of the diagnostics which may be more relevant for the ITER bolometry project. The characteristics of the plasma and of the machine are briefly described. A more detailed description of the instrument can be found in the paper by Martin et al. (1997).

RFX is a large (minor radius a=0.46 m, major radius R=2 m, pulse duration up to 0.25 s), air-core experiment designed to operate in high plasma current regimes (up to 2 MA). Despite the size of the machine the access to the plasma is limited by portholes which are kept small to reduce magnetic field errors and this requires substantial miniaturisation of the diagnostic equipment. A careful engineering of the project was necessary because of: (1) the highest temperature (300 C) at which the first wall can be kept during operation; (2) the UHV conditions of the vessel (base pressure $\approx 10^{-9}$ mbar); (3) the need for full remote control of the diagnostic and for decoupling it from the machine without breaking the main vacuum; (4) the fast timing of the configuration set-up and of the evolution of some important physical phenomena which characterise the plasma (sub-ms time scale) and (5) the relatively high level of electromagnetic noise around the machine.

THE DIAGNOSTIC

Several units compose the diagnostic: the insertion and the detection systems, the support structure, the signal processing electronics, the data acquisition system, the head

[1]University of Padova, Department of Physics

temperature conditioning circuit, the control and safety system and the in-situ detectors and electronic calibration systems.

The insertion subsystem includes eight versatile and multi-purpose bellow drive manipulators (seven of them used simultaneously) which bring each photocamera to its operating position with ±0.1 mm positioning error. Two versions of the manipulators have been produced, which differ only in size: that coupled to vertical portholes (six units, stroke of 800 mm, length 2345 mm) and that coupled to the horizontal, shown in Fig. 1 (two units, stroke of 500 mm, length 1524 mm). Four manipulators (3 vertical and 1 horizontal) are dedicated to bolometric tomography (48 lines of sight) and the remaining four to SXR tomography (78 lines of sight). The manipulators are supported by three independent mechanical structures, anchored to the floor to decouple the diagnostic from machine vibrations.

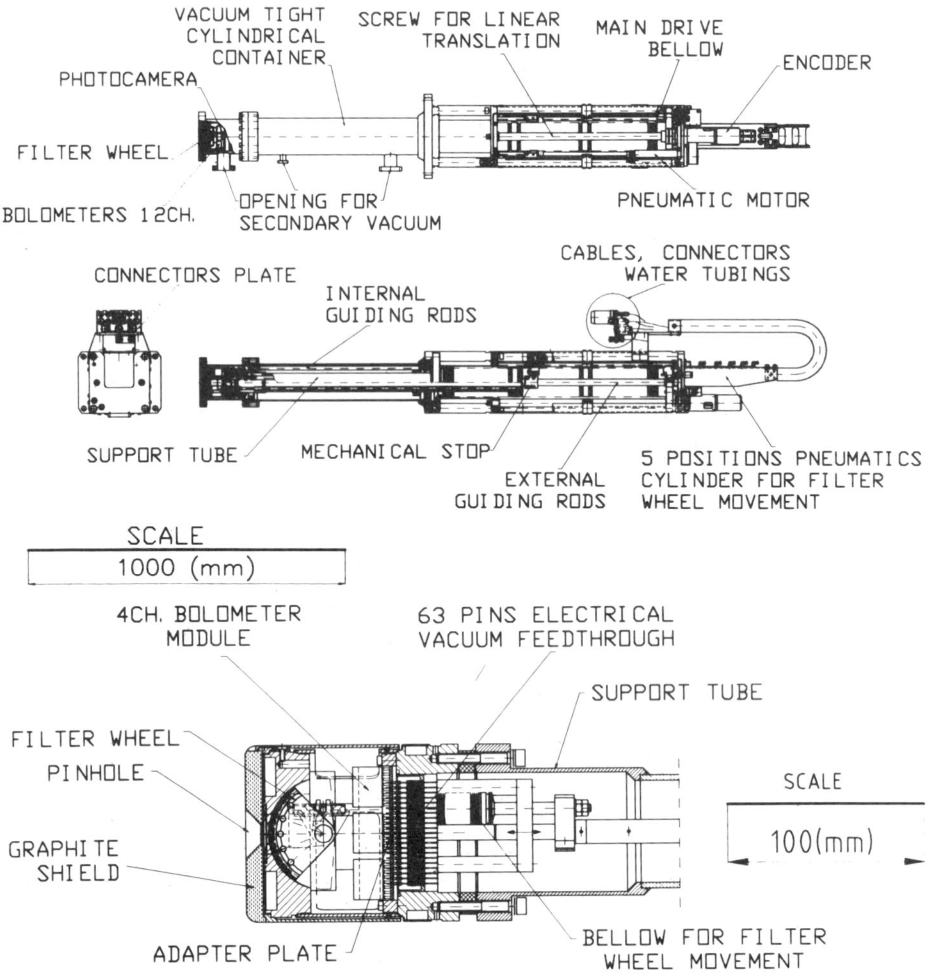

Figure 1: Cross-sectional view of one manipulator (top) and of one bolometric photocamera-horizontal version (bottom)

The bolometric and SXR photocameras are identical and have cylindrical shape, with a diameter of 83 mm for the vertical version and of 89 mm for the horizontal (see in Fig. 1 a cross-sectional view of one bolometric photocamera). To mount both the SXR detectors (Si photodiodes) and the bolometers without any resort to wiring in vacuum, they are equipped with built-in female connectors fitting to corresponding pins on a ceramic adapter plate, which in turn can be directly plugged into specially made 63-pin electrical vacuum feedthrough welded in the head rear flange. This solution makes also easy to interchange detectors for test purposes. Each head can be kept at a selected steady temperature in the range 8-90 C by means of a water circuit.

Four-channels miniaturized low-noise metal resistor bolometers, with an active area of 3.8x1.3 mm^2, are used for total radiation measurements (Müller and Mast, 1984; Mast et al., 1991). Small high power remotely controlled light emitting diodes are installed in each photocamera and provide a system for fast and on-line testing of the detector correct operation. Absolute and in-situ calibration of the bolometers can be made by means of remotely controlled electronic systems. Four different material foils can be interposed between the pinhole and the detectors by means of a 5 position filter-wheel (the fifth is left open), driven by an air pressed stepping cylinder and a small bellow.

The status display, the remote operation of all the diagnostic, all the alarms and the overall safety of the diagnostics are managed by the control subsystem.

The signal processing analog electronics, fully remotely controlled, (Murari et al., 1997) is located in cubicles close to the machine. The signals are amplified and digitised in the same board with 1 MHz sampling rate and 12 bit resolution to be then optically converted and transmitted via TAXI-link to the VME units, located in a different room, along 60 m long fibre optics. The bolometer amplifier is a 50 kHz carrier frequency amplifier with amplitude modulation technique. With hardware filtering the overall bandwidth of the electronics can be selected between 1, 4 and 10 kHz. With specially developed calibrator circuits the gain of SXR and bolometric amplifiers can be absolutely calibrated with an accuracy of 1%.

DATA ACQUISITION AND ON-LINE ANALYSIS

Data acquisition and on-line analysis are carried out by a distributed system based on a configurable number of VME CPUs which perform local data preprocessing, as software filtering and resampling, and exchange pre-elaborated data to perform on-line tomographic analysis. The system resides on three different platforms: a VAX/VMS cluster, where the main RFX control and data acquisition system is installed, VME CPUs for data acquisition and on-line tomographic inversion and a UNIX Alpha station for support facilities and local storage of raw data. The system architecture allows a number of inversion routines (even with the use of different algorithms) to be assigned to any CPU of the system. It is therefore possible to balance workload among CPUs and to increase the computing power of the system by adding new CPUs. The communication protocol has been defined to support any number of participating CPUs. Communication is achieved on a LAN using the IP protocol. The inversion is performed after the shot when all data have been acquired and the results are then sent to the central control and data acquisition system to be stored in the pulse file.

Most of the existing architecture can be used in a real-time system, as that needed by thermonuclear reactor, to perform on-line tomographic inversion for feedback applications. In a real-time system, in fact, preprocessed data for all chords must be exchanged among the CPUs at the time they have been acquired and tomographic inversion must be completed before the next acquisition. For this reason the overhead introduced by the LAN would be unacceptable, and CPU faster than that currently used would be required.

Therefore other hardware configurations should be considered and two solutions could be adopted: i) the use of a reflective memory, which guarantees that pre-elaborated data stored in the memory are available with a very short delay to the other sites; ii) the use of a DSP with dedicated high speed communication ports for data exchange, such as for example the C40 by Texas Instruments.

EXAMPLES OF MEASUREMENTS AND CONCLUSIONS

The bolometric tomography has proved to be very effective for the study of many phenomena. Two quite different examples are shown in Fig.s 2(A) and 2(B), where a maximum entropy approach (Buck, 1991) has been used. Fig. 2(A) shows a typical distribution, with radiation emissivity concentrated in an outer layer extending over the last 15-30% of minor radius, whereas in the case of Fig. 2(B) a much more poloidally localised radiative structure, due to the interaction of a locked mode with the first wall, is present.

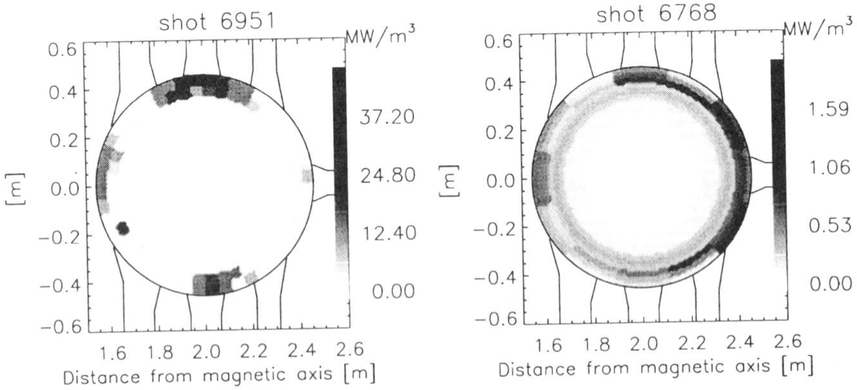

Figure 2: Emissivity profiles in RFX. Left, (A), typical case; right,(B), locked mode case.

These examples, together with the technological characteristics of the diagnostics, with its versatility and with the features of the RFX experiment and of its plasmas, should indicate that this system may be suitable to host some specific test dedicated to the ITER bolometry project. For example various versions of the detectors can be interchanged in a relatively easy way in the photocameras. The existing electronic set-up, with a bandwidth up to 10 kHz, could allow their test to monitor high frequency phenomena, with possible developments of a prototype of a real time system. Tests of the detectors in various temperature conditions or dedicated to the study of electromagnetic compatibility in rapidly varying magnetic fields may also be planned.

REFERENCES

Bagatin M. et al., 1995, *Fusion Eng. Des* **25,** 425

Buck B., Macaulay V., 1994 "Maximum entropy in action", Oxford Science Publications

Martin P. et al., 1997, *Rev. Sci. Instrum.* **68**, 1256

Mast K.F. et al, 1991 *Rev. Sci. Instrum* **62**, 774

Murari A. et al., 1997, *Rev. Sci. Instrum* **68**, 951

Müller E. R., 1984, and Mast K.F., *J. Appl. Phys* **55**, 1984

Rostagni G., 1995, *Fusion Eng. Des* **25,** 301

ABOUT THE EFFECTS OF THE MOVEMENTS OF THE CRITICAL REFLECTING LAYER ON SWEPT REFLECTOMETRIC MEASUREMENTS

Roberto Cavazzana, Maurizio Moresco, and Emanuele Spada

Consorzio RFX - Corso Stati Uniti 4, I-35127 Padova - Italy

INTRODUCTION

The impact of density fluctuations on reflectometry measurements has been studied extensively and is still matter of discussion. Detailed numerical models [1-4] (either in WKB or full-wave simulations, 1-D and 2-D) gave insight on many effects that the fluctuations may cause to the reflectometric measurements. However some fundamental requirements which have to be taken into account in designing swept reflectometric systems can be recovered from a simple 1D model with analytical considerations.

BASIC REQUIREMENTS FOR HOMODYNE DETECTION

The typical signal from an homodyne reflectometer is:

$$s(t) = A(t)\cos(\Phi(t)) \tag{1}$$

Different methods have been proposed to recover the phase law from the reflectometric cosine signal, such as sliding FFT [4], digital complex demodulation [5], digital frequency discrimination [6] and Hilbert transform. These advanced methods give typically a good phase resolution, and have the ability to filter out moderate or sporadic phase distortion induced by fluctuations; however in presence of a strongly distorted signal also these smart methods will fail. For the purpose of this work the basic fringe count method (the phase is increased by π every time a fringe is found) will be used, because of its predictable and clear behaviour.

Time Domain Analysis

The simplest case to discuss is a linear phase law perturbed by a sinusoidal phase modulation of frequency ω_f and amplitude α:

$$s(t) = \cos\left(\omega_0 t + \alpha \cdot \cos(\omega_f t)\right) \tag{2}$$

Fringes can be found searching the minima and maxima in the signal; to find them just look for zeroes in the signal derivative:

$$\omega_0 t + \alpha \cdot \cos(\omega_f t) = k\pi \tag{3a}$$

$$\omega_0 - \omega_f \alpha \cdot \sin(\omega_f t) = 0 \tag{3b}$$

In the unperturbed case (α=0) only eq. (3a) has solutions at positions $t_k = k\pi/\omega_0$, which can be considered as the proper extremals. The phase can be correctly recovered until:

$$\boxed{\omega_0 > \alpha\,\omega_f} \tag{4}$$

Diagnostics for Experimental Thermonuclear Fusion Reactors 2
Edited by Stott *et al.*, Plenum Press, New York, 1998

If this relation is not satisfied spurious fringes appear from eq. (3b) and the total number of extremals is higher than expected. This is the major trouble of homodyne detection scheme: in presence of disturbances it is very difficult and often not possible to detect if the phase function is increasing or decreasing.

In the case of reflectometry the IF beat frequency ω_0 depends on the average target distance d and microwave source sweep rate v_{sw}:

$$\omega_0 = (4\pi/c)\cdot(v_{sw}\, d) \tag{5}$$

while the parameter α on the oscillation amplitude h and the microwave wavelength λ:

$$\alpha = 4\pi \cdot h / \lambda \tag{6}$$

The condition (4) appears to be symmetric regarding fluctuation amplitude and frequency, but the effects of these two parameters on the signal are quite different. As can be deduced from Fig. 1, the distortion of the signal at a fixed product amplitude by frequency, is stronger in the case of fluctuation with high frequency and low amplitude. Thus high frequency fluctuations are likely to be dangerous even if small.

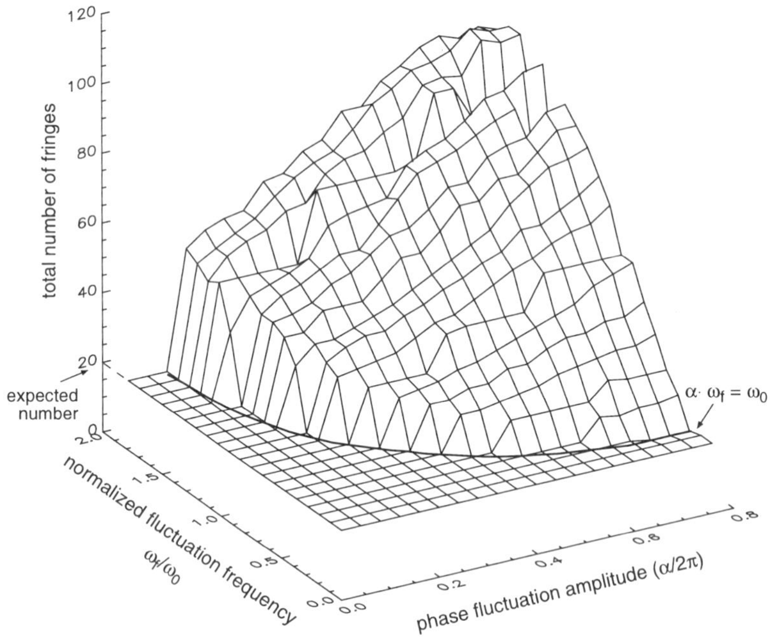

Fig. 1: number of counted fringes in a homodyne perturbed signal as function of the fluctuation amplitude and frequency.

Frequency Domain Analysis

A similar analysis can be applied in the frequency domain and easily extended to more general cases. In the previously considered case of the sinusoidal perturbation (eq. (2)) the amplitude spectrum of the signal contains a series of harmonics of frequency $\omega_0 \pm n\cdot\omega_f$. The amplitude of the n-th harmonic is given by the value of Bessel functions:

$$A_n = |J_n(\alpha)| \tag{7}$$

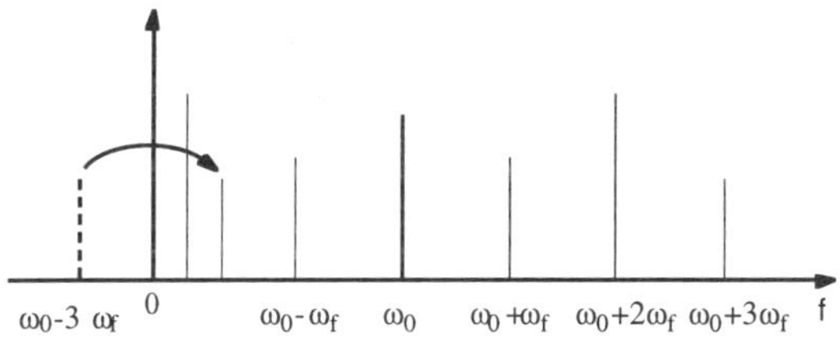

Fig. 2: example of an FM spectrum with aliasing.

The obvious method to find the frequency ω_0 is to find the central frequency in the spectrum. In the considered case with a real signal if the frequency is negative, it will appear in the positive frequency domain, because the imaginary part of the signal is missed.

So it is possible to find the correct central frequency only if there is no negative frequency aliasing. Recalling the Bessel function property:

$$J_n(\alpha) \approx 0 \text{ if } \alpha > n \tag{8}$$

the condition found in the time domain case (4) can be recovered again:

$$\omega_0 > \alpha\, \omega_f \tag{9}$$

Now this rule can be extended to the general case of an arbitrary perturbing function applying the Carson's law [7], which states that the bandwidth of the modulated signal Bw_{mod} depends on the typical peak amplitude of the phase variation α^* and on the total bandwidth of the modulating signal, which in this case is the fluctuation bandwidth Bw_f:

$$\boxed{\omega_0 > Bw_{mod}; \quad Bw_{mod} = \alpha^* \, Bw_f} \tag{10}$$

Typical high frequency density fluctuations have a noise spectrum of the form $f^{-\beta}$; $\beta > 1$. For this cases the parameters can be chosen as the bandwidth with 90% of energy and 3 times the r.m.s. amplitude as peak amplitude.

Since two sidebands of width Bw_{mod} appear around the central IF frequency, even in the case of heterodyne detection the total IF bandwidth of the system Bw_{IF} should be at least $2 \cdot Bw_{mod}$ to properly recover the phase information on the density fluctuations.

EFFECTS OF FLUCTUATIONS ON PROFILE RECONSTRUCTION

If the target is not stationary during the measurement time, the profile reconstruction can be erroneous, even if the phase can be correctly recovered from the reflectometric signal (homodyne or heterodyne). This effect can be easily outlined in the case of an oscillating metallic mirror, but it is still present in any plasma profile reconstruction [8]. The phase law written as a function of the linear varying probe frequency $F = \nu_{sw} t$ becomes:

$$\Phi(F) = \frac{4\pi F}{c}\left(d + h\cos(\omega_f F / \nu_{sw})\right) \tag{11}$$

Reflectometric density profiles are reconstructed by means of Abel inversion using the formula:

$$x_c = \frac{c}{2\pi^2} \int_0^F \frac{\Phi'(F)dF}{\sqrt{F^2 - F^2}} . \tag{12}$$

In the considered case:

$$x_c = \frac{2}{\pi}\left(\int_0^F \frac{d}{\sqrt{F^2 - F^2}} dF + \int_0^F \frac{h\cos(\omega_f F/\nu_{sw})}{\sqrt{F^2 - F^2}} dF + \int_0^F \frac{-(F\omega_f/\nu_{sw})h \,\mathrm{sen}(F\omega_f/\nu_{sw})}{\sqrt{F^2 - F^2}} dF \right). \tag{13}$$

The first two terms contain the correct physical information about the mirror position and displacement, while the third which is introduced by the phase derivative is a spurious effect that typically tends to amplify the measured oscillation amplitude.
In principle this effect may be corrected, but a very clean and precise phase signal is needed or it can be attenuated by averaging on different measurements. In any case the condition to reduce this effect is:

$$\boxed{v_{sw} >> \omega_f \Delta F} \tag{14}$$

It does not depend on the average target distance *d* and emphasises the role of the sweep rate v_{SW}. This result applies both to homodyne and heterodyne detection systems. An example of a profile reconstruction on a metallic mirror with $v_{SW} \approx \omega_f \Delta F$ is shown in Fig. 3.

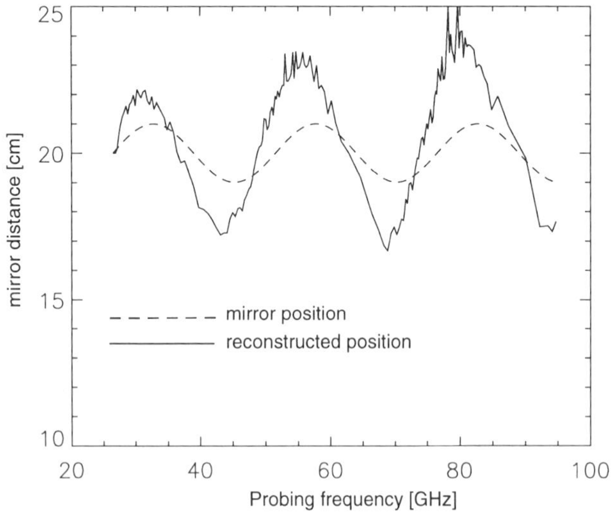

Fig. 3: simulation of profile reconstruction on a moving mirror.

EXPERIMENTAL MEASUREMENTS

The discussed statements for homodyne systems have been experimentally verified on RFX [9]. In RFP plasma the fluctuations of the cut-off layer seen by the reflectometer are mainly caused by plasma rotation. Because of the configuration properties, higher modes are present and this gives raise to a wide frequency spectrum.

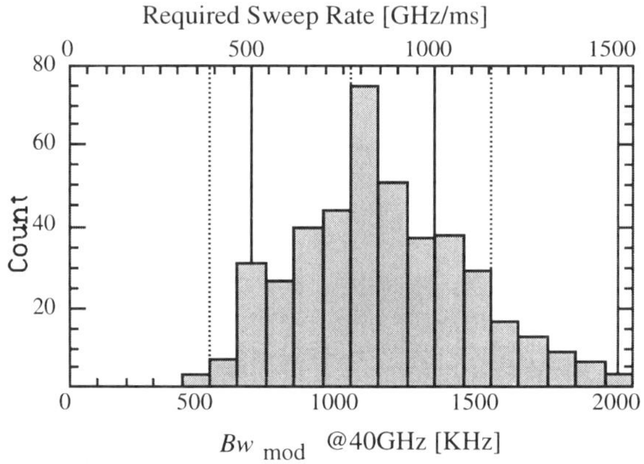

Fig. 4: typical modulated bandwidths of IF signals measured on RFX by the homodyne reflectometer.

The bandwidth Bw_{mod} has been measured directly by operating the RFX reflectometer at fixed frequency and values up to 2 MHz have been found (Fig. 4).In these plasma conditions we were able to successful recover the phase from the signal using a sweep rate of 1 GHz/μs or higher. A further raising of the sweep rate to 4 GHz/μs produced a better S/N ratio (Fig. 5) and increased the fraction of signals on which the phase could be correctly recovered from 35% to more than 70% .

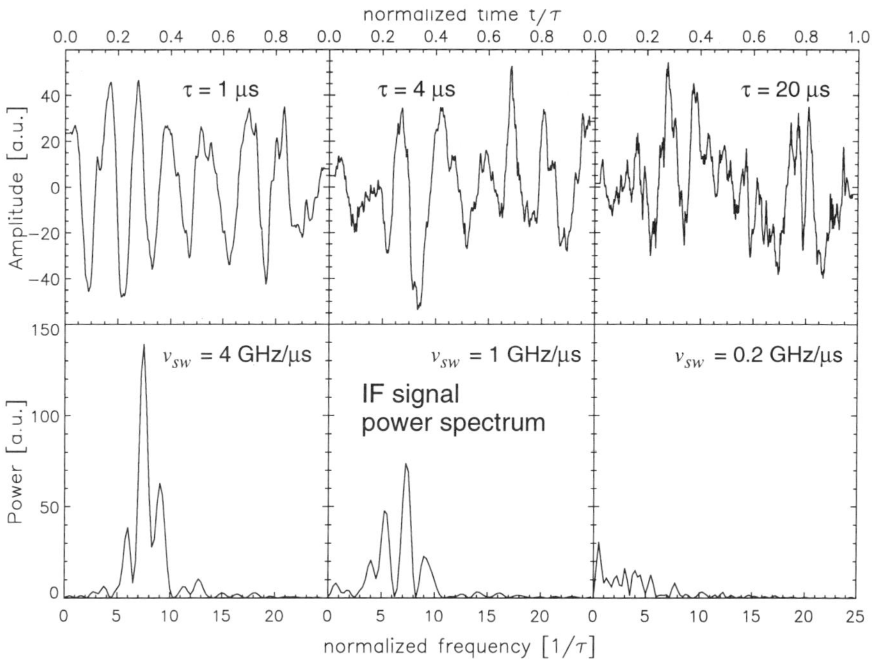

Fig. 5: IF signals from RFX plasma measurement obtained at different sweep rates.

CONCLUSIONS

From a simple 1D model it is possible to outline some fundamental requirements that are to be taken into account while designing swept reflectometric systems.
For homodyne reflectometers it is possible to extract the phase from the IF signal only if the product of sweep rate v_{sw} by average plasma apparent distance d satisfies the relation:

$$\boxed{(v_{sw} \cdot d) > Bw_{mod} = h/\lambda \cdot Bw_f} \qquad (15)$$

where h is the typical peak amplitude of the fluctuation, λ the microwave probing wavelength and Bw_f the frequency bandwidth of fluctuations.
In both homodyne and heterodyne systems the sweep rate should be keep as high as possible to reduce the errors introduced by plasma movements in density profile reconstruction:

$$\boxed{v_{sw} >> \omega_f \Delta F}. \qquad (17)$$

Furthermore a high sweep rate gives also a better S/N ratio.
To obtain a correct information about fluctuation the bandwidth of the IF section must be wide enough to accomodate the two sidebands Bw_{mod} around the central frequency:

$$Bw_{IF} > 2 \cdot Bw_{mod}. \qquad (16)$$

REFERENCES

[1] I.H. Hutchinson, *Plasma Phys. Control. Fus.* **34**, 1225 (1992).
[2] R. Nazikian, E. Mazzuccato, *Rev. Sci. Instr* **66**, 392 (1995).
[3] B.B. Afeyan et al., *Plasma Phys. Control. Fus.* **37**, 315 (1995).
[4] V. Zhuravlev, J. Sanchez, E. De la Luna, *Plasma Phys. Control. Fus.* **38**, 2231 (1996).
[5] K.W. Kim et al., *Rev. Sci. Instr.* **66**, 1229 (1995).
[6] P. Varela et al., *Rev. Sci. Instr.* **66**, 4937 (1995).
[7] A. B. Carlson, *Communication Systems*, McGrawHill (1975).
[8] R. Cavazzana, E. Spada, *RFX internal report FB/71* (1996).
[9] R. Cavazzana et al., *24th EPS Conf.,* Vol. 21C (1997).

INDEX